A BOOK

I0394398

Manuel Pagan

A BOOK

A Series of Essays

Copyright © 2017 by Manuel Pagan.

Library of Congress Control Number: 2017916890

ISBN:	Hardcover	978-1-5434-6319-4
	Softcover	978-1-5434-6318-7
	eBook	978-1-5434-6317-0

All rights reserved. No part of this book may be reproduced or transmitted in any form or by any means, electronic or mechanical, including photocopying, recording, or by any information storage and retrieval system, without permission in writing from the copyright owner.

Any people depicted in stock imagery provided by Thinkstock are models, and such images are being used for illustrative purposes only.
Certain stock imagery © Thinkstock.

Print information available on the last page.

Rev. date: 11/07/2017

To order additional copies of this book, contact:
Xlibris
1-888-795-4274
www.Xlibris.com
Orders@Xlibris.com
769503

Letters

$_{1}$Do$_{2}$

$_{2}$Not$_{5}$

$_{3}$Edit$_{9}$

$_{4}$in$_{11}$

$_{5}$this$_{15}$

$_{6}$book$_{19}$

Contents

¶|||24s.0 28lm.s. R.S.C. {Reading Secondths Count : in (.milli) : secondths-{whole numerical value apose millisecondths // subterior values minutes}: minutes-{{:}whole numerical value apose secondths}}|||§ For as we have known literary-interpretation-(s), I know how their exist-an deviation of vocal-vernacular-voiced-competent-interpretation-(s), amid my typed context, (that not have been) commonly-existent yet, (for whom I am to) insight you -{an Reader}-(maybe amid other competent-individual-(s')), (in this) Predicate-Directions for Reading...

◊ A. {¶|||29s. 62m.s.||| § - Essay-Paragraph-Subsection-interpretation}; indicates each ("0")-indication-(s), prior to each essay. This continue per Article-(s) (as each) Essay, indicate-(s) at-page-deviation, (as how this) book inclinate an Month-Dynamic-Interpretation-Cycling-(s), thinking how pacing those subterior-indication-(s), apose sporadic inductive-indication-(s),| 1 | continue literary-progression through (1)/(2)/(3)/(4)/(5)/(6)+; (when at an end) per essay, an interpreting from redact-paragraph, juxtaposing, affirm subsection-(s), through

¶|||22s. 61m.s.||| §| 2 |§[a{Count-Preposition-of} / b{Lines-Count} / c{Article-Essay-Count-(s)} / *d{Book-Essay-Count-(s), after (a the) initial-essay} / {An redact-paragraph offset Essay-Paragraph-Subsection-context, interposed [1]interpretation-(s) / [2]indication-(s) / [3]intellect-jargon-(s) / [4]term-(s) / [5]time-calendar-mathematical-inclination-(s) / [6]paragraph-juxtaposed-literary-neutral-indication-(s) / §[7+]etcetera...}]

while asserting an initial-premise, (for my) reader-(s') blogging refining, each excerpt-inclin-ation-(s) by omitted numerical-indication-(s),| 3 |and predicate-dictionary-word-form-interpretation-(s), et cetera...

◊ B. ¶|||08s. 41m.s.||| §{- - dash-interpretation}; indicates {[1]word-conjunction-(s) / [2]word-interpretation-(s) {-(s) / -(suffix-participle)} / 3lines-edge-paging-refinement / 4word-(s)-an-(particular-preposition-interpretation}...

◊ C. ¶|||07s.95m.s.|||§{/ - virgule-interpretation}; indicates [synonym / antonym / empirical-executive / et cetera offset... interpretation-(s)] (by an) [bracket / numerical / interpretation-extent-(s)]...

◊ D. ¶|||07s.60m.s.|||§{& - and-interpretation}; indicates comparative, contradicting, or et cetera preposition-offset... two-fixed-word-(s)-interpretation-juxtaposition-(s)...

◊ E. ¶|||09s.97m.s.|||§{[a. / b. / et cetera...];| 4 |indicates [brackets-virgule-word-(s)-posit-extenuation-underline-dynamic-interpretation}; indicates various interpretation-(s), scattered throughout the English-diction-lexicon...

◊ F. ¶|||02s.50m.s.|||§{, - comma-interpretation}; indicates spacing in read phrasing...

◊ G. ¶|||05s.53m.s.|||§{; - semi-colon-interpretation}; indicates break in essay-syntax-interpretation-(s), in syntax-extemporaneous-fashion...

◊ H. ¶|||10s.96m.s.|||§{. - period-interpretation};| 5 |indicates end of paragraph-redact-subsection-interpretation-(s), as well an reiteration of past-historical-usage-period-sentence-break-interpretation-(s)...

◊ I. ¶|||08s.81m.s.|||§{... - ellipsis-interpretation} - indicates an transition of essay-paragraph-subsection-(s)-{¶§}, by paragraph-redact-dynamic; as well (as an) indication similar to semi-colon, (but in) synonym-style...

◊ J. ¶|||29s.15m.s.|||§{(word-(s)) - (predicate-subdue)-parenthesis-interpretation}; indicates isolation of [two-plus-preposition-(s) / being-positions' / auxiliary conjunct-predicate], amid context, as well (as an) isolation of suffix-{(s)}, when also incorporating an word or so, relevant upon each preposition-over-usage-extenuation-(s), used for subduing those preposition-vernacular-interpretation-(s)-over-usages, while inclinating-an offset-interpretation-(s), upon second-person-subversive-dynamic, responding when in word-initial-parenthesis-interpretation-(s)...

- ◊ K. ¶|||04s.15m.s.|||§{({word-(s)})} - eagle-brackets-interpretation-(s)}; indicates [inductive / deductive offset-word-(s)-interpretation-(s)]...
- ◊ L. ¶|||04s.53m.s.|||§{: - ration-interpretation}; indicates an measure of {minute-(s) : secondths : millisecondths}...
- ◊ M.¶|||05s.85m.s.|||§ {> | < - greater-than / less-than-interpretation-(s)}; indicates an disparity of [higher / lower numerical-observational-comparative-inclination-(s)]...
- ◊ N.¶|||22s.15m.s|||§ {1 / 2 / 3 / 4 / 5 / 6 / 7 / 8 / 9 / 10 / etcetera...-Numerical-inductive-offset-synapses-book-encyclopedia-redact-inclination-(s)}; indicates an point-(s) of synapse-spacing-(s), for bordering the blanks of-a-the book-page-(s)-sequential-inductive-reasoning, mode-time-period-redacting, deductively, juxtaposing how each interval-period, combines those point-(s) for reasoning each moded-interval-period-(s), juxtaposing [Timing // dating]; upon physiological-offset-visual-tonal-vocal-reverberated-perspective-(s')...
- ◊ O.¶|||26s.91m.s.|||§ {¶|||07m.:53s.91m.s.|||§-stopwatch-interval-period /// ellipses- ¶-Paragraph-§-Subsection-interpretation-(s)}; indicates those focused-concerted-time-period-(s), for extending how many time-(s) their period-reduction-interpretation-usage-(s), incur while those basis-book-learning-function-(s), direct how to comprehend, an fluent-competent-offset-genetic-reader-divided-ordering-(s), from Commercial-Market-Literary-Education & Continuum-Genetic-Seperation-Ecological-Geological-Labor-Wage-Rocky-Mountain-Plain-(s)-Hudson-Bay-West-Central-Canada-Political-Boundary-directing-(s'), from Saint-Paul & Montana-(north of a-the Missouri), by-an Literary-Revolution.

*Footnote; first read a-the paragraph on page 491, under Gregorian-Calendar-Schedule
Circulation-Survey-360°...

...^{1}Redact$_{6}$

^{2}in$_{8}$·

3·your·$_{12}$

$_{4}$·own$_{15}$

$_{5}$note-pads$_{23}$...

<u>Numbers</u>

Prolegomena

¶[Selective-Service in [[0]National-Guard / [1]Army / [2]Navy / [3]Air-force military-service-(s)], (do not) comprehend (how to) influence sociological-genetic-offset-enlistment], (for an) abrupt deviation offset-at lethal-circumstance-(s), interrupting sociological-introverted-subconscious-perspective, yet understood how literary-document-(s), amid celestial-physics-offset, affect introverted-individual-perspectives’, amid tectonic-plate-centripetal-force-revolving (conun-drum), disrupting how death-approach, fear independent-sociological-individual-American-enlistment-(s’), not comprehending-an international-pressure from Germanic-Hierarchy, affecting each outcome from mass-population-offset-introverted-circumstantial-perceptive-interpretational-thought-(s’) (engineering-enlistment-error, amid infantry-thinking-process {(military form, DD-214 (directive-discharge)-2.1.4(count)))}; forgetting how surrounding perceptive-religious-matter, affect each [[1]object / [2]subject / [3]article / [4]kin / [5]species / [6]physiological-interaction-(s’) / [7]geological-labor-effort-(s) / [8+]et cetera...] affirming why I write this book, and type print it, where centripetal-force-continuum-revolving-parameter-pressure, (not yet) confirm who’s self-pride, is supposed to segregate from one-another, (from how) [[1]eye-color / [2]skin-type / [3]hair-color-texture / [4]head-shape / [5]nose-style / [6]head-size / [7]neck-width / [8]physiological-component-(s) offset-characteristics’ / et cetera...], affirm when natural-mundane-motion-(s), affect how every-motion-(s), characterize where inclinated-notion-(s), conjure what defined tangible-interpretational-focus-(es), affirm individual-(s’)-subconscious-perspective-existing, intermittent deviated-monologue-(s’), because how vision-offset, continue along-an axis where tectonic-hourly-timed-dating, another reason I create an book, to reduce the market of ambient-competition, to an more directly-lively-entertaining-book; scarcely-recollecting resources, documented per Lineage-dating-(s), because each individual not force-an direct feminine-proglamated-repopulation-extent-(s), when every-whatever-individuals', matter default, when existence, simply go pass intersection-(s), without each momentum-indication-(s'), pertaining voluntary-listening-(second-person-posit), (when youth not be) directed reserve-(s)-training, adults' differing university-accredited-debt, and military-communion, subjunct terribly without-an understanding, (for how many) children, women are supposed (to be) directed-to, male-strict-totalitarian-directing, and weak-inhabitant-(s) executively-eliminated, (from an) legislative-executive-perceptive-focus, over legal-vanguard-senate-legislational-remarks', yet working physiological-physics-male-sweat-effort-(s), where each physiological-motion-(s) not affect-an direct impact of physics, unless other inhabitants' apply-an direct-focus for architectural-geological-developmental-purpose-(s), circumstanced-outcome-(s'), affecting direct developmental-conjured visual-depth-vocal-ear-canal-tonal-perspective-(s'), where each direct-segmented-hourly-time-posit-moment-(s), continuum offset, affirming in quick-allegiance, (why an) individual matter-(s), upon our limited county-city-Federal-commercial-residential-traded-lifestyle-reliances'; (when in) high-school, children choose to dick-around, youth-sexual-activities, (as I be 26-continuum-sexless (hopefully before thirty), (when each) offset individual-(s) [[(*[ignore] my parents' saving their money, uncertain those aging marks for monetary-physiological-survival, (amid this) contextual-book, not with editing-economic-support *ignore priori, please for my honesty]], amid historical-progress in though conventional-circumstances", (that I can not even) officially-directly-explain, (as what) proglamate this initial-context (not having youth-opposite-sexual-reproductive-interaction-(s)); | 6 |The prior from my Los Angeles-Post-Fort-Benning-Ethnic-Schizophrenic-Surroundings, amid indirect social-surrounding-circumstance-(s) still not stopping, in the shadows (as I am) subconsicously-pressuming from Los Angeles, continually amid the prior, sarcastically awaiting the neurological-subdue (affirming language is not clearly vocally shitty, but somewhat internally-imposing), not seeing much how human-extraspection is, amid their terrible-extraspective-effect, or genetic-repopulating, (as I teeter with the) Literary-Macmillan-Blonde-Wives-perspective (though Macmillan may not pan through, in regard of experience, or so), through-an blue-eyes-inquiry per dozens of children-sociological-forms, amid family-literary-generational-encyclopedia-book-per-child-breeding, resuming (from my) youth-birth-silence, before my subconscious-silence, (After Los Angeles, after my Army-Fort-Benning-Non-Enlistment (initial-white-national-*billions-literary-continuum-assumption-(s)-inquiry, when thinking quick, (how to erect an) nation, not knowing my subconscious was shown from birth, amid my silence from sounds in my subconscious, I either beforehand or, require now, an [Book-Editor / Literary-Agent / Publishing-Company], neutral-approval for such context, (so I in my) initial-prolegomena, making clear my intention-(s) from literature, when expecting how Americans (are so) economically-literary-ignorantly-dry, to not ever work-an historical-past age of common-80-year-(s)-of-age, ignoring at offset-Genetic-tens-billionths-two-decades-to-century-national-population, my [{Military}-Fort-Benning-(expected)-non-enlistment / {Sociological}-Los-Angeles-adventure-ambitious-false-expectation-purpose-(s)], not exactly knowing (if their) subconscious-Subconscious-neurological-American-effects, for why blacks are black, religiously-morally, not controlling ““schizophrenia” [*I believe my Father, (with my mother, not help my initial-comprehensive-{N.Y.-Book-Editors, (residing in Connecticut)-{{$5,732-comprehensive edit}-out my labor-pocket-{May-2016-June-2017}, maybe proglamated 2-months, to 1.5-year-17-months, including the Streets of Los Angeles bothering-me from indirectly-lying, not worth delving into*]||||||||||; (as how I) nearly display (why my) initial-context, (show how each) [American / International / Indifferent inhabitant-(s')], (why their is an flaw upon) American-Historical-Dow-Jones-stock-market-trade-discourse, conducting how independent-individual-perceptive-bias, choose to ignore (Americas'-Military) Montana-Geological-physics-genetic-seperated-planned-developing, (not from) militarial-recognition, north (of the) Missouri, (because how) individual-purpose-(s) affect (a the discourse from) human-activity-(ies), when developmental-sociological-traded-purpose-(s), affirm (why each task-(s), are managed at an) approximate perimeter-diameter-distance-developing, because silent-introverted-individualistic-independent-inhabitant-(s'), shy (from how) mute-introverted-illiterate-individual-(s'), (yet are capable of referencing in an) simultaneous-fashion, when common-poverty-exist, not considering how other individual-(s'), can affect off-time-work-activities, from religious-militarial-Vanguard-economical-debt, working how resourcing is still relied on, (from an) international-trade-debt-disparity, practicing exchange-trade-funds-investing (from those) objective-item-(s)-tradition-(s), Military-Engineering-Dynamic into socialism, not infantry-{A.R.40-D.D.214}, per geological-developmental-discourse, (because where) families offset-genetic, not consider seriously, (why their is an) evaluative-discourse-(s), male-commercially-constantly-building, those urban-housing-sector-(s), an genetic-offset-allegiance, (where their not be an consideration for those) refined-factory-production-valuable-(s), (which would pertain to an) general-compliance-dynamic, (not affecting much of those) property-perimeter-boundary-(ies), (that should be covered when) human discourse, run-off election-(s), not considering selective-service-national-guard-engineering,| 7 | (before age [35 / Thirty-Five] service cut-off), why extraspective-community-territory-developments, have-an consideration-process per birth-repopulation-(s), which affect geological-title-property-impact-(s), when extraspective-individual-(s'), (are to interact at) various elemental-condition-(s'), following independent-commercial-labor-wage-operation-(s), not communicating from high-school-(s)-county-offset, (whom is to enlist into) [military offset branches reserves /// sociological-Labor-pool], graduating (into an) singular-communicated-offset [ARMY / NAVY / AIR FORCE classification-(s)], (for how) commercial-operation-(s) (are to be performed by) independent-inhabitant-(s'), (for whom) develop [roadway-(s) / commercial-entities / block-housing / convention-center-(s) / et cetera extemporaneous-required-development-(s)],| 8 |in creative-resource-patent-developmental-discourse, by inhabitant-commercial-developmental-discourse-(s), (which would matter where) physics-geological-ecological-entity-(ies)-developmental-discourse, signify (what we all not ever are in an) congruent-fashion, conducting each political-affair, (where their exist-an) inconsistency in common-community-literary-extraspective-religious-family-trust, affirming in my opinion, why Offset-Political-Genre-Books, (not yet exist, when more) important (than the) Bible, (for how my near) article-(s)-literary-context, simplify-continuum, an greater article-(s)-genre-tier-deviation, expository-exposing-explaining,| 9 |(how their exist-an) numerous erroneous-facts, (military 62-years-old-enlistment-entry-cut-off-limit-(s)), which I believe should (be in to) ninety-five-years-old, (as an) O.T.H.-four-year-non-enlistment-(s)-passoff-period-(s)-{3}, amid-an Federal-government-monitoring, [Vanguard-(exchange trade)-fund-(s)-$3,000 / $10,000-yearly // seasonal // monthly-(respecting)-religious-communion-payment-period-(s)]| 10 |funding American-Private-non-currency-(ies), because (their not ever exist an nation that) remain self-sufficient, from community-relied-resources, when existence continue-an neglectful-disregarded integration-human-neutral-blank-extraspective-interaction-(s)| 11 |{why I believe the military should practice [national-ethnic-separation // white-genetic-characteristic-offset-separation-(s)], or else life exist blank-existing, until (their be an) American-World-War-national-extinction; not understanding an nice-lifestyle existing, because how human-offset-visual-interpretation-(s), interpose [Family / Co-workers / Community extra-spective-interactive-discourse].¶|||09m.:00s.29m.s.|||§ I from birth, would simply be thinking how existence came to offset-formulate, encompassing-an neutral-deviation of notion-(s)-circumstance-(s), impeding-how development-(s) of terrain exist, through various-discourse-(s), conjuring what inclination-(s), affirm how dating continue amid voluntary-human-impasse-discourse-(s), with an allegiance to common-human-moral, affirming those indication-(s), that objectively afflict every single inclination-(s) from offspring-morale, amid centripetal-force-continuum-deviation-posit-visual-tonal-perceptive-discourse, persisting-an varied impasse-competent-objective-usage-circumstance-(s), influencing (how our) subconscious-visual-awake-retentive-impasse-recollection-(s), affirm which indication-(s), consider (why their) remain-an deviation of those commercial-product-factory-item-(s)-production-(s), traded at market-(s), (from when) offset visual-dating, pertain-an multiplicit-conundrum, considering (why their are) physiological-being-(s), that continue to accede what may incline, where dating is supposed to pertain objective [1collecting // 2refining // 3shipping // 4market-retaining-(re-shipping)-on-site-placing // 5longevity of circumstanced-notion-(s)], counting under hour, those posit-circumstance-(s), (that allude to each) (milli)-secondths-minute-(s)-count-deviation, for review of wage-operation-process-(es), per commercial-market-labor-procedure-(s); practicing on denominating Each & Every circumstance-(s), affirming why customary-development-(s), physiologically (make how) dating-year-continuum-reset-(s), are orchestrated under-an capacity of reasoning, aforemention, what object, whom be, why process, when to [place / communicate / interact / sociologically-schedule-affirm continuum-hourly-minute-(s)-timed-dating-procedure-(s)], when daily-day-hourly-dating, affirm what individual-(s'), pertain-an minor-educated-purpose-(s), where perimeter-sectored-parcel-governmental-territory, remain barren at rural-terrain, because when each individual continue circumvention-process-(es), that inhibit those formal operational-procedure-(s), processing at impasse-payment-(s)-work-wage-moment-(s), what shipped-product, (is to) circulate, what intermittent-deviation-trading-(s), per spaced-parcel-moment-exist, because when dating is supposed to account-an community-purposes, trading-(s) amid-an colloquial-bias, having men work geological-climate-condition-(s), while women are supposed to [mansion-rural-housing / transit-conventional-reclusive] prestige physiological-male-characteristic-(s), when dating by security-measures',| 12 |pertain [stock-(s) / bond-(s) / (exchange-trade)-fund-(s) / securities] circulating those industrial-industries, which formulate-document,| 13 | commercial-operation-(s), not yearly-documented, amid I.R.S-yearly-trade-routine-(s),| 14 | (which consider why their be an) area per city-planning-police-policy-coding, that routine-an formal-function-(s), when performing transit-trade-operation-(s), because which dating-periods, is monetarily-considered, when-what individual-bias, is formally considered, (due to) ordering-circulated-commercial-type-procedure-(s), operating those mechanical-function-(s), (due to) common-practice, which consider where dating review those affect-(s), per hour, (as when an) geological-task-(s) is intend completion, (under those) [managerial // supervisor procedure-(s)], looking (about at) those balance-(s) per budget, apposing [common-labor // employees'], when culminated circumstantial-focus-(es), are suppose to wage, an directive-mass-effort-(s), for developing-an extraspective-sociological-industrial-development-(s),| 15 |amid discussed-subterior-geological-degree-(s), understanding how tectonic-plate-developmental-measure-(s), remain limited from reproductive-cycled-matter,| 16 |predicate each circumstance per moment, frictioning matter.¶|||12m.:19s.$_{26m.s.|||}$§ (So now I am) taking minuté-time, diminuendo my prolegomena-manuscript, and delving (into an) randomization (by my) finer-written book-pages, establishing the grounds of my book, (the priori-requisite) frustration.....¶|||12m.:3 0s.95m.s.|||§

...The main idea of this book, is on page-one-hundred-fifteen, {February}.¶|||03s.65m.s.|||§

.....How many Century-(ies), may it take until the next finest book could have been personally-adjusted, from offset current-American-English-notion-inclination-(s) ¶|||07s.50m.s.|||§

...Requisite my body essay-(s), I do not fully intend to complete all of those essay-(s), in paragraph-formal-response...¶|||05s.75m.s.|||§

...Individualism is how we adapt away from our children, not counting those moment-(s), as the most precious moment-(s), we ever have with our raising of self-offspring...¶|||07s.62m.s.|||§

...[Peach / White / Pink] Woman are amazingly attractive...¶|||02s.37m.s.|||§

...I haven't aged like I would like to...¶|||02s.03m.s.|||§

...Work place, is quite difficult, when relying the subjective effort-(s) of an [singular / several] individuals...¶|||05s.62m.s.|||§

...Peers can only be looking out from their best means', not in the vague interests, from other individuals', amid family-lifestyle-discourse...¶|||06s.69m.s.|||§

...Simply read through those lines, recollect interposed-proportionally, how matter is timed, and figure on an scale of 100%, what may be my April-scaling, not yet logirhythmn each [mathematical / grammatical / punctuation-timbre-(s)], so I not know how to edit my book alone, but with the monetary-deviation for time, their exist an similar-process, for how editing work by publisher, rather than word, but in format, I remain inconclusive to say, as how type-convey, amidst an written-cursive-dynamic, for how tapestry-weight, after the measured-fact, from numerical-incremental-circumstance-(s), that remain out the lexicon of any tribal-accent-language, colloquial per momentary-reflexive-motion-(s), as amid through contextual-fact-(s), granted honor, when formal historical-word-referencing, expanse those bordered-edges, as to incorporate an total-jargon, apose each dictionary-reference-(s), relevant-motivational-access...¶|||52s.66m.s.|||§

...Lines should reflect word per second, while letters are in an millisecondths-reading-rate-interpretation-logic, per letter-influctuated-frequency, juxtaposed [1letters // 2words // 3terms // 4syntax // 5definition // 6Obituary // 7Poem // 8Article // 9Essay // 10Paragraph // 11news-article-(s) // 12passage-(s) // 13Psalms]...¶|||14s.78m.s.|||§

...I am trying to sell the walls-reprinting's, off this book, through scanned paging... ... Imagine how pages could make an room-wall-reprints, upon ethical-walls (paper)... ¶|||09s.06m.s.|||§

...Finally, if you see any - (dashes), out of place from the border of the paper-(s), pardon the perimeter-dash-incoordination, nearly 500 pages, and I can see their be an difficulty in precisioning such edit, taking it as an *participle-break, please, amid those later flaws of my virgule-syntax, by brackets and underline, the editor was not going to punctuate, as much as grammar-edit my book pardon those punctuation-error-(s)....¶|||24s.12m.s.|||§

...If any individual becomes too self-dependent, or if the sociological-system continues to lie about their introverted-behavior, it constantly makes me worry about an World War-Scenario, due to each outcome of customs amid physics...¶|||10.97m.s.|||§

Prolegomena

...Try not reading each essay simultaneous, if too long, remember only thirty-minutes an day reading is my commercial-literary-attempt, to make the reader understand not too long on the article, and more intelligent intricating by those means of literary-entertainment...¶|||12s.38m.s.|||§

...Please read consciously-moderate-aloud, it would give the reader an gauge for how much to read in an day, proportional other activities...¶|||06s.10m.s.|||§

...I devised this book, so the reader can read during the workweek, each essay, intermittent daily work-effort-(s), to structure an reading pattern, that is not difficult to follow...¶|||07s.94m.s.|||§

...[Letters /// Words /// Punctuation] marks are how symbols made logical-interpretation...¶|||04s.00m.s.|||§

[Article:] [January]

¶1:Date-(s) (of a the) [day / night] sequence-(s), (are of an) superlative grandiose (in)-tangible perspective-offset by concepted-view-(s'), (of the) reasoning per hourly-time, (by an) wage-numerical-calibre (of those) facts of perspectives, stemmed from being by several natural-force-(s), Solar-Fission-Combustion-Expansion-Infinitive, afloat plasma-temperate-molten-fusion-substantial, (by an) configuration of Earth-Celestial-Body, (due to an) default of Rotating-Force, acting upon tectonic-plate-(s), happening (for an) continuum of celestial-energy, (that is) propelling-force upon human-being-perspective, while rotational-pressure act upon perspective during centralized-aspects, physics of [[1]Rock / [2]Water / [3]Soil / [4]Plain / [5]Prairie / [6]Desert / [7]Forest / [8]Mountain Composites], for which (are of an) tangible-mass of matters, remaining an part of existence (for how) moments become perceived (through an) voluntary-physique-efforting-(s) of survival, that ingrain oneself (upon a the) place of Terrain, that forebode (how an) being is willed (by an) elect of individual-self-needs, ordering survival-piece-(s) of composited-matters from mass-family-(ies), adjunct sedentary-refinement, for ever partaking (in an) primary-cycles of self, which brings imagery of synapses to be perceived (by an) Interval-Lapses', (by of) blood-pulsed-intervalizing-perspective, in [view / sight / vision], (as an) Intangible-fulcrum of physical-being, (by an) requirement to [[1]eat / [2]drink / [3]defecate / [4]urinate / [5]collect / [6]place / [7]clock / [8]plan / [9]budget / [10]order / [11]monitor / [12]harbor / [13]write / [14]type / [15]weigh / [16]examine / [17]patrol / [18]man / [19]serve-(ice) / [20]ingrain / [21]formulate / [22]regulate / [23]locate / [24]count / [25]clean / [26]factify / [27]proof / [28](en)-act / [29]rate / [30]procure / [31]deliberate / [32]debate / [33]affair-(s) / [34]model / [35]develop / [36]intercede / [37]Transit / [38]Trade / [39]habitat / [40]adopt / [41]birth / [42]ship / [43]average / [44]compile / [45]Ticket / [46]observe elapses of time / [47]play / [48]Lead / [49]Operate / [50]Sort / [51]event / [52]clear / [53]progress], (how an being-(s) are) posited (in to) focus, by rigor of character-selection, in attempt to factify Each & Every notion of daily-government-value-(s), (by) through seconds-interval-act, to foretell-efforts of noted-observation-(s), (by of) physical-timbre-objective-use-frequency, (to those) forms of act, that preclude-activity-(ies) upon physics-of-Terrain, (to an) expectancy of Locale-of-participation, in virtue (to an) abundant-configuration-(s), (per (of)) individual-selves, using tentative-objective-focus, (from of) attention-span-efforts, elapsed generalized-segments per date, affecting (a the) manner of character-skill-(s), for who-(m) impasse familiar (visual-evident) exterior-relevance, (in an) configuration by depth-compass-circumference-bend-surface-area-perimeter-boundary, (in so that), influence of formal-conduct, (can be) conduced, (in the) nature of Approximate-Locale & Payment, in dues for purchasing unit-object-(ive)-(s), per political-matters (from of) taxed-state-(s), (while in of an) accessory to usage, via [individualism // religious] genetic-mass (-efforts') during affairs of districted-housing-city-planning, for enabling developmental habitant-mass, while decisions of offspring, are brushed 32L.aside, for adulterated-living, uncommunicative Location & Citizen-requirements' to state the needs, per self-voluntary-existent-inhabitant-citizens'...¶|||02m.:57s.34m.s.|||§

§0...Count-[38 / Thirty-Eight]-preposition-of…| 17 |..Lines-[33.48678 / Thirty-Three-whole.1fourtenths-2eighthundredths-3sixthousandths-4sevent-en-thousandths-5eighthundred-thousandths]..| 18 | ..Article-Essay-[1 / One].. What was the lesson? ¶|||08s.84m.s.|||§

¶2:If marking being-(s), (is preferred to be from an) ["(wo)-man" // man genitalia observing)], (what can be concluded from such) dull assumptions', when beings perceive (through an) spacial-transit, blood-pulse-synapses', ventricle-pressurized, exterior-luminescence-transfiguration-air-vapor-oxygen, surrounding upon physics-status, as eye-plasma-ventricle-visual-perspective-(s) data-interpret "dating", (from an) parental-conscious-lineage-recognition-(s), traditional from perspective, amid compass-plane-depth-exterior-visual-presence-(s) (by of) existence, formally supratangible, apose interior-self-physical-tense-(s), (by a the) notions', [blood-pulse-invigoration / sight / sound hyper-tense-(s)], (from an) variation of [density-(ies) / buoyancy-(cies) / centripetal-vapor-pressure-(air)], tensing types of [substance-(s) / Liquid-(s) / Kinetic-Wave-Energy-Pulsation-Infinitive], affecting (a the) being per blood-pulse-synapse-(s) relayed, pending upon date-rotation-(s)-inclination-(s), (from an) pre-historical-unconscious-lapse-cycle-(ing-(s)) (through of) Solar-Plasma-Combustion-Infinitive-Kinetic-Radiation, emanating an constant-infinitive-emissions,| 19 |from Volume-Density-Substantial-offset-perceptive-Pressures, (for of an) vague-ambiguation, (not to be) counted through nature;$_{1}$| 20 |(for awhile) tenses' (come in to) awareness by common-effect.$_{1}$¶|||01m.:11s.$_{61m.s.|||}$§ Time not be configured, when blood-pulsed-synapse-being-(s) [refer / affirm], (from a the) linear-accrue (by of) tangible-(s), upon supratangible-existence, producing objective-(s), posed (to an status of) interval-(ization)-(s), from commercial-production, at Federal-Commercial-offset-locations, per Metropolitan-Civilization-(s), depicting how [woman / man]| 21 |agenda-materialize, encompassing form-(s), (for an) nominative-flex-(ive)-reaction-(s), (by size of) minuté-spacing-millisecond-second-blip-sight-depth-perspective-responses', under influence of hour-state, as [Sedentary / Reflexive / Stanced / Motion-(s) / Sat position-(s)] proceeding-imagery of [Density-(ies) / Volume-(s) / Fabricate-(tions) / Dimension-(s)], emanating synapse-blood-pulsed-being-(s), whom adapt an usage of repetitious-daily-pattern-(s), (through an variety of) capabilities, under an affected-outcome-(s), interact-(ing)-confirmation-(s), per objective-(s), through reset-offset-(s) of day-(ily)-effort-(s), tended tensed-placings,| 22 |torqued human-population-count-(s), (from (by of)) manned-placed-being-(s), entailing various account-(s) from hold at [[1]Warehouse-(s) // [2]Construction Site-(s) / [3]Station-(s) // [4]Facility-(ies) // [5]Market-(s) // [6]Repair Shop-(s) // [7]Free Standing Entity-(ies) // [8]Hospital-(s) / [9]Legal Office-(s) / [10]Justice Center-(s) / [11]Supreme court-(s) / [12]Bank-(s), [13]et Cetera...], enable-(ing) individual-(s'), (at an) assignment, character-incremental-continuum-interpretation-(s), flowing by sensational-wind-pressurized-force-(s), interweaving hourly-time-posit-(s), (to be an focus (by of)) factor-(s), that believe in possessed-object-(s) from tangible-clasp, while [reference / retention / memory / place] are unconfigured (by of) elemental-succession, as Individuals'-Perceive (at those facts of) comfort, never "being" (at an) exert per effort-(or)-(s), retaining tentative thought subconscious-imagery-recollection-(s), experiencing "free will" enabling soul, or note-placing-(s), calibrating thought, (from an depiction of) purpose-(s), monitored by bank-account-(s), (due to) focus of Date & Day, in medium of {Time} & [Calendar], (as an) fixation per act-(ion)-(s), that must be constantly scaled, (from by) unbiased-inhabitant-(s'), during those discourse-(s) (by of) perceptive-observational-existing, enabling [man / being-(s)], remaining a-(n) part of (red)-action-(s), significant by movement-(s), (for of the) nature per seconds'-elapsing-(s'), ticking along an actual-rate of process-(es), continued through an revise-(ing)-(s) of reasoning-(s), (from how) (in)-habitant-(s) (citizens of state (not national referendum // policy)), serve oneself, to come up through belief by characterization-(s), (from vigor of a the) previous-generation-(s) subjective-period-decade-(s)-influences, intolerable adjacent-congruent extraspective-social-behavior-(s), imagining aging-being-(s), through lapse-(d) presence-(d)-effort-(s), involving others whom skill implicitly, effectively partaken at rated-wage-act-elapsing-(s), sustaining frequency-(ies), envision-(ing) perceivable-circumstance-discourse-(s), (from of) deviation-(s) offset sun, (as of an) unaccounted-continuum-pendulum-earth-rotation-effect-(s), projecting kinetic-energy, (through an) superfluous-pleather of medium-(s), impacting youth-principle-rules, observing scale-(s) of limit-(s), understand how Individual-being-(s), bias upon existence, to appear in cause of day, an figure by Metropolitan-Planned-design;[2] (awhile an) fallacy-linear-normal-lifestyle-(s), take parental-child-lineage-familiarity, light, (by self) belief, causing awareness in self-act-(s), Scheduling & Date Interval-(s), freely convened administration development-(s), fancily collective by populants', thinking who can implement direction-(s) by schedule, proportional per approximate-location-(s)-input, regard-(ed) in adjustment-(s), order-(ed)-(s) from Vehicle-identification-number-accounting, to [fix / repair / maintain] at areas of squared-off-perimeter-space-(ing-(s)) adjourning's' while presence-cause, signify through possession-(s), upheld constructive-development-(s) of daily-securing, freely-objecting, as those cause-(s) of Individualism, conform voluntary-credit, work-position work-character-(s'), becoming full of option-(ed)-choice-(s), (from a the) domestic-tranquility at peace, in fact by time, for how lapse-(s) hover (in through) attention-span, allotting [purpose // value-(s)], (from the concept of) time.$_{2}$¶04m.:47s.$_{74m.s.|||}$§ "Today", repeat-(s) (at an) familiarity apose time, formed at place, through hourly-paper-identifying, date-(ing-(s)) driven bank-adjacent-adjunct, each passing hour, progressively elapse-(ing), constantly resetting [form-(s) / entity-(ies) / place-(s) / activity-(ies)], (from an)) [tense / synapse / tension-(s)], congruent, complementary-supplementary-efforts', referenced (by an) variation of [objective / subjective / substantial / auxiliary maneuver-(s)] for physique, interpreted by customary-methods, inclusive in step from [tone / pitch / vibration-(s) / wave-(s)], (as an) [thought-process // perplex-(s)] by character-(istic-(s)), respiring cardiovascular-respiratory-blood-pulse-synapse-tension-(s), (e)-affected by perpetual-centripetal-force, acted-through air-pressure forms of pressurized-wind-(s), intermittent reflex-(es), in order at [day / night earth-centripetal-force-rotation-(s) / Earth-orbit-around-the-Sun // Revolution].$_{3}$¶|||05m.:32.$_{09m.s.|||}$§ Breath (as an) annotation [(to)-inhale / (from)-exhale], which proof or emit life, (from a the) human-cardiovascular-organ-ligament-ventricle-skin-substantial-tension-(s), as [liquid-blood-pulse // respiration] of oxygen, friction sensational-perspective-(s), cause-(<) what constantly stabilize-(s) bone-posit-(s), natural per occupant-impasse-causes', [breathing // wind-cardiovascular-timed-limit-(s)], motioning-(s) response-(s), to [practice / repeat / discourse] at logic of form-documenting, living natural-discourse, requisite those multiple-numerative-complex-(es), independently-form-(ed), adjust-(ing) hourly-time-posit-(s), (due to) act-(s), being coherently-competent, required for [land-spacing / cultivating-crop-(s) / grazing-horticultural-mammal-(s) / apiculture-caging / maricultural-industrial-fishing], [[1]Foul / [2]Hen / [3]Chicken / [4]Sow / [5]Hog / [6]Pig / [7]Cow / [8]Bull / [9]Taurus / [10]Horse / [11]Calf / [12]Rooster / [13]Peanut / [14]Peas / [15]Tomato / [16]Potato / [17]Carrot / [18]Yucca / [19]Grain / [20]Rice / [21]Oat (flour) / [22]Celery / [23]Barley & Malt / [24]Molasses from Maple-Tree-(s) / [25]Bee-(s) / [26]Wasp-(s) / [27]Orchid-(s) / [28]Rose-(s) / [29]Tulip-(s) / [30]Carnation-(s) / [31]Gardenia-(s) / [32]Avocado-(s) / [33]Garbanzo-bean-(s) / [34]Red-Kidney-Bean-(s) / [35]Black-Bean-(s) / [36]Lentil-Split-(s) / [37]Vine-(s) / [38]Pepper-Kernel-(s) / [39]Red-Pepper / [40]Yellow-Pepper / [41]Green-Pepper / [42]Onion-Patch-(s) / [43]Garlic-Clove-(s) / [44]Orange-Tree-(s) / [45]Pomegranate-Tree-(s) / [46]Grape-(s) / [47]Cranberry-(ies) / [48]Blue-Berry-(ies) / [49]Black-Berry-(ies) / [50]Cherry-(ies)-Trees / [51]Lemon-Tree-(s) / [52]Lime-tree-(s) / [53]Almond-Tree-(s) / [54]Walnut-Tree-(s) / [55]Pecan-Tree-(s) / [56]Pistachio-(s) / [57]Oysters / [58]Anchovies / [59]Sardines / [60]Octopus / [61]Tuna / [62]Shrimp / [63]Crab / [64]Lobster / [65]Clams / [66]Kipper / [67]Olives / [68]Eggplant / [69]Sunflower (Seed-(s)) / [70]et cetera....], out for prose, (by of) continental-shelf-topography-terrain-offset, amid factified-title-territory-offset, where blood-pulsed-synapse-being-(s) require such piece-(s) of [produce // agrarian]-value-(s), (due to) basic-human-right-(s) for lifestyle, administering whom inhibit [women / men / children], requiring [metropolitan / geological / ecological existent-act-(s)], per day;[3] awhile secondths-continuum, offset count-(s), (have an) definite-limit (per of) Intangible-year-lifetime-ambiguous, labeling each and every crop, by individual-classification-(s), requiring an intervening (by of) International-Federal-government-trade-policy-(ies), (from an) cause [to // from] resource-cultivation, amidst commercial-product-sale-(s) at free-enterprise-sociological-trade-(ing)-post-(s), (where when) farmer-(s), whom primary focus (is on) mass-singular-crop-rotation-(s), draw (on an) call for 94L. lifestyle-(s), in solace of territory-boundary, (from an) registering of Metropolitan-Being-(s) purchase-rates' per year, circulate-cycling season-(s), interval-(ed) where market-shelf-stock-sales occur...¶|||07m.:49s. 06m.s.||| §

§0 ...Count-[47 / Fourty-Seven]-preposition-of... ..Lines-[95.52875 / Ninety-Seven. 1.five -tenths- 2.two -hundredths- 3.eight -thousandths- 4.seven-ten-thousandths-5.five-hundred-thousandths].. ...Article-Essay-[2 / Two]… .. How to schedule what to eat?! ¶|||09s.75m.s.|||§

¶3:(A the) dynamic of Socialism, sees from enabling ecological-effort-(s), through unreferred-bias (of a the) citizen-class, from juvenile-intersexual-affair-(s), residing in approximate-locale, at urban-suburban-property-(ies), by vague-desires', cultivating fertile-grounds', preferred focusing (on a the fact of) fraternizing, (as how) dipitchipated-pulsed-being-(s'), would laze (by of) operation-(s), (by an) preference through velocity-depth-abstraction-(s), ambiguous (by at) character-role-position-(ing-(s)), cycling intermittent, exterior-skin, way of life, (by an) needed-succession, for ego becoming an bolster-piece, amid congenial-matter-(s), assent-acceded, (an) lineage (by of) offspring-being-(s), which conceive through sensational-perspective-(s), sequenced-method-(s), personal-independent-individualism, being left to believe, notion-routine-(s), simultaneous character-family-peer-(s), whom enact self-motivations', in prerogative from achieving-material-success, awhile subconscious-passing-(s), omit vocal-visual-imagery-tonal-volume-pitch-wave-vibration-(s), intermittent social-lifestyle-(s(')).4¶|||58s.28m.s.|||§ To ever draw out effort-(s) from being-(s) soly monetary-wage-hourly-budgeted-check-payment-(s), motivate (an) occupant (by of) habitant-living, cause belief enables soul-(s) of bigotry, which forces inadequate-character-development-(s)-residential-cubic-space-(ing-(s)), per visual-surface-area, formulating, objective-space-storage-title-perspective, yet through possession, at weight, for unit-object-(s), post-day velocity-individuating, secular-parameter-(s), (at of) ones' limited-boundary, [1type / 2kind / 3character-(istic-(s)) / 4piece-(s) / 5label / 6form / 7sort], applicating through the nature of affair-(s), traded-commercial-reliance on| 23 ||[other-(s) // national-constitutional-state-constitutional-extenuation-(s)-{extraspective-physical-wage-act-requirements', governmen-tal}, in fashion from pent up culture-(s), bemused by sedentary-existence, when human-activity, engross natural-parameters, from Biloxi-tradition, felt from steadfast-complacency, from self, visual-referencing, attain-accrue-passing-moment-(s), measuring-through, impasse-environment-entity-variation-(s), [enter / exit-(ing)] entities, at blatant-dismissal from temporality, systemizing, personal-vocal-affair-(s), pursued by [“belief” // “idea”],| 24 |over-than, [“fact” / vocal-vernacular-subjective-conjure-interposing / reactive-interacting-tonal-mutation-(s)], enjoying, sitting (in of a the) excess at place-(ing-(s)), moded when, [appeasement // entertainment], is too little, or too much, not asking, (for an) deed of title-pronoun-responsibility, by [house-(s) // vehicle-(s))], (for how) man title-vin-vehicle-velocity, tempt fate (from of) lineage-destinies, under constitutional-formation-(s)-intrication-(s)-deviation-(s), constructing act-(s), to guide an basic-common-metropolitan-ecological-effort-(s), from visual-vocal-tonal-motion-(s), [force / velocity], through challenge-(s) of geological-metric-(s), (for of) (kilo)-meter-(s) of mile-(s), vague ambiguate journeying, each road-(s), repetitively transiting in trust, circulating ones’ habits’, (at / to), an ease of task-(s), which linger temporarily, evading traded-principle-(s), serving [being-(s) // person-(s) offset-ration-deviation-(s)-interaction-(s)], integer none-reasoning-(s), repetitively simultan-eously-occurred, from auxiliary-transit-device-(s), along an history of plagued-process-(es), typing through nature, an superfluous city-population-(s), as each tenure grow-(s) accustomed from age-(s)-(ing)-work-transgression-(s), childless, when Older & Older, abiding constructive-developmental age-(d)-characteristic-(s), log-value-(s), engrossed by mentality of ambiguity, (due to how) “mind”, is clearly the only way, any being enforce-(s) personal-decision-(s), when default [geological // metropolitan offset-(s)], deviate how personal-effort-observation-(s), maintain an basis per [article / topic / issue-(s)], because practice at mute-physical-individual-self-(ves), factor sexual-affair-(s), relayed at youth sustained-interval-interaction-(s), marking perceptive-genetic-being-(s), whom conjure logic, as tales’ (by of) act-(s), debated under pursuit-by vocal-word-vernacular-hoc-note-(s), defining an United-global-achievements, {Globalization}, at [[1]reason / [2]cause /

3nationalism / 4ethnic-identity / 5resourcing / 6infrastructure / 7parameter-boundary-(ies) / 8forced act-(ion-(s))], mutually-patterning, register-control, necessary for substantial [dry / wet // (meal) value-(s)], (at an push of) Grande-hypothesis, sustaining heard-sound-limit-elapse-counting-(s), hyper-tensing, negative-recidivism (for in those) belly-(ies) from men, proprietary-theatre, retain without an rigor, (of a the) prior-inclination-(s'), for men are dormant through street-air-hall-way-mode-moment-temporal-luke-warm-passing-(s), drifting-off inadvertent destiny, considering false [1taboo / 2mysticism / 3tradition / 4habit / 5homage / 6customs / 7rituals / 8infinitive possessive-belief], accustomed to castigate-prerogative-(s), never admitting [1faults / 2error-(s) / 3contention-(s) / 4issue-(s) / 5dilemma-(s) / 6topic-(s) / 7conundrum...], intermittent-at daily-performing, free-will Social-Requirements' & Placing-(s), where personal-role-(s), liberating-freedom, idiosyncratic common-belief, emit from men 53L. electorally-leading woman-(en), sanctifying oneself, (to those) idle bemusements,| 25 |contest-(ed) from callous-bias-possessive-collective-habit-(s')...¶|||04m.:30s.42m.s.|||§

§0...Count-[22 / Twenty-Two]-preposition-of... ..Lines-[54.261/Fifty-Six.1.two-tenths-2.six-hundredths-3.one-thousandths].. ...Article-Essay-[3 / Three]... ...how does habit affect subjective-vernacular, past-read-interpretation-(s)?¶|||09s.72m.s.|||§

¶4:Why is "legal-judgment" implemented by law, (for the) longest recording of history, known to [basic-human-rights // reason-(ed) existence], backward-(s)-physiological-continuum-documenting-cycling-system-(s), intrapersonal [common bias // hoc affair-(s)], sustemming [1educational / 2personality / 3collective-mass / 4soluble-(s) / 5resource-(s) / 6technique-(s) / 7application-(s) / 8momentary-attention / 9etcetera...], diagonal through government, (as an) socialism-at-mutual-petition-prerogative-will-methods, [conducting // behaving] by natural-voluntary-will, at attendance location-(s), cycling longevity, from inhabitant-(s), [place-(ing-(s)) / objective-(s) / schedule-(s) per registered-entity / person-being-relationship-(s) / valuing crop-(s) per citizen], (amid an) rush through secondths-continuum-mode-count-interval-offset, redacting-(s) article-(s)-essays-period-activity-(ies), (by at) state, holding obligations', (due by) ethical-fate, practicing-familiarity upon spacial-objective-proportioning-effort-(s), impassing metric-distance-(s), by cardiovascular-reason-(ing-(s)), transient-maneuvering placed, (through / by) routine daily, when introvert-(in)-habitant-(s), practice believing [self-retained-trade-debt-purchase-pronoun-objective-possession-(s) // property-(ies)], proceeded per monthly-recreation-cycling-(s), licensed at affair-(s) (by of) mutual-worship, not romanticized, emitting thought at exponent-repopulation-aspiration-desire.$_{5}$¶|||01m.:08s.$_{50m.s.|||}$§ (For how bias alone by) possessive-objective-property-individuation goes, dipitchipational-being-(s), [maintain / obtain tangible unit-(s) usage-(s)], performing biased-act-(s), mirror-reflective, perceptive-subconscious-visual-cerebral-interior-head-(memory), affiliated when, congested-claustrophobic-unilateral-policy-planning-(s), parallel transient-commercial-municipal-order, (for how) transit-velocity, rely residential-metropolitan-city-(ies), fatiguing human-limits', (at an) singular-spacing-location, given intermittent by ambient-enjoyment-motion-(s), proportioning through physical-efforted-action-(s), serving procedural-agenda-command-(s), instructed via dismissive-competency, wrangling inhabitant-being-(s), (from of) [[1]tax / [2]tariff / [3]fee / [4]bill / [5]bond / [6]grant / [7]deed / [8]order / [9]stock / [10]surcharge / [11]ticket / [12]voucher paper-(dollar) // [13]metal-(cents)-note-currency-offset-denominating-elemental-commercial-unit-(s)], demanding called ordered-purchase-cycling-(s), amid common-barbarism, (that stem from a the) mass-offset-origin-(s) of being-(s), whom continually-operate-mutely, yet believing (in an) dark-historical-aging-presence-morale, idealizing youth-breeding, as contemporary-adult-(s), conceive by hospital-offspring, gossiping love, not genre-political-party-Totalitarian-Federalism-tier-decimal-book-shelf-coordinated-choreographizing, in categorical-alliance, with regard to, superlative-ethnic-aboding, adequate sufficed-daily-traded-purchase-survival, an notch above mammal-creatures, yet coherent to apply off-work-time-inhibited-prose, observing those parity-(ies) of [market-objective // substantial-nutrition // literary-interpretations', as monetary-circulation-progression-(s)], generate stagflated motive-mode-development-(s), not alluded to, when, hourly-time-posit-(s) gradually-recede, repetitive-intermittent-metropolitan-dynamic, apose geological-rural-terrain-engineering, grandeur amidst off-time-transit-activity-(ies), limited secular-distance-(s), parallel plane-(s) at visual surface-area-measurement-(s), requiring-subjective-document-(s), to fill or interpose, scheduled-timed-hour-dating-interval-(s), naturally remaining rested, ensuring [{Law} / {Government} / {Politics} factor-(s) (of) commercial-enterprise], mechanize how lifestyle, meander algorhythmn, basic-human-rights-infinitive-belief, by modern-religious-regard, interpretation-(s), for communication-(s) (as the way of) [volume / mass-density-(ies) / air-centripetal-pressure], scaling mass-elemental-substantial-weight-(s), (for which are) balanced by juxtaposition-(s) per mundane-temporal-being-perceptive-physical-registering-(s), upon millibar-scaling-(s);[4] awhile centripetal-force encompass breath-(s), (by of) being-blood-pulse-imagery-impasse-(s), reasoning how measuring-evaluation, increment-locale-limit-(s), mathematical-grammatical-documents-supplementing, (in view of) Land-surface-area-perimeter-distancing, where bias cannot qualify, stationary-objective-posit-(s), adjusting (from an) common-routine-impasse-(s)-method, simply reasoning, or no skill upon detailed-elaborated-truths, conjure elicit-explanation-(s), periodically-simultaneous, [subjective / article-(s) value-(s)], (for when) repetitious-practice, accentuate-articulated-impasse-emphasis, placed through familiar-maneuver-(s), repulsing (while at) muscle-physiological-reflex-(es)-(ed)-hyper-tension-(s)-sight-sound-reverberated-vibrational-kinetic-stagnant-tounge-taste-nose-fragrance-blood-pulse-synapse-(s)-(subconscious-memory), placed at focus, through motion-(s)-timbre, offset-notion-(s), from objective-wage-usage-(s),| 26 |pertaining

[Article:] [January]

[Identification / Key & Lock / bend-of-pan // pot], accumulating live-commercial-living-product-(s), reflexive-vocal-imagery-impassing, apart from [past-motion-(s) // future-motion-(s) / after-life-bone-placing-(s)]?¶04m:29s.42m.s.|||§ Surrounding-surface-area, where alive-movement-(s), forgo act-(ion-(s)), a-laterally-perplex, ordained ascertained-right-(s), during locational-stance-(s), deciding choice-(s), by absurd-abstracted-reflex-(es), in ardor from serviced-duty-(ies), ulterior subjective-letters-word-(s)-syntax, occurring from lifestyles' (by of an) human-being affixed-causes, premising bias-improper-inductive-reasoning-(s),| 27 |amidst sensual-tempo-motion-(s)-timbre, similar as, beef-grain-rice-cream-water-substantial-physical-tensions', interval hourly-time-posit-(s), deviating interior perceptive-reflective-unseen-skull-posit (mind), or blood-pulse-synapses-elapse-(s), remaining simultaneous per day.6¶|||05m.:02s.60m.s.|||§ "Relevant-relativity" or momentum, pronoun-conjure-(presently), from off-time-conceptive-popularity-era, (if ever to be from), work-place-career-fatigue, significantly abounded, by formal visual-physical-impasse-fatigue-investigation-(s), when awhile military-brother-bi-proxy-religious-belief, aptitude an bastardized-national-weapon-affair-(s), confused-confident in arm-(s), demonstrating an coordination from living-stature, at civilization-sufficing elemental-environment-temperate-condition-(s), abbreviated (from an) Largo-Length-cubic-vocal-resonance-subjective-retaining, temporarily-calm, (by of an) way-(s'), enjoying self-physical-idling, [amidst // interval],| 28 |voluntary-extraction-(s) (by at) monetary-task-(s), from involuntary-instinctual-human-behavior, in focus from depth-perceptive-elapse-(s), metering at limited-peripheral-dimension-(s), [adjacently-paralleled // blood-flow calculated per mode-minute-(s)-count-reset-(s)], spacing an balance,|29||per [[1]integer-progress-process-impasse-subjective-interpretations'-limit-extent / [2]recalling requisite-term-(s) // [3]word-noted-observation-(s) /// [3.2]annotation-(s) //// [3.3]conjuring for where location require what objective // [3.4]subjective-timed-interval-directive-(s)], focusing visual-vocal-tonal-physi-

cal-perspective, (through an view at)| 30 |depth-perceptive-visual-im-
agery-bias-determination-offset, intertwining | 31 ||[[1]due-(s) / [2]obliga-
tion-(s) / [3]process-(es) / [4]procedure-(s) / [5]protocol / [6]agenda / [7]schedule
/ [8]talk-(s) / [9]et cetera..], by visual-imagery-vocal-tonal-letter-word-sen-
tence-comma-syntax-mathematical-weights-measures-scales-gram-
matical-punctuation-ontological-etymological-topic-literary-synthe-
sis-proof-(s)-vibrational-synapse-interpretation-(s')-elapse-devia-
tion-reasoning, appraising cubical-square-height-maintained-sur-
face-area-limit-(s), perceived by independent-(hyper)-tense-(s), ob-
long, spherical-circumference-distancing-(s), fathoming stature-(s),
(for how presence upon) integrity, remain reflexive, at moded con-
fluence, in ulterior sedentary-elapse-(s), intermittent continu-
um-elapse-secondths, valuing from self, (where at an) sight-visu-
al-imagery (perceptive-reception), remain [kinetic as / of motions //
at force-(s)],| 32 |when physics-secondthsless-distance-(s), geologi-
cal-generalizing,||||per [hour // thirty-seven-secondth-(s)-view-ten-
mile-maxim-sight hourly-time-posit-(s)], elapse passing hours, en-
dure axis-revolving-matter-(s), not moveable from physiological-lim-
it-(s)-dynamic, physiological-muscular-movement-(s)-motion-(s), (as
how) bias affect circulated non-fiction-reasoning, mass-presumed-ti-
tle-proprietary-interacting, rewarding objective-(s), by man of men,
aged in use, believing in debt, accumulated massively, historical per
extent notion-(s), analogous antagonistic juxtaposition-(s), clearly
chosen, from substantial-free-time-activity-(ies), pertained upon
[traded-waste // water // electric current-cycling-matter-(s)],| 33 |while
superlative-mass-motion-(s), exponent-offset, amid perpetual-cen-
tripetal-force-momentum, by [self / extraspective influence-(s)], not
expressing personal-bias;[5] awhile managing commercial-mathemat-
ical-budget-market-trade-system-(s), center by [fact-(s) // proof-(s) //
truth-(s) // unit-(s)],||||||at concrete-metal-lock, motions' per wage-
work-reflex-(es), labeled upon paper-ink-documented-vin-part-(s),
proved various offset-commercial-sociological-demand-market-(s),
not subservient before ever tabling [count-(s) / lapse-(s) / Grammar

Choice-(s)], when Vocal-Term-Definition-Interpretation-Tempo-Rate-(s), subject-predicate-visual-tonal-volume-pitch-perceive-interpret, designated-parameter-(s), for whom-(s')-tenure, inductive [inclinate // indicate thoughts'], relaying meters-depth-perspective-(s'), valuing (through an) ethnic-default-perceptive-matter-(s)-offset, where documentation-note-(s) by metropolitan-civilization-mean-(s'), juxtaposed| 34 |[book-reference-(s) // commercial-operating-instruction-(s) // political-bill-policy-process-procedural-act-(s)], congressional-referenced-when, [weight-unit-metering, solubles // densities, dimension-(s) / tangible-(s) / parameter-(s) / Luminescent-highway-discourse], affluently simultaneous, age-genetic-generation-(s), (for whom have been) skewing [commercial // residential // congregation-market-trading-factor-(s)], in relative-regard, (by when) [age / self // family calendar-interval-(s)], forgone present-conclusion-(s), predicate nominative-reflexes, per-tense-(s), apposing secondths-tic-deviation-interval-(s), directing how, accumulated-market-unit-(s), askew from Language-hoc-tension-(s), word-note-(s), through tier-incremental-pricing-(s), upon millisecondths-biological-geological-reasoning, when sedentary-place-(s) become-an center for human-prerogative, by [group-city // county-planning effort-(s)], default by road-managing, what surrounding perimeter-area-point-(s), upkeep an interior-fitting-(s)-upgrade-(s), (from what) circumventable| 35 | [residential // commercial city-map-infrastructure-documenting], county-transit-injunctions, (by what) others believe upon counting in rhetoric-vocational-vocal-tone-reverberation-(s), reflexive-hyper-tense-motion-(s), since before youth-post-presence-daily-numerative-denominating, impassing at subjective-interpretation-function-(s), focused on natural-selection-process, identifying finite-value-(s), allude-(ed)-mind, not yet scheduling romantic-opposite-sexual-affair-(s), [validify-(ing) / exchange-(ing)-(s) / purpose-(s) / trade-(ing)-(s) / range-(ing)-(s)], offset where confided community-repopulation-task-(s), withdraw having prominent-interaction-(s), meriting passed [stint-(s) // dating period-(s)],| 36 |becoming adept through work-experience, getting access (to those) product-(s) passed elemental-compositional-extracting-(s), as tangible-physical-presence, comradery, not imbue, accustomed manner-(ism)-(s), basing bias-idle-self-(ves), upon individuation title-surface-area-perimeter-boundary-(ies), (as for what) depict proper document-date-(ing)-(s)-procedure-(s), revealing currency-traded-exchange-note-(s), directly at objective-placing-(s), surrounding constituted government-incorporated-metropolitan-area-variation-(s), context-(ed), sizing cubic-surface-area-reasoning, fixated (from an) duty-lapse-presence-(s), reel-(ed) in part, adjoining-hour-time, with calendar-dating-period-(s) observing possessive timbre work-mode-continuum, registering manual-method-readings, (from by) humanity memo-manuscript-perspective, attention-span reminded, by municipal-nature, piecing millisecondths-offset-matter-dynamic, by formation-(s) (at / through),| 37 |secondths-continuum-interval-aging-juxtaposition-offset, from self-perspective, (where / when) unit-usage-(s), sustem an [prerogative-cause / retainable-cause / application-cause / proprietary-cause], integral periodical-moment-referencing-(s), from [being // inhabitant / mass-physical-basic-vocal-functioning], amongst perimeter-peripheral-tense-(s), paralleling interpretative-focus, sort-order-labeling, various physique-type-(s), moded at area-operation-(s), in place by educated-man, (at an) tactile-response, from county-dynamic-offset-notion-(s), accumulating-independent-calorie-(s), where time-trade-deviation-(s), metre under solar-setting, [water-tubing / industrial-wire-engineering-current transitioning], minutes through hour-(s) per utilities, [input / maintaince / output timbre-1/4hour-time-posit-(s)-instance-(s)-count-(s)], delayed objective recollect-(ed)-hardware-imagery-(ies), intermittent daily-hourly-inset-deductive-set-(s), balancing juxta-position-(s) per [secondths-letter-word-incremental-impasse-synapse-interpretation-rate-(s) /38/ minute-(s)-referring-rate-deductive-interpretation-elapses-deviation], having mentality, craven per [week / month / season], at reference-point-(s), (once an) [year //

season], tier-reference-arrange, how temporal-hyper-tense-(s), vary per auto-biographical-fiction-hoc-existent, reflexing emasculatory-motion-(s), superlative voluntary-impasse-(s), defining quasi-location-entity-objective-chapter-subjective-balance-conjuring, where familiarized self-perspective-conceiving, factor through perceived-locational-sight-depth-wave-reverberation-(s), [kinetic / sedentary], physical-compositional-stature, (as where each) objective-piecing-(s), emote hyper-tense-reflex-(es), conglomerated attention, [where // when], set area-(s), approximately-gauging, [to / from / upon], visual-tonal-hyper-tense-(s), in vocal "de facto", from medium-focus, by tangible-tense-(s), perceiving-population, affirming at tangible-trade-post-(s), qualification-(s) where terrestrial-continental-shelf-exist.[7]¶||||12m.:22s.[63m.s.|||]§ "Can", is restricted by fatigue-reflexes, depth-perspective-conjure-affirming, systematic-parameter-(s), intuitive physical-motion-tense-(s), tangible-surrounding-focus, conducting conscious-interaction-(s), never pieced while [presence // afterlife], (as method of accruing stored-data), differ hourly-rate-time-intervals, affinity-perplex, per [proof / fact / truth / a-(n) / tangent, elapse-objective-data-posit-(s)], rate per planned-city-county-perimeter-surface-area-developments, perimeter-visual-valuing, natural by objective-duration-(s), intermittent parcel-ambient-effect-(s), [[1]hot / [2]warm / [3]lukewarm / [4]tepid / [5]cool / [6]cold / [7]scorching / [8]freezing condition-(s)], (set at) environmental-influence, fortified, yet untamed at municipal-code-offset, for skewing normality-trade-bi-proxy-reliance, upon linear-parameter-(s), currency trading, commercial-note-(s), as ruse by [content / context writing letter-form-(s)], creating objective-usage-complex-effort-(s), when fluent-manuscript-parameter-motion-(s), focus revision-(s), as will 401k-plans, [disclaimer-(s) / term-(s) / condition-(s) / clause-(s)], disseminate ambiguous-stretched-hour-time-elapse-(s), for [[1]fact / [2]reason / [3]virtue / [4]place / [5]dimension-(s) / [6]parameter-(s)], emitted from self-temporal-sensation-(s) (not yet hyper-tense-(s)), from [seeing / hearing / tasting / smelling] (revising touch (as an) sensation);[6] invokable by common [sight / sound cognitive-registering-(s)], hoarding, viced-object-(s), locked-grip-holding-(s), from individual-right-(s),| 39 |(not an part through) physiological-free-will-movement-motivation-(s), inclined extraspective-genetic-blood-pulsed-being-interaction-(s), implementing from existing mechan-ical-ingenuity, [where / when],|||||||||passing-moment-(s), attire aging-historical-prestige, where elapse-(s) form-iterate-observation-(s), "read", not correctly-interpretable, [what / when] prerogative align [person-(s) / being-(s) / activity-(ies) / alliance / negotiation-(s)], affirm past-day, requisite-impasse-compulsion-(s), deviating [vehicle / blood-pulsed-being-(s) / Centralized-housing / (Sub)-Urban-commercial-entity-(ies)], dating each faceted-value-(s), in discourse upon tangible-depth-perspective-elapse-(s), eschew, from content intermittent secondths-tic-continuum, timing style of effort-(s), commerce-denominating, enumerated-crop-count-(s), aligned on property, for [rural / metropolitan consumption-agenda-(s)] apposed (sub)-urban-parcel-planning-(s), [warehouse-(s) // land-(s)], in market-trade-conjunctive-yield-(s), learning from common-error-(s), form-subjunctive-documenting (via receipt), reliant upon farmer produce-collection-(s), (as an) method intricating, referred-routine-recipe-tier-effort-(s), governmentalized by familiar-affair-(s), amongst content-social-frat-askew, autobiographically-defaulted, by physique-complexion-(s), monitored because an simplistically placed-competency, intermittent, term-presence-rate-limit-interval-deviation-(s) (interval sleep), post-refer, an honest joint-(s)-reflexive-boneless-tense-observational-set-visual-imagery-perceptive-compositional-posit-(s)-posture-(s'), non-conventional, through [[1]weekly-article-reading-(s) / [2]grammatical-daily-hour-(s)-time-reset-coordinated-ordering / [3]seasonal-objective-purpose-evaluating / [4]monthly-subjecting // topic // article-(s), recoursing [5]year // [5.25]season /// [5.083]month], date-hour-impasse-intricating, planned [event-(s) // scheduling-(s)], (for an) confluence (by of) inductive-reference-hyper-tense-eye-sights-imagery-ear-sounds-reverberations,| 40 |enticed by annotational-book-cited-redacting, com-

[Article:] [January]

piling day to year-aging-documenting, conjuring human-mass-offset-territory-environment-discourse, (by an purpose for) schedule-reset-(s), planning calendarized-interval-(s)-impasse-(s'), offset at [work // sleep // transit // market-purchase-(s)],| 41 |Method & Collection, concerting closed-infrastructure-corridor-(s) surface-area-(s), from primordial-intimacy, where feminine-interval-participation-act-(s), constantly-circulate, placed-objective-instruction-(s), observing depth-perceptive-measured-parameter-(s), deviating-space-(s), because how self-interior-hyper-tense-(s), qualify constantly, emphasized collection-(s) (per of object-(s), at) [mansion-entity // land-parcel // yacht-vessel // fine-vehicle-(s)], "up to now", not incorporating, rural-terrain (an flaw) offset, range-county-estate-rural-spacing-(s) concerting repopulation-constitution-method-development-(s), cluing physical-reflexive-maneuvering-(s), through each house, (as for whom) derive-commercial-objective-patent-objective-possession, urgent (from those) non-factor-(s), servicing human-need, remaining in adjunct-juxtaposition-(s), apose compositional-data, collected at elemental-substantial-supplementary-Federal-state-offset-requirement-(s),| 42 | inductive-notes-referencing form-document-(s), compartmentalized-scheduled-memory-(ies)-(visual-imagery), at [[1]church / [2]parish / [3]insured-mortuary / [4]funeral / [5]cremation / [6]memorial site-(s)], piecing past-dating, blood-pulsed-physiological-reflexes', as sensational-physical-fluid-direction, remaining ignored present-continuum, genetic-mass-feminine-apose-sexual-reproductive-spacial-offset-(s), male incapable, common-compromise, by physiological-title-territorial-effort-(s), sported off (in an) turn-off per subjective-attraction-(s), forgetting (there is an) non-fiction-parameter [([a]Macmillan Corporation) / [b]McGaw-Hill / [c]Glencoe / [d]Harper-Collins / [e]Harcourt / [f]et cetera...], insighting an opposite-sexual-non-desire, from common-work-place-wage-complacency, affecting repopulated [livelihood // existence], (at an) exponential-community-(ies), quick through subjective-lesson-prose, while governmental-commercial-currency-denominating, accede (through an) infinite-possibility-(ies), disregarded male-commercial-interracial-commercial-game-(s), before serious-commercial-subjective-genetic-religious-diligence, fortifying feminine-repopulation-physique, [dating / deviation-count-proportioning / hourly-time-retrograding],| 43 | remaining amidst, progressive-finite-continuum, apprehending temporality, as intuitive-factoring, at-will-perceiving, whom signify importance through calendar-dating, where present-lifestyle-impasse, present historical-lineage-offspring-existence,| 44 |(from an) anticipation per historical-superlative-free-developmental-excess-resources-developing, not yet retrodating, concerted-planned rural-geological-territory, subject-interval-period-documenting (metropolitan county-(ies))-article-lesson-tier-ordering-(s), apose natural-lazy-free-idle-sensation-(s), enticed by male-victory-entertainment, (by of) gay-game-(s), deliberately-incompetent, from male [race / ethnic] idle-interactive-common-preference, when male-superiority, infringe, white-(pink)-feminine-repopulating, title-housing-city-(county)-filed-extension-(s), (as when) women abide to such chide, (by an) sweatless-nature-of-human-(s).[8]¶|||18m.:05s.[20m.s.|||]§ Whom be fit to lead, not fully coherently conceive how hypothetical-future-instance-(s), proglamate past-historical-established-plan-(s), through pre-requisite-interaction-(s), fully incapable by alternative-governance-(s), apose A The United State-(s) of America-trade-methods', (for I continually-inquiry otherwise), not collecting those ancestry-(ies) bones, from [[1]Nepal // Nagpur[1] / [2]Mongolia // Manchuria[2] / [3]Israel // Saudi Arabia[3] / [4]Jordan // Kazakhstan[4] / [5]England // Northern-Historical-Colonies[5] [6]Spain // Mexíco[6] [7]Greece // Germany[7] / [8]Egypt / Congo[8]], elapsing (by an) historical-promenade, secularly-shrouding, physical-spacing-(s), culminating in America {U.S.A.} at social-security-birth-reset-limit-(s), alluding humanity-dynamic, being religious by exterior-housing-exerted-effort-(s), dividing boundary-spacing-(s), (as in the) Age of Discovery, from fundamental-monetary-trade-nation-(s), western-front,| 45 |[Netherlands-Western-European-North America) / Spain-(Western-European-Latin America) / Morocco-(Native-Americans / Mexicans / Slaves)], before the China-Knowledge & Indian-Gene-pool-(s)-split-Assyrian-Kabul-River, through knowledge from calendar history, upon hourly-time-reference-(s), visualizing [data // memory // imagery], adjunct sea-board-human-developmental-economies, constructed, superseded-governmental-boundary-block-parcel-limited-development-(s), [where / when] set spaced [residing / driving / commercial marketing], [to / through / from], mode-effort-(s)-impasse-time-posit-(s), presiding-at present-secondths-rating, (for an) [attempt / practice / repeat / behavior], affecting [Physical-Perceptive-Being-(s) & Objective-Auxiliary-Warehouse // Storage-Collection-(s)], particulate, when person-national-reference-belief-(s'), emphasize secular independent-present-aging-origin-power, burdening land-being-(s), modified by personal-mass-extent-mathematical-engineering-conceptive-(s)-generating, relevant throughout routine-schedule-(s), understanding why populational-mass, exceed over an [thousandths / millionths, city // county // state transit-population-labor-wage-circulating], creating constitutional-measure-(s), prior free-enumerated-view, relative-superlative, general-mass-conforming, at psychological-inquisition-(s), yet never accumulated by political-matter-(s), per [topic // tangent of bias], asserting cellular-possessive type-from, "typewriter", accounting letter-characteristic-(s), bended-trade-reliance, in plain view of natural-selection, where auxiliary-device-(s), (can not be needed by) mass-common-human-being-(s), but when want (by of) desire, rather than entity-title-development-(s), proceed ([1]by [2]what [3]has [4]been [5]from [6]past) dated governmental-patent-commercial-conceptive-general-population-mechanical-trading, investment-stock-funded, patent-product-common-investment-(s), vocally-accounted by [[1]race / [2]Ethnicity / [3]county / [4]city / [5]church / [6]market / [7]store / [8]station / [9]individualism], understanding Modern-Being-(s) & Contemporary-Political-Auto-Biography-(ies'), signify who validify, historical-visualized-lifestyle-(s), with regard where broad-account-(s), become billed by metropolitan-being-(s)-bank-account-(s), responsible by claustrophobic-possessive-area, mental-cognitive-interpreting, where disorder of mind, affect physical-perceptive-presence-being-(s'), (for where) tangent becomes "too much", from self-idol-behavior-(s'), (more than others are not) challenged at individualism-bias-dipitchipation, as254L. Age & Will-Transgression-(s), settle-by voluntary-free-will-impasse-efforts... ...continuously revolving in centripetal force rotation-latitude perpendicular of tectonic-plates-three-dimensional-axis...¶|||21m.:13s.17m.s.|||§

§0...Count-[27 / Twenty-Seven]-preposition-of...||||||||.Lines-[254.573 / Two-Hundred-Fifty-Four-whole.1fivetenths-2sevenhundredths-3threethousandths].. ..Article-Essay-[4 / Four]... ..Four-Sixty-Year-Generation-Cycle-Sequences, since year Seventeen-Seventy-Six-{1776}, continuum deviation 0.0166-0.0352.-matter-rate-documenting.¶|||15s.62m.s.|||§

¶5:[How / When / Who / Where]!, are those first being-(s) at technological-development-(s), (from an) configured representation, by live-tense-(s), scaled [physique / dry // wet substantial-(s)] in [gram-(s) // ounce-(s) /// pound-(s) American-measure-increment-(s)],| 46 |(where when) instrument-objection (by of) act-(s), not configure Schedule & Square-Feet-Surface-Area, placing belief, (in an) conjunct physical-mentality,| 47 |as individualized-self-obligation-(s), task constitutional-geological-population-boundary-(ies), (as where) objective-auxiliary-usage direct-post-posit-(s), (in an) conscious-retardation-offset, that exist reluctant by commercial-education (commonly-preferred [youth-high-school-diploma / university associates or Bachelors'-diploma]) (with what) context innuendo, induced by kin, or iteration, direct subjective-lesson-(s)-interpretation-(s), contexting (what has been for what thing-(s)), one will require, timely amongst maximum-elapse-hourly-time-posit-(s), as four-minutes-earth-centripetal-force, unconscious perpetual tectonic-plate-continental-shelf-horizon-degree-(s), every twenty-four-hour-(s), per [sixteen: eight-hour-interval-(s)-(work // sleep-mode)], elapse by count-population-status, for One Hundred and Ninety-Eight-hour-(s), apose Eighty-hour-(s) (population-association), while substantial-packaged-product-(s), accumulate per week, by conscious-hour-presence-reset-(s), posture at space-(s) [sat // stanced act-(s)], (by an) One-hundred & Twenty-hour-(s)-work-week-denominative-posit, apose Forty-Eight-Hour-(s)-weekend-denominative-posit, formulating an gauge by work-cycle-perspective-fatigue, enumerating-inducting upon temporal-designated-location-(s), as sleep-junction-(s), factually-interpret, year to month-reset-(s), focused on direct-syntax-referencing, at [[1]redact // [2]review // [3]revise // [4]reiterate // [5]reedit // [6]increment // [7]inclinate // [8]indicate // [9]intercede // [10]Etymological-ontological-proof-reasoning // [11]minor-highlight-methodical-work-mode-(s)-exponent-act-offset], elapsing year, intercorrelating, considered visual-imagery-variable-factor-(s), [(to) be / (that // this) it / (been) is / (being) as / (at an-location // placed) at], when dating remain posited, [as / is, tense-relationship], (present / past), with no future anticipated documenting-means, as resolute-receipt, via transaction-(s), requisite presence-present, (for how past) objective-accumulations, make (how we would) think, referencing from book-(s), those commercial-objective-purchase-(s), requiring human-being-(s), to configure Industrial-commercial-objective-trade-variations, because national-tectonic-plate-mass-default, (at an use from) past-date-present-time-systematic-processing-(s), affix periodical-interval-(s), by "aging-year-age-offset";[7|] 48 |ignoring age-tier-offset-development-deviation-(s), juxtaposed deciphered conceptive-display-(s), at modern-age-presence, interval four-minute-(s)-degree-rotation-revolution-deviation-(s), denominative-offset-inductive, two-hundred-forty-secondths-elapse-reset-continuum-(presence)-count, moded (to // from) variating reference-imagery-motion-(s)-(a)-effects', climate cerebral, twenty-four-hour-day-interval-(s), calendar-dating, {[0—100-years{[8.]3,155,760,000sths. / [7.]52,596,000m. // [6.]876,600h. /// [5.]36,525d. //// [4.]5,217.85w. ///// [3.]1,200m. ////// [2.]400sea. ///// [1.]100y.(including leap-year-deviation)] one-hundred-percent-median-offset-count-deviation, time-denominating // dating-numerating-denominating // month-denominating /// seasonal-denominating //// yearly-denominating, time hour-integer-posit-(s) // minutes-integer-mode-gamma-interval-motion-sequence-(s) |/4/9/| secondths-continuum-count-offset-denominating, input calendar-dating-notions, into time-dating-continuum], secondths-micro-interpretation-matter-offset-observing, contraposed variable-count-deductive-reasoning, extemporaneous, Physical-Limit & Act-(s), upon hourly-time-posit-(s)-counts, (by of) Centripetal-force-drift-rotational-revolution-posit-perceiving, observant through exterior-limit-existence, as [minute-(s) // subjective-deductive-impassed-notion-(s)],42L.| 50 |continuously revolving, to proceed an continuum-notions-offset-dynamic... ¶|||03m.:49s.15m.s.|||§

§0...Count-[2 / Two]-preposition-of…| 51 |..Lines-[42.177 / Fourty-Two.1.one tenths-2.seven. hundredths-3.seven thousandths]..| 52 |..Article-Essay-[5 / Five]..| 53 |..How do City-Mayor-(s') plan the commercial-community-operations?.. ..How can one, schedule geological-psychological-sociological-ecological-architectural-economical-commercial-repopulational-agenda, (by our) present commercial-state, of literary-affairs; upon 1-100-years, Time & Calendar-Dating-Scaling?! ..From perceptive-hyper-tenses'-{4}, upon exterior tectonic exponent-miles-distances-impact, through [measuring / weighing / gauging], [solid-densities /// liquid (subtract (-)) from dry-density-vials // bottle // beaker // et cetera…measuring)] / [dry-substances /// atmosphere-air-vapor-miles-gauging]… ...requiring an survey of humanity-subjective-competency… …letter-(s)-word-(s)-semi-colon-(present common period)-paragraph-sentence-(s)-period-essay-(s)-article-(s)… ...for later essay-(s)-article-(s)-communicated-reordering, (for how) House-Congressional-Legislational-Politics, should have always been handled as Federalist-Totalitarian-Party-Rule, under John Quincy Adams, since 1800.¶|||57s.91m.s.|||§

[Article:] [January]

¶6:(As for when those) mass-(es) offset humanity, time function interval calendar-year tabling, yet proceeding visual-imagery, (yet to be by) blood-pulsed-synapse-being-(s), comprehending box-unit-nominative-value-(s), (without an) mass-extraspective-composite-elemental-collection-influence, producing-data, differentiating systemized-dynamic-(s) of commercialism, creating an persuasion by interaction-(s), reasoned-from western-central-state-nation-state-continental-measure {Texas // Montana per-twenty-million-initial-national-reproduction-(s)-(minimum)}, scripting [context // constitutional population-industrial-development-(s)], delving physical-compositional-form-(s), by natural-offset-(s), surveying perimeter-spacing-(s), fixate-focus on mutual-public-metropolitan-congested-area-(s), apose soly personal-family-lineage-housing-ten-acre-range-extension-(s), pertaining-literary-prose;8 awhile residence-(s) (from an) American-white-male-dominated-mariner-majority-waste-body-civilization-Treasury-populating-era-(s), thus far barren-deviate, approximate-designated-metropolitan-city-title-surface-area-parameter-(s), throughout continental-shelf-infrastructure-(s), factoring from population-census, dimensional-constitutional-public-construction-(s), factoring inhabitant-labor-effort-(s), exponentially-radicalized by tectonic-plate-elemental-extraction-(s), proportionally-rationalized apose, Tenure-Hierarchical-managing,|54|variating-entity-objective-cycling-agenda-monetary-circulation-hourly-wage-task-(s), budgeting Electricity & Water, positing-physical-motion-act-(s), numerative objective-use-(s), sporadic tentative-act-(s), enable-ing-peer-(s), false, greater-than family, methodically paid, when engineering metropolitan-(sub)-urban-geology, publicly effecting housing, where ordered motion-(s) impact general-locational-impassing-(s'), supposed (to be) [inquired // evoked natural-ambient-purpose-(s)], where county-city-limited-district-(s)-placings, should request factory-ordered-cycling-(s), in vicinity from population-property-objective-substantial-limit-requirement-(s), mass-proportioning secondths-objective-unit-notion-elapses,| 55 |inductive-denominating community-deviating-offset-task-(s);[9] awhile mass-interaction-(s)-in-set-formation-(s), counting objective-juxtaposition-(s), waged (as for of what) [window-pane / brick / cement / pavement-slab / barrier / et cetera… count-(s)] (by an) tempo-timbre-blood-pulse-breath-respiration-(s)-synapses'-metronome-timbre-secondths-degree-minute-cent-influctuation-(s)-offset-align, discoursing human-being-(s), unfolded by differences from work-routine-pattern-(s), pace-timbre, metropolitan-spacing-(s), depth-perspective-physical-motioning-timbre physique, apose, relevant-proportional-city-planning-federal-book-ration-rate-deviation-(s), rationalizing, various secondths-elapse-notion-(s), refer-encing calendar-reasoning, not from theoretical-intangible-(s), (but when) [time-motion-(s)-elapse // empirical-subjective-denominative-topic-verifying] velocity-influctuate-impasses', through centripetal-force-tectonic-plate-continental-shelf-drift-observation-(s), incurred-awhile [day // night temperate-effect-(s)],|56|correlate breath-respiration-blood-pulse-friction-tension-count-(s), syncopating simultaneous-fashion, denominative-(long-term) apose numerative-(short-term)-simultaneous-act-(s) syncopated linear-fact-(s), along scheduled-basis, from ante-meridian-waking-cycle-perspective-(s), progressively-exerting concerted-force crescent-interval-intermittent prime-meridian-passing-(s)-(American-Hemisphere);[10] (as when) [tense-of / act-of / place-of], sit-in orator-manner, fatiguing from factoring, bias-possessive-grip, rather converse-(ive) (auxiliary) objective-article-(s), debating [what // why], each solid-compilated-product-patent-tangible-(s), sectionalize (at an) location, for calendar-scheduling, from [offset-commercial-market-trading // task-(s)], patent-pending-(clause) in part [who // when // where population-communal-extraspection], affect market-pricing-(s), per product-(s), (not purchased communally), settling-circulation-inquiry-(ies), often offsetting economic-commercial-price-(s), per market stimulus-cycling;[11] trusted (from when) work-progressive-payment-(s), encourage false-congressional-enticement, unconscious, (for no human ever) formulate non-fiction-population-observation-literature, prior commercial-dime-out-excess-purchasing-(s), intermittent each intangible-population-offset-continuum-observation-(s), metric-distance affecting [manner-(s) // matter-(s)], when personal-possessive-commercial-collection-(s), possessively believe Mass-Population-Deductive-Object-(s) & Individual-Personal-Partial-Bias-Self-Habitant-Article-Usage-(s), reviewed where one-(s')-personal-vision-posit, apose tectonic-plate-topographic-climate-variation-(s), upon continental-contour-shelf, populate-persuading, national-repopulating-subjective-existence, studying (learning), subjective-periods-numerative-impasses', contraposed numerative-factory-box-market-shelf-unit-piece-(s), currency-I.R.S.-denominative-yearly-dating, psychologically-pertained objective-subjective-motion-(s)-capacity-inquiry-(ies), elapsed-extent-(s), from scaling [human-conduct // behavior],| 57 |elapsing in [time-secondths-continuum-juxtaposition-numerative-count-(s), per [1]sleeping /////// [2]eating /////// [3]transiting /////// [4]reading /////// [5]writing /////// [6]working /////// [7]redacting /////// [8]conversing // [8.1]dialoguing // [8.2]talking // [8.3]vocalizing /////// [9]debating /////// [10]architectural-planning-(seasonal-hourly-schedule-posit-(s)) /////// [11]constituting-(monthly-hourly-schedule-posit-(s)) /////// [12]psychological-counting-(weekly-hours-schedule-posit-(s)) /////// [13]theatre / [14]philharmonic / [15]ensemble-performance-(s) / [16]festival-(s) / [17](literary)-soirée (not as like party-(ies) more book-conversive-(monthly) // [17.1]seasonal-(ly) /// [17.2]yearly-denominative-referencing)];12 gauging how conducted-operation-(s), serve an major-denominative-location-purpose-(s'), communally-numerative-subterior, operational-hour-(s), scheduling apose yearly-impasse-denominating, where religious-communion-mundane-examin-ation-(s), (our theological-community off-time) juxtapose minuté [transit // 62L.commercial-work-place // sleep / eating], (as the) judgement-(s) /or/ legislating of human-objective-market-executive-excess-usage...¶|||05m.:37s.16m.s.|||§

§0...Count-[6 / Six]-preposition-of... ..Lines-[63.37475 / Sixty-Three-Whole.1.three-tenths-2.seven-hundredths-3.four-thousandths-4.sev-

$_{en}$-ten-thousandths-$_{5.five}$-hundred-thousandths].....Article-Essay-[6 / Six].. ..How many motions' can humanity-offset, concerting effort-(s), pertaining [[1]eating / [2]sleeping // [3]clothing /// [4]family-converse //// [5]reading ///// [6]transit ////// [7]work-operating ///// [8]purchase //// [9]communal-converse /// [10]attendance // [11]participate / [12]plan-future-present-amicable-day-interaction-(s), (for self-lineage-offspring-historical-documenting, amongst-an state-territory-maxim-genetic-population-appreciation]?¶|||25s. $_{81m.s.|||}$§

$_{¶7:}$Conjuring visual-direct-peripheral-perceptive-fact-(s), phenomenally intercede physical-blood-pulse-synapse-being-(s), visually-vocal-tonal-perceiving, fixated non-sequitur, non sequenced, non-fiction-reference-ordering-(s), apose, generational-lineage-reset-repopulation-focus-orchestration-(s), periodicalizing monetary-elapse-abstraction-(s), configuring age, listless from determining act-(s), reproduced to extenuate mode-(s) at placed-objective-living, servicing an offspring-imagery-objective-temporal-subconscious-recollecting, after-self-life, upon presence in live-lineage-transgression-imagery-physical-perspective-(memory), note-(written-document-ing) noted-(circulated-currency)-notion-(s), (along an) simultaneous-deviation-continuum-secondths-tic-estimate-documenting.$_{9}$¶|||39s.$_{34m.s.|||}$§ Harboring participation, cordially characterize-(s), what men-stand-for, as blood-pulsed-synapse-individual-perceptive-being-(s), elapse cycling-(s), every [Four-Minute-(s) // Two-Hundred & Forty-Secondths, (celestial-centripetal-force-time-cartography-degree-latitude-360°-tectonic-light-water-vapor-kinetic-friction-impact)], posed for value-verifying, mere-maintainable commercial-patent-unit-productions, (from discourse at of) road, training population-thought, brainwashing effort-(s), (along an) centralized-secondths-intangible-observation-(s)-denominating-natural-constant-(s)-posit-(s), while year, sustain Jeffersonian-Gregorian-Calendar-Year, as account-(s) (by of) being-(s), regulate religious-enabling, without [cultivation // constitution], founding title-grounds, from written-discourse, affirming reasoning-(s), when blood-pulsed-synapse-being-(s), perceive intangible gauge-evaluation-(s), efforting-act-(s), intangibly discoursed, documenting, where being-(s) should participate in part of consensus, per year, amid decade, (every ten day-(s)), flux accompanying, [[1]food-produce / [2]clothing / [3]shelter / [4]bowel-discourse / [5]age-mutation / [6]skin-tense-(s) / [7]perimeter-surface-area-deviation-(s)],| 58 |in self-extraspective-belief-commercial-market-shelf-accumulation-(s), apose bias-extraspective-interaction-(s), (for how vehicle-(s) have been) intended to affect-daily-impasse-(s), above terrain, perimeter-lot-spaces, affecting commercial-objective-outcome-(s), where accustoming self, at wage-company-policy-instructed-pay-directive-(s), provide family-congenial-living, accruing cultural-visual-interceding-(s), while men of man, (be in cue) (by an) default-existence from state-legislation, clasping post-Jefferson-Era-default-national-constitutional-influence, persuaded constitutional-denominating, no further than extenuated [secular-county-(ies) // city-(ies)],| 59 |because managerial-appraisal, not volunteer literary-lifestyle-documenting, cycled-by congressional-population-documenting, as nature shift momentum-grandiose-impasse-(s), simultaneous visual-imagery-(ies), at tangible-clasp-focus-(es), table-count-coordinated, [body-(ies) // entity-(ies)], executed by senate-congressional-bill-policy-referendum, whom share an analytical-mutual-discourse, analyzing commercial-aging, by Military-retirement-62-maxim-notion, scheduling [agenda / sociological-ecological-schedules];[13] serving avid-communion, by inhabitant-(s), requiring loyal-allegiance through mass-age-deviation-(s), reproducing industrial-populational-developmental-engineered-objective-constructive-effort-(s), foreseeing domestic-unit-product-shelf-deviation-value-(s), present-prior planning, through counting caucuses, reasoning-hyper-tense-(s)-subjective-tier-time-hour-dating-sequential-ordering,| 60 |common by jointed-reaction-(s)-reflex-(es), Scholarly due to [Associates /[2]/ Bachelors' /[4]/ Graduate /[5] // [6]/ Master-(s) /[8]/ Philosophical Degree-(s) ([10]years at an singular-institution)], denominating-mass-observation-(s) upon era per relevant-state.[10]¶|||03m.:00s.[84m.s.|||]§ Accounting individual-honor-(s'), from [commercial-work-hourly-(numerative) // salary-(denominative)-wage-payment-method-(s)], associate-(s')-effort, trivial natural self-reliance, affecting human-disposition, periodically-continued to interposed blood-pulse-synapse-air-respiration-(s)-motion-(s)-reflex-(s)-action-(s)-pace-tempo-timbre, affluent-interacting-(s), under Municipal-Code-Rule-Intuitive-guidance, piece eschewing, pre-intangible-interpretation-(s), con-certing repetitive-interval, age-relevant-tier-capability-(ies)-offset, sectoring community-effort-(s), (1in 2of 3an 4form 5of) transition (by6) Genetic-State if ever considered, requiring conclusive-content, retaining at naïve-allegiance-method, conferring debate-method-function, [1tooled / 2objective / 3Scheduling / 4Subjective-lesson-Topic-(s)-Commercial-Article-(s)-essay-ordering / 5Physique / 6Activity-(ies) / 7Hourly-visual-imagery-upkeeping], apose manager-empirical-reset-offset-(s), reinforced by Common-Literacy-Boundary-(ies)-Limit-(s), inconclusive, when interpreting-context, scheduled at mathematical-faux-paus-errors, never morally-introvert, amid free-(sub)-conscious-thought, where-when-whom observe objective-value-(s), (at an) State of mind, basing 51L. reliance on mass-developmental-population-function-(s)-counting-(s), (from those requisite-(s), by)| 61 |blood-pulse-synapse-work-wage-terms-existence-usage-count...¶|||04m.:20s.76m.s.|||§266.84%21=12.70667-syntax-deviation-rate||Two-hundred-sixty-six-decimal-1.eighttenths-2.fourhundredths-divide-twenty-one-equivalent-twelve-decimal.sixtenths-one-hundredths-ninethousandths-syntax-deviation-rate||

§0...Count-[7 / Seven]-preposition-of...| 62 |..Lines-[54.4725 / Fifty–three.1.four-tenths-2.seven-hundredths-3.two-thousandths-4.five-hundredths].. ...Article-Essay-[7 / Seven]... ..University- Graduation-deviation. [Celestial-solar // Earth-Cartography-latitude-degree-(numerative-count), Four-minutes-time-denominating each rotating-revolutionary-revolving-degree-(s)].¶|||15s.57m.s.|||§

¶8:I persistently (have an) problem with commencing my books'-predicate initial-discourse, for obtained-focus, [(un)-bias /

hoc-(ed) non-fiction-literature], (for I) constantly consider, visual-vo-
cal-fact-(s), attempting proof, by personal-observation-(s), (as how)
non-fiction, (is under an) national-basis, from opinionated-excerpt-(s),
when educational-degree-(s), fulfill assertive-obligation-(s), superla-
tive extent-contortion-(s), devising methods-of-trade, emphasizing
juxta-position-(s) (by of) live-matter-(s), from metropolitan-com-
mon-offset-repetitive-mass-population-national-motion-influctua-
tion-(s), continually corps circumstantial deductive-referencing, sort-
ing associative-consumeristic-(ism)-causes', not incorporated, at
offspring-dating-(s), (for how) [auto-biography // biography]
genres-personality-dynamic per-hourly-time-elapse-(s)-posit, de-
mon-strate individualistic-effect-(s), partaking sociological-tran-
sit-idle-enjoyment, (from how) reclusive-individualism, believe vehi-
cles are required, when government not direct (why an) vehicle should
never (have been) bailed-out, because when each aspect (by of)
perspective, tier (by an) independent-individual-(s) drive (purport-
ed-directive-motivation), observing (how without) finite-document-
ing, (what can any) [citizen // inhabitant] expect from hourly-time-im-
passe-(s), (without an) calendar-dating-subjective-continuum-verify-
ing, (for how we) recollect data, by hours per book-article-(s), with
sleep intermittent, suggesting (from the) [Person // Human dynamic],
(for how) documenting (is like if) Global-population-7,036,259,814-pop-
ulation-count, (have an) total-lifespan-average-living, of 80-years per
individual, suggestion from [estimate /apose/ deviation] age-
114.0876712328-(32 days)-denominating-millionth-hour, depicting
those stages of aging perspective, perceive each date [9,566-days /
hours-notion-holiday-objective-resets], emphasizing an deviation of
political-systems, functioning from [[1]Nation // [2]State-(s)-Constitution-
(s)-(for only America has state-constitution-(s)) /// [3]County ////
[4]City-governments, requiring documenting (to be as) [1.1]National-En-
cyclopedia // [2.1]State-Dictionary /// [3.1]County-Book-Volume-(s) ////
[4.1]City-Book-Volume-(s) / (Independent)-Sociological-Population
-Compilation-Book-(s)], (where the) superfluous-billionth-word, (can

only) offset, by human-perspective-word-competency-maxim, to proceed (at an) deviation of national-offset-political-dynamic-population-purposes, 473,040,000-secondths-individual-words-fifteen-year-consecutive-count, continuing (to be) spaced through Dating & Timing, (for how) common-political-candidate-perceptive, falls well short of this mark, individualistically sleeping proportional (for an) international-watch, mimic-superlative, (as any) American-Industrial-Patent-(pending)-Development, probably depict (why our) lord shipped American-ingenuity, overseas (china), outsourcing component-elemental-resources-balancing, not constitutionally-segregating-formally ethnic-bible-literary-subjective-physiological-speicies, when congressional-committee-(s), over-budget, irresponsible common-population-educational-deviation-(s) per monetary-circulated-bank-currency-(ies), when literary-Valedictorian-processions not characterize, those Problems & Issues, concerting human-contact, (for how they prefer to be driving and reclusive) among common-extraspective-denominative-referencing, having how qualified those upper [15% / 10% / 5%, category-graduate-(s)-competency-(ies)], ratio-rate introverted-construing, similar (to all) common-inhabitant-(s)-(bottom 85%) impasse-quality-(ies), where mass-offset-perspective-(s), affirm non bias-interpretation-(s), when action-(s) simultaneous| 63 |[written-read-redacted-subjective / article-(s)-inquiries], served upon physiological-constraint,| 64 |an sedentary-livelihood-physical-population-documenting-dynamic, subversive geological-metropolitan-rural-terrain-(s).11¶|||03m.:26s.80m.s.|||§ As (for // to), whom it would concern, whoever is responsible by act-(s), would than table secondths-elapse-(s)-counting-motion-(s),| 65 |verifying [constitutional-planned-surface-area-dimensional-spacing-(s) /// patent-product-conceptive-objective-(s) /// subjective-empirical deductive-reference-interactive-reasoning-(s)],| 66 |secondths-mode-motions-count-supplementing-schedule-location-timed-act-(s)-theatre, from physiological-fatigue-limit-(s), (from by those) capable-competency-(ies), not yet depicting conventional-Non-Fiction-genre, as Totalitarian-Federalism-genre,| 67 |preserve participle-preposition-presence [in /// out / of / for / from // to], what bring-forth, [to // from], an momentum (by of) shared-discourse-(s), local-in-motion-(s), reconstruing biography, (from an) state of primary-historical-fiction, [tempo-pressure-vibration-(s) // liquid-bounced-blood-pulsed-resonance // air-intake-vibration-(s) influctuation-press], perceiving hourly-time-interval-(s), during calendar-dating-period-(s)-offset, alluding how repetition, build an more intelligent-individual-(s), when living amid concerted-existence, as who-when-where-what, place under consideration, interval [interior / exterior trade-value-(s)], for commercial-product-material-(s), impassing [action-(s) // motion-(s)] from responsive-focus mundane-reporting, following-an daily-reprieve, (as for whom) quickly-assert, visual-tangible-notion-(s)-observation-(s'), considering an average-mean-(s), to rely dependent-self, by inquiring [time-period-(s) / time-frame-(s)], momentum subterior attuned, forming annotation-(s), per [metal-wire-stripping / wire-length-seventeen-inch-(es)-layered-serie-(s) / kilowatt-(s)-frequency-107,000-gama-up-station-deviation // Hertz-frequency-105,000-gama-down-station-deviation], relating mass-production-default-synchronizing, exampling meticulated dynamic-(s), visualizing [effort-(s)-(input) // effect-(s)-(output)], upon unit-objective-possession-(s), gamma-adjunct, perceptive-parameter-(s), iterating intervoluntary-effort-(s), as kudos-bias, retaining key-possession-(s), centralized where, city-boundary-(ies) limit physical-creativity, remain probable because, rate-less-inhabitant-subjective-physical-offset-input, show how book-shelf-(ves), remain [dormant / vacant], time-tempo-beat-count-(s), without gauge for calendar-history, continuum intercorrelating at-will, an façade-wall-foyer-discourse, integral at defaulted-value-(s), placed-upon [boundary-(ies) // limit-(s)], rectangular upon land-proprietary-jostle, basing parcel-territory-land-block-(s), by personal-ethical-(sub)-urban-lordship-adjustment-(")equivalency("), adjunct-parallel, [Urban / Suburban / Rural County-government-spacing-(s)], where [road / house planning-(s)], interpret property remark-(s), [self-bias-work-effort-(s) / bank relationship-(s)], (by an belief in) free-mass-market-shelf-unit-volume-(s), unweighted per, from sensational-thought-(s), rather thinking-visual-imagery-organ-tissue-(s) emitting compositional-configuring, from ventricle-juxtaposed-fitted (like bone // river-bed), acceding synapse-syntax-observation-(s),| 68 |temporal intangible-perceived ordered-documented-pre-concepted-paper-article-referencing-(s), remaining-centralized in city-county-(tribe)-metropolitan-nation-state-(s)-trade-influence-(s), determining competency-(ies) in error by [post-retirement // youth-energetic-physical-activities], sustaining those unplanned revolving-year-(s)-aging-(s'), allocating commercial-religious-genetic-factory-discourse, (through an) post-revolutionary-genetic-call to arm-(s), yet assembled through an militia-entrenchment, (due to those) population-vernacular-vocal-confirmation-motions', their of theatre.12¶|||06m.:31s.61m.s.|||§ No individual has summon an mass, for ordering commercial-industrial-agenda-schedule-material-(s), dictating confided-perspective-(s), from self, amongst subterior-(county // city—(districts))-(80-meters by 35-meters-blocks))-population-(s), characterizing-space-(s), due in part, (none) note-(s)-inquiry, per (quire // ream), (for how) technological-commercial-engineering, sustem past philosophies, (detailed-analytical-theorizing) amid sociological-governmental-commercial-geological-ecological-mass-population-system-(s), efforting when placing-(s), amid faint-sleep, perceive imagery-register-focus, affirming-lifestyle, by will, [defaulted-to // by-government-(s)] ecologically-globally-offset, when elapsed-precedence, (not having an) origin-belief-(s), from ethnic-conglomeration-(s), rather suggesting etymological-ontological-proof-documenting, (yet) direct how using object-(s), at regular-behavior, ignore (a the) pattern-(s) (by of) language-(s), confirming an Estezzuavarian-existence, (as from how) mechanics-discourse, (an error of human-resourcing upon genetic-identification), amount superego, balancing Past-Fatherly-Subconscious, featured-fact-(s), contexed at influence, delving personality, only intermittent-mentality, from humanity relying person-instructive-enabling, functioning literate-directive-(s), complacently-relied on, (by an) false-proportional-constitutional-metre-reasoning, nationally discoursed, | 69 |evident common-school-educational-grade-point-average-hedonism-sociological-hoc-enabling, featuring shelf-life-cycled-fact-(s), callous by context, continuously dating various-(s)-complex-(s), shorted by [belief / idea], rather grandiose-ambivalent-discourse, unplanned (by children though) adult-aging-succession-(s), juxtaposing ontological-grammatical-punctuational-mathematical-algebraic-chemistraic-geometric-biological-trigonomic-anatomic-calculus-pathological-psychological-phyiological-philosophical-geometrical-physics-etymology, reciprocal [numerative // denominative variable-datum] (through an pleather of medium-(s)), [[1]picture / [2]film / [3]book-context / [4]textbook / [5]manual / [6]dictionary in Medical / [7]Legal / [8]Mechanical / [9]University-Institution {Oxford} / [9.1]{American Heritage} / [9.2]{Merriam Webster's} / [10]Encyclopedia-(s) / [11]Press / [12]bible / [13]redacted-note-(s) /// et cetera…], (as for who can report in accordance), [educational-words // syntax rate-interval-(s)], (post-date-reset-(s)-examination-(s)) depicting-rate-conjugation-competency, identifying land-surface-area-(s), when objective-tool-discourse-survey-inquiry-(ies), (are supposed to) influence ordered-product-(s), expecting an future identity, (as through) individual-voluntary-surveyed-tense-(s), routine dollar-currency-transaction-(s), interval sleep-elapsed-fallowing, holding degree-merit-credential-(s), sharing measure-(s), inactive by commercial-transit-sleep-eating-showering-dispelling-affair-aging, amidst exponent-Federal-unit-notion-offset-county-state-responsibility-reproduction, (when where) spouse-(s)-alone, conglomerate-affair-(s) (in of) state, hither fro & to, laze by (sub)-urban-location-(s), titled at development-(s), reserved away from barren-territory,| 70 |revolving-revolutionary-centripetal-force-tectonic-plate-continental-shelf-existence, reminded electric-currented-outpost-(s), following those boundary-limit-(s), assuming [requisite // prior past value-(s)], radical by blood-pulsed-synapse-being-(eu)-genetic-(s), interlaced by bastardized-transit, intersected imagery-registering-cognition-(s), never commonly-noting, rural-spacing-(s), proceeding physical-role-(s), ([1]by) spouse-(s)-dynamic,| 71 |([2]via) individual-age-communications', reoccurring ([3]at) habit-repetitive-posit-(s), comfortable by, variablizing linear-measure-(s), [4]awhile [color-primary / tone-primary], not invigorate prac-tice of sweat by men, amidst geological-exterior-climate-environment, for issues at response, policed when time-periodical-interval-(s), account-(s) (from an) pleather-associative-talked-mass-(es), assessing per materials-projecting-consumption-(s), alluding package-discard-recycled-output, administered by person-(s)-rationing-proportion-(s), surface-impassing, spatial-constitutional-quadrant-(s), thinking how population, merit retrodated-consumeristic-auxiliary-compost,| 72 | economically-interchanged, (dialed in from) political-person-(s), ($_{1}$at) mode-(s) ($_{2}$by) commencing-fact-(s)? Would stancing not configurate, or may timbre-tempo-blood-pulse-vibrational-time-rate-count-interval-(s), (in / of) acquiring title-property, affect affluent rural-metre, (for an) International-Home-Federal-debt, (at of) developmental-continuum, set aging, (yearly) with ($_{0}$no) rebellion ($_{1}$of) wealth, cause M.I.A.-timbre, Retire & Archive, ($_{2}$an) decimalized-rubric, ($_{3}$while) presence declare statement-(s), ($_{4}$as) monotonic-motion-(s), ream-document, voluntary-effort-(s), reading ($_{5}$from) bias-personality, ($_{6}$not) genetic-perspective, generating auto-belief, close ($_{7}$to) personal-ideal-(s), gazing technological-inclination-(s), surging population-(s);$_{14}$ (at an) direct free-will-duty, ethically inclining human-being-figure-(s), upon configurative-census, where ethnic-adjustment-(s), compensate equivalent-attended-non-effort-(s), resided amongst intercorrelated-sect-(s), from county-elapsing-maxims ($_{8}$through) calendar-article-planned-dating-(s), ($_{9}$at) view, simultaneous hour-offset-time-posits, [four-minute-(s) /// Two-Hundred & Forty-Secondths, syntax // synapse-elapses],| 73 |(memory=attention-span=(caution)=(warning)=(beware)-notion-(s)), offset-perceived visual-imagery-ten-miles-limit-(s),| 74 |(not to) case-caused-orchestrated-schedule-effort-(s), point-(s) per day schedule, balancing genetic-state-objective-(s), while youth-rubric-lessoning-(s), allude mathematical-literary-prose-reference-experience, formal-decorum, inlay (by whom) comprehend (a the) House-Representative-(s)-function-(s), by Pronoun-interacting, (social-reclusive-issues'), identifying population-(s), from conceptualizing genetic-discourse-(s), where state-development-(s), collect-data, oxygenated amid physique-respiration-(s), interpreting bias street-transit-discourses', when communal-impasse-(s), secularize sedentary-sensational-intimacy-impasse-(s), aging pass ecological-elapse-(s)-discourse, (my suggestion, we impasse daily-hour-(s), (without an) common-literary-decimal-numerative-ordering-purposes) (for how) primary-commercial-consumption, by-an auxiliary-enabling,| 75 | signify where locale rely an national-Federal-government-monetary-trade-policy-(ies), (for how) 143L. inadequate common-behavior apply subjective-educational-international-multiple-choice-examination-(s), without thinking how mathematics, remain perceptive-peripheral-subterior, each voluntary-moment, (as when) hours-lifetime-denominative-expectancy, is generally at 701,280-hours, (when at) Age Year, Eighty-(80)-census-bureau...¶|||12m.:25s.53m.s.|||§

§0...Count-[19 / Nineteen]-preposition-of... ..Lines-[145.85 / One-Hundred-Fourty-Five.1.eight-tenths-2.five-hundredths].. ...Article-Essay-[8 / Eight]... ..How can we survey population, (for an) formal-denotative-Congressional-Library-subterior-political-affairs-documenting, amid discourse of political-state-constitutional-politics, (from an) illiterate-perspective, per commonly-educational-85%-lower-E-classification-(s), where committee-planned-resources, (are not centralized to an)| 76 |state-furnished-architectural-engineering-commercial-scheduled-production-repopulation-inductive-offset-methods, from literary-continuum-non-fiction-Commercial-documenting?¶|||32s.03m.s.||||§

¶9: If patterns as such confine an being, does coagulated-inductive-reference, form idle-seldom-scheduling, budgeting populational difference-content, deriving familiarity-(ies) at bias-placing-(s), decimal-compartmentalizing, visual-tonal-vocal-perceiving-offset, or does morality-embody idle-tense-(s'), (as where an) task, is supposed (to be) (red)-acted-(ing), during minutes-mode-elapse-competency-posit-interval-(s), impassing [day // night interval-period-(s)], ordering secondths, motioning secular schedule-difference-(s),| 77 |proportional volume-visual-perspective-(s), elapsing wind-air-respiration-centripetal-force-kinetic-expanse-transference, for Land & Territory metropolitan-offset-(s), sufficed (by an) national-order, variated valid contextual-truths, fluent, consecutive-secondths-calendared-maxim-(s), aboding effort-(s), dating by [{43,200-subjective-cardiovascular-active-awake-secondths}-forty-three-thousand, two-hundred-secondths // {12 hour-(s)} Twelve-Hour-(s)], (for as) free-will-duty, industrial command (super)-ego-motion-action-(s), attempt the most difficult-geological-movement-(s), amid theatrical-timbre-ambient,| 78 |(when where) genetic-pronoun-reorder-survey-psychological-accounting-(s'), abash abstract-ambiguation, unsaturated at celestial-hour-ten-minute-(s)-formed-interaction-(s), following interval-(s')-reset-(s), [{24 Hour-(s)} Twenty-four-Hour-(s) // {86,400 secondths} Eighty-Six-Thousandths & Four-Hundred-Secondths earth-centripetal-force-rotation-inclinations], programming sequential-schedule-elapse-(s)-mode-action-(s), subterior {Three-Hour-(s)-Reset-Periods-Virtue-Etiquette-physical-exertion // literary-competency, maxim-extent-counts}, balancing fatigue when mass-live-continuum-effort-(s), offset-live-perceptive-elapses, every four-minute-(s), per [celestial-horizon-east-west-centripetal-force-temperature-drag-impact // effects'], planning motion-interval-(s), awake [[1]reading-(s) // [2]writing-(s) // [3]exercising // [4]interacting-(s) // [5]eating-(s) // [6]dispelling-(s)-synapse-elapse-(s)-minute-(s)-impasse-(s), apose [0.1]sleeping /// [0.2]working /// [0.3]transiting-extension-(s)-subterior-elapse-(s)-hour-(s)-impasse-(s)], conceiving how denoting-action-(s), interpose population-numerative-purpose-(s),| 79 |(amid an) never-ending-present-offspring-lineage-perspective-denominative-infinitive, curbing [month-ly / seasonal // yearly legislational-impasse-(s)], prosing an ambiguity-continuum, unscheduled offset-population-personal-prerogatives', (for how to orchestrate an) county-city-housing-population-living, when their remain those default-discourse-(s), by executive-government-dynamic, (for humanity is too non subjective (from an)) legislational-cursive-writing-educational-communications-dynamic, since 1912-forward, through periodical-hours, subjective-offset, non-article-extents.$_{13}$¶|||02m.:00s.$_{78m.s.|||}$§ Can bank-(s) loan general-inhabitant-(s') money? No! Not quite so;$_{15}$ (but for whom can) modify-language-(s) through inadvertent borrowed-letter-(Franco-Ger-manic)-participle-word-syntax-literary-(Western Hemisphere)-culture-synapse-(s),| 80 |denote-numerative-act-(s)-method-(s), denominate-(ing) written-tangible-book-composition-(s), as tempo-one-hundred-twenty-eight-beat-(s)-per-minute-participle-beat-influctuation-blood-pulse-harmonious-rhythm, apose syntax-deviation-measure-(s), reasoning [[1]housing /// [2]Industry /// [3]tool-(s) /// [4]currency-note-(s) /// [5]subjective-tangent-refined-datum /// [6]population-perceptive-discourse /// [7]comestible-substantial-reproduction-(s) /// [8]clothing-resource-(s) /// [9]elemental-psychological-mass-population-objective-evaluation-(s) /// [10]statehood-climate-territory-denominating // [10.1]constitutional-deductive-confirming, mass-population-(s)]| 81 |throughout [formal-objective-developing-extraction-edible-substantial-compositional-planned-housing-acreage-offset-(s)-clothing-pattern-transit-routine-square-mile-per-capitia-documented-constitutional-hundredths // thousandths, notions-influctuation-(s)],| 82 | amidst human-being-(s)-Political-Person-(s)-mentality-complex-continuum.$_{14}$¶|||02m.:45s.$_{07m.s.|||}$§ Fundamental-requirement-(s) of state-taxes, mix able competent-objective-instructed-article-(s)-wage-workers', referencing in-ductive-document-(s), toggled by notion-(s), extenuating nation, for example, (a the populous of) [California {36, 567,

493 million inhabitant-(s)-estimate-continuum}-{November-2015} // Thirty-Six-Million, Five-Hundred & Sixty-Seven-Thousand, Four-Hundred-Ninety-Three populants'], case how, House-Representative-(s), per capita (by of) city-ordinance-(s), suffice state-developmental-discourse-(s), [regionalized / metropolitan-county-sector]| 83 |exceeding population-demographic-proportion-(s) of geological-land-(s), which amount inhabitant-(s) preference-bias, from historical-biological-idle-state of affair-(s), offset implicit relied-development-(s), daily-dated, when reset circumstance, receipt-(s)-market-shelf-transaction-(s), instead (of an) confluential-mass, confirming conceptive-architectural-effort-(s), because how common-idle-dynamic, absolute, traded consideration-(s), from Director-(C.E.O.-Management-Operations) & Follower-(employee) where Competency & Ordering, effect those (re)-population-effort-(s)-statehoodship-(s'), (where when), desire is seldomly-regarded, because how terrible (a the nature by) idle-independent-individual-(s), believe two-child-thirty-years-neutered-offspring-enabling, influence those formal-creative-transit-impasse-conceptive-imagery-(ies), surrounding lifestyle-decision-(s).$_{15}$¶|||03m.:55s.$_{57m.s.}$§ Date elapse-(s) (in an) fashion, conjunct (by how) [will // could], "Lend", confide Possession & Trade, awhile accustomed-habit-(s), father insurance, (as an) [belief // idea // habit], through Man & Economy, where method-(s) (by of) equal-writ, consensus-belief, from Iteration per men, before national-identity, prior Age of Enumerated-Trade amid luminescent lighting kinetic-hemispherical-charge, posit understanding from study-course-work, proxy date, apose composite-view, forced by natural entity-denominative-accumulation-secondths-effort-(s)-offset, {currency} juxtaposing [Day / Week / Month / Season posit-denominative-deductive-configuration-(s)], (by-at) (red)-action-(s), from [Sight / Sound / Taste / Smell compartmentalized-(reasoning)-hyper-tense-(s)]| 84 |Seven-million, seven-hundred-seventy-six-thousandth-Secondths, ration per season-(s), flagrant from omitting [Solstice // Equinox temperate-impasse-celestial-effect-(s)],| 85 |for

petty-entertainment-relief, non-sequitur (by an) limited (pro)-noun-extents, stemmed at realm surrounding, leading upon tasked-performance-order-(s), formulating social-requisite proceeding-(s), presence per dating, concert day-primary-function-(s), (with where) night complementary fixated-conjure-progression-(s), inductively-denotes, formal-order-(s) factoring count-deviation-(s), from population independent-observation-survey, (by of) exponent-objective-usage-deviation-(s), from hundredths-count-denominative-percent-focus, (apose an) stock-like-numerative-influctuated-deviation-(s), relevant apose, investment-(s) (that have enhanced our presence by) [continuum day // night offset-reset-dating], 73L. persuading voluntary-mode-(s), dating per [day // season // year // several-year-period-(s) /// decade-(s)], traded-confidence-investment-(s), centralizing two-centuries (by of) American-English-Transgression-(s), not thought, when emancipated-revolution-(s), form such past-requisite-(s) (by of) human-blood-pulse-pitch-tempo-vocal-visual-reverberated-vibration-(s)-resonance, when visual-imagery-vocal-tonal-vernacular-perceptive-interpretations'-timbre, direct how existence is not etymological,| 86 |(for of) Physical-Blood-Pulse-Synapse-Visual-Imagery-Vocal-Tonal-Perceptive-Vibrations & Grammatical-Punctuation-Ontological-Proof offsets...¶|||05m.:48s. 51m.s.||| §

§0...Count-[12 / Twelve]-preposition-of…|87|..Lines-[78.83973 / Seventy-Eight. 1.eight- tenths- 2.three -hundredths- 3.nine -thousandths- 4.seven -ten-thousandths- 5.three -hundredths-thousandths].. ...Article-Essay-[9 /Nine]… ..I have never referenced an book of mass-Federal-offset-city-county-state-location-population-surveys.¶|||11s. 91m.s. §

¶10. I've thought pink-white-women, an personal-exotic, (as an) theatre of teacher-(s), at various Classroom-Settings (not public), (yet) being-formed, yearly-impregnated, for physical-offspring-exponent-presence, yearly lessoning, scheduled-offspring-mode-(s), affecting tension-(s), adjunct-hourly-time, lessoning redundant-synapse-birth-age-equivalent-lesson-(s), juxtaposed uninclined offsetage-(s)-posit-(s), adept aptitude, (dissecting-analysis), Taxonomic-Classificat-ion-(s), studying (by an) method per location-environment-practice, during individual-perspective-offset-(s), concluding various physiological-anatomical-biological-masses,| 88 |{10 – 100 Colleague-(s)-Citizen-(s), (at an) dated-occasion} defining along parallels for class-fraternalized-affinity, aging-hyper-tense-(s), (like physical-beard-growth-elapse, or geological rock collecting) dissolving amongst intermittent-effort-(s), when conscious-capable-limit-(s), (sleep-pattern-mundane-inefficiency), concisely pass daily-impasse-extent-(s), verifying-limit-(s)-reset-political-boundary-(ies), conferred discretion, (where an) awake-impasse-juxtaposition-fatigue, vary women-fatigue (an woman (or so)), harmonized (by when) male-counter-part, think how retirement-idle-impasse-(s), upon present-day-(s)-physical-hourly-wage-impasse-(s), conceive purposes at house-posit-location-(s), considering independent-subjective-bookshelf-processes, through feminine-reproduction, apose (a the) muscular-physiological-organic-impasse-(s), for White-male-subservient-white-feminine-superiority, intended enhancing,| 89 |orchestrated continental-shelf-purpose-(s), construing feminine-family-community-subject-four-day-off-spring-periods-teaching-(s), surveying youth-interpretation-(s),| 90 |apose present-adult-hood-confluence-contextual-continual-congressional-literary-upkeep, (when whom matter by) visual-imagery-vernacular-vocal-tonal-perspective, remained significant between one-another, (as where), Syntax & Motion-(s), are not rural, for stint-(s) of labor-fact-(s), ordering industrial-products, as commercial-entity-solved-solution-(s), constantly testing experiment-(s), (by where) [exponent-cubic-supratangible-geological-matter // product-unit-deviation, 0.1 / Zero-whole-decimal-One-Tenths // 0.01 / Zero-whole-decimal-One-Hundredths /// 0.001 / Zero-whole-decimal-One-Thousandths Count-Extents],| 91 |observe [sociological-expenditures // geological-matter], (genetic-preference-experiment-surveying), elongated senile-age-disparity, naturally ideological, by clueless-faith, based from monotheistic-principle-(s), relying prosed post multiple-choice-imagery-contextual-interpretation-(s), temporal physiological-fatigue-hyper-tense-(s)-visual-imagery-awake-consciousness,| 92 |upon [ecological-constructive-spacing-(s) / topographical / primary-bodies], (at an) centripetal-force-existence, [investigation-(judicial) / prose-inquiry-(legislative)-legal-form-focused / instated-(executive)],| 93 |by literary-case-purposes, prosing bookshelf-proprietary-competency (metropolitan-centralized-city-retardation), concerting non-fiction-deductive-context-referencing, seasonally redacting, observed generalized-physical-character-feature-(s), reflexive, visual-tonal-taste-smell-hyper-tense-(s)-conjuring, en-abled designated perimeter-surface-area-offset-perspective, at [reviewed / redacted] consum-eristic-advertisement-consumption, being handled (on an) bridled [balance-(s) / scale-(s) / mea-sure-(s)], through year-Gregorian-millennia, singular-century-birth-offset-posit-(s), retrodating visual-imagery-hyper-tense-(s), interposed live-referring, past historical-competency-limit-(s), retrograding [syntax // synapses four-minutes-celestial-horizontal-horizon-secondths-rotation-impasse-cycling-(s)-count-(s)],| 94 |(educational through literature), contraposed [age / senile-elder-(s) / Jeffersonian-Era / (Solar-Fission-Four-Hundred-Eighty-Eight-Year-Maxim-presumption) // Protestantism, presence-offset-staged-phases (by of) beings], furthering varied “commercial-industrial-Choice-(s)”, in computational subjunctive-hourly-centripetal-tense-(s), [past-present-continuum-reference-documenting-deviation / present-presence-acting /// documenting // future-calendar-dating], defining parameter-depth-perspective-distancing-(s), distinctly-intermittent from age-year date-simultaneous, past-historical-abstract, dating [generation-(s) / era-(s) / decade-(s) / lineage transitional-periodicalizing-(s)], amid Two-Hundred-Forty-(American)-Year-(s) (Decade Y-Current-Present)) {July 4, 2016};[16] [when / where] physics is not {Federalist-{Alexander Hamilton / John Adams}-Democratic-Republican-{Thomas Jefferson}, per visual-direct-peripheral-observation-count-(s), from defined term-(s), continually abstracted, through transient-destination-(s), amidst demonstrating activity-(ies), physical crescendo, offset incremental-notion-(s), amid-at visual-surface-area-tense-(s), volumized, approximate-ambient surface-area-title-space-(s), vibrating congested repressing-ideal-(s), (for of no) contingent belief-confluence-communication-(s), by| 95 |[Whigs-dynamic |in // out|, an Montana-State-Geological-political-literary-redevelopment-party-ten / twenty-millionths-(white-pink)- population], awhile mental-vehicle-memory-cognition,| 96 |(psychological-discourse) preferred factifying-featured-vehicle-car-insurance-drivers'-license-operator-(s), copper-topped by requisite-traded-acquisition-(s), regardless commercial-factoring-(s), Psuedomatricgeoplaused, when age-fatigue-elapse-deterioration, adjunct sleep, before depth-force-anterior-psuedo-personality groom ecogeocomsteering, aligned [[1]avenues / [2]boulevards // [3]courts // [4]drives /// [5]expressways //// [6]highways ///// [7]Interstates ////// [8]places /////// [9]roadways //////// [10]streets //////// [11]terrences], class-aging-subjective-clause, characterized (where when) live-voice-halogen-belief, emit repetitive-natural-vocal-vibrations-transient-base-hypotenuse-perceptive-tangent-deviation-(s),| 97 |not from those absolute soluble-elemental-tangible-(s), where physical-perceptive-timbre-characteristic-(s), pace intermediate regeneration, calendar-hour-dating,| 98 |year-comprehension-contention-impasse-inset-focus, (by an) opposite-sex-intimacy, communicated through religious-extraspective-communion-interaction-(s),| 99 |per currency-note-denominative-method-(s) enhancing (through) lifestyle-quality-(ies), per impasse through written [manuscript / cursive // typed Federal-documented-dating], upon $_{\text{65L.}}$those hourly time-posit-(s), for scheduling dating-posit-point-(s), (by of an) visual-depth-perspective-tangible-qualities-offset... ..not commonly-copyrighted...¶05m.:34s.$_{\text{19m.s.}}$§

$_{\text{§0}}$...Count-[5 / Five]-preposition-of…| 100 |..Lines-[66.3974 / Sixty-Six.$_{\text{1.three}}$-tenths-$_{\text{2.nine}}$-hundredths-$_{\text{3.seven}}$-thousandths-$_{\text{4.four}}$-ten-thousandths].. ...Article-Essay-[10 // Ten]… ..Can an dozens-mil-

[Article:] [January]

lionths-white-population (ever have men whom) worship voluptuous-white-pink-blue-eye-women, (transitional-blonde) for educating an genetic-communion-children, by women-housing, male-transit-geological-subjective-payment-efforts, trading like an labor-civilization (by of) men, learning [exponent-feminine-birth-educational // masculine-labor-title-subjective-constructive-repopulation-dynamic], yearly populating from Federal-Whig-dynamic, an Political-Party-Currency-Literary-Congressional-Language, for (or sooner) retirement-aging-national-constitutional-erecting, circulating an formal-printed-currency (from a the) European-Union-Dynamic-retrograde (amid those) various tangents of self-provisional-personal-visual-possessive-perspectives', at transit along exponent-tons of Federal-Interstate-highway, unmeasured from lazy-subjective-historical-labor-documenting, yet essay-semi-colon-syntax-articlized per volume-blood-pulse-respirating-inhabitant-(s')?¶|||58s.62m.s.|||§

¶||For words-inductive-referencing syntax, breve rhetoric limit-(s), present modern-placing-posit-(s), rated at perceptive secondths-limit-(s)-continuum-age-tier-historical-decade-centuries-offset, present when, modern day-impasse-(s), intersect present-day-denominative-value-(s), whole day-retrograde-zero-reset-data-reasoning, juxtaposing calendar [denominative-integer-offset /// numerator-integer-offset], reciprocal [seventeen / eighteen-hour-(s)-awake-per-day // seven / six-hour-(s)-night-deviation], counter, rate-deviated hour-(s), off-work-time-schedules, where decimal-placing-(s), as zero-whole-decimal-nine-tenths-eight-hundredths, awake-pattern, upon time-primary-operating sequence-(s), per secondths-notion-(s)-offset, wandering when calendar-period-interval-(s), anticipate year, (secondary), following objective-placing-(s), historically posited, where relative-elongation, topic-extenuate, Stem & Leaf-Plot-two-dimensional-graph-reasoning, pertaining light of day, amid centripetal-force-spherical-circumference, [rotational // revolutionary drift], dropping clue-(s), (as what) facts formally affirm-(s) literary-book-offset-purchase-rate-deviation-(s), referencing apposed, rated-circuit, spurted physical-live-tense-effort-(s), not pulled in subjective-note-(s), subjective-tier-evaluation-(s'), noting which perceivable off-site (balance-(s) pertain mass-population, in basic-human-rights-objective-deviation) capable-effort-(s),| 101 |log amid daily-secondths-mode-interval-count-denomination-(s), [reading / writing // log documenting], language-construes, from when [velocity-depth-force resonation / tone // pitch // volume // vibrational range], construe dipitchipational-physical-work-being-(s), requisite live-blood-pulsed-synapse-being-(s) say by modern-contemporary-act-(s), Renault-process, adult per [decade-year-juxtaposition-(s) / child-age-comprehension-deviation-output-(s)], resolute salary-tenure, conducting personal-age-id-experience, amidst conjured interval day-(s), while [hour-(s) / instance-(s)], forebathe in petrowashymn-loveational-proper-tense-(s), mood through meander claritive-reliant perspective, when received-payment-(s) (bi-weekly // weekly // daily), methodize calendar-effort-(s), live accustomed from accounting vague aging, ancestrally-defaulted, one birth-day primordially forever, year-interacting (lifetime-memory), (not) aboded amidst [park-ambiance / Canyon Hill-(s) / Mountain / Desert / Forest perceivable-geometric-surface-area-(s)], adapting vocal-mutation, perceptive-offset, creative (by where) visual conduct, securely-maintaining work-entity-commercial-effort-(s), affixing [volume-petroleum / iron-ore-density-(ies) / luminaire live-saturation-intangibles], corporately sponsored, where locational objective-placing-(s), bemuse habitual-inhabitant-ruse, impassing past-day sequenced commanded-movement-(s), unconsciously emitting analytical-purpose, in focus from other [inhabitants' / object-item-(s) / storm-drain / sink / toilet-tub-outlet // inlet-flow-(s) of water], placing instructed-object-(s), at scheduled-perimeter-surface-area-limits', (by an) cubic-perimeter-area-dimension-(s)-set, where placed parameter-(s), identify [surface-area / interior-square // rectangular perimeter-limit-space-(s)], objecting metropolitan-city-planned-block-housing-lock, Populated & (Un)-Affirmed through where configuration-(s) (by of) tangible-uncollected-ton-(s), best set mode, contemplating conclusive-operation-(s), resulted when age, review through age-limitation-(s), requiring physical-act-(s), undaunted when fate by purchase-price-objective-unit-deviations, deviate-at locational-hourly-time-posit-(s)-elapse-impasse-(s), (for how) calendar-impasse-influx-continuum, (has an) multiple-count-(s)-process, progressing upon present locational-circumstances, what result is required for deviating each notion-(s) (at an) Federal-Mysticism-production-offset, 39L.common when offset-below-average-intelligence-query, premise how thinking is perceived at location, from [self-interior-bias // hoc-perspective-offset]...¶03m.:13s.34m.s.|||§

§0...Count-[3 / Three]-preposition-of...| 102 |..Lines-[40.38425 / Fourty.1.three-tenths-2.eight-hundredths-3.four-thousandths-4.two-ten-thousandths-5.five-hundred-thousandths].. ...Article-Essay-[11 // Eleven]... ..If dozens (of millionths of) inhabitants, not know (how to document through) [reader // writer] article-(s)-books-publishing,|103|political-genres of non-fiction- (auto)-biography-documenting, (could never have my prior) Essay-Redact vision, (for how) lifestyle affects commercial-sociological-patent-copyright-productions, from registered trademarks, over copyrights', relying on [musicals, serials, photos, et cetera...]| 104 |to transit lineages away into retirement, [dying // not repopulating], (as I) believe, (earlier then) physical-economic-circulated-capacity, (for not) literary-communion-planning, housing, children, [eating periods / preparation], resourcing, patent-product-entity-Qr-cycling, clothing for topography-environment-conditions, sociological-lifestyle-festive-celebrative-interacting, Appreciating & Cherishing visual-imagery-vocal-tonal-vernacular-perspective.¶|||53s.12m.s.|||§

¶12(Are their enough) perceptive-blood-pulsed-synapse-being-(s), affirming commercial unit-value-(s), affording (relevant costs' of living) variable-(s), developing case-(s), reasoning [Land //// Water //// Electrical-Current-(s) //// Wind-pressure], purporting-mass, collecting-accrued-unilateral-debt, following nationalism-faith, periodical-perpetual, perimeter-square-surface-area, denoting count-(s),| 105 |from visual-pupil-eye-plasma-depth-vocal-tonal-respiration-organs-offset-continuum-interposed-perspective, secondths-elapse-reset-(s), propelling crust-superior-mass-perceptive-posit-(s)-offset, breathing constantly, affirming how voluntary-movement-(s), surplus-incentive-(s), while object-(s) remain (due through) intervoluntary individualism-belief, conducting ways' of extraspective-brother-(ly)-bias, using trade-off-(s), for meriting-educational-degree-(s), individually-possessed, accumulating-subjective-credit-(s), circulating [bond-(s) / grant-(s) / scholarship-(s) / bailout-(s) / trust-(s) / loan-(s)], developing [Governmental-I.R.S.-earnings-currency-year-denominating // Commercial-Enterprise-Enumerated-Trade], slighted in parameter-view, participated where expert trading advice, suggest frequency-factor-(s), [weight-(s) / volume-(s) / natural resourcing], presuming (a the) toll of debt, frequently apposing economic-outlook-(s), affirmed by consumeristic-spending, (for how) allegiance define position where communal-interaction-(s), mean current-note foretelling, promising-land, elucidated accumulated product-(s), mass-produced, factory commercial-conveyer-item-unit-output, simultaneous religious-function-(s), generating generalized-parameter-(s), lacking House-Populus-political-reasoning-overview, from conscious-Federal-vied-act-(s), pertaining work-week, [hourly-placed / minutes continuum-instructed-objecting], greater-task-(s), by presided-documentary-spacing-(s), asserting territorial-title-claim-(s), apartment-sufficed when greater than blind-convention, legislatively-evaluate, contort apose associate-career-rates, when collected timed-elapse-synapse-period-(s), lose hyper-tense-fixation-(s), factoring measurement-(s), per individual;[17] accounted by each being, fortifying independent person accounting-(s), per individual-(s')-character-role-(s), by communication-(s), per human-basic-rights'-resourcing, validifying-fact-(s), sustaining [[1]analytical-blood-pulsed-being-(s) / [2]ingestible-soluble-(s) / [3]household-appliance-(s) / [4]household-spacial-appraising / [5]work-cycle-effort-(s) / [6]procedure / [7]produce / [8]methodical-transportation-practice / [9]clothing-production / [10]Territorial-Boundary-(ies)], applicable Concord-blood-pulsed-being-(s), who-(m) perceive understanding, from comprehending basic-human-comradery, affluently discoursing, genetic-aligned-allegiance, gathering natural-resource-(s), (due in) [storage // transit weight-part-(s)], [creating // developing], extraspective-31L.objective-collection-development-(s), consulting constitutional-territory-(ies), (from when) mean-(s)-offset, per locale, balancing visual-tonal-depth-perspective-measure-(s), [pertain // obtain // sustain], intermittent [room-property / weigh-(ing)], [commercial-yield / Consumer-packaging / balance per personal-saving-(s)], intricating approximate-territorial-objective (goals), aligning [date / age / year ordering-(s)], by interactive-impasse-continuum-offset, always requiring subjective-intelligence, when serving developmental-fact-(s), by visual-imagery-voice-vocal-tonal-vibrational-reverberations-deviation-continuum-perspective...¶|||02m.:35s.12m.s.|||§

§0...Count-[3 / Three]-preposition-of... ..Lines-[35.62 / Thirty-Six-.1.four-tenths-2.three-hundred-ths].. ...Article-Essay-[12 // Twelve]... ..[international / national debts], relying Federal-American-Monetary-Department (from of) State-Commercial-Currency, from Senate-Congressional-Calendar-Acts-Denominating, from introverted-trade-policy-(ies), I wonder through literature, how human-ethics-works', amid present trade policy-(ies), during daily-poverty-mentality-millennium-dynamic.¶|||18s.60m.s.|||§

¶13How would one syncopate, Liberal-Art-(s), juxtaposed technical-engineering, symmetrical where tempo-syncopated-timing-(s), calibrate calendar-date-rate-(s), [occasion / interaction / happening / event-(s)] perceived amongst blood-pulsed-synapse-being-perceptive-compulsive-attentative-actions', aforementioned New Gideon's'-lifestyle, influencing an timbre, per American-Liberty-Abigail-Roosevelt-being-(s), when individual-(s'), act per [physique / stature of age / constant-up-keep (per of an) objective-entity // auxiliary manual-instructed-posit-(s)], infrastructured input beam-spacing-(s), cemented by general-upkeep, accolading past day compositional-dating-conceptive-patent-products, interpreting at present-location, adrift, contin-ental-perpetual-momentum-centripetal-force-volume-speed-(s), surpassing visual-imagery-per-spective, aging amid grandiose-superfluous existence, anterior west-hour-time-dating, in part independent-application-(s), entrance mechanical-ingenuity, psychological-timbre, maturing visual-imagery-tonal-vocal-vernacular-referencing-decimal-physical-motion-(s), adjacent sec-ondths characterizing-(s'), (at an) global autobiographical-dysfunctional-scheduling-dues'-progress, through Renault-biographical-ecological-discourse, roller-coaster living-virtue-(less)-(s), grid spherical physiological-being-elapse-time-drift-(s)-micro-subterior-impassing-(s), during impasse-posit-(s)-offset-juxtaposition, characterizing surface-area, where plane-circumference-depth-perceptive-existence, suffice without adjourning molecular-momentum, approximating [from / at] an mathematic-timbre, conjuring (from through) visual-perceptive-imagery-(ies), calculating geometrical-calculus-algebra-mathematical-processing-(s), proportioning periodical-monetary-method-(s), algorithmic composite-accumulated-note-(s), formally (not commonly) participated, when substantial-nutrition-requirements, receipt urspurated indication-(s), apose past-thought-documenting-(s), extended from [vent / steering / placed-condition-(s) / automatic-transmission],| 106 |for methodical-metallurgical-maneuver, by fine-black-ink-posit-thinking, contraposed fine-blue-ink-posit-thinking, onto white-blue-lined-paper-medium-(s), congruently inked, physically from perceptive-motions', visualizing-hyper-tensed-perceptive-interpretation-(s),| 107 |effecting location-objective-physical-being-sociological-interaction-outcome-(s), where relevant locational impasse, vary independent-form-(s), summing [fraction-(s) // percentage-(s) // equation-(s)], depth-largo-hour-day-year-elapsing's', for nothing simultaneous contemporary-maneuver-reading-(s), (due at) magazine-(s)-{military-auxiliary-M-{I / II}, transitional infantry-line, apose say, contemporary-shelves-(super-(market-(s)}, from sensational-hobby, not discoursing ["physics" // geology], while metrics abstract, [in // by] [shape / dimension-(s) / ecological-response-(s)], debate-argumented-offset, where road [turn / straight balance / stop & go-incur / character-identify], modern-transit-engineering-technological-commemoration-(s), per [physical / medical / cardiovascular-response], monitoring timbre-characteristic-(s), not Theatrically-Communicated, when, where objective-use, intermittent peripheral-sight, surround environment-physical-depth-visual-imagery-tonal-vocal-vernacular-perceptive-response-(s), relaying-an delayed-response-(s), via V.I.N.-1-A-{Vehicle Identification Number}-presence, amidst millennium-dating, accounting intrication-(s), (un)-calculatable, factifing sight-standard-deviation-(s), from abled-passing-(s), as tempo-blood-pulse, inductive-numerative-mode-continuum-fraction-(s)-dating-mathematicalcal-calculating-deviation, presume presence-(s), as past-prelude upon tomorrow-foresight, alluded from physical-motion-(s)-tensings', following week-context, next where present-day-stress-anxiety, hold writ, serving [Self & Table // Desk-ration-(s)], date impassing guided personal-practice, ready where [perspective / sensation / thinking / feeling psychological-compass-notion-(s)], mode where feature-production-(s), [[1]define / [2]depict / [3]derive / [4]determine / [5]deviate / [6]deter / [7]date / [8]distinct / [9]denote], incremental-crescendo [inclination-(s) // indication-(s)], affecting human-being-subjective-inductive-perspective-(s), from homage for American-Fathers'-caffeination-allegiance, heir-generated, chronologically, sip-eat-drink-ingesting, whatever pass down ones'-throat, forgetting [word / act / place], influencing secondths-continuum-offset-millionth-hour-age-114.4738, apose those [minutes-(60 / 24) // month-(s)-(30.41667)], [secondths-(60-reset-deviation) / season-(s)-(91.25)], [milliliter-(1000) /// secondths-(60-reset)];[18] because F.D.A.-regulated-confirmation-(s), amount blood-pulsed-being-(s)-inductive-term-reference-competencies, losing present-tense, modernly affirmed, ruling belief, selling from solicited-wage-physiological-tense-(s)-effort-(s), communal by family-independent-receipt-payment-(s), yet subjected with transitive-movement-(s), mounting-belief, for auxiliary-use [in / of / to // from], repeating physique-being-quality-(ies), through managerial-subjecting task-(s), (from how) adult-individualism, case-(s) product-unit-amounts, (as an) lazy-bemusement, inculpable for never habitually acting, (by an) coordinated etymological-ontological-mathe-matical-grammatical-punctuational-chemistraic-biological-anatomical-physiological-geometric-algebraic-trigonomic-psychological-calculated-date-progression-retrograde-impasse-interposed-processing-(independent)-prerogative-retrodating-method-(s), surrounding genetic-pre-dispos-ition-(s)-offset, sociological-survey-inquiring independent-individual-(s'), at [physics-metric-miles-perceptive-limit-offset /108/ geological-boundary-title-land-property-housing-offset], blocked per limit-(s) population perspective-(s), upon national-denominative-genetic-creative-reformation-(s)-referendum, proceeding calendar-date-(s) ranged [day-kinetic-white // night-vapor-no-pressure-space], in distance, [discussing // personally-envisioning // formulating], how personal-semantic-(s), pertain bank-account-(s'), through Physical-Day & Individualism, interceded intuition, setting condition-(s);[19] awhile day-night-centripetal-force-impact, effect barrier-exterior surrounding-placing-(s), predominated by male-proprietary-limit-(s') in hold of [form / title / payment / lease agreement-contract-(s)], I work Literary-Contracting, rather [Sports-Contracting / Music-Contracting / *Managerial-Contract / Acting-Contract / *Military-Contracting / *Commercial-Auxiliary-Contracting], from hourly-time-posit-(s), age-offset-expiry-season-year-millennium-offset, often voluntarily configured, (due where) monetary-payment-(s), select amid hover-celestial-impasse, above circular-tectonic-plate-(s)-centripetal-force-drift-offset, suffice-an physical-interactive-influencing rotation-impasse-continuum-sleep-offset, thinking rate-count-offset, juxtaposed in return when day revolve-"return", amid solar-kinetic-energy, propel molten-plasma-fission-force, [corollary (physics) / celestial offset], latitude-longitude-trajectory, grandiose default-mundane-nature, affecting physiological-cycle-(s), (not Taxonomic mammal-(s)) oxygenated physical-mouth-nasal-esophagus-throat-aorta-ventricle-blood-inlet-respiration-circulation-flow, formal at posit-east-to-west-revolutionary-centripetal-force (compass North), directing humanity pass grandiose-effervescence motions', per Tectonic-Plate-(s)-offset in [Sevenths-Displacing-(s) / Ocean-Bodies in Fourth-(s)-Anterior-Conjoining / Humanity in Two-Hundredth & Five-National-State-Civilizations-Developing-State-(s) / Animalia-Specie-(s)], (not) Edible-Intangible an primary-horticultural-maricultural-agricultural-thirdths-consumption-cycling, universally divided, fragmented beyond Two-Hundred & Fourteenth-Boundary-Limit-(s), sparsely-populated upon millennium-driving, (without) motivate-ion, (not reproduction) when [day / week / month / season limit-(s)], define our listless-endeavor, (through a the) great-undefined-abstraction-continuum-unexplored-existence, daily raising, having errors serve unanimous-reasoning, responsive when colloquial-ability-(ies), denominate educational-visual-imagery-book-context-redact-percentage-inept-count-(s), aging historical auto-biographical-obituary-memorial-funeral-layings', ambiguous perceptive-bias-documenting-offset, interacting conscious-receptive-interactive-criticism, upon natural-physics, (as an) reproductive-silence, subconsciously motivated, when rule direct deliberate-debate, in Legislational-Parameter-(s) & Tier, per genetic-human-population-order-(s)-count, stated past political-party-denominative-decade-count-(s)-interval-(s),| 109 |Free-Democratic-Republican where enabled-resourcing, never consider objective-proportions, embarking literary-congress-ional-discourse, contingently defining human-behavior, before, classical-prose.$_{16}$¶|||06m.:52s. $_{40mz.s.s|||}$§ Lesson & Lecture remain incapable at silent-static-solitary-introverted-score-(s)-posit-(s), | 110 |audience-decibel-audible, [mutual-redact-denoting / Individual-interactive-discussion / consensus-opinion-(s)], setting physical-sake, from [down-syndrome // autistic common-mute-behavior], which emits', obligation by faith-based-trade-(s), market-entity-(ies)-secured, serving duty-(ies), at land (owner) (lord)-title-housing-block-proprietorship, where barrack-style-city-planned-construction-based-development-(s), form aging-fatigue-physical-posit-(s), interceding generational-preparatory-routine, accustomed without adult [peer-scholar / co-worker interaction-(s)], fatigued during formal-interactive [Senate // House] congressional-debating, inquired at opposite-personal-perceptive-blood-pulsed-being-(s), variated through point-(s) at passing-existence, intercontorting characteristics' per individuals', not formally (by of) social-personality, misconstrued by preference-bias-mute-text-talk, abstracting existence timbre discourse-path, (when an) squabble (by an) subjective-prose, blissfully ignore physical-reproductive-purpose-(s), principle-lessoning criticized work-place-commercial-jockeying, accustomed at mute-introverted-wage-monetary-mentality-impasse-(s), shyly recluse past [time / date posit-(s)], reacting (as an) onomatopoeia-cordialities-bit, retarded by both-sex-(es), functioning reflexive [Politics / Religion / History] of Treasury-suffrage, where personal-book-shelf-reference not matter, inductive-redact-reasoning-rule, executively-watched,| 111 |by idle house-representative-literary-article-(s)-cursive-observational-documenting,| 112 |streamlining congressional-senate-house-legislating, along with executive-state-defense-commercial-urban-labor-transportation-agricultural-health-interior-education-cabinet-sociological-literary-congressional-senate-calendar-monetary-policy-documenting, producing conclusive methodical human-preference,| 113 |for vision-(s) (by of physics at) still-posit-(s).17 ¶|||08m.:25s.61m.s.|||§ Way of ego, mutual-decade-year-(s)-complacency, affect reproduction upon human-being-(s) (The Chinese (outsourcing) Philosophy), while male-(genitalia), remain adrift varied daily-child-raising, having women bear offspring, believing physical-acts-only, concept commercial-resourcing better [Bearing / Raising / Offspring-rating (per trimester) aging], passed per perceptive-blood-pulsed-synapse-being-(s), outsourcing quicker presumed than many technological-article-(s)-offset, clausing human-behavior from misconduct by objective-elemental-labor-(physical-effort)-interior-(subjective-documenting)-extraction-(s), deriving hour-time-offset-impasse-notions-exponent-deviation, from one-million-hours-notion-maxim-perceptive-count-offset,| 114 |per monthly-mass-handling, in [hours /// minutes // secondths offset], juxtaposing, when instated judiciary-review, deteriorate-age, taking advantage, (by the) construct-(s) of executive-police-fueled distance-(s), incomparable where introverted-mute-cordial-social-requisite, formulate Artistic-Language-inquiry, at [1Write-(ing) / 2Read-(ing) / 3Redact-(ing) / 4Question-(ing) / 5Response-(ing) / 6Dialogue-(ing) / 7Lesson-(ing) / 8Interact-(ing) / 9Theatrical-Piece-(s)-past-milestones-documenting], acquired (when at) present multiple-choice-response-memory, forgetting why municipal-depth-perspective-offset, primarily define, living-civilizations-county-population-offset, timbre-mode-locational-environment-(s), setting-an genetic-population-(s) furthering lineage-(s), from self-family-function-(s), (while amongst activities per) mass-130L. collective-citizen-(s),| 115 |serving political-militarial-religious-sociological-economical-statehoodship...¶|||09m.:45s.33m.s.|||§

§0...Count-[9 / Nine]-preposition-of…|83|..Lines-[127.43 / One-Hundred-Twenty-Seven.1.four-tenths.-2.two-hundredths].. …Article-Essay-[13 // Thirteen]… ..What effort-(s) of living, help (any of us) achieve (a the) millionth-hours of life, (amid an) global-common-offset-thinking, ritual by customary-traditions' for general-lifestyle?¶|||16s.00m.s.|||§

¶14No pre-reflective-cogito, align with perceptive-referencing, from taxonomic-classification-(s)-scientific-water-weight-count-(s)-offset, while mathematical-proportion-(s), tier photosynthesized-category-(ies), (as an part per through) enveloping-focus, [resourcing / mass-populating / cleaning-activity-(ies) / lawn-care / space-rural-hydrant-covering-(s)], demonstrating [cause-(s) // reason-(s)], per offset schedule [job / career] by style-class, observing ethical-prose, in relation amid partial-act-(s), referring from book-(s) [Dictionary / Encyclopedia / Chronological-Book-Series], juxtaposed Time-elapse-(s)-continuum-offset, observing apose timeline twenty-four-hour-simultaneous-day-offset-[time-dating-variables // calendar-dating-variables], series-(s)-perceptive-offset, calendar-seasonal-(monthly)-median-rated-interval-(s)-{a.90-days / b.2,160-hours // c.129,600-minutes /// d.7,776,000-secondths} denominative-thinking-point-(s) apposed title-property-effort-(s);20 when non-fiction-impasse-offset, exist amid travel, transiting, while never subjectively having been perceived properly, interpreting approximate-physical-depth-visual-tonal-perceptive-proof-(s), visual-imagery-subjective-prose-perspective-conjunction-offset, per individual-approximate-location-formal-competency-interpretation-cap-acity-handling.18¶|||53s.47m.s.|||§ Consumeristic-commercial-hobby-activity-(ies), concept patent-product-(s), intermittent [industrial-commercial, factory-objective / plant-auxiliary heavy-development-(s)] (my literary-goal to juxtapose, by strict diligent, documenting-purpose), marketing consumer supply and demand, when resource-(s), are [Federally China / et cetera.. Outsourced], (for as a the) Federal-Common-commercialistic-method, per Commercial-Productivity & Human-Wage, lack extraspective-at-work-off-time-interaction-(s), upon off-time-communal-activities, (over seen through) international-government-expiry-elapses', wayward-discourse-(s), free-will meandering national-referendum, matriculating confirmed-voluntary-will-(s), (by an) basic-birth-component-(s), where intercede-(ing-(s)) proceed mathematically-subversive-thought, before how [Grammatical-Thirty-Six-Letters-(Ill-(90%)-literate-(10%)-population-offset)-Word-rate-hour-(time)-Day-(calendar)-cursive // manuscript documenting], observe responsive-act-maneuver-check-confirmation-(s), piecing data together, (without an) incremental-common-minor-career-subjecting-documenting-period-(s),| 116 |evidently-observing congressional-library-educational-documenting-interior-rural-integration, (this book elapse-(s), an full-four-day-read of [Fifteen-Hour-(s) per day // an estimate of Sixty-Hour-(s)-to-Seventy-Hour-(s)-total-book-476-pages], naturally two-hour-lessoning-(s), one-month-elapse-(s)-denominative-hours-(730-Hour-(s)-per-month);[21] competently survey-referring, how human-conduct through yearly-ages', weather impertinent sub-par-behavior,| 117 |before-at voluntary-free-will-motions-imagery-peripheral-visual-tonal-subconscious-offset, omitting conscious-unconscious subjective-interactings', upon surrounding-setting-(s'), evoking location-timing-purpose-(s'), [by / from] desire for women, allured away, (an pleather by of) offspring, "fine rural-estate", coordinating scheduling-(s), motivation-(s), segregated-genetic-pertinent-industrial-factory-commercial-market-interactions, (as when) monetary-literary-denominative-security, family-communal-entity-bi-weekly-payment-check-interact-(ions), motion-location-notion-offset-effort-(s), serving for other individuals', (an worshipping by of) elemental-compositional-objective-preference-usage-offset-posit-(s), visual from community-interaction-(s);[22] difficult (for how) habitual-continuing homage-custom-tradition-hobby-habit-personal-independent-guidance, requires' convincing conscious-visual-pupil-respiration-blood-circulation-perceptive-blood-pulsed-synapse-being-(s), containing-circulating ventricle-blood-pulse-pupil-synapses, by [[1]head / [2]neck / [3]torso // [4]arms / [5]hips / [6]legs /// [7]feet /// [8]hands component-compositional-beings'], jointed by [[1]wrists / [2]knuckles / [3]elbows / [4]shoulder / [5]Trachea / [6]Larynx / [7]groin / [8]knees / [9]ankles], accustomed-ancillary-suburban-metropolitan-rural-terrain, (not yet) redact-tiering, taxonomic-classification-(s), or non-fiction-genres, when city-county-metropolitan-block-estates-(s) per family, by millennium progress, from human-prerogative, serve-duties, per decade-year-hour-focus, per bias-hobby-effect ambiguating, perceived political-wage-act-(s), where metropolitan-boundary-perimeter-posit-(s), extreme before self-limit-interior-visual-perceptive, defining tensable-existence, (by an) majority of populant-(s'), whom intercede (at any) given moment, idling 50L.rather than studying, by reference-owned-extents of book-(s), understanding why reasoning by communal-religious-political-house-subjective-transit-trade-means', interactively interact-(s'), physiologically..¶|||03m.:10s.69m.s.|||§

§0...Count-[6 / Six]-preposition-of… ..Lines-[49.575 / Fifty-One.1.five-tenths-2.seven-hundredths-3.five-thousandths].. ...Article-Essay-[14 // Fourteen]… ..Generally all human-beings', are introverted, over 50%-70%-percent, (affecting literary-competency, physical-effort-(s)). So how can it may be further, when voluntary-efforts', are seldom and few, by inhabitants'-economical-belief-discourse,| 118 |amid [extraspective-work-wage-site-circulating // personal-collective-wage-retaining-stagnating], by inhabitants' circulated-trade-methods'?¶|||24s.22m.s.|||§

¶15If "human-being-(s)",| 119 |enable compulsive commercial-C.E.O.-General-Major-Manager-locational-tier-way-(s), thinking may continue to disregard, how birthed-creation-bias, hoc male-calendared-count-(s), physiologically-imposing, sat-weighted-stature-(s), (as how) hour-time-interval-deductive-discourse, has time-calendar-mathematical-definite-numerative-inductive-minutes-secondths-point-(s),| 120 |which would suggest cycling government-election-cycling, (where an) $1,000,000-commercial-position-cutoff-salary-exemption, question those literary-credentials', (for how to) [generate // circulate], sociological-commercial-literary-(not subjective)-purposes, rather cardiovascular-effort-(s), amid $^{5}/_{100}$-20-whole-category-classification-(s) modifying geological-political-territory, referencing amongst isolation-(s) (at of) book-data-article-genre-decimal-ordering, determining (how to) reason denominative-monetary-educational-debt-analysis, (for how) commercial-systems [change / transpose] proof-fact-(s), by mundane-attention (from a the) Two-Hundred & Sixty-One-day-(s), at work-place-merit-(s), five-day-(s) per week, not including Holiday-(s)), conjuring One-Hundred & Four-Day-(s)-weekend-off-time-period-(s), offsetting hobby-activities, (when where) identifying surrounding parameter-(s), space (where in at) local-city-approximate-distance-development-(s), adjourn clues per house-residence-(s), private-displays impassing-hour-(s), during [exercise / Observation & Write / tabling water / scheduling inhabitant-(s')], per physiological-perceptive-blood-pulse-synapse-circulation-human-being-(s), organizing visual-tonal-vocal-perspective-(s), abounded where posit-(s) amid existence, encompass cardiovascular-response, intuitive physical-invigoration, fatiguing at [[1]environment / [2]temperature / [3]climate / [4]terrain / [5]topography / [6]setting];[23] contrasting inhabitant-worker-lotto-conversation-(s), interluded micro-measurement-(s), Subterior,| 121 | [fragmented-compartmentalized-incremental-component-unit-(s) /122/ Commercial-article-(s)-paragraph-syntax-deviation-extent-(s)] elapsed substaining subterior progressive-deductive-mode-perspective-elapse-points-referencing-(s), while where abstract-momentum, rotate simul-taneous, abound tectonic-plate-drift-physics-offset, remaining uncompiled, uncoordinated, amidst commercial-labor-expert-(s), whom mundanely isolate, subterior-lesson-commercial-subject-documenting, not tiered (by an) literary-genres-subject-(s), extenuating genetic-neighboring, politically-warned, reasoning confirmation-(s), not past-account-(s), per free-will-independent-individuals', self-independent-liberated-individual-(s)-partial-bias, (as what) define total-generalizing, America-(s')-free-practice-method-(s), deducing order, until individual-competent-inductive-component-(s), derive vast-mass-territorial-state-(s), which identify existence, (through an belief (by of)) auxiliary-developmental-housing-market-labor-literary-reader-syntax-redacting-wage-payent-circulating-national-sociological-militarial-industrical-commercial-governmental-worshiping, working (away from) legal-judiciary-case-practicing, (from how) paper & ink, affect much (of those) sparse-county-(ies)-population-intersexual-discourse-(s), for hygiene not brim motivation, per inhabitant, by feminine-dress-code, instilling security, (for an) documenting-constitutional-writer-reader-order-(s), deviating default-religious-authoritarian-relaying, superseded, Western-Hemisphere-Mythological-presence, never interceded (by at) hour dating, mutual-attained-order-dictating-sequential-progressive-function-(s),| 123 |(as when) tectonic-plate-continental-shelf-topographic-climate-terrain-surface-interior-cubic-area-spacing-(s), (television rule enabled perspective-(s), idle-movement-(s), (from an) lacked free-will-discipline-procedural-purpose-(s), mundane upon lifestyle-aging-interaction-(s)).$_{19}$¶|||03m :05s.$_{46m.s.|||}$§ [(Pink)-White-Woman / (en)], (come so) sparse and infrequent, encountering (a the) majority-mass total-population-tectonic-deviation-offset, demonstrating male-mute-youth-interactive-dynamic-(s), innately-reclusive, by physiological-blood-pulse-synapses', apose written-observation-(s), generating growth, naturally inclined, overdeveloped early-age-intimacy, unaware (by those) monetary-deviation-(s), by textbook-data-prosing, for actively-referencing, Literary-Book-shelf-Commercial-Circulating-Data, [collecting / circulating] monetary-unit-box-values, regenerating Genetic-Classification-(s), practicing extraspective-written-documenting, (as an) way to work an religious-congressional-county-literary-forum, (to affect how) reader-creativity, has (yet to) exist unclassified, where retirement recalibrate independent-individualism-(auto)-biographical-copyright-obituary-newspaper-serial-perspective-(s), not refocusing class-characteristic-(s), from [[1]face / [2]waist / [3]hips / [4]jaw-line / [5]skin-tone / [6]hair-color / [7]nose / [8]ears / [9]eye-color / [10]lips / [11]forehead-shape / [12]chin / [13]cheek-(s) / [14]head-size / [15]hair-texture-(s) / [16]gluteus-maximus / [17]Breast-Size / [18]Pectoral / [19]genitalia // [20.1]Vaginal-Wall // [20.2]Penis];$_{24}$ annotating why accounting constitution-(s), (has yet) accompanied population-(s)-supplementary-exponent-item-accounting-method-(s), denominating from commercial-production-unit-count-(s), superlative populations at National-Constitutional-Tier;[25] having dime-out-(s), per [commercial-operating-entity-(ies) // excessive-constructive-stagnant-ordering-(s)], simultaneous cronyized-friendship-(s), without future-geological-excavating-architectural-resourcing-market-incorporated-company-product-location-schedule-agenda-monetary-time-circulation-conceptive-dynamic, serving at barren-terrain, remaining by three-decade-offspring-three-century-generational-repopulation-physiological-effort-(s'), inclined (for how) objective-retaining-dynamic, apply perceptive-blood-pulsed-being-effort-(s), systemized by [numerical-indication-(s)-0 / 1 / 2 / 3 / 4 / 5 / 6 / 7 / 8 / 9, count-(s)-(10-999) /reset/ count/tens/hundred/order// notion-(s)-(1000-1,000,000) //// motion-(s)-superlative-offset, over 1,000,000-exponent-division-standard-deviation-reducing, by excessive-inclinations', (such as leaves of an) tree, sand-grains, agricultural-seeds, et cetera... (presuming (that we do not go over our) lifetime-awake-visual-nature-deviation of average-age-eighty, 42,048,000-minutes)], [upon / of / in] an expository visual-interpretation-influctuation-(s), mutually obligated, revising lifestyle-(s), by juxtaposition at communal-extraspective-natural-impasse-(s), reclusive like animal-(s), vocalizing interaction-(s), at voluntary motion-posit-(s)-impasse-(s), when considering social-schedule-aligning, intermittent national-Renault-development, resourcing (from an) ambiguous geological-physics-development, not asserted governmental-agenda-demand-(s), defining intelligence, (from of a the) basis population-common-objective-perceptive-conceptive-tangible-compilation-(s), when [PATENT-(S) // COMMERCIAL-FACTORY-PRODUCTION-(S)],| 124 |sustem an balance, heavier than simple workload, as constitution elucidate specific-territorial-populational-monetary-balances, outlining physical-efforts-limited-condition-(s), apose dating-period-(s), (from an) mass-consensus, [when // where] constitutional-interpretations-(s), (is fragile for such) belief, because-an nearly Five-Hundred-Year-(s)-physics-presence, amid existence, yet not confer daily an common assertion-(s),| 125 |at objective-item-unit-(s)-product-(s), defining peer-totalitarian-national-documenting through constant-conscious-interaction-(s), [sustaining /// obtaining // maintaining], clasp-affirmations', per [effort / case / amendment / article / schedule / Agenda], as participated-guideline-(s), to persuade citizen-(s), (at an) offset-relevant-juxtaposition-continuum, enumerating free-will-purpose-(s), (for an) millionths-mass-constitutional-furthering-exchange-cutoff, proposed, elucidated human-being-(s)-behavior, from subjective-feudalism,| 126 |not exceeding Twenty to Twenty-Five-Million-inhabitant-(s), per National-State-of-Affair-(s), (for how) reproduction occur-(s) by lineage-repopulation-direction-(s), impassing voluntary-existence, where grandiose reformation, not require only being, (but a the) vision Emit & Succeed, meticulous physical-psychological-subjective-diligence, (from all that) surround general-perspective-(s),| 127 |(in through to push for an) new-protestant-congressional-religious-feminine-yearly-repopulation-sects, by attempts' per fifty-year-range, emphasizing literary-article-(s), (in an) present-consistency, outlining each mathematical-88L. geometrical-dimension-subjective-inclination-(s), factoring mass-population-elucidated-ontological-etymological-documenting, considering how power is suppose (to be) ambitious, not reclusive...¶|||06m.:58s.38m.s.|||§

§0...Count-[8 / Eight]-preposition-of... ..Lines-[89.983 / Ninety-One.1.nini-tenths-2.eight-hundredths-3.three-thousandths].. ...Article-Essay-[15 // Fifteen]… ..I proudly adore pink-white-women, whom have particularly Blue-Eyes & Blonde-Hair, (as I have hazel-green-eyes, tan-white-pink-skin, [interior shirt / pants white-pink], facial-arms-lower-legs-tan, from parental-birth), (tanned unreasonably from birth) (from an) variation of Hittite-offset-genetic, having circumference [26-inches-waist / 29-inches-hips], slender [arms // legs], (by an) delicate-soprano-alto-vocal-tone-voice, (from say my competency!), having circumcised-genitalia, and petite-nipples-(one-inch-wide x one centimeter-high), without caring (too much for the) ears, (with an preference for an) [gentle-bend // straight-nose], 50-56-centimeter-head-size, (myself 53-centimeters), size 7.5-inches-feet, six-and-an-half-inches-hand-lengths, (having an) slight muscular-gluteus-maximus-bend, (learning / understanding) (as how) numerical-counts-notions-exponent-offset-dynamic, applies through existence, through physiological-components', as well miles-physics-distance-(s), from meter-meticulated-measures, gauging how [solid-densities /// liquid-substantial-(s) /// dry-substantial-(s)], are calculating why literary-documenting, (is intended to verify each) physiological-voluntary-visual-hearing-perceptive-inhabitant-(s'), (by an) notions-exponent-appreciation, for Living & Lifestyle, when respirating by [visual-vocal-tonal-tangible // supratangible-ontological-etymological-existence].¶|||01m.:17s.$_{06m.s.|||}$§

¶[16]As previous, I personally adore [Blonde-hair / Crystal-blue-Corona-eye / Blush-Pink-red-white-skin-tone], gloriously-radiating, pore-glitter-shimmer-shining from skin-tone, characterizing adoration, by classicism-personality, apose per behavioral-nature, where traded-affairs-(s), interconnect, correlated Suffice & Expanse, Show & Tell, Supply & Demand role-development-(s), interweaved from continental-shelf-expanse,| 128 |grandiose through commercial-entity-work-wage-processed-perspective-(s), operating (at an) [stance // sat motion-(s)], continuing commercial-governmental-ingestible-social-consumer-developments, all awhile hourly-time-posit-(s), at day-dating, elucidate our temporal-existence.$_{16}$|||36s.$_{79fm.s.|||}$§At this moment, I do not partake (in an) emotional-intimate-relationship (beginning of paragraph indication), for account not budget-subterior my control, (where I tender at) monetary-commercial-circulation-confidence, noticing sale-(s) compliance, when receipt-dating, (on an) market-fluent-success, prior-note-(s'),| 129 |how I.R.S., affect commercial-monetary-Federal-recollection-reproduction, (hour and / or minutes-mode-count-deviating), literary-reference-dating intermittent formal-procured tender-subtle-note-accent-(s), by organic-physiological-composition-(s'), tending offspring (as an duty of) dimensional-development, as well, enjoyed impasse-trade bemusement, accumulating-tenure as commercial-experience, balance lifestyle-(s), upon common-disposition, illiterate per [yearly // seasonal non-fiction-phenomenal-book], gandering commercial-wealth, juxtaposed interceded-cycle-duty-(ies), (from an) call of citizen-(s'), by economic-outlook, thinking, what [formal inclination-(s) // indication-(s)],| 130 |origin-spouse-reset-(s), consistent mass-elemental-material-production, cooperative at person-in-place, visual-analysis, (by those) tentative-focus-(es), set daily, before physiological parameter-(s), offspring-tabling hospitals, claustrophobic metropolitan archive-(s)-data, because mutual moral-obligation-(s), not met at, locational-deadline-(s), sustem from elder-(s), assisting youth-proceeded naïve-interpretation-(s), present duties to ontological-visual-vocal-tonal-proofs, considering conceived interval-(s), at-location by physiological-presence, (that for) now, remain reclusively-idle, by parental-parameter-(s), serving duty-(ies), collecting compositional-food-(s), (as an) method for practice upon, consistent-daily-reproductive-generalized-nation-status.$_{20}$¶|||02m.:06s.$_{51m.s.|||}$§ Discern not be evoked, by voluntary-common-living, handling hardware, pray-swear amid [physics / beings' / subjectives / article-(s)-schedule-order-(ing)-(s) / Tenure-act-(s)], where / how / what / when / who] performance, absolutely-require, visual encapturing, formal documenting motion-(s), | 131 |developing-an tier-subjective-word-definition-(s)-article-(s)-rationing, at modern-presence, where (inter)-voluntary-interaction-(s), generally underperform each [content / context focus-(es)], historically configuring-purposes, equilibrium extraspective-daily-topical-issue-(s)-diatribe-(s), emphasizing physiological-stage-(s), where per intersexual-affairs-(s), implement whom conjure elemental-creation, defining what surrounding physiological-covered-capacities', subjectively-elaborate, congenial interactions', for commercial-production-purpose, tasking [water /

soil-condition-(s) / sand // rock // stone densities], to remain relevant where genetic-being-(s), formulate-document-discerning, particular grande-interpretation-ambiguous, per person-character-(istic-(s)), mutual (by at an) aptitude-guidance, severely incrementing [syntax // paragraph // chapter // article-(s) capacity-(ies)], (upon an) month-year-age-progression-rate-(s), offset birth-season-(s)-week-of-progression-(s), [age-tier-formulating interval-season-(s) / interval-month-(s) / interval-week-(s)], corrugated by discretion of inhabitant-(s), articulated seasonal-orator-directive-(s), manipulating [metal-(s) /// plastic-(s) /// glass-(s) /// wood-(s) /// twine / cotton], (by those) competent-inhabitant-(s) [1observation-(s) / 2protocol / 3procedure-(s) / 4task-(s) / 5process-(s) / 6funded-effort-(s) / 7communal-project-(s) / 8instructions / 9directives / 10agendas], pertaining individual-causes', per Inhabitant & Community-Offset-Discourse, abstracted where ambiguous-state-(s), settle purport by day-hour-(s), idle (through an) aging-longevity, per [Passing-Hour-(s) & Physics-Earth-Centripetal-Force-Continental-Shelf-Hour-Elapses / Minute-(s)-Continental-crust-passing-(s) // revolutionary-year-dating-fifths-second-(s)-tic-(s)-placing-(s), equivalent-one-minute-impasse //// day-denominative-minute-(s) deductive centripetal-force-synapse-period-(s), elapse two-hundred-forty-numerative-secondths-sum, apose One-thousand, Four-Hundred-Forty-Minute-(s)-denominative-sum, equating syntax-deviations per day-night-rotation-period-count-(s), six-minute-(s) // three-hundred-sixty-secondths-synapse-interval-(s)],| 132 |impassing schedule-purposes', through full twenty-hour-(s)-sequencings', apose mode-objective-momentary-elapse-period-(s), centripetal-force-continental-shelf-posit-(s), depositing, [self-perspective / being extraverted-ordering-(s)], | 133 |when relevant objective-units, abide policy-trade-requirements', instinctually surviving yearly healthy senate-legislational-committee-president-cabinet-executive-branch-philosophical-calendar-reasoning-(s), (from how) free-will, affects' personal-visual-interpretation-(s), capable-at stance-lapse-(s), visual-imagery-perceptive-retaining-of, acre-(-

less) per [hour-daily-manning / walk of residence-mode-terrain], (yet) affecting how offspring-interpretation-(s), impasse defined, geological-topographical-constitutional-political-boundary-parallel-(s), formulating various operational-location-(s), trading monetary-unit-debt, (where at), physical-documenting-boundary-(ies), perceive from market-item-shelf-unit-value-(s), (as when) sociological-free-trade-method, (not) schedule individual-(s) per entity-developing-genetic-state, (from where) title-moral-terrain-parcel-curtails, present national-identity (upon an state of) relevant-repopulation-inhabitant-documenting-archive-input-(s), circulating when [women // men], could physiologically-redevelop, but allude to my primary-theme, A Book;26 An Serie-(s) of Essay-(s), human introverted-reclusion, identifying (how sale-(s) can only be freely-expected, amid post-commercial-congressional-monetary-circulational-purpose-(s), supplementing income, when individual-family-differentiation-(s) of topography-boundary-limit-(s), consider entertainment, as value-(s'), introverted by [market // store] merchandizing trade-(s), merely-surviving, (for no) past mass-religious-offset-consensus, assisting political-agreement-(s), [when / where], developmental-juxtaposition-(s), feeder away, consistent pursuit, (as whom) volunteer, participated trading fashion, pre-ordering, 70L.elemental-compositional-continental-shelf-tectonic-plate-extraction-(s), relevant-at [psychological /134/ confessional inquiry-(ies)] through moral-characteristic-(s), identifying agenda-proglamation-(s), (for how) theatre role-play, cycle (as well as) festival-(s), where conjuring individual-character-role-(s) subterior society, remain-an difficult-challenge, interposing mass-fidgeting, non astute elucidated formal-honest-reformation-reasoning, to [respond // interpose appropriate-critique], | 135 |from self-cursive-written-fundamental-interaction-(s), where studying proglamated human-mind superfluous-abyss, sleep, accumulate, hourly-experience, numerative-inductive, an interceding years-denominating-experience, amid each hourly-minute-(s)-time-dating-impasse-continuum-progresssion-(s)...¶|||06m.:18s.36.m.s.|||§

[Article:] [January]

§0...Count-[10 / Ten]-preposition-of.….Lines-[77.21 / Seventy-Six-•1.two-tenths-2.one-hundredths].. ...Article-Essay-[16 // sixteen]… ..(there are) human-beings'-population-purposes, which coordinate with [metal / wood / glass / plastic / cotton / coal], from elemental-extraction-refining, from [trees / shores / excavation-sites / fields / mines], for consideration, (on how to schedule an) calendar-agenda, (by the) political-genetic-party-dynamic, establishing an nation, (by those means' of)||||||||||sociological-psychological-geological-ecological-commercial-educational-reproductive-repopulating-county-city-zone-district-classification-(s),|136|requiring hard-data, hand-written, in cursive, along with retrodated-typed-documents, variating [live // historical data], by proceeding presence, before (their ever can be) free-time, to commiserate, celebrated-festive-lifestyles, balancing how [military / economic-trade / political / sociological denominating], remain subversive, by political-ties, held up, (by the) most common-meager-effort-(s), of human-voluntary-extraspective-impasse-effort-(s').¶|||52s.75m.s.|||§

¶17Human-Being-(s), (as we) commonly vocally-mundanely-refer, are averagely normal, adoring mediocre-effort-(s), (by what corners can be cut), relying congressional-calendar-committee-executive-cabinet-reasoning, upon person-(s), whom evaluate equivalent-commercial-volume-units-proportion-(s), through male-aging mass, as peer-(s) interact vocally-mutated, at commercial-corridor-(s), census per decade-aging-(s), (if any) [time // date], consensus [men / ethnicity / women], render incapable, physiological-cardiovascular-influence-(s), unpersuasive from free-will-direction, believing identity, defining practical-impasse-effort-(s),| 137 |serving duty, from military-obligational-defense-mechanism-(s), (personally pursued), when Men & Property slip conscious-title-deed-(s)-extension-planning-purpose, per mundane-city-watch, deterred avidly, from governmental-periodical-interval-examination-(s), per decade, since World War, signaled presence-at statehood-nationalship, rather (than an) Montana-state-nation-political-boundary-tiering,| 138 |conglomerated, shared facilities obsessive-auxiliary-objective-constructive-market-use-(s), resourced, retributed terrain, estate (when where) commercial-theatre, venue various-decade-period-(s),| 139 |similarly refined, by Montana-Totalitarian-Federalist-national-identity, shifting American-conceptive-resourcing, from Chinese-outsourcing, (not to have an) creative-adjustment,| 140 |from entity geometric-genetic-centralized-national-population-constant-denominative-documenting, mass-impact-(s) from Paper & Ink parched, local-county-state-developing, rather commonly-preferred, (by an) manuscript-Federal-comme-rcial-reserved-centralized-documenting-(s), simplistically commercially developing Industrial-Militarial-Treasury-currency-savings' (from an) congressionally-historical-illiterate-basis, noting [[1]geological-books // [2]atlas // [3]encyclopedia // [4]dictionary-(ies) // [5]auto-biography // [6]fiction-story-(s) // [7]judiciary-case-(ing-(s)) // [8]political treaties / [9]constitution-(s) / et cetera...] enabling common-belief, where offset sequence-(s), rely resource-(s), sustain (far more than) celestial-evident-etymological-past-history,| 141 |expedited (from an) unknown-European-political-kingdom-hierarchy-monetary-origin-continuum-directing, guised behind secular-privacy, affirm-ing trade adequacy, by human-mere-effort-(s), confirming poverty-belief, unconceptive by objective-requirement-(s), comprehending intermittent sequentially-simultaneous, level schedule-progression-(s), an ambiguous-parameters, ongoing theatrical-act-(s), following whose order-(s) instruct where person-debated-population-agenda-schedule-cycling, enable depleted recycled-traded-goods, temporally observing, structured system-(s), affirming monthly-payment-(s),| 142 | impassed, where [influctuated [paper // metal] currency-balance-determination-(s)], (are ignored from an) common daily-cursive-context-analyzing, determining method-(s) of payment-(s),| 143 | dispersed apart, [typed-manuscript-communal-grandiose-developmental-county // city-govern-ment-reasoning-(s)], remain undisclosed per [[1]entity-spacing-(s) /

[2]cursive-documenting-(s) / [3]lecturing-(s) / [4]debate-(s) / [5]day-secondths-empirical-denominative-(re)-counting-maxim-denoting / [6]theatre-performance-hall-(s) / [7]orchestral-structuring / [8]reference-oversight], defining Geology & Genetic-Offset-Population-(s), where physical-being-(s), proportion-offset,| 144 |when cardiovascular-exerting, affects' respiratory-muscle-ligament-tension-(s)-extent-(s), (undercoor-dinated) (where at) physical-visual-tonal-perspective-(s), by human-development-(s), offset [[1]secondths // [2]cent-(s) /// [3]breath-(s) per minute // [4]ventricle-musclutory-skin-blood-pulse-synapses-(dialysis)-blood-pulse-(s) per minute /// [5]centripetal-force-continental-shelf-physics-celestial-techtonic-plate-(s)-geology-mobius-pressure-fission-impact], posit-(s), per [four-minute-(s) // two-hundred-forty-secondths-elapse-(s)-shift-(s)], east to west, fixed by North direction [Arctic-Ocean // Antarctica-Physics-Tectonic-Dynamic], empirically-ordered between hourly-time-interval-(s)-motion-(s)-impasse-sort-referencing-(s), juxtaposed-an| 144 |day-century-calendar-dating-impasse-(s)-agenda-interval-(s)-hourly-effort-(s)-period-(s)-citizen-schedule-wage-political-labor-periodicalizing, defining population (as an) six-parallel, ration-offset, ranging ongoing [Physics-Metrics-Abstraction-(s) & Oceanic-Bodies / Tectonic-Plates-Metrics-Political-Offset], (uncontrolled) by literary-prose, (in an) denominative-Federal-governmental-control, weighing [volume // dry-collected-proportion-(s) of logic], historically documenting, constitutional-human-observation-guidance, directing commercial-entity-operation-(s), by hu-man-impasse-structural-development-(s), present per [year // census period-(s)], refining focus, amid religious-mass, relevant by physical-subjective-visual-depth-color-shape-tangible-vocal-tonal-vibrations-interpretation-period-(s), stratagem (from how) mass-perceptive-population-count-(s), (are in an) position, from conscious-effort-(s), per elapse-degree.[21]¶||| 04m.:09s.[82m.s.|]§ Attempting to identify human-being-(s)-boundary-fatigue-limit-(s), (is an fancy by my) visual-letters-words-syntax-imagery-vocal-tonal-sound-perspective, comprehending [literary-im-

passe-inclinations], when others mark proportional, calendar-dating-documenting-aging-capacity-(ies), fluently-conjecturing, commercial-factor-(s), by physiological-stature-presence-(s), for personal population-tangible-mass, swayed away from wilting with pastor-persons', (as when) [political-state / nation offset-boundary-grandiose-parameter-(s)], require present thinking, (for how to incorporate through) daily-dating, minuté-stagnant-increment-(s) of objective-(s), (amidst such dynamic of) [Auxiliary // Entity-(facility), title-operator-(s) security-protection], for oneself impasse by vehicle, intersecting metric-territory-(ies), while mile-(s) per individual [unsituated // unsettled // uncirculated terrain], (not ever Stop & Consider letter-numerical-inductive-inclinating why) Genetic-Segregated-Repopulation & Subjective-Commercial-Non-Fiction-Book-Docu-menting, are mid-points currenting-an direction (by of) [Self-Family // Hierarchy-Lineage & Community-Congregation], for national-political-party-Federal-State-currency, based (on no) presidential-attempt directly, abiding political-literary-developed-state, (to extract those) Treasury-Trillions-Savings, by literary-congressional-documenting, (for how) uncoordinated humanity remain by belief from title-possession, physical rather sustainable-estate-document-ing.$_{22}$¶|||05m.:14 s.$_{60m.s.}$|||§ Whomever touch-think-sensation-sense-(s)-(5), (for I) differ, suggesting we offset four-hyper-tense-(s), by wind pressure affecting [smell // taste // hearing], diminuendo-blood-pulse-synapse-pupil-air-circulating-eye-sight, while [tongue // interior-nose // interior-ear], offset ear and nose, by bone-cerebral-pressure-vibration-(s), while eye contain male-sperm-plasma, or pupil, corona, plasma, offset spider-veins, subterior ventricle-system-physique-aorta-air-oxygen-sensation-intake, (not to be consciously aware of "one" or the "other" consequence-(s)), (for how) objective-unit-(s)-commercial-succession-(s), rent apose title-(s), all depending Federal-interior-state-offset-geological-tectonic-compositional-resource-(s), (by an) equipment-requirement, storing subjective-conceptive-schedule-theory, when "will", (be through an) range of constitutional-claimed-documented-territory, comparative offset, estranged family-(ies), unaware globalization-conglomeration-(s)-(errors), per bastardized-77L.physical-trade-(s),| 145 |(due to an) competent-inept-observational-interposed-cited-documenting-referencing-(s),| 146 |(yet upon) day-calendar-hourly-time-dating-populant-community-(sequential-(per-feet))-ordering-(s), cycled governmental-currency-item-unit-patent-commercial-documenting…¶|||06m.:10s.94m.s.|||§

§0...Count-[10 / Ten]-preposition-of… ..Lines-[79.7875 / Fourty.1.seven-tenths-2.eight-hundredths-3.seven-thousandths-4.five-ten-thousandths]..| 147 |..Article-Essay-[17 // Seventeen].. ..[Military / Government / Currency-Note-(s) / Sociological Economic-Circulation], should be indicators (for an) different national-state-sized-currency-generating-political-party, working-an nation, away from (bastardized free-democracy-republican)-policing, (by our) perceptive-hyper-tense-(s)-genetic-reformation-reasoning, building centralized public-population-entities, proportional by survey-political-development-(s), [funding / investing] (into an) construction-corporations, developing barren-Montana-terrain, by labor-effort-(s), (for how) political-population-survey, must intertwine (in-between an)| 148 |select-genetic-warehouse-(Costco-entity-corporation, you (need an) identification-card-corpor-ation), centralizing developing estate-(s), working away from metropolitan-congested-slums, (for then) orchestrating [agricultural / horticultural resources], (amongst an) lineage-numerative-county-state-estate-offset, of [cultivating / grazing resources], (as well an) corporation-denominative-stockpile-resources, working along-an basis of elemental-extracting, refining ordered-literary-scheduled-product-(s), eliminating production-waste-(toy)-product-(s), streamlining how intersexual-repopulation, [male / female effort-(s)] are required, amid [sociological-work / festive effort-(s)] .¶|||01.m.:09s.18m.s.|||§

¶18Depth-perspective elicit visual-focus, often near approximate-depth-perceptive-sight-(ing-(s)), sedentary-relapsing-(s'), [shape / form / color / structure-(s) / ground-(s) erected-nature], disambiguous depth-visual-imagery-perspective-view-(s), amid natural-human-motions-notion-(s)-offset-impasse-(s), encountering-incurrence-(s) moment to moment, methodically self-voluntary, skewing routine-repetitious-axis-motions' through [day / food / clothing / shelter // housing], when human-nature, not educationally be guided to think, (for how) [motion-(s) // letter-(s) // word-(s) // grab // wipe], are collective-notion-(s) for thought, by [walking / bicycling / auxiliary-transit-(s)], (to an) designated,| 149 |objective-coordinated-entity-offset, relevant-instructed-directive-usage-(s), centripetally-shifted, amidst visual-perceptive-elapsed-climate-condition-effect-(s), acceding-interaction-(s), through various pattern-(s), primordially protag-onist-parameter-imprinting, motivating rural-conceptive territory, Designed & Taxed by propri-etary-act-(s), [accredited // funded], for raising inhabitant-(s) perceptive-nature, where rural-inhabitant-(s), space-off each factual-physiological-perceptive-vocal-blood-pulsed-synapse-being-(s)-fatigue, scheduled by linear-wage-merit-trade-processing-(s),| 150 |[{||Coordinate-(misplaced-offset)-Latitude // Longitude // altitude degrees'||}] placing sequential-procession per each commercial-entity-objective-(s)-instructive-dynamic,| 151 |retaining-hungry-accounting (by those) objective-date-cycle-ordering-(s), timed per consumer-demand-(s), through impassive inhabitant-educational-reproductive-theatrical-festive-self-feeding-agricultural-group-yield-sharing-(s), sustaining daily-tasked-objective-(s), from sociological-fervor, (having an) common human-personality-error, believing oneself (as an) political-person, while rural-terrain remain barren (upon of) geological-topographical-title-deed-infrastructure-appropriated-abstraction-(s), | 152 |administered relied-commercial-market-Federal-headquarters-warehouse-shipping-schedule-(s)-operation-(s), base-(s) priori present-day, routine-interval-(s),| 153 |fulfilling [day-secondths-maximum-limit-(s) = Eighty-Six-Thousand, Four-Hundred-Seconds' // 86,400 S.P.D.-reset], where continuum complete-rotation-$^{1}/_{365}$-day-revolution-drift, revolve gently afloat amid common-offset-visual-imagery-perceptive-deviational-adjustment-(s), executing [agenda // schedule-(s) // strategy-(ies) coordination-(s')], while weekly-act-(s), support secondths-contin-uum-miniscule-micro-offset-(s)-reset-counting, [going / gone] (time)-focus, apose year-denomin-ative-revolutionary-impasse, revising elapse-(s), moded where population focus (on an) micro-cosm-{numerative}-system-(s), amid ecological-political-boundary-dynamic-{denominating}, catalyzing mutual-citizen-informing-(s'),| 154 |(gossiping lifestyle unreproductive), offspring-live-creating, (non) utopic-reordering-(s), egalitarian directed-punctually, if ever amassing totalitarian-sect-(s), rurally-inclined, diligently counting [citizen-(s) // person-(s)], whom caucus conscious-rate-interaction-(s), default-maneuvering consistent letters-words-rubric-comprehen-sion, amid metrological-etymological-ontological-proofing-(s), proportional one-month per representative-(s)-{2,700 citizen-person-(s) per month minute-interaction-(s)}, rurally-identifying, why terrain should be [observed // constituted] per thousandths-mass-(total family, not soly men), no more than fifty-thousand-citizen-(s) per civilization-(s)-book-constitution-(s)-denominating, creating an new millionths-nation-documenting-currency-dynamic, similar (to the) European Union, yet else notion-documenting, (from the) 50,000-population-interactive-dynamic, (into an) [five-nine-millionths-population-50-100-volume-365-pages-per-book-documenting / one-nine-tenths-millionths-population-100-1,000-volume-365-pages-per-book-documenting /155/ one-nine-hundredths-millionths-population-1,001-10,000-volume-365-pages-per-book-documenting, not exceeding an library-room-formal-read-written-denominating-perspective, more than one-billionths-population, (for of) national-shifting, from such untangible-communications];$_{27}$ redacting repetitively, for minuté-inclinated-indication-(s), by vocal-conscious-vernacular-inductive-reasoning-impasse-(s).$_{23}$¶|||03m.:19s.$_{22m.s.|||}$§

Their (is no) singular personal-view, perceiving ontological-proof, so what define thought, without [(live)-writing // read-participating-competent-(s)], (from an) ongoing process-(es), (by whom can) interpose [etymological-(physical-perceptive // production-tangible)-ontological-(subjective)-proofs (Merriam-Webster-dating)],| 156 |deriving past-pertinent-historical-genre-library-decimal-order-(s), (I ignore, and trust live-documenting), while centripetal-force, spherical-burst-rescind, molten-pressure-kinetic-energy upon tectonic-plate-(s)-metrics-mile-(s)-spherical-proportional-drift-deviation-(s), mathematically-shaped, continental-shelf-political-country-state-(s)-territor-ial-boundary-(ies), evoked by human-blood-pulse-synapse-rate-(s')-deviation, mobius-afloat-awhile kinetic-molten-spacial-centripetal-drift-tectonic-continental-shelf-offset, mundane-emit, thinly-cooled-temperature-water-drift-vapor-(s), upon physiological-respirational-oxygen-[inhale // exhale]-intake-presence, visual-perceptive-imagery, evident display-(s) from inhabitant-(s)-perspective-metropolitan-conjunction-(s), causing-an individual-(s') [bias // hoc array], complicated from reset-(s) (by of) day-order-cycling-(s), meriting sleep at night, well-comforted at self-family-stable-anticipation, differing comforts from taxes, partaking among hourly-wage-method-(s) of payment, practiced [when // where] independent-being-(s), enable work-effort-(s), creativeless by surrounding barren-area-space-(ing-(s)), documenting commercial-reproduction-(s), (by an) genetic-mass-identifying, (from an) protestant-sect-default-subjective-competency-offset, defining self-genetic-characteristic-preference-(s), "intimately premised", appreciating [visual interaction-(s) // effort-(s)] through constitutional-terrestrial-action-(s), intermittent-interposed-boundary-focus-(es), unilateral (by an) genetic-agenda, psychological census, sustain-ing participating citizen-(s), whom elucidate patent-conceptive-requirement-(s), understanding objective-time-frame-interval-(s), periodically-dated, developing prose per genre-referenced-[fact / proof]-(s), control designating totalitarian-genetic-government-currency-hourly-time-interval-dating,sequentially-periodicalizing, denomin-ative-documentation-(s), juxtaposing| 157 | [government // psychological-firm // university-structure-(s)], systemizing-an [checking // saving-(s) // cash // emergency-funds of survival common-human-being-daily-exist-ence].24 ¶|||05m.:15s.43m.s.|||§ Is time ever completed? Or will animated-physique continue [contenting // contexting], abstract-spacial-existence, indirectly-focused, from Vision & Hearing how far (into a the) future can anyone see? Whom do we know, have note-(s)-concerted, cited, from (felt)-consumption, where purpose of objective-(s), remain dormant, where interior-perimeter-surface-area-currency-circulating-entity-(ies), identify individual-character-presumption-(s), customary comforted appliance-(s), by auxiliary-effect-(s), 74L.determined by blank-perceptive-moment-(s), unreasoned through near-sighted commercial-calendar-schedule-effort-(s), (by an) classical-consistent-socialism..?¶|05m.:46s.88m.s.||||§

§0...Count-[6 / Six]-preposition-of… ..Lines-[75.13 / Seventy-six-•1.one-tenths-2.three hundredths].. ...Article-Essay-[18 // Eighteen]… ..Perspective & Genetic… lifestyle would (be an goal for) National-Federal-Montana-Political-Party-(other name-abridgement), or chest rating commercial-literary-structures, encompassing lifestyle, (by those) means' per trade, comfort, classification-effort-purpose-(s), embellishing lifestyle-hours-dynamic, amid Earth-Centripetal-Force & Physiological-Visual-Vocal-Tonal-Vibrational Perspect-ive, Genetically-offset, from political-religious-observational-characteristic-(s').¶||26s. 06m.s.|||§

¶19 Winter initially commences by [December-Twenty-First // December 21], as each year, be perceptive, (by an) grandiose-celestial-transit, while earth-centripetal-continental-revolutionary-axis, revolves around English-conjuring-observation-(s) through January-primary-month-inclination-(s), climate affecting, ethnic-disrespect-(s) interceded holidays [New Year's Day-{1st} / Martin-Luther King Day-{17th}], object-swearing under American-colloquial-commercial-unit-objective-trade-usage-(s), royally partaking-task-(s), at placed [objective-(s) / currency-(ies)], conclusive ideological-freedom, practicing-preferred task-workload-(s), as [plumber / technician schedule-40-usage-belief] where imbued physiological-embodying, unaware [economic-turn-over / government-legislate], permit [House-Representative-(s) // Senator-(s) Calendar-Bill-Permission-(s)], defining expectation-(s) (for of) Mass-Population Supply & Demand Order-Cycling-(s), with-holding general-inhabitant-(s)-decision-(s), unratified national-constitution-(s), where state-(s)-offset-territory-legislation-(s), confirm emancipated Federal-congressional-order-(s), differentiated apose bastardized-genetic-discretion-process-(es), never considering (how each) notions-(s) offset, affect intermission-progression-(s), valued because how constitution verify Population & Designated-Boundary-Territory-(ies),| 158 | variablizing, tierology-organization-(s),| 159 |apposing document-(s)-continuum-perfection-ordering-(s), (from of) tectonic-centripetal-continental-geological-nature, where commercial-product-objective-posit-(s), abstract from defining-point-(s), terminologically-inept, where term-definition-(s), confirm past-ontological-affirmed subjective-interpretation-(s), (when an) fixation per article-(s), consider how, explained-definition-(s) (from of) point-term-(s), exemplifying term-usage-(s)-non-point-definition-competency-ambiguous-points, not intercontort per identified-article-(s), existing at particular-location-posit-(s), offset per independent-inhabitant-(s) spacial-premise, affecting instructed-auxiliary-manual-usage-(s), intermittent undefined [where // when], parameter-(s) remain barren, by intersexual-repopulation-reproduction-purpose-(s), lacking cause, (from an) primordially-personable recollecting, impasse-visual-imagery-perspective-(s), exper-iencing contingent subjective-focus-(es), askewed through metropolitan metrological constitutional-review,| 160 |internationally-Federally-trade-tolerant amicable with [ethnic // (racial)], being-inhabitant-(s)-background-(s'), when basic-human-rights-belief, fear-(s) lower-class-ification-starvation, (which would be an) better way at handling objective-instructed-technique-(s)),| 161 |assumed through retaliation-weapon-occasional-assault-(s), posited mass-life-influctuation-(s), remaining offset, tectonic-grandeur-awe.$_{25}$¶|||02m.:41s.$_{68m.s.|||}$§ Cold frigid spacings apose solar-temperate-continuum-impact-effect-(s), juxtapose solstice-weather-pattern-(s), when revolutionary-pattern-(s), inset fourths-season-(s)-impasse-(s), adrift thirdths-temperature-effect-(s), gently-bending north (a the) equator-temperature-extreme, curiously drifting such physics-phenomena, unlike physical-extension-(s), physiologically disregarding personable-extension-(s), where social-reclusion-(s'), mind possessive-self, pertaining monetary currency interchange, while mobius vast-momentum, accept broad-trade, free-bias-persuading, ridicule per objective-requirement-(s), identifying territorial-motion-act-(s), disclosed where local-estranging-(s), affect commercial-factoring-(s), (by when) tectonic-elemental-compos-itional-fact-(s), not yet have architectural-individualism, develop mathematical-proof-(s), congruent per patent-objective-construe-(s), proportioning visual-proof-(s), intending-an [[1]calendar-patent-cycling // [2]industrial-commercial-objective-production // [3]market-substantial-box-unit-volatility // [4]housing-lifestyle-(s) // [5]sociological-extraspective-personal-time-interaction-(s) // [6]sleep // [7]Auxiliary-Mechanics-Corporation, all by Genetic-Congressional-Literary-copyright-inductive-periodicalizing-(s)-numerative-continuum-effort-denominating, (suggesting we need congressional-legislation, of judiciary-casing-(s), (for how to) denominate-humanity, [subjective-(s) // article-(s)] (from evident-educational-economics-observation-(s), (for why) humanity is illiterate, by written-cursive-manuscript-means, relying type-"manuscript"-case-contexting, verified identified vocal-capable-inhabitant-(s)), amid executive-branch-total-always-physical-requirement-(s)], maintaining commercial-item-posit-(s), where human-blood-pulse-synapse-timbre-(s'), monitor physical-offset-interpretation-(s) per instructed-commercial-shelf-unit-objective-(s), abiding selfish-lifestyle-(s), affecting circumstantial-outcome-(s), without-an tectonic-developmental-purpose-(s), (vaguely) understanding extraspective visual-image-perspective-(s), conceded elucidated evident-impasse-requisite-action-(s).[26]¶|||03m:57s.[12m.s.|||]§ (So how would) winter affect inhabitant-(s)| 162 |[eating-routine-(s) / clothing-routine-(s) / social-timed-interaction-(s)], planned-ahead commercial-scheduled-designation-(s) of housing (electricity / water), sleeping repetitively at interior-title-tile-posit-(s), for self-security, defining loyal-abiding-inhabitant-(s'), when voluntary-mass-syntax, fashion synapse-(s), elapse-(s) within limited-boundary-competency, per moded-minutes, [age // perspective-(s) juxtaposition-(s)], derived maintained debt commercial-unit-purchased-objective-(s), using compositional-product-part-(s), by mass-expectation-demand-purchase-parity-(ies), conveyer, factory-commercial-production-(s), with governmental-commercial-elemental-geological-shared-load-(s), supplying industrial-stock-purchasing-(s),| 163 |from Federal-Reserve-denominating-guidance, | 164 |Internal-Revenue-Service, past-earn labor internal-interior expected-hourly-wage-year-output-(s), when service-(s) monitor educational-(family)-granted-accreditation-(s), per [citizen-(s) // commercial-corporation-(s) // credit-bureau-(s)], [patent // historical-biographical-commercial-component-objective-factory-(ies)-composition-(s)].[27]¶|||04m.:53s.[35m.s.|||]§ Much time dissolve past human-cognition-(s'), faulting inhabitant-sensation-(s) "cogito ergo sum", simply pulse-(ing-(s)) pass-(es), at [[1]work / [2]clothing / [3]showering / [4]fraternizing / [5]eating / [6]driving / [7]drinking / [8]bowel-dispelling // et cetera…], where those public-county-commercial-reliance-(s), | 165 |surround daily-life, ignorant formal-character-identification-(s), interposing infrastructural-motive-(s), (by of an) lifestyle-(s), causing [[1]housing / [2]agricultural-yield-(ing-(s)) / [3](super)-market-trade-(ing) / [4]terminological-review / [5]reproductive-concern-(s) / [6]sophisticated-socializing / [7]psychological-documenting / [8]government-branches // [8.1]powers // [8.2]jurisdiction-denotative-commercial-objective-Federal-balancing-(s)], requiring mass-population-locational-repopulating-offset-(s), resourcing elemental-value-(s), exemplified seasonal-report-(s), where various-efforts, impact perceptive grade-rate-(s)-point-(s), representing constitutional establish-ment-(s), rurally resided, when subjective-systemized motion-(s), impasse daily-physics-regard, implicating thought, (by an) mass-trade-method-ignorance, representing politically-enabled-trading, (not) personally-reasoning person-ration-conceptive-patent-after-effect-(s),|166|con-traposed, commercial-box-unit-production-(s).₂ ₈¶|||05m.:52s.₄₈m.s.|||§ Neurological-synapse-tense-blood-pulse-(s)-inhabitant-(s), tension-₇₆L.blood-pulse-ventricle-friction, sense-decay, from [sleep // fatigue], interposed centripetal-force-molten-tectonic-revolving-rotating-air-respiration-vapor-(s), adrift, tilt (of a the) globe-(s')-axis, slant intermittent centripetal-force-impact-effect-(s)...¶|||06m.:0 8s.₉₈m.s.|||§

§0...Count-[7 / Seven]-preposition-of.... ..Lines-[77.92 / Seventy-Eight.1.nine-tenths-2.two-hun-dredths].. ...Article-Essay-[19 // Nineteen]... ..Scheduling requires' an addressing (by of) physical-health-count-(s), interceding physics-work-act-(s), orchestrating-an formal-develop-mental-genetic-classification-(s)-society-(ies), per lineage-communal-voluntary-respects'. [...240-Count-Preposition-of... Start ...Fourty.170731....]| 166 |redact-deviation-median-offset-page-range [Fourty-Two / 0.048780.... // 0.170731....]¶|||26s.79m.s.|||§

¶20...Node-winter (is of an) freezing-state, empty where physiological-nature, perceive blood-pulsed-synapse-tensed-being-(s'), cooled-down, where kinetic-friction-tense-(ing)-(s), decay intermittent natural-circumstance-(s), exterior celestial-boundary-(ies), remaining vacant, per distance-(s), unmeasured, (by those) inhabitant-(s) measuring-(s), (for whom) consider how articulation by vocal-vibration-(s), modify letter-(s)-word-(s)-term-(s)-participle-(s)-syntax-(es)-sentence-(s)-definition-(s)-elapse-(s)-debate-(s)-paragraph-(s)-phrasing-essay-(s), upon [chapter-(s)-(historically) /or/ article-(s)-(present-comprehension-concertion-attempt)],| 167 |book-copyright-article-(s)-(chapter-subject-maxim-extents, suggesting [youth // collegiate subjective-extension-(s)] ramble on, without-an clear benchmark-constitutional-articlizing,| 168 |essay-paragraph-phrasing, those proof-(s) of location-objective-currency-weight-(s), proportioning-article-reasoning-(s),| 169 |identifying how inductive-incremental [inclinations // indication-(s)], progressively-interpret, depicted conscientious perspective-(s), defining interposed-interval-concerted-motion-time-period-(s), counting by visual-imagery-effect-(s), determining-(s) how developments, remain elementally-resourced, retained for, dictatorial-repopulation-impact-(s), by participated-being-(s')-effort-(s), or simple-standard-(s), pervading persistent consistency-(ies), because (a the lack by of) free-voluntary-will, upon elemental-honest-objective-extraction-(s), inductive-denominating, (by those) dimensional-spacing-(s).[29]¶|||01m.:13s.[75m.s.]|||§ Infertile-field-(s), positing along the northern-hemisphere, geologically-ranging intermittent latitude-degree-offset-(s), apposed The United-State-(s) of America-North-South-Gama-parallel-(s), per [Twenty-Four-degree-(s)-Thirty-Three-Minute-(s)-North-84°-W.-(Key-West-Florida-Straits-Southeastern-Point);[28] by Forty-Seven-Degree-(s)-Twenty-One-Minute-(s)-North-74°-W.-(Maine-Canadian-Border-Atlantic-Northeastern-Point);[29] by Seventy-One-Degree-(s)-Seventeen-Minute-(s)-North-72°-W-Barrow, Alaska-North-Western-Point);[30] by Thirty-Two-Degrees-North-118°-West-Imperial Beach San Diego-Southwestern-Point-(Pacific-Ocean-Mexico-American-Border)], defining various State-Constitutional-Continental-Shelf-State-(s), abstract by Population-Designation-(s) & Tectonic-Continental-Coverage;[31] spacing-off [agricultural-development-(s) // intersexual-repopulation-(s) // psychological-scheduling-(s) // governmental-objective-inductive-denominative-debate-(s) // university-educational-production-(rather commercial-factory-output-(s))];[32] Producing & Decaying through frost, because intangible-field-surface-area-(s), relaying stagnant-energy-(s), (un)-populated interval kinetic-photosynthesis-aquatic-bodies-currenting, plant-crop-process-nutrition-posit-retention-(s), balancing diet intermittent with maricultural-comestibles, individualize like plant-algae-particular-type-(s)-{taxonomic-classification-(s)}, physically-arranging soil by proportional [title-property / commercial-governmental-county-state-Federal-Contract-(s)], verifying-matter-(s), where numerous [plant-(s) // crop-(s)], offset juxtaposition-(s) per plant-species-(s), for various hypertension-(s), reset motion-objective-usage-(s), relevant to [décor // edible-(s) // shading // medicine-(s) // ambient // condition-(s) per crust-continental-shelf-political-boundary-(ies)],|170|seasonally breaking observation-(s), from [space-solidified // molten-solstice-(s)], apposed equinox-lukewarm-medium-temperate-impasse-rotation-impacts, variating offset north-south-hemisphere-centralized-fixed-posit-(s), (from when) solar-celestial-physics-distancing-(s), impact terrestrial-taxonomic-order-(s), at tectonic-mountain-mass-east-west-seismic-centripetal-drift-continuum, | 171 |constantly-pressurized, supratangible immovable physics-matter-(s), physiologically inclined, differing [hundred-(s) // thousand-(s) spherical-cubic-miles-(s)], offset physiological-limit-(s)-quality-(ies);[33] yearly intervalizing visual-imagery-observation-(s), per fifteen-day-(s)-interval-(s), (like day-hour upon yearly-transit), observing calendar-seasonal-passing-(s), celestially-abstract, while inertia, transit-an pleather of matter, as celestial-four-minutes-degree-(s), [latitude-rotation // longitude-level-trajectory], altitude-centripetal-depth-perspective, where personal-individual-height, posit-perceive-offset, juxtapose physics-distance-challenge-(s), by| 172 |[climate // temperature // topography // geographical-political-statehood-origin-birth-offset-livelihood-(s')].[30]¶|||03m.:44s. [35m.s.|||]§ We are capable of receiving message-(s), preparing from seasonal-program-preparation-(s), Preparing & Stocking crop-(s), per [[1]Wheat / [2]Grain / [3]Oat / [4]Rice / [5]Corn / [6]Potato / [7]bean-(s) / [8]lentil-(s) / [9]carrot-(s) / [10]yuca], during offseason-development-(s), having-an order continuum nutritional-routine, from edible-delectable-(s),| 173 |favoring-an clean smooth-digestion-processing's, planning preparation-(s) during Summer-Solstice, anticipating regards', apose balancing [mean-(s) // meal-(s)],| 174 |from documenting-tangible-weight-(s), proportional ounce-comestible-proportion-(s), cycling food-group-(s), relating growth-pound-(s), juxtaposed packaged-ounce-(s), eating comestible gram-(s), from [fork / spoon / knife], [[1]cutting / [2]lifting / [3]inserting / [4]chewing / [5]swallowing / [6]waiting / [7]digesting], pieced-increment-(s) of digestible-taste-consumption-matter-(s), considering by weight-digestion, proportional equal-rationing-(s'), offset economic-developmental-fact-(s), as [muscle-mass / organ-tissue-lather-remineralization-(s) / skin-tissue-remineralizations / ventricle-system-remineralizations / blood-pulse-mineralization cycle-process-(es)], proceeding avid constantly human-political-limited-parameter-(s).31 ¶|||04m.: 41s.10m.s.|||§ For now, abiding municipal-metropolitan-corporate-(restaurant)-order-(s)-trade-method-(s), base-premis corporate-commercial-Federal-Headquarter-system-(s) of relied-trading-technique-(s), not fluent in cursive-written-documenting per individual-(s)-(ism)-morality, referred by independent-self-(ves),|||||||||||||||||||interacting amongst conjured-flavored-note-(s), infused intelligent-book-collection-(s), from market-resource-(s), interpreting accoutrement soluble-edible-(s), significantly incrementing lifestyle-interaction-(s), Fatigued & Replenish-(ed), (not thorough), genetic-ethical-survey-(s) (by of) socialism-requisite-(s), considering objective-factor-(s), by collected-quality-nutrition-supplement-(s),| 175 |tamed through constructed-infrastructure, (for how) recipe-crop-regenerating-cultivation-hoarding {One-foot-Four-Inch-(es)-Height // by Length // Width per field-development-(s)}, coordinate population congregation-segregation-congruent-effort-(s), [when // where] year-round geological-crop-rotation-yield-(s), overgrow in concentrated-cultural-yield-(s), imbalanced practical-tool-(s), farmer-hoarded, resourced-soil-composite-(s), (bricklaying) (as an) community-ecological-effect-(s')-impact, influenced from mass mundane motivational-movement-(s), where physical tangible-object-placing-(s), perceive through visual-depth-perspective-distance-(s), among approximate-terrain, thinking how voluntary mass-consumption-effort-(s), restrict physical-exertion-fatigue-limit-(s), without restraint from legal-hemispherical-trust, perpetuated by personal-habitual-behavior-(s'), commonly endeavorless, by geological-tectonic-political-party-pursuit-(s), (by of) social-population-(s), under auxiliary-restraint-circumstance-(s), Time & Calendar rating, economic-stimulus per season-(s), for accounting tax-(es), incurring repetitive-directive-(s), setting agenda-purpose-(s), by market-demand-resource-balance-proportion-(s), instilled from Past-Requisite-(s') & Future-Budgeted-Requirement-(s) upon global-affairs-(s), proportional commercial-market-traded-goods, acquired by social-service-(s), centralized at [ambiguous-city // county-(not state / Federal-Nation), planned-housing-system-(s)], funded from Federal-developed-circumstantial-commercial-discourse-(s), requiring [[1]land-purchase-(s) / [2]concrete-brick-count-(s) / [3]mechanism-(s) / [4]mutual-citizen-(s)-terrain-wage-developing-(s) / [5]wire-production-purchasing-(s) / [6]water-fortifying / [7]current-(s) of electricity coordinating county-governmental-water-plant-(s) (why I see nationalism the difficulty of inhabitant-(s)], (in regard where at no such moment), an fancy of mine, if ever Sociologically-Surveyed & Nationally-Militarily-Engaged formally, work-alleviated, ritual-work-burden-(s), present by faith, engineering-entity-market-housing-development-(s), by discourse per timid-pre-requisite-period-(s), mapping ecological-system-(s), for sociological-sophistication.[32]¶|||06m.:55.[81m.s.|||]§ I understand the excruciating-burden very well, not applicable in any decade of timing, (but as for the) analysis through competency (by an) dozen-(s)-millionths-humanity, node-(ed) Non-Fiction, (if ever our) education-system succeed expectation-(s) (from those) neglected-burden-(s) of subjective-physics, by free-will-individualism, intended (from a the) discourse of engineering, in [[1]extraspective-factory-production / [2]spacing-(s) per governmental-infrastructure bridge-(s) / [3]draw-bridge-(s) / [4]ditch-(es) / [5]dam-(s) / [6]water & power-facility-(ies) / [7]field-crop-collecting / [8]et cetera...], default by entertained-inhabitants', which ridicule rigor from focus of dimension-(ed)-conceptive-visual-imagery-vernacular-vocal-tonal-perspective-(thinking), impassing locational-routine, time-(ing)-observation-(s), [(a) / picture-envision-(ing);[34] (b) / principality-resource-concept-developing;[35] (c) / survey-(ing) inhabitant-(s) // citizen-(s) of state;[36] (d) / constitutional-population-political-party-ordering-(s);[37] (e) / fund-(ed)-raising-effort-(s) by commercial-advertisement

& book-sales;[38] (f) / developmental-monetary-dissolve of national-congressional-order;[39] not (h) / further];[40] for attention-span, is dulled by alert-sense, not in functional-response through innuendo-logic-(s), from notion-(ing) nationalism, (or the) World War-(s) (of the) twentieth-century, purpose-Significance upon extraspective-posit-juxtaposition-(ing)-(s), interacting amongst competent-(s) that market how trade is applied, from mass-educational-ignorance, continued by inconsistencies, as reclusive-introverted-inhabitant-(s), whom not understand psychological-genetic-honesty, affect all that we retain, (from an) simultaneous-commercial-market-debt-access-system-(s).[33]¶|||08m.:12s.[2] [8m.s.|||]§ So how ever does anybody think, would not intuitively register, [budget / religious virtue-dynamic-(s)], persisting incompetent-aptitude-(s), saving from [wage / annuity / fund-(s) / income], self endeavoring retirement-planning, in cause of collect-(ing)-object-(ive)-(s), in relation per calendar-dating-(s), surveying citizen-(s), from inquiry amid American-national-product-trade-reliance-longevity-discourse, deciding either modern-individual-lifestyle-(s), past retirement, until death looms, poor for instating State-National-Cycle-Reform-(s), demarking territorial-metropolitan-posits, from constitution analysis of Population & Objective-Mission-Discourse-(s'), (as thought for) lifestyle-objective-effort-(s), rather than, stagnant-work-place-trade-consistency-(ies), affirm where American-state-(s), offset settlements, primarily (by the) Nineteenth-Twentieth-Century State-Constitutional-Transition-(s), nationally constituted, under Federal-Union, for methodical-thinking, past those requisite-(s) of public-commercial-development-(s), transiting pass-beyond millennium-boundary-(ies), yet tensing-an lineage from existence, well under thousandths of year-(s), illiterate by aged-work-place-(s), limited by creativity, per commercial-market-production-(s), minimally-value-(ing) product-(s), self-introverted-affirming cause-(s) of patent-concept-development-(s), by land-proprietorship, owned in allegiance through [constant-activity-(ies) // renovation-(s)] stable by rural-site-(s), affecting [[1]Tradition-(s) / [2]Custom-(s) / [3]Homage / [4]Habit-(s) / [5]Hobby-(ies) / [6]Ritual-(s) / [7]Native

/ 8Vaules / 9Historical-Archive-(ing)-(s)], mattering (upon an) university-method-distancing-(s),| 176 |differentiating from present commercial-mean-(s), periodically-discoursing act-(s), [walking / driving / entity-renovating upkeep], 123L.enveloping-an cognitive-purpose with terrain, defining religious custom-(s), (from due-(s) by) parental-citizenship, for exponent-offspring-populating, by local principality-resourced-development-(s)…¶|||09m.:47s.79m.s.|||§

§0...Count-[25 / Twenty-Five]-preposition-of...|177|..Lines-[124.95 / Eighty-Five.1.nine-tenths-2.five-hundredths].. ...Article-Essay-[20 // Twenty]… ..How from America & International-Territory, (would their be an) mass-congressional-population, to collect those formal-soluble-elemental-resources, for practicing proportional-estate-(s), (which would space off), each of those factor-(s) per lineage-communal-interactive-repopulating, (at an) century-decade-rate-(s), gandering (how to) cultivate-customary-development-(s), ahead per present-day-dating, (for an) seasonal-yearly-crops-field-development-preparation-(s), while industrial-development-(s), require patent-literary-concertion, (to affect each) participating-competent-citizen, whom comprehend (how to resource), serving cycling (by at those) proper cultivation-(s) offset tectonic-plate-(s)-statehood-territory, when lifestyle should (be the) ultimate-goal, (for understanding how to) develop geological-boundary-(ies), (amid an) present-tense, (by of) work [wage // salary] payment-method-sufficed-living-(s), ecological-development.¶|||49s.78m.s.|||§

¶2LIt is impossible to “expect” an effort of loyalty, (as any) free-will-inhabitant-(s’), (is due to) genetic-pre-disposal, operating from stubborn-mentality-(ies), input while waged-hour-(s)-pass-by, and human-existence merely suffice upon relevant-condition-(s) under State, surviving empirical-monetary-trading-method-(s), as numerical-observation-(ing-(s)), remain dormant by linear fashion driving pass [numerative sign-(s) / road-(s)], habitually-impassed, never documenting ones'-lifestyle by self, intermittent [co-worker-(s) // colleague-(s) state-interaction-(s)], coagulating-compound-(s) by surrounding-rural-inhabitant-(s'), viewing city position-(s), per individual-self-(ves), following along with surrounding-temperature-condition-(s), fatiguing citizen-(s) from city-offset-residence-(s), micro-work-career-settling, practicing various obligation-(s) (by agenda as an) goal of existence, from notion-(s) per [Year-(s) // 365-(6)-90-(1)-(2) // Season-(s) // 90-(1)-(2)-30-(1)-(28-(9)) // Month-(s) // 30-(1)-(28-(9))-7 // week dating-(s)] per Dating & Hourly-Time-Posit-(s)-interval-(s).[34]¶|||50s:59m.s.|||§ Why is non-fiction, (so abstract from a the) discourse of human-historical-documenting? For existence thus far, the globe circulate-revolve around the sun, solidified from tectonic-plate-lunar-exo-solidification-(s)-revolving, persisting of seven (Six, one ice-rock) tectonic-plate-(s), revolving celestial-ambient, (by those) independent-state-(s), sovereign by international-matter-(s), (when those) territory-(ies) of state, populate land, conjuring through terrain-impasse-(s), where land-surface-area-proprietary-space-(ing-(s)), remain always too remain under governmental-intervention-(s), by calendar-(ed)-infrastructure-developmental-maintenance-planned-scheduling-(s), citied by work-trade-interval-(s), tensed when interjection conjure juxtaposed-perspective-(s'), conferring method-(s) of entity-routine-(s), maintaining those resolve-(s) by [water-pressure-currenting / electric-metal-wire-currenting / Cement & Rebar infrastructure / wood-beams // metal-frame-dry-wall-(gypsum-board)-interior-parameterizing / storm-drain-system-(s)],| 178 |as taxed-effort-(s) by Federal-state, facilitate placed-infrastructure from free-enterprise-trade, [entrance // exit], commercial-development-(s);[41] awhile discourse by [bias / hoc partial-opinionated-decision-(s)], remain undictated because human-being-inhabitant-(s), prefer individual common [apartment / (town)-house-(s)] residential-parcel-(s), living for patterned-work-wage-routine-(s), unacquainted (by the purpose of) [racial // ethnic // genetic] interaction-(s),| 179 |requisite formal-visual-perceptive-functioning, counting population-(s) upon terrain for physical-existence, encouraging effort-(s) by [House-Representative-(s) // Senators],| 180 |residing by [family-tier-accounting / water-currenting / electric-currenting] at residential-parameter-(s), factifying unit-count-(s) through individual-concept-(ed)-creativity, defining constitutional-population-(s), determining public-market-spaced-area-(s), at mass-interpretation, comprehending surface-area-spacing-(s), valid-ifying collected-mass-objective-(s), allocated through freedom-disposition-(s), for private-research where natural-weather-condition-(s), compare discourses' from scheduled-intervention-(s), enacting focus per composited-matter-(s), with surface-area-(s), fulfilling those self-physiological-limit-(s), meriting individual-(ism)-(s) by aligning fund-(s) through Commercial-Work & Literary-Library-Congressional-Non-Fiction-Purpose-(s), in observation-(s) present modern-year, thinking how future discourse sustem ideal-(s), continuum-time-offset, mass-genetic-identification-(s'), through presence-perspective, while (in the realm at) self-cubic-feet-spherical-fathom-area-(s), posited upon infinitive-reset-present-year-(s), as calendar-(less)-usage from aligned-human-being-(s), prefer factor-(s) at various distance per day, traditionally [resided / transited / worked / interacted], encompassing [day // night] physiological-perceptive-blood-pulsed-synapse-being-(s'), sleep-regimented, during surrounding-religious-impasses upon state-developments, when referendum, define factoring, miniscule-fragmented-compound-(s), apose medium-alternative-timing-(s), aligned where calendar-dating, pertain orchestrated-entity-operation-act-(ion-(s)),| 181 |reference-experiment [[1]compound-density-(ies) / [2]distanced-range-approximation-(s) / [3]decay-(ed)-life-count-(s) / [4]water-mass-level-count-(s) / [5]color-mineral-phosphorus-testing-(s) / [6]transitive-usage-(s) / [7]sodium-chloride-test-(ing-(s))], circulating constantly impassing circulated variable-(s), [which // that] surround (us by) frequent-activity-(ies), deserving tallied-mark-(s), reasoning count-(s)-adjustment-(s), incorporating mutual-relay, from experiment-count-(ing)-(s), (used for) defining-parameter-(s) at commune-junction-(s), casing material-objective-development-(s), (by an) aged prerogative-genetic-community-(ies), from release of possession-(s), viewing-object-(s), through distancing tangible-(s), approximate weight-density-(ies), for-visual-thought-reasoning-(s), tiering| 182 | [Day // Week // Month Hour-(s)], apposed [seasonal /// yearly Date-(s)], proportioned by median-acreage-spacing-(s)-designation-(s), positioning limit-(s) conjecture, determining purpose for objective-placing-(s), when citizen-discourse-(s), abode land upon terrain, while focus rely coordinated-condition-(s), by governmental-person-(s)-role-(s), defining Incompetent-Being-(s) & Undeveloped-Surface-Area-(s), because outcomes per live-review, interval Solar-impact, emitting kinetic-aquatic-vapor-horizontal-latitude-rotational-revolutionary-axis-spin, [extraction // instilling], kinetic-energy-discharge-offset-(s), regenerating plant-algae-composition-(s), [crop-(s) / Mammals' / tree-(s)], discoursing hunger-survival-motivation-(s), actively setting propose-(d)-effort-(s), (due to a the) nature of human-consumption-(s'), devouring edible-goods;[42] awhile date correlate constant-decay, apose human-free-will-rebirth-effect-(s), hence comparing age-(s) per tier-(d)-impasse-deviation-(s), to day-(te) progressing retrodated-grade-point-(s)-posited-qualities, juxtaposing child-(ren) by [[1.a.]dawn / [b.]teen-(s) /// [2.a.]mouning // [b.]young adult-(s) /// [3.a.]day //// [b.]mid-aged-adult-(s) /// [4.a.]mid-day ///// [b.]pre-retirement-adult-(s) /// [5.a.]afternoon ////// [b.]retired-adult-(s) /// [6.a.]evening /////// [b.]elder-(s) /// [7.a.]dusk //////// [b.]Grand-Mother-(s) / Father-(s) /// [8.a.]midnight ///////// [9.b.]Great-Grand-Mother-(s) / Father-(s) /// [10.a.]early-mourning ////////// [b.].Great-Great-Grand-Mother-(s) / Father-(s) aging-discourse-(s)], remaining (at an) position, amidst nature, too grandiose for absolute-specification-(s), amid those hour-(s)-interval-(s), expecting extraspective-discourse, retrieving focus, [tasking / preparing / serving],| 183 |Recipe-Comestible-Substance-(s) & Age-Progressive-Deteriorating-Posit-(s), (I be of the young-adult-variety), interposing period-(s) per time-posit-(s), directing location, rigor Extracting & Extrapolating, each subtle

nuisance as secondths-continuum-minute-(s)-mode-count-(s)-offset,
scale sole-purpose conjuring non-fiction-method-(s) (by of) living,
furthering hyper-tense-schedule-progressive-prerogative-(s), by iso-
lating those requirement-(s) in harmony where want-(s) secure com-
mercial-pricing, adjusted in relevant-regard, an bastardized-elemen-
tal-extraction-(s), producing commercial-factory-currency-note-(s)-
denominative-verified-adjusted-configurations.$_{35}$¶|||05m.:49s.$_{24m.s.}$§
Prior position-(s) pass past aged-citizen-(s), control (from an) auto-bi-
ographical-pattern, thinking when cursive-etiquette, never acceding
genetic-referendum, while isolated focus, remain from Children &
Retirement, for educational-prose, where article-(s), inset purpose
(far further than) discourse-conjuring, alluding impact reproduc-
tion-exponentiation, setting-an schedule-dynamic for de-
cade-time-posit-(s)-anticipating, stating purpose-(s) by presence,| 184
|an introverted-expectation-(s), amid political-extraspective-expecta-
tion-(s), (as how) survey conduct (a the) purpose (by of) present-his-
torical-progression-(s), unable to let go, past-error-developmen-
tal-fault-(s) at workplace, deconstructing through labor, those vacant
[office / residence-(s), edifice-(s) // entity-(s)], defining city-plan-
ning-effort-(s), by 401k-pension(s), never too be enough, while pro-
portional-literary-discourse, (has yet) examined (a the) excruciat-
ing-effect-(s) of Mass-Population-(s) & Legislational-Government,
(the error of Executive-Person-(s)-Observation-(s) & Com-
mon-Law-Judiciary-Review), International-Legal-Binding-Mone-
tary-Trading-States-Theatre, define decade-aging-judgement-(s),
still-idle from monetary proportioning, by frequency (per of) [mass-
act-(ion-(s)) / length of day-interval-discourse / yield of human-re-
sourcing-effort-(s)], which incorporate "time-deviation-(s)", attempting at elucidating literature, in concerting Accurate & Explicit On-
tological-description-(s) of territory-(ies), upon human-popula-
tion-statehoodship-discourse, | 185 |envisioning how acre-(s) become
pre-concepted, from self [weekend // off-time-religious-inquiry-(ies)]
(sport retardation-brainwashing-expending), while requirement-(s) of

technological-unit-objective-usages', concert commercial-production-(s), visualizing those acre-(s) per [forestral // pasture spacing-(s)-natural-arrangement-(s)], by aesthetic-narcissistic-body-conscious-view-(s), (as of) subduction-zone-ambient-vapor-centripetal-force-impact, imbue (as like) grandeur of perceivable-existence, always attempting by complete-work-week-material-(s), infusing work-week with weekend-scheduling-(s), arranging sleep-pattern-(s), adjusting lovely-characteristic-(s), often disregarded where cycling mutual-counterpart-(s) upon perceive-able-exterior-surface-area-handled-distancing-(s), securing [live-presence-dating-schedule-(s) / future-impasse-calendar-interval-presence-dating-(s) / book-present-reference-retrograding (think-(ing) inbetween [here // at]], an live-dating by grammatical-data, perceiving requisite visual-tonal-vocal-interpretation-(s), sociological-living-routine-(s)), ordering historical-Dewey-decimal-past-referencing-(s), [understanding / comprehending] formal etymological-purpose per [life // existence], apposed memory-post-existence-purpose-(s), thoroughly executing action-(s) (upon those) nature of affair-(s), (a the efforts per Woman-(en)) in lifestyle by discourse of sedentary-reproduction, continually disregarded by men, for preference of Americas' (celebrity)-white-collar-work-routine-(s), existing (greater than) religious-Treasury-communal-offspring-intracontorting genetic-identity) (due to) pension-(s) per character, rather Expansion & Passion, that would come from say an R.V. (Road Vehicle)-excursion per week intermittent an county-state-tour,| 186 |apose monthly state-national-coverage-impasses, interceding-an surrounding bucolic-extraspective-genetic-nation-state-socialism,| 187 |working 114L.offset-sleep-schedule-calendar-agenda-constantly, on geological-ecological-economical-territory-architectural-develop-ment-(s),| 188 |at rural-genetic-estate-(s)-offset...¶|||08m.:51s.65m.s.|||§

0...Count-[24 / Twenty-Four]-preposition-of... ..Lines-[115.173 / One-Hundred-fifteen.1.one-tenths-2.seven-hundredths-3.three-thousandths]..| 189 |...Article-Essay-[21 // Twenty-One]...|190| ..I simply wonder weather any mass-mutual-genetic-society-motivation exist, (when out in the) world, not rationalize which measures, are formal in requiring an formal-aging, (through the age (by of)) [One-Hundred-Thirty-Five-Year-(s) / 135-Y.+ or older], acceding expectation-(s) of retirement, enjoying [1Literature / 2Labor / 3Festivities / 4Drinking / 5Smoking / 6Resourcing / 7Constructing / 8Numerical-Constitutional-Ordering / 9Subterior-Calendar-Population-Directive-Ordering / 10New-National-Currency-Drawing-(s) / 11Interactive-Bill-Debate-(s) / 12Statehood-Transiting / 13Weekly-Feminine-Impregnating-(an goal of mine by Thirty-Two or so) / 14Sociological-Genetic-Meticulated-Literary-Documenting / 15Comparing Genetic-Characteristics (from an) basis of Blonde-Pink-blue-Repopulating-Goal, tiering, 15.1head-size // 15.2nose-structure // 15.3ear-size-shape // 15.4hips // 15.5waist // 15.6eyebrows // 15.7lips-size // 15.8cheeks // 15.9jaw line // 15.10neck-size // 15.11hair-texture-color juxtaposed-physiological-char-acteristic-(s')]. (I do not care about) political-ethical-ethnic-racial-jargon, I rather focus (on my) singular-genetic-ambition, (as how I **boldly** assume in my lifetime), establishing an nation, (awhile those) work-fatigues, get old when subjective-interpretation-ordering, remain terribly faulted human-history, (as I opposite) sexually-adore-women sensuously.¶|||01m.:05s.38m.s.|||§

¶§22.Manuscript & Cursive exist (as an method for) subjective-off-set-tier-lessoning, designated for expert-lector-(s)' usage-(s), (upon an) focus of [finger-(s)-tips / hand / eye posit-coordination], persuading extraspective-effort-(s),| 191 |where exterior-visual-value-(s), [table-(s)-(surface) / paper-composition-(s) / ink-plastic-rubber-dispensing / et cetera...], from alliteration-prose-written-execution-(s), affirming logic notion-(ing)-(s) voluntary-participating-competent-inhabitant-(s).36¶|||22s.78m.s.|||§ Generally inhabitant-(s) prefer (to be by) oneself-(ves')-individual-identity, while auxiliary-vehicle-identification-(s'), not amount [dipitchipational comprehension // effort-(s)], by [Cubic / Square Area], perspective-(less) at perimeter-measure-(s), ratio cubic-feet-interior-spacing-(s), (for how) Weight-(s) & Extraspective Purpose-(s), apply upon existence.37¶|||38s.31m.s.|||§ How does human-inhabitant-(s) own property, remarking intervoluntary-fraternalized-family-interaction-(s), in debt from national statehoodship, when choice-(s) from status, are pertained upon theatrical-effort-(s), populated at constraint, where [boundary-(ies) // limit-(s)],| 192 |define human-driven-experience (petroleum station boundary-(ies)) by Friday-effect (00:08:52:48 P.M. / week-denominative-comparison-into-hours-of-day($^5/_7$), superimposed day-Twenty-Four-hour-denominative-fraction-percentage-cycling) (by of) mental-timing, passing-mode-(s) of interaction, forgetting-as human-(s)-(synthetic-personality), (a the) organic-vocal-quality-(ies), interacting basic-presiding, appreciating human-extraspective-competent-interaction-(s).38¶|||01m.:14s.37m.s.|||§ Without guidance of rating-(s), presumed note-(s) per [count / memory-biological-limit-(s) / infinitive-rotational-expanse], juxtapose molten-kinetic-plasma, amid centripetal-force-revolutionary-posit, featuring fact-(s), when [date / time / calendar], affix central-superior-case-(s), contorted by [vision / conscious-pitch-tone / parameter-square-octagonal-resonance-reason], by vernacular-vocal-pitch-timbre, variating various individ-ual-(s)-subjective-past-read-present-conducted-competent-interpretation-perspective-(s)-(In-dividuation), for weather mutual-ethical-pragmatic-right-(s') matter, (due in part) idle-worshipping, which remain greater than vehicle or housing-abode, as hypoactive-attention-deficit-disorder-(superego), ignoring momentum (from of) centripetal-force, influencing continental-shelf, in rotation of season-(s), approximate (grammatical)-numerical-mathematical-posit-(s) from visual-interior-skull-subconscious-perspective-(s), when [picture / painting / vision / sight / video in V.H.S. / D.V.D. / Blu-ray / byte-(s)], proof tangent-(s) of mute-abstract-expanse, rather [writing // reading reasonings], from self upon population-discourse, topically resettled per [syntax // elapsing], passing each [fourteen-minute-(s)-twenty-four-secondths /// eight-hundred-sixty-four-secondths],| 193 |per reading-synapse-impasse-(s), ([two-page-paragraph timbre of this book-(s)-essay-(s) /194/ eight-decimal-six-tenths-four-hundredths-centripetal-force-continental-shelf-degree-elapse-secondths-point-(s), superimposed mundane-vision, from day one [hundred // thousandth] percent-(age-(s))-denominative-impasse-(s)]), incrementing per [whole // tenth interval-(s)], impassing hourly-time-posit-(s);[43] awhile [[1]sitting // [2]standing // [3]working // [4]lifting // [5]carrying // [6]writing // [7]conversing // [8]talking // [9]idling], aid from human-automatic-driven-control, (not Congressionally-Literary-Controlled), relying patent-invention-help, rather manual-effort-principle, blue-collar-work-effort-(s), meaning more inhabitant-(s') lifestyles', apose white-collar-managerial-helm-scheduling, in control (by of) budget,| 195 |(from use of) [Federal-Reserve // Stock-Market], toileted sink-work-auxiliary-placing-(s) at developing-entity-(ies), practicing (hobby)-craft-(s), from perplexes-perspective-sociological-purpose-(s), featuring product-objective-item-fact-(s), before without subjective-proof, soliciting effort-(s) through no constant-communication-outlet, (from an) mass-hemisphere-belief of bastardized-ethics, pace-timbre court-legal-judiciary-government-county-state-Federal-count-(s)-casing-(s'), (for of) competent-reform, ignoring [fluent-frequency-word-tempo-(ing-(s))-vernacular-essay-syntax-elapse-(s)-extent-(s) // $^{1}/_{20}$-redacting], (by an) mute-pitch-read-style [male-voice / female-voice / child-voice], rather commonly amounting relief of sound-(s), (instead writing), when not believing participle-chord-vocal-discourse, vocally modifying each variation of vernacular-conscious-individuation, affirming each fact of live-presence, upon other person-(s)-opinion-proof-(s), furthering contexted-article-subject-(s), yet without elongated-syntax-interpretation-patience, comprehend-ing redact-intermezzo-written-interaction-intonation-timbre, sustaining written-reasoning-(s), emphasized (from focus (by of) Population & Territory-(ies), [from / at / for / cause / reason], (not to be) alone, [in / when] affirming logic, apparently occurring, by topics-offset-construe-(s), continuum apposed centralizing perspective-range-imagery-thought-recollection-impasse-(s)-construe-(s),|196|(for how) time-continuum, progress-intermediate-variable-(s), that over pressure self-used-collection-(s), per product-objective-(s), having topic-(s) (be rushed by an) figurative-cause, from colloquial-hoc, amidst [position / juxtaposition-(s)] as [inhabitant / employee / human-being], assert-confidence, by awareness from instructed-self-infinitive-blank-lapse-(s), decimal-ized (from a the) zero-split-infinitive-continuum-offset, by [1sliver / 2grind / 3brush / 4push / 5sit / 6balance / 7place], never (in an posit of) "seeing", but, (by an) visual-perceptive-physical-posit, simultaneous when anew focus tangibly apose skin-tension-pressing-(s), visually-perceiving, intangible-reasoning-(s), ascertaining-focus, simultaneous where fact-(s) meander matter-(s), (from an) referenced-routine-progression, per calendar-agenda-scheduling, correlated by aptitude pertaining [word-(s) / term-(s) definition-extent-(s)], (as where one would believe to) supersede analytical-reason-(ing), by person trademark-registering, amid being-census-count-(s), instilling "consciousness"-brainwashing per inhabitant-lifestyle-type-(s),| 197 |by market-trade-limit-competency-maxim-life-act-(s), basing complacency of matter-(s), by motivated-ability-(ies), not accrued from requisite national-genetic-resourcing-monetary-reform, (as whom be qualified to) errand task-(s) of engineering, while dimension-budget-tabling, premise metrics-operation-(s), orderly imbuing, interpreting how place settled conjoin land by physique, compose ideal-(s), self-idolized without a-(n) total-land of populant-(s), whom follow house-associates'-diploma-thirty-thousand-being-(s)-handling, presumed allowing permission from representative-(s), under Senate-Calendar-Legislative-Law, free-will state-legislative-incrementing, historical-unincorp-orated-mass-(es), deviating decade-exact-increment-book-deductive-reference-(s), from factor-(s) of tangible-mass, under debt, (to an) social-ecological-affluent-debt, by bastardized-national-genetic-"superiority"-default, accustoming individualism to account (auto)-biographical-literary-genre-logic, through transgression age-retirement-future-wanted-propensity-401K-plan-kick-back-(s), forgetting how [theory / hypothesis] (is to) survey-scheduled-population-practice-(s), progressing live-reason, by [presumption / presence], [ob / con / main / sub / at, tain-(ing)] [product / resource / tangible trade-(s)], intertwined per calendar-aging-infinitive-population, (from those curtails by) age-constant-live-decay-deterioration, intermittent physical-mass-presence, affecting place, in attaining stable-property (by of) [urban / suburban / rural territories] per governmental-public-retention, compiling 78L.compounded-resource-(s), infinitively presuming managerial-helm-effect-(s'), controlling possession of object-(s), where placed-central-affirmation-adequacy, define Will & Development-(s)...¶|||05m.48s.75m.s.|||§

§0...Count-[23 / Twenty-Three]-Preposition-of… ..Lines-[80.575 / Eighty.1.five-tenths- 2.seven-hundredths-3.five-thousandths].. ...Article-Essay-[22 // Twenty-Two]... ..*It is important to write-cursive when documenting*... ...312-Count-Preposition-of, total thus far…¶|||11s.25m.s.||||§

¶23What guideline-(ing)-(s) (would be for use), amid roadway-transit, except for yield-signage, generally adrift, for course to place, having human-being-(s)-motive, sustained mere identified self, repeating routine congruent-transit-limit-(s), (around those) apparatus of reproduction, working reasoning by Government & Private-Entity-(ies), serving community-needs per employment, upon house-method-conversation-(s),| 198 |epoxy by sedative-virtue, viewed as [analytical-physic-(s) / analytical-representing / analytical-commercial-resourcing], dissolved per bias upon law, for developmental latter industrial-technology-(ies), anew separational-method-(s) per state, factoring inhabitant-(s), not to state blood-pulse-synapse-tension, upon daily-contravene, daily proceeding from sleep, afloat Supratangible-tectonic-plate-(s), posit [density friction-(s) //

spacial ambient], requiring career-(s) by hourly-wage-check-payment-(s), maintaining-an title-housing-objective-place, subterior state-constitution-offset, when focus of dating, confine around general-parameter-(s), retaining relevant objective-topical-issue-(s),| 199 |held in bank-account-(s)-Federal-monetary-subjunction-circulating, proof proprietary-posit-perspective-mobius-logi-rhythmn-(s), intonated, interval-physical-passing-(s), extenuated-feverously, for timbre in concise secondths-posit-deviation-(s), range [gauge / scale-(s) / balance-(s)], monitoring compositional-unit-developmental-matter-(s), apose weight-(s)-(ed)-designated-value-(s), pressurized subterior decimalizing, date-maxim-self-age-expectancy, alert attention-span, not extrapolated concern from political-disposition-(s), year-to-date-maxim interpretable-reasoning-(s), based from free-will-individualism, (not apparent through) communal-interaction-(s), (as how) recent physics actually exist, confirmed by Gregorian-Calendar-year-Reset, January-New-Year 2018 Years, when English-Oxford-term-(s)-count-(s), oversee all existence (Trivial & Significant), approximating term-(s)-existence, as exterior visually-evident.$_{39}$¶|||01m.:35s.$_{97m.s.|||}$§ Followed by intermediary-guideline-(s), parental-order conceiving at movement-(s) by note-(d)-trade-(s), when metropolitan-Federal-Government-trust, confidently modernize superlative-birth-(s), deeming perceptible-offspring-workplace-value-(s), as inhabitant-social-human-characteristic-(s), encap-sulate tangible-physical-act-(s), burdening faith by self-evident-emancipation, freely-reigning national-extent, [per week // two-thousand-hour-(s)]], sufficed cubical-2,156-feet-mean-median-deviation-space-(s), in rate of hour-(s) per year-currency-method, providing practical, independent-task-(s), substantial-inhabitant-(s), appointed at posit commercial-trade-efficiency, ignoring reality, askewed developmental-act-(s), amid seasonal-passing-(s), timbre-pace-youth-fore-shadowed-adult-capabilities,| 200 |pursuant abound city-district-(s), methodizing free-will-individualism (as an) pragmatic belief, when blank-blood-pulse-evocation-(s), demonstrate how lifestyle pertain Governmental-Commercial-Treasury-Circulating, relied mass-population-(s) having only existed for under 500-years.40¶|||02m.:24s.25m.s.|||§ [What / Who], (is an) dipitchipational-being, in lieu of growth (at an) entity hourly-work-effort-(s), sustaining no foresight from historical Jefferson-Constitutional-Order-(s), cherishing volume-petroleum-being-(s), (for an) collection of metal-panes (flat-hydrahedron / tubular-circumference-(s)), concentrate-focus at Stop & Go-motion-(s), shortening attention-span, by odometer-fuel-senses', elapsing reasoning, (by no analysis from) House-Represent-ative-Common-Reasoning-(s) (30,000 B.p.R.), dipitchipational-being {(Di)-e / Pitch / I / Occupation-(al)};44 (not) marking wage-deviation-(s), awhile receipt-purchasing, denominative-total-population-(s), yearly as [adolescence / child not count-deviate from House-representative-(s) adult eight-thousand-nine-hundred & Sixty-Eight-count], from modern-year, [Two-Thousand & Sixteen / Seventeen] (December-(estimating from [book agent /201/ corporation-contracting])), reasoning housing from [city-planning / county-transit-planning] circulating metropolitan-geological-Federal-housing-reason-ing, regenerating population, for objective-unit-circulating-offset total-moment-(s), impassing population-exponent-extent-reduction-reasoning, per [form / stature / parameter] adjust governmental-metropolitan-spacing-(s), formulating [pavement-cement / brick-rebar infrastructure], positing land-development, concerting focus, local in Federal-Product-Unit-(s)-objection, [private / public] free-will-bias, liberating conjure per state, in view, not to study principle-guideline-(s), of (per) [State-Constitution / Federal Constitution-debt-accumulation-(s)], ignoring international-reliance, when chronizing calculated-budget-managing-method-(s), religiously believing| 202 |[eight-hour-(s) per [day // forty-hour-(s) Metropolitan-dipitchipational-inhabitant-(s')], should bore through living, enjoying entertainment, after-work-hours, without conversing those visually-attractive-opposite-sexual-citizens', 52L. (whom are) appreciated simply (from an) Adorable-Appearance-(s) & Conventional-Discussion-(s'), (by those) economic-systems, which require what survey of extraspective-population-effort-(s) from labor, directly appreciating common government State-Developmental-Territory-Lifestyle-(s')...¶|||04m.:04s. 35m.s.|||§

§0...Count-[7 / Seven]-preposition-of… ..Lines-[54.235 / Fifty-Four.1.two-tenths-2.three-hundredth-3.five-thousandths].. ...Article-Essay-[23 / Twenty-Three]... ..(When is it that) dipitchipational-human-(s'), would follow their ritual-place, by communication-(s) of transit, extenuating effortless-reset-mile-(s), inestimably placed, with depth focus of Road-Way & Velocity-Effect-Dynamic. To pet-peeve-oneself, transiting-existence, arid metropolitan-interval-progression-(s), presuming product-accumulation-(s) per sub-stan-tial-ingestible-(s) at market-(s)-retention-(s), effort only when compulsive-transitive-impasses', enable virtue of belief, from [residence-living / human-scheduling] auxiliary-modern-contemporary-effort-(s),| 203 |teetering between Automatic-Transmission & Steering-Balancing, continuum-theatre, intersecting-value-(s), by preference from individual-(s), executing voluntary-efforts, cause family-purpose, via governmental-debt-relief, order various historical-unit-accumulations, by market-tensable-act-(s), never reasoning Transit & Physics, for metropolitan-driving, superlative-use-(s'), compulsively-driving, as aging act-(s) with enjoyment, luxury senile retirement-aging each inhabitant-off-set-shoulder-objective-continuum-usage-(s), amid continual-human-physical-contin-uum-handle-auxiliary-objective-location-tense-legal-usage coordination-(s).¶|||58s.91m.s.|||§

{January-(31)-Month-Day-Hour-Minute-(s)-Secondths-Count}-{2,678,400 D.H.M.S. = 60S.× 60M.× H.× D.}-{744 H.}-{44,640 M.}-{31 D.}

[Article:] {February}

¶24What circumference is Supratangible by mass, (but not of an) dimensional-equivalent upon physics-metrological-centripetal-drift? However anyone has counted each lunar-cycle, (is still not of an) factual observation-(s'), (for by an) rationale of count-(s), not continue aware, persistently documenting live-posit-node-tense-vision-(s),| 204 |juxtaposing solar-centralized-posit, apose lunar-fixed [rotation // centripetal revolution], exemplified when cold-season-nature, (from an) default-condition-(s), by Earth-molten-core, tilt from Tectonic-Lunar-perceptive-axis, positing Three-Hundred-Sixty-Five-Day-Drift, apose passing-continental-terrain, amid celestial-North-South-Degree-(s)-Thirty-two-Degree-(s)-Parallel-(s)-Exterior-Extreme-(s), continuum-offset revolutionary-drift, during each passing-momentum-degrees, drift hovering along physics-grandiose-molten-tectonic-vapor-offset-pressure.38 ¶|||43s.94m.s.|||§ Lunar-cycling, disposition tectonic-plate-(s), (for how we) centripetally-exist, solidified by density-matter, supratangible, mundane (from our) recognition, by extremes of [rotation // revolution-(s)], emitting kinetic-vapor-(s), oxygenating human-being-(s), [vapor-flow // oceanic-bodies-water-centripetal-physics-motions], contraposed solar-gaseous-molten-fission-excess-impact, estimated Four-Hundred & Eighty-Six-Year-(s)-{486-Years}-(From 2016), retrodated February-14L. waning Tenth-Full-Moon-Miami-Day-calendar-impasse-celestial-molten-combustion-continuum-plasma-initial-(d)-rip, apose October-Tenth-discourse, by one-hundred and twenty-one-and-an-half-days... ¶|||01m.:10s.41m.s.|||§

§0...Count-[4 /Four]-preposition-of... ..Lines-[16 / Sixteen]... ..Article-Essay-[1 / One].. // ..February.. / ..Book-Essay-[24 / Twenty-Four].. ..Physics is simply [Solar-Combustion-Infinitive-Magma-Molten-Fission-Continuum-Density-Liquid-Matter-Plasma /// Lunar-Total-Tectonic-Centripetal-Force-Revolving-Force-Dry-Matter /// Earth-Tectonic-Plate-(s)-Aquatic-Oceanic-Rivers-Lakes-Channel-Canal-Straights-Terracultural-Crust-Layer-Molten-Magma-Combustional-Infinitive-Continuum-Fission-Density-Liquid-Core-Matter-O-Zone-Layer-Aquatic-Vapor-Lunar-Tectonic-Maxim-Galaxy-Layer-Blue-Spacing]-{Warden-Jung-Freeman-copyright-publish-recollection, not library-minor-note-(s), per. integer.at-f-grade-primary-subjective-deviation}.¶|||37s.63m.s.|||§

¶25Has anyone bothered to count each lunar-cycle-(s) from tectonic-offset-visual-impact-effect? Or does generic passed interpretation-(s), [when // where] objective [resourcing / ingestible-soluble-(s) / cotton-grown-resourcing or clothing-production-weight-analysis] proportion-ordering-(s), from census population-set-interval-goal-(s), skewing [responsibility-(ies) // project-(s) // task-(s)], upon communal-free-enterprise-development-(s), commercially produced, appraised intermittent quarterly-annual-interval-(s), per crop-yield-rotation-(s), (from an) accumulated-measure per inhabitant-(s) year to date, revolving revolutionary-communal-need-(s),|205|(Three-Hundred-Sixty-Five-Day-(s)-impasse, multiplied by exponent-reset-offset-count-(s), serving qualities per [1age / 2crop-yield-count-(s) / 3boundary-limit-coverage / 4denominative-currency-count-(s) / 5population-boundary-extent-(s) / 6commercial-production-count-(s) / 7industrial-production-part-(s)-account-(ing)]| 206 |from [volume-(s) / substantial weighted-pound-(s), compound-(s) / ounce-(s) // (milli)-liter-(s) per soluble-product-(s)], amid faith in market-volatility, regarding attained-unit-good-(s), retained at isolated-private-collected tangible-good-(s), creatively-conjured, by article-lesson-historical-application-usage-(s), administrated from conglomerated-trade, instructing commercial-unit-value-(s), offset per [farmer-substantial-sustain-able-cultivation-(s) / government-horticultural-agricultural-regeneration-facilities], actively (relying Federal-population-constitutional-principality-(City-District-Parameter-Area-(s))-surveyed-account-equivalent-(s)), adapting [horticultural / agricultural] yield-(s), compiling commercial-systems, which produce basic-human-attire, abiding commercial-product-usage-(s);45 awhile commercial-at-will-employment, develop citizen-based-trade-system-(s), extracting lured-operation-(s), industrializing commercial [head-quarter-(s) / market-(s) / Shipping & Handling-Facilities / Factory-Production-(s) / Warehouse-(s) / Shipping-Truck-(s)], serving human-being-(s), (by per an) scheduled Gregorian-pre-sent-year-reset-observation-(s), approximately posit-(ed), having auxiliary-transit-objective-(s), direct security-box-unit-mean-(s), through shipping-way-(s), logistically-budgeted, ordering unit-object-(s), per cubic-surface-area-(s)-dimension-(s)-boundary-(ies), defining career-objective-directive-(s), ecologically-juxtaposed, instilling instructed-directive-focus,| 207 |at monetary-circulating-entity-(ies), comfortably-modified, vernacular-literary-economy, to [concept / model / build // construct / mode], county-legislational-reasoning-(s), (as an) delegation of state-deductive-denominative-filed-data, retrograding settled-stock-composite-construe-(s), from Federal-resource-composition-(s), (one example, (for why) literature-denominating-non-fiction, is incumbent), limit-(ed) market-shelf-life-expectation-(s), determining tenure per manager-(s), not cycling diploma-degree-credential-(s'), [communally // politically] extraspective, amid globalization-Federal-incorporation-(s), intricating traded-good-(s), noted at freely-demanded, (under an strict-analysis of fate), inconsistent metropolitan-market-shelf-unit-relaying, when field-cultivating-effort-(s), define why evaluation per cubic-surface-area-prism-(s), not partake blue-collar-effort-(s), requiring awareness at [educational // institutional // Contextual // commercial purpose-(s)];46 per rural-movement-(s), working away, (from those) centralized-development-instate-(ing-(s)), default, human-mass-population-existence, (as an basis for of) survival, not sophistication.40¶|||02m.:34s.09m.s.|||§ (If these are not) pre-instated, (how ever would those façade-(s) per) construction-site, plan scheduled spacial-impasse-(s)-continuum-requisites', pre-imposed commercial capable traditional-development-(s), swaying influence-(s), requisite free-persuasion-(s), being regulated (by from) (political)-person-perspective-(s), present per {{County-City-Government-Official-(s)}-{State-County-House-Representative-(s)}-{County-Mayor-Councilmen-Populations} // {Federal-Senate-{Two per State}-{Federal-State-House-Representative-(s)}-{adjusted from State-population-(s)}}.41 ¶|||02m.:56s.87m.s.||| § Reluctant focus toward [word / term / definition / language-skill-(s) / literature], sustained socialistic [passing // enabling], requisite precept-(s) of human-being-(s), needed for vehicle-trade-(s), by geological-metropolitan-order-(s), amassing daily-effort-(s), from govern-mental-road-construction-(s), directing focus (by those) mass-(es) afloat [supratangible / intangible] existence, transitioning planned-coordinated-development-(s)-subjective-currency-circulating, managed by citizen-requirement-(s), handled from [individual // communal] impasse-sect-impasse-surrounding-(s), needed for objecting resource-(s), relatively-49L.proportional objective-unit-fact-(s), per concepted-conveyed [objective // subjective] retained-property-collecti on-(s)...¶|||03m.:26s..02m.s.|||§

§0...Count-[5 / Five]-preposition-of.. ..Lines-[49.715 / Fifty-1.aevne-tenths-2.one-hundredths-3.five-thousandths]....Article-Essay-[2 / Two].. //..Book-Essay-[25 / Twenty-Five].. ..From who-(s') perspective do we ever [see / hear / feel / talk // tangent focus] subversive motive-(s), (by of an) national human-being-(s), formulated at first-person-posit-(s'), defining archival-Auto-Biography-documenting-reason-(s), from human-perspective, adjusted where constitutional-detailing, awaited educational-order, tier Taxonomic-Classification-Deviation-(s), learning how intangible-tectonic-spacing-(s), offset-from personal-observation-(s), to articlize inhabitant-(s)-surrounding-nature, (unaware (of a the purpose for) psychological-evaluation), requiring tools-use-(age-in presence-performance-action-(s), to affirm each tangible-focus-applicated-unit-objective-usage-article-(s)? ¶|||27s.91m.s.|||§

¶26As for consciousness, reasoned-purpose, not exist.41¶|||02s.75m.s.|||§ Life consisting (of an) various array by manner-(s), offset at location under flag per Governmental-Planned-Developmental-Resourcing-Center-(s) & Subjective-Feudalism.42¶|||10s.40m.s.|||§ No symbolic-sharing define personal-possessive-responsible-independent-objecting, when socialistic [turmoil // freedom // requisite-domestic-tranquility // revolutionary-era], estrange local-community-(ies), cause deliberating-character-role-(s) in one fashion or another, isolate human-being-(s), peripheral (from the) method-(s) of routine-(d)-day, steadfast at tense-(s) per presence-(s), resetting reclusive upon natural-off-time-period-(s), a-moded from behavior, (unconducted after academic-reiteration-(s)),| 208 |allude through present millennium-Protestant-infinitive, unaware of Christianity-Ultimate-Purpose, by economical-commercial-budget mechanism-(s), Gestalt amid fact-(s) by laze, accrediting intermittent exterior-perceivable-value-(s), from posited procrastinated-effort-(s), energized by age interaction-(s) upon city-county-state-swarming, accountable Momentum & Life, upon present-future-live-interval-decay-requisite-move-ment-(s);[47] influctuated (as when) momentum not be count-(ed), from mathematical-timbre, or age be sworn-retarded, because how retirement-decai-planning, remain unrenovated from early-20th-Century-Government-Planned-Architecture, tenure upon present-instruct-(ed)-act-(s) at work-place, for self-indulgence while aging, having remain creeped-up, undaunted (from a the) impossibility of mobile-stabilization, affecting invariably [visual-tonal-vocal-vernacular-perspective / mind], (by an) Balance & Blend per extraverted-interaction-(s), forgotten by introverted-independent-perspective-(s'), posterior-reflex-(ed), synapsing evocation-(s), by prerogative from [vision / hearing-cerebral-sensation-(s)], [elapsed / synapsed], for construing unstated, sound-exterior-resonance,| 209 |volume-pitch-tone-vibration-(s)-count-participle-word-term-extent-document-(ing-(s)), scaling tone (by of) blood-pulse-cardiopulmonary-breath-intake-count-impasse-(s), imputed from ear-canal-vibrational-focus (upon an) minuté-minuscule muscular-ligamental-tension-(s), reverberated from blood-skin-sat-refreshing-(s), tentatively eaten when reflexive-vibration-(s), emit effort, for objective-focus, (from an) unsubjective-socialism-nature, tense-ventricle-system-blood-respirational-friction-Mobius-pressure-(s), [near / far / close-depth-perspective juxtaposition-(s)], confined by physiological-perceptive-limits', affecting (a the premise for how all) human-pulsed-fatigue-tense-(s), focal-pointed, timbre vague form of tempo-rhythm, perceiving (forgetting) [beat-(s) // note-(s), per minute /// blood-pulse-rate-interval-(s)-pre-anticipation-calculating-(s)], in tier-deviation-(s) from complicit-rule-(s), in sort from verbatim-reason-(s), identifying commercial-product-unit-fact-(s), ranged from title-possessed-property, (upon an) type-manuscript-governmental [title / form / lease / contract / agreement / term] documentation-default-demand-register-directive-(s)] afforded by individual-(s') identifying-storage-parameter-(s), without-an homecoming by longevity, for influencing up-keep, during at [communal-development / receive-appreciation], per intangible-mass-goods of state, extraspectively-prioritized, from hardware-objection-(s), omitted pass, how vehicle-(s), moderate (a the) expanse, (in)-sweatable in lieu of intermediate [self / others], comparing [default / goal-(s)], pre oriented while factifying [(in)-tangible [solid-(s) / liquid-(s) / centripetal-pressure-(s)], requiring Religious-Government-ordered-developments, watched (at from an) offset-(s) of trial-error-(s), (by the) origin-physiological-existent-state-(s), not documenting,| 210 |hypothesis-trial-experiment-reasoning, present at offspring-perspective-community-effort-(s)-deviation-(s), (by of an) expectancy of death, (at an) unconscious-moderate-aging, apose covering, (a the) dynamic of physics, (for an) view of reproductive-subjective-simultaneous, [remodeling // renovating-up-keep-socialistic-methods of government-living].•43¶|||03m.:03.•34m.s.|||§ Methodical obedi-ence, is relied by ratio person-count-(s), apose religious-being-mass-accumulative-accounting-(s), rather deducing socialism-individuals-housing-offset-reasoning-(s), constituting intermittent historical-archive-(s), for not soly relying auto-biographical-effort-(s), from mimic auto-biographical-literary-response-(s), year per year;48 year per decade;49 year per century;50 year per season;51 year per month;52 year per week;53 unscheduled where day-at-will-employment-guarantee-(s), exclaim through trial-fault-(s), which remain in population-deviation-(s), apose (rivalry-(ies)) conscious-reality, visual by color-tone-(s), as physique-composition, tangibly limit, maxim-approximate-output, [brick / metal] idly fused in contact at perimeter-surface-area-space-(s), amounting property-limit-objective-retention-accumulation-(s), per personally-unscheduled-objective-(s), (as an) unconscious-mission per objective-commercial-trade.44¶||||04m.25s.99m.s.|||§ For none have examined such fact, self-ignorance, be delighted daily, enjoying co-worker-limit-tasked-expectancy-effort-(s), sipped from coffee (water-pressurized-refined-soil) caffeinating, sub-par remaining estranged where perspective of peripheral-surrounding-place-(s), technically-pass offsetted blank-creative-vocational-written-artistic-tectonic-documenting, [in-to / in-at / of-a-the vehicle / house / entity / exterior-sociological-posit-(s)-community-dynamic];[54] ambient without direct-inclination-(s), for crafting unit-title-property-[copyright / patent]-verification-(s) of religious-message, through commercial-production, under-an government-constitution per genetic-pre-dis-position-passage-(s), counter ethical-developmental-nation-state, (from when) ten-thousandths of book-(s), equivalent, my in-depth-literary-developing, for architectural-visionary-impasse, (at an) sociological-governized-system-(s), (yet from a the) dynamic of common-housing, (or else retard-(s) just simply relying on government-traded-handling-(s)) during state-political-affairs), per social-sustainable-identity, ["reflective" / "memmorium" / "apprehended"], short-elapsed, thinking from unit-thing-(s), unwritten proportionally per [word-count-(s) // letter-(s)-count-(s)] per unit-objective-weight-(s), (by an) scaled-system, from proportion-weight-(s)-mineral-type adjusting, how elemental-resource-(s) of continental-shelf-crust-layer, (have not been) planned by person grandiose-effort-(s), along elemental-statehood-article-(s)-extraction-resourcing-(s), (by an) genetic-fixation-(s) (on an) decade-offspring-simultaneous-reproduction-(s) per relative-genetic-identity, defining each day, from dedication-(s), upon state of diplomatic-boundary-affair-(s), commonly-morally-participated, when genetic-primary-identification-(s), separate, (from an) pre-genetic-detailing, (a the) requisite-(s) of 67L.factory-unit-factor-(s)-production, considered, (by an) survey-communication-(s), constantly-up-kept, because how dating is upon hourly-time-posit-interval-(s), when date remain fixed (on an) consuming-ingestion-practice-balance, by timbre-focus-twenty-four-hour-elapse-cycle-period-(s)...¶|||05m.:18s.84m.s.|||§

§0...Count-[22 / Twenty-Two]-preposition-of... ..Lines-[69.85 / Sixty-Nine.1.eighttenths-2.fivehundredths].. ...Article-Essay-[3 / Three].. // ..Book-Essay-[26 / Twenty-Six].. ..Appraisal of history, does not enhance our future-vestige-(s), for ever modifying by physical-presence, present-day-tense, upon our requisite-(s) by of) offspring-continuum-humanity. ..Pagan-Psalms-Two-thousand, what rhetoric-rate, would subbcome *notary-mark-January?¶|||16s.10m.s.|||§

¶27Common-reasoning not consist truth-(s'), by natural-discourse of Federal-State-Affair-(s), bread swear circumstantial-influence, having depth-perspective-way define relied-ideal-(s), managing-directive-(s), focused (from an) (un) educational-preference, basis (on), physical-motion-(s), motivating order-(s), aided upon reason per surface-area-space-(s), from senate-house-congressional-representative-(s), by common-sedentary-effort-(s), displaying required-work-effort-(s), reclusive by adequate-pay-ment-(s), (as the) case, of Subjective-feudalism, subcon-sciously-sustaining youth-stigma-(s), sexually chided, for offset-(s) of repetitive individualistic social-belief-(s), awed time and time-again, forgetting as implementation, yet group-genetic-resources-requirement-(s), by constant grammatical-punctual-mathematical-medium-canvas-characterized-inductive-introverted-perceptive-deductive-perspective-defining, from construct-ive-creativity, to transpose, from two-dimensional-conception, upon grandiose-elaboration-(s), by terrain of physic-(s), (out of an) question-(ing) of those cycle-(s) by prerogative-reasoning-(s'), forming individual-lease-mentality, upon titled-proprietary-contract-spacing-listening-enacting-discourse-(s), complicating personal-family-lineage, apose an surrounding-characteristic-(s), (as a the) primary-job of [labor // commercialism].45¶|||59s.50m.s.|||§ Context of inductive reasoning, misconstrue endeavor, enabled from requisite-effort-(s), act-(ed), by birthed belief, upon adult-sweating, with no male-chivalry of women, to focus-energy-(ies), by community, considering how what [alliterational-device // method of payment] can filter subjects upon amidst an extent of daily-instinctual-passing-(s), empherally revealing physiological-form-(s), functional-capability-(ies), conterposed sedentary-self-(ves), (as an clue at) extraspective-timbre, intercontorting conclusive-cause-(s), (for why) individual-task-routine-(s), should matter (from an) extraspective-human-interaction-(s), in regard (from of perspective by) Competent-Individual-(s) & Physics-Default-Mundane-Existence.46¶|||01m.:36s.80m.s.|||§ Rigor at participated-presence, affirm-dating, cluing insight-(s) from inhabitant-disposition juxtaposition-(s), fundraising aimless of tectonic-plate-strategizing, when navel-contouring, stretch those vast extent-(s) of veteran-mentality,| 211 |adjusted from state-referendum, sole-physiologically [personal-denominating / bill-(ing) / annotate-(ing)] those development-(s) at political-militarial-discovered-boundary-(ies), changing accumulation, from civilizational-deviation-(s), taxed-religious-duty-(ies), ([1]by [2]an [3]body [4]per [5]of) [inhabitants // citizen-(s)] whom practice [seasonal // yearly // monthly centralized-continuum-oscillation-fact-(s)], comparing affirmed juxtaposition-(s) of interpretation-(s), (as this such book competency expect-(s) to copyright-monitor) disclosure-(s) per article-(s)-evidence-(s), after facting [Book-Agent /212/ Publishing-Company-Books-publishing confirmed-Contracted-Trusts], (*path-ologically-pre-meticulated (obituary)), by sociological-non-read-(uhhh) evaluation-(s), incapable of legitimate educational-redacting-(s), from commercial-professional, requiring decorum, for initiating constructive-(s)-physics, when natural constitutional-idling, believe (for an) millennium of populating, to create offspring, where treasury not truly (be of an goal for life), regardless from commercial-production-shelf-shipment-product-objective-value-(s), per [object-(s) / element-(s) / place-(s) / environment-(s') / climate-(s) / condition-(s)];$_{55}$ upon topographic-demo-graphic-juxtaposition-(s), where I have noticed, from Military-Enlistment & Miami-Lifestyle, an secret | 213 |abided by free-democracy;$_{56}$ while [white // black-race-(s) /// international-offset-ethnicities], range (as an) total-ethnicity, per ethnicity=(ies), not speaking (from of such) (juxta)$^{-}$position-(s), where classism tolerate poverty, for not having interjection-(s) of [objective-(s) // sociological-ideal-(s) // mission-(s)] of centralized-government, ambiguating as more & more American-subjunction-(s), confirm-an religious-parent-commercial-political-enabling-posit-(s'), uninfluenced for [obtaining // sustaining // maintaining, property /// family-heirloom-(s)], centralizing [governmental-physiological-universal // book-commercial-(esce)-constitutional-population-denoting-(s)], impassed by housed-inhabitant-(s) for view, as live-discourse, past-an array intersymposium, per Moral-Awareness & Reasoning, for making each objective count-(s) reciprocally, upon residual-act-(s), duplicating motion-(s),| 214 |from location-objective-instructive-directive-continuum-resourced-commercial-production-circulated-usage-obser-vation-(s)-confirmation-(s), when$_{48L.}$ commercial-manager-(s) think how spaced-out-budgeting, create-an greater mean-(s) per commercial, influencing how employee-mentality, require-(s)-an numerous amount of story-(ies), as present-impasse-exist, (for not being an) fluent-avid-citizen, being possessive, from commercial-objective-instating, determining-choice-(s), anointed by annotated-interpretation-(s)...¶| ||03m.:49s.$_{20m.s.|||}$§

$_{§0}$...Count-[23 / Twenty-Three]-preposition-of..|215|..Lines-[51.167 / Fifty-Three.$_{1.one}$-tenths-$_{2.six}$-hundredths-$_{3.seven-}$thousandths].....Article-Essay-[4 / Four].. //..Book-Essay-[27 / Twenty-Seven].. ..How do [any / we] us, operate [60$_{Sths.}$· 60$_{Mins.}$· 24$^{Hs..}$ 365$^{Ds.}$ Multiplication-Deviation-Offset], [micro / Macro] year-offset-count-(s), continually undocumented, simultaneous per visual-distance-pressure-effort-(s)-continuum-offset.¶|||17s.53m.s.|||§

¶28When deserved-recognition (is an) aspect of fluent-contingent-successive-comprehensive-functional-proper-work, what naturally inhibit inhabit-(s') at posit-(s) for duty, cause by oneself (upon an) serie-(s) of religious-gesture-(s), customizing historical-incurrence, stated (from a the) following past-day-interpretation-(s), (from how) human-behavior, (can never have been as fast through) vision by thought, to have landscaped perceptive-surrounding-(s), from second-person-error-(s), unlistened at posit-(s), to consider how auxiliary-machine-functioning apply (to those) development-(s), (from an) rushed-hedonism-mass-routine, juxtaposed managerial-purpose, in no clear fashion, (as what) others may inquire in to, for validifying-tense-(s), prior objective, when substantial-comestible-(s) suffice human-doctrine, through natural-analytical-discourse, unsuccessful mutual-youth-community-peer-(s)-collegiate-co-worker-retire-ment-competent-inteligent-genetic-communion-succession-(s),following-an comprehend-sive-surrounding-primary-genetic-feature-(s)-background, (for then) tier-(ing) each refinement of physiological-characteristic-(s), apose-an consecutive-progression-(s) by day-night-govern-mental-social-loyalty-developing, abated money, (in an) personal-grasp, believing to perceive-reason, by no intent for mutual-interaction-(s), as does colloquial-vocal-vernacular, remain rampant, value-(ing) [opinion-(s) / literal-observational-fact-(s)];57|216 |linearly observing discoursed-action-(s), aforementioned from present-tense-continuum-perspective, anticipating [event-(s) // scheduled-interaction-(s) // friend-(s')-interaction-(s) // routine of payment-(s)] (in an) constant-reset-future, ideal for present-communication-(s) differing location-(s), (un)-constituted when [objective-constitutional-(esce)-amendment-article-(s)-documenting // subjective-consti-tutional-(esce)-amendment-article-(s)-documenting // population-constitutional-(esce)-amend-ment-article-(s)-documenting // designated-territory-proportional-weighing-constitutional-(esce)-article-(s)-documenting (proportional to population physiological [male // female] impasse manual-capable-coverage apose [transit-mass-impasse-purpose-(s)) // schedule-arranged-constitutional-(esce)-amendment-article-(s)-documenting // linguistics-colloquial-vocal-vernacu-lar-archaic-finalize-decade // generational-era-penultimate-constitutional-(esce)-article-denomin-ative-inductive-documenting // transit-function-constitutional-(esce)-amendment-documenting // road-way-empirical-deductive-increment-article-(s)-repair-(s)-constitutional-(esce)-amendment-documenting-(s) /217/ entity-planning-developmental-impasse-constitutional-(esce)-amendment-documenting /218/ population-sociological-psychological-physiological-schedule-dating-agenda-act-(s)-constitutional / (esce)-amendment-article-documenting-(s) // commercial-venue-account-(s)-constitutional-(esce)-amendment-article-(s)-documenting]] periodicalizing everything (from an) university-(ies)-literary-denominating-position-(s), juxtaposed-an psychological-literary-denominating-position-(s), transitioning auto-biographical-entry-(ies), consisting graphic-(s) per architectural-planning-offset, to display discourse-(s) of creative-common-input, building (to an) utopic-sociological-class, together on trade, entering [discourse / leaving] waning off, those focal-point-(s), [hampering / improving] select average-(s) per individual-being-(s), fixating local-effort-(s) prior national-contin-ental-expansion, [fundraising / scheduling / budgeting / maintaining / implementing], attempt-(s) to consider how exponent-reproduction, (can be an) genetic-objective-(s), aware (of a the purpose for) scheduling mass (upon an) dating-schedule-(s), where national-revise, (has had an reliance on) initial-retirement-state-influence-impact, (to sustem from an) natural-embellishment, mutual-offset-discussed-discourse-(s), asserting-prerogative-(s), (in open) avid-effort-(s)-prejudice, when ones' action-(s) supersede (a the) bias-(es) of idle-responsive-spoken (out of tounge-(s') / turn-(ed) / beings'), urspurated from vanity, understanding objective-resourcing-dynamic, (which should be based at an) count-(s)-central-community-entity-usage-(s), rather than [garage // house // storage // thrown-away // attic // shed // out-house // et cetera...] carefully-planning, each factoring-(s) of objection, (that have an) awareness of time, by year-longevity, tasking common-interaction-(s), prior commercial-shelf-unit-product-transit-shipped-objective-(s), Resourced & Produced, interposed-centralized-studying, subjective-purpose-(s), tasking-objective-(s), from individual-place-(s) per direct-visual-tonal-perceptive-focus, expanding entity-isolating-particular-(s)-task-(s), rather-an conveyer-constructive-pattern, instilled written-analytical-conceptive-denominative-factoring-(s), waged numerically-standardizing-an, total-revolving-revolutionary-genetic-methodical-scheduling, administering dues (through an) association-premise-basis, [in // upon] perimeter-surface-area-offset-limit-(s), focusing value-(s) per Fund-(s) & Prerogative-Census, (not in part of why we are ever capable of developing an call to reason-(ing)), without-an basis (of an) collective-societal-mutual-subjective-comprehension-(s), tallying adjourned primary-beneficiary-(ies) of will, unclucidated from Federal-Monetary-note-(s), denoting trade-pact-(s), (by an) inductive-practice-(s)-offset lineage-belief-longevity, through each intention of action-(s), guiding ones'-personal-method of leased-development (per $^{1}/_{8}$ /deviation/ $^{1}/_{6}$ block-housing), (from an) physiological-maximum of purpose, juxtaposing-an surrounding-livelihood impact-(s) upon continental-tectonic-elemental-extraction-(s), living life, from deviation-(s) apposed [syntax // synapse // elapse mode-count-(s)-impasse-(s)], concerting location when activity-(ies), dwindle without extraspective-competent-(s'), interacting-determination-(s), voiced at local-approximate-mute-concern, defining self-auxiliary-command-(s), scheduled interposed centripetal-force, corollary mass-production-(s), amid view at Exterior-Existence & Self-Proportional objective-input-(s), (upon such) condition-(s) of requisite-$_{65L.}$historical-competency, surveying human-disposition, in regard of unfaltering human-progress-sion...¶|||04m.:27s.$_{56m.s.|||}$§

$_{§0}$...Count-[19 / Nineteen]-preposition-of...| 219 |..Lines-[63.14 / Sixty-Five.$_{1.zero}$-tenths-$_{2.nine}$-ninths].. ..Article-Essay-[5 / Five].. // ..Essay-[28 / Twenty-Eight].. ..Whom commonly understand vernacular-vocal-tonal-vibrational-letters-words-European-off-set-syntax-deviation?¶|||10s.$_{72m.s.||f|}$§

¶$_{29}$How does cycling pertain "day", upon year, meander calendarized-periodical-grand-event-happening-interval-(s), acted past, budgeted-cycling-(s), (to have an) cause of reasoning, when being-(s) (un)-caliper natural-order, free for fact-(s), never commonly understanding budgeted-expense-(s) (as an) post-day-return, from present-input, (to return in an) fashion, corrugated population-mass? Is non-fiction-ordering, not considered past textbook-deviation-ration-(ing)? Or does inhabitant-(s) (have an) extemporaneous-compulsion, socializing market-product-(s), past subjective-purposes, acquiring factory-direct-object-(ive)-(s), askewed amongst mass-ecological-reproductive-discourse, tiering progressive-physiological-letter-word-syntax-elapse-paragraph-synapse-perceptive-imagery-retention-(s), by self-worth, rather compulsive-behavior, incessant by the belief of individuation, possessively claiming independent objective-product-(s), through commercial-entity-(ies), cycling-an superfluous-account-ing-(s'), disposing human-production-need, perceptively-limited by physiological-effort-(s), not invigorated by physiological-cause though variation-(s) of discourse (by of) existence.$_{47¶|||-}$41s.$_{82m.a.|||§}$ Focusing Legislational-government, imposes-an impact of will, to contravene reasoning of development-(s) for maintenance, when adjoined physical-posit-personal-will, passed family-blood, imbued causes' directing voluntary-brainwashed-civilian-circumstances, by use of tangible-development-(s), upon numerous-factoring-(s), [Governmental / Commercial / Economic / Industrial / Social-empirical-deductive-reducing-order-(s)], sufficing technical-off-set-(s), alerted commercial-productive-volume-mass-trade, ordered into percentile-dynamic, for variation-(s) of mass-trade-(s), never to referendum merited workplace-entity-(ies), percentile conclusive when constructive-development-(s), amount integrated municipal-unilateral-bipartisan-manual-work-documentation, directing each day, for no formal-calendar-subjecting-intermission-(s), at monetary-capable, and genetic-unilateral-developmental-populating, not having an conception, for how limited those bias-(es), remain upon personal-objective-belief, gifted as present date-conjunct, without invigoration for journey, but movie-tangent-(ed)-dis-course, preference limit-(s) per off-time-factory-direct-ordering-(s), by product-purchasable-material-(s), (un)-engineered by note of direct-mass-prerogative-(s), characterized by at-will-employment;58 awhile time-elapse past dating, to factor perimeter-surface-area-(s) in transmission by id-limit-perspective-(s').48¶|||01m.:44s.51m.s.|||§ Id not identify affluent-palpatory-posit-(s), repetitious (from the) manner of movement-(s), deviated while [use-in / use-to // their-of], physical-transitive-passing, experience redundantly, Aptitude-(com)-place-(ency) & Limited-Territory, (for of the) feminine-masculine-ethnic-racial-connotational-dynamic, affecting how treasured each inhibited view of "mind", idealize [idea / faith / belief / theory / figurative / inanimate thought-(s)] illiterate of physical-ordering, as how effort-(s) continue to remain, relied upon an unequal-nature, by human-basic-rights-protest-extortion-(s), due in part vehicle-(dis) order, by view of Odometer & Time, auxiliary-preferenced, rather than subjective-order-incremental-perceptive-reasoning, juxtaposing transit-trade upon substantial-intake, rated when scheduled-discourse, never involve daily-interval-exact-signed-location-(s) (officer-passing-(s)-method), hypertensed compressed-metro-politan-unobserved-space-(ing)-(s), hypo-perceiving at religiously-individualized-memorial-contact-(s), interval, time-date-total-elapse-non-count-(s), abstract-ed, through calendar-year-(s), (as what) aspect solidify past present-tool-usage-(s), before one can conjure an formal-extrapolation for factory-commercial-product-patent-conceptive-usage-(s), meandered present-time-passing-(s), affixed for mundane-nature, instinctually-inclined, by bias, amassed [hourly / minute-(s) // secondths interval-(s)] per time-posit-(s), measuring along "scale", those perceivable-weight-(s), affirmed from year of day, {year to date commercial-collected-product-mass-trade}, uncoordinated interposed-hourly-segmented-timing-(s), at location and sign, from letting free-will, act upon work, factor-(ing) proportion-(s) per substantial-pressurized non observation-(s), while quality remain competent by grab use as grip, transiting-limit-(s), intersected where prevalent municipal-count-(s), not mathematically-pre-intrigue perspective, apposing calendar-time-rate-posit-(s) of data, as perceivable-impasse-routine-reasoning, abounded physical-terrain, tensing fixture-(s) of objective-(s), surrounding conceptive-focus, present (at an) direct governmental-square-foot-price-rate-protection, generically bastardized, utilizing historical-legal-history, remained vacant by blank-observational-mass-deviational-memory-(ies), unequipped for seeing rural-territory, enveloping contemporary-vision-(s), citing-refer-ence-(s) of action-(s), upon daily-intervention-(s) (86,400 S.P.D.) (infer-daily-offset-(s)) // (31,536,000 S.P.Y.) (referencing deductive-calendar-period-ambiguous-extent-(s)), (as for when) act-(ion)-(s) will be extracted out of "at-will-employment", creatively directed upon physics-notion-(s) (primary-element-(s);[59]| 220 |intricated from detailed-elaboration-(s)), [(un)-formention-(ed) during variation-(s) [to-where / from-a-the-who (m) / as-of-why / for-of-how]] conjured at [competency-discoursed-aptitude-reference-logic // interactive-inferencing] applied (to what is figured by an) synapse of elapse-article-deductive-reference-application-(s), instructed (as when tenses' from) sight-approximation, perimeter property-place-(ing)-(s) (upon a (n)) designated-space-(s), in square-surface-area for assessing in square-feet, reset-(s) of plane-surface-area-surrounding-space, for when language (hoc / bias), not prove an base of merit-(s) formulate-(d) formal per hourly-locational-timed-sequencing-(s), unordered-perfectly, segmented-minute-(s)-act-analyzation-(s), to form-merit, per hour-(s)-setting-(s)-minute-(s)-elapsing-(s), variating sequence-(s), upon temporal-condition-(s), (for when) day-tier-placing-(s), eventually inhibit-passed-market-space-(s), unthought for style, residential-blue-collar-purpose (blue-collar-effort-(s), abiding metropolitan-limit-(s)), by serie-(d) methods per Protista-taxonomic-space-(ing-(s))-area-proxy-(s), where suburban-area-server-(s), act by fatigue through sweating, cause combustion-kinetic-pressure-flammable-continuum-friction-(s), move in estimated-area-approximation-(s), grid-graph-matrixed-mapped, by road generated petroleum-data, fuming future, for Gregorian-housing-sequence-(ing), to tangent-road from house-distance-(ing)-(s), fixed at forefront, in signature by [write / talk hoc voluntary-abstraction], while [sat-in / stance-of / about / central-of / exterior-nature], perceive through secondths-time-elapse-effect, (in an) expiry, from momentum-condition-(s), primary (by at of) tectonic-control, free by conscious-subjective-feudalism, excerpting context-senator-legislating, by executive-enforcement, requisite judiciary-review, (for how) advised-legal-counseling, discretion when act-(s), move [through // upon parameter-complex-(s)], time-constraint-(ed), directing method-(s), defining receipt-tempo-data, by calendar-ottoman-planned-collection-(s), Hourly-Fatiguing & Clothing-Worn, elapse-(s) where planned-spacing-(s) per objective-pre-set-place-(ing-(s)), posit-(s)-conjunction [Season & Hour / Month & Minute / Week & Secondths] inadvertent adventuring an lifetime of [fact-(s) / proof-(s)], (upon an) primary-locale, settle-(ing) document-(s) where little is competently-applied, revolutionizing momentum, as when prerogative-bias, not move from hovered-effect-(s), elating ventricle-idolization, (un)-focused by tense-(s), at center per parameter-feature, greater-than lecture-article-commercial-trade-subjective-discourse.49¶|||05 m.:33s.51m.s.|||§ Market eviscerate public-commercial-spacing-(s) per communal-distance-accumulation-ratio-proportioning, spaced {Headquarter-(s) // Chief Executive Officer / Chief Financial Officer-placing-(s)} numerating, commercial-entity-(ies)-order-ing,| 221 |centralizing stable-Federal-transit-resourcing-(s), at Workermens'-challenge-(s), not reasonably discussed discretion-(s), from instruction-(s), designated in present-pursuit-perfection, by [House-Speaker-first-person / floor-speaker-third-person-overview], defining-tangible-objective-(s), from mysticism, historically auto-biographical-fact-(s), referred at present-literary-sat-posit-(s), set in past after-interrogative, labeling [person-(s) / place-(s) / thing-(s)], from substantial-pressure-carbon-dated-conjunct-marking-(s), drifted dexterity-movement-(s), affirm-ing repetitious-routine-(s), pertaining-placings per designed-feature-(s), defining offset internal-eye-plasma-imagery-perceptive-interpre-tation-(s).50¶|||05m.:16s.45 m.s.|||§ Would factor-(s) be temporal when clasping objective-commercial-possession-(s), or is soothing solidarity, from (sub)-urban-residence-(s), vague gauging choice-decision-(s), determining access per road, mystified from payment-(s) earned, (as an) enumerated-offset-(s) (not) desire ethnic-mass-pay-belief, simul-taneous upon construction-(s) where free-inhibition, maintain possessive-item-acquire-(ing-(s)), crediting lease-objective-(s), unthought-of, bastardized-elemental-patent-commercial-demand-extraction-(s), expiry in comparison, social-lifestyle-deviation-(s) at property-posit-(s) (by of) Mass & Payment-(s), occasionally-reading until temporal-understanding empirical-tasking, factoring four-primary-operation-(s) of budget;60 [Infrastructure / Leasing / Order-(s) / Marketing], for which, Government tax goods of factory-direct-mass-trade-(s), for International-transcend-ence of Illuminated [mile / knot / measure / weight-(s) / kilometer], scaling-(s) human-transit-micro-milli-measure by temperature-resetting-(s), progressive from hemispherical-comparative-county-weight-responsibility-(ies), transient (from how life has been) construed of tangible-physics, where fact, self-data-modify, (to) each lapsed-hour by latitude-hour-origin-offset-(s), when parameter-(s) offset, metropolitan-year-miniscule-fragment-(ed)-data-inhabitation, bound-aried where ecological-duty persist from workers'-performance-(s), by scheduled-interval-(s), passing momentum, not evocated from studying human-being-(s'), preferring self-physical-property-objective-(s), voluntary to possessive-dictation, aimlessly remaining intermittent greater-construct-(s), meandering governmental-technical-need-(s), uncreative for handling tectonic-plate-crust-continental-shelf-plane-adjustment, apposing mutual-inhabitant-(s), Pronoun-documenting, cause perspective drift from molten-impact-(s), paved-away, by metal-frame-rivet-mount-part-tube-valve-piston-fuel-contraption-motor-aluminum-frame-door-(s)-w/lock-(s)-rivet-space-tiering, inset tube-fed [pipe-(s) // wire-(s) // p.v.c.-(schedule-40) // et cetera...], deviating objective-conceptive-2-dimensional-planned-perimeter-spacing, valuing measured-weight-(s), per product-package-(d)-trade, embodying (a the) physical-three-dimensional-creative-living-conceptive-entity-(ies), while physical-terminological-reasoning, (not be of) Taxonomic-inhabitant-(s)-confidential-worker-classification-(s) {Protista}, concepted commercial-govern-mental-hold, (for how) patent-objective-(s)-pre-concepting, is both [two-dimensional-pronoun-individual-concepting // three-dimensional-corporate-clay-molded-pre-concepting]| 222 | multitude-difficultly-intricating, minuté Spherical-Space & Perimeter-Surface-Area, adjustment-(s) per each mass-product-ion, proportional Federal-Corporate-afforded-objective-value-(r)-(s), whom demand consumeristic-spending-mysticism, having note, general-impasse, serve "mind" (I believe [mind // memory], (is to be) after life mortem-state, not live-head-posit-organism-composition-(s)), as placed-program-method-(s), space perimeter-surface-area, apose [physical-manned // objective-dating-(s)], upheld (not afoot) interceding-parameter-(s), routine [note-interchange / value-appraisal-(s) / insurance-appraise-(s)], sedentary upon approximate-lifestyle-location-(s), as Dating & Reproduction, (are denominated from an) basis of numerative-focus, when millennium never be fully counted per [secondths // minutes // hours deviation-offset], timed (as how a the curb of Population & Reproduction) century-live-duplicate, impending legislative-Senate-Executive-Federal-monetary-policy-(ies), tentative tenant-road-shuffle-reasoning, his-torically-perplexed, comforted in century-(ies), not conceiving time-dynamic before millennial-belief, (from of) [1street / 2road / 3avenue / 4place / 5court / 6boulevard / 7lane] as impasse-(s') continue apose centripetal-force-impact, when reasoning has yet been formally-contingent in labeling, millennium-inferencing-Gregorian-Calendar-Progress, impeding-offset day-week-month-season-year-decade-time-interpose-population-reasoning, deducting (to a the) inductive-fact that, human-(s) are 136L.without-an common-genetic-structure-civilization-self-extraspective-motivated-pur-pose-(s) amid vernacular-vocal-vibrational-volume-pitch-comprehension...¶||| 09m.:37s.20m.s.|||§

§0...Count-[39 / Thirty-Nine]-preposition-of... ..Lines-[132.76 / One-Hundred-Thirty-Six.1.seven -tenths-2.six-hundredths].. ...Article-Essay-[6 / Six].. // ..Book-Essay-[29 / Twenty-Nine].. ..Secondths-continuum-effect, (is a the) soly-primary-inductive-reasoning-premise-principle[0], by

a.[[**60**ths / Sixtieths] [1]minute-secondths-reset /

b.[**3,600ths** / three-thousandths-six-hundredths] [2]hour-secondths-reset //

c.[**86,400** / Eighty-Six-thousandths-four-hundredths] 3day-secondths-reset ///

d.[**604,800** / Six-Hundred-Four-Thousandths-Eight-Hundredths]| 223 |4week-secondths-reset ////

e.[**2,628,000** / Two-Million-Six-Hundred-Twenty-Eight-thousand]| 224 |5months-deviation-secondths-reset /////

f.[**7,884,000** / Seven-Million, Eight-Hundred-Eighty-Four-Thousand]| 225 |6seasons-deviation-secondths-reset //////g.[**31,536,000** / Thirty-One-Million, Five-Hundred-Six-Thousand] 7year-deviation-(leap-year)-secondths-reset; pertaining no [ªdecade-(exponent 10 from basis year) / ᵇcentury-exponent 100 from basis year) // ᶜmillennium-(exponent 1,000 from basis year)], as when secondths-continuum-effect, introverted-individual-offset-hypertense-(s)-exist, (for of how) physiological-tension, evoke (by an) approximate surrounding-physical-registering-hyper-tension-(s)-offset amid those [facilities / vehicles] at sustainable-commercial-resources, only pertainable from physiological-tensable-visual-reflexive-neuro-functional-being-(s).¶||||58s.94m.s.|||§

¶30If dating require-(s) effort-(s), (why is it) humans-prefer effort by week, or year of calendar, apose time tasking-interval-(s), poignant for punctual-timed-minute-(s)-segment-(s)-effort-(s), revising placed-entity-location-(s), acceded while physical-effort-(s) extenuate further at planned-incorporated-community-impasse-(s), thinking how populus-resident-(cies), are supposed to survey land, per [county // city planning-(s)], (where from) National-Federal-House-Representative-(s)-Account-(ing), per offset mass-population.51¶|||21s.50m.s.|||§ [Animal-Mammal-(s) / Pet-(s)] remain (un) train-(ed), (by an) nature of inhibition, untamed by vision to think, so why possess such an creature, or foot an piece of land-surface-area, (not thinking how to coordinate animal-control, with an field for animal-hoarding, or so) to plan future from tensed-possession-(ed)-areas-(s), directing latter-presence-memory, when land, [never / not / no / none] from Veteran's-affair-(s)-purpose, recognize formal-dating, (by an) basis of time-interval-(s), (to be) fully-evaluated, by free-will-examination, [philosophically / philanthropically / genetic-mass-effort-(s)], ascertained where context by factual-data, (upon an) visual-audible-inter-ior-response-(s), innuendo colloquial-physique-composition-(s'), while perspective adrift intermittent hyper-tense-(s), affecting transitive figurative-axis, littoral revolving momentum [in / upon], lapsed-metropolitan-square-capitia-aging, expiry while presence of being, (where when) form requisite object-mode-(ing), through day Asset & Attaining (s), where (re)-action-(s), square-off self-bias-moment-(s), in place [where / when / what / whom] be rated by view of object-(s), (as when) house feature-focus, stasis complexed-fervor, by inhabitant-humanistic-need-(s), enjoying interceding day, from timed [cycle-(s) / effort-(s) / elapse-(s) / mode-(s) / interchange-(s) / transit-(s) / tense-(s)] acting while date, demonstrate Physiological-Structure & Physique-Tense-(s), from genitalia-nature-being-(s), whom accompany-pronoun-age-tension-(s), affluently noting, rated variable-(s) upon [space-(s) / genetic-inhabitant-(s)] stylized (by an) singular-perceptive-affirming, second-person-tense-interchange-(s) upon hourly-time-calendar-dating-inclination-(s), factoring presence where [urban / suburban / rural-spaced-placing-(s)] coincide adaptive inductive-incremental-words-functional-interpretation-(s), affiliating tangible-observation-(s) (from held-understanding) by ethical-synapse-(s), (as how) voluntary-independent-variable-(ed)-being-(s), count upon prelate-dated-calendar-progression-(s), juxtaposed daily-time-interval-(s), to [be / are / were / their] in volume, by blood-pulsed-being-(s'), well supplemented for solid [density / viscosity], factoring erected-vertical-altitude, variating [cause // purpose] by feminine-masculine-relationship-libido, mingling dated-hourly-time-(s), (at an) relationship present by date-(s)-(ing), (for how count through) census-education, elapsing through present-tense-(s), when day per valley-pasture-plain-forestral-desert-mountain-canyon-terrain, dynamic physics, variating reverberating wave-tectonic-impact-fixations, as [oxygen-(H2O) / solid-density-(ies) / centripetal-force-continuum (for no base fact of origins)] intermittent bias possession of [tool-(s) / pen-calibre] (un)-configured from population-district-city-county-interval-dating-communication-(s), when calendar influence being-locational-directive-(s), accounted hourly per time-frame, produced affluently where human-offset-dipitchipation, (is not in an obligation from) [human-being-(s) / inhabitant-physiological-form-(s)], once at experiencing-moment-(s), (by what) perspective-(s) motivates' inhabitant-(s)-attentive-percep-tive-focus, affording stock-shelf-priced-value-(s), from comestible-substantial-cause-(s), self-indemnified, (from an) religious-genetic-order-(ing-(s)) by individual-(s), lurid-bias, objective-title-property-location-holding-possessing, amongst-mass-mutual-production-trade, rather-an particular-resourcing-recycling-developing, intertwining [physical / physique / stance posture-(s)], [sitting-in / sitting-at] designating marked-place-(ing)-(s), juxtaposing age-sex-purpose-(s) by acts-per-time-elapsed-interval-(s)-tool-usage-sequencing, amounting experience, by documenting incremental-numerical-inductive-deci-mal-deviation-interpretation-(s), per various-physio-forms, either [being / creature / inhabitating] physique, winding cause, through daily-person-contact-reasoning-(s), sufficing residence-(s), (un)-literary (from an) timbre of paper-back-illustrating, reading note-(s), arrayed-an range-plane-surface-area-perimeter-spacing-(s), per article-(s), thought at revising-(s), proportional educational-redacting, causing (hyper)-tense-(s), unclear (for how focus by) religious-extraspective-congregation, modify commercial-development, adjusting (to an) practical-housing, relevance by balancing ratio-(s) per genetic-survey-mass-(es), conjuring weather their exist an objective-purpose through commercial-aging, instilling thoughts' at, per factory-direct-production-offset, (from an) warehouse-welding-compositional-drinking-resources-recycling, conscious-direct-contact-procrastination, when inhabitants, are among state, for focus of tangible-(s)-value-(s), transiting to site, developing article-(s) relevant to community-objective-(s), rather national-commercial-production-reliance, vague from mathematical-elemental-objective-usage of resource-(s);[61] awhile product-commercial-economic-free-enterprise-trade, distort western-free-setting-(s), from unloyal-inhabitant-(s') reclusive-effort-(s), divided by national-state, as solemn-mutual-monetary-recognition, from topic-(s), relevant from word-understanding, due to passing-time, unconducively focused on longevity (by of) physique, along with retentive-balance, moding way of [currency / object // objective-population-mission-elemental-extraction-method-(s)] balancing power, needed by mundane-reasoning, affecting our basis, at parameter-limit-(s), dictating affluent-dialogue, when timbre-tempo, influctuate current-frequency-(ies) by perceptive-wave-vibration-(s), [[1]mechanical /// [2]Transverse /// [3]ocean /// [4]sound /// [5]electromagnetic /// [6]longitudinal /// [7]surface /// [8]physical];62 for how listening apply an feature of rationale, for interval-subterior fourteen-minute-(s)-twenty-four-secondths-syntax-elapsing-(s), characterizing [numerator / fract-ion-numerator / integer-numerator / numerical-numerator // percentage-numerical-denominating], operating perceptive-class-ified-article-(s)-tangible-(s), relative by deviation-(s) from live-tensed-blood-pulsed-being-(s), affecting [produce / element-(s) / substance-(s) / climate-territory / ethnic // genetic-inhabitant-residential-divide], [adjacent / hypotenuse-integer-variablizing], featuring per dimension-(s)-offset, [perimeter-square-feet [length / width / height], base-posit per [weight // measuring-(s)] deviating72L. denominating-personal-property, inductively reasoned communal-perceptive-denominative-constitution-(esce)-numerical-document-filing, observing-inhabitant-(s) dimensioned-abstract-spacing-(s), where of terrain-citizen-populating...¶|||05m.:21s.83m.s.|||§

§0...Count-[15 / Fifteen]-preposition-of...|226|..Lines-[71.425 / Seventy-Three.1.four-tenths-2.two-hundredths-3.five-thousandths].. ..Article-Essay-[7 / Seven].. /227/ ..Book-Essay-[30 / Thirty].. ..Minutes follow an rate, (for how) [we / us / you], English-manuscript-interpret-comprehend, visual-letter-word-(s)-syntax-vocal-vernacular-dictionary-documented-observation-(s).¶12s.49m.s.|||§

¶31[[1]Inductive / [2]Conjure / [3]referencing / [4]letter-word / [5]deductive / [6]formulation / [7]calculating / [8]conceiving / [9]redact-(s) / [10]perceiving / [11]addition / [12]subtraction / [13]division / [14]multiplication / [15]deviation / [16]greatest-common-denominator / [17]reading / [18]write / [19]syntax / [20]grammar / [21]punctuation] by [sight / sound] decrescendo tension-shape-distance-dimension-color-form-peripheral-incande-sary-(in)-[direct-depth // reverberated-vibration-(s)]-tone-count-per-spective-(s'), consider where observation (not memorize), demon-stration-(s) by will, [showing // displaying] natural-timbre, (by an) analytical-perspective, from inhabitant-being-(s), memorizing eight-hour-five-day-(s)-work-routine, Fourty-hours-present, while day-fo-cus, accrued [financial debt / credit], from excessive [substantial // objective] possession-(s), financed by personal-account-(s), approving currency, rated (by an) consensus-mass-purchase-compliance, (for an) lifestyle-objective-(s), percentage-(s) of Federal-commercial-pur-chase-trading, by verbal-interaction-(s), askew from commercial-fac-tor-(s), observing dormant-transitional-rural-develop-ment-(s), un-observed in offset-physics-comparison per metropolitan-technologi-cal-conjuring, thinking at all time-(s), (a the) hypothesis of reproduc-tion, placing-self-responsibility-(ies), (upon an) pronoun-habitant-(s), methodizing locational-development, juxtaposed rural-terrain, when municipal-warning-caution-advisory-no-trespassing-signage, de-fine metropolitan-boundary-notion-(s), ruling governmental-Feder-al-power, advising hierarchal-comparison, holding designated-ar-ea-(s), (as an rationalizing per of) fact-(s), having [upper-middle-class // upper-class] maintain-sustain-disparity, amid Middle-Class-Life-style-(s) & Poverty, not conveying reliable-thinking, broad abound at-tention per Space & Tangible-Objective-Item-Variable-(s), adjusting proportion-(s) (by of) possessive-hyper-tense-(s), designating con-trol where perimeter-surface-area, designate parcel-(s) of land-plan-ning-(s), governmentally-denominating required post-mainte-nance-development-(s), where terrain by voluntary-inhabitant-group-ing-(s), segment state-planned-space-(s), focusing [[1]who / [2]what / [3]where / [4]when / [5]why / [6]how reasoning matter-(s)], come in to effect, through physiological-subjective-directive-documenting-effort-(s);[63] when historical daily physical-terminological-factifying, define tan-gible-physique, from commercial-develop-ment-(s), for rural-con-ception, remaining dormant without-an consideration of how unincor-porated-territory remain cheaper-than incorporated-territory, from substantial-sustain-able-trading-access-posit-(s), maneuvering through imagery-sighting-(s), denominating day-night-hour-visioning-sleep-reset-adaptation, inductive tangible-development-(s), offset-per state hardware-infra-structure-purchase-transition, (for why) [1fee-(s) // 2tax-(es) // 3permit-(s) // 4fine-(s) // 5ticket-(s) // 6bill-(s)] follow legal-judiciary-executive-warning-(s), process [objective // land-space-territory // physiological-substantial-nutrition] affecting our mundane-natural-impasse-(s)-decision-making, when solid-hold, have [patent-(s) // copyright-(s')] maintain-trade, per budgeted-operation-(s) under governmental-trade, comparative historical-obituary-primary-inhabitant-(s), where developmental-metropolitan-placing-(s),351L. grammatical-punctuation-syntax-aptitude, soly (from an basis from) national-commercial-trade-reliance, (comfortably-numb) unfiled county-city-building-fees-code-compliance-(s), (one or the other), pending Residence or Commercial-Planning, under [county // city infrastructural-development-(s)]...¶|||02m.37s.75m.s.|||§

§0...Count-[6 / Six]-preposition-of... ..Lines-[36.167 / Thirty-Seven.1.one-tenths-2.six-hundred-ths-3.seven-thousandths].. ..Article-Essay-[8 / Eight].. // ..Book-Essay-[31 / Thirty-One /// §02m. : 37s. : 75m.s. / two-minute-(s) : thirty-seven-secondths-essay-reading-rate].. .follow along the essay-reading-rate-deviation.¶|||09s.03m.s.|||§

¶32What study-(ies) of whom reside upon government [city // county planning], intermittent state-Federal-government-constitution, not think while momentum-drift, (as how) community is suppose to formally disband requisite-(s) task-(s), if ever to save money, while creating-an genetic-union, basing reproductive-quality-(ies), (away from) recipe-receipted-trade, fluently-interpreting subjective-conjured-consideration-(s), flowing physio-psychological-progression-(s), for act-(s) per intuitive-state-(s), upon contingent-restructuring, (as how) government-function, (but when) climate would take part in decisions from military-statistic-(s), ignoring Texas-Instrument-(s)-mechanism-function-(s), balancing how [constructive // engineer development-(s)], are [defined // documented // denominated] where place-per-individual, naturally compulse, (but should be eliminated from by of) applause-fanatic-presence-driven-historical-census-count-(s), [happening / accrual / glottal-stop-go] amused impasse transitive-momentum-physics-presence-impassing, entailing [nature / mammal-(s) / insect-(s) / body-(ied)-waterway-(s) / road-(s)], naturally disambiguous;$_{64}$ awhile human-physiological-characteristic-feature-(s) focus international-commercial-production-factor-(s), collecting present aging past-memento, defining career-isolated-experimental-field-(s), benign-at-an physiological-continuum-prelate, (as where) value of area is put to perimeter-surface-area, (when a the) repetition of [edifice // constructive-renovated-residence-(s) // commercial-governmental-constructive-entity-(ies)] conjure-$_a$-$_b$-inductive-success-ion-effect-(s), (as when) order-tallied-count-(s) per effort-(s), pertain (un)-inquisitive moved [a]-circa-era, (by then) $_b$-perplex-momentum, (as where) $_c$-resumed-composition-(s), perceive why referencing deviate (from an) [c]-thesaurus-error-(s), (due to an) lack of literary-auto-biographical-communal-purchase-denominating-purpose-(s), investing further into [movie-(s) // sport-(s) /// idle activities];$_{65}$ awhile visible-physics perspective-rate-(s), not concert directed-focus, extraneous in place of force, rampant with exponent-peril-factoring, cause (for how each) force of physical-nature, posit-physiological-composition by calendar-year-(s)-deviation-(s), apose exterior-perceptive-calendar-hour-density-time-temporality-rate-(s), consistently moding upon bias, without-an index of matrices, examining how movement-(s) rationalize reproductive-offspring-exponent-causes', during mass-bias-retardation, pulsated-passed posit-(ed)-compositional-movement-(s), anticipating will-attention-span-belief, amid physiological circumference-drift inertia, not confirmed by vague-obligation-(s), ambiguated (by those whom perceive a the) pertinence of logical-(person-(s));$_{66}$ awhile others accelerate-dormant-Gideon's'-reader-yield, [fault // fruit] manual-context, physically-blank-thinking, legacy-interval-(izing)-memory, decomposing fact-(s), at placed-goings, with flow of time, (as when) perspective is formed by tense-(s), validify-(ied) evident physical-human-interaction-(s), disregarded by self-infetesimal-numerical-para-meter-(s), where visual-depth-perspective, see through momentum-vapor-(s) from belief, amid kinetic-temperate-dispersion, harmonized solar-molten-combustion-impacts;[67] awhile earth-terrestrial-centripetal-force, molten-mobius-drift, offset-an [solar-infinitive-combustion // earth-molten-aquatic-tectonic-plate-(s)-continental-shelf-terrestrial-rock-soil-sand-deposit-(s)-moment-um], impacted (from how our) perspective afloat continent, where [due-in // due-of act-(ion-(s))], affiliate elemental-modification-(s), schemed-agenda, (future-as-when) scheduled-human-inter-section-passing-(s), live in affinity-status, by area to moving amidst perspective-(s'), offset [to // through area-(s)], oxygenated (earth-molten-yellow-star-vapor-cooling-effect-(s)), seclusive reclusive-introverted-isolations', infetesimalized synthetic-activity-(ies), by taxonomic-day-secondths-word-maxim-gammas, yet exponenting elemental-extraction-population-ration-devia-tion-(s), during exact-minute-(s)-hour-(s), exponenting physiological-intelligence-effort-(s), when hour-(s) (have not been) juxtaposed from introverted-livelihood, juxtaposing intersected age-(s), when working on experience per task-(s), mattering varied subjective-gamma-(s)-sociological-extraspective-interpretational-congressional-documenting, vast (out where) territory, engineer-tranquility, upon tectonic-governmental-boundary-(ies);[68] awhile metropolitan-longitude-hour-reset-(s), focus latitude-minute-(s)-north-south-political-variation-(s), graph-deviating x-axis-force-horizontal-spherical-momentum, incandescently-perceiving per [balance / measure / scale] subterior-labeling, tangible [commercial-object-(s) ///// physiological-person-(s) ///// populational-being-(s) ///// governmental-perimeter-boundary-parcel-(s) ///// personal-title-action-(s) physiological-maneuvering-(s)] scaling [hour-(s) /// minute-(s) /// secondth-(s)] apposing subjective-inter-pretation-focus-(es), specified by [pronoun / particular-patent-object-(s) / movement-(s) / reading / writing-impasse-elapses // syntax-(es)-synapse-(s)], tiering [weight-(s) // dimension-(s)], articulating those form-(s) of conceptive-patent-product-development-(s), per [percentile / fraction percentage-(s)] for article-extent-refinement-(s).•52¶||||03m.:55s.•72m.s.|||§ [Percent-age-(s) // fraction-(s)], define factoring-method-(s), mounting posit-(s) at particular-develop-mental-point-(s) upon denominating light-contin-uum-property-offset, per daily-routine-repet-itive-impasse-(s), tangible (when where) commercial-shelf-item-object-(s)-cycle-(s), per entering-entity-purpose-(s), maintain-operation-(s), intermittent operational-procedural-function-(s), fathoming heightened-direct-cubical-surface-area-depth-visual-tonal-vo-cal-perspective, reason-ing hyper-tense-(s), in plain view for tangible-output-(s), conjuring method-(s) from adver-tisements, persuading employed-at-will-inhabitant-(s'), balancing control, commercial directive-(s), (by an) religious-perspective, resumed (when / where) resourced-manufactured-commercial-based-ordering-(s), [process // ship] on hourly-time, commercially-shipped electric-water-govern-mental-based-location-(s), awaiting National-Agenda (America-(s')) tectonic plate), [persuading // influencing physical-movement-(s)], coordinating purchase-value-(s), for physique-fatigue-recuperation, relying self, (from the) mass-composite-production-method-(s), modifying surrounding development-(s) at governmental-constitutional-count-(ing)-(s), day-population-(s), variating offset-career-task-(s), when national-debt, sustem [Totalitarian // Communistic // Sociological society-(ies)] apposed America-(s') [Treasury-posit // Federal-extraction-(s)] historically commercially cronyized, by deliberate-ignorance, offset-per [family-(ies) // familiarities // friend-(s)] per Constituting-Territory & Political-Discourse, affirming Gideon's'-hillbilly-country-sociological-lifestyle-(s), at metropolitan-(sub)-urban-movement-(s), tabling-counts, massed at Federal-paper-print-form-count-currency, without common-written-cursive-typed-document-inquiry-(ies), denominating perspective-(s), from operation-(s) delicately sustained (from those) competent-bosses superior-value-(s), unreferenced literary-purpose-(s), rationalizing proportion-(s) per free-will-offset, furthering (a the) discourse of family-community-reclusion from national-object-(s), per period-(s) from blank-thinking, not perceiving passing-physics-momentum, concisely-conjuring-commercial-factor-(s), mutual per inhabitants'-prefer-red-responsibility-(ies), detailing revised nation-(s), making Rich & Poor, sustain mutual-Stockholm-effect, uninquisitive mutual-literary-interacting-(s), offset genetic-class-(es), then maintaining commercial-unit-production-objective-(s), volumizing-upon, impasse-(s')-transit-totalitarian-method-(s), pertaining up-beat-activity-(ies).53¶|||05m.:37s.70m.s.|||§ Metal continued (to be) collected, while among brick-infra-structure-(s), entranced in conventional-discourse, acting upon those requisite-impulse-(s), sharing cause-(s) from introverted-possession-(s), continue require-ment-(s), per common-legal-discourse-(s), affirming purpose-(s) handling matter-(s), unused by House-Representative-Dynamic & Supreme-Judiciary-Federal-legal-authority, regarding [finding / collecting] free-will-citizen-(s'), whom share-an mutual-interest, upon [[1]mutual-interaction-(s) / [2]territory-(ies) / [3]climate-(s) / [4]terrain-(s) / [5]constitutionalizing / [6]national-governmental-referendum / [7]calendar-planning / [8]restructuring entity-(ies) at territorial-claim by mutual-genetic-mass-effort-(s) / [9]manufacturing / [10]resourcing / [11]infrastructuring / [12]denominating / [13]studying-existence], intermittent residence-hourly-living per day (overwhelming any contemplatable-dating, when formal-physical-action-(s), reduce my belief of non-fiction, (to be an) much more dull-linear-mass-accounting (babysitting), common-mass-(es)-offset (including sleep at home)), presiding at posit-(s), human-sedentary-discourse.54¶|||06m.:16s.20m.s.|||§ Stature of character, define self-individual-(s), reactive commercial-trade-volumes-objective-item-requirement-(s), (to eat for an) living, not in creamed-echelon per indulgence, being founded on new-barren-rural-terrain, unconstituted post-retirement-anticipating-planning-discourse, (by an) lack of future-thinking, in Day & Age, (un)-attempted to resolve the unscheduled-socialism-genetic-discourse-dysfunction-(s), not elegantly-aligned, physical-cursive-effort-(s) elder-generational-mentality-(s), vacant-minded, thought-process-(es), [concept-(ed)-city-offset-planning-(s)], passing those physical-fatigue-(s), wilting-away, deterred lust from primordial-existence, historically where [[1]word-(s) / [2]work / [3]schedule / [4]discourse / [5]time / [6]calendar / [7]intervalized-experience / [8]interaction-(s)], discourse whatever invigorate-(s') inhabitant-lifestyle-(s), from prerogative-requisites' by human-modern-census-count-(s), pre-occupied by per personal-limit-(s') per individual-choice-(s), upon Existence & Repopulation, not communally-sweating, the irritated nature from political-theatre, conditioned (by the) assertive, whom have had pushed-effort-(s), not rescinded heavenly-direction, understandable by limit-(ed)-way, of personality from individual-(s), whom not intricate-concise-claritive-conjured-understanding, (from those) land [lord-(s) / individual proprietary-title-(s)], maintained leased-mentality, pre-requisite-ideal-(s), default-repopulating, (not in an) titled-regard, for momentum-incremental-counting, denominating mass-commercial-act-(s), transiting at offset-succession-(s), past those boundary-limit-(s), surrounding aged-fatigue, past the decade-intensity, when incapable present-elapse-(s), per practice intermittent, focused-aptitude-intelligent-individual-(s), writing interposed [physiological-component-(s) / act-(s) upon limit-(s') of boundary / variation-(s) of texture-(s) / depth-perspective / vibrational-kinetic-transitive-wave-(s)], grandiose-superfluous, common-protest-labor-reasoning, pertained-from compartmentalized group-approximation-(s), collecting tangible-matter, examining minute-(s)-segment-interval-composited-mass-matter-note-(s), pointed-apposed intangible-air-vapor-(s), writing ink onto [paper /// parchment /// velum / text / type] fundamental conceptive-creating, passionate-purpose-(s), reasoning vision upon hardware-tense-(ing-(s)), which evoke hyper-tense-(s'), interval-interposed [centripetal-force-momentum-deviation-(s) / respiration-breath-vapor-blood-pulse-tension-(s')], that apply secondths-posit-(s) amid Motion-(s) & Action-(s), invigorated where [agricultural / horticultural mean-(s)], clue voluntary-effort-(s), writing Power & Water pre-concepting territory, as elder-life-basing, offspring rendering [style / model / readjustment-(s)], per subjective-reasoning, exposed written-context, meriting [mathematical-inductive-reasoning / literary-grammatical-punctuation-development-(s)] denoting scientific-observation-(s), when qualities per observable-scaling, discourse commercial-experiment-(s), lineage-passing article-(s), for post-life subjective-directing, competently-moding methodical-population-enhanced-contin-uum-purpose-(s).55¶||08m.:39 s.60m.s.|||§ Parental-economic-deeming, fluster (for how) naturally partial work-method-(s), affect attainable-objective-(s), (when either no cause through) subjective-comprehending, persist, for being-bias, displaying aged-stillness, conjunct from constitutional-manual-biblical-textbook-literary-competency, abounded by Schedule & Planning, yet modifiable compartmentalized-imagery, silhouetted unconsciously, where legislation remain from Federal-state-constitutional-order-(s), [claused / termed / conditioned / sanctioned] in confluence, (from the) style of political-boundary-traded-resourcing-affair-(s), geologically-sustemming (sustained-stemmed) county-city-[commercial // residential] Federal-trade-planning-location-(s), enveloping acre-environmental-purpose-(s) per lifestyle, garnishing governmental-default-boundary-(ies), when genetic-segregational-sociological-discourse, denominate all expenditure-(s), transitive biological-analyzation-(s-(’)), believing [1self-individual / 2family / 3friend-(s) / 4partner-(s) / 5group-(s) / 6co-worker-(s) / 7religious-sect-(s) / 8school-fraternity] (as how) sixteen-candles-post-aging-after-effect-(s’), present future [inclination-(s) / indication-(s)] conditioned offset-per physiological-tense-(s), dating unclear-fact-(s), from set-placings, retiring aged mental-focus, from adult-youth-effort-(s), efforting into retirement, (an non passion for) experiencing the globe.56¶|||09m.:37s.10m.s.|||§Attention-span affiliate physiological-segmented-minute-(s)-elapse-(s)-physics-offset, intriguing-personal-study, fielding-out, (un)-syntaxed-perceptive-focus-context, (in-of) entity-parameter-(s), [sustaining / pertaining / maintaining / containing] (person-(s) / human-decision-(s')), reasoning-how physiological-discourse impasse mile-(s) from perimeter-planned-boundaries, where [two-dimensional-concepting // surface-area-exterior-shape-(d)-(s)] figure [three-dimensional-mental-reason-(ing)] accruing-variable-(s) by consumeristic-objective-count-(s), trading monetary-act-(s), affinitive-influx, forbidden-boundary-city-county-state-governmental-access, by locked-aged-interaction-(s), crescendo-acceding "involuntary" inhab-itant-human-male-female-superego-enigma-individual-(ism), enchanted self-idolization, before (un)-reproductive-self-bias-hoc-dynamic, symbiosis 150L. grand-parental-mysticism, abyss those way-(s) of fortune, defining-classicism, in-thought-regard, depth-perspective-discourse, defining humanity-perceptive-focus-post-life-document-ation-transitional-lineage-generational-error-(s)...¶|||10m.:30s.86m.s.|||§

§0...Count-[22 / Twenty-Two]-preposition-of...| 228 |..Lines-[147.675 / One-Hundred-Fourty-Eight.1.six-tenths-2.seven-hundredths-3.five-thousandths].. ..Article-Essay-[9 Nine].. // ..Book-Essay-[32 / Thirty-Two /// §10m. : 30s. : 86m.s. / Ten-minute-(s) : thirty-secondths-essay-reading-rate].. ..Human-interaction-(s'), (are an) slow-process, when work-effort-(s) not comprehend how to apply article-subjective-essay-lesson-(s), per visual-momentum-interaction-interpretation-(s). ¶|||19s.91m.s.|||§

¶33 Snide aging, not believe self-mentality, treasuring western-hemisphere-status, not instilled monetary-commercial-objective-proprietary-value-(s) through aging, while passing-time, [fatigue // experience aging], becoming adapted by subject-(s), along period-(s) of hour-(s), (from an) ethnic-laziness, affecting total-monetary-economic-geological-outcome-(s), producing American-factory-production-(s), formulating idle-push-(es), from illustrated-luminescent-graph-ing, voluntary acting through operational-mode-(s), featuring age-driving-fatigue-(s), by daily-incompetency, access-(ing) free-enterprise-entity-(ies), not capable of expanding surround-ing boundary-(ies) of constitutional-limit-(s), without religious-off-time-communion, practice-(ing) national-group-discourse, by usage of proprietary-requirement-(s), (from when) recollecting needed item-(s), (are an) part of transfiguration using personal-physical-self, soluble-physical-ventricle-respiration-(s), emitting-response-(s) (from of) air-vapor-centripetal-tension-(s), yet factifying literate-genetic-recodifications, modifying Schedule-(s) & Territory, reasoning thought from international-commercial-trade-dynamics, Thirty-Thousand-Titled-citizen-(s)-per-House-Representative-(30,000C.-1H.R.), following constitutional-orders, familiarized by appraised-estimate-(s') following Five-Hundred-Twenty-House-Representative-(s), multiplied apposed commercial-volume-count-(s) per Fifteen-Million, Six-Hundred-Thousand-Citizen-(s), (male) covered by family-population estimating Sixty-Two-Million-Four-Hundred-Thousand-Citizen-(s), abiding Federal-Representation.[57¶||||]01m.14s.[97m.s.|||§] Decades mass-(es), age-simultaneous, contra-dict personal-virtues' upon hourly-timing-(s), tensing physical-motion-serie-(s), physio-logical-reflex-(es), trading-product-(s), from off-work-time-moment-(s), sufficing life-style-(s), (by an) type of none-discretion, disclaimed physiological-perceptive-tense-(s), reminding conclusive work place effort-(s), candidly avid, managerial-monitoring, placed-planned-area-(s), upon placed-factor-(s), inhabitating nominative-focus, bolstering traded-factor-(s), featuring traded-commercial-production-(s), dating lost-property-focus, at aged-juxtaposition-(s), sufficing activity-(ies), dollar-per square-foot, each inch-incremental-notion-work-week, tabling-down [monthly-expenditure-(s) / comestible-(s) // seasonal clothing // yearly-solid-objecting by partiality of fact-(s)] verifying family-(ies), caring amidst modern-dating, when habitual-individual-(s) kin work-parameter-(s), collecting apartment-belief-independent-object-(s), Sharing & Listening, discoursed at separate-living-pattern-physiological-group-pattern-(s), recalling introverted two-dimensional-parameter-graphing-(s), (not) factoring-place where domicile call to order, enforced-physical-effort-(s), separating genetic-aligning, making decade-(s)-impasse-(s), amid perspective-visual-impact, (from when) focus (by of) Lifestyle & Schedule (are from an), experiencing age stabilized-perplex-(s), figuring 31L. factor-(s) at commercial-rental-present-tense, past set-exampling-(s), defining future-schedule-(d)-inhabitant-(s), congregating familiar-datum, because where commercial-purchase-objective-placing-(s), maintain-an responsibility at designated-traded-activity-(ies), mundane Creature & Instinct, sub-par luxurious-human-exist-ence...¶|||02m.:00s.40m.s.|||§

§0...Count-[9 / Nine]-preposition-of... ..Lines-[33.9189 / Thirty-Three.1.nine-tenths-2.one-hun dredths-3.eight-thousandths-4.nine-ten-thousandths].. ..Article-Essay-[10 / Ten].. // ..Book-Essay-[33 / Thirty-Three /// §02m. : 00s. : 40m.s / two-minutes-essay-reading-rate].. ..what motivates' interactive-mutual-vocal-interpretation-appreciation, historically by English-American- historical-word-vernacular-interpretational-usage, amid discourse of American-Federal-monetary-debt? ¶21s.06m.s.|||§

¶34How obsessive are Bride & Groom, by youth-mode-(ing) per subject-(s)? (If for not) default-development-(s) inspire-being-(s) (with an) car, (for an) approximate-circulating [government-offset-inhabited city // county spacing-(s)], Limited-wage-Circumstance-(s) & Boundary-(ied)-Controlled-Territory, affecting those Germanic-Trait-(s), mysticisized by contact at [commercial / residential governmental-order-(ing)-(s)], (by an) pertained voluntary-genetic-defining, by tectonic-mass, having metropolitan-congestion, forecast where street-(s), traffic transited-traffic-passing-(s), showing metric-mile-(s), forcing-physics-impact-(s), qualifying module-act-(s), viewing pass human-reasoning, where surface-area, define aging per independent-being-(s), by child-development-(s), valuing self-default customs, acquainted-commercial-interaction-(s), finding newer-avenue-(s), where sociological-impact-(s), permit how interaction-(s) (is upon) daily-impasse-(s) per lifestyle-(s).58¶|||43s.60m.s.|||§ What might occur, if continuum of human-dissolve, developless-character-skills, generationally-transitioning, morphed life from physique, maintaining mental-capacity, designating populant-(s), (un)-astute by "rationale", (as when) variated Elapse-(s) & Cognitive Repetition demonstration-(s), affirm conclusive-truth-(s), where human-(s) (are by an pattern for of) valuing-commercial-context, when natural-cause, resolve new-measure-(s), implying work, where focus at market-posit-(s), impasse-conjure Population & Ingestible-Soluble-(s), listing agenda, presently 16L.dating [elapsed // syntaxed] passed-interactive-recognition-(s), (age-decaying-continuum-history-calendar-years [season-(s) // month-(s)]) defining per month-elapse-recipe-listing-(s), by what motivated human-behavior, construe objective-(s)-conduct, substantial planned-mutual-mass-resourced-land, acquiring fluently conversed trade tense-rationale... ¶|||01m:26s.35m.s.|||§

§0...Count-[2 / two]-preposition-of.. ...Lines-[19.1579 / Nineteen.1.one-tenths-2.five-hundredths-3.seven-thousandths-4.ten-thousandths].. ..Article-Essay-[11 / eleven].. // ..Book-Essay-[34 / thirty-four /// §01m. : 26s. : 35m.s. / one-minute : fifty-seven-secondths-essay-reading-rate].. ..Life is short, conduct formal effort.¶09s.25m.s.|||§

¶35[Arriving / departing], affect how people presume responsibility-(ies) by state-planning, having person-(s) designate, class-posit-(s), [from / range / development / construe] state-ordered-operation-(s), defining county-Federal-logical-ordering-(s), by human-population-need-(s), (when those) elder-(s) past generations, Sat & Re-Posit-(ed), those pre-stigmas, of transient-trade, in awe for administering redundant-claim-(s), excessive where populus-mass, believe power, abiding State-Federal-Legislating, having how major-metropolitan-county-square-mile-(s), become-an captivity of power, without tenure, capable conscious-dialogue, at positional-status, affecting (how each) inhabitant-(s) seasonal one on one reasoning, interact-(s').59¶|||32s.56m.s.|||§ Hence, no reform can file under collective-grandiose-mass, (due to) ethnic-subjunction, when legislative-law, proceed power-(s) by House-Representative & Senate-Member-(s), upon Speaker-floor-monitoring, confluential-population-political-practice-(s), (by an) scheduled-budgeted-agenda, from Goods & Services, attempting budget, flowing economic-system-(s), where free-will-being-personality, (not have an) extraspective-direction, planning for retirement, with other [co-worker-(s) // peer-(s)], from thought of resource-(ing) per decade-census-coordinat-ing.^60¶|||56s.^75m.s.|||§ In obtaining piece-(s) of matter, [man / woman / child], partake along-an accompany-(ing)-purpose-(s), patent-marketing, traded-matter-(s), (from an) default-of empirical-deductive-corporation-offset-action-(s), obligated by responsibility-(ies), when Federal-State-Constitutional-Established-method-(s), practice through daily-upkeep, maintaining population-count-guideline-(s), balancing ruled observation-(s), perceiving inhabitant-(s') order, during mundane-purpose-(s) when physiological-blood-pulsed-being-(s), derive parameter-(s) from mechanically-gilded-contexted-constructing, converting fact-(s) from methodical-effort-(s), following instruction-(s), to conjure from physical-perspective, method-(s) of societal-development, (while those) physical-boundary-(ies), surrounding littoral-area-(s), ensure purpose-(s) at territorial-22L. land-development-(s), by vocal-vernacular-inter-active-constant-upkeep, (while upon an) area of possessive-development-(s), resourcing-product-(s), near around Federal-Commercial-Denominative-accounting... ¶|||01m.: 38s.15m.s.|||§

§0...Count-[7 / Seven]-preposition-of...| 229 |..Lines-[23.78 / Twenty-Four.1.seven-tenths-2.eight-hundredths].. ..Article-Essay-[12 / Twelve].. // ..Book-Essay-[35 / Thirty-Five /// §01m. : 38s. : 15m.s. / one-minute : thirty-eight-secondths-essay-reading-rate].. ..Why has humanity not thought intuitively... working on developing terrestrial-state-perimeter-boundary-(ies), reordering-an genetic-political-party-classification-(s), by tier of Execut-ive-Senate-Cabinet-Committee-Labor-rural-housing-commercial-district-(s)-develop-mental-commercial-educational-article-subjecting, (furthering) how practice of living affect-(s') each moment of physiological-tense-(s).¶24s.06m.s.|||§

¶35*1·The prior elaborates' (why the rich are wealthy, and the poor are sufficed. Each day, the partiality that inhibit-(s) family-individualism, yet congruently learn how file-ordering mass-genetic-characteristic-(s), enhances community-genetic-configuration-(s) from methods of population-survey-(ing), efficiently influencing State-Territorial-Reform, while account-(s) of reasoning, order commercial-production, set at residential-creative-purpose-(s), constituting-an architectural-graphed-planned-developmental-territorial-claim-(s), amounting-an effective-act-ing-practice, documenting commercial-objective-technique-(s), (legally-denominating), re-fer-enced citizen-author-stories, (by whom) calendarize-scheduling, during hour-time-interval-(s), by pre-mass-survey-planning, accommodating8L. mass-action-(s), motioning past momentary-segment-(s) interval decade to century-death-existing... ¶|||39s.37m.s.|||§

§0...Count-[<u>2 / Two</u>]-preposition-of...| 230 |..Lines-[<u>9.09 / Eight.1.zero-tenths-2.nine-hundred-ths</u>].. ..Article-Essay-[<u>13 / Thirteen</u>].. // .. Book-Essay-[<u>36 / Thirty-Five /// §39s. : 37m.s. / thirty-nine-secondths-essay-reading-rate</u>].. ..As poor-inhabitants' don't work, so do the rich under-labor (sweat), those empirical-developmental-city-minor-county-major-master-planning-constructed-ordering-ing-(s), (for how intellect is not in) political-literary-congressional-genre-(s), (because of an) lack of physical-preferential-effort-(s), to erect such development-(s), amid lifestyle-psychological-bias-hoc-individualism.¶||||26s.15m.s.|||§

¶36How kin work commercial-operational-service-(s), define-(s) how momentum juxtapose-(s) individual-property-hold-(s), believing in voluntary-possessive-objective-claim-(s), when shipped-ordering-(s), reveal-fact-(s), apposed stated-worded-sequence-(s), (by an) Age & Definable-Reasoning-(s), influencing-inductive-evocation-(s), by movement-(s), [for / from / of / to-awhile-from], “observational-statistical-percentage-rate-deviation-(s), rated-by singular-posit-(s), (at an) hourly-time-basis, juxtaposed impassed consistent-percentile-deviation-(s), upon present-day, when Senate-Shepard-(ed)-being-(s), interact by social-gossip-work-style, bored prior value-(s) per silent-concerted-workplace-individual-(s’), (cleaning) effort-(s), centrally metro, without rural-characteristic-(s), defining perspective upon physic-(s), from dating commercial-subjection, Religiously-obligated standard-practices’, from date-year-variation-(s), per reflexive-median-average-disambiguous, wage-rated relied need-(s);69 unscheduled when [bias / hoc], affect inhabitant-opinionated-decision-effort-(s), affirming [expenditure-(s) / saving-(s) / checking-debt-purchases / trust / yield / bond / stock / scholarship-age-validification-(s)] defining precedence by education, affirming populant-fact-(s), when perspective in surrounding-imagery-gamma, reside, human-behavior, (by an) non-loyalty-free-will-fate-(s), priori true-conceiving, natural commercial-objective-title-property-possession-(s), early by parameter-proximity, (for how) physics-lock, accumulate-suburban-road-deviation-(s), becoming-attached-to common-daily-average-thought-process, per passing hour-(s), atoned turning onto physics [solar / terrestrial / lunar-untangible-(s)] (from an) vocal-timbre-concerted-focus, dialogued political-literature-genre, (by an) vernacular-vocabulary-glossary-dictionary-purposes, communal as Gideon’s-impact, affect rate-(s) per word-(re)-synapse-(s), intricated by impassing-time-interval-counting-(s), tiering [age / year / physics-ontological-proof-(s)].61¶||| 01m.:39s.13m.s.|||§ Introverts’ not cater extraspective-interaction-(s), by habit of men, consuming commercial-inspired-recipe-meal-(s), forgetting time-interval-(s), pass calendar-year-(s), unaware how invigoration per interaction-(s), fixate on personal-ideal-(s), encompassing invigoration-(s) per lifestyle-(s’), through act-(ion-(s)) in vehicle-module-(s), where required place-(d)-resourcing-objective-(s), supplement-account-(ing)-(s), congested where designated-city-county-centralized-empircal-road-way-entity-(ies), | 231 |submerge [market-(s) / garage / domestic-residence-(s) / warehouse / centralized-government-department-(s)] factifying monetary-citizen-Federal-currency-market-purchase-trade-(s), impassing momental-means', centripetal-force-continuum, east-to-west-latitude-trajectory-metric-force, impassing metro-bias-deviation-(s), sufficing-currency-means', observing simultaneous-motion-(s), (by an) productive-self-prerogative-(s), intermittent subconscious-thought-(s), (for an) infetesimal-variable-(ization), defining Federal-patent-tangible-(s), affirming belief per matter-(s), exterior before self-physical-anatomical-composition-(s'), ventricle-eye-plasma-ligament-reflexive,| 232 |bones-muscular-skin-posit-offset, using piece-(s) of receipt-data, discoursing interpretational-competent-capability-(ies), (from an) empirical-belief-subjunct-(ion), vocatively accounting, Pen & Paper, personal-objection, through [letter /// word /// paragraph /// syntax-imagery-passing-(s)], in turn voluntary-agenda-(s), maintaining family-individualism, where lifestyle-plateau-(s), retiring locked in on day-dating-discourse, affirming in juxtaposition of hourly-time-posit-impassing-moment-um, tense-(s) per-secondths-time-continuum, by youth-routine-order-(ing-(s)), instating directives' from activity-(ies), daily-counting, acted-effort-(s), deviating pronoun-compositional-inhabitant-(s'), devising product-(s), pertaining perimeter-surface-area, in approximation by volume-substantial-modern-recollect-(ing-(s)), perceived through visual-scope when aging at hoc-affinity, without-an aptitude for lecture-(ing), parole-(ing) scale human-density-tension-(s), (as a the) fundamental-comparison-(s), inductive-numerating [pound-(s) / ounce-(s) / gram-(s)], invigorating Personal-Classical-Intelligence & Customary-Community-Class-(ification-(s)), perceiving each [objective // substantial-quality-value-(s)], apposed currency-denominating-tangible-confirmation-purpose-(s). 62¶||| 03m.:28s. 19m.s.|||§ Priori befud-dles sighted-view-(s), at off-time-interaction-(s), reclusive where residence-(s), not participate with community-approximate-location-(ing)-(s), serving colloquial-act-(s), at

[Article:] {February}

Locale & Day, [set / reset posit-(s)] at proportioned-distance-(ing-(s)), from mathematical-prose, emitting-an observational-dynamic upon terrestrial-physics-cartography-expedition-exploring, while common-mass-physical-hoc, not pertain [grammar / rubric / literature] (accelerated reader not youthfully applied), aged-yearly-monthly-dating-aging, (at an) basis for context-comprehension-interpretating, per book-(s) applied balance-(s) of work-activity-(ies), passing season-(s), qualifying read-extent-extenuation-(s), perceiving (a the) fact of studying, sustemming [teacher / professor maxim-(s)] concluding quality of merit-(s), equating terminological-reason-(ing), juxtaposed, season-week-cycle-period-(s), facting value-(s), from census-accounted-place-(s), noted-perceptive-qualified-credit-(s), compliant where consistent-persistency-(ies), constantly-charge bias, idling-state of dating, fulfilling how human-(s) thrive (from the) ignorance of life, that enable-(s) those dull surround-(ing-(s)), that pertain-location-(s), by default-luxury-(ies), that not account [story-(ies) // discourse of community-event-(s)], persuading ability-(ies), from at-will, conceptive-communal-constructive-thinking, why location-(s), used (as for when each and) every act, apply-an stance at future-planning, (like retirement) designating discourse for entity-(ies), upon fluent-literature-per-book, thinking when method-(s) of collecting-objective-matter-(s), institutionalize-order-(s), that have been "sooo blessed", (by the) American-Tresurigal-Probing, without-an face to declare-an absolute total-influence, when communal-action-(s), abide to law, (for how) stilly-sweatless, inhabitants' work, and contextually-illiterate they remain by traditional-standard-(s').₆₃¶||||04m.:51s.₄₇m.s.||§ Accustoming age, settles, (by an) location of period-(s), contingent of [hoc / bias] passing time when centripetal-force, blank-memories, from purpose of self, upon methodical-documenting, confirming compositional-decay {bone-(s)};₇₀ perceived amidst existence, through second-(s)-cardiovascular-circulated-blood-pulse-(s), simultaneous-offset-(s'), preferred by numb-stillness, passing-off, equal-decay, surrounding common-nature, unclear (for why) existence matter, because (of an lack of) thought, (through of) self, upon lineage-tiering.64¶|||05m.:14s.56m.s.|||§ The superego will is stubborn of documenting, significantly inclined for accounting, (by when) will, pass along effort-(s), alluring degree-(s) of internal-blank-minded-72L.innocence, claiming introversion, not enforced, by post-humorous-life-matter-(s), where regard define, (a the) true-way of living, Pure & Honest, upon those motion-(s), per live-blood-pulsing, by when self-skull-interior quiet-passing-(s), remain undaunted of bodily-interior reproductive-motivation-(s)...¶|||05m.:35s.66m.s.|||§

§0...Count-[20 / Twenty]-preposition-of... ..Lines-[74.41 / Seventy-Five.1.four-tenths-2.one-hundredths].. ..Article-Essay-[14 / Fourteen].. // ..Book-Essay-[37 / Thirty-Seven /// §05m. : 35s. : 66m.s / eight-minute-(s) : twelve-secondths-essay-reading-rate].. ..Congruent-simultaneous-centripetal-force, revolving physics-three-dimensional-of, developmental-rotational-terrestrial-crust-surface-tectonic-plate-(s)-deductive-offset, when anatomical-physiological-perceptive-motion-(s), incur-inductive, (amid an) superfluous of residential-circumstance-(s) (inductive), apose those [commercial / governmental] (deductive) entity-(ies)-currency-circulated-development-(s), (due to) factor-(s) (out of) [shipment-processing / patent-factory-develop-ment-(s) / Federal-Commercial-Resourcing / Market-trading-outlet-(s) / Consumer-purchase-parity], cycling Goods & Service-(s), for comprehending (how to) [maintain / sustain / contain / pertain / retain], purchased-trade-object-(s), per family-title-proprietorship, amid continental-statehood-resourcing-development-(s).¶|||42s.78m.s.|||§

¶37Time operates, (by an) tic-secondths-inductive-thinking, paralleled to day, (by an) year-denominative-dynamic, rotational revolutionary-centripetal-force, using account-(s) for collecting-objective-(s), from managerial-observation-(s)-on-site-observation-(s), effectively moving, uncharacterized feature-(s) at sedentary-place for equivalent-perimeter-parameter-(s), in direct factoring, instating-transitional-passage-(s), as duty-(ies) are responded to, apprehending existence, where volume-blood-pulse-observation-(s), sustain-an placed living, after subjective-centralized-(re)-comprehending-(s), balanced age-perceptive-interpretation-(s), from private-expenditure-(s), [scaling / trade-(d)-measure-(ing) / equivalent-total-accumulate-(d)-elapse-(s)] requisite antiquity, (as what) beseeches fortitude-(s) per place, for property-maintenance, in relay of defense, featuring-content, partially-influenced, by topical-attitude, from body of constitution-(al)-discourse-(s), by social-illiterate-dynamic-(s), affirming bias-friendship-(s), where [monetary-convincing / humble-living / self-worker-(s')-post-space-cubical-(izing)], singular-degree [temper-ature // educational-degree-(s)-extent-(s)], where movement-(s) from physical-motion-tense-(s), program progression jovial-relationship-(s), early-by adulterated-mentality, empirically-entran-ced, impassing constantly, human-default-physical-effort-(s), from [*eating / clothing / domestic-cating* commercial-residential-parameter-(s) of government-objective-parcel-planned-space-developing], apposed [self-limit-(s) / religious-group-guidance] from educational-passing-(s), by individual-bachelor-(s')-senior-senile-timbre, 17L.evading extraspective-interaction-(s), understand-ing elemental-extraction, per factory commercially-produced-objective-(s), placed-by tranquil-domestication-(ing), hourly-time-posit-(s) (by of) physical-pragmatic-grouping-(s), specifying hour-(s) upon [locale / longitude foothold] perceiving sight where reset-development-(s), antedote-(less)-(s)-tangible-written-oral-redact, possessing personal-property, content by common-citizen-(s')... ¶|||01m:34s.04m.s.|||§

§0...Count-[4 / Four]...| 233 |..Lines-[21.156 / Twenty-One.1.one-tenths-2.five-hundredths-3.six-thousandths].. ..Essay-[15 / Fifteen].. // ..Essay-[38 / Thirty-Eight / §01m. : 34s. : 04m.s. / two-minutes : thirteen-secondths-essay-reading-rate].. ..If condition-(s) of nature affect our mundane-effort-(s), why has globalization-modernization, affected humanity (in such an) null fashion, from historical-lecture, amid Air-Conditioning & Showering? ¶||||15s.72m.s.|||§

¶38Human Being-(s) ("are" from an) variety of geological-circumstantial-sequence-(s), emitted from physics, when solar-pressure-fission-friction, place existence underneath ones'-personal-physics-perspective, tangible from self-perceptive-tense-(s), defining our unsynch-ronized local-domestication-(s'), (as an) existent, from juxtaposition-(s) by segmented-minute-(s)-period-(s) of transit, offset by human-objective-resourcing-duty-(ies), accounting colloquial-discourse, focusing hand-eye-coordination, timbre-motioning [straight-going-(s) / intersection-(s) // thoroughfare stop-(s)], (upon an) cardiovascular-response-(s), (from an) [individuals' word-inquiry // construe-(s)], variablizing-focus (due to) inhabitant-being-blood-pulse-ingestion-visual-imagery-(ies), Gestalt when self-individual-act-(s), define time-proportional-focus, intermittent digestion, segmenting variable-value-(s) interval [crop-(s) / fruit-(s) / meat-(s)] proportion-(ing), physique-blood-pulse-cardiovascular-aorta-ventricle-organ-blood-pulse-pressure-impasse-(s), (as how) physiological-presence-(s'), motion physiological-hyper-tense-focus, datum from [cursive // manuscript-routine-subjective-parameter-(s)] through cul-de-sac-posit-(s), lineage-visualizing-perceptive-trust-conjuring, congested surface-area-presentation-(s), where resident-(s') interlink, yearly-governmental-default-denominative-documenting.65¶01m.:05s.97m.s.§ How do inhabitants' perceive through visual-blood-pulse-organ-ventricle-tissue-wave-pulmonary-response-(s'), (as when) [one / we] visualize pass, clear-effervescent-chilled-molten-vapor-reverberation-(s) (as seen from a the ground, mid-day) present dating-hourly-time-posit-(s), in ordered adjustment-(s) from industrial-(ism) wave-frequency-(ies), kinetic-energy-after-effect-(s), variated celestial-centripetal-degree-(s), afloat topography-territorial-effect-(s), offset anatomical-physiological-human-blood-pulse-elapse-frequency-(ies), eight-decimal-one-tenths-four-hundredths-five-thou-sandths-secondths-offset-influx, motion-(s) progressing, balanced around decade-tense-frequency-(s), enumerative-functioning, [from // upon // at // by // of] century-pos-it-(s), (as an) mark of future upon motive of status.$_{66¶}$01m.:45s.$_{44m.s.§}$ Person-(s) [is / has] not applicated perfectly, an status of political-party-literary-commercial-genetic [measure-(s) /// scale-(s) /// balance-(s)] for evocation by common-beings'-handling, remaining dipitchipated during daily-eight-hourly-wage-payment-menial-routine-(s), not-participating through literary-vocal-voicing-interaction-method-(s), hourly-time-posits-(s), currented during physiological-motion-(s')-impasse-(s'), amongst natural-settings', Collecting & Holding, grifting (un)-superior commercial-modification-(s) complex composite-matter-(s), per individual-self-(ves), without first, creatively-concepting, Objective-Physiological-Limit-(s) & Geological-Ecological-City-Block-Transit-boundary-(ies), second, literary-weekly-circulated-leadership, directing executive-refined-agenda-(s), surveying elapse-synapses from mass-voluntary-disposition-(s), motivating impassing-year-(s), per decade-century-exponent-deviation-dynamic, seeing if deducing daily-perceptive-hourly-offset-rotation-revolutionary-impasse-(s), affect millennium-term-(s)-ideal-(s), defining how definition-(s) abstract visual-vocal-vibrational-boundary-parameter-(s), pertaining day-currency-note-rate-(s), median by reason, (for an) maxim-county-parameter-limit-(s), per relevant-parameter-(s), (not in an) moveable commercial-referendum, (from those) chain-(s) per article-(s)-literature, yet surveyed, limited-human-capability-(ies), per offset-circumvention-boundary-(ies), questioning weather genetic-communion, document proglamated-matrix-(cies), remained [(un) graph-(ed) / plot-(ed) / grid-(ed)] perimeter-boundary-development-(s), learning initially, the dynamic of [place // subjecting], from motivated-survey-participant-(s'), thinking whom comprehend why inhabitant-intuitive-interposed-reasoning, matter when affirming colloquial-effort-(s), (from a the) historical-birth-approximate-community-offset-(s), understanding how suffering under-an physics-security-Federal-Governmental-Complacency, yet interpret how decai approach through living, each sleep, until death arrive, where living is quick to think how to pass unto Lineage, what object-(s), defend-an title-property-terrain, informing remaining-placed, never fully architect-turally-planned, retirement-locational-defining, from communal-extraspective-genetic-religious-prerogative-(s), (by of) [family / neighborhood-developmental-dynamic-(s)] given-worship, during (as when) inhabitant-(s')-compensation, become intensively more still, (with each) passing-age, birthing intermittent, [date /// calendar // time], [fixate / focus / access-(ing)] physiological-anatomical-timbre-(s'), by individual-personal-bias-hoc-characteristics, permitting coverage at approximate-city-metropolitan-perimeter-designated-spacing's, in vehicle-voluntary-acts', reset-(ed) by blah-blank-out-talk-gossip-inhabitant-memory, imaging undocumented tangible-directive-(s), Philharmonic-Score(s)}-{Drama-Script(s)}-{Literature}-{Entity-Construction-Composition-Interior-acoú-tre-mónt}-{Realism}-{Impressionism}-{Constitutional-Survey}, depicting visual-median-(s) from dead-memóir, defending defined-apartment-housing-block-title-lease-resid-ential-boundary-(ies), ignoring literary-rural-interior-cursive-wallpaper-exterior-painting-de-velopmental-upkeep, (as when) inhabitant-role-(s), juxtapose group-serie-(s)-perspective-(s), [influence // persuasion], genetic-community-intention-(s).67¶|||04m.:31s.01m.s.|||§ Mathematical-reason, upon Literary-Prose, is intended to educate subjective-tense-(ing-(s)), when order of planned-peers', gather around-an [central-town / village / city-district] (as an) community, (upon an) county-state-parameter-boundary-trading, that intend reasoning-article-(s)-commercial-class-ification-(s), (as the) dynamic of individualism, remaining in comparative-contrast-examination, intermittent religious-sect-dominium-(s), longing physics existence, apposed how physical-limit-(s), yet vocally-enable grandiose-celestial-galaxy-universe-boundaries, repopulating offspring-embedded petroleum-stationing, (as how) uncultivated-terrain, pertain Undetermined-Perspective-(s') & Metropolitan-Boundary-(ies), while reading (from an) subservience, incapable for defining superego, balance what is intended, to account each visual-imagery-tangible-excess-exponent-weight-(s), apposed individualism-prerogative-(s), by family-posit-(s), fluent in [literature // writing // redacting] defining literate-leader-(ship), whom [direct / dictate] [objective-(s)-agenda // population-work-trade-agenda // scheduled-developmental-motion-(s)] (by of) [[1]human behavior / [2]discourse / [3]pattern / [4]action-(s) / [5]motive-(s) / [6]conduct / [7]effort / [8]academic-incremental-interval-(s)].68¶|||05 m.:29s.14m.s.|||§ Government-subjective-feudalistic-population-(s), direct daily-hourly-time-period-(s), first planning work-type-(s)-function-progression, by university-governmental-documenting, using psychological-inquiry, to order developmental-population-(s), from character-function-capability-rate-(s), minding [military // resourcing // factory-consensuses-objective-production-yearly-obligation-(s)] amongst [sustainable // maintainable-nation-status-continuum] competently varied item-objective-pertaining, selection-(s) in literary-prose, held-at purchase-trade-unit-volume-(s), (due to the) dynamic of reference.69¶|||05m.:56s.45m.s.|||§ Context (has an limit upon the) mode of day, offset apose ingestion, for survival by mass-proportiated-means, (un)-thought how interactive-inhabitant-(s) differ social-interaction-(s), collecting-together, scheduled-effort-(s), from [agenda-objective-(s) // substance-(s)] pre-planning retirement-aging.70 ¶|||06m:09s.16m.s.|||§ Going against the law is [forbidden // prohibited] as such constitutional-referendum;71 remained incapable (since those) fore-father-(s) (of this) majestic-nation, once challenged, heralded, (such) requisite, (yet to be) reduplicated, (for it can not), by litergary-reference-extent-(s), congressionally-monitored, suggest-an Germanic-Vocal-Vernacular-perceptive-modification-shift-(s), juxtaposed, non chronicalized mutual-bookshelf-prerogative-(s), in finance per (bi)-weekly-payment-discourse-(s), as morale guide human-reliance, pledged placed-configuration-(s), at trade-(d)-order-(s), dormant Political-Party-genre-(s), practicing before (into an) Off-set-National-Ecological-Geological-Military-Governmental-Economical-Sociological-Commercial-Factory-Manufacturing-Religious-properties-of-thinking;[72] when civil-basic-human-right-(s'), require bioluminescence, guiding paved-corridor-(s) differed at location-offset-discourse, without orchestrated-requisite, continuum-present, directing human-guidance, when [parental-birth / youth-raising / peer-fraternizing / co-worker-plateau] through years of century-millennium, not concerted-awareness per hourly-time-frame-location-transit-dynamic, transit-scheduling, planned-entity-(s)-location-impasse-(s), placing efficiency (within those) organ-response-(s), individual-base-value-(s), appreciating-existence, through surrounding-parameter-(s), per-offset development, influenced per se, voluntary-persuaded-progression-(s), yet discipline during year, thinking of how decai transition upon [[1]age-deviation-(s) / [2]ethnicity / [3]city / [4]vehicle-modal-(s) / [5]obituary-discourse / [6]birth / [7]workplace / [8]university // [9]college / [10]day-care / [11]elementary-school / [12]middle-school / [13]high-school-(s)], arranging unpopulated schedule-(s)-inhabitant-(s')-impasse-(s), per [year / season / month / week], as day intonate perspective upon each hour (dating), for putting woman in housing, by needs of commercial-reproduction, insighting respectable-men, (whom are to) [serve / treat / appreciate women], [understanding / comprehending] (how their is an) unconscious-governmental-municipal-mentality-code, persistent per offspring per parent, factoring cargo, discoursing means (by an) [birthed-family / coworker / neighbor / religious-member-(s) / friend-(s) / shopping-aurora], evoking checking-style-method-(s'), [pose / apose / juxtaposed human-comfort-(s)] primary-personal-prerogative-(s');[73] awhile celestial-theatre, revolve-continuum-align-(ed), millennium-dating-dynamic, under review at legal-parameter-(s), evaluating default-municipal-code-(s), apposed legal-guideline-(s), when human-customs-tradition-ritual-dynamic-(s), forget how contexting human-offset-lifestyle-impasse-denominating-discourse, denote auto-biographical-enabled-context, omitting (in lieu of) commercial-worker-competency, encyclopedia-deviated convergence, (not ever explored by an) pre-planning, two-dimensional-pre-planning-concepted-effort-(s).71¶||||08m.:30s.06m.s.|||§ Maintaining industrial commercial-production allot demand-responsibility-(ies), where market-coordination, volume-deviate stock-market-confidence, from false-currency-(ies), believing manuscript-documentation, superlative intercoordinated-ordering-(s), affecting stock-market investors, investing-in commercial-market-corporation-(s) like [[1]Kroger / [2]Ralph-(s) / [3]Pavillion-(s) / [4]Albertson-(s) / [5]Aldi / [6]Haggan / [7]Costco / [8]Kmart / [9]Target / [10]Walmart / [11]Publix / [12]Winn-dixie / [13]Piggly-wiggly / [14]Sprout-(s) / [15]Trader Joe-(s') / [16]Whole Food-(s) / [17]Pavillion-(s) / [18]Vons / et cetera... (some under parent companies;74 not including 0Presidente / [0]Navarro's / 0I-G-A-latin-mom-&-pop-etcetera...)] [when // where] communal insight-(s'), offset-physiological-movement-(s), conditioning rural-territorial-living, interceding-voluntary-motive-effort-(s), written at subject-ivity, claiming how lifestyle-(s'), transcend through vision, temporal-visual-imagery-condition-(s), (from of) weather-blue-white-collar-effort-(s), balance perceptive binder-documentation-context-(s'), offset uniform default-acquisitions, abided under government-denominum-dow-(futures)-($20,975.09 // $20,910.00) stock-volume-currency-units-deviation-{04 / 26 / 2017}), influencing how American-Federal-neutral-bonds, offset each incorporated-stock-investment-group-(s), financial-historical-advise, when market-trade-volatility-neutrality, simply view stock, under-an guise of presumed ethical-interactive-circumstance-(s), premised from individual-bias, conjured (from the) historical-religious-movement-perplexes', per population-discourse, reducing mass-territorial-cycling, (amid those) basic-resourced-substance-(s), sustainable governmental-repossession-(s), at each and every incorporated-intersecting-impasse-(s), publicly-relied on, not from private-commercial-enterprise-separation, via international-religious-community-fund-(s), when Private-Vanguard-{Dow Jones // S & P 500}-Dutch-(Neatherlands)-investment-groups-stock-market-variation-(s),| 234 |pertain (retain) [129L.]

commercial-Federal-public-neutral-investment-trade-funding, per International-Stock-Market-Centre...¶||||10m.:15s.$_{11m.s.|||}$§

§0...Count-[21 / Twenty-One]-preposition-of... ..Lines-[129.57 / Ninety.$_{1.five}$-tenths-$_{2.seven}$-hundredths-$_{3.seven}$-thousandths-$_{4.four}$-ten-thousandths-$_{5.two}$-hundred-thousandths].. ..Article-Essay-[16 / Sixteen].. // ..Book-Essay-[39 / Thirty-Nine /// §10 : 15 : 11 / Fourteen-minute-(s) : sixteen-secondths-essay-reading-rate].. ..Where can human-purpose pertain, contain-collected-resourced-elemental-object-(s).¶||||11s.$_{25m.s.|||}$§

¶39 Weathered down from each day, no common-effort-(s') per hour-(s), extrapolate extra-spective-subservience, concluded where presence, not elongate, utterly abstracted personal-prefer-ential-preference-(s'), based (from when) personal-introverted-prerogative-(s') at physical-sensation-(s), ward pastor-guidance-behavior, intermittent [1village / 2community / 3district / 4city / 5county / 6state / 7country-national agenda-population-operation-(s)], implementing function-(s), before-center metropolitan-limit-(s), spaced close, from approximation-(s), unthought from motive-(s), calculate excessive-count-(s) per accessible [resources // elemental-compound-extraction-(s)] per exponent-reproductive-repopulating, means of existence, in part (for how) perceptive-characteristic-(s), focus motion-(s) when human-hand-eye-coordination, define [two-dimensional-conceptive-blue-print // master-plan-(s)-matter-(s)] defining-magnificent-harmon-iously, mass-discourse, interacting (by those) formal-subservient-interaction-(s), accentuated, [developmental // deconstructive // dismantled discourse-(s)] affirming commercial-decision-(s), where live-thinking, observe surrounding others', (by an) coordinating-patent-concept-(s), resorted (in lieu of), [1climate-territory-boundary-mass-preference-redeveloping / 2residential-mass-housing-dynamic / 3transit-development-dynamic / 4economic-denominative-discourse / 5ingest-ion-substantial-resourcing / 6clay-soil-composite-brick-resourcing / 7metal-refinery-resourcing / [8]educational-entity-volume-ordering / [9]adult-lecture-center-(s) (subjective-emphasis-discourse) / [10]boundary-limit-rationalizing (by of) individual-(s) interactively-apart mass-population-reasoning-(s) why objective-placed-discourse / [11]factory-production-ordering / [12]refinery-(ies) / [13]manufacturing auxiliary-objective-concepting // [14]developing / [15]coal-mine-(s) / [16]water-infrastructure-(ing) / [17]electric-current-(ing) // [18]wiring / [19]mass-festivity-center-(s) / [20]paper-documenting-production // [21]toilet-paper // [22]paper-towel-(s) // [23]cardboard // [24]instruction-(s) // [25]manual-(s) // [26]receipt-(s) // [27]shipping-receipt-(s) // [28]books // [29]post-it-note-(s) // [30]agenda-bill-referendum // [31]currency-printing // [32]printing-paper // [33]product-packaging / [34]Hickory-pignut-Shagbark-Shellbark-harvesting /// [35]Oak-Blackjack-Bur-Cherrybark-Chestnut-chinkapin-over-cup-pin-post-red-scarlet-shingle-Shumard-swamp-chestnut-swamp-white-harvesting /// [36]Red-Cedar-Eastern-harvesting /// [37]Long-regular-milled-white-rice-harvesting /// [38]Long-par-boiled-rice-harvesting /// [39]Whole-Grain-harvesting /// [40]Oat-Bran-Germ-Sorghum-Harvesting /// [41]Barley-Grain-harvesting /// [42]Farro-Grain-Harvesting /// [43]Red-White-Yellow-Onion-Harvesting /// [44]Garlic-variety-harvesting /// [45]Apple-tree-harvesting /// [46]Orange-tree-harvesting /// [47]Peach-tree-harvesting /// [48]Pear-tree-harvesting /// [49]American-Milking-Devon-cattle-cultivating /// [50]American-white-park-cattle-cultivating /// [51]American-Yorkshire-cattle-cultivating /// [52]Leghorn-cattle-cultivating /// [53]Delaware-cattle-cultivating / et cetera.....];[75] [thinking // considering] (how expensive our) lifestyle-(s) intertwine, locational-objective-commercial-item-(s)-examination-(s), developing cultivated-entity-construct-(s'), apose-upon pound-rate-(s), eating per individual, an mass-proportional-substantial-consumption-cycling-(s), intermittent scheduled-agenda-routine, inputting resourced-energy, by consumption upon [physiological-anatomical-digestive-nutri-tional-retaining // decomposing-timing-(s)], intervened, later consumption-balancing-(s'), because weight-cultivating, proportion-out, mass-quantity-equivalent-individual-(s), thinking how day upon [season // year] remain unproportioned, pronoun-cattle-harvesting, contingent (by an) community, where scientific-citizen-(s), [[1]schedule // [2]agenda-coordinate // [3]literary-document-(s)-reference // [4]book-(s)-articlize // [5]shelf-collection-(s)-coordinate // [6]Constitutionally-think // [7]infrastructure territorial-boundary-(ies) // [8]diction-vernacular-decade-extent-(s) // [9]psycho-logically-meet // [10]continuum-developmental-engineer-housing-infrastructure // et cetera...] concluding-interceded-thinking, by common-standard-(s), limited offset-perceptive-focus, then furthering expanding rationale-competency, (through modification-(s) of an) open-routine, present (by of) self, upon spouse, upon family, upon neighbor, upon community, upon competent-political-population-45L. discourse, centralizing-an Federal-mass-population-collective,-individual-(s)-fam-ilies-inhabitant-(s'), understanding militarial-reserves-engineering-functional-pur-pose, inter-mittent free-will-sociological-purpose-(s), (as how) living is blood-pulsed (from an) mobius-centripetal-pressure, responding upon those tense-(d)-purpose-(s), motivating lifestyle-individualism-calling...¶|||02m.:56s.60m.s.|||§

§0...Count-[5 / Five]-preposition-of... ..Line-(s)-[48.205 / Fourty-Eight.1.two-tenths-2.zero-hun-dredths3.five--thousandths].. ..Article-Essay-[17 / Seventeen].. // ..Book-Essay-[40 / Fourty /// §02m. : 56s. :60m.s. / five-minute-(s) : three-secondths-essay-reading-rate].. ..Extraspective-individual-perspective-{myself, longing for an 75%-ration-25%-introverted-work-ethic}, pertaining with-an light-eye-color-customary-inhabitants, tying in those holes, centered at Federal-district-(s)-offset-deviation-(s), because how colloquial-community-routine-(s), can only be practiced at hand, requiring [lineage // genetic] interactive juxtaposed-neighborly-appreciation-(s'), for sustaining those qualities of patent-commercial-production-product-(s), along with human-community-individual-juxta-position-(s), continually-rating, each and every piece of matter, on perceptive-scale-(s), that (have an) purpose-function, in delving in to each and every Executive-topic, offset literary-educational-reference-reasoning, understanding that we should be carrying an [Bible / Book] around, at all times', fashioning our method-(s) of trade, with-an shoulder-bag, similar (of an) tool-belt, (due to) referenceable-interaction, moding year-season-dynamic, (to day of the) week, saturating the intensive-focus, (from how) dating is passing so perceptively-rapid, that we should draw (upon an) optimistic-method of living, when performing any task (we are to be) [operating / processing / functioning et cetera...], always consciously-awake, rate-appreciating everything, conventional similar-offset-genetic-individual-(s), mattering (from an) commercial-product-(s)-production-(s), amid [racial / ethnic-offset-bias-(feminine) / hoc-(masculine), sexual / lifestyle / workload / geological-ecological-economical-estate-title-retained-objective-product-(s)-deviation-fiscal-monetary-currency-exchanged-interpretations']. ¶||||01m.:30s.19m.s.|||§

¶40Did ever reproduction occur through group-function? No.72¶|||03s.18m.s.|||§ Did soly individualism, have an impact (by of) offspring-interval-ordering, hourly-time-posit-(s), of centripetal-force-continuum-crust-surface-area, styling basis from residence, separate, inhabitant-(s)-discourse, by coordinated-responses', fortifying commercial-factor-(s), where word-inductive-reasoning, group-adjust-(ing), physiological-impasse-(s), from retained-subjective-lesson-(s)-flaw-(s), (as how) existence pertain, through physiological-cardiovascular-pulmonary-circulation, when anatomical-conjuring path-an purpose, from self-function, making intertwined-subjective-intrication-(s), from empirical-deductive-(s) of content-(s), passed where the latter, supersede interval-period-(s)-influctuation-(s), rendered Reading & Redacting (characteristic-(s) of religion, (un) pertained from (wo) men (thus far), discoursing-an historical-physiological-continuum-male-directive-non-responsive-documenting-Lineage, entertaining to proceed-ones' lineage, along with-an national-dispertion of similar-genetic-document-(s)-appreciation);76 charging context, abbreviated per point-(s) of informed-data, spaced off at Season & Time-Frame Proportion-Logging, interdialouging peer at peer-interaction-(s), conferring aptitude-rate-(s), apposed year-semester-(s), of [vertical-longitude / horizontal-latitude / height-altitude] cubic-space-reasoning-deviation-(s), [height-tangible // altitude-tectonic-spacing, air-spacing] upon visual-depth-perspective, plane-two-dimensional-infinitive-aligning-discourse, commonly-resetted, basically-modifying, mass-repassed-discourse-(s), continually bemused by approximate-constructive-dynamic, (un) emphasized at local-spacing-(s)-accentuation-dynamic, principle-align, Centripetal-Force-Pressure-Drift-Distancing & Tectonic-Plate-Tier-Continental-Shelf-Topographic-Climate-Plane-Artistic-Architecture-Community-Planning, adjunct at secondth-(s)-prevalence, where metre per cubic-tension-(s), weighing patent-commercial-packaged-prepared-unit-(s)-objective-volume-(s) in [dry-measure / substantial-dense-quantity-(ies) / substantial-plasma-density-(ies)] mass-conglomerated-homogenized-perspective-(s), (not be) properly formed, per place, among timed-elapse-interval-(s), sequestrated when syntax-impasse-(s), align objective-purpose-(s), motivating-an harmony with opposite-sex-reproductive-desire, (as an) condition, for calibre, [scaled / ruled / parameter / taped / dated] means of measuring tangible-composite-quality-(ies), that compare against weight-type of matter-cubic-surface-area-denominative-appraising, balancing [visual-electro-magnetic-color-solid-density-wave-(s) // sound-reverberational-vibra-tional-wave-(s)] apose impassing [year / decade expiry-objective-effect-(s)] evaluating longevity of century-denominative-3,153,600,000-secondths-tics, (not including leap-year) from [objective-deterioration // fatigue-aging], remeasuring each fact, (from an) juxtaposing continental-climate-topographic-offset-(s), defining each individuals', (by such an) extent-(s), that general-vision, proglamate from millennium-physics-illiterate-interpretation-(s), against commonly accepted, actual-Gregorian-calendar, based economic-turmoil-merit-(s), during physiological-primary-presence of humanity, conjuring choice-(s), formal by indicating, why ambiguity be [difficult // required impassing] adjusted requisite-presence, when natural-continuum-centripetal-force-discourse, program timbre, not collect-an consistent average of book-(s), per relative-function-(s), [when // where] entity-schedule-functions, require competent-average-interpretation-consistency-(ies), that in my favor, prefer ignoring the completely-poor, enabling-an variation of [rich // wealthy // moderate-competent-voluntary-citizen-(s)] and specifying how unless literature go (through an) massive transformations', remain (in an) terrible-dormant state of constitutional-verifying, planned by local-county-city-living, awhile federal-regulation-(s), not oversee congressional-library, revising how condition-(s) are conducted (by those) position-(s) at political-power, [Executive /// Legislational /// Judiciary Branches of Federal-State-Government] ...¶|||03m.:25s.78m.s.|||§

§0...Count-[15 / Fifteen]-preposition-of...| 235 |..Lines-[42.31 / Fourty-Three.1.three-tenths-2.one-hundredths].. ..Article-Essay-[18 / Eighteen].. // ..Book-Essay-[41 / Fourty-One /// §03m. : 25s. : 78m.s. / four-minute-(s) : thirty-two-secondths-essay-reading-rate].. ..Checks & Balance-(s), are formed per under an international-trade-bonds-bias.¶|||12s.59m.s.|||§

¶41Chronological-street-road-rural-ordering, not exist (as an) literary-genre, positing-perception of mass, primarily-along natural-unoccupied-spacing-(s), for modern-centralized-comfort-(s'), by means through access-survival-sufficiency.73¶|||12s.59m.s.|||§ Ecological-effect, is approximately-reseted, when [day // night] rotation-revolution-three-hundredths-sixtieth-five-whole-denominative-percentage-fraction-proportioning, offset those confide-(s) where enabled-construction-point-(s), quality-rate mass-individual-listening-juxtaposition-(s), incredulous-by verbose verbal-skill-(s), impacting dialogue-interposed-effect-(s), when mathematical-grammatical-punctuational-discourse, survive amongst scientific-hourly-time-interval-calculating, [measuring /// scaling /// balancing] organism-variable-(s), adjunct human-perspective-inter-pretation-(s), labeling collection-(s) sorted-subjecting-material-(s), by ameri-

can-government-document-bailout-monetary-bonds-denominative-intervention, stabilizing those checking-com-munal-saving-(s), thinking why currency, deviate retirement-saving-(s)-impact, when terrain-remodification-(s), are planned at formal-perimeter-limit-(s), effective-effort-(s), concerting preliminary default-mass-scale-accounting, sociological-parameter-(s'), through inquiry from voluntary-capable-tense-(s), competent their-at climate-surrounding-observation-(s), retained by focus (by of) mutual-person-work-equivalent-documenting-(s), simultaneous-air-pressure water-bodies-vapor-drift-(s), according to latitude-east-west-perspective-(s), (upon an) latitude-parallel, by blank-idle-interaction-(s), communicating language,| 236 |(as an) common-mass-comprehension-code, intended to instill-motivation-(s), impassed-at, present-simultaneous-percentage-fraction-deviation-(s), working apose continuum-numerative-quantities, from mini-scule-possessive-juxtaposition-(s), cause human-listening-seeing-tasting-smelling-sensational-quality-valuing-discourse, qualify-(s')-matter, accumulated (to be) re) cycled, from the nature of [ideal-(s) // belief // faith] in aging-confirmation-annotating, reinvigorating those fact-(s) (while at an) approximate-definable-observable-feature-(s)-sighting-(s).74¶|||01m.:46s. 38m.s.|||§
Distance should be measured by height from surface-level-notion-(s), in depth, simultaneous collected-objective-interval-(s), developing an ecological-routine, sociological-physical-requirement-(s), where hourly-parameter-(s),| 237 |[document / adjust / procure] 25L..tangible-quality-(ies), compartmentalized-input-motion-(s), from exterior-variable-output-(s), by human-mass-deviation-motion-(s)-unconventional-discourse, working force, comparative schedule-agenda-methodical-ordering-(s) extracting-matter,| 238 |(for an) government-entity-dynamic, away from treasury-method-handling... ¶|||02m.:10s.04m.s.|||§

§0...Count-[3 / Three]-preposition-of...| 239 |..Lines-[28.115 / Twenty-Seven.1.one-tenths-2.one-hundredths-3.five-thousandths]... ..Article-Essay-[19 / Nineteen].. // ..Book-Essay-[42 / Fourty-Two

[Article:] {February}

/// §02m. : 10s. : 04m.s. / three-minute-(s) : ten-secondths-essay-reading-rate].. ..I inquiry what the formal lines-minutes-into-secondths-deviation may be, from the various-juxtaposition-(s) of randomized-article-(s)-to-refined-essay-topic-order-tiering may be, into each period-semi-colon-offset-deviation, amid my paragraph-redact-inquisitioning-refining offset geological-location-(s) [1]department of defense / [2]department of commerce / [3]department of housing and urban-development / [4]department of state / [5]department of labor / [6]department of interior / [7]department of health / [8]department of justice / [9]department of homeland-security / [10]department of transportation / [11]department of agriculture / [12]department of Treasury / [13]department of education / [14]department of energy / [15]department of Veteran-Affairs, in contrast of Senate-Budget-Committee denomination comparison, from house-general-ecological-sociological-psychological-economical-bill-inquiring, (from how) military-budgeting, expropriate an socialism-lifestyle-deviation-(s) (from such) institutional-financial-systems, amid an reprieve to thinking, (as it is too) difficult for youth to learn educational-predicate, amidst adult-commercial-lifestyle-(s), ignoring why retirement is instilled, insighting those factors from Labor & Trade, how Supply & Demand, pertain-an function for learning, after youth-education-(the minor-league- (s), apose adult-commercial-labor-(big leagues), economical-commercial-cycling-(s)].¶|||01m.:07s.88m.s.|||§

¶42Roaming approximate-locale, impact congestion, vehicular-disorder, keeping occupant-(s), apprehended by sedentary-comfort-refinement-(s), from bi-monthly-payment-(s), adhered by routine, Back & Forth*, workplace-environment-(s) (labor pool), affecting [individuals' / religious-fervor] (by of) territorial-claim, conjuring methods of elemental-resourcing, where area, at various-configuration-(s) of terrain, pertain-an collection per [tool-(s) / produce / scale-(s) / utensil-(s) / cooking-supply-(ies) / provision-(s)] using minded habitant-(s), for methodical-scheduled-purpose, (in lieu of) practical-belief, developing-trust, by observable-discourse, where participated being-(s), whom can reason House-Representative-Discourse, congregate communal-direction-(s), by senate-commercial-resourced-goods, serviced when mass-discourse, remain extremely-limit-(ed), requisite historical hour-per-spective-offset-(s), repeating each synapse-elapse-time-posit-(s)-impasse, because how thought pertain discourse of development-(s) amid perspective (as how) value is interpreted, upon Physics & Physiological-Blood-Pulse-Perceptive-Visual-Hearing-Reverberations-Wave-Vibration-(s)-Kinetic-Chilled-Vapor-Temperature-Variance-Influctuation-(s).75¶|||52s.41m.s.|||§| 240 |Industrial-commercial-governmental-militarial-belief, not yet instill those mutually-competent-motivation-(s)), supplementing collected-products, centralized-climate-topographic-shift, from state of birth, {I.e. Miami, Florida}*, continually examining those [measure // metric distance-(s)] defining article-(s)-offset, from dynamic-(s) per mass-peripheral-pupil-perspective amid continental-shelf-human-discourse,| 241 |(not) [1]intonating-[2]intuitive-[3]intelligent-[4]interposed-[5]informational-[6]interdialouge-{[1]transitive-present-continuum-tense / [2]intransitive-verb / [3]pronoun / [4]transitive-in-past-tense / [5]adjective / [6]transitive-verb, if my editor or readers, could work on an blog, for tiering my books predicate article-progression-process, [We / I] could then study intertwined-data to further elaborate, (how the) formal word-resets of predicate-thinking (are to transpose the) vernacular-vocal-voice-interpretation-dynamic, (for why their exist an purpose from those) idle nature-(s) of thinking, when vison-hearing-dynamic, (have an) function of enjoyed-invigoration, (for then) [possibly / probably], enticing an nation, developing-an formal-functional-ecological-predicate-society-psychological-scheduled-circulated-cycled-citizen-function-(s)}, as continuum-silent-stillness, inhibit men, whimsically passing woman, not viewed for pertaining children, (or natural-physiological-voluptuous-arousal) registering those dynamic resource-judgements', where waiting during passing time, voluntarily examine, (if so), working away from the faith-debt-base, trading by action-(s), genetic-intelligent-mass-classification-(s), (from whom) concern physiological-head-facial-characteristic-(s), tiering-data, from physiological-nature;[77] when mass-inquiry (have yet an) relevant-disposition (at the) moment, working toward an bastardized-genetic-memory, uninquisitive, by survey redacted-critique, considering how perspective is mattered, (where at an) impasse of lifestyle, (when in my case), extraspective-interaction-(s), have been severely under construed, apose natural-interaction-(s), by celestial-perspective, upon general-posit-(s) from historical-topic-(s), affecting our present-discourse, where we Live & Domesticate... ¶|||02m. 31s.84m.s.|||§35.595L.

§...Count-[10 / Ten]-preposition-of... ..Lines-[35.595 / Thirty-six-1.five-tenths-2.nine-hun-dredths-3.five-thousandths].. ..Article-Essay-[20 / Twenty].. // ..Book-Essay-[43 / Fourty-Three /// §02m. : 31s. : 84m.s. / three-minute-(s) : thirty-one-secondths-essay-reading-rate |||milli-secondths-rate-1,000 = 10-secondths|| 21,344-millisecondths|.. ..A-the priori millisecondths-inclination, I intend to reason-an denominative-method-(s) of Time-{millisecondths : secondths-matter-reason}, apose {secondths : minutes-matter-reason-ing}, apose {minutes : hours-matter-reasoning}, proportioning {[d]Hours : [c]Minutes : [b]Secondths : [a]millisecondths continuum-count-(s)}, apose-juxtaposed an Calendar-{[1]day : [2]week : [3]month : [4]season : [5]year day-dating-percentage-(s)-deviation} converting those inclinations of Time, into Exterior-Compartmentalized-Advanced-Date-Aging-Categor-ical-Decimal-Enumerated-Year-Denominative-Value-Quantity-(s) deviating Article-Essay-period-semi-colon-Column-Offset-Order-reading-interpretation-comprehension-Classification-(s), practicing an system for [weighing / measuring / calculating] all the (In)-Tangible-quality-quantities, juxtaposing how untangible-peripheral-physics-metrics-measuring, deviate tectonic-plate-(s), water bodies, apose human perspective, apposed industrial-commercial-development-(s), juxtaposing incorporated-Federal-Perimeter-boundaries-offset, furthering how those effect-(s) of intelligence, not require education, exemplifying why individual-perspective, not ever contain exterior-visible-matter, discoursing-an ecological-food-chain-cycling-dynamic, continuing to motion for invig-orating existing-life-style.¶|||01m.:10s.40m.s.|||§

¶43Date meander passed minutes-timed-interval-(s), acting where workplace-decorum, motion where designated-property, survey means' amidst elemental-condition-(s), maintaining [hygiene / clothing / eating / recovering / idling / et cetera..] (from those) familiarity-(ies) of state, trading at national-settling, serving-an duty of man, far greater (than those) role-(s) of woman, omitting [dialogue / conversation / analytical-revisiting] how unit-items-objection pertain-(s'), monitoring-effort-(s) per, [1genetic-classification-(s) / 2reproduction / 3resourcing-elemental-compound-(s) / 4construction of housing, in comparison of provisional-entity-(ies) / 5financial-entity-(ies) / 6political-mansion / 7department-(s) of nation / 8center-(s) / 9garden-emporium-(s) / 10operational-center-(s) / 11+etcetera...] abided by financial-objective-development-(s), auxiliary through [territory / sociological-parameter-(s)] defining characteristic-(s), when average-mode denominate [1produced crop-(s) / 2mammal-hearding / 3educational-infrastructuring / 4textbook-volume-(izing) / 5continual-dialogue form mass-analytical-discourse / 6Territorial-monitoring / 7age to date // 7.1month // 7.2season-proportion-examining], upon confide-(s)| 242 |at timed-calendar-centripetal-force-impassing-(s)-dynamic, in place per hourly-time-upkeeping where kinetic-effect-(s), (such as) [photosynthesis anatomical-blood-pulse / temperature] tangibly-qualify territorial-matter-(s), constantly upkeeping-an local-prevalence, from social-class, individual-istically coordinating organized-file-structuring, upon work-place-emphasis, constructive reno-vation-impasse-living.76¶|||01m.:17s.84m.s.|||§ Orchestrating notion-(s), incur while dating is feasible at limited-physiological-impasse-(s) (transit-included), identifying why extraspective-interpretation-competent-comprehension, matter physiologically-inclined from agenda, pertaining those arduous-burden-(s), per daily-effort, motioning while adult-youth-reading-curb, not contain-an count per [mammal-(s) / plant-(s)] seeing through visual-observational-discourse, ordering physiological-motion-(s)-secondth-(s)-transition-(s), when hyper-tense-observational-confirming, tier-comparative incremental-cycling, Political Matter-(s) & Affair-(s), hand in hand, with day-succession-(s), rather than proxy-year-simultaneous-effect-(s), resetting juxtaposition-(s) offset per age-limit-(s), surveyed-analytical-analysis, determining learning, interposed amidst date-day-time-impassing, developmentally-conjure-(ing), periodicalized interval-reasoning-(s), considering how Water & Electric-Metal-cylinder-Wiring-total-density-Piping-tubing-Conductive-current-(s), are directed thoroughly, denominating Political-Commercial-Sociological-trade-impact-(s), by County-Federal-Treasuring, when mutual-inhabitant-interaction-(s)*, forget to pay their Royal-Vanguard-Fiscal-Funds, Dow-Jones-S-&-P-500-Dutch-religious-war-international-evaluation-debt-(s)-{3000 / 10,000-dollars-deviation, per city-district-church-house-subterior-pastor-repre-sentative, upon financial-North-Eastern-pilgrimage-journey-visitation (but while artists like Paul Van Dyke & Tiesto, signify (why the Dutch do not) exactly spend (all their) exchange-trade-funds correctly, but still require (Federal)-funds > E.T.F.s (by an) deviation of circulated-Federal-Currency, (why I) prefer this literary-piece (to be with) Macmillan-international-neutrality},|243| offset English-French-Nasdaq-trading-(I have not yet figured out whom they rely their funds /or?/ exchange-trade-fund-(s) upon, but observe the F.T.S.E.-(London Stock Exchange Deviation C.A.C.-Paris Stock Exchange-{|PAGAN|{[3]C. = [19]S. phonetic in Francis-(frán[0]sh[0]wa)-(is, [9]i. shifts into a third-first-vowel-(s)-priori-opposite with [1]a. median (=)equivalent 5-numerical-inclination-([5]e), adding-an [23]w. before, signifying "wa"-deviation = 11.5 k.= l. inbetween [23]w. before-after (in the) [5 / 6 letters], while [19]s. shifts-an fifteen-consonant-letter-posit, shift into [8]h. median (=)equivalent 21-numerical-inclination-(v), amid "sh"-deviation = 13.5 m.=n.} lesson}-(from a the) solar-fission-impact-infinitive-continuum, regenerating [plant-(s) / Protista / grass / tree-(s) / edible-mammal-(s)], incrementing-static-kinetic-particle-(s) amid ambient-nature, as chore for labor, timing growth-accounting), crescendo, mass-personal-isolating, marking those requirement-(s), by reusable-data, fundamental from basic blood-pulse-facts, awhile [citizen-(s) / parcel-living-cycle-(s) / development-(s)] order (due to) physical-terrain.77 In succession of act-(s) (has their ever been an) default of perceivable-existence, cycling [mammal-(s) / bird-(s) / cow-(s) / bull-(s) / chicken-(s) / rooster-(s)] determinant per efforts, meticulously-intricating, complexity-(ies), central from astute-study-(ing), [matter-(s) / mass-(es) / mean-(s) / way-(s) / balance-(s)] per [weight-(s) / scaling-(s)] as population-deviational-discourse, inhibit mutual-citizen-prerogative-(s), while state-traded-fund-(s'), understand America-(s')-Federal-Government (to have an) strain of reason, by European-dictionary-literary-non-fiction-purpose-(s), contexting partial-fact-(s), surveying perceivable-potential-maximum, where Territory & Physics, administer further fiscal-complexities, rating (non)-rural reasoning, where voluntary-impasse-discourse, mass-analyze-reasoning, aforementioned, at-will-dynamic, when blue collar-budgeting,54L. form social-sophis-tication, (as an) goal of determined-citizen-inhabitant-(s)-vocal-interacting...¶|||04m.: 29s.64m.s.|||§

§...Count-[11 / Eleven]-preposition-of... ..Lines-[54.9825 / Fifty-Four.1.nine-tenths-2.eight-hun-dredths-3.two-thousandths-4.five-ten-thousandths].. ..Article-Essay-[21 / Twenty-One] // ..Book-Essay-[44 / Fourty-Four /// §04m. : 29s. : 64m.s. / four-minutes : twenty-nine-secondths-essay- reading-rate ||33.958m.s.].. ..I enjoy those European-cultures economic-systems, wondering if they can ever be more explicit than I? Or rather I simply continue interpreting each economic-minuté, grammatical-literary-cultural-interpretation-(s)-reasoning. ¶|||18s. 47m.s.|||§

¶44Cycling nature, gamma-spectrum, series-sequenced by intangible-respiratory-breathing-in/out-ventricle-blood-pulse-friction-motion-(s), emitting blood-muscle-pressure intertwined bone-system-stabilizing, amid ligament-tensions of soluble-density-mass, (like the moon & earth-terrestrial-tectonic-plate-(s)), upon organ-cartilage-digestive-tract-system-(s), atop muscle-ligament-filing, (as like the crust of the earth), to emit from skin-sensation-(s), juxtaposed earth-tectonic-plate-continental-shelf-fission-revolution-rotation-impasse, depicting centripetal-force, (of an) primary-subterior-cycling-(s), kinetic-molten-vapor-friction, [water // sweat-perceived] while spaced-off-civilization-(s), matter, {John Freeman / Carl Gustaf Jung // Sigmund Freud /// Herman-Rorschach //// Meyers'-Briggs psychologies offset} (due to the) unconscious-nature of extraspective-interaction-(s).78 How existence came to be-(ing), (is not my) natural-discern, when revolutionary-period, not deviate too far (from a the) initial-ultimate-commencement of existence, falsely-hypothesized, ((for the) last Four-Hundred & Eighty-Eight-Year-(s), recurring-theme), afloat Default-Presence & Future-Day-Year-Infinitive-Continuum, offset core-progression-momentum, upon movement-(s) through passing-force, soliciting movement-(s), monitored (by those) progression-(s) per individual-(s'), (from an) idle-view, as terrain hovering grandiose-calm, remain greatly uninhabited, from developmental social-cause, at location lifestyle handling.77¶|||01m.:02s.96m.s.|||§ Solid-density-(ies), can only "be of", cycling by [industrial-jack-hammering / excavating-weight-(s)] offset organism-water-nutrient-enrichment-(s), per volume-development, aboded by citizen-voluntary-participation-after-census-influence, formulating from perspective supplementary-influence-(s), particulate measuring molecule-(s), (by an) mathe-matical-scale-denominative-rate-influctuation-subjective-perceptive-applicated-comprehension. 80¶|||01m.:24s.24m.s.|||§ I not know fully, anybody (whom can flourish an) entire botany-garden, | 244 |by self will alone.81¶|||01m.:29s.77m.s.|||§ Discern for control, vary regeneration, awhile numerous-tally-(s), are calendarized by civilization-(s)-cultivation-(s), (at an) average-means simultaneous of routine-(d)-lifestyle.82¶|||01m.:36s.:89m.s.|||§ Determination of measure, (can only be) aligned, if trusted-surveyed-citizen-(s), take initial-progressive-competent-intelligent-impasse-step-(s), incrementing data, (by of) present-affair-(s), Studying & Experimenting, [replacement // workplace] infrastructure-piece-(ing-(s)), soluble dated-analysis, from hourly-time-posit-(s)-labeled-grandiose-population-offset-document-counting, intended for ordering an progressive-constant-momentum-crescendo-continuum-aligning, [sprout / flourish rotation-(s)] documenting-materials', without focus of dues, cultivating land, when figurative-dreaming, remain unweighted, per [dry-measures / liquid-measure-(s)] enumerating mass-grouping-common-objecting, unfocused, per timbre of independent-character-description-(s), intricated step per step-purpose, breathing in relation to observable-surface-area, requiring documenting, discoursed venerable cause, preliminary supplementary-fact-(s), formally-revised, counting cultivation, in adjustment to physiological-mass-documented-eating-mode, still (at an) common-ambiguous-fashion, (for how) human-free-preference, yet comprehend how expenditure-pertain human-deterioration, [permeate // recycling] climate-condition-(s) offset citizen-(s), concerning daily-objective-(s), ordering factory-product-objective-(s), reproducing organic-matter, where various-facet-(s), [transmit / receive] tangible-scientific-elemental-properties, pertaining effective-effort-(s), accounting Community-Spacing-(s) & Building-(s), [measured / physique-composition / dry-weighed / liquid-weighed / pre-conceived-objecting of cubical-room-spacing-(s) / et cetera...] not collecting citizen-hypothesis-determination-(s), acted for attaining matter, servicing planned-territory-goal-objective-(s), [claused // conditioned // biased] from title-property-purpose, only 42L. appreciating lifestyle, when imagery-perceptive-objective-(s), incur past visual-vocal-tonal-perceptive-live-lineage-memory...¶|||02m.:56s.49m..s.|||§

§...Count-[12 / Twelve]... ..Lines-[42.80709 / Fourty-Two.1.eight-tenths-2.zero-hundredths-3.seven-thousandths-4.zero-ten-thousandths-5.nine-hundred-thousandths].. ..Article-Essay-[22 / Twenty-Two].. //

[Article:] {February}

..Book-Essay-[45 / Fourty-Five /// §02m. : 56s. : 49m.s. / two-minutes : fifty-six-secondths-essay-reading-rate || 17,649m.s.]..
...Counting is significant for comprehension, interposed those reasons for accessing land-territory-property-perimeter-parcel-(s), considering (how to) excavate below geological-surface-density-(ies)-composition-matter-(s), refining (each of those) properties developed (amid an) perceptive-conjured-motivation-(s), (due to) variablized-major-position-focus-quality-(ies), concerted from extraspective-surveyed-population-personal-morale-effort-extent-bias, affecting, each momentum-impasse of celestial-solar-physics, apose-(d) physiological-ethnic-mass-continental-specimen-(s) (due to) consideration-(s) by voluntary-(yes /or/ no) percentage-calculation-(s), from effective-evaluation-(s) per county-terrain-resourcing-population-error-dynamic.¶|||54s.65m.s.|||§

¶45 Whom perceive for another presence, past or future present-occasion? What exterior-observation-(s), can be "presenced", by daily-effort-(s')? Reset-(s) incur awhile institution set focus, denominum-currency, where residential-citizen-(s), familiarize oneself, at setting, disavowed-perimeter-surface-area-(s), by parental-guidance, affecting early-age, quality-lifestyle, development-(s), pertain each individual-(s)-perspective, amid daily-hourly-geological-cent-ripetal-force-deviation.¶|||22s.19m.s|||§

{January-(31)-February-(28 / 9)-Month-(s)-Day-Hour-Minute-(s)-Secondths-Count}-{5,097,600 D.H.M.S. = 60S. × 60M. × 24H. × D.}-{1,416 H.}-{84,960 M.}-{59 D.}

{Article:} [March]

¶47How does monotone-chromatic-memory, view imagery intersecting human-being-(s), planning calendar-(ed)-residence-(s), agenda-constitutional-(esce)-ordering, formal-capacity, per year-to-date primarily, juxtaposed daily-week-hourly-discourse, variating contextual-posit-(s), offset per genre, (heavily-primarily-fiction-mystery // secondary-auto-biography);[78] historically-referenced, not seen, upon mutual-basis, deducing clue-(s) from body-text, affirming blank-purpose, discoursing Story & Discourse literal when technological-development-(s), occur simultaneous, (where their are) barrier-(s), county-boundary-(ies), viewing tangible-possessive-matter-(s), where common intangible-mass-commercial-production, tangent apose feature-(s) at lecture, discoursing [subject-review / psychology / redact-(ing)] coordinating Populant & Agenda, offset method-(s) from schedule-(s), affecting inducting elemental-extraction-(s), Measuring & Balancing extracted elemental-weight-(s), for stone-pillar-development-(s),| 245 |defining visual-subjective-imagery, contexting tangible-lift-able-construction-(s).[84¶||||45s.94m.s.||§] Counting being-(s), matter-(s), (for of a the) responsibility, upon governmental-denominating, verifying (by an) extra-document, per patent-commercially-taxed-facts, determinant by historical-relevance, modifying present-day-year-discourse, affecting temporary action-(s) through-an rapid decade-year-(s)-impasse, from spacial-areas-(s), passed by lineage, settled from offspring around commercial-silver-spooning, as how Age & Millennium intercede-(d), mode-(ed)-routine, from mean-volume-unit-average-(s), statistical per percentage-numerical-factoring, inductive-equating-progressing day, familiar by nature, affirming fact-(s), by interval-(s) at date-(ing), [hot // cold] temporal-adjustment-(s), participated in part, by age-reset-(s), rationing year-(s) of productions, adjacent near-passage-reset-deductive-cubical-zone-(s), family-oriented, when commercial-value-(s), remain residential, where community require, human-interaction-(s), sustaining conversation-(s), about how conducting neighborhood-affair-(s), in regard to timbre of territory, differ [tangent / objective

/ development] from governmental-commercial-boundary-limit-(s), covered-divided, at appoint-(ed)-parameter-(s), coming from representative-(s)-extraspective-effort-(s), (as why) placid-complacency, figuring mind of mass-(es), by case from discussion (of how to) order tangent-(ed) continental-hemisphere, parcels-by western-hemisphere-trade-method-(s), having how free-will require an Good-(s) & Service-(s) demand-(-ed) from purchase-parity-influctuate-(d)-trade-rate-(s), no further than state-ethnic-genetic-regrouping-(s), when tectonic-mass, secure-parameter-(s) (by-of) trade-uses, objecting intermittent comestible-intake, having weight-(s), be unproportioned, where parameter-(s), apose neighbor-conjure-development-(s), because decorum-post-construct, compare activity-(ies), as central-distancing-(s), transiting passed an multitude of entity-(ies), where scheduled-activity, elapse through discourse, [entering / exiting] where limited-perspective, at various construction-(s), overwhelm perceptive-imagery-interpretation-(s),| 246 |incompetent effecting, amid hourly-time, categorized when calendar-inquiry, is suppose to assess those tangible-factor-(s), from effort-(s) impassed, awhile common-disposition, should legislate, where orderings sectionalize social-cause-(s), for inhabitant-discourse, during daily-impasse-continuum.85¶||||02m.:20s.66m.s.|||§ Reasoning unincorporated spacing-(s), (have yet to exist, for how) defining place, interval lifestyle-(s), anticipate amidst [offspring / neighbor] interaction-(s)] anticipating after-life-continuum-recollecting-(s), performing novice-being-theatrical-purpose-(s), mundane an extraspection-fear, incapable, where tense-minute-memorizing-impact, comfort-obligation-(s), in thought, from children-debt, impending theatre, assessing commercialism, from action-(s) [form-(ed) / made / concept-(ed) / develop-(ed)], an regard of propertied-territory-(ies), suppose (to be) deviating constitutional-ground-(s), comparative (to a the) House-Representative-(s)-mass-consensus-legislating-count-(s), leasing, body-mass exemplified through motivation-(s) for title-(d)-fact-(s), prefer-(ed) by senile-aging diminuendo syntax-insurance-assurance, not expanding library-reference-context, proportional at live-perspective-(s), expositorially surveying population, (about a the) capability-(ies) by [[1]water-treatment-operator // [2]electrician // [3]police-officer-(s) // [4]military-member-(s)-{mechanism per operation-count-(s)} // [5]farming // [6]construct-ion-worker // [7]engineer // [8]designer // [9]labor-manager-(s) // [10]store-manager-(s) // [11]sales associate // [12]government-office-documenting // [13]employment tasking juxtaposed managerial-monitoring, chaining by district // branch // regional // national-manager-(s), commercial-usage, for the fact of trade] as when mathematical-margin, affect tool-usage, where square [feet / yard / meter] granulate-increment Water-Pressure & Electric-Current-Kilowatt-(s) equivalent by converting data, inputting hardware-intangible-usage, confirming conduct, denominating substantial-progres-sive-interval-(ization), juxtaposed human-ingestion-nutrition-day-time-interval, snack-intake, defining an consistent digestive-pattern, by ingestion-intake-rate-(s), compared substantial-nutritional-rate-resourcing-(s), (as how) every facility, be accessed by educational-guideline, logically-grading, calendar-period-(s), rather having an [constant-year // off-time-period-(s)] for grading efficient season-(s) in thirdths, to juxtapose conceptive-model-manual-context, uninfluen-tial, by thirty-days-retrograde-period-(s), [58L.] interval one-season-impasse-period, practicing [Eat-ing-Habit-(s) /// Resourcing-Method-(s) // Imaginary-Inspired-Auxiliary-Sphere-Sculpture-(s)] for [yard // park // ambient-perimeter-planning-surface-area(s)-factory-population-concensused-objective-production-(s)] as housing-renovation-(s)-developmental-period-(s), scheduling human-effort-(s), more methodically, focus purpose on after-life-lineage-transitional-specimen-role-(s) disciplining... ¶|||04m.: 29s.$_{97m.s.|}$§

§...Count-[9 / Nine]-preposition-of.. ..Lines-[62.12375 / Sixty-Four.$_{1.one}$-tenths-$_{2.two}$-hun-dredths-$_{3.three}$-thousandths-$_{4.seven}$-tens-thousandths-$_{5.five}$-hundred-thousandths]... ..Article-Essay-[1 / One].. // ..Book-Essay-[47 / Fourty-Seven].. ..[scale-(s) // rate-(s) // measure-(s) // war-(s) // Count-(s)...]¶|||10s$_{59m.s.|||}$§

{Article:} [March]

¶48Timbre of characteristic-trait-(s), [step-(s)-count-(s) / tempo of musical-beat-(s)] upon [[1]measure-elapse-interval-(s) / [2]time-infinitesimal-calendar-rationing / [3]respirational-kinetic-vapor-blood-pulse-beat-tensing-(s) / [4]secondths motion-(s) // [5]action-(s) // [6]maneuver-(s) // [7]transiting // [8]dialogue // [9]reading // [10]writing // [11]mathematical-documenting // [12]idle-blank-impasse-(s) // [13]first-person-conscious-dialogue /// [14]second-person-subconscious-wait-listening] for [response /// third-person-subconscious-written-observation of topical-debate of minute-(s)-gamma-continuum-impasse-effect-(s) / note-(s) of denominative-tangible-mass-population-unit-commercial-production-operating // human-basic-effort-(s) / product of shelf-demand-margin-(s)] appraising rate-(s) of population-demand-purchases, per dated-period-(s);[79] relevant to expiry-impact, defining margins of objective transit, commercialized, pass governance-agenda, when inhabitant-(s') enable, date accreditation, dissertation usage conjuring in theory, sociological-observation-(s), formulating an infinitive-(un)-hypothesis-theory, perceiving passed population-persuasion-(s), pertaining ability-(ies) where infrastructure-reset-order-directive-(s), are relied by visual-imagery-experience-impasse-competency, engineering objective-(s), in harmony of [continental-shelf // human-modification-(s)] for [continental-shelf /// entity-(ies)-structure-(s) /// hardware /// technology] pertaining [software-operation-(s)-denominating /// cursive-written-documenting, object-(s) /// typed-print-manuscript-text-(s), basic-human-rights'-operation-(s)-maintaining] influenced vocal-confirmation-(s) from dimension-(s)-omission-(s), not defining point-(s) at point-blank-documenting, simultaneous per [breath-respiration /// blood-pulse-rate per minute-(s) /// cent-dollar-deviation-(s) per human-task //// objective //// account-(ing) through human-mass-population-work-order-classification-(s)-ordering-(s)] appraised by insurance-agent-(s), whom order-objective-(s), for where usage-configuration, table area upon [apose / juxtaposed] constant-rotation-(s), per employee-(s), whom posit objective-(s), from managerial-directive-(s), per calendar-period-designation-(s), placing monetary-volume-unit-(s), perspective to juxtapose, work-location, separate from ecological-purpose.[86] [When / where / why] citizen-(s) exist, adapting per language, abided by subjective-relevance, not quantitively-defining tangible-fact-(s), mutual for commercial-production-acquisition-(s), surveying communal-mass, from-an [formal-written // manuscript-documenting-(s)], (as when) time-elapses, monitoring live-tense-(s), memorizing pronoun-requisite, amidst presence, syntaxing period-(s) from tensable-character / (istic)-(s), putting self physical-composition, (in an) zero-person-self-bias-perspective, subconscious by reference, intended for guidance, where discourse, meander fair-practice, margining rule-(s), percentage, checking those data-bit-(s), (upon how) etymological-nature, verify ontological-proof, from unperfected-incremental-word-term-definition-subjunct-order-sequencing-(s), conjuncting weight-(s) of affair-(s), from physical-composition-(s), whom are (un) referenced [reading / difference / inference / affirming] cause, reviewed in example-(s) per volume unit-proportion-(s), dating subterior hour-(s)-tense-(s), (as an) body of territorial-longevity, varied those ways by existence not reintervalized, timed-motion-(s)-coordination-posit-(s), focusing pass (a the) zero-person-self-bias-perspective-(s), defining life, by the four-secondths-deviation-synapse-(s), aptitude subjective-invigoration-(s), unintellectual for construction-(s), juxtaposing common-purpose, emitted by the continuum-offset, intertwining residential-perimeter-boundary-citizen-activity-hour-minute-(s)-architectural-furnishing-objective-appliance-(s)-agenda-(s'), serving genetic effort-(s), working from routine-schedule-(s), cause mentality-evocation (thought), yet believed-by log of scheduled-voluntary-moment-(s), tensing interval-age-(s), periodical-(ized) obituary-routine, radical driver-operation-(s), where-when, credited account-(s), [bring / place / move] mass-production-object-(ive)-(s), by self-bias, tensing presided-observation-(s), when year-(s) of presence, slip perceptive future-reset-(s), intermittent weekly-offset, daily-sleeping, calendar-reasoning, remaining

{Article:} [March]

[conceptual / hypothesized / uninhabited / precept] hard-copy-past-day-presence-area-duty-(ies)] comprehending extensive-converse-(s), date-facting-products, contently-considering, why recipe, [retrograde act-(s) // self-requirement-(s)-articling] crop-growth-collection-(s), ordering classification-(s) where [study / practice / congrugated aging] concepting cause, impasse person-mass-comprehension, trial abroad, those locale-(s), not efforting-community-(ies)-interaction-(s), where territory, remain affected (from of) solar-kinetic-molten-combustion-energy, responded by civilization-(s), mattering elapsed-fact-(s), where parameter-(s), rest by night-sleep, margining unit-(s) of shelf-data, crescendo by capable-trust-(s), ideologically-inclined, cycled-sequence-(s), pertaining presence by territorial-characterizing, when subjective-competency is defined when Time & Date are elaborated by [entity-transit-sociological-schedule-objective-impasse-ordering // elemental-extraction-population-objective-periodical-psychological-ordering;[80] genetic-median-average-age, mode interceded activity-(ies) // motion-(s) // directive-(s)] cycling retrograde-article-(s)-ordering-(s), impassing time, when secondths-natural-passing, daunt perceptive-citizen-(s'), impassing date-(s), affirming-data, factoring-unit-(s)-equivalent-(cy), apose time, focusing objective-use, deviated in miniscule-incremental-(s), reducing elapse-expiration, per syntax-synapse-(s), offset individual-citizen-(s), process-(es) greater than, cubic-objective-deviation-(s), denominating [elapses / measures / calculations / places / date-(s)] upon two-dimensional-perimeter-boundary-map-planning-(s), an generally-defined-surrounding-mile-(s)-extent-population-civilization-cultivating, apose surface-area-square-foot-parcel-measuring, finite, balancing territory (twenty-foot-underground-space // two-thousand-foot-air-space-civilization-pattern-process-(es)), upon population-voluntary-psychological-perspective, juxtaposed substantial-nutrition-comestible-consumption-weight-incremental-(s), posing motion-(s) at [work-place-action-(s) // community-impasse-converse] along an means of intangible-process, secular-dependent, where defining [schedule // agenda] in fluent-process (by an) genetic-mass, presume greater through operation-(s) for life, always considering how to [cultivate-soil // entity-terrain-posit-(s)] from requirement-(s), (in to an) elegant-bemuse of Lifestyle & Class.[87¶|||]05m.:23s.[63|||§] This operation, I [deem / qualify];[81] as our limit of perspective, cause exterior-extent, remain limited, by spacial-property-(ies), from metropolitan-dynamic,| 247 |pursueding [public / private-funded-account-practice-(s)], affirming referenced-account-(s), present future-passing-(s), requisite impasse effort-(s), clarifying pragmatic-conjure, rural-offset-(ed), numerative-operating, juxtaposing rich-rural-territory apose metropolitan-poverty, past, pronoun-physiological-decay-denominative, simultaneous by word-logic-referencing, initially colloquial-local-thought, basing perspective (from an) relied-fatherhood, of trade-(s), because premise of Lifestyle & Impasse, remain ignored by all human-organism-(s), because of basic-human-motion-(s) [[1]eating /// [2]sleeping /// [3]studying /// [4]working /// [5]transiting /// [6]sexual-intercourse /// [7]walking /// [8]sweating /// [9]interacting /// [10]communion /// [11]dispelling /// [12]cooking /// [13]purchasing /// [14]talking /// [15]idling;[82] defining existence, forgetting commonly to write /// read /// redact /// debate] article-(s) for exceeding present-infinitive-human-disposition.[88] At where, can note permit longevity, for others' partaking comfort, in post-mortem-lifestyle-(s'), apose present-understanding, lively in clear resourcing-routine, upon majority-mass-(es)-effort-limit-(s), inconclusive population-adjusted-result-(s) fervor for national-political-religious-communion, defining all object-(s) acquired in life, by thinking how proportion-(s), interlinking simultaneous (time) mass-perceptive-existence, where at an offset of Physique & View elicit what can be done, if genetic-swearing, implement an rigid-regimen, avidly succession-elemental-population-objective-count-extracting, subjecting each [entity // roadway], by an literary-purpose, upon those standards of life, contingent in reforming each visual-imagery,| 248 | denominative all [weight-(s) // measure-(s) // distancing] balanced (by of) human-physique.89¶|||06m.:49s.01m.s.|||§ Religious-fervor;83 payment-method-(s) reclusive by individual-(s'), prior-base self-interest-(s) by singular-zero-person-introverted-bias-bigotry-character, subcon-scious habitat, where-when human-routine, intercede co-worker-peer-interaction-(s), (un) astute by age-bigotry, impassing human-mass-subjective-routine, characterizing Ethnic & Race-Rate-Deviation-(s), from family, (un)-pertain-(ed) (by of) matter, (for how) national-state-agenda, recognize each proprietary-citizen-motion-(s), upon calendar-infinitive, uncounted from effort-(s), interval-(izing) an order of chain-sequential-operation-(s), vocational at conjure (by of) human-limit-(s) cordiality-greeted, disclaimer-offset, interacting from focus of material-limit-(s), synthetic blood-pulse-exert-effect, timed by clause, characterized by time of date-(s), assessing elapses upon routine-effort-(s), elaborating ecological-climate, adjunct sociological-routine.90¶|||07m.:30s.73m.s.|||§|| 249 |Inept ten-(s)-tense-(s)-exponent-population-offset-reset-deviation-(s), unattested-to, hundredths-exponent-mass-reset-deviation, uninspired for thousand-th-(s)-character-trait-identification-base-person-posit-set, from youth-age-offset-(s), reset-(s) interval-(s) per time-frame, characterizing those default-posit-mass-extent-(s), percentile-perceived, upon an fraction-incremental-crescendo, from integer-whole-count-(s), decimal-reset-(s), coordinating-an subjective-empirical-denominative, decimal-miniscule-matter-incremental-progression-(s), balanced apose versive [<u>agenda // schedule // regimen, calendar-periodical-denominative-action-(s)-count-(s)-affirmation-process</u>].91¶|||08m.:03s.76m.s|||§ Commercial-faith, under-study spacing-(s), yet upon an reflexive monitoring, notion-(s) inexplicably simultaneous, relevant freedom-(s), charming national-identity,| 250 |deriving state-affairs, (by an) constraint, from mass-live-effort-(s)-prerogative.92¶|||08m.:16s.42m.s.|||§ Senile-perspective, rationalize prerogative, from identifying free-will, by basic-self .93¶|||08m.:20s.77m.s.|||§ No progression exist, unless an mass, instruct object-(s), impassing serie-(s) of disclaim-Gestalt-conjunctive-action-(s), elongating attention-span (<u>warning / caution / advisory / alert</u>

/ yield / stop / attention);[84] shock-notice-(s), as thought of after-effect-extent, while present, reset-subterior-cycling-(s), not space, for mass-title-purpose, requisite national-male-maxim-housing-space-limit-(s), spacing metropolitan, upon continental-shelf, uncultivated their of grandiose-ambient.[94] None have marveled upon the boundary-limit-(s), characterizing factor-(s), untangible where spacing-(s), has alluded an finite-demarcating, what procedure-(s), are methodically implement-(ed), assessing responsibility, (to an) leadership-median-(ten)-thousandths-tier, form-documenting realty;[85] by mental-spacing-(s), that distract, or not pertain, formal-subjective-referendum, from colloquial-hillbilly-dialogue, incoherent per-a-the [file-cabinet-(s) / book-shelf / (personal) / library] posit-(s) as cycle-ordering classification-(s) for live-population;[86] where mass, not convey fluent-vocal-chorus-tone-syntax-denoting-data, by means of non-fiction-literary-ways, persuading directive-outcome-(s), from point-A-abstraction, regardless educational-modern-age-acting, simultaneous various-extreme-(s) of rationale, pervade metropolitan-limit-(s) of boundary, (when whom) physiological-concerting, proportion conscious-interaction-(s), defining [self / tenth-men-conscious-impasse / hundredth-men-driven-impasse / thousandth-men-contravene / ten-thousandth-men-governing] variation-(s) of structure, yet fluently conveying transient-auxiliary-objective-(s),| 251 |by word-tangent-accountability, (from an) [singular-individual-first-person-vocalizing // second-person-individual-listening] seldom when third-person-listening-affirm-reforming, as I attempt to offset-advantage, such reasoning, from [literary-survey / house-representative-contacting (once I move from Florida) / senate-member] as particle-factor-(s),| 252 |assessing member-bill-documenting, historical-mass, upon population-subjective-interpretation-educational-fair-evaluation-(s), upon linear-means of mathematical-succession-(s), rating mass-error-(s) from relied-trading, meandering monetary-production, by those credit-cause-(s), uncontrolled of human-conduct, where [politi-cian-(s) // writer-(s) // policemen //

military-men // common-socialism] share an common-thread, of inductive-performance, inconclusively referencing article-(s)-data, (from an) denominative-decade-theatre-passion, of grandiose-abstraction, pertaining ethical-means, where Mass & Motive, are suppose to reveal relevant-free-will-survey-accountability, in nature of synthetic-assertive-refinement, focusing factor-exponentiation, from population-survey, at territory-offset-(s), hypothetical state-reform-dynamic, inquisitioning independent-singular-effort-denominative-denoting,| 253 |upon mass-individual-(ism)-reading-writing-reset-deviation-count-numerative-denoting-competency-(ies), understanding nature, upon hard-data, (by an) denomin-ative-English-centralized-language-accessing, evaluating personal-soft-data-clarity, assessing population-inhabitant-(s), by deviation-(s) per (of) rate-(s), upon time-calendar-spacing-(s), where an genetic can enact upon intervention of competent-discourse, prevalent personal-spacing-crop-retirement-assessment-(s), upon Will & Way, efforting, awhile impasse, wilt-introverted-nature, passing-elapsed-notion-(s), before any recognition of such dismay, regard those ignorant-(s'), whom evade realty, discoursing mass-product-composite-concept-developmental-trading, Germanically not formally processed, as dynamic of physiological-characteristic-(s), word syntax-deviational-refinement, alterior-current-interpose-syntax, taxing commercial-item-unit-tangible-(s), by Ink & Paper, refining parameter-limit-(s), efforting from self, an state-extraspective-exterior-value-(s)-dynamic, relating data, mundanely, referencing sentence-(s), as underlying fact, upon those piece-(s) of non-fiction-articling, conveying to recipient, how credit-responsibility, pertain from contexting transit-limit-usage, variating residential-sleep-work-dynamic, influctuated routine-confidence, by human-emotion-(s), series-(ed)-parallel, programmed for reading, but when collective-means are not fulfilled, by each and every individual, from either non-fiction-expertise-(accredited from cursive-context) or auto-biography,| 254 |upon retirement-impasse.95¶|||11m.:48s.11m.s.|||§ As when simultaneous-intercede through hour subterior day, customs' are checked, as for whom one is, in-charge of placing, an offset-structure-resourcing-order-(ed)-responsibility-routine, familiarizing approximate-local-sequence-(s), for (religious) guidance, of perimeter-spacing-(s), pertaining creative-instrument-(s) (as such this program of chapter-paragraph-comma-sentence-syntax) from regiment of human-population-auxiliary-prerogative, not established from blood-pulse-response upon rectangular-perimeter-feet per hour, or moded transit;[87] at observation of ambient, (in an) symphony of sound-(s), that are moved, upon factor-deviation-period-(s), temporal when blood-pulsed-perceptive-tense-response-sight-wave-frequency-flux-sound-vibration-(s)-reverberation-expon-ent-tenths-lapses', emit by back-forth-tension-(s), through second-retrograde-progressive-motion-(s)-offset-(s)-accounting,| 255 | blood-pulse-rate-frequency-(ies), upon day-offset comparison [week-offset // month-offset /// season-offset //// year-offset;[88] when motion-(s) // instruction-(s) // effort-(s) // action-(s) // reading-(s) // transiting-impasse-(s) // et cetera…] are intended for documenting [secondths-elapse-count-deviation-(s) // minute-(s)-elapse-count-deviations] time retrograded from mass-quantity as-purpose, for calendar-whole-integer-hundred-(ths)-count-day-ordering-(s), by Thousandths-Day-Cycling & Decade-Day-Cycling-(s), categorizing ten-thousandths-day-cycling-(s), by limit-(s) of human-fatigue, and reproductive-concern, for when mass-proportional-rationale-effort-(s), exhibit fact-(s) of area, discerning order, elaborating acre-(s) [per / of] [cubical-air-space-limit-(s) / horizontal-length-width-square-mile-(s)-boundary-objective-designated-intervalizing] perimeter-surface-area objective-weighing-(s)-measure-(s) assessing object-(s)-posit juxtaposed [nature / synthetic-existence] adjusting habitat by trade, or governmental-ordering, from how commercial-ism-dynamic, pertain, faster, refined, perceptive-infrastructure, eliminating physiological-trade-effort-(s), not orderable, by [sorted food-choice-(s) // clothing-preference // housing-needs] exploring exemplary cause for Self & After-

{Article:} [March]

Life-Lineage-Mass$_{174L.|}$ 256 |existence, hopefully due (to an) totalitarian-genetic-literary-population... ¶|||13m.27s.$_{63m.s.|||}$§

§...Count-[55 / Fifty-Five]-preposition-of.. ..Lines-[174.4807 / One-Hundred-Seventy-Eight.$_{1.four-}$tenths-$_{2.eight}$-hundredths-$_{3.zero}$-thousandths-$_{4.seven}$-ten-thousandths]...Article-Essay-[2 / Two].... Book-Essay-[48 / Fourty-Eight].. ..Keep words short now, with minimal-response-reminders.¶|||09s.$_{97m.s.|||}$§

$_{¶49}$During, reoccurs by awake-time-posit-(s), durating [lapse-(s) // synapse-(s) // particular-motion-(s)];$_{89}$ interval-upon, blood-pulse-set-tense-prerogative-(s), living pass various inhabitant(s'), for cause of documenting-(tion) in order to consider elder-aging-process, from-an youth physiological-purpose, practicing population-communion, off-set government-reliance, (by an) requisite male-front-line-effort-desire, blue-minded, impassing-existence, tranquilly-thinking, in line of objective-duty, as when mass-cultivation-directive-(s), agenda [biological / geological / sociological / historical] accounting topic-(s), articlized population-subjectivity, articulating objective-agenda, as governmental-process, commercially-instated, energy-parameter-(s), tolling tax-(es), to understand the nature of denominative-denoting (as this book intend plus ten-thousand-individual-word-intrication-(s));$_{90}$ balanced upon perceptive-competent-person-blood-pulse-(s), where inquiry-maxim, offset time, entertaining worded compound-conjunction, for light of reading-exponent-(s)-logic, (as I backtrack-reference comparative the books read of accelerated-reader-program) that clue (re)-spending, mass-prerogative-purchase-(s), defining (a the) commercial-cause from leadership, incapable of Tresurigal-mass-developing, population, Vast & Empty, (un)-communicated in youth-relationship, from the retardation of intersexual-relation-(s'), that inhibit no reason from those three-primary-deviation-(s) of second-sibling, upon the individual-sort, [girl-boy / girl-girl / boy-boy offspring-pattern] that have an simultaneous-simulation, of spacing-deviation-(s), verifying perimeter boundary-capacity by [state // county // city-district-(s), parameter-(s)] abstracted from how hour-(s) of adult-coverage, can verify territory, juxtaposed-offset, population, planned from walking-distance-maxim-rate-impasse, post-engineered, industrial-com-ponent-(s), as commercial-neutrality, confirm from government, discourse of mathematical-denoting, solid-resourced-chemical-(s), denominated by,| 257 |per quality-equivalent-word, upon pre-mass-documentation-effort-(s), characterizing fact-(s) upon self-use-limit, before involving other inhabitant-cause, from the traded-effort-(s), sustaining (a the) inclination of involved-objective-effort-(s) upon territory-permanent-posit-(s), (for thinking what is a the) most effective-method, for using objective-(s), upon territory-development, spacing entity-(ies), by walking-impasse-effort-(s), practical from historical-present-outcome, applying restriction-(s) upon transit-coverage, involving daily-age-dating-discourse, that offset the sake of ignorance, uncounted by exponent-personal, outward-maxim-average-vernacular-syntax-accounting, affirming mass-effort-(s), that ensue property, upon territorial-claim-(s), that exist, from [Christian / Judaism / Muslim military-faith-(s)] trading since slavery, awhile consuming food, (from a the) mass-mute-physical-metropolitan-existing, (un)-categorized, by common-perspective.96¶||| 02m.:17s.10m.s.|||§ Catholic is out of such formation, (but of an) Roman-Maxim, whom partake in season, from monthly-week-lecture-(s), unnoticed under the full-limit of [day // night] [unscheduled /// unregimented] by an focus of mass, apose century-millennium-discourse, pertaining period-(s) of year, by dating-age, corollary technological-default-offset-(s), piecing each & every tangible-obstruct;91 variating general-feature-(s), from manual-dexterity, affirmed, when each task, involve living, for state-denominative-objection, circulating federal-commercial-excess-educational-inclination-(s), yet to extrapolate the longevity of traffic-(ed) (solicit-(ed)-matter-(s), denoting time apose trade, loophole-(d) for basic-human-right-(s), involve an incredulous pattern of Territory & Power, presumed inhibited-space-(s), without formal-inquiry, for those component-function-(s), because

{Article:} [March]

Perimeter-Space & Formal-Engineering-Discourse, [clause / common-law / supreme-court / condition / term / definition / temporal] defining individualism, by reclusive social-identity, not communicating empirical-ordering, amidst human-theatre.97¶|||03m.:06s.79m.s.|||§ Textbook-(s) of educated-informing, elicit object-(s), from auxiliary-effect-(s), having physiological-identity, state premise, meriting self-blood-pulse-being, collar of career-mode, defining order-(s), rescind incremental-fundamentalizing, an singular-territory, pertaining men, by no capable-conception, lacking an desire for insider-purpose, in to [mansion / private-housing] categorizing surface-area, in cause, centralized-fundamentalism, exterior, greater extraspective-limits, exploring housing per thirty-thousand-citizen-(s), per two-year-conscious-leader, directive-limit.98¶|||03m.:31s.76m.s.|||§ So I do not know how to extend such [objective-(s) // subjective-(s)] for balancing word-mark-(s), associating information, commercially fact-(ed), conveying why maintaince, weight balance of factoring, from how competency, remain heavily out of the order of documented-faith, read-ceaselessly, delivering no purpose of non-fiction-entertainment, pass time instead from movie or lotto-purchasing, (un)-incremental of hard-cover-non-fiction-articling-(s), effervescent, quality year-developmental-fatigue-(s), practicing Wait & Effort-(s) intent economic-routine, drifted upon untangible-quality-(ies), of [tectonic-plate-(s) // molten-lava-core-heat-friction-plasma] inaudible understood by totalitarianism, because epitome, remain, cortical, awhile Egalitarianism & Totalitarianism inept of magnitude, rifted refined scaling, perceived by subjective-weight-quantity-(ies), that superlative-reasoning, functioning from [lease / loan prerogative-(s)] for observing those-limit-(s) of perplexive [condominium / apartment / duplex / trailer / townhome / business-room / private-tower / mansion / house] as parcel-(s) of citizenship, that store responsibility, between [market-(s) / workplace-environment / transit-impasse]| 258 |recuperating from physiological-fatigue, from governmental-developmental-construction-default-feeder-pattern-(s), systemized by sign-restricted-obstructive-(s), reminding human, upon habitat, the sensational-way, flawed of praise, tense-(d) from being, with indication, plateau extent-(s), deviate classification-limit-(s), collecting for an requisite-posit-(s), intending an after-life-longevity-remembrance.99¶|||04m.:43s.64m.s.|||§ Methods daily construes', [commercial / governmental / economic / industrial / militarial / social / transcendental trusted-order-(s)] formulate religious-cause for how contraceptive-method, define masculine individual-character, exacerbated by mass-run-of-the-mill-effort-reflexive-discourse, preferred brotherism, before exponent-repopulating, focused (from an) posit of pre-dated-matters', loyal to subjective-feudalism-ordering, of statehood-maintenance, because the awareness of technology-warning-(s),|259|when Life & Action-(s) are male-introverted-restrained, because what is acquired for appeasing, rather sensuous-interaction-(s), afraid of female-genetalia, because how individual-(s)-male-response, perceive from objective, before Physic-(s) & Perspective due to generational-language-brainwashing-adaptation-comfort-preference,72L. enjoying fantasy before chore, as how setting affect intersexual-relation-(s')...¶|||05m.:20s.21m.s.|||§

§...Count-[37 / Thirty-Seven]-preposition-of.. ..Article-Essay-[3 / Three].. // ..Book-Essay-[49 / Fourty-Nine].. ..Lines-[72.798 / Seventy-Two.1.seven-tenths-2.nine-hundtredths-3.eight thousandths..] ..Swifting out every moment not posit cause for entity-development, amid an sociological-cause, so where retirement live, not save over $100,000, per decade circulating expenses of bills, verifying cause, due to certain aged qualities by entity objective wage moments, offset from scheduling, where location can only identify which language skills can be candid due to area at trading posts, reminding which instruments of cooking matter where elaboration of restorm, share an restraint usage of developed objective, instilled at an estate of bias, rather centralized interpretational government, per circa life, infringed at aged passages, under an revolving temperature-layers, geometrically offset uninhabitable when retirement not collet genetic, from formal documenting efforts under library-verification-(s).¶|||43s.76m.s.|||§

¶50.Male-inhabitant-(s), prefer their personality-error, protecting Treasury & All, at the detriment of female-counterpart, perceiving as common-being-(s'), rather mutual-person-(s),|260|undirectable for ordered-reasoning, by subjective-empirical-merit-(s), acted under population-psychological-order-communicating, sequentially periodicalizing, pronoun-individ-ual-(s') to objective-residence, by elemental-proportioning, conscious interaction-(s), apose consensus-effort-(s), labored for effective water-current-engineering,| 261 |powering [residence // commercial] entity-posit-(s), abiding governmental-fundamentalism, followed by state-dip-lomatic-affair-(s), for legal-bill-judiciary-senate-committee, affirmation-(s), from population-protest-pressure, scaring politician-(s), whom lack those literary-skill-(s), to exemplify why human-(s'), condition affect each effort-(s), at minimum of living, an auto-biography-literature-example, per decade-impasse, implying that humans simply Idle & Enable one another;92 rather avid Expansion & Repopulating to define those parameter-boundary-(ies), receiving by interaction-(s), a-the pleasure, interaction, with fine-looking being-(s), whom are defined (from an) critique of characteristic-(s), as well intellect-capability-(ies);93 in defining-existence, where motion-(s) pertain upon perspective of individual-characteristic-(s).100¶|||01m.:06s.35m.s.|||§ For example, citizens are perceptive-physiological, from-an ideological-view, count-offset [military / men / idle-commercial-account-accumulating] minimal industrial-performance-(s);94 sarcastically relied on fundamentalism, pre-instilled credit-faith-based-trading, hurt from those topographic-condition-(s), when commercial-prerogative-(s), balance apose Calendar-History & Mass-Family-Populous-Effort-(s)-Reasoning, held under affair-(s), by person-(s), whom nationally factor, factory-mass-dynamic, reasoning sociological-interaction-(s) of theatre, reasoned by the free-will, person-barrier, believed effective, while an complete failure of repopulation;95 incoherent to subbcome to such fact that sociological-living is far too bastardized, to permit going in such foul-direction, filthy by the deliberant human-responsibility, sharing attained objective-factor-(s), pertinent in determining word-constitutional-reasoning, repetitively-aware, (from an) largo-comprehension, incomparable, by height-cut-off, upon surface-area-length-width-distancing-(s), which would define territorial-discourse, if historical-perspective, pertain, prevalent-present, by dating elite-status, embellishing [territory / auxiliary-transit-objective-(s)] by passes of title, for being at, an location for further-repopulation-subjecting, opposite-sex, entity-purpose-(s), labyrinth, [interior-structure // exterior hedges-(s) // tree-(s) // plant-(s)] décor-positing, various type-(s) per [taxonomic /// agricultural /// terrestrial or terracultrual /// stone //// wood //// plastic-statue-(s)] detailing how dating, should have long been expedited, but when now Tresurigal-control, clearly be for basic-human-right-(s), and nothing more.101¶||||02m.:29s.20m.s.|||§ Collecting require-(s) space, for maintaining numerative-matter-(s), because weight-proportion-(s), are perceivably involved, upon practicing letter-mark-numerical-tally-documenting, upon an [subjective-documenting / physical-outlet-(s)] signifying purpose from self, through an genetic-communal-mental-mass, aware by-those limit-(s), by the stasis of organism-(s), rendered wed, (re)-cycling matter-(s), intermittent depreciating-data, constantly progressing mind, pass obituary-day, as perceptive-limit-(s), in review of objective, for how humans partake intermittent, [bingo / lotto / betting / gasoline / comestible-(s)-shelf-life / fund-(s) / checking expenditure-(s)-gambling] pertained storage allot-amount-(s), cause not in thorough concertion, for why we are not based in an origin-genetic, but elapse, by miniscule of lease-style-thinking, influctuated by Retirement & Crop-Collecting decaying by continuum, as for how recipe affect lineage by either an confided-reliance,| 262 |or rural mass-reasoning, equivalent from subjective-effort-(s), proportional Tresurigal-discourse-denominative-fate, {Two-decimal-Seven-Tenths-One-hundredth-thousand-th-nine-ten-thousandth-five-hundred-thousandth-count} variated by Person-Intelligence & Common-Incompetency -(ies).102¶|||03m.:33s.63m.s.|||§ Post-millennial-reason, pre-base-conceive focus, through attention-span, rather natural-observation, trading matter, reduced from depreciation-parity, by the mass-average-deviation-(s), perceivable, productive-matter-(s), counted by state, after-model-formal-use, reparent as figurative-reason-(ing), for faith in mentality, by the bias-superego, yet savvy terrestrial-population, arguing that objective, be dissuaded from educational-background, affecting off-time-spouse-relationship, reclusive from work, until corridor-(s), from adherence to tax-(es), to [allocate / insure / toll / bill / tariff] constitutional-faux-paus, because how, day intermittent seconds-omission-(s'), select pronoun-reason, simultaneous millisecondths-fourth-sixteenth-quarter-note-maxim, as a the gamma-deviation-(s) of {cent-reasoning;[96] (Indust-rialism Commence) {Abraham Lincoln;[97] (1860-1864)} /{Five-cent-reasoning;[98] {Thomas Jefferson;[99] (1776-1806)} // Ten-Cent-Reasoning;[100] {Franklin Delano Roosevelt;[101] (1933-1945)} / Twenty-Five-Cent-Reasoning;[102] {George Washington;[103] (Pre-American-Conception-1783)}};[104] privy denominative-fact-(s), gesturing appropriate-discourse, by auto-biographical-history, memoir by motive-(s), which inspires sociological-discourse, from off-time-diligence, interceded-place-(s), at focus,| 263 |[subjective-impasse // present-input // future-dating-interposing] for an genetic-longevity, through governmental-trust-confidence, defining affair-(s) by mass-human-appropriation, billed-note-(s),60L.|264|from basic-human-rights'-count-cycling, modifying mundane-order, in lieu of cultivating an entirely new process of Repopulating & Denominating, per [Territory // Objective-(s) // Population-work-effort-(s) // Housing // Comestible-(s) // Socio-logical-activities]...¶|||04m.:59s.80m.s.|||§

§...Count-[13 / Thirteen]-preposition-of.. ..Article-Essay-[4 / Four].. // ..Book-Essay-[50 / Fifty]....Lines-[62.1987 / Sixty-Two.1.one-tenths-2.nine-hundredths-3.eight-thousandths-4.seven-tenths-thousandths].. ..Whomever concern retentive-subjective-interpretation-thought-extents, comprehend posture-prose of surrounding area for objective retentive-directive-agenda, relevant upon conditions per [exterior / interior surface area], pertaining how educational-lesson, is suppose to function subversive vocal-grammatical-mathematical-visual-tonal-vibrations, from when dating is counted at random superlative interpretation offset-(s), without offset-inductive-prose of millisecondths-motive-interpretation-(s).¶|||29s.91m.s.|||§

¶51.Velocity agglomerate progress of living-passing-habitant-(s), impassing occupancy, interval academic-gathering, by percentile-deviation-(s), from inferencing percentage-numerative-fraction-(s), as [respiration / writing / redacting context / vernacular-voice-reading-(s) / debate of resource-(s) / population-account-denominating of genetic / creative-term-literature-intervalizing] retrograde calendar-interval-(s), dated in time-deviation-(s), of elapsed-effort-(s), thinking how grade, affect each matter, by expiry-effect-(s), configuring citizen-mass-discourse, through condition-(s) of temperature, by moral-guideline-(s), tangent apose, degrees-(s) where [climate / location / terrain / season / entity] impact perceptive-effect-(s) of exterior-posit-(s),| 265 | observable from tense-(s), when mass-humanity-commonality, modify territory, uncoordinated by scheduled-agenda, isolating-data, primarily-significant, by contort-directive, devised mysticism from physique, image-viewing, through kinetic-vapor-pressure, after aquatic-pressure, offset tectonic-plate-rock-density-pressure, sanded of aquatic-divide, as state-(s) of volume, part of an [ecological // physic-(s) perspective-offset-(s)] greater than, whole deducing-observable-moment-(s), from time, in despair of sex, without time for children-lessoning, when their be an basis of live-matter, tiered from interaction-(s) of progressive-motion-(s), where neighbor-community, are for affirming, conclusive discourse of men, by mode of creative-input.103¶01m.:16s.37m.s.|||§ Through blood-pulse-invigoration, tense-exist, before synthetic-material-(s), theatrically-aware [matter / mass / material-(s)] ordering from act-(s), how to conjure surrounding surface-area, by an [Federal // Tresurigal budget] dormant Modification-(s) & Observation-(s) limited from restricted-resource-(s), advised formal-proportion-(s), pertaining subjective-comprehension, past-documented, by an Tresurigal-survey-effect, aging modern-presence, unclear for whom define national-boundary-(ies), when state individual-territory-offset-(s), offset population-(s), defining after-life-prevalence upon temporal-existence.104¶|||01m.:42s.31 m.s.|||§ What indicate-(s) purpose in life? Various method-(s) of county, measure through syntax [reading / writing / date-data-documenting / table-referencing / pronoun-attendance (deviation of freedom) / subjective-account-judging] appreciating objective-transit, as level-(s) of parameter [in / of] deviational-pre-concept-development-perimeter-area, climate-influence an simultaneous-offset, by hypermatic-reactive-reflex-(es'), matric-posited commemoration, passed along character-timbre, hyper-tensively responding, from-by Trust & Men, whom validify variable-(s) for testing discourse-(s), of Territory & Family-Mass-Population, defining means for handling sustained affair-(s) of state, blood-pulse-active, transitive-genetic-structuring, conducting duty-(ies) by-an centralized-living, which encompass Date & Age characterizing experience, [crescendo // plateau] from individual-(s') juxtaposed tense-sort-organized-temporal-ordering-(s), from-of per market-data, under operation-(s)32L.| 266 |spaced at district-controlled-limited-boundary-(ies)...¶|||02m.:25s.46m.s.|||§

§...Count-[21 / Twenty-One]-preposition-of.. ..Article-Essay-[5 / Five].. // ..Book-Essay-[51 / Fifty-One].. ..Lines-[32.18 / Thirty-Two.1.one-tenths-2.eight-hundredths].. ..diminuendo of redact paragraphs, due to the context complexities.¶|||10s.62m.s.|||§

¶52How does consideration vary from individual-(s'), while relying on resources, from one-another? Passing-terrestrial-centripetal-drift, never impact-(s), human-(s) bone-physiological-trajectory-coordinate-ordered-mile-(s)-elapse-passing-(s), because vehicle-approximate-distanc-ing-(s), following an egalitarian-metropolitan-compact-order-(s), conscious by presided tabling, when time-frame, will have parental-generational-passing-(s), census-era-aging, [path / journey / endeavor, (ac)-count-(able)-reference-(ing)] historically-lessoning, present-secondths, by none of a the minute-(s)-elapse-(ing)-retrograde-(s), saying [game / set / match / period / quarter] comparative-act-(s) abstract-consensus, impassing after, hour-(s), barrier-less, abstract-mile-coordinate-horizontal-breathing-centripetal-pulse-space, subservient-under, moon-tectonic-drift, asleep-exhausted, harmonious-routine, where passing-(s), define a-the approximate-vicinity, of factors, mutual by employee-(s), by square-feet, upon an fixed-set-placed-location-(ing), permitted [lease / loan / payment] alone at area, moding psychological-type-(s), insured by vehicle-weight-accounting, from a the physiological-structure, in self-reputational-posit, by mode of transit, upon metropolitan-governmental-public-living, spacing out, from comforting, centralized-constructive-mass-living, as when date, is abided by schedule, congruently fashioned, because those required-product-marketed-discourse-(s), that include an relationship, based by expenditure-merit-(s), passing-day, from the present-year-offset-(s), deriving [Gregorian-Two-Thousand-Sixteen-year // Actual-Four-Hundred-Eighty-Eight-Year-(s)-continuum]| 267 |of tectonic-plate-continental-shelf-blood-pulsed-transit-being-(s), that accumulate-matter-(s), intermittent interior-warehouse-constructive-lot-(s), for an livelihood, based on Commercial-Trade, & Demand-Elemental-Extraction-Production-(s).$_{105}$¶||||01m.:35s.$_{14m.s.}$|||§ Rate be factored, from dating, as two-discourse-(s), one, is from the calendar-incremental-count-(ing), past-day-(s), upon present, then retrograded interval-(s) by day-(s), upon year-denoting, of the day-cycling-impasse-(s), seasonally facting, revolving-rotating-increment-(s), of centripetal-force-drift;$_{105}$ second, is the date-hour-cycling, approximately-blood-pulse-rate-(d), by those factoring-(s)-simultaneous, [dimension-(s) / pulse-elapse-count-(s) / exterior-wave-frequency-(ies)-posit-juxtaposition-(s) / solid-objective-data-(m)] pertaining custom-(s) of physical-muscular-response, upon visual-imagery-mark-(s) as word-term-definition-sentence-paragraph-language-reference-(ing) (playback), of those composite-matter-reformation-factor-(s), that are in lieu with of personal-responsibility, apart from ontological-etymological-terminological-optological-literary-grammatical-punctuational-proof-ing, determining payment-wage-disparity, requisite actual-equal-right(s'), when [realty / work-cycling-(s) / off-time-abode-(s)] parameter state-national-ordering, alluded parameter-(s), from area, during time-posit-(s)-qualifying, from third-person-observation-over-sight-offset-posit, [listening / mediating] brief-interpretation-(s')-adjustment-(s), learning of boundary-extent from how [population // reproduction], affect dispersed-disposition-(s), requiring destination-discourse, at all moment-(s), for transit amidst metropolitan-offset, objecting character-traits, [perpendicular motion-(s) // sustaining-(s)] at posit-(s) for interpreting what surrounding-agenda, is required, | 268 |for [future-developmental // renovation-inquiry].106¶|||02m.:51s.59m.s.|||§ Depth-perspective, sustem by three-dimensional-density-substancial-(un)-tangible-(s)-respiration-blood-pulse-perspective-offset-(s), pertaining two-dimension-marking-(s), by height-density-(ies), substan-tially-space-(ing-(s)), those deviation-(s) of matter, intermittent three-dimension-zero-person-perspective-(s), retrograded, two-dimension-whole-zoning-(s), pertained when individual-(s), hamper one-hundred-percent-fluctuation-deviation-incrementing-age-percentile-drift, commonly-offsprung, not deviating from yearly-capability, motivating documentation, far too enjoyed, by subjective-listening, referencing article-(s), for practicing-scheduled-effort-(s),| 269 |to understand Hierarchical-Genealogical-Geographical-Encyclopedia-Fathom-Deviation-(s), affirm-ing papier-fiat currency, through method-(s) of Population-Work-Effort-(s) & Commercial-Growth to place alterior-focus, during day, sleeping, juxtaposed night-resting, interval-moding conscious-passing-(s), unlabeled signage upon parameter-approximate-surface-area-(s), scaled by date, in decimal-whole calendar-age-growth-simultaneous-percentile-deviation-(s), because of individuals'-perceptive-offset, when day-(s) in ten-thousandths-whole-degree-reasoning-syntax-(s), juxtapose daily-hourly-minute-(s)-label-motion-(s)-act-(s), crescendo-incremental-elapse-(s) present, offset-(s) at city-county-civilization-juxtaposition-(s)-posit-(s), reading-article-(ized)-syntax-posit-method-(s), pertained through seconds-elapse-deviation-(s), not memory, for live-intervalizing, pertained mathematical-rigor, by grammatical-prose-elapse-(s), paragraphed-sentence-(ing-(s)), referencing [extent / re-(view)-(dact) // observational-mutual-competency] measuring median-means, from comprehension-thousandths-word-extent-(s), divided in to whole-denominator, observing exterior-matter from-an natural-perceptive-posit-base-offset, (un) equivalent total-denominating, by day-elapse-(s), upon merchant-factoring-(s), periodical-serie-(s)-discoursing, ensuing declaritive-dialouge, stated at merchant-mathematical-product-tangible-approximate-weight-mass-specialty-effect-sale-price-not-ion-(ing)-(s), pitched from purchasing, maintaining-operation-(s) while space, requisite [electric-current // water-current / product-ordering-(s) / default-trade-activity-(ies)] reset how elemental-extraction, intrigue perspective, when physiological-attaining, can fluently, transpose factory-production, and objective-attaining, by an process of Extraspective-Purpose & Gifted-Requirement-(s) emitting an aurora of fact-(or-(ing-(s))), implicit of requisite-history, because of an lack of educational-preference.107-¶|||04m.:52s.93m.s.|||§ Perspective is present at approximate-placing-(s) by individualism, when mass-accounting is progressed, accumulating-data, by each moment impasse, furthering thought awhile continuum of house-representative-documenting-proportioning, taxing populant-(s'), from bill-approval-(s), when resourced-data, pertain yearly-(s)-passing-(s), where human-origin-sourcing-ambiguity, partake upon planned-monitored-spacing-(s), by individual-appraisal, creating-motives of trade, upon limit-(s) of common-man, titling property, as the evaluating of family-directed-activities', as bank-account-prerogative, is incapable of sharing latter-age-ten-thousand-day-schedule-agenda-reformation, when organizing etymological-data, defining ontological-observation-(s), where those genetic-inhabitant-(s), disperse throughout physiological-human-mass, not intelligent, defining required-effort-(s) of [elemental-extraction // agricultural-development // horticultural-development-(s)] thought by department of commerce-documenting, when post-life-offspring, order mass-congregation, retaining denominative [objective-(s) / substantial resource-parcel-(s)-(far more intricate)] maintaining land-(s), refined because single-state-order, not populate further out of metropolitan-city-county-(s)-mass-development, interdialogue-perspective, mindful of the boundary-limit-(s), instated, by mass-perspective-dynamic, when tectonic-plate-(s), partake in metropolitan-discourse, because linear-survival, accede basic human needs, from how development of entity-(ies), are an part of space-(ing)-(s)-juxtaposition-(s), for impassing during an mode of time, upon an calendar-routine-impasse, affecting those emotion-(s) of individual(s'), awhile [daily // nightly mode-(s)] not balance concurrent of population enumeration.108¶|||06m.:20s.34m.s.|||§ How rural-practice of repopulation, ever work, would require women at home, and educated-economically, informing peers at [home / neighbor / pregnancy / tabling of motion-(s)-count-posit-juxtaposition-(s)] requisite ordering tangible-resourcing, as male-labor-pool-dynamic, practicing territorial-claim-infrastructuring, abide loyal-survey-constitution-mass, following formal-step-(s), initial process, when exponent-classroom-confirmation-documenting, where adult-decade-interval-reasoning, prepare-future, upon perspective-capacity, when Man & Mass exist, because of an linguistically-influence, persuading repopulation, awhile human-nature remain mute for care of repopulating.109 ¶|||06m.:57s.94m.s.|||§ Pronoun, is an position by which we are addressed by defining blood-pulse-motion-tense, placed awhile posit-limit-(s), pre-instate-focus, from blood-pulsed-being-(s'), where fact rely from pre-dating routine, upon spaced-hypothesis, for nature of Title & Monthly-Payment-(s), remind why temporal-existence, axis adrift, tectonic-plate-molten-lava-density-plasma-drift, featuring momentum, cause how timbre-tempo-time-note-(s), affirm reference from past, by religious-locale-divide-(s), through action-(s) by-of temporal-limit-(s), confided [eating / dispel, of excess-mass] untested quality-(ies) of monitored-juxtaposition-(s), because dynamic of cult, not reason literature, for median-focus, when surveyed-mass, embody [parent-(s) / elder-(s) / specimen-(s)] of sociological-taxonomic-transient-psychological-doctorate-legal-Ground-(s);[106] conterposed [Representative / member / Engineer / operator / officer] means of maintaining domestic-tranquility, contracting rural-space-(s), from the mass-survey-act-execution-(ing-(s)), efforting labor, distanced of crust-untangiblility, [testing / probing / examining / learning / listening / measuring / weighing / refining / applying] syntax-term-aptitude-individual-retention-deviation-(s), showing still motion-(s), upon word-retention, by gist of act, factoring in objective, consulting from contemporary-education, apose child-yearly-offspring-communal-interaction, building upon decade-(s), parental-guidance-perspective, by those fact-(s), that are communicated between an confluence by reputable-citizens-(s), whom have passed mutual-examination, with term-(s), concerted by off-time-prevalence, perceiving numerical-context, accustoming-tense-(s), for-of littoral-purpose, from tepid-response-(s), temporally scheduled, coordinated-reliance on retirement-insurance, enabling our basis-pulse-limit-(s), around an initial, post-world-war-period-(s), that consisted introduction-(s) to Retirement & Labor-Wage-Increment-(s), continued from mid-age-commercial-purchase-theatre, defining why mentality, mathematically-incline, idle state-(s), (in)-capable of nation-(s)-population-(s) over one-hundred-million-inhabitant-(s), per state-(s) [[1]Florida / [2]Texas / [3]California / [4]Arizona / [5]New Mexico / [6]Louisiana / [7]Mississippi] that remain incredibly hot, of [summer // spring forecast], evaluated for developmental-prerogative-(s), awaiting-command, directing latter-passing-(s), (from an) sociological-genetic-motion-(s)-effect, operating-perspective-mode-(s), from the vertigo of transit, that isolate routine time-fully upon day, from census-reasoning, not interpreting data, in age-pocket-(s), uncalculated from term-(s), independently male-race-responsibility-proxy, of (sub)-urban-parcel-yard-leisure, (un)-rural for neighbor(s')-effect, upon acre-(s) of transit-responsibility, [unconcepted / unhypothesized / unscheduled / unenvisioned / uncapable / unexponet-populated] an natural-limit, that demonstrate denominate-professionalism, for accessing default-basic-human-written-space-(s),| 270 | which require firm-practice, sorting from how inferred-urspuration-(s) of transcendentalism, partake continuum ancillary-talk, without an finite-end, to constitutional-period, that is slept in good-faith, by such Longevity & Defense, not comprehending from the national-bastardization, the global-impact, by offset-war-(s), to engage in population-reduction, leaving those brother-(s)-in-arm-(s), teetered by no age-compensation, retiring physiological-time-aging, of geological-topography-approximation-(s), to define the dynamic of metropolitan-environment, grossly-populated, where climate resume day, and omit-night-period-(s), for thought of any such mean(s'), from post-swear-period-(s) of educational-discourse, aligned by event-(s), which shape the work week, by drove-(s) of men, whom are over-stressed, fatiguing, age-insighted-pause;[107] when human-common-preference, prefer-sensational-feeling-(s), rather regard of national-mass-genetic-offset, having how I view my economy, as the way of national-objective, positing those [absolute-truth-(s) // fact-(s) (my book be fact, not judiciary-review)] when I remain unclear of sales-(s) via publishing-company, and have an requisite perspective, to believe that from deviation of those [month-(s) /// season-(s) /// week-(s)] those none sale-(s), build by case further, that I claim a the initial-American-English-observation, of mass-motion-(s)-function-(s)-inadequacy, sch-eduling how Time & Thought, can be incredibly short, while mass-population-interaction-(s), along with those impasse-(s) of livelihood, define those secular-space-(s), [unratified // unconstituted] in defining an denominative-accounting-method, of mass-effort-(s), upon those basic-human-requirement-(s), which instill motivation, for motion-(s) during [revolving //

rotating-vapor-pressure-hover-impasse].110¶||10m.:41s.29m.s.|||§ Insighting-pause, rests' reading, affect-ing cardiovascular-respiratory-blood-pulse, as [physique-posit-(s) / physical-act-(s) / subject-article-context-lessoning-discourse / literature-progression-(s) / developmental-terrain] through macro-composite-micro-producing, purchase-depreciated objective-means, while shelf-life, not be demanded, from free-will, perceptive-quality-(ies), expiry of outcome-(s), by commercial-production ignorantly abiding to boundary, through consensus-constitutional-review, leading guidance by-of human-discourse-(s), variating those total-mass-(s) apose an total-rationalization of matter, not weighed in juxtaposed comparison-(s), for-an total-untangible-tectonic-mass-composite-(s), reasoning-unit-(s), electronically-characterizing land, incapable from self-sole-survey, alone, (even forever) [conjuring // considering] coursework intention, examining efforts of human-disposition-(s'), by the dynamic of youth-study-(ies), proportioning those time-(s) of [fraternizing // socializing]146L. in preference-anticipation of adult-work-mode-(s), modernly-defining, why movement-(s) of driven-mass, are accustomed by daily-work-effort-(s)-impasse...¶|||11m.:36s.45m.s.|||§

§...Count-[71 / Seventy-One]-preposition-of.. ..Article-Essay-[6 / Six].. // ..Book-Essay-[52 / Fifty-Two].. ..Lines-[147.42 / One-Hundred-Fourty-Seven.1four-tenths-2.two-hundredths].. ..Marching be iterated, yet conceived, from latitude-tectonic-visual-offset-cubic-meter-distance-exterior-area, so modification of variables, juxtapose-physics via tectonic-plates, amid perceptive-offset-tangible-matter.¶|||15s.66m.s.|||§

¶53[Options // decision-(s) exist]. What [effect / impact] may pertain historical-confluence, through [fan / church-member / civilian-citizen-infinitive-input], never redacting, assessing [what // who // where // when] motion-(s) of [male /// female /// children] are remodifying, for male-being-responsibility-(ies), having how women-yearly-reproduction, require teaching [offspring // communion-children, in comparison of cooking-recipe-(s) // house-keeping // furnishing // repopulating // lessoning // instructing] trialing prerogative-sensible-awareness, percentile aver-age-comprehensive-unit-(s), from commercial-live-population-resourcing-production, when daily-dated-variable-(s), automatically-reference, mute, probable-bias, appealed legal-monitoring-review, while temporal-reflexive-response-(s), in-fluctuate-continually, when legal-review, has not been abolished, by an genetic-nation-(s)-juxtaposition-offset-(s), relevant to those preferences', by legal-nation-(s), because where climate-county-range, b-inserted, [road / residence / destination lot-entity] define means, by payment-methods, deducing from those grammatical-point-(s), labels responsibility-(ies), for every person per thousandth-(s)-word-act-(s), semestral-Fourty-five-commercial-week-day-(s), juxtapose [residential-homework / entity-presence-identification] when cell-work, classify taxonomic-posit-specificity, uncharacterized of organism-(s)-deviation-(s), tangible upon tectonic-plate-feature, when pulse-rate, emit hour-place-posit-perspective, receiving an momentary-data-setting, mattering by how hour-(s) are involved thorough each impasse of [day // night], only prevalent is if an consensus can perceive of such data, concerting means', while developmental-encounter-(ing-(s)), incrementally-deduct, by an one-hundred-year-(s), from birth-juxtaposition-(s), affirmed logic, amidst interpretation {For Example;[108] as I [type // edit] this portion of perennial-aging, the date-hour is {September Ninth, Two-Thousand-Sixteen // 8:32 A.M.}-{North-Miami, Florida} so my birth-transient-hour is in an influctuation of [progressive-hour-incremental-impasse // anticipated-hour-death-hour-presumption] by {April 25, 1991 // 3:42 A.M.} having this hour be at, Hour 219,679 continuum, apose one-hundred-year-age-denominative at, 876,600;[109] intervalizing life-span-expectancy-continuum, by 656,321, hour-(s)-expected-remaining construed when continuum affect how perspective is conducted, because progress of secondth(-s), enhance word-rationale, by syntax-synapse-(s), focusing simultaneous-offset-(s), from where popu-

lation-census-article-factifying, remain without an core common-developmental-schedule, conjuring-feature-(s) of routine, semester-(s) through year, undictated, in cycle-progressive-time-posit-(s), underscoring age, with simultaneous-offset-(s) of dating [1book // 2motion-(s) // 3sleep // 4eating // 5transiting // 6working // 7interacting-articlizing-reference-impasse-count-(s) //// 8hour-central-mode-syntax-thinking-impasse //// 9minute-(s)-motion-(s)-rate-deviation-(s)-expectancy //// 10secondths-retrograde per cent-pertinent-motion-(s)-output] passing modern-continuum, tiering a the dissolve of men, abided at location-(s), by quasi-year-lapse-formulation-processing, stagnantly-developing, ascertained [constitutional-settlement / resolve of articled-educational-discourse / complete-(ing) of task-(s), intervalizing dissent of population-housing-offset-(s), upon scientific-observing;110 ordering from table-(s), task / errand / objective / effort / cashier / financial-officer / stock, by efforted-motion-(s), directed // concerted] if an genetic-representation, denominate those enumerated-numerative-product-(s), quantifying commercial-activity, by-an over-production, of indirect-cause-(s), effecting [expiry // shelf-life // stock // transportation // production] over-lapping, perceptive-balancing, of [economic-disparity // turmoil] convincing inhabitant-(s), unrepresented by Federal-trade, promised-trust-(s), for acquiring, limited-holding-(s), stressing issue-(s), in lieu of hospital-repopulating of mass-population-(s), discoursing ecological-county-limit-activity-(ies), requisite formal-participation, upon those terrestrial-boundary-(ies), uncharacterized when continental-taxonomic-shelf-class-ification-(s), are unperceivable by posit-(s) at still recognition of such [modification-(s) // formulation-(s)]44L. of elemental-resourcing-(s)...¶|
||03m.:42s.81m.s.|||§

§...Count-[19 / Nineteen]-preposition-of.. ...Article-Essay-[7 / Seven].. // ..Book-Essay-[53 / Fifty-Three].. ..Lines-[44.19 / Fourty-Four.1.one-tenths-2.nine-hundredths].. ..Timing is decaying awhile dating is living preserved, how else to conjure visual-images than, using ones' personal-vocal-tone, (offset) other living *functional-organisms.¶|||14s.79m.s.|||§

¶[54]As I proceed though this book, I fully expect to make those formal-adjustment-(s), accounting a the denominative-etymological-year-hour-(s)-notion-(s)-extent-(s), labeling each particular-inclination-(s), set on [explaining // defining] how existence can not have existed more than Five-Hundred-Year-(s).111¶|||13s.91m.s.|||§ Extraspective-observational-discourse, iterate notion-(s), when age-simultaneous-offset-(s), label observation-(s), partaken in affirming frequent pass-(ing-(s)) of [genetic // bastardized-national-inhabitant-(s)] awhile moment-(s)-(um) [rotate // revolve] as mass-public-road-interdialogue-effect, establishing historical-civilization-(s)-tectonic-galaxy, interweaved each and every governmental-accounting-specimen, by an denominating-influctuation curb, consumeristically-expenditured, ordered adherence by-of municipal-code, after incurrence of instance, be maintained purpose, cause introverted mathematical-momentary-observing, not beat musical-note, denoting as instrument-tone, through pressure-gauge, winded those walled-limit-(s), insetted, direct population, from place to place, sustaining no pre-agenda-schedule, for determining those spacing-(s) impassed during time-rotating-day-night-offset-(s) upon dating-revolving-impasse, resetting rotated-hour-(s), apose revolving-dating;111 influctuating mass-trade, accessing [objective-(s) // substantial-(s) // comestible-(s)] from basic-human-right-(s)-indication-(s), embodying power, as the [exterior / interior] time-posit-(s) influencing subjective-governmental-article-(s), dispatched by-the path of common-human-effort, never to interpose thinking when greater-rural-existing, plot through [residential / commercial / public / private-domain-(s)] posturing-surrounding-surface-area, of public-access, for encountering parameter-(s), circulating radical-organized-formation-(s), by [length // width-perceivable-perimeter-parameter-two-dimensional-distance-spacing(s)] sequential of habit, not concurrent table-self, accumulating-fact-(s), relevant legislational-psychological-sociological-annetational-mathematical-articling-(s)-documenting, deducing tangent-(s) of continuum-incremental-data, superlative syntax-redact-written-refinement, as why [tonal-vocal-pitch-day-hour-subterior-minute-(s)-secondths-elapse-(s) // synapse-(s)-frequency-(ies)],| 271 |deviate [word / letter] count-(s), independently understood, mathematical-annotational-articling, (un)-congruent, where data proportion population-(s), apose universal-pop-ulation-common-sense-inquiry, requiring person-(s) to factor base perspective-(s), upon exterior-dating, in temperate-adjustment, contin-ental-shelf-statehood-territory-claim, adapting lifestyle upon condition-(s) of land, abstract-ambiguity, possessive-claim-(s'), possessive subjected-objective, archaic, primordial-conjuring, our Founding-Father-(s), initial-disbanding of monarchy, for an free-empire, Nouveau Sans Frontieres, human-belief, for living by an motivation for basic-necessity-(ies)-soly.112¶|||02m.:09s.25m.s.|||§ Aspecting from this method, have how conscious-individual-order, factor (un)-schedulable-off-time-prerogative-(s), when self-individual-monetary-interval-(s), centralize living-abode, where solid-place-(ing), line those pocket-(s) of men, uninterested in (hyper) tense-prevalence, to ascertain a the personal-appeal of extraspective-location-observation-constant-conversing, inclined those tense-(s) of other-citizen-(s)', learning from eras-period-(s), [Slavery / Holocaust / Western-wars-dispositional-reference-(ing) from-of-those article-(s) of religious-ordering-(s), powering place / person-citizen-dynamic / economic-resource-genetic-century-year-decade-reasoning-(s), in fifth-(s)-reconstructive-direction-(s)]| 272 |in case of regenerating-genetic-identity-mass, by an intellectual-examination-(s), upkeeping timbre, discussing, identity of schedule, upon date-effect-time-deviation-(s), through an [season-denominating / month-denominating] those week-day-passing-(s) of [hour-(s)-interval-(s) / minute-(s)-interval-(s) / secondth-(s)-interval-(s)] calculating calibre, by act-rate-individual-effort-(s)-intrication, spacing surface-acre, in comparison-(s) [feet / yard / meter / acre-measuring-(s)] as a the goal-(s) of aging, assessing those fact-(s) of discourse, by [measure-aligning // mile-distancing-(s)] defining time-ordering-(s), as the attempt to instill-formal-subjective-motion-(s)-intrication-(s), when conver-sive-discourse, elapse so rapid, juxtaposed syntax-impasse, for synapsing read-article-(s), in congruent-fashion, working from a the ambiguity-simultaneous-impasse-effect, to centralize an tens-millionth-mass-population-perspective, intent on an genetic-reformation-procession-(s), merited upon ordering at all moment-(s) of existence, effort-(s) to fulfill lifestyle, from year, upon dating, through hour-(s), acquainting minute-(s) by an gamma per impasse, affirming each [indication-(s) /// inclination-(s) /// innuendo-(s)] of aging-notion-(s), for intrication,50L. to apply in ever becoming alive for [one-million-hour-(s) // One-Hundred-Fourteen-Year-(s) & Fifty-Six-day-(s) / Four-Prime-Meridian-Exact]...¶|||04m.:02s.22m.s.|||§

§...Count-[21 / Twenty-One]-preposition-of..| 273 |...Article-Essay-[8 / Eight].. // ..Book-Essay-[54 / Fifty-Four].. ..Lines-[51.3789 / Fifty-One.1.three-tenths-2.seven-hundredths-3.eight-thousandths-4.nine-tens-thousandths].. ..Posterior our spine pose, apose anterior-torso, motioning physiologically, at an cardiovascular-millisecondths-tension-respiration-rate, yet tangible upon measuring scaled-tectonic-air-impasse-inclinated-notions.¶|||20s.56m.s.|||§

¶55-For example, while I [write // type] no tangible-belief, correct visual-perspective, complementary as inhibition, seeing from prose-less-conjure, solar-impact, for guideline-(s), directing individuals of mass, living by contrary-community, physical-conjure, without fluently using [1term-(s) / 2time / 3definition / 4condition / 5pattern-(s) // 6motion-(s)-synapse-ordering-(s) // 7map-(s) // 8dictionary // 9book-(s) /// 10determinable-document-directive-(s)] temporal-sensation, amid environment as interval-lapse-(s), evoking an form of thinking meriting Sport-(s) & Lotto-Bingo-luck-continuum, greater-than book-written-typed-context-hard-cover-purchasing, binding legion-(s) of independent-populant-individuals'), when-as-of reset-galaxy-universe-planetary-system, offset-residential-city-limit-(s), holy by perspective set interior, conscious-metropolitan-square-perimeter-fencing-lease-mentality, festering amidst the individualism-dynamic, upon decade-century-dynamic, falsifying-millennium-clause-count, that exist in an year-millennium-memory-assumption-perspective, from life-time-elapse, that require-focus of [woman-primary-inhabitant // man-primary-citizen] as occupant by-of planned-designated-housing, in free-will-decision-specimen-(s), by the national-merit, overemphasized, from vocational-degree-(s), preferred by coffee drinking-off-time, not [tasting // seeing] display by solitary, impasse calendar-dating, affirming word-multiple-simultaneous, not examined, from states of biological-memory, existing from requisite past-data, not aware for future-present offspring-data-expanding, ones' physiological-effects', upon continental-shelf-existence.[113] ¶|||01m.:31s.[04m.s.|||]§ Style exist upon the physiological-present, moving upon memory, associative-motion-(s), awhile present-tense, apose Cent & Age, from past-requisite-letter-word-concept-notion-(ing), timing motion-(s), for work on motion-analyzation-(s), unordered by text-documenting, variablizing elemental-extraction-(s), incandescently from-of plasma-combustion-kinetic-energy-origin-present-offset-density-matter-mass, denominating comparison-(s), by measuring objective-(s), surrounding an [plane // cubical-interior-spacing-measure-(s)], for cubical-space, feet per cubic-feet analysis, where entity-(ies)-development, have those piece-(s) of denominating, upon latitude-degree-equator, as longitude-parcel-placing-(s), distance awhile land pertain under, lunar-equivalent-denominative-matter,| 274 |untangible-mass-density-matter-(s), unconfirmed from [writing // reading observation-(s)] by perennial present-inhabitant-(s), which bemuse passing existence, unthought for why mass-group-method-(s), are implemented, because of how grandiose physic-(s) remain, apose [physiological-self // population-(s)] when requiring an composite-bone-(s)-weight-density, reasoning year, by [self // directive-(ed)-agenda-action-(s)] denominating elemental-weights-extraction, comparative, factory-production-excess-cutoff, by then packaging-data, interval-historical-resourcing-content, intended for mass-individual-objective-preference-abstraction, not denominating an numerative-degree, upon rationing of greater-numerative-individual-degree-(s)-denominating, by ruling amendment-article-predicate, factoring [currency-note-(s) // cent-(s)] from family-mass-population-deviation-(s), existing by tectonic-divide-(s), ranging [statehood-(nation) // county (state) // city (county) // city-district (city) // religious-church-house-(s) (principality-(ies))] restricted from how stance upon reproductive-dynamic, rely counterpart, intermingling, dynamic-(s) of [Man / Woman / Child-/-(ren)] that are in an conjunction by [race-(s) / ethnicity-(ies) / nation-(s)] not denominating illiterate-retarded-matter-(s), that remain (un)-emphasized, present, as modern-future, clarify matter-(s), by an etymological-ontological-thinking, defying how offspring, enhance such experience, by those guideline-(s), unparticular to those individual-(s'), whom determine action-(s), by planning project-(s), in light of governmental-development, because their not exist offset-English-language-(s), from mass-impasse, that define cultivating of continental-shelf, [top-soil // rock-(s) // sand // deposit-(s) of matter] embellishing how life is to plan by hourly-focus, from those minute-(s), that exaggerate what can be done by requisite-mass-survey-planning-inquiry, from notion-(s) of methodical-territory-excavating-living-production-planning.$_{114}$¶|||03m.:49s.$_{89m.s.|||}$§ The language English;$_{112}$ has an religious-passing, set (from an) approximate out of body experience, where trade-post-civilization-(s), yet focus [present-elapse-(s) // synapse-(s)] from youth-educational-physiological-developing-comprehension, scheduling limit-(s) of physio-logical-limited-boundary-(ies), upon live-present-observational-discourse, existing from those mundane observation-(s) [[1]Sun / [2]Moon / [3]Planet / [4]air / [5]ground / [6]fire / [7]tree / [8]space / [9]human] as fundamental-impasse-(s), when literary-intrication, not equate [solar-fission // lunar-orbit // tectonic-plate-(s) // ocean-body-(ies)-distance-surface-area-(s)] inanimate, from cement-tile-(s), explosion-(s) from butane-flint-flame, amongst thickets of forset-surface-area-length-width-dimension-boundary-(ies), when information remain an part for how conjuring of language, influence literary-reading-(s), but when writing remain uncommon, for how to instill motivation-(s), for inhabitant-focus, upon each impasse of daily-photosynthesis, for organism-extraction-(s), intermittent night settling of territorial-space-(s).$_{115}$¶|||04m.:43s.$_{43m.s.|||}$§ Respite freely wish-(es), the language riddle, not yet understanding from mass-perspective-civilization-exponent-deviation-(s), independently pertained political-matter-(s), from [Federal // State-constitutional-order-(s)] never protesting in mass-population-(s), about how to conduct those affairs of mass, by an governmental-system, influctuated, (lunar-two-hundred-mile-clearance-limit) territory-offset, those [density / substantial / productive, matter-(s)-production-maintaining-mean-(s)] provisioning mass-population-fundamental-requirement-(s), demanding common-understanding of those Spacial-Measure-(s) & Spacial-Acre-Boundary-Entity-Count-Dynamic-(s) defining physical-blood-pulse-beat-moment-(s), tense-(d) by worded-competency, unled abound parameter-(s), of differentiated-weight-(s), [measuring // distancing] entity-interval-placing-(s), for entity-developmental-progression-(s), congruent, balancing compositional-matter-(s), in form of unit-(s), requiring balancing of such object-(s), apose parameter-three-dimensional-cubic-feet-spacing-(s), anti-cipating engineering, for reasoning, construct-(s) of [sea-border / mountain-contour] at matter from the daily-impasse-(s), basing from substantial-perspective-(s), emitting [blood-pulse // sweat] intermittent kinetic-solar-molten-vapor-impact, effervesantly-invisible, clear of direct-recognition, while surrounding-concertion-(s) of [rubber // metal] isolate such friction, as pressure-gauge-comprehending of such matter.$_{116}$¶|||05m.:54s.$_{09m.s.|||}$§ Library-reference, is not medical-subterior-auto-biographical-ex-amining;$_{113}$ coronating average development, from extraspective-community-collecting-examination-(s), periodical technological-development-(s), which deviate when, year denominate season-(s), commercially-progressive, without an maxim-potential-outlook, from requisite-experience, pertaining trade of volume-unit-(s), by population-psychological-inquiry-dating-time-cycling-(s), streamlining those factory-production-(s), from elemental-weight-(s)-resourcing-(s), timing those factor-(s), from concepting how to construe an corporate-schedule-(s), based on an twenty-four-hour-cycling, of [human-trading // objective-production-(s)], as when commercial-development, should base (from an) variation-(s) of inhabitant-participation-(s), working on an national-agenda-outlook, because how development, perform by voluntary-trading, because of how bias affect those impasse-(s) of perspective, when applicable, from direct-analyzation-(s) of trading, segregating an preferred-genetic-mass, through such commercial-mean-(s), and paying whatever [governmental-fine-(s) // tax-(es)] implemented, to affect such outcome, when nationalism, would be best by commercial-effort-(s), simply around the corner, of fined-worth, because of a the working environment-(s), that pertain upon an confidence-development-production-(s), pertaining stock-market-investment-(s), awhile each piece of trade, would be well to conceive for an participating mass (weather [Blonde-hair-blue-eye-pink-skin-nationalism // Blue-eyed-rosy-yellow-skin / et cetera by white characteristics], working towards becoming blonde-hair-blue-eyed-pink-skin;[114] attempting to find inhabitant-(s') from U.S.A.-primarily, and secondary Europe & Latin America);[115] creating an Walmart-(esce)-corporation, incorporating stock-market, having all voluntary-national-participant-(s) (if ever Surveyed & Mundanely agreed), investing in to a the stock-market, Saving & Investing, an aspect of mixed-block-development, fixated, (by trading on an) standard-parallel per American-Free-Trade, basing motive-incentive-trade, not on [chief-executive-officer // branch // store (assistant // department) // regional-manager-(s)-economic-preference] corporate-dynamic [collecting /// resourcing /// relocating-community-(ies) of genetic-petition-survey-population-(s)] (from an) communicating of national-prevalence, focusing on an initial rural-market, for then constructing community-boundary-(ies), by an purpose of nationalism-transition-(s), mundanely differing inhabitant-(s'), (from an) marketplace, upon either [common Latin // black-ethnic-racial-background] focusing sale-(s), by customer-ambient, for an [white // pink] purchasing-center, segregating (not racially profiling) those inhabitant-(s) by commercial-religious-company-clause, having how each inhabitant of ethnic-background, affect white community-trade, as for how the price-(s) are an sliver higher than those of cheap-market-scheme, for diverting-traffic, and not permitting those inhabitant-(s) of color, to affect sales-flow, by an utopian-sale-confidence;[116] swearing on an 0.00% loss-prevention-sales-dynamic, from segregational-picture-non-permitted-customer-performance, differing those ethnic-customer-(s), or profiling them as theft-prone-poverty-inhabitant-(s), while also working on legislational-loop-hole, of civil-right-(s)-act of 1964, by affirming the national-identity-aspect-(s) of trade, not ethnic-profiling or segregation, but juxtaposed those national-commercial-market-outlet-(s), that have an ethnic-congestion, and abuse their right-(s), for not working on an national-constitutional-disbanding, in role of national-identity, not segregation-commercial-stagnation, as well as using non-fiction-context-literature & methodical-survey, to offset those fact-(s) of discrimination, swearing on copyright-context-usage (my copyright-(s)-impasse directive-intent, rather than visceral-feeling, from slave-encampment, to divert those inhabitant-(s')-preference-(s), of trade;[117] isolating focus of a the 1964 Act, by Title VII Paragraph 2, stating "In very narrowly defined situations, an employer is permitted to discriminate on the basis of a protected trait, where the trait is a bona fide occupational qualification (BFOQ) reasonably necessary to the normal operation of that particular business or enterprise.[117]¶||| 09m.:37s.[32m.s.|||]§ To prove the bona fide occupational qualifications defense, an employer must prove three elements: a direct relationship between the protected trait and the ability to perform the duties of the job, the BFOQ relates to the "essence" or "central mission of the employer's business", and there is no less-restrictive or reasonable alternative (United Automobile Workers v. Johnson Controls,

Inc., 499 U.S. 187 (1991) 111 S.Ct. 1196). The Bona Fide Occupational Qualification exception is an extremely narrow exception to the general prohibition of discrimination based on protected traits (Douthard v. Rawlinson, 433 U.S. 321 (1977) 97 Sc.D.. 2720).118¶|||10m.:21s.85m.s.|||§ An employer or customer's preference for an individual of a particular religion is not sufficient to establish a Bona Fide Occupational Qualification (Equal Employment Opportunity Commission v. Kamehameha School — Bishop Estate, 990 F.2d 458 (9th Cir. 1993))" ;118 focusing on transitioning trade from various corporate-outlet-(s), in to an company-developing, by an population-customer-regular-interacting, meriting an millionths-population, for grounds of company-dynamic, not segregation from when there are numerous [national // regional] trade-outlet-(s), for an similar-product-(s), not affecting a the total-trade-market-dynamic, while working on an [Twenty // Twenty-Five] million-white-blue-eyed-trading-population, on an national-scale, but swearing upon an interior-nationalism-effort, to constitutionally-transition an national-impact, as a the company-policy, for governmental-awareness, in transition of market-scheme, and Cleaner & Better, ambient-premise-(s), for conducting trade of an upper-middle-class-impact, for separating from religious-lies, and working on an company, based upon an boutique-effect, of trade, by renovating factory-production, by Investment-Effort-(s) & Operation-(s)-Dynamic, having no individual profit in to a the million-(s) per year, from trade, focused on factory-production-development, as how to keep the government off company trading impact, in other word-(s), if one-million-inhabitant-(s), are purchasing at $100, per customer, per season, then the C.E.O. makes proportional, an hundreds-thousandths-dollar-rate, tiering an simultaneous-hour-(s)-responsibility, when the one-million-dollar-(s) per year dynamic is achieved, by an ten-million-card-member-market, efficiently-producing those need-(s) of an white-customer-service.119¶|||11m.:56s.39m.s.|||§ (I think college sports, have dissuaded humans-thinking, from understanding governmental-law, for discourse of a the Civil-Right-(s)-Act of 1964, not seeing how Title-7,141L. apply by an commercial-proportional-observation-(s) of Market-Trade & National-Transition, as from how our Founding Fathers', had acceded from The British-Union...¶ |||12m.:12s.73m.s.|||§

§...Count-[68 / Sixty-Eight]-preposition-of.. ...Article-Essay-[9 / Nine].. // ..Book-Essay-[55 / Fifty-Five].. ..Lines-[142.31 / One-Hundred-Fourty-Two.1.three-tenths-2.one-hundredths].. ..I script this book for the reader to understand why an long term entity would be required, reinvigorizing why reading such context is important to circulating an ecological-terrain, by commercial-wage-agglomeration-(s), when dating past grandiose-elemental-objective-(s) for lifestyle-comfort-(s).¶|||24s.68m.s.|||§

¶56-[Server / Accountant / Intern / Nurse / Apprentice / Subservient], not exist intermittent of one another by work-place-conducted-affair-(s).120¶|||07s.19m.s.|||§ Why is sociology the observation yet psychological? This is the purpose of social-scrutiny, contradicting free-will-action-(s), for insight-(s) of psychological-scheduling, when free-will partake in psychotherapy, yet rationale, for conceived-coher-ent-fluent-comprehended-apprehended-interpreted-understood-grasped-iterated-conjured-apparent-proficient-visual-hearing-vibrational-thinking-proportional-indicated-inclinated-imagery-interpose-documenting;119 as cognitive-offset, reflect cogito ergo sum-mental-ity-belief.121¶|||36s.81m.s.|||§ Yield may never be comprehended, juxtaposed those effect-(s) of aging, [uncollected // unsaved] for constitutional-territory & transit-crop-objective, which impact our perspective-(s), when surrounding construction, similar-model era, equipped by limit-(s), emphasize boundary-(ies), covered (from an) literary-median-prognosis, expected by-an [mass-non-fiction-cursive // manuscript-written-confirmation-(s)], from a the way human-cycling, process by governmental-default-biological-means', [partial / possessive] beyond comparable-reason, when extraspective-community, recluse to an introverted-family-standard, apparent through Constitutional-Spacing-(s) & Self, founding grounds, extemporaneous, objective-piecing-inferencing, Time-Intervalization & Date-Time-Ordering-(s), developing an extra-effort, vague by population-agenda-scheduling, for mundane-order, define logic, [barren // listless // blank // inept] upon future retirement-advertisement-commercial-flow-entertainment, post age Sixty-two-year-file-ordering, for capable-act-(s), in still-adult-requisite, idol for of those effect-(s), of solar-parallel, juxtaposed [winter-summer-solstice //// spring-autumn-equinox-celestial-tectonic-rotation-cycle-(s)] perpendicular sequential-order, cause parallel naming, matter by the observation-simultaneous-elapse-ordering-(s),| 275 |for practice greater than study;$_{120}$ monolithic, commercial-monotheistic-religious-view, product-crafted, from commercial-expert-collective, designated at particular-limit-(s), emitting light, requisite general-visual-presence, for objective-purpose, in elder-prerogative-reset-anticipation-(s), that lead present from past-requisite, for styling [eating / clothing-method / sequential-routine] of work-mode, maintaining-housing;$_{121}$ following the nature of belief, instilled in axis, ontological perspective, yet asking from self, amongst other-(s), cause for census, their of period-(s) savvy belief in subjunctive–peripheral-view, acted in [of / interior] mundane-theatrical-role, driven by extent, from exterior-boundary-(ies), limited by the county to county two-degree-circumference-mile-(s)-boundary-limit-(s), of family-offspring-reproduction, delivering pass [[1]lane / [2]road / [3]alley / [4]court / [5]street / [6]avenue / [7]boulevard / [8]place] congesting passing memmorium, virtuing-value, by national-bastardization [concept / hypothesis] transiting auxiliary-intuition, by glory of perceivable-sensation-(s), not in written-thought, simultaneous of peer-(s), determining circumstance, by an extension of dialogue-discourse, amongst [genetic / ethnic / race / populated-locale-(s), of unranged territory] not constitutionally-proportional, since The Louisiana-Purchase;$_{122}$ ranging county-(ies), when planned-development, intersect-parameter-diameter-(s), revolving-centripetal-infinitive-coach-locale, thinking from voluntary-belief, setting free-will, by local-point-impassing-(s)-motion-(s), as planned-genetic-viewable-bias.$_{122}$¶|||03m.:00s.$_{03m.s..|||}$§ [To / from] consume parental-growth, deciding by currency, tax-deviation-(s), their of human-behavior, cause having stem (from an) sacrilegious-mass-documentation-deviation-(s), yearn by idle-belief, susceptible to none-else, yet vernacularly-coherent, fluidly-marking, conclusive-sedentary-paging-(s), adjoining creative-alliteration-proficient-literature, for appeasing tense-(s), explicitly conveying participating-reader, to enjoy the discourse of work upon monthly-reading-(s), from [weekend // week-impasse-incremental-paging-(s) // paragraph-reading-(s)] per entry, to define the dynamic of [Physiology / Subjective-Coursing] from effort-(s) of national-preference, or disinterest of non-fiction, as much of life exist in an fog, for ambervision-effort-(s), [parallel // perpendicular] relevant-focus, upon article-objective [sedentary // motion-(s)-focus-(es)] instilling reasoning through all motion-(s) we act upon.$_{123}$¶|||03m.:45s.$_{50m.s.|||}$§ Autonomous present, is hour-centripetal-day, upon month logic, (un)-incorporated of subjunct-form-(s), where defining live-weight-scale-data, persist past-data-collecting, for finalizing the classroom-purpose, reordering through those hall-(s) of [private-commercial // housing-sector-(s)] as the way of Reconstruction & Reproduction of Offspring, from how a the celestial-offset, involve festive-motivation-(s), apposing tense-(s), as the physiological-take, intermittent sat entity work-cause, [thought / endeavor / present-life / monetary-denominating-matters], by Formation & Storing Data-Information communally typed, of postunned-inteligence, through {I.S.B.N.} International Standard Business Number {subjective-commercialism};$_{123}$ {L.C.C.N.} Library Congressional Control Number {govern-mental-oversight};$_{124}$ clasping objective, by the limit of influence, as how entertainment settle, upon read, and consecutive-serie-(s), as my route of analyzation upon existence, balancing weather human-intelligence exist, or weather human-life remain far too idle, to value even trillionths-dollar-debt.$_{124}$¶|||04m.:39s.$_{84m.s.|||}$§ Interceding diatribe, can not influence common-mass, for self-hyperactive-nature, not stillfully thinking, from (hyper)-tense-(s) upon objective, ensuing from census-accounting, deviated, [government-labor // elemental-resourcing] apose an multitude of other denominating-(s), from-of citizen-social-timbre, upon various physiological-demand-(s), | 276 |set by default-push,61L. upon terrestrial-day-rotation-revolution-drift... ¶|||05m.:01s.15m.s.|||§

> §...Count-[30 / Thirty]-preposition-of.. ..Article-Essay-[10 / Ten].. //..Book-Essay-[56 / Fifty-Six].. ..Lines-[61.1167 / Sixty-One.1.one-tenths-2.one-hundredths-3.six-thousandths-4.seven-tens-thousandths].. ..Formal etiquette, have yet found an forum for genetic-dialogue, due to educational-reclusion of homework, by religious-commercial-function, practicing filing as an technique for linguistics-skills, topic-basis, for endelving upon deductive-redact-article-(s), from each paper-deviation-fiat, amid colloquial-living.¶|||23s.56m.s.|||§

¶57Hello how are you? Response, pause;125 for how nature of physicality remain, the human-genome, has merely learned that cordialities, have an impact of those passing-(s), bastardized applied from-focus from the millennium-effect, rating cent-(s), intermittent, tectonic-nation-articling, denominating receipt-(s) of purchase-(s), by flat-rate-stock, per dollar, yet adjourned by an mass-group, from individual-family-religious-effect-(s), estranged their of city-commercial-trade-(s), merit-(ing) past-action-(s), during present day, gossiped in to level-(s) of story-(ies), remaining barred by approximate surrounding-(s), sufficing night-sleep-revolution-effect-(s), resetting hour-(s) of day, by the aging-process, impacted, from how hour-(s) of unconscious-cycling, complacent-sedentary-effort-(s), from how vague-referencing, materialize objection, persist sedentary posit-(s), at daily-parameter-impasse, ensued hyper-tense-blood-pulse-soft-friction-sensation-(s), rhythmed by (mute) voice(-d)-interpreted-observation-(s), configured where tangible-(s), form physiological-stature-weight-minor-offset-(s), comprehending how extent of existence, formulate human-physiological-transient-being-(s), better understanding the nature of (un) exponent-reproduction, where market-structuring-(s), are supposed to be strategized, campaigning influential-marking-(s) by statehood-tectonic-continental-shelf-national-posit-(s);[126] apose eastern-tectonic-offset-(s), demonstrating surface-area, [to / from] individual-work-(er)-(s), in faith of routine, demonstrated when perspective of matter-(s), commonly-receive, an non conceptive-product, as dynamic of commercial-market, where at hand, accompanied through rate-(s) of past-year-decade-century influence-(s), classroom-schedule-(ing), deter [mass-race-(s) / ethnicity-(cies)] upon modern-state-trading, placing the genetic-diplomatic-affair-(s), by an new frame of mind, grading Term-(s) & Condition-(s),| 277 |upon territorial-governmental-expanse-coverage, publically-domained, by private-commercial-offset-(s), defining length-width-surface-area, by spacing per rated-word, when omitted literature, is instrumental, by sport-complex-ideological-system-(s) of university-ecology, forgetting how to juxtapose, syntax-tangent-compartmentalized [Term-(s) // Definition-(s)] not accounted from the largo-ambient-default [memory / time-secondths] per-of minutes-passing-(s), reducing incremented-impasse-obser-vation at calendar-point-posit-(s), without an historical-purpose, in to present-day (major / minor-subterior-tier-reasoning-(s)),| 278 |[elapsing // synapsing] revolutionary-cycling-(s), that proceed awhile present-day-secondth-(s), ground how [sound / sight / vocalized-vernacular-vibration-(s)] define dated-affairs, upon an gathering-continuum, collecting resourced-elemental-tangible-(s), comprehended by staffing-extraspection, where those setting-(s) by an difficult-environment, modify, (from an) incur by video-minded-static-curenting-pulsed-inhabitant-(s), whom bear offspring in an ambiguous-fashion, from hospital, differing thinking of each pulsed-form, for interacting, from self, for sufficing mere-condition-(s), at an surrounding territory-topographic-environment;[127] aging at workplace-reasoning, locally-comfortable, by fraction-numerative-motion-(s), impassing timed-hour-(s), when day-mode affix-hour-(s), from the mentality of being-(s), pertaining limit-(s) of life, plentiful [whole / integer] margin-conglomerate-unit-(s), ensuing, proportional [measure-(s) // weighing-(s)] fatigued by matter, which is [Supply-(ed) & Demand-(ed) (national) / Good-(s) & Service-(s) (northern-mentality) / Gun-(s) & Butter (southern-mentality)] equivalent-comparative, cognitive-focus, living measured, as parameter-placing-(s), from property-objective-(s), having thing-(s) be there, not in splendid specified-documenting, entailing, posit-(s) of natural-observation-(s), of grand-matter, understanding topic-(s) of [transit / forest / tree / mammal / housing / social-service-(s) / public-decorum / etcetera...] not formulating measure of lot-spacing-(s), outside discourse of year-(s), patterning favor, not by an monthly-denominative-interval, apose day-week-dynamic, when [hour-(s) / minute-(s) / secondth-(s)], [to / though] voluntary-discourse, log different extent-(s), from [motion-(s) // action-(s) // command-(s) // instruction-(s) // demand-(s) // order-(s) // contemporary-piece-(s) // objective-deterioration-(s) // machinery-life-span // book // comestible-(s)-shelf-life // disposable-(s) // et cetera...], mass-tangible-denominative-accounting, noting of periodical-age-(s), influctuated various [hour-(s) // minute-(s)], to [wake / sleep / eat / dispel / walk / collect / prepare / trade / appraise / think / interact / transit / study] default-factor-(s),48L. by-of territorial-development, methodizing existence, from perceived-limit-(s), of a the overwhelming-effect-(s), of surface-measure-cultivation... ¶||||04m.: 26s.63m.s.|||§

§...Count-[28 / Twenty-Eight]-preposition-of.. ..Article-Essay-[11 / Eleven].. //..Book-Essay-[57 / Fifty-Seven].. ..Lines-[49.46 / Fourty-Nine.1.four-tenths-2.six-hundredths].. ..deposits of sedimentary rock, layer tectonic compositions of terrain, for mechanical-engineer-ing compositional-extraction, amid an revolving title-trade-forces, that demonstrate how individuals are to be supplemented refined compositional-terrain, unto mechanical-engineered-factory-refinement, due to commercial-market-traded-means, produced per box-units, relevant upon circulated-seasonal-yearly-consumer-purchasing-parity, per metropolitan-capitia, at-will from wage, currenting each variable upon free-will-individual-wage-hoc-circulation. ¶|||33s. 68m.s.|||§

¶58Colloquial-Verbial-Act-(s), base from human-need, those objective-possession-(s), thus vindicated, default-governmental-religious-commercial-pressure-(s)-act-(s), upon humanistic-existence, loyal to individualism, not comprehending, why there is an European-pressure, when the dynamic of freedom, insist an psuedo-extraspection of mutual-passing-inhabitant-(s), disregarding purpose of non-fiction-literature, analyzing objective-jubilee, where mass not instantaneously-perceive, the constitutional-mass, extraspective-interaction-dynamic, haphazard-ly-studying, from [bible / textbook / historical-context-lessoning] uninquisitive Vehicle-Itinerant & Schedule, furthering phenomenal-effort-(s), where surrounding-surface-area-spacing-(s), collect by an objective, matter, for an extraspective-mean-(s), practicing standards, from how Federal-Reserve & Competency, align by an ignorance of homework, for romanticism, unstill, to focus on study-(ies), greater than, sported-motion-(s).125¶|||50s.54m.s.|||§ Remaining in vehicle, is powerful, in keeping up the morale of inhabitant-(s), due to those idle-inclination-(s), unaware, theatrical-interlay, which render the fanatic-mentality-mass, idle, askew tangible-(s), as how engineer-worshiping-approach,| 279 |show how literature has been dull, preferring [spectacle // stands // fanatic-ideal-(s)] loyal to idling, not contorting how existence is suppose to observed objective-collection-(s) of matter, discoursing applicated unit-(s) per volume, rendered unclear, for after purchase usage, when contract-data, periodical commercial-parameter-(s), inset where objective, with governmental-legal-filing-(s), being upon the extraspective-default, commercially-required, mathematical-maneuvering, each interlaying piecing-(s) of hard-data, conjunct, paper-comprehension-subjunction, not intuitively conjured, weighing, in comparison, other exterior-value-(s), administering directive-(s), by class-staffed, routine-continuum, executing act-(s), still lost from mass-intuition, because literature, upon legislative-issue-delving, receive feedback, when discourse of human-event-(s), see the progressive conceptive-legislative-ecological-bastardized-genetic-mass, handling matter-(s) of production, by setting-place of affair-(s),| 280 | upon envisioning an preferred-routine, by housing-approximate-progression-(s), subdued upon state, for no appreciating those interaction-(s) of characterized-blood-pulse-being-individual-(s), rating interaction-(s), ensued scale of hour-time-posit-(s), denoting [mathematical / Grammatical subjective-tier-ordering-focus]26L. by those relevant issue-(s), defining prerogative through passing observation-(s), where perspective-pertain, apposed the interior-view, labeling each subterior-notion-(s), by an empirical-relevance, affirming from each [person // being] passing-purpose, upon objective-ordering, invigorating-existence, in degree-order, of revolving-force... ¶|||02m.:28s.83m.s.|||§

§...Count-[13 / Thirteen]-preposition-of.. ..Article-Essay-[12 / Twelve].. // ..Book-Essay-[58 / Fifty-Eight].. ..Lines-[29.12 / Twenty-Nine.1.one-tenths-two-hundredths] ..Retirement come as an reminder for genetic-redevelopment, yet individuals demonstrate an evident laziness, not cooperating on developing ecological-geological-terrain, as an proportion per factor-(s) understanding how consumerism is reasoned by daily-developments, ordering how an entity-product-dynamic, exist from shelf-life, rationed apose ecological-transit-circulating objective-means.¶|||21s.78m.s.|||§

¶59What time will (not) consist calendar-date-counting, cycling human ordered-routine, microcosm minor-studies-effect-(s), subterior individual-accounting, affirming work-presence, where subjunct-processes, undervalue familied-governmental-entity-(ies), thought up, from only basic-human-right-(s), for no steps-progression-directive, persist upon mathematical-prose, or canvas-painting-labeling, by an [magistrate / monarch-social-intrigue] entailing parameter-(s), (from an) stadium-access, confirming duty-(ies) by service-(s)-obligation, when invigoration (not entertainment), interact by common-interaction-perspective, having tense-(s), pulse, without an count of rate, upon life, being defied, when moment-(s) of celestial-latitude-drift, vary upon longitude-posit-(s), where ones'-place, exist amongst, age modifying, in regard of those interval-decade-year-after-life-memoir-(s'), when live-act, code-effort-(s), in lieu of documenting-self-purpose, for how communal-population, tier along an juxtaposition-(s'), when constructive-purpose, enhance imagery, Measured & Distanced, by payment-(s), in view-sound-resonated-scene, setting an daily-acting, thought upon spacial-bound-(s), because schedule-arrange-tense-(s), passing-(s) hour-(s), habitual physical-destination-(s), repeating [when / where] time is pertinent upon hour-denominating, by those secondths of minute-(s)-action-(s), motion for activity-(ies), specifying task-(s), subjunct according perceptive-mean-(s'), when feature(-s) of commercial-production, be through margin-(s) of blood-pulsed-being-work-demand-effort-(s'), intangible by one another, aware that perceptive-limit-(s), instate law,17L. directing an basis of living, by residence, as-why,| 281 |religious-persecution, be due in part of physiological-perceptive, when idle-observation-(s), conceive by, swearing-discourse.¶|||01m.:40s.81m.s.|||§

§...Count-[9 / Nine]-preposition-of.. ...Article-Essay-[13 / Thirteen].. //..Book-Essay-[59 / Fifty-Nine].. ..Lines-[18.5215 / Eighteen.1.five-tenths-2.two-hundredths-3.one-thousandths-4.five-tenths-thousandths].. ..dating is counted by an compositional-dynamic, retrieving exterior-tangibles, upon visual-receptive-sight, vocally-observing how each topic can be produced, from an heard-input-ear-reception, timed through secondths of human-housing-impasse, offset an location-dynamic, providing how each inhabitant is conducted through tensible-effort-(s), at an designated-location, for of volume per unit, amid pounds per unit, measures, at human-coverage-capable-competency, normally commonly uninquisitive for which security is verifiable per individual-family-dynamic, for of offset, variablizing an constant-subjunction, under entity-copyright-(s)-registering-(s), ordering from trade-marked-factory-productions, as market-registering, settle under an heated-climate for common-adequate-living.¶|||50s.70m.s.|||§

¶60Pre-instated, modern-date, does daily-parameter, have an capacity in each passing moment of schedule, applied awhile mutual-purpose, maintain each inhabitant-(s)-discourse, interceded [Water / Power / electric-appliance-(s)] by utilities, that motivate each being, by-an concerted-effort-(s), of mutual-co-worker, when public-ground-(s), take step-(s), during weekly focus, deriving infrastructure, in proximity of lifestyle, for contingently progressing, pass an sequential-impasse, of motion-(s), in an intent, on completing article-focus-(es), because interposed-interpretation-referencing.[126]¶|||28s.50m.s.|||§ Whom is capable of annotating the movement-(s) of an individual, when those secondth-(s) are quite brief, for ordering objective-(s), by fact of placing-(s), in shelf-product-place-(ing-(s)), as the call of command, upon temporal-reaction-(s), discoursed through gastronomic-preparation, discharging excess-comestible-(s), from kinetic condition-(s), restricted by limit-(s), when there are tasted-food-(s), that apply upon our capable-intake, of digestion, better balanced, by an balance of bowel-discharge, apose our compulsive-reaction-(s).[127]¶|||53s.06m.s.|||§ Glossary factor, after context relay of data, post-defining, posit-(s) of researched-information, lessoning the purpose of an subjective-direction-(s), for delving in to vibration-(s)-relay, measure-(d) by note-(s), cycled by frequency-(ies), in various tangent-(s), upon the [cause // case] of order-reference, persuaded from effort-(s)-researched-impassed, directed, upon inquiry by-of [grammatical-punctuation / mathematical-prose / scientific-count-table-labeling / physiological-reaction / exterior-mass-sociological-discourse-lessoning], [series-sequential height / weight / ventricle-system-circumference];[128] working awhile aging-deviation-decay, deter self, from conveying those point-(s) of invigoration, through an routine of work, incrementing scale-(s) of [density-pressure // viscosity // buoncy // air-pressure] from zero-whole-percent-median, radiant one-whole-percent-denominative-deviated-factoring;[129] incrementing simultaneously by-an numerative-decimal-objective-scale-offset-(s), anatomical-tension-(s) apose biological-destination, surveying usage, as a the primary-deviation,| 282 |for mathematical-prose, intermittent elemental-table-characteristic-(s).128¶|||01m.: 51s.37m.s.|||§ Second come, daily-temporal-sequencing, which appear (from an) inset among the prior-cycle, reasoning how minute-(s) of hour-(s)-motion-(s)-impasse, input an biological-age-anticipation-day-count, for case of article-(s), verifying constitutional-fashion, for reference by currency-numerical-notion 8.10213243546576, ordering-article-(s), of paragraph-essay-book-mass-matter-(s)-notion 8.10.21.32.43.54.65.76 (.10 = day-week-discourse-filing)(.21 = day-month-discourse-filing)(.32 = day-season-discourse-filing)(.43 = day-year-discourse-filing)(.54 = day-four-year-discourse-filing)(.65 = day-decade-discourse-filing)(.76 = day-century-discourse-filing);130 filing each ecological-denominative-cycle-(s) of matter;131 recognized upon observational-prose, from substantial, human-aging, comestible-(s), in to solid-tectonic-deposit-(s), upon factory-developed object-(s), discoursed where issue-(s) are from Population-Denominative-Count-Method-Practice & Tectonic-Expanse-Resource-Ordering, delving deeper in to an forms-superfluous, during daily-hour-rate-(s), when awake [sixteen-hour-denominative-elapse-(ing) // synapse-sequence-(s)] differed, when individual-(s') remain rampant of freedom, not understanding a the mathematical-subjective-physics-method, evident by all clear perspective-(s'), but when treasury-characterizing, is not formally ethical, because how development-(s) of those inhabitant-(s), are to work effort-(s) in an [physiological // subjective-parallel] validating existence, by two-median-(s), formal accreditation, by when human-(s) are impassed regular-living, trading of those product-(s), but by individualism, not an national-interaction, persuading an genetic-identity, avidly working to overcome those other ethnic-identity-(ies), defining parameter-development-(s), and enhancing an sociological-class.129¶|||04m.:02s.43m.s.|||§ Filing objective-count-(s), denominate, how we impasse an repetitive-process of Dimension-(s) & Usage, making an majority of a the currency up, from labor-effort-(s), circulating, subjective-observation-(s);[132] [product-objective-(s) / spacing-(s) / substance-(s) / metropolitan-boundary-(ies) / state-constitutional-limit-(s) / mass-analytical-yearly-observation-(s)] of event-discourse, [ecological-discourse / website-.org-.gov-.mil-.com-.net-significant-data-filing / taxonomic-classification-(s) / et cetera...] for non-fiction-inquiry, late or never even of conceivable comprehension, when perspective is dormant of documenting such fact-(s), upon blood-pulse-perspective, when real-observation-limit-(s), exceed physiological-gastronomic-anatomical-non-fiction-reasoning, forgetting internet-non-fiction-book-document-ing, relevant apose mass-survey-competency, by an nature of affair-(s), instilled from mass-prerogative-(s').[130]¶|||04m.: 48s.15m.s.|||§ I intend to test the discourse of such thought, but as for soly myself, I would see that I initiate such turn of culture, or maxim thė notion-plane, upon individual-prose, remaining subservient, contrite-skepticism, for my critique be only from "what-else-incur", [furthering // stagnant] constitutional-limit-(s), tier-periodical-reordering, ecological-origin-continuum-input, resetting means, where of an existing-presence, among objective-product-trading.[131]¶|||05m.:13s.46m.s.|||§ Geological-scaling I do not fully believe be ordered correctly, I prefer to disagree with much of the dating before the year {1536};[133] (for the) solar-fission-impact, we exist, breath-interior-cartilage-reflexive, (by an) contingent-order per creative-circulation-national-genetic-communion-effort-(s), from those inhabitant-(s) of national-individualism, "O.K.(ed)", parental-discourse, that involve an adult-extraspective-debt-accumulation-(s), Federally Deposited, persistently disregarded trade, insured documenting fiction, believed an fantasy of corporation-merchant-trading, rather hypothetical-prose, conveying observation-(s), from self-action-(s), in discourse, by mass-human-bias, not under [Federal // State-Constitutional Discourses] when age-scaling, offset from say my perspective of decade in to millennium, omitting focus per day upon calendar-interval-(s), as false-point-(s), impassed at mass-negated-effort-(s), when [time // calendar] posit-(s), not intertwine grandiose millennium seasonal-festival-perspective-(s), planning festival-(s), in homage to industrial-object-(s), from entity-outlook-perspective, through ink-flow-paper-documenting, per secondths apposing each state of matter, for an variation of secondths-matter-milliliter-exponent-grandiose-matter-miniscule-extent-scaling, first identifying posit-(s) of tangibility, relevant per each matter-posit-(s), then practicing weather their (be an) rate of [time // dating] properly pertained upon such tangible-(s), redefining (a the) periodic-table of element-(s), scientifically-mute by year, comprehending how solar-kinetic-impact, affect each still-cooled-solidified-element, velocity-propelled-weight-comparison, amid centripetal-force, intermittent lunar-impasse, which is felt by perceivable-competent-matter-(s'), temporally-possessing, an accustoming of [governmental-boundary-subjective-space-documenting //// university-limit-subjective-objective-space-bloom-flourishing] visually effecting [Flat-Ground & Taxonomic // Objective –Placing-(s)], when hourly-time-posit-(s)-interval-(s) {Hour}-{Minute}-{Secondth-(s)};134 affirm matter miniscule of human-composition, ratable by motion-(s) their at physiological-impasse, not requiring machinery, in [measuring // dimension-(s) // weight-(s) // balance-accumulated-factor-(s)], awhile [housing // entity // vehicle-(s) // tectonic-plate-(s) // appliance-(s) // installation-piece-(s) // water-body-(ies) // machinery // subjective-prose] are of calendar-dating-posit-(s), requiring an advise of minor-common-usage, from major-expert-analysis, of those robust factor-(s) of greater than singular-person-effort-(s), because how {Day}-{Week}-{Month}-{Season}-{Year}-{Educational-Period(s)}-{Decade}-{Century};135| 283 |pertain focus of [distancing // metrics // mile-(s) // ton-(s)-superfluous-tangible-(s)] greater than self-physical-recognition.132¶|||07m.:35s.90m.s.|||§85L. How else would we walk upon life? Do we rush step-(s) pass surface-area, elapsing discourse of ordering, taken place, to effectively-task, mass-inhabitant-(s), from peer-(s)-dynamic, ensued well thinking, governmentally-resourced, for commercial-product-(s), in constant-upkeep, identifying the ground below us, [1excavating / 2mine-shafting / 3stabilizing / 4ordering / 5collecting / 6shipping / 7creative-producing / 8forming / 9marketing of data] thinking per hourly-time-frame, how to partake in activity-(ies), for meriting impasse at posit-(s) enhancing life...¶|||08m.:02s.77m.s.|||§

§...Count-[52 / Fifty-Two]-preposition-of.. ..Article-Essay-[14 / Fourteen].. //..Book-Essay-[60 / Sixty].. ..Lines-[89.925 / Eighty-Nine.1.nine-tenths-2.two-hundredths-3.five-thousandths].. ..Where individuals not agree at neighborly-yields, no field can be demarcated, per individual, for of extraspective-shared-resources, apose community-program-perimeters, for adjusting individual-perception, per lifestyle at family-handling, per maintained-waged-entity-work-effort-(s) -cycling, continuum-afloat hovering percept-ive-peripheral-distances, not thought subversive, awhile visually-awake above tectonic-plate-density-distancing-subversive.¶30s.38m.s.|||§

¶61If cohesive mutual-group-interaction-(s), are how we obligate our focus, upon one another, while incurring commercial-trade-requirement-(s), conceiving demand, by-of paper-dollar, trading tangible-good-(s), scheduled of [substantial / solid-density-circumference-diameter-measure / wind-spacing-centripetal-force-velocity-hour-minute-second-drift] affirming continental-shelf, atop of tectonic-plate-(s)-revolving-force, impassing centripetal-force-celestial-circumference-trajectory, as theory, scale-documenting, from visual-observation-(s) of inanimate-matter, alluding day-operation-(s), [maintaining // manipulating], [object-(s) / form-(s) / ticket-(s) / bill-(s) / product-(s) / etcetera...] as-an Gestalt-ordering-(s), continuum, by default-economic-(s)-reliance, only thinking when physical-subject-(ing), command live-act, at entity-objective-case, operating awhile date, upon year, intercede bi-weekly-payment-(s), default by employment, from power of Federal-Engineering, valuing those passing-variable-(s), [in // at], living-examination of religious-communion, shepherd through transit statehood-responsibility-(ies), spacing where neighbor-socialism, not conceive dynamic-(s) for interest, from physiological-elapse-non-time-count-ordering-(s), requiring comprehension for [supplement // complement] calendar-dating-interval-(s), dating mass [impact / motion-(s)] chronologically listing formal inquisition, adjacent astute-acquaintance-(s), during off-time-interaction-(s'), happening at momentary-dialogue, upon-an pre-studied-basis, by notion-ration-(s), sustemming aging-population-(s), by date-interval-(s), considering, state-mass-genetic-reform-(s), orchestrated instrument-(ed)-citizen-(s), thinking of aging, for society-obligation, to resource those means by modern-day-government, rather sufficing "awe" of product-collection-(s), that partake in an debt-method-(s), not efforted in more work-field-(s), by natural-curb in motion-(s), having environmental-impact, requisite Each & Every focus by-of [1.medical / 2.legal / 3.governmental / 4.commercial / 5.industrial / 6.educational-cause-reasoning-(s)], of [1.territory / 2.mass-population / 3.resourcing / 4.production / 5.transit / 6.comestible-(s) / horticulture-regeneration / agriculture-regeneration / boundary-limit-monitoring] as those factor-(s) for constant-conceiving, in rotation of daily-momentum-passing, as while individualism, validify citizen-(s), from developed-account-(s), responsible for ordering, Service-(s) & Product-(s), designating work-effort-(s) in place of generational-aging, pertaining longevity from act-(ion-(s)), for solid-elemental-construction-(ing), amidst grandiose-territorial-spacing-(s), by those peripheral-expanse-glance-(s), of visual-appreciation, when dating is not applied to those visual-tangible-matter-(s),| 284 |along with [objective // subjective // product-timing-(s)] ordering from [weight-(s) /// measure-(s) /// dimension-(s)] balancing all those movable piece-(s), amongst an surrounding-premise-(s).133¶|||02m.:28s.25m.s.|||§ Not mechanical-device, or [inanimate-object // flash-memory] can [1observe / 2touch / 3reason / 4maneuver / 5conjure / 6conceive / 7evoke / 8perceive] cause of human-voluntary-act, (as an) soulless-fiat-trade, requiring human-daily-drive-act-(s), primarily-consuming the majority of human-activity, in absolute understanding of "cash";136 resolving those fact-(s) of fair-practice, not simply of dues, because love of accompany-(ing), one-another, [sibling-(s) / friend-(s) / co-worker-(s) / acquaintance-(s)], define [physic-(s)-extraction / vocal-vernacular-conscious / constitutional-territory-mass-population-claim / dimension-(s)] upon tangible-measure;137 posit-(ed) requisite present-dating, resetting conjecture from data-piece-(s), not formally-tiered for ordering from population-impasse, an progressive-marketing-strategy, cycling examinable-focus, by self and other-(s), in succinct interjection, for segregational-private-market-place-(s), different from public-domain, where perimeter-parameter-(s), by citizen-inhabitant-(s), have had constant-attention, pertaining dimension-(s), adjourn-reasoning, offset from metric-(s)-distancing-(s), by what creative-direction, has been put in to attention of physics, for how land, posit an tectonic-mapping-celestial-time-interval-mass-metric-distancing-(s), horizontal-latitude, [revolution // rotation] centripetal-force;138 (at a the) greatest deviation of metric-mile-(s) per secondths-circumference, apose minute-(s)-latitude-mapping-degree-mark-(s), four-minute-(s), per mapping-degree, designating, elapse-synapse-time-posit-(s), for an vague-ambiguation, cause physical-tension, sight, in plain-observation, an unilateral-bi-partisan-trade-reasoning, temporally-sufficed-living-area, by whom review those cause-(s) of day, amidst entity-day-hour-passing-dating, subterior-accumulated-deductive-hour-time-interval-development-(s), impassing [Input / output, trade-purpose], (from an) tectonic-plate-(s)-drift-impact, (un) prevalent to human-posit-(s)-offset-(s), because of self-bias, upon [physic-(s) // motion-(s)]50L. yet reordered from how midnight-timing, should exist at Five-ante-meridian;139 Miami, Florida;140 from Greenwich, England...¶|||04m.:13s.07m.s.|||§

§...Count-[23 / Twenty-Three]-preposition-of.. ..Article-Essay-[15 / Fifteen].. //..Book-Essay-[61 / Sixty-One].. ..Lines-[51.38 / Fifty-One.1.three-tenths-2.eight-hundredths].. ..I inquiry where individuals-work and plan to retire at, examining how commercial-production, is fragmented serving-(s), shared upon daily continental-development-(s), per moment at developmental-configuration-(s), when dating-periods, are reset weekly, from an lack of genetic-tectonic-mail-language-commodore, amid celestial-solar-fission-drift, axis at offset visual-living-cognitive-purpose, only relevant by the community of cooperative-objective-entity-obligation-(s) too.¶31s.31m.s.||| §

¶[62]How is pertinent-fact significant, when populant-(s) are free to behave however millennium-aging-method, affect bias-introversion, where familied-command-(ed)-act-(s), formulate temper, at workplace-ordering, relied soly on gift-(s) of time, uncalculated while local-timbre, alleviate self, dulled interaction-(s), awhile being remain limit-(ed), from voluntary-lance-effort-(s), in place, during interval-(s) [at / in / near / situated] an [[1]area / [2]place / [3]car / [4]stadium / [5]park / [6]entity-ies) / [7]residence / [8]market / [9]stock-exchange / [11]factory / [12+]etcetera...] remained under citizen-duty, for whom focus posit-(s), intricating-identity, fatigued from centripetal-drift, unclear kinetic-vapor-pressure-rate-(s), passing by, in an familiar-clue-(s), for how brain is biological, perspective is psychological, cerebellum is medical, but common unspecified-characteristic-(s), are sensational, working soly from physiological-impasse, relaying synapse-(s), upon latitude-degree-(s), for international-unilateral-policing-align-monitoring, of trade from eastern-hemisphere-origin-(s), consisting edible-matter, remaining inclined hemispherical-tectonic-plate-(s)-national-boundary-(ies)-offset-(ed), reasoning of day, by plane cause, encountering event-(s), due to extraspective-sustain-(ing), pinpointing seasonal-equinox-transition-(s), awhile subtle-day-hour-elapse-(s)-movement-(s), emit kinetic-energy, transiting solstice-maximum-shift-(s), amidst [solar / lunar-placing-(s)]15L. continuum way-developmental-progression-(s), upon surface interval-(s), when year by linear-limit-(s), affect each adjustment of symmetrical-ordering, upon [terrain / territory]...¶|||01m.:20s.22m.s.|||§

§...Count-[4 / Four]-preposition-of.. ..Article-Essay-[16 / Sixteen].. //..Book-Essay-[62 / Sixty-Two].. ..Lines-[16.37 / Sixteen.1.three-tenths-2.seven-hundredths].. ..cubic area-impasse, remain from an clearance of feet, upon meter-approximate-two-dimensional-area, comparing impasse by an two-degree-compositional-dynamic, apose physiological-14-cubic-feet, for comparitive-etymological-existence-observation-(s).¶|||17s.78m.s.|||§

¶63Weather one can be trusted with [1.allegiance / 2.loyalty / 3.obligation / 4.duty / 5.responsibil-ity / 6.service / 7.leadership-role-(s) / 8.task / 9.list / 10.errand / 11.requirement / 12.energy / 13.follower / 14.objective],| 285 |dating, while we are living-cardiovascular-respirational-perspective-hyper-tense-reasoning-being-(s), which act, from familiarity, at local-proximity, hoarded to an place of central storage, [residence / public-road-space (via-vehicle) / constructed-entity / church / station / market / etcetera] enjoying throughout commercialism, hobbied-activities, familiarizing motion-(s)-interval-(s), accustoming routine of vehicle, through station of gasoline-purchase-handling, at designated location-destination-(s), independently believing, how passing-hour-(s), form communication-(s), to [recite / converse / gossip / light-physical-interaction-(s)] ambiguous while abstract-extemporaneous-observation-(s), revere [local-familiar-habitant / object-appreciation / feeling-(s) of television-article-(s) / past-encounter-(s)] when unclear-decision-(s), in omission of topical-affair-(s), affect how ones' mood upon an practice by-of day, [resolve / affirm / guarantee / absolute / emphasize / sum] each moment, through an passing-(s) of live-memory, tolerating simple-naivety, merely-appreciating existence, from-an parental-physical-dynamic, by Birth & Age, oral-perceived, by independent-tangible-composition-form, in lieu [comprehension / study / understanding;141 1word-point-(s), 2defining article-(s) / 3document-(s) / 4form-(s) / 5application-(s) / 6press

/ 7novel-(ing) / 8etcetera...] influctuated perceivable-subject-(ing-(s)), tiering from untangible-solid-density-abstract-terrain, notion-(s) for modifying compound-collection-(s) of composite-form-matter, refining tangible-mass-objective-proportioned-mean-(s), article-(s)-documenting, at factory-production,| 286 |in transition of shipping-plant-(s).134¶|||01m.:33s.81m.s.|||§ Why is live-discussion, not by an, voluntary discourse, when influence of bias, affirm [physique stance / sat interaction-(s)] from Sight-Eye-Ventricle-Pulse-Elapse & Sounded-Ear-Canal-Ventricle-Pulse-Elapse;142 awhile presence at live-impasse, past before (any / our / your / I, physiological-position-posit-(s)) Eye-(s) & Ear-(s), determining subjective-comprehension, by physical-sensation-(s), [to / through] an hour-(s) per minute-(s), duplicative date-centripetal-rotational-revolutionary-cycling, where-when physiological-tangible-mass-force, suggest physics upon the conclusive fact of mass, but when one can not cover an entire individual-effort, from method of trade, not directly concerted-effort-(s), by inhabitant-mass, concepting those acre-(s) where territory be covered, as the effect of mass-preferred-poverty, idling by entertainment, rather diligent constitutional-commercial-religious-activity-(ies), developing an state of artistic-architecture-infrastructure, upon those observational-passing-(s),| 287 |intermittent,29L. [solar-fission / earth-tectonic-plate-(s) / posit-perspective of universal-molten-gamma].¶|||02m.:32s.63m.s.|||§

§...Count-[15 / Fifteen]-preposition-of.. ..Article-Essay-[17 / Seventeen].. // ..Book-Essay-[63 / Sixty-Three].. ..Lines-[29.84 / Twenty-Nine.1.eight-tenths-2.four-hundredths].. ..allow millisecondths to simultaneous along secondths, and compare compositional-meter-(s), apose exterior-visual-distance-observation-(s), amid fatigue through aging, decaying along an time-frame of payment-(s), waged in succession per dating, when arrangement-(s) of commercialism, periodicalize familiarity, by synapse-mode-offset-aging, from an comparative-body-species-rate-(s), comparative-offset revolving-revolutionary-mode-circular-spherical-circumference amid oblong-visual-sedentary-voice-existing-human-posit-(s).¶|||31s.09m.s.|||§

{Article:} [March]

¶[64]Creative-perspective, timbre simultaneous-notion-(s), invigorated from time-posit-(s)-tiering, arise expansional-force, aligning [fact-(s) / (objective) upon notion-(s) / (subjective)], by either [day-calibre / year-calibre] of taxonomic-classification-state, finding [mammal-(s) / plant-(s)] inanimate from cognitive-nature, sparse way-(s), as pulsed-organism-(s),| 288 |variating domestic-inhabitant-(s)-environment, from particular identifying characteristic-(s) of Mass & Dimension-(s), temporally perceiving, physiological-day, for which conjuring, impassing through momentum-shift-(s), bypass mass-independent-individualism, count-rate-(s), for aligning those notion-(s) in simultaneous of aging-decision-progression-method-(s), from either [ventricle-blood-pulse / organ-muscle-cartilage-tissue], definable-presiding [mass / matter],| 289 |defining through kinetic-vapor-intangible-vibrational-rate-deviational-vernacular-vocal-subjective-per-spective, familiarizing [factor-(s) // notion-(s)] creatively-aligned, by an technical directed-purpose of discourse.135¶|||49s.31m.s.|||§ Now can only exist while present, interceding age-conscious-estimate-(s) of time, apose age-interval-life-expectancy-(ies), difficult for how present-year-juxtaposition-(s), from Three-Hundred-Fifty-Thousand, Four-Hundred-Minute-(s) per [year / Twenty-Seven-Million, Thirty-Two-Thousand-minute-(s), per citizen-life-expectancy-cycle / (estimate), deviate motion-(s)-impasse, every hours-twelfth-interval-(s), operating through-an practical interacting-consciousness, for life-expectancy-method, pertaining how decay-maximum-expectancy-limit-(s), intercede hour-(s), date calendar-(s)-function-(s), two-hundred-seventy-six-thousand-hour-(s) of daily-ten-hour-simultaneous-extraspection, remaining significantly less than two-thirdths, million-minute-(s)-calculation], virtue estranged-reclusion, [routine / entertainment / work / unwind] general-unclear-average-common-mass-(es), inept stable-effort-(s) indepen-dently-developed, because land, require mass-configuring-act-(s), in bias-reluctance, of simple-task-(s), generating motive-(s), for acquiring-object-(s), by an notion-(s)-evaluating, prior each objective-production.136¶|||01m.:

39s.28m.s.|||§ Intervalizing mass-infinitive-notion-(s)-extent, pertained when time, not ever tense other-inhabitant-(s) (hyper) tense-(s), but under an requirement to persuade perspective, order retrieval objective-(s), defining purpose, upon an grading-routine, (from an) imagery-tangible-(s)-ambient-distance-perspective, ascertaining reference-data, by present-life-motion-(s), in light, live-future-discourse, modifying [sight / sound-impasse-recollection-(s)],| 290 |organically-reflexive, when fact-(s) of exterior-observation-(s), pattern an path of anatomical-muscle-memory-routine, persuaded through non-verbal-communication-(s), way of monetary-discourse, unparticular [reading / redacting / article-referencing / writing] referencing influenced-familiarity, where comfortable-act-(s), enammer-cause, for how feeling-(s) affect sociological-sensation-(s) interacting by-an individual-character-(ization)-(s), as cause of reasonable-acquiring, commercial-market, product-commerce,32L. containing nutrition, for maintaining physiological-matter...¶|||02m.:24s.85m.s.|||§

§...Count-[10 / Ten]-preposition-of.....Article-Essay-[18 / Eighteen].. // ..Book-Essay-[64 / Sixty-Four].. ..Lines-[32.367 / Thirty-Two.1.three-tenths-2.six-hundredths-3.seven-thousand-ths].. ..Matter of fact!! if decimal-measure of cubic-distance-visual-offset from human-cubic-meter be, what distance can be covered, if title-measure, appropriate-pounds-per-inches, agglomerated-shapes-commerce. Valuing an unit-%-percentage-per-volume-box-parcel-(s), at warehouse-deductive-interior-clearance-proportion-dating-month-weekly-date-volume-(s), for generation-(s) of year-(s), as goal for Lifestyle & Transit; because how proportions of perception remained cased, under an lack of formal-paper-usage, per individual-documenting, from personal-written-invigoration.¶|||34s.10m.s.|||§

¶65What time-zone-(s) parallel latitude national-conjunction, marching cross-intersected, from year-denominative, metric-miles-(s) tectonic-latitude-vertical-degree, for an circumfer-

ence revolutionary-transit-nation-state? Earth, contain various
elemental-deposit-(s), existing, through an continuum, revolving
circumference-centripetal-force, effervesce of passing-hour, under
influence of solar-sun-combustion-heat-friction-plasma-core-la-
va-expansion-explosion-pressure-infinitive, emitting an vapor-par-
allel-extreme, of developmental-kinetic-energy, unclear how phys-
ic-(s)-supratangible-grande-posit-(s), remain inanimate, by min-
ute-tension-(s).137¶|||30s.65m.s.|||§ Immutable fact-(s), of human-perspective,
emit such an align of peripheral-awareness, for when sensation-(s) of
kinetic-brim, power us as living-organism-(s), because how our ca-
pable-existence, cue day-interaction-(s), repetitive in daily-human-fa-
tigue, scorn stressors, from action-(s) monitoring objective-develop-
ment-(s), relevant to those feature-(s), developing by an mass-con-
sensus,| 291 |for customizing terrestrial-impasse-(s), from elemen-
tal-resourcing.138¶|||47s.44m.s.|||§ Scientific-venture, not origin-being-(s),
askew virtue, viewing human-behavior, lineage-transgression,
fact-(s) of hierarchy, whom comprehend religious-fervor, aboding
directive-itinerant, expanded disperse Governmental-Influence &
Territorial-Placing-(s) upon tectonic-plate-(s), internationally balancing good-(s), requisite European-trade-route-(s), when social-pres-
ence, envelop-issue-(s) of state-(s), cause population not have much
to say, believing such is worthy of grounds of bias-trading, when
travesty because idol-petulance, whittle drift of physics, during each
celestial-impasse, daily-three-hundred-sixtieth-degree-(s), Twen-
ty-Four-hour-rotation-(s), intercede, no determined ground-work,
completing task-(s), for calculating-bias, by human-subjective-feu-
dalism-perspective, cause balancing Orator & Audience, complicit
received-information, constantly-upkept, apose [listening // persua-
sive-vernacular-vocal-tone-vibrational-subjective-skills] conveying
how kinetic-vapor-intangible-vibrational-pressure-rate-(s), fore-
sha-dow event-(s) of future, alluding interaction-(s), as dignitary-(ies)
of social-citizenship, following marginal-order-(s) at workplace-in-
stance-(s), if ever being able to perceive how objective-held-posses-

sion-(s), can equivalent resourcing-output, where state-federal-balancing, deter empirical-company-requested-order-(s), by citizen-natural-tendency-(ies), of tranquil-domestic-living.[139] ¶|||01m.:51s.[15m.s.|||]§| 292 |As solar-parallel-earth-year-rotation-exist, does those origin-(s') of humanity, be Raped & Mutated, unclear from religious-dating, as-why, those presumption-(s) by-of [language // literature] why deliberate-faltering, exist in an dating physics-elemental-extraction-evaluation, ontologically-tested, since African-(s')-Israel-Gaza-Strip-Genetic-origin-offset illiter-ate from ontological-proof, how entomological-competency, due to [choice-(s) // bias // decision-(s)] lacked by discipline of trade, building an nation of inhabitant-(s), deducing instinct-plane, deviating civilization-(s), impassing an genetic-population-basis, per Year & Millennium, by dynamic of group-modifying, where Land & Perspective, is not cultivated completely, by even an billionths-mass-population, but when inhabitant-(s), remain free of political-concern, and ideal blank-impasse-cerebral-perspective, by bias of content-trading, enabled from community-communion-belief in lifestyle. Context has never elicited lifestyle, abiding mundane residence order, following daily-weekly-limited-mentality, (from an) natural-effort-(s), intermittent attention-span-limit-(s), for of common-temperate-passing-(s),[38L.] conditional where environment, fluctuate by an flurry of [day-seventy-four-degree-(s)-median-average / night-sixty-three-degree-(s)-median-average] (from how I am born & raised in Miami, Florida);[143] occasionally rain-shower-(ing), and overcasting cloudy-forecast-condition-(s), (from an) method-(s) of living, reciprocal, human-constitutional-continuum-disposition... ¶|||02m.: 20s.[21m.s.|||]§

§...Count-[24 / Twenty-Four]-preposition-of.. ..Article-Essay-[19 / Nineteen].. //..Book-Essay-[65 / Sixty-Five].. ..Lines-[41 / Fourty-One].. ..location have an ambiguous-distancing-(s), per individual-location-capacity-conjunction-(s), from brick-80-lb-counts, per individual-interpretation-(s), amid colloquial-etiquette-talk, falsing millennium-belief of population, by an historical-solar-belief, off formal rotation-revolutionary-count, from year per season, amid month-weekly-transit, for visualizing how physical-effort, proportion in an clearance-cubic-spacing, each moment encountered at an entity, as human-introversion, not learn how to agree with one-another, on an national-basis.¶34s.04m.s.|||§

¶66If we are not careful, from those variation-(s) of tangible-mean-(s), upon what consensus will character-building, persist amid religious-fervor, having how direct-effort-(s), formulate citizenship, measuring, foreseeable distance-(s), impassing experience, while collecting [package-(s) // packet-(s)] of resource-(s), ensued spacing-(s) measured-distance-(s), formed adjunct neighborhood-discourse, building atop those centralized-facet-(s), [central-county-sewer-system / national-road-terrain], while under state-electric-wire-grid-(s), elongating spacing-(s), currenting-friction from water-metal-wire-conducing, taxing [proprietary-lender-loan-(s) / leased-parcel-(s) / outright-ownership] permitting the introverted-dynamic of metropolitan-development-(s), to annex capable-interceding-(s),| 293 |stably-controlled, by operation-(s)-conducting, traditional-affair-(s) of statehood-belief.140¶|||39s.44m.s.|||§ Forever associating life, from male-direction, has been instilled by those forefront-(s), from physical-perspective, "getting done", pre-sequenced-work-order-file-placing-(s), instructed so habitant-(s') can precluded labor, by demand from market-impasse, denoting those ideal-(s), based on merit of supply, to accredit from Federal-deposit-(s), insured by an routine [cash // check] influctuation-payment-method-(s)-orders, obligated quota-exchange-(s),| 294 |by deadline-(s) of accredited-payment-(s). Blue-collar-tasking, gleefully ignore concerted-maneuver,| 295 |manipulating continental-shelf, for entity-development, because mathematics, not inspire architectural-planning.141¶|||01m.:09s.09m.s.|||§ This focus, (has an) jading of human-conduct, behaving by mute-thought, incensed from nothing, omitting why interaction is a the fundamental-impasse of [living // existence], moding physical-self, away-from listening of vocal-vernacular-vibration-(s)-intonation-pitch-tone-quality,|

296 |metronome intonating tone-volume-tempo-vibration-(s), synchronizing data, for order-referencing of word-(s), by elapse-(s) of pitch, as secondths-elapse-(s), due to various-focus-(es), by participle-tone-form-(s), [ritardando-baritone // largo-tenor / andante-alto // allegro-soprano] letter-tone-inter-pretation-(s)-shift], intermittent word-tone-frequency-syntax-conjuring, thinking how order-(s) by grammatical-tabling, sort focus for tone-interpretation-(s), vibrating from secondths, by an millisecondths-letter-scaling, creating-an denominative-scale, counterpoising word-article-form-(s), when an diction-practice, should be implemented, for constitutional-grammatical-punctuation-tonal-vibration-(s)-graphing.142¶|||01m.:55s.53m.s.|||§ If order not denominate sequential-routine-(s), what define synapses, by an redundant-fashion, driven for opinion, amidst continuum-friction-belief, scared to inquiry why inhabitant-(s') are an issue of existence, from psychological-annotation? Population of inhabitant-(s), proceed an limit of individual-humanity, not including teenager or younger, for annotation-conjure,30L. yet experiencing, impassing-[date-(s) / years], when review of each (in) tangible-precept, affirm connotational-awareness of value-(s), when Date & Time, segment [(re)-set / cycling / sorting] offsetting along simultaneous-moment-(s), year-date-scaling-elapse-(s), simultaneous tier-motion-(s)-continuum-deviation-(s)...¶|||02m.:27s.00m.s.|||§

§...Count-[16 / Sixteen]-preposition-of.. ..Article-Essay-[20 / Twenty].. // ..Book-Essay-[66 / Sixty-Six].. ..Lines-[33 / Thirty-Three].. ..how to count from staff, vocal-harmonic-throat-pitch-notes, upon an harmony of measure-(s), succinct per extent of interpretation-reset, apose millisecondths-notion-continuum-count-(s), as usage of human-letters-words-numbers-interpretational-denominating, define literary-remark, proportional [serial // copyright-extent-(s)] by via means for primitive-commercial-trade, bogged by genetic-tectonic-continental-dispertion, not understanding how genetic-territory-(Montana)-reformation, require an continuum of commercial-schedule, in literary-thousandths-extents-literature, per individual-occupant, from how after-life-lineage, affect an offset calendar-period-(s), questioning neighbor-commercial-shared-bastardized-statehood-communities, not considering why land is important to have an collective-religious-investment-group, for appreciating how developing is characteristically going to be pursued, amid an lifetime of existing, and passing of lineage-quality-locational-characteristic- s.¶|||50s.00m.s.|||§

¶67Left in awe, while human-discourse, inhibit touch, around general-parameter-(s), land remain claimed, because how individuals', rebuttal success, requisite affirmation, occupying governmental-taxed-territory, as topical-issue, by sequestration, for daily-living, when commercial-effort-(s), require [containing / maintaining] fact of order, while attempting to rein in an population, focusing methodical-ordering, by practice-(s) of [1food / 2cultivation / 3crop-recipe / 4mammal-regeneration / 5elemental-resourcing / 6water-reservoir-stream-currenting / 7electric-wire-conducing / 8artistic-architectural-infrastructuring / 9field-rotation / 10land-valuating-upkeep / 11+etcetera...] readjusting-pace, of daily-method-(s), foddered along those curtail-(s) during daily-payment, encaptured by activity-(ies), roused of daily-accounting, from socializing.143¶|||41s.18m.s.|||§ Introverted bias, in my belief, characterizes, an mentality-flow, impassing affair-(s) of physics, | 297 |by Western-Hindu-Muslim-African-Caucasian-Tradition, old-school-easy-living, without angst-antagonism, of juxtaposed-ethnic-posit-(s), passing day, for family-structuring, nestling total-population, when schedule of dating, impasse presence, undeniable tangible-fact-(s), weighing in observation-(s), upon measure of cubic-spacing-(s), review from tense-sighting, sounding out still-focus-analysis where [place / objective / order-(s) / maintaince], experience when past effort-(s), allude at an dry-self-presence, incapable of assuring fact, for sole-perspective, is with an personality-error, as how deviation-(s) of being-belief, consider to be person, but not scaling those factor-(s) that affect the outcome-(s) of politic-(s) of industrial-commercialism, subsided parameters, by self-interior-weight, requiring antedote-relief, measuring increment-(s) of distance-impact, upon self-tangible-mass, comparatively visualizing how human-subjective-grading, not be in an commercial-home-work-pattern, post educational-institution-(s), brigaded noun-maintaining, from pronoun-response, for whom presume duty-(ies) of natural-appliance, possessed so that stock can be collected from natural-resource-(s), developing ecological-posit-(s), when causes of self, input production, (as an) credit, for instilling an motivation in to inhabitant-(s)-will, that go as far as to imagine in idle-theory, no ecological-sociological-hypothesis, concept-directing, population-mentality-offset, away from a the heard, of steer, whom yet understand an common-subjective-society,25L. intelligent through those listened-peer-(s), whom understand a the process of listening, greater than, talking in jargon...¶|||02m.:12s.21m.s.|||§

§...Count-[17 / Seventeen]-preposition-of.. ...Article-Essay-[21 / Twenty-One].. ..Book-Essay-[67 / Sixty-Seven].. ..Lines-[26.14 / Twenty-Six.1.one-tenths-2.four-hundredths].. ..sequencing motions, has yet been possible, through book-scheduling-observation, annotating length of motion-(s), in secondths-count, comparing how physiological-compositional-stature, competently-function, for developing geological-terrain, from commercial-genetic-investment-resourcing, figuring out in to meticulated-specification, who pertain to what matter circulating economy by an product-purchase-parity, reliant on basic-living, not specified for an purpose of documenting housing-entity, from military-mission-restless-upkeep, forgetting what objects to pass along our lineage-repopulation, for future aging, by general-means of parameters, per moment by what method of designations are required from an individual-posit, for determining each point of terrain, due to physiological-developmental-effort-(s).¶|||46s.21m.s.|||§

{Article:} [March]

¶[68]Inuit-(Un)-Incorporator-(s) not live filing under conscious-colloquial-Ebonics-vocal-inter-action-discourse, for illiterate manuscript-instruction-(s)-purpose, not interpret Consti-tutional-sworn-cursive, rating writing-university-locational-discourse-scale-proportion-(ing) weight-(s), by consumer-socialism, defined by e-tier-geo-eco-impact, by-an circumstantial-outcome-(s), dissolved individual-independent-right-(s) from men, in an political-male-natural-omission of reason, from celestial-observation-(s), gay by female-component-(s), when sport-mundane-allegiance, put black-being ahead of women, before considering female-counterpart, in an earlier-precedence, nationally-subjectively-transgression-(ed), centralizing national-tectonic-allegiance-state-order, with how international-politic-(s) are with an ethnic-offset-(s), of genetic-disposal, incapable yet, for how Tenurial-comprehension, where integer-(s), sustem from population purchase-demand, upon variation from blood-pulse-inhabitant-(s'), aging in an ignorant-succession, through educational-study-(ies), focusing longevity, at perceptive-abstract-impasse-(s), not prosed, without an extraspective-non-fiction-genre, factoring from proprietary-share-(s)-inhabitant-(s), consciousness, with family-religious-local-year-decorum, offset where governmental-residential-credit-title-deed-parcel-payment-resource-(s)-responsibility, are Pros & Cons, amidst Observation & Detract Data, put upon individual-perspective, mass-extent, continually inputting data, without an relay of order, pre-instilled by the discourse of schedule, to have an focus, by perspective in hour-ordering, when why is data not tense-(d) compartmentalized-stock-market-ordering, emphasizing year-date-scaling, upon time-interval-analysis-posit-(s), alluding from a the rule earlier, instilled, when hourly-time-offset, range thoughts', apose [calendar-grandiose-objective // tectonic-plate-(s) // continental-shelf // elemental-extraction-(s) // excessive-weighted-object-(s)] comparative elapse-expiry-matter [interval / decay / resourced / ordered constant-(s)], amid per population-mass, during basic objective-concept-shelf-development, throughout era-(s) impassing momentum, plateau consensus by-of [individual-perspective-(s) / educational-perspective / co-worker-perspective / retirement-perspective] routine by hour-day-year, elemental-resourcing-product-concepted-basic-human-right-(s)-impasse-ordering-(s), etching how lifetime should partake free-will-directive, diminutive depreciation, imbue from substantial means, reducing the amount of shit, by substantial-receipt-material-purchase-intake, having an variation of checking-expenditure-(s), offset balance, temporal daily-activity-(ies), for church not be an scientific-center, and remain claused along with metropolitan-infrastructure, taxing-affairs, reminded to resource, away from housing-developmental-center-(s), as well as, factory-developmental-component-resource-center-(s), affirming team-effort-(s), taken part, where-awhile, year-discourse, yet define century-juxtaposition-perspective-offset-(s), while closer than appearing, because, decade is rationed in to season-(s), by an whole-numerical-sum of Fourty {40}, apose season in to decade as zero-decimal-two-tenths-five-hundredths {0.25}, upon those pre-instated-limit-(s) of extraspective-extent, having age be offset where pre-archaic-cause-(s), priori equivalent-deviation-(s), rating-material-(s), as market package-food,| 298 |during day-week-month-expiry-live-refrigerated-preservation, calculating analytical-discourse, sum of [percentage // fraction-integer-sum-(s)] temporary where solid-data, (ob) (sus) (main) tain-(ing), particulate each piece of observation, to be render-(ed) inanimate, fostering American [reading / writing] hypothetical conceptive-interior-perceptive, analyzing prime data, furthering [understanding // comprehending] upon those impossible-national-total-spacing-(s), sum-amounting [men / women / child-(ren)] genetically-sect-offset, decade-millennium-time-impasse-interval, emphasizing calendar, apose offset-circumstance-(s), deduced cordial-attention, elongating live-physiological-breathe-blood-pulse-being-mass-count-offset, unintegrated the location of designation, not designating schedule for control by any particular-matter.144¶|||03m.:50s.49m.s.|||§ [Solid-(s) / composition-(s) / substance-(s) /

liquid-(s)] pressurized per square-inch, when tire-p.s.i.-ton-complex, incorporates day-temporal-discourse, not in an forum of discourse, because, interacting upon form-(s) tensing being-(s), corroborating mutual-colleague,| 299 |intermittent commercial-community-individual-Workermens'-discourse, influences family, at an maxim, where hour-(s)-routine-open-time, concluding day, from [sleep / transit / market / housing / affair / church-outing] uncentralized, motive where free-will, evoke physical-blood-pulse-mean-(s), subjecting, rate-note-(s), vibrational, [note-beat-(s) // rest-(s) / pulse-beat / secondths] per [minute-(s)-elapse-(d)-time / letter-tone-word-second-elongated-spoken-vocal-tone-elapse-count(s)] resetting intonation during cycling instrument-symphony-ensemble, tone, beat-quarter-note-sixty-largo-tempo, per second, as when rate-(s) intonation tonal-instrument-(s)-reverberational, one-second per cycle-intonation-mode-(s), amidst ensemble-note-range-count per second, offset, woodwind-(s), [piccolo = 1{D4-C7} / flute = 2{Bb3-D7} / clarinet = 3{E3-C7;[144] P. Bb} / oboe = 4{Bb1-Eb5} / bassoon = 5{EBb1-C5} / French horn = 6{F#2-Bb5} / soprano-saxophone = 6{G3-A5} / alto-saxophone = 7{Bb3-F#5;[145] P.Eb} / tenor-saxophone = 8{F#2-D4} / baritone-saxophone = 9{D2-A4}] exemplifying vibration-tonal-range-(s), boasting sound-intonation-blending, layering-time, denominative per-se,| 300 |entity-resonance, confiding, timed-elapse-(s), [influencing / persuade inclination-(s)] during effect of spacing, first when three-fourth-(s)-measure = 15-second-(ths), be interluded instrument-tone-(s), upon vocal-dialect-(s), harmonizing survey-mass-citizen-(s'), comparative-denominative, brick-count-(s), defining dimension-spacing-(s), per [room / hall / theatre / philharmonic-stage / stadium] measuring distance, for self-spacing solid-density-resonance;[146] filing order where word-vocal-minute-syntax-mood-frequency(-ies), synchronize-scaling, [interior-entity / interior-breathe-blood-pulse-circumference-height-being / exterior-interior-cubic-spacing-(s) / breath-blood-pulse-being-tone (mute syntax usage)-vibration-(s)] identifying within interior-entity-setting-(s), tangible-historical-account-(s),| 301 |affirming data, as, posit-date, objective-comprehension-deterioration, seasonal [hammer // nail-(s) // gypsum-board // dry-wall // et-cetera…] from order date;147 starting textbook-glossary-reference, as article-(s) object-comprehension, for Bachelors'-pre-set, preliminary military-male-condition-(s), implied, decade-act-fervor, continuing the ancillary-trivia-remarking-(s), fulfilling goal-(s) for perspective, in regard of anatomical-blood-pulse, upon the perspective-observation-dynamic, for activity-(ies) inset by physical-act-(s), instead of subjective-text-non-fiction-literature-survey.145¶|||06m.:43s.79m.s.|||§ Consideration by extraspection-error, is why my context at times', will apparently be perceived as redundant, because how each article-essay, remain in an inconsistency of interpretation, when stressing how physiological-vocal-vernacular-tone-reverberation-vibration-resonance, affect how each notion-(s) of action, remain in an non-fiction, critique of human-interaction-characteristic-quality-(ies), designating objective-(s), in an metropolitan-population-conjunction, overcrowded, by [fault // fallacy], from men distancing-metric-gamma, obeying an national-military-allegiance, following an unconscious-international genocide, of human-mode, unclear of what trigger such an offset, awhile an national-credit-extortion, from political-metre, not measure from founding-father-free-will, how such error, proglamate human-trade, from how subjective-free-will-expression, can perceive such opinionated-interpretation-(s), impassing [secondths /// minute-(s) //// hour-(s), posit-(s)] for further interpretation-(s), by subterior-subjective-notion-synchronizing, how millionths-mass-population, live without an extraspective-interactive-communion-timbre, illiterate from such focus, due to ideological preference, upon etymological-subjective-studies.146¶|||07m.: 47s.31m.s.|||§ Writing-technique, can not instill-program of logic, upon requisite incoherent-operation-(s), [diagram /// graph // grid / mapped] merit discourse, intermittent affair-(s), balancing operation-(s), with independent-preference-hobby, consistent-strict-writing-(s), conceptual-value, upon physiological-capability-(ies), relevant from individual-push, (from an) genetic-offset, because of ethical-self-bias, pertaining life-median-deviation-end-(s), not in an simultaneous-common-writing, defining visual-imagery-interpretation-(s), understanding diligent-interaction-discourse, from competent, conceived data, construing constitutional-denominating, [requisite // present // future] human-legal-value-(s), extracting from [1space / 2composition / 3chemical / 4metal / 5alloy / 6ion / 7soil-clay-composite-brick / 8gravel-brick / 9refined-mountain-sand-(s)] dimensional by constraint-(s) where human-limit-(s), measure in-of thousandths-day-concept-discoursing, proportioned an scaling of [weight-(s) //// measure-(s) ///// time in hour-(s) // minute-(s) / secondths] for subterior-defining, purpose for constitutional-population-mass-political-territorial-distancing-cultural-development-(s), constructed when schedule-year-usage, amount community-purpose, where constitutional-mean-(s'), theatrically-sustain, dormant passage, having each entity require an hour-(s) of operation-(s), for dating e-tier-sch-ego-geo-purpose, without an life-median-ten-thousandths-day-scaling, discoursing analytical-prose, offset perspective-(s'), [1understanding /// 2comprehending /// 3interpreting /// 4apprehending //// 5communicating //// 6concepting //// 7calculating //// 8pertain-applying] location-hour-planning, intermittent [retirement-introversion / extraspective-survey-inquiry] enabling conceived generation-decade-action-offset, built-in, from text-educational-limit-(s), effect-(ed) [where // when] peer-reading-attention-span, cannot be-an part, for fact-(s) of day, determining logic, during at, locational-work-product-reliance, resourcing from government-trust, vernacular-defined-limit-(s), reduced from resource-depreciation, affecting influence by credit-ordering, major-auxiliary-objective-(s), where residence, contrast development, yet considering how accounts', verify social-security of-an mass-population-consensus, as individualism discourse, interlude tiding-(s) for trade deduced offset individual-perspective, upon topographic-metropolitan-environment-deviation-(s), elaborating-daily-hour-count-(s), uncompared [decade / year

/ season] contrast integer-variable-(s), spacing built median-interval-trust, from Federal concise-collection-(s), of taxonomic-matter-(s), as [horticulture / agriculture mean-(s')] resourcing, community-management, serving means, when life is perceptively-survived,112L. for no formal-mass-population-voluntary-planning of state-constitutional-measure, from elemental-extracting of resourced-matter, exist by means' of complex-trading, by our systems of reliance on other economic-commercial-systems of monetary-trading..¶|||10m.:19s.50m.s.|||§

§...Count-[33 / Thirty-Three]-preposition-of.. ..Article-Essay-[22 / Twenty-Two].. // ..Book-Essay-[68 / Sixty-Eight].. ..Lines-[114.411 / One-Hundred-Fourteen.1.four-tenths-2.one-hundredths-3.one-thousandths].. ..Preposition is reseted, various times per task, amid an force per visual-description-(s), lived at century-yields, variating terrain and road, where interpretations, force past visual-vocal-tone-describing, lackluster per individual-aging through retirement under an East / West Coastal-tectonic-development, never resourcing from an centralized-water-post-error, not recognized by foundry-funding, federally-separated from land-posit-monetary-development-effort-(s), when land cognition from each metropolitan-ten-miles-meter-straights-dynamic, impassing an lifetime of secondths estimate 14, 836, 207-secondths in vehicle-active-rolling-motion; ranging 122,489-miles-vehicle-petroleum-impasse, for how land require-resourcing from land, but by an population-mass, per 11 square meters, per individual, amid development of a the terrestrial-land, apose population-objective-resourcing. ¶|||01m.: 00s.16m.s.|||§

¶69If day-elapse, objective-response, require Pay & Retirement, for corridors by cabin-interior, unplanned where tapered-focus, analyze scheduling, as [id / ego / superego / instinct] involuntary-independent-objective-inhabitant-(s), adequate by [residence / housing / market-scheduling] place mass-first-person-bias-hoc-offset, where [house

/ city / state-ground-(s)] sustain dialogue, constitute, space-request, per citizen, deliberate where mean-(s) of work,| 302 | cultivate-terrain, in off-time-hour-(s)-scheduling, that place those commercial-object-(ive-(s)), upon daily work-load, interpreting time for each fact of enumerated-discourse, balancing [weight-(s) // measure-(s)] during governmental-extraspection accounting denominating-data, as their rationale literary-denominating, for objective-collecting, from material-(s), engineering [terrain // objective-(s)] in comparison, for private-instrument-(s), continuing dismay, for any one who know-(s), weather graduated high-school-inhabitant-(s'), [discourse // refer] validated-notion-(s), by survey when physiological-inhabitant-(s),| 303 |point-(s) [1.spacing / 2.writing / 3.dimensioning / 4.contracting / 5.carving / 6.statue / 7.book-discourse-three-hundred-sixty per entity-dynamic-documenting] yearly populating, capacity-written-ordering-(s), defined in an present-historical-documenting, verifying cohesive conclusive-analytical-discourse, from logical-output-origin-(s'), affirming relative-influence, persuaded ideas, thinking how physiological-group-effort-(s), concur natural-cause, delaying developing-schedule, for incremental-date-ordering-offset-posit-juxta-positional-impasse-definition-interpreting, affirming those notion-(s), for grasping how idea, refine each of those impasse logical-interpretation-(s), because daily-inductive-interpretation-(s) impact, where data is matter-(ed) because application where at an impasse-effect from population, is to influence by persuasion, conscientiousness, validating population-juxtapositions, to create an ground-(s) of cultivation, when article-(s) matter by usage (from an) present-learning-application-(s), for-of determinant-conjure.147¶|||01m.:48s.03m.s.|||§ Friend-(s) learn in youth, associate-interaction;148 hence, why not acquaint-associate-(s), because interaction-(s) of neighbor, by group-trust-saving-(s), when off-time-activity-(ies), affirm through an repetitive-physiological-impasse-interpretation-(s), because of daily-existence, where offset-communion, matter for how existence pertain per day of living when of an existing survival, by notion-(s) upon lifestyle-routine.[148]¶|||02m.:10s.[22m.s.|||]§ Individualism & Religion, confirm this notion, not noticed by road-sign-deviation-(s), of metropolitan-offset-count-limit-(s), when tasking of off-time-effort-(s), insist eight-hour-work-effort-(s), upon taxonomic-prerogatives', not counted as documenting, for each piece of matter, verified perspective, left upon visual-hearing-ordering, reproductive-activity-(ies), margin daily-overture, emphatic hour, thought by-of rural-territory, because means-concepted, take an formal-discourse, uninfluential during governmental-international-study-(ies), operating those decision-making-process-(es), proportional-mean-(s'), which effect each other, competent, by an unknown-subversive, when context pertain upon each notion-(s) of data, that collect affirmation-(s) of visual-sound-kinetic-vapor-reverberational-vibration-(s)-pitch-tone-gamma-resonance, pertaining tense, [where // when] an anatomical-component-(s)-mass, particular individual-movement-(s), reflexive-joint-reaction-(s), that can not be in interpret-(ed) data, unless study be redact-contexted, focus-emphasized, surrounding territorial-claim, reasoning information, by an attempt to subject-time-ordering-(s), elucidated educated-voluntary-participant-(s), whom can finally figure an age or class, for informing-inhabitant-(s), desire by mutational language, literature, from national-survey, adjusting Tresurigal-trade, (from an) private-genetic-subversive-act-(ing-(s)), Federal Deposit Income Corporation;[149] employed directive, deriving mass-function, by presence of act, prevalent commercial-form-document-scheduling, intermittent astringed, neutral-workplace-trade-citizen-(s), whom [follow // obey] every dictation of constitutional-guidance, requisite immigrant-history-compilation-subjunction, intricating-contorting-notion-(s), rendering date by act-(s), in submission at payment-voluntary-interaction-(s), for idle-freedom,| 304 |conveying way of men upon family-developmental-survey-impasse [prepare-(d) / contain-(ed) / package-(d) / ship-(ped) / process-(ed)] ordering sequential-discourse, to determine, purpose of fact-(s), identifying each and every piece of matter, from government-receipt-W-2-401K-planning, pre-determined mentality-maximum-competency-capacity, mutually-offsetted, for thinking, concluding effort, before an impasse be motivated, cause moment (um), perceives' as [day // night // morning // evening-temperature-(s)] apose centripetal-force, as seasonal-impasse, affect climate, in lieu of the physiological-mass, not literary-perceived-future;150 remain afloat metropolitan-method, trading at post-(s), designated, from Protestant-heart, that be too entranced by transit, to communicate those subjective-discourse-(s), for entity-dynamic-development-(s), while character-(s) of socialism, become more and more disenchanted, individually-progressing, per daily-inquiry, subjective-mean-(s'), affecting how trade platform, direct each inhabitant by an motivation of pulse-living, for benign-cause, simply wanting object-(s), rather seeing the active-participation, requiring fully coherent-document-(s)-existence, by the Physiological-Blood-Pulse-Respiration-Being-(s') & Subjective-Perceptive-Marked-Article-(s), referencing each citizen-quality, informing analyzation, through an succession of [space-measure-concept-remodifying / object-dimension-composition-(ing)] of [abstract-resource-matter / place-time-date-order-designating / mass-fundraising-survey-effort-(s) / distan-ce-population-coverage-impasse-date-ordering] conceiving formal-discourse, for motion-(s), incorporating Trade & Repopulating, upon [aging / location / objective / composition / perceptive-limit / genetic-characteristic-(s)] intended to be uninfluenced, not pertaining time, relevant relative-national-discourse, visualizing how the international-binding that requisite-proceed, have an formal-pattern, cognitive human-prerogative, by fact, our preconception of time, is expiry by organic-matter, upon eating-drinking-non-documenting, perceiving [pound-(s) / pint-(s)] transferred-of [ounce(-s) / gram-(s)] micromanaging-increment-(s), aligned when [milligram / milliliter measure], organic-nutrition, piece each matter, (as an) lesson pertaining documenting, fluently ordering [elapse-four-second-cycling-count-deviation / synapse-four-minute-(s) // Two-Hundred-Fourty-Secondths cycling-count-deviation] [emitting // inking subterior-count-(s)] per secondths-conjunct-periodical-logic, intending conceptive-activity-discourse, working from our omission of [time by hours / minutes // secondths //// date by day / week / month / season / year / decade / century, milestone mark-(s)] by aging, impassing movement-(s), for mid-late-age ordering, me;151 twenty-six, as of April 25, 2017{SSN 595-11-6217}, missing those [9,496 days /// 227,904 hours /// 13,674,240 minutes /// 820,450,400 secondths] sequencing (from an) requisite of what can be done, from proportion of minute-(s)-impasse, for denominating upon secondths, how to schedule dating, by [hour-(s) // day-(s)]76L. per motion, thinking of repopulation-reproduction, awhile accounting for how we cycle our indigestion, and pulsate-impasse... ¶|||06m.:49s.03m.s.|||§

§...Count-[32 / Thirty-Two]-preposition-of.. ..Article-Essay-[23 / Twenty-Three].. // ..Book-Essay-[69 / Sixty-Nine].. ..Lines-[77.39 / Seventy-Seven.1.three-tenths-2.nine-hundredths].. ..aging fatigue with how experience increment pass our [awake / sleeping hours], per moment at an awake subconscious-representation, influencing how each outcome pertain movements from men on land, as women are suppose to document lifestyle-effort-(s), pertaining upon terrestrial-land, as each day revolve pass solar-drift for an influctuation of tectonic-land-matter, from sociological-physiological-space, pertained at each impasse of visual-vocal-living.¶|||30s.44m.s.|||§

¶70.By outlining an Fourty-minute-acreage-terrain, a-the fundamental-collecting-(s) of data, proceed revolving sedentary-outcome-(s), where concerted-effort-(s), pertain per day, because individual-purpose, trade (from an) [import // domestic aspect] affecting genetic-extraspection, when guide for those action-(s) of effort, are due to balancing human-morale, with intention of act-(s), forming each purpose-(s) of [daily-family-time / field-work-time / product-recipe-concept-resourcing-time / adult-decade-aging-subjective-entity-(ies)-timing-(s) / market-trading-timing-(s) / theatre-entity-(ies)-timing-(s)], as fundamental-(s) by-of day-hour-scheduling-common-deviation, daily discoursing, an robust purpose upon transited-affair-(s), by communicating with an circle of direct-friend-(s), (three-hundred-sixty-degree-(s) another example of human-limit, pertaining purpose from each of those factor-(s) of date, from reasoning life in an approximate gauge-(ing) of 36,500-day(s) {Thirty-Six-Thousand, Five-Hundred-Day-(s)},| 305 |by the age-one-hundred-maxim-median-deviation-primary-mean-rate-deviation, perceiving simultan-eous-aging, as deviated from each alive-being, ordered governmental-currency-enumerated-trading, from Federal I.R.S {Internal Revenue Service}-post-year-documenting, W-2-form-credential-(s),| 306 |positing denominative company-(ies)-resourced-development-budget-operation-(s), along those credit-(s) conjured, [resourcing / outsourcing-data] brought in from plant-station-operation-(s), for refinery-development, putting each piece of personal-data, abound tectonic-trade-application, never formal local-resourcing, concluding each piece of elemental-matter, designate by concept-trust, (from an) past-historical-pronoun-expert-prestige-reliance, apose an modern-educational-active-decade-aging-community-concept-planning, from cycle-(s) by [[1]genetic / [2]state / [3]population-territory-constitutional-articling / [4]resource-construction / [5]crane / [6]Cat 390F L-(excavator) / [7]Cat 793F(mining truck) / [8]Hb 68(brick forming stationary-machine / [9]122 SD-(cement truck) / [10]Vertical Shaft Kiln Cement Plant / [11]Flour mill / [12]grinding mill / [13]mineral grinding plant / [14]pneumatic bag packing machine / [15]ball mill / [16]vibrating screen / [17]hammer mill / [18]belt conveyer // [19]bucket elevator / [20]material feeder / [21]metal furnace-plant / [22]et cetera] for the resourcing means, basing natural-chemical-(s), for mold-concept-modification-(s) of matter, from directive at constitutional-survey-cursive-documenting, physiological-being-count upon sparse-spaced-territory, appropriating land,26L.| 307 |with service-(s) intended to cycle ecological-reproductive-repopulating-affair-(s), by an means of Physiological-Perspective & Voluntary-Motion-Effort-Discourse...¶|||02m.:29s.32m.s.|||§

§...Count-[14 / Fourteen]-preposition-of.. ..Article-Essay-[24 / Twenty-Four].. // ..Book-Essay-[70 / Seventy].. ..Lines-[27.531 / Twenty-Seven.1.five.tenths-2.three.hundredths-3.one.thousandths].. ..intermittent continuum revolution-x-rotation, each latitude-degree, revolve-offset each individual-human-being, amid an circumference-discourse, per independent-thought-process, born each dating, of twenty-four-hour-daily-pressure, in motion at axis awhile housing shelter perspective of distance where at east-to-west-pressure comprehension at mode-continuum.¶|||27s. 03m.s.|||§

¶71.For date proceed by momentum, passing of tectonic-plate, never inspiring mathematical-intuition, for nature of Mass & Ate, refer-from [sight / sound / touch, interior-blood-pulse-exterior-skin-sensory-cognition] delving human-behavior-discourse, as-an [blood-pulse-rate / timbre of pace / beat-(s) per second-tone-observation-(s) / tempo-measure-piece-frequency-(ies)] funda-mentally-rationed, by each citizen, [when // where] means because educational-limits', construe commercial-trade, away from socializing-limit-(s), leased on cause, due to introverted-individualism, culturing ones' lifestyle, from means' based believing minimalistic-prerogative, for trading and perspective, upon methodical-mass, existing during impassing-discourse, extracting-element-(s), for intent per day, in-an awareness, spiraling long-term-goal-(s), aging juxtaposed an simultaneous hundredths-thousandths-day-effect, cause-(s) self-extraspective-genetic-verbal-interaction, build on intuition, by creative-concept-(s), lessoned-interacting, by peer-(s) of equal-age, (not incest-breeding, deploying determined-individuals', whom base thought, not perspective, as how schedule of location-(s), are interpreted, interval-(ed) monthly-evaluated psychological-discourse, [influencing / persuading], an mass to enact an motive of longevity-existence, affirming [theatre / festival destination-goal-(s)] through discourse of event-(s), pertaining living-character-development-(s'), emphasizing offspring-preference-memory, determining prerogative of discourse, simply routine, when extraspective-genetic-simple-appreciation, influence Interaction & Ecological Discourse, during community-event-scheduling, awhile dating, not juxtapose, each individual, from birth to [present-dating-age-year // season // month // week // day // hours // minutes // secondths].149¶||||01m.:29s.28m.s.|||§ Intimate relationships, are [hard // difficult] to come by, awhile commercialistic-scheduling, [design // designate], interaction-(s) at sitting-entity-(ies), primarily consumerist, at an motion-adjusted-transit, without sense of awareness for extraction-effort-(s), on timed-schedule, by discourse of government-constitutional-cursive, relayed an inter-voluntary-purpose, per day, affecting sleep, by night smoothly, not invigorated by being at night-impasse, or effect of earth-centripetal-force-drift, when maneuvering-perspective, focus day-returning-force, by night-exerted-force, balanced (by of)-a-the [revolutionary-circumference-drift // rotation-influctuation-(s)] partaking in human-discourse, (from an) mean-(s) of humanity, whom conjure purpose by alive-intricating, incrementing blood-pulse-being-(s), which rate secondths-deviation, from the tangible-discourse, of interest, counted, by [interaction-moment // eating-impasse-substantial-nutrition-influence] in proportion during day-sequential-offset [matter-(s) /// issue-(s) /// topic-(s)], ordering, upon set-parameter-(s) awhile day, define a-the quadrant-perimeter-(s) designated for idle-success, where active-participation-interaction, further development, (from an) [weight // measure // counting] at population-Federal-needs', apose formal-territory-clause-resourcing.150¶||||02m.:29s.57m.s.|||§ Now I do not know per-se, but I am at heart, extraspective, in my demeanor for existence.151¶||||02m.:36s.04m.s.|||§ Finding others', is not inclined naturally, for how one thrives' from interaction-(s);152 but when I get such an rush of energy, when active, and verbal for-how observational-existence, [elapse // synapse // cycle]35L. before our very visual-reverberational-sound-wave-perspective-(s')...¶||||02m.:51s.19m.s.|||§

§...Count-[15 / Fifteen]-preposition-of.. ..Essay-[25 / twenty-Five].. // ..Essay-[71 / Seventy-One].. ..Lines-[35.59 / Thirty-Five.1.five-tenths-2.nine-hundredths].. ..Interacting fervor, revolve around each perspective, as count offset an compositional-issue per parameter, as individualism, square-off-meters, without an denominative-determinant-cause, per individual at premise-meter-(s), exclusive at cause, when at-will-employment exist with no prerogative of individual-purpose, at sedentary-dating, each revolution of combustion-solar-molten, apose human-genetic-dispertion, apose tectonic-plate-populat-ion-genetic-characteristics, motivating action-(s) when numerical-inclinations, occur intermittent-sporadically, at off-time-prerogative-(s), from individual-motion-(s), denominative-accounting, from sedimentary-physics-still-exitance.¶|||40s.50m.s.|||§

¶72. Being-(s) are Born & Raised, in-an systematic-order, where introverted-affair-(s'), obtain-item-(s), by [leased-abode / title-(d)-property-spacing-dynamic] while each way-(s') of life, bring in-to perspective, an hold of objective, identifying-individualism, characterized by natural-will, lacking human-being-mass-subservience, (as an) code of arms', for handling velocity in observance of Physics & Territorial-Surface-Area-Spacing-(s).152¶|||21s.16m.s.|||§ Are woman any different from men? Or are their only fact-(s) by-per-of Persecution & Petulance, instigating a-the dynamic of average-discourse, simultaneous-amidst [movement-(s) // motion-(s)] from-an [Hallow-Listening & Seeing], understanding through touch-possession, enveloped-routine, by way of local-insuring, having an piece-(ing)-(ed) lived-interaction-(s), without confidently pursuing an goal of sensation, as-an impassionate-cause, reasoning, reproductive-habitant-lifestyle in physiological-fact, by cardiovascular-discourse, because observed sight, define from-an formal-documenting-verifying of data, with usage, in cause (from an) anatomical-physiological-cardiovascular-response, pertained upon each effect-(s) of extraspective-mass-influence,13L. impassing-universe, stable upon soil-terrain, for resourcing-material-(s), from concept-present, relative-relay of objective-collective-usage...¶|||01m.:07s.57m.s.|||§

{Article:} [March]

§...Count-[11 / Eleven]-preposition-of.. ..Article-Essay-[26 / Twenty-Six].. // ..Book-Essay-[72 / Seventy-Two].. ..Lines-[13.695 / Thirteen.$_{1.six-}$tenths-$_{2.nine-}$hundredths-$_{3.five-}$thousandths]....ritardando to normal shading upon staff-fluent-note-intonation-progression, as when intervoluntary-synapses react from voluntary-discourse when posterior-individuals, stand for an scheduled-objective, amid entity-title-trusting-(s), from an square-feet-two-dimensional-ego-perspect-ive, apose three-dimensional-superego-mobius-perspective, through physics revolving-gap. ¶|||25s.$_{78m.s.|||}$§

$_{¶73}$Munching on memory, is an part of how, live-silence, be-an effect upon those senses'-inhibited, due-to historical-discourse, without scaling [read-data / measure of spaced-terrain];$_{153}$ defining from subjective-perspective, live-physical-act-(s), pertinent though process, which instill notion-(s) of state, through men-mass-influence, per hour of day, passing in-as-an part of say, Eleven-Ante-Meridian {11:00A.M.}, for brunch, apose interval-prior-period-(s), eating intermittent lunch-from period-(s), or Five-Prime-Meridian {5:00P.M.}, for Supper & Drinking, after work, not in communication at night by religious-chaste-interaction-(s), persuaded amongst mass-action-(s), influenced no directed tangible-fact-(s), upon vision at physiological-anatomical-discourse, focusing daily-discourse, through hourly-accounted, entity-transit-cycling, settled, in-an census-population, having manner-(s), to reach out to those act-(s), yet conceived for [Hypothesized / Rudimental Purpose], being, motivated by human-(s')-physical-similar-presence, apose-posit, women from men, whom remain amidst commercial-production, by-an based-discourse, not in thought, for how power, or superior-resourcing, materialized, due to each piece of inhabitant, equivalent matter, from-of perspective, unconfirmed due-to thought-rates, influx off-time-moment-(s), idle, vacant-minded-mentality, cause for each day, be too short for engineering-formal-thinking, but [where // when] day not refine mass upon presence, applicating compositional-matter-reformation, in lieu of conceptive-directive, offspringing from-an spouse-schedule, intuitive, from tense, for discourse of mass-schedule-directive-(s), as a-the millennium-transit-continuum-impasse, where each notion-(s), not react of tense-(s), for how every formal-interaction of community, pertain [minute-(s) // secondths // dating-anticipation-(s) of hour-(s)], as locational-obligation-(s), by nation-state-(s);154 define where mineral-deposit-collecting-(s'), sustem-from commercial-conceptive-governmental-denominating, those data-find-(s), when visual-conscious-reasoning, elaborate each inhabitant-discourse, economically-considered, through year-mass-fundamental-development-(s'), differing non-thought of data, because present-state-(s) of affair-(s), sustem-from a-the mass-individualism-bias, that ensue, cause tangible-defining, not ever affirm reliable-trust of data, because, live-temporal-day, upon calendar-interval-(s), examine each piece-(s) of information, for-an calculated-code-(s), where word-interpretation-(s), [synapse // elapse] syntax-sentence per-an paragraph-discourse, cause mode of humanity, pre-set living, by confide-(s) of Public-Space & Housing-Posit-(s),| 307 |(not) (for-an) through-thought (as of) simultaneous-person-analyzation, leaving blank-memory, intermittent-upon tectonic-plate-history, never upon an theory of ecology, emphasizing those composite-material-(s) of trade, from state-redevelopment-locale,| 308 |awhile America-(s')-Federal-resourcing-abstract-state-deviation-(s), deter (from an) international-analysis of transit, upon mass-population, because inhabitant-order, not formulate amidst fluent-dialogue of colloquial-thought, pertaining [mass / persons'] order-contrast, waiting by [month / year calendar-count-(s)] depicting industrial-product-(s), as conjure for formal-developing of territory, [where // when] material-(s) of space, require active-thinking, of-how, (re) order-(ing), those assorted-variations of hardware, remain in lieu of fact / (oring) by-each [day / week-interval-deviation / month-documenting-offset / seasonal-comparing day-effort-(s) // weekly-mass-picture-discourse-interval-(s)] as method for effective reapplying, direct-concerted-effort-(s), in case where [weight-(s) / measure-(s) / distance-(s) / dimension-(s)] affect resource-(s) of nation-state, from denominative-measuring, of-each enumerated-discourse, from North-American-Quadrant-Hemisphere-Tresurigal-discourse, where civilization-offset-(s), differentiate social-security-dynamic, per introverted-disguise, adhering privacy of character, in form of state-objective-cause, pertained when each development of physiological-dating, in lieu of order, influence those base-notion-(s), where all human-(s) are of-an family-mute-introverted-thinking-process, not coherent for contrast at daily-simultaneous-rotation-synapse-impasse-(s), during revolutionary-physics, drifting off, each elapse-(s), amidst drift of day-oxygenation, pertaining night-solidification-(s) per [substantial / solid matter-(s)],46L. in line of dating, because mass-developmental-millennium-discourse, relax thoughts' for from mute-impasse, constructing-an default-directive, by inhabitant-mass-existence, for an default-intervoluntary-motions', of male-political-existence...¶|||04m.:11s.41m.s.|||§

§...Count-[38 / Thirty-Eight]-preposition-of.. ..Article-Essay-[27 / Twenty-Seven].. //..Book-Essay-[73 / Seventy-Three].. ..Lines-[47.82 / Fourty-Seven.1.eight-tenths-2.two-hundred-ths].. ..Without scaling an acreage of living, purpose by those metropolitan-miles, not intercede human-introverted-individualism-bias, leaving how government is to handle politics, different from each individual-handling, as when an climate of terrain, not computate what is significant for handling each surrounding inhabitable-parameters, amid an solar-daily-impasse-discourse, under ultra-violent-rays upon tectonic-taxonomic-plant-crop-regenerations, simultaneous-human-racial-ethnic-population-(s).¶|||33s.53m.s.|||§

¶[74]When [[1]does / [2]will / [3]could / [4]has / [5]for-of / [6]pronoun / [7]act for natural-yield-(s) / [8]effort of law / [9]medical-deoxyribonucleic-blood / [10]governing of land-territorial-count-(s)] in thirdths-power-current-charge, branch-(es)-simultaneous, affirmative-action-(s), by executive branch [{White-House} / legislative branch {Capitol-Hill} / judicial-system-tier (legislative-grounds) {supreme court {Federal}-{State}-{Superior-Court-County-System}] not affirm city-district-residential-usage, by executive-limit-(s), limited defined papered humanity, to confirm each and every factual-usage of article-(s), in-an case trade, scaling matter-(s), by-an time-interval-(s)-precedence, maintaining an form-documenting-offset-instruction-manual-application-contract-W-2-401K-plan-discourse, set from present-impasse, for pre-future-instatement, requisite analytical-ruse, following each prerogative due-from, past-requisite under Eisenhower-Madison-(s)-progression-present-perspective, referencing amongst comparisons' of data, upon scheduled-day-(te)-hour-routine-designation-repetition-pattern-(s), pre-set on random-normal-discourse, pertain-ing religious-obligation, in order-lessoning-function, fulfilling free-will-effort-(s), prerogative for Man & Data by the dynamic-discourse, existing from each repetitive-act-(s), by-an developmental surface-area, designating surface-area-approximate-recourse, motioning each-fatigue, from man-hold, obtaining hardware-data, by-an conceptive-diagramming, pre-acted, for-of resourcing;[155] constraint, because-of term-(s), claused by [condition-(s) // definition-(s) of nature] by movement-(s) intermittent surface-area, harbored at each case of directive, efforted by mute-social-security-interaction-(s), [where // when] trust by aging, sustem from independent-individual-(s'), not in-an exponential-simultaneous, cause physique-offset, not in live-active-tense, per [passing / acting / interacting deviation-(s)] succinct interposed interval-integer-notion-even-odd-reasoning-(s), upon surrounding-observation-(s), [[1]prefix / [2]participle / [3]word / [4]term / [5]definition / [6]paragraph / [7]sentence / [8]suffix] from literary-opto-grammatical-punctuation-hearing-vibrational-resonance, by constant-setting-offset-(s), amidst [height-width-perimeter-parameter-surface-area / tone-instru-ment-hour / intonation-vocal-logic-harmony] lesson-(ing) age-synchronous-gap, existed in present from past-video-requisite, as how, saving-data, slip past, each inhabitant, during denominative-currency-transaction-(s), as while each populant-(s), guarantee, non conclusive-determined-momentum, when impasse study-(ies), where human-biological-directive, requisite directive-work-order, consequential from directive-discourse, as free-will-gauge, not scale live-data, apposing archaic-data, recalibrating reference at present, to be incapable retaining produce-(s), by male-aging-pre-instated-directive, for marine-military-guidance, by mass-population-common-discourse, affecting how mutual-genetic-intelligent-(s'), not exist, for religious-pursuance, sit-listened, continuing [1Custom-(s) // 2Habit-(s) // 3Hobby-(s) // 4Homage // 5Method-(s) // 6Mysticism // 7Native // 8Ritual-(s) // 9Tradition-(s) // 10Value-(s)] by present-mass-balancing, future-elongation from past-trust-expectancy, upon present-impasse-comprehending, because finite-able, pre-requisite-present, distanced by weight-(s), stocked by shelf-(ves), modified from perspective-elapse-(s), not by-an peripheral-direct-sight, through passing, intermittent every day, where location-(s) for motivation, are requisite brute-idolizing, presiding-(s) where-at discourse, (board) meeting-(ed), in-to schedule, by federal-transaction, amused passed minute-(s)-secondths-retrograde-counting, upon revolving hour-elapse-three-thousand,| 309 |six-hundred-secondths, by simultaneous-mass-distance-perspective-offset, not denominative (from an) singular-day, while those [vacant / empty / blank / abstract / unclear, elapse-(s) of space-(ing)-(s)], remain apart, titled-data, interposed position-(s) by cubical-methodical-trading, influencing each surface-area, by-an human-limit-(s), not free from coverage of territorial-ground-(s), where offsets of various perimeter-surface-area-(s), affect how conjured observation-(s), requisite-sustem tense-(s)-impasse-visual-sound-wave-vibrations-reverberation-(s),43L. as how evocation, imbue present-existence...¶|||03m.:46s.58m.s.|||§

§...Count-[12 / Twelve]-preposition-of.. ..Article-Essay-[28 / Twenty-Eight].. // ..Book-Essay-[74 / Seventy-Four].. ..Lines-[43.53 / Fourty-Three.1.five-tenths-2.three-hundredths].. ..While circulating, each individual serves an role upon schedule, for an set amount of hours, working an wage per instrument-product, circulating the entity-premises, for whom inhabit an entity at any given time of designated-working.¶|||18s.77m.s.|||§

¶[75.]Pitch-vocal-vernacular-median-volume-decibel-frequency-offset, not harmonize tone-vibration-decibel-beat-note-mark-instrument-toning,| 310 |upon interior-tone-space-resonance-exterior-reverberation-(s), upon parameter-cubic-interior-square-feet, as for defining constraint-area, not as philharmonic-stadium-intonation-development, affecting tangible-overlay, where [federal-roadway-development-resourcing / planned-constructed-space-(ing-(s)) / object-stance-statute-(s)] have an mathematical-timbre, reasoning blood-pulse-elapse-rate-second-tic-median, by word-sentence-paragraph-rate-observation-offset, for pre-tensing-concept-focus, by goal-(s) per [week / month / season / year / two-year / four-year / decade -reset-(s)] for gauging-perspective, from depth-perspective, upon revolutionary-rotational-centripetal-force-tectonic-plate-drift, ordering by-an fluent-periodical-sequential-progressive-sequence-(ing), year-scaling, elapse-expiry-matter, sorting-data, by storage-spacing-site-(s), by mass-count-variable-dimension-comparisons, distancing constructed-entity-(ies), in thought of common-motion-impasse-(s), so then much of-those philosophical-theory-(s), from present-elapse, sustain an date-maxim, while-at location, resetting [year-W-2-form / day-receipt-(s)] casing each fact, from date-inclinations, that if not sustained, what can be verified, for [confirming // verifying fact-(s)] except| 311 | [governmental-Federal-State-county-legal-discourse / library-author-fiction-mystery-auto-biographical-historical-account-denominating / newspaper / encyclopedia-documenting] in-ept of [dictionary-purpose, by diameter / circumference / *radius-dimension-volume-solid-substantial-organ-composition-proportional-density-pressure-reasoning, per-of matter] by [blood-pulse-respirational-observation / subjective-documenting / post-present-interval-reasoning] by method of dating, upon live-presence, not act before-day, or past-day, as continuum-infinitive-reset-(s), incur per hour, claused by aging, ordering gross-object-(s) proportional parameter-pre-instated-guideline-intention, because blood-pulse-sensation, not influence each constraint, where still-idle-human-blood-pulse-cardiovascular-respitorial-muscular-motion-(s)-reflex-(ing), pertain hyper-tense-(s') from [eye-sight / ear-canal-hearing / nose-organ-smelling / tounge-ventricle-skin-(Na)-tasting] substantial-matter, conceiving appreciation, intermittent elapse-deviation-(s) period-icalizing day-denominative-offset-roundup, not in-an year per decade comprehension-update-tier-ordering-mass-offset-(s), concluding season enough to be in year, but not of an collective-aging-effort-(s), due-to via-identified-individualism, self-coherent, by personal-possessive-perspective, in lieu for no vocal-vernacular-literary-loyalty, reading facts of dating, from present-date,| 312 | impassing opposite-sexual-offset-(s), incapable of exponent-matter-reproduction, per mass-repopulation, construing conceptive-anticipation, for Documenting & Counting-Data, Supply & Demanding, those mean-(s') from physiological-blood-pulse, verifying value-(s), informing from fact-(s), those factor-(s) of mass-production-depreciation, to have an influence from physiological-documenting, without false-critique (for which I continue to open-ground-(s) for thinking of), human-discourse, remaining intermittent mobius movement-stop-go-continuum, wanting idle-trust, commercial-component-(s), where trade of state, sustain sworn obligational-contract-trading, yet for intended-conceptive-state-mass-matter-schedule-development, cause of the means of Individual & Mass, intersexual-exponent-reproductive-resourcing, attempting to fulfill (a the) formal-proportional-deviation-(s),38L.| 313 |per tectonic-plate-surface-area-coverage-population-loyalty-deviation-(s)...¶|||03m.: 29s.22m.s.|||§

§...Count-[13 / Thirteen]-preposition-of.. ..Article-Essay-[29 / Twenty-Nine].. // ..Book- Essay-[75 / Seventy-Five].. ..Lines-[38.07 / Thirty-Eight.1.zero-tenths-2.seven-hundredths].. ..As for surface-area, if individuals do not perceive of such surrounding surface-area, then perception remain flat-omitted, not interpreting each function of surrounding-parameters, when dating timed surface-area, per titled-premises, as when individuals not cooperate from high-school, in an community-function, per activity, when processing each daily-function of titled-land.¶|||25s.59m.s.|||§

¶[76]Normally, human-(s') pass off centripetal-force, ignoring ecological-trading-routine-limit-(s), ordering-discourse for-of movement-(s), present, when age impasse during year-decade-offspring-friend-resetting, identifying Age & Generation, affluent pre-interpretation, pertaining subjective-reference, by the legal-paper-dynamic, insurrected mundane-truth, (un)-factored from judiciary-review, but commercially-executed,| 314 |conjuncting constitutional-belief, for awe by-of mass-equivalent-information, while order of routine-sequential-placing-(s), pattern [posit / transit / maneuver / reflect / contemplate-intangible-perspective / sleep] as method of [mind / body / soul], fas est et ab hoste doceri future-belief, of Western-freedom, when wilting-away, hold regard of aging-effort-purpose, as invigorating individual-self, by life-cycling, influence-(d)-(ing), each passing of date, amongst [family // neighborhood // community-(ies)] while momentum-elapse-four-minute-(s), passing-an degree-per-elapse, while thought under frequency-measure-cutoff-point-(s), subscribe extraspective-degree, not in congruent-angle-(s), denominative per measure-balancing, integer-variable-(s), per degree incremental reset-(s) of [density / substance / liquid-tangible-setting-offset-(s)] not detail-elaborate, individual-human-autistic-hyper-active-self-extraspective-offset, defining perimeter-(s) of distance-(s), from tectonic-repopulation-dynamic, not in scheduled-form, maintaining those product-material-(s), that enable [safe / successful discourse of action-(s)] by pre-written-anticipation-method, for accounting [notion-(s) // inclination-(s) // indication-(s)] as blood-pulse, conjure-comprehend data, in-an focus of subjective-reference-redact-analysis, through upon objective-interval-use-place-action-(s), pre-sent-reset-requisite-notion-simultaneous momentum, not individually-affected, as legal-boundary, executively-act, without either the legislational-discourse, or literary-discourse, for empirical-reset-entity-order-(s), requiring restroom-distance, approximate from bowel-digestion-effect, which all human-being-(s), are in-an objective-limit-compliance, due to [timed-locational motion-(s) // subjective application] that yet be practiced of centralized-governmental-action-(s), surveying-mass, due in part, each day meander-puzzle-effect-impact, piecing-together, detracting-interaction-(s), yet feasibly in temperature-stable-offset-(s), inclined daily history, not in an live-actual-era, of lifestyle, for-when each inhibition is indicatable, directing scheduled-will wilted-off, because living, be massively-inclined, through forefront, at-those abode-(s),| 315 |emboss-reflecting, remembered-anatomical-memory, singled from numerated-thinking, free not to comprehend, denominative-discourse, aging individualism, without an documenting per individual, for intricate-observational-impasse-interaction-(s), defining dictionary-data-limit-(s), having encyclopedia-effect, lie fiction-(ed)-genre-(s), comprehending each rule of entity-(ies)-development, as clue for data-collecting, when each date can be meter-(ed) for count-(s) per data-information-bit, interacted when reasoning, affect mass physical-effort-(s), never to partake in subjective-documenting-observation-motion-(s), affirmed through [data-hour // minute-(s), increment-(s)] in spacing-written-reflex, upon [territorial placing-(s) / Universe-(al)-(ity) (-(al) city in Los Angeles) / tectonic plate-(s) / oceanic body-(ies) / human-mass-(es)-{race-(s)}-{ethnicity-(ies)} / mass-objective-denominating / planet / galaxy / university / city / county / state / federal / housing / bank-(s) / market-(s) / lot-(s) / road-transit-reset / field / park-(s) / crop-rotation-perimeter-parameter / empty-barren-terrain / government-station / city-planning / empty-cubicle-spacing-(s)] apart of-each-momentum-clause, condition-(ed)-discourse by existence, less-referenced, each act, from more referenced motion-(s), incremental-interval-act-(ed)-(ion-(s)), from self-focus (myself upon the constraint-(s) per data-usage, from government-restriction-(s), unvocalized, perceptive-motive, when arrangement cause, an developmental-diagram-concepting abound presence, for post-life-aging-value-(s), in comprehensive-usage, preferred conjure, over game-control-idling, rendering human-perspective, incapable for most of-those scenario-(s), pertaining twenty-four-hours-per-day, simultaneous year-denominative-aspect-focus, remember-ing attention-span, not present active-day-hour-sequential-order-deviation-referencing, from "read-mind",| 316 |for thinking {Reading}-{Writing}-{Redacting}-{Comprehending}, clause-circumstantial-temporal-conditional-advisory-data, as empirical-offset-inclined receipt-(ed)-purchase, prior, tabled-agenda-operation-(s), what define dues' upon Daily-Liberal-Maintaining & Handling commercial-production amidst, lifestyle communal-neighborhood-(s), as-an basis from every-moment, deviated-by, Shipping & Handling, from Sociological-Demand & Supply-Production.[153]¶|||04m.:18s.44m.s.|||§ Budgeted-staff from spontaneous-customer-interaction-(s), pertain merchanted Shipping & Handling, sovereign-domestic-trade, in hold where formulate-(tion-(s)), sustem fact of receipt-(ed)-purchase, from-prior antedote, table-(d) intermittent Maintain-(ing) & Handle-(ing), while agenda-operation-(s), sustain daily-liberal-production, developing due-upon [stock / market / receive-(ing) requisite-merchandising] where-at an area, communal continual-boutique, upkept sale-(s)-presence, performing custom-(s) by repetition, not denominated [by // from] History & Present, depended-simultaneous for each coherent-thought-process, cause each-individual-offset, interpret posit, without solar-time-implication, proportioning each-day, by [morning-future-present // mid-day-action-present // afternoon-past-present-continuum-offset-reset-repetition-continuum-influx-operations'-aspect].[154]¶|||05m.:04s.29m.s.|||§ Every day is presented an incumbent-process, illiterate present-juxtapositions, because, introverted-perspective, present an presence at an impasse per-moment, that influctuate-focus, because how motion-(s), motivate-fact, where notion-(s), make an physiological-perspective, pertain upon an tectonic-surrounding-(s) implemented at an default-operations', (un)-preferred of manipulation, because how, progression-impasse, affect each of those purpose-(s), at an notion-perspective, where-at, motivation, for delving subjection, apose construction, requires objective-mass-population-engineering, to affirm, how reference-article-incremental-deductive-indicated-interpretational-induct-ive-understanding, applicate-(s) where each dating of human-subjunction, remain without affirmation at-of communal-religious-et-cetera-progression, upon an communal-perspective, per individual-appli-cation...¶|||05m.:45s.02m.s.|||§

§...Count-[22 / Twenty-Two]-preposition-of.. ..Article-Essay-[30 / Thirty].. // ..Book-Essay-[76 / Seventy-Six].. ..Lines-[68.17 / Sixty-Eight.1.one-tenths-2.seven-hundredths].. ..How to schedule your child from birth until adulthood, is an conundrum I encounter, that through literature, I would hope to develop book-forums, for different decade-aging-stages, when understanding how we are dying, and require wage-savings, to be hoarded by an group of individuals, whom cooperate through family-events, funding money for an nation, by first buying mid-west-spaced-terrain, for millions of dollars, for then commercially-developing constructive-premises, for later engineering-directives, practicing how to resource-elemental-compositions, for then studying how commercial-governmental-patents, are to be factory-produced and refined, on an basis of unit-counts, proportioning the ecological-entity-(ies)-routine-scheduling, per individual at hand for helping construct each notion of individual-group-motions-function, pending the ecological-environment circumstances, due to each fact of circulated-motions.¶|||48s.25m.s.|||§

¶77Thousandths-deductive, not intertwine, cause of value-(s), where-when [1government-center-(s) / 2superior-court / 3senate-congressional-hall-{In God We Trust} / 4E Pluribus Unum-House-Hall / 5County-Court / 6State-Supreme-Court / 7Attorney-(bail-(s)-bond)-office-(s) / 8Doctor-Office-(s) / 9Independent-New-Auto-Center-(s) / 10Mall-Parcel-dynamic-open-space / 11Private-plumbing-office-(s) / 12Electric-Plant-(s) / 13Open-air-routed-current-(s)-space-(ing-(s)) / 14Militarial-Base-(s) // 15Operation-(s) /// 16{M.E.P.S.} Military-Enlistment-Processing-Station / [16]Auto-Part-(s)-Store-center-(s) / [17]The-Home-Depot / [18]University-build-(ing-(s) / [19]Gymnasium / [20]Arena / [21]Stadium-(s) / [22]Elemental-resource-Objective-Factory-Plant-(s) / [23]Workout-facility-(ies) / [24]Park-Center-(s) / [25]fundamental-Elementary-School-(s) / [26]approximate-Middle-School-(s) / [27]local-High-School-(s) / [28]College-(s) / [29]Institute-(s) / [30]Plant-Center-(s) / [31]City-Hall-(s) / [32]County-Hall-(s) / [33]Bank-(s) / [34]Concert-Hall-(s) / [35]Laundry-Mat-(s) / [36]Water-Treatment-Plant-(s) / [37]Firefighter-Station-(s) / [38]Office-Supply-Store-(s) / [39]Pet-Supply-Store-(s) / [40]Hardware-Store-(s) / [41]Golf-Course-edifice / [42]Funeral-Home-(s) / [43]Mortuary / [44]Insurance-Center-(s) / [45]Dorm-Facility-(ies) / [46]Museum-(s) / [47]Coffee-Shop-(s) / [48]Strip-Mall-(s) / [49]Federal-Bureau of Investigation / [50]Pawn-Shop-(s) / [51]Disney-Land // [51.1]World // [51.2]Resort-(s) / [51.3]Hotel-(s) // [52]Motel-(s) / [53]Post-Office / [54]Private-Shipping-Store-(s)-warehouse-(s) / [55]Shipping & Handling plant-(s) / [56]Marine-Warehouses / [57]Tire-depot-(s) / [58]Smog-Check-Center-(s) / [59]Sporting-Apparel-center-(s) / [60]Smithsonian / [61]Edison-Factory / [62]Capitol-Hill / [63]State-Capital-Build-(ing-(s)) / [64]Barber-Shop-(s) / [65]Bicycle-Shop-(s) / [66]Lounge-(s) / [67]Theatre-(s) / [68]Federal-Reserves / [69]Treasury-(s) / [70]Post-Office-sequential-orchestrated-entity-operation-(s)], municipal-coded by county-jail-city-operation-(s), per civilian-citizen-establishment-(s), active-day-hour-sequential-order-deviation-referencing, from "read-mind",| 317 |for thinking {Reading}-{Writing}-{Redacting}-{Comprehending}, clause-circumstantial-temporal-conditional-advisory-data, as empirical-offset-reminder-cycling, having alert-reminding-(s), for not verifying human-characteristic-(s), naturally accepted, upon fact of reasoning, through day in existence, [to / from] mass-offset-(s), perceivable-composition-data, where rotational-revolutionary-impact, amass by mundane-default-condition-(s), applied, just for the sake-of, natural-perceptive-existence, not explained from human-perspective, for illiterate-fact-(s), inanimate, of study, because preferred-ignorance, affirming how tempo, is deviated, presence-tense, when human-(s), prefer, illiterate-alliteration, rather, variation of genetic-preference, insisted, from perspective at birth, yet to realize genetic-intrication, because, how individualism, pertain upon each impasse of perspective, where-at-focus, from individualism, because, what motivate, priori, whom motivate each synapse of physiological-existence, from how [academia /// effort /// conduct] affect each outcome of logic, (from an) pre-requisite, upon present-tense-individual-population-offset-continuum, [where /// when // why] each article, is of an preference for each fact of life, because how focus pertain, where influence, [means // ways] each fact of notion-(s), developing construction, at an impasse, from will-trusted-trade;[156] emphasized, from lineage-transition-trade, because how [fact-(s) / proof-(s)], pertain per day, awhile article-usage, is rather preferenced, idiomatic-ideological-introverted-reflection, upon human-mass-instinct-cogito...¶|||02m.:40s.82m.s.|||§

§...Count-[12 / Twelve]-preposition-of.. ..Article-Essay-[31 / Thirty-One].. // ..Book-Essay-[77 / Seventy-Seven].. ..Lines-[35.43 / Thirty-Five.1.four-tenths-2.three-hundredths].. ..How to access each entity-(ies) objective needs, is an issue I encounter, for understanding how factory-resource-ordering, affect market refined-product-box-unit-count-(s), scheduled along an weekly-monthly-basis, circulating interior-entity-(ies)-development, practicing on saving items, to embellish the entity-(s') quality, depending on centered-designation, pending on the circulated-scheduled-motivation, of a the social-mass and circulating of entity-operators.¶|||28s.28m.s.|||§

¶[78]March please! If not for devout-cause, can perspective perspire, yet of due-(s), required in lieu of action-(s), unprepped by literary-discoursing, for abstract-ambient, not sustain cohesion, gauge-perceiving from [Ma-(e)-n / Woma-(e)-n / Children / Elder-(s) / dead-sentimental-relative-mass] an impact per density-volume-pressure;[157] from pronoun-responsibility-(ies), per mass-equivalent-direction, due to those ideal-(s) at an figurative-perspective, not factoring-space, as government documentation for observational-concept-diagram-model-planning, untrue per fact, impassed-inspiration, that is desired, inhabitant-mode, causing idle-memory, sustained-unclear-reason-relay, as-why commercial-production, be rated by extraspective-gauge, sustaining matter-development, per collection of present-circumstance, from no confirmation-(s) from living-discourse, perennially-present simultaneous, short-intrication, only matter-(ed)-affair-(s), from mammalistic-regard, Documenting & Individual-Impact-Identifying, state-political-discourse, when Arrangement & Order, suffice-upon present tempo-conjecture yet abbreviated each repetitive-reference, from each day, affluent anno-domini-count, from-past-requisite, before fact at-of-upon difference-(ing),| 318 |display past-mass-year-deviation-ration-(ing-(s)).155¶|||01m.:-09s.97m.s.|||§ What second be interrogative, from how fraction-(s) per truth, not balance value, present-focus-elapse, due-to what moment per season, invest-adjecture, for modification at archaic-moment-(um), present-mass-deviation-(s), where-when prerogative, visualize factual-basis, before each state-thesis, prestated-instate-diploma-appreciation-value, merited upon, reaction-(s) from material-object-scheduling, as matter-itinerant, simultaneous symmetrical-reaction, segregate-second-(ths), as-due from process, developing each moment controlled, from command, an perceptive-value-adjunction, by surface-area in common-perspective, philosophically-inclined, instating-fact, from per thought, by [budget / receipt / title / currency / trade] intended schedule-agenda-simultaneous-note-currency-enumerated-freedom-(s), notion-extended, per reflexive-tense, grandeur granule-increment-article-substaining;158 at-an domestic-trade, formulated from each fact, inclined-receipt-(ed)-purchase, prior, tabled-agenda-operation-(s), due-to Daily-Liberal-Maintaining & Handling established residential-(ce-(s)), moded-daily-act-(s),| 319 |entranced road-transit-commercial-operation-(s), at-an general-dating, (in)-formal where interpretation-(s), increment-interaction-(s)-requisite-comprehension, interacting-at formal composite individual-youth-comprehension-study-(ies)| 320

|(blood-article-subterior-reemphasis), from when extraspective-elemental-modification-(s),| 321 |balance.29L. article-scale-balance-population-modification-focus...¶|||02m.:23s.28m.s.|||§

§...Count-[4 / Four]-preposition-of.. ..Article-Essay-[32 / Thirty-Two].. // ..Book-Essay-[78 / Seventy-Eight].. ..Lines-[29.4 / Twenty-Nine.1.four-tenths] ..As would I would reason, you should be informed, that the punctuation has fallen off, from the preposition-deviation.¶|||12s.84m.s.|||§

¶79Religion depict, international-American-broad-record, traded by at an mass-routine-adjust-(ing-(s)), per sort of time, when-where memorial-action, believe work at-an tense-(s), clivity clause, by [where // when // why // whom] state-(sis) event-(s), per superego, apose persuasion, incremental per step-(s), at an [denominative-tense-(s)-lapse-physique-stretch / forum-talk] unvocalized by pronoun-mode, touch-character-(ized)-(s), from respirational-synapse-blood-pulse-tension-oblong-skin-bone-muscle-organ-ligament-ventricle-compositional-beings', before-breath in nature acclivity respirate, (in) cardiovascular interaction, presumed deviation-(s), at place, when self-view, still (un)-cause-(d) at effort-(s), during fatigue-endure, from tranquility-peace, fair for denominative-reason-(ing), while education pertain exterior-occupation, scheduling referred-object-(s), [where // when-fact-(s)] interpose present-day, intermittent-voluntary physiological-impasse-silent-perspective-offset, as like an [duck // cow // bull // dog // pig // hog // sow // et cetera of mammal-(s)] responding upon surroundings, not assertively-conquered, acceded incremental-deductive-thinking,| 322 |because how tension-imagery-respirational-blood-pulse-perspective (time), juxtapose-offset, physics-earth-global-aquatic-tectonic-nation-state-county-city-district-acre-objective-posit-(s)-purchase-aspect|323|(calendar).156¶|||01m.:19s. 09m.s.|||§ Government-Periodical, is not fashion-(ed) by-an accrue, intended through, pragmatic-effort-(s), expensing infinitive-article-clivity-increment-(s), piecing from pronoun-(s), how matters' have per realm, blind sum-(s) of data, from solar-fusion-greater-superfluous-matters', while focus pertain, attuned-present-work-(er)-man-(s')-sterile-yield, percentage-fraction,| 324 |class-act-present-spatial-normal-effect-(s), awhile attuned objective-article usage-repetition, attune-present, at entity-space-(ing-(s)), upon individually-compartmentalized ids', from self, as-when bias-hoc-id-routine, affirm future-present, from-during English-present-memory, featuring, present familiar-affinity for [family // neighbors // readers' at still-solace-place] intermittent loitered-moment-(s), when each fatigue-(s), from muscle, interpose subduction-zone-respiration, underneath crust-posit-density, apose ways' of resourced-means', before, simultaneous-system-offset-function,| 325 |space pronoun-objective, awhile self-centered-independent-confidence, be implicated during-intermittent, respirational-synapse-blood-pulse-beat-rhythm -(s), averaged by individual-independent-dependent-spouse-(s), personally-handled, when affair-(s) rely (from an) natural-physiological-dependency, based on day, to order-fact-(s), when-where-by, [[1]decibel-sound-reverberation-(s) // [2]resonance /// [3]tone-pitch-intonation //// [4]tempo // [5]Key-Signature-(s) ///// [6]Measure-(s) ////// [7]beat-(s) /////// [8]note-(s) //////// [9]vocal-vernacular-variation / [10]sight-color / [11]shape / [12]depth-perspective / [13]object-(s) of shaped-matter (form-perceptive) / [14]subject-discourse] weight-(ed) micro-measure, discoursed when tense-(s), suture between voluntary-diligent-(s)-affair-(s), [[1]forming // [2]manipulating // [3]cultivating // [4]developing // [5]conjuring // [6]assenting // [7]aspecting // [8]affirming // [9]enhancing] Each & Every;159 of those piece-(s) of data, for objective-juxtaposition, | 326 |upon respirational-synapse-ventricle-blood-pulse-muscle-ligament-organ-tension-pressure-skin-compositional-continuum-bone-nail-posit-offset-(s), by-of, personal-perceptive-charact-er... ¶|||02m.:52s.25m.s.|||§

§...Count-[8 / eight]-preposition-of.. ..Article-Essay-[33 / Thirty-three].. // ..Book-Essay-[79 / Seventy-Nine]. ..Lines-[34.78 / Thirty-Four.1.seven-tenths-2.eight-hundredths].. ..Common-human-beings, introversion shows, not understanding the significance of language, from book-context, as for how to convey each piece of objective-worth, from when housing should simply require beds, furnishing those cooridors of inhabiting, amid an proportional-reasoning of each unit-object, upon an [residential / commercial-entity-circulation], of each factored-unit, for ecological-scaling, measuring those distances of level-proportions, for cycling the geological-terrain, for resourcing from compositional-matter, those refinements of directed-factory-objective-(s), upon an surrounding parameters, defining how terrain is suppose to balance human-genetic-being, housing entity-(ies), commercial-entities, resourcing-sites, factory-production-ware-houses, et cetera, for objective-placing-storage, scheduled-functional-usage-(s).¶|||42s.44m.s.|||§

¶80. Discourse of rubric-cause, intermittent [year / child-(hood) / physical-mass] believing compulsive-pulse-beating-being-(s'), objective-error, developing-cause, perceiving communal-aging, instinctually-instated, when physiological-discourse, elapse-focus, during time-frame, due to presence, amongst-presence, informing ungrounded-affinity, by-at-an time-analytical-discourse, for calendar-date-observation-(s), foresworn during character-letter-theatre, mundane physical-fact, reverberating an rhythm-routine, as tense-confirmation-(s), account dry-substantial-composite-paper-page-sheeting, per date pleather, gamut in to area, as reaction-(s), place from state-subterior-infinitive-locale-(s), for introversion, by family-routine, influctuating-interaction-(s), as-when date elapse-(d), [[1]audit / [2]receipt / [3]car-note / [4]insurance-note / [5]city-water & power / [6]state-appliance-usage-electric-bill / [7]personal-food-income] trading product-(s), in Law & Nature, for role in conjure-(ing), climate synapse-elapse-(s) from each date-interval, while second-(ths) pass in role for dating-object-(ive-(s)), living from character-role-play, not directable at mutual-comprehension, in-of-an [fatigue / stress / elapse / synapse] accounting denominative-count-(ing)-(s), from video-motion-imagery-perspective, not counted, in present longevity-reset-(s), defining article-progression-(s), remained ambiguous defining-data, as what will-be, for truth, [value-(ing) / appreciation / determining] informed-data, while present-aptitude, fact-(or)-(s) data-hardware, by-an current-mysticism, routed at [entity / road-way / tectonic-barren-terrain] claused [where // when] fact-(s) of information, pertain physical-comprehension, upon physiological-capacity, viewing-matter throughout life, in routine per mode, acutely-accentuate-incrementally-present-form, where visualizing perceptive-tangible-mass-matter, per impasse of individual-posit-perspective, juxtapose-offset, from one another...¶|||01m.:36s.47m.s.|||§

§...Count-[4 / Four]-preposition-of.. ..Article-Essay-[34 / Thirty-Four].. // ..Book-Essay-[80 / Eighty].. ..Lines-[19.37 / Nineteen.1.three-tenths-2.seven-hundredths].. ..Still Counting those Essays their.¶|||06s.19m.s.|||§

¶81March indicated centralized-fact-(s), from-of perceptive-seasonal-revolutionary-rotational-celestial-pattern-(s), defined by basis of individual-perspective, upon visual-parcel-miniscule-matter-(s), pertaining at an relevant perimeter-parameter-impassing-(s), which order simultaneous military-routine, for command-base-operation-(s), defending those contexed logic-(s), implementing order, from free-will, instilling characteristic-(s), by pulse-(d)-physiological-anatomical-matter, following those command-(s), in progressive above-average-disparity, [1measuring // 2weighing // 3timing // 4dating // 5sequencing // 6scheduling // 7instructing // 8contexting, social-order], to maneuver objective-posit-(s), for factoring-(s), their-of, post-industrial-revolution ingenuity, relying auxiliary-vehicle-comprehension-(s), instated due [to // at] balance, obligated-trust, currenting return-payment-(s), (in an) economical-cycling, [when // where] citizen-(s')-prerogative, constraint-(s) decorum, persuaded by command-ordering-(s), where commercial-hard-data, pertain [to // from] each individual, whom abide by the military-discourse, cycling citizen-(s)-schedule-(s), (as an) juxtaposition-process, through commercial-religious-industrial-mass-trading, throughout those period-(s) where troop-count-(s), simultaneous-mass-ordering-(s), along with [labor // commercial production-(s)] sequencing act-order-process, defining each regiment, by regalia-output, timing impasse-(s), of-those circumstance-(s) of juxtaposition, by sociological-freedom, are uncapable of an national-ethnic-juxtaposition, upon western-hemisphere, genetic-population, totalitarian-reform.[157]¶|||01m.: 16s.84m.s.|||§ Account is conceived, by the basic-training-merit, which has always involved, military, from-an multi-diverse-bunch of citizen-(s), whom deem those quality-(ies) of training-routine, to man militia-equipment, vaguely incrementing, tier-order-discourse, involving each individual, whom have an base-competency-trust, by operation-(s) occurring smoothly, through the general-confiding, from governmental-entity-standing incurring population-discourse, affecting outcome-(s) of each and every inhabitant, from whom pertain-oneself, upon group-development, as nature of common-law, systemizing our subjective-order, (from an) legal ramification-(s), for no cohesive written-comprehension-order apose our constitutional-inset.[158]¶|||01m.:47s.89m.s.|||§ What can any one individual do, if learning is not in an simultaneous-conjuring, by intensity of [hour /// minute-(s) /// secondth-(s)] upon year? Instead, could all of us be restraint by the year-decade-century-millennium-dating-effect not inputting our ecological-concepting, by an literary-graphic-thousandths-word-point-(s)-mean-(s), concept-defining from socialism, an further identity, for survey by-of those competent-individual-(s), whom conceive objective-purpose;[160] awhile physical-labor, be the [gift // blessing] of terrain, by a-the space-(s)-constituting, of each pieced matter, [articlizing / composition] cause in turn of metropolitan-discourse, where lie-(s) of the rural-terrain, remain barren of engineering, for trading resource-ordering, persuading influence, by perspective-(s), from each cooperating-inhabitant, whom attempt to Interact & Formally-Agree with principle-(s), set by average-inhabitant-(s)-interpretation-(s), attempting to persuade daily-parameter-motion-(s)-function-(s), from requisite-article-designated-space-referencing;[161] noting space-variable-documenting, upon hour-(s) of passing-time, ridding entertainment, for-of-a-the fact-(or-(ing-(s))) of data, outputted, from no clear means, documenting the hardware-tangible-(s) (governmentally present), by literary-year-date-scaling, elapse-expiry-matter-(s) ordering daily-intentional-taxonomic-document-(s), proportional [soil cycle-(s) / tree-cycle-(s) / plant-(s)-cycle-(s) / animal-(s) / measure-(s) of territory-surface-area-composition-distance-interval-(s) / mammal-(s) / agriculture-cycle-(s) / horticulture-cycle-(s) / human-population-cycling-(s) / psychological-encyclopedia-documenting of mass-inanimate-movement / entity-development-schedule-designated-purpose / road-way-deterioration-impact] as those factor-(s) per pulse-perceptive-being-(s), interdigitated discourse of affair-(s), designating by decade-cycling-(s), pertinent from material-(s), ruled by-an directive at live-analysis, rendering observation-(s), as either present, or ineffective, from what attempt-(s) to reform, are required, while noticing inconsistency-(ies), from human-introverted-bias, comforting oneself,47L. through conformed circumstance-(s) of international-trade, juxtaposing religious-effort-(s), by-of Eastern-Orthodox-Nationalism, apose Western-Commercial-Individualism...¶|||03m.: 37s.51m.s.|||§

§...Count-[26 / Twenty-Six]-preposition-of.. ..Article-Essay-[35 / Thirty-Five].. // ..Book-Essay-[81 / Eighty-One].. ..Lines-[47.75 / Fourty-Seven.1.seven-tenths-2.five-hundredths].. ..diligence upon ecological-terrain, is impossible without an soveirgn-genetic-society, working on series scheduled-issues, per inhabitant, required at an capacity-entity-functional-operation-(s).¶|||15s.54m.s|||§

{January-(31)-February-(28 / 9)-March-(31)-Month-(s)-Day-Hour-Minute-(s)-Secondths-Count}-{7,776,000 D.H.M.S. = 60S. × 60M. × 24H. × D.}-{2,160 H.}-{129,600 M.}-{90 D.}

{Article:} {April}

¶82Chalice-objective, befuddle-mysticism, for individualism, remain practicing basic-human-need-(s'), governmentally-protected,| 327 |for population-trade-impact, ordering-data, by the hardware-capability-(ies), inactively pre-instated, before approaching an [entity / citizen] about those factor-(s) for discussion, upon commercial-directive-limit-(s), that affect those [formulation-(s) / conception-(s) / inception-(s) / conjuring-(s)] discoursing globalization-process, to inform those inhabitant-(s) of laze-pulse-thinking, of an future-ongoing-direction, sustaining effort-(s), during present-day, in transition for process, timing motion-(s), upon calendar-literary-ordering, offsetting dismay, for those fault-(s), family-commercial-consumer-individualism, religiously-coursed, instilling understanding, for why Lesson & Warning, are insignia, interval-entity-directive-reference, to remind the public, how each piece of data, is in an aversion of fact, [stock-(ed) / shelf-(ved) / order-(ed) / resource-(d) / Elapse-(s) // Synapse-(s) // Syntax] emit individual-perspective, voluntary momentum-interval-(s), for how self, pertain family, when bias-incur, affirming logic,| 328 |by Age & Temperate-Time-Impasse-Interval-(s);162 awhile-amidst physique, perceive, physic-(s)-drift, interceded variable-variance-gamma, requiring human-act-(s), attuned-present, typically casting human-temporal-calendar-conduit-method, from synthetic-effort-(s), at location of resourced-needs'. upon;163

> Age statistical-rate-(s), by-of timbre-synapse-effect-(s'), by human-census, aged-awhile, year-two-thousand & fifteen; are apose past-historical-decade-(s)-impasse-(s), capturing timbre-antedote, from year-(s) of

[Nineteen fifteen]; year past present age maximum interval of {1915 Y}

{P.Y. = A.D.}

- *[Nineteen-Fifteen]=One-Hundred-year-denominative-age.. {1915 = 100}*
- *[Nineteen-Twenty]=Ninety-Five-year-denominative-age... {1920 = 95}*

- [Nineteen-Twenty-Five]=Ninety-year-denominative-age... {1925 = 90}
- [Nineteen-Thirty]=Eighty-Five-year-denominative-age..... {1930 = 85}
- [Nineteen-Thirty-Five]=Eighty-year-denominative-age..... {1935 = 80}
- [Nineteen-Forty]=Seventy-Five-year-denominative-age.. {1940 = 75}
- [Nineteen-Forty-Five]=Seventy-year-denominative-age.. {1945 = 70}
- [Nineteen-Fifty]=Sixty-Five-year-denominative-age........ {1950 = 65}
- [Nineteen-Fifty-Five]=Sixty-year-denominative-age........ {1955 = 60}
- [Nineteen-Sixty]=Fifty-five-year-denominative-age......... {1960 = 55}
- Nineteen-Sixty-Five=Fifty-year-denominative-age........ {1965 = 50}
- Nineteen-Seventy=Forty-Five-year-denominative-age... {1970 = 45}
- Nineteen-Seventy-Five=Fourty-year-denominative-age... {1975 = 40}
- Nineteen-Eighty=Thirty-Five-year-denominative-age...... {1980 = 35}
- Nineteen-Eighty-Five=Thirty-year-denominative-age...... {1985 = 30}
- Nineteen-Ninety=Twenty-Five-year-denominative-age.... {1990 = 25}
- Nineteen-Ninety-Five=Twenty-year-denominative-age..... {1995 = 20}

[0.322580645 file day-month-clivity / 0.010989011 file day-season-clivity / 0.002739726 file day-year] by rotation-year-cycling-(s), balancing-humanity, due to manner upon physical-perceptive-area, from [technical / vocational bind-(ing-(s))] presumed when word-clause, age-process, individual-being-(s), enfranchised by-an incorporated-governmental-survey-social-count-(s), where Mass & Matter-(s), affair from state-physics-mundane-prerogative, persuading perspective-(s) of mass-population, subjunct from-an family-developmental-ideology,| 329 |where individual-opinions', tense-vision, allocating self-synapse-blood-pulse-respiration-tension, interposed, [spouse / child / deed / act / tax] enabling voluntary-will, by-an conception perceivable, vibrational-intonation being capable to act, deducing reason from-of mathematic-(s)-numerical-analyzation, conducting common-law, as-an legal-practice, to-analyze grammatical-fact-(or-(s)), produced from commercial-tax-act-(s), affecting pre-militarial-governmental-resourcing-motion-(s), as-an allegiance of reliance, upon debt-based-purchase-(s), which agglomerate globalization, to abode at denominative-capitol-venture, in ordained sustain-Tresurigal-inflation, reserved by way of Nationalism & Education extemporized abstract-deviation, for exampling tense, apose [time-rate // tempo-rate /// frequency-lapse-(s)] which [allude / agglomerate], [human-sedentary // motion-(s)-deposit-matter], celestially passing, syncopated-orchestrated-population, interceding formal tectonic-territorial-reproductive-time-effort-(s), religiously-examined-weekly, seeing if human-(s) are capable to study-rate-(s), competent content-denominating-constitutional-survey-(ing), median-proportional, directive-matter-(s), per [solid-objective / dry-substantial // wet-substantial / auxiliary-device / liquid-(s)] hovering ambient pressure deviation-(s), scaling dimension-density, where-when visual-distance-approximate-requirement-measure-expert-(s), apose-juxtaposition, factor-(s) by-of literal-material-(s), cause case when variable-documenting, receipt year-communal-intention, will-potential, as gauge for what any one individual can [persuade // effect // materialize] upon an surrounding-mass-(es) with data-material, pertaining day-elaboration, from physics, apose year-governmental-curb, identifying [why / what-is important] by domestic-trade-meter, publicly [two-dimension-distance-space-offset-tax-para-meter // private-two-dimensional-blue-print-constructed-posit-offset]61L. comprehending-matter-(s) (of state-constitutional-residential-responsibility)...¶|||04m.:32s.78m.s.|||§

§...Count-[14 / Fourteen]-preposition-of.. ..Article-Essay-[1 / One].. // ..Book-Essay-[82 / Eighty-Two].. ..Lines-[61.9 / Sixty-One.1.nine-tenths].. ..Aging deviation, I find to be an preset-date-year-assertion, to compare those variables of five-year-deviation, to compare how every-season circulate-money, with when daily-wage, compare an superlative-economy, not formally ethical, by collection alone, for how parameters are to be discussed amongst other individuals, whom comprehend why an entity-utilities, function for human-population-circulation-function, when dating mode an median of counted-circumstances, amid an process of numerical-plus-one-thousand-count, in comparison of integers both Greater & Lesser, in due part from how decimal-extension, vary from plus trillionth-column-exponent-comparison.¶|||33s.97m.s.|||§

¶83.Imagine an approximate century-(ies), focusing Time & Date, upon an intermittent-present-day, perceptive-vision, sleeping physiological-fatigue, by measure, vowel [distance / time-elapse] [period-(s) / weight-(s)] depicting dimension-comprehension, affirming tangible-value-(s), required by subservient-action-(s), that have their gauge per reluctance, identify mass-movement-(s), per trade, due to customary-tradition, pleasantly efforted, commercial-goods, for retention of objective-article-value-(s), upon an personal-lock-spacing-location.159¶|||29s.22m.s.|||§ Thought-(s) react upon such reflex-(es) by-of receipt-budget-order-(ing-(s)), directing work-week-influence, to confide each routine-simultaneous, reliance of ecological-trust, administering feature-goods-production, by-an default-nature, from mass-correlated-subjunct, identifying fact-(s) of matter, to weight-scale, intuitive physical-mass-goods-comparison, value-gauging, tangible-product-(s), without formal-collecting, mass-production-means, from equivalent-time-purchase-(s), amidst communal-cooperation, to-assess-currency, by-an national-denominative idle-blissed, through contact, at defined-analyzed-vehicle-continuum-progression-(s), Public-Government-Access-Road-Space & Private-Commercial-Revenue-Residential-Objective-Procedure-(s), fluctuating monetary-obtain-(ing)-(s), per inhabitant, whom input commercial-private-production-(s), performing managerial-operation-(s) difficult upon stated-affair-(s), for no employee-voluntary-effort-extent.160¶|||01m.:07s.97m.s.|||§ Climate & Territorial theatre, are ignored by terracultrual-impasse-effect-effort-(s), applying tense-listening, upon exterior-vibrational-order-(s), pre-surveyed, where Self & Genetic, collect through-an age-adult-decade-generation-reforming, writing each facet per ink-droplet-(s),| 330 |from-an organized-chromatic-social-governing-method, blend-(ed) paper-composition-(s), maneuvering [major / minor] subjecting-timbre-(s), manipulating inhabitant-focus, by contact-vocal-vernacular-predicate, conscious among impasse interaction-(s), plotting a-(n) culture of specimen, for psychological-schedule-ordering revealing minor-payment-account-(s), clearly past-interval-motion-(s), self-perceivable, as major-calendar-past-referencing-systems, collecting data from when, calendar-interval-(s), per voluntary-constitutional-patient-(s), restructure religious-message-(s), awhile physics deviate communal-active-(re)-development-(s), because architect-directive-progression-cycle-(s), are not an total-peer-common-application, from derived-effort-(s), pertaining how impasse, [influence // persuade // direct] communal-territory-structures.161¶|||01m.:54s.63m.s.|||§ [Debit // Credit] not persuade inhabitant-housing-objective,| 331 |for how [inflation // payment] program western-development-(s), methodically-modified, from plastic-checking-payment-(s), when societal-parameter-effort-(s), cubically-appointed, forward-depth-public-plane-two-dimensional-offset-present-continuum, ordering pertinent-interval-development-(s), at-an, consistent cultural-commodore, [sight / sound / ventricle blood-pulse-timbre-offset] seen from, [being-(s) // inhabitant-(s) /// citizen-(s) placing-matter-(s)] which stabilize substantial-material-(s), upon solid-composite-requisite, pressurized, pertainable-fact-(oring-(s)), surrounding perceptive-view, determining [route / process] documenting-data, verifying paper-default-systemizing, appointing citizen-subjective-objection, by metre of distance-proportioning. [Creed / feature / book-articling-(s)], affirm [day / night sequence-(ing)-(s)] upon centripetal-force, as Earth, perceive from-an tectonic-crust-surface-posit, periodicalized-documenting, each hour by centripetal-drift-report-(ing), unfathomable during matter-(s), customary temperate-effecting, factoring from shelf-life-purchase-expectancy, per perceivable-spacing-(s), dues to formal-material-(s), per note-of human-inheritance-behavior reconciling, recalibrated applicable-physical-effort-(s), by stretch-physical-cardiovascular-reasoning, prevalent Physical-Being & Tectonic-Nature, elaborating-duty-(ies), by word-conjure-mass-spread-civilization-(s)-infinitive, exponent-deviated, where proper discourse, affect common-simultaneous-offset-(s), determining perspective where human-means, limit trade-pursuit, deteriorating point-determination-(s'), interval hour-decimal-tenth-(s)-notion-reset-(s), along with rate-whole-note-referencing-(s), cycled among physical-synapse-blood-pulse-respiration-awareness, in line, where cultivating per physical-operations'-rate-deviation-(s), stabilize thinking, from prior-requisite-generation-(s) thinking through routine, how to [plan // order // routine] reclaimed-territory, found-established, when religious-communion, agree on an genetic-directive, pertinent perimeter-parameter-dimension-(s)-impact, awhile factor-(ing-(s)), visualize-data, by genetic-place-(ing-(s)), scheduled mass-rate-impasse-article-influctuation-(s), when individual-(s), as well as psychological-practice, premised, one-on-one-recollecting-(s) per [sight-(s) / sound-(s) / activity-(ies) / act-(s) / work-routine / eating / clothing / interacting] defining local-proximity denominating-currency, upon dimension-comparable-limit-(s), asserting [product-(s) / good-(s) / service-(s)] with retrospect, by physiological-presence, upon parallel date, collecting data-reference-(s), at voluntary-impasse-style, skewing human-application, hypothe-sized receipt-examination-(s), post-articlized, from matter-(s) defining entity-location-timing-(s), viewable at-of-an territory 56L. [approximate-acre-(age) / physical-cardiovascular-respitorial-muscular-reflexive-synapse-respiration-blood-pulse-tense-(s)] in [worship / objection / right-(s) / duty / value] per territorial-default-exploring-process-extent...¶|||04m.:20s.63m.s.|||§

§...Count-[8 / eight]-preposition-of.. ..Article-Essay-[2 / Two].. // ..Book-Essay-[83 / Eighty-Three].. ..Lines-[57.83 / Fifty-Seven.1.eight-tenths-2.three-hundredths].. ..Money is an process of wage, amid an shared-historical-objective-entity-usage, not thinking to collect an national-genetic-agenda, from literary-table-non-fiction-prosing, of various facets of existence, upon ecological-terrain, evaluating in secondths, how much an moment matters, in view at depth-perspective-distance, from various spacings of land, by an governmental-control, observing how to proportion a the land by directive-agenda, from when dating is supposed to be orchestrated, by circumstantial-proportioning, making swift-decisions, upon (entity)-scheduled-(population)-wage, of programmed-economy, resourcing from compositional-terrain, at barren-development, for identifying why processes function, and require updating or refining, pending upon issue handled at such moment.¶||||41s.38m.s.|||§

¶84Religion (has an) militarial-biblical-Chaplin-corps premise, from three-primary-religious-sect-(s), communally-gathering, awhile belief-faith-sect-(s), defining mass-quantity-(ies), per [Christianity / Judaism / Muslim ethnic-cultural-order-(s)] contain mental-limit-offset-(s), from-an post-dating articlizing, pursuant inhabitant-motivated-preference.162¶||||16s.10m.s.|||§ Pursuit unite racial-hobby-group-(s), formulating humanity, unclear as for how conjuring united-effect-(s), requisite past-present-middle-point-stillness, (by of an) [American-Protestantism // Germanic-Catholic-Christian-Protestantism /// Italian-Catholicism //// General-Reclusion] faithful by currency-based-trade, enveloping customary-familiarity, set by self-(ves), among product-collection-(s), deliberating daily-value-(s), by-an largo interpretation-(s) per fact-(s), civilizing human-will while daily-act-(s), concert motion-(s), past incapable Germanic-sole-philosophical-retort, passing day by Christianized-filing, due-to solar-combustion-default-cycling-(s), from perceptive-being-(s), identifying daily-mass-mental-capacity-interaction-discourse, when mini-scule-denominative, per data-article-(s), fact-inform, medium-(s) of surface-acre, similar book-sheet-flip, comparative flat-width-height, upon cubic-feet-numerative-measure-(ing), perceivable scale space-differentiation-(s), clarifying space, emitting from daily-vocal-vernacular-rebuttal, cause how moment-(s) impasse interior-perspective, as response-limit-(s), calculate-deviation-(s), enabled where default-mass-handling, reciprocal-repercussion-(s), scholastic scholar, sustaining-subconscious inaccurate-literary-device, internalizing fact-(s), when extraspective-mass-requirement, incandesantly affect purpose, from peripheral-conceptive-impasse, [direct / peripheral] sight-discourse, spacing surface-area, before composition, where reoccurring-presence, affect dimension-diagram-concept-material-(s), in routine, weigh-(ing) conception-development, required-from a-the pattern, of individual-participant-(s), whom schedule-agenda, | 332 |directing on-site-directive-(s), constructive-by, those elemental-refined-quality-(ies), engineering part-(s), needed from matter, to then reapply (con)-current-(ing) matter-(s), historical-product-purpose, as prerogative-data, commercially-clasped, when weekly-intercept, guideline writing, by documenting spoken-psalm, in an, word-deviation per daily-impasse, interpreting, information accumulating apprehended multitude fact-(s), by independent-city-limit-(s);[164] simultaneous federal-denominating, affirming Federal Bureau of Investigation-method of compliance, by congenial general generational-offspring, not in-an claritive-discussing, hardware, at fact, prevalent parameter-guideline-(s), that imply on-site-factor-scheduling, when survey per human-will, distance boundary-(ies) upon annexing, intended residential-cause, when-where moment-(s) of commercial-cause, performing routine- interval-day-sequence-(s), interval observational-collected-limit-(s), not participated by [each / all] individuals'-honest-effort-(s);[165] perceiving data, (instead information), as house-process, persuading survey-count-(s), yet collected in simultaneous-order, from when national-distancing, define inhabitant-(s), not understood, resourcing-clause, by mundane-limit-(s), where hardware-application, is during facilitating surface-area, for industrial-engineering-mass-weight-examination, by mundane-fact, cause purpose amidst millennium-examination, pulling information from those curtail-(s), at past-reference-discourse, awhile date, persuade woman, amidst simultaneous-conjoining, along pregnancy-age-cycle-capability, juxtaposed [Twenty-(ies) / Thirty-(ies) / Forty-(ies) /

Fifty-(ies)] retiring ones' Sixty-(ies), (as an) daily-routine-prerogative, along offspring, amassed data, through discourse, preferred where interval-upkeep, quid momentum, as cycle-comparative-whole-interval-piece-(s), perceptive-decimal-sequence-(s), offset-comparative-perspective-(s), determin-ing each value per data-article-(s), as conjecture from mind, incrementing-date, impasse [collected // stored] tangent-infer-article-sort-ordering-(s),| 333 |paragraph-reset-(ed), because mode-measuring, note synapse-respiration-blood-pulse-beat-count-(s), tangible perspective-offset, working on confirming value-(s), in-an routine where pursuance by active-trading, per article-material-(s)-interrogative, visualize-impression [memory // thought-(s)] as preference, from mass-blank-idle-still-substantial-reliance apose owning crop-(s), worked on territorial-land, as-an part, by living-impasse-discourse, understanding agricultural-kinetic-energy-cultivation-transition-dating, juxtaposed eating-cycle-practice, balance handling, horticulture, based upon introverted-nature, where humanity, recultivate horticultural-routine-maintaining, due to matter-decay-rate, impacted through [[1]day // [2]mid-day // [3]afternoon // [4]evening // [5]dusk // [6]night // [7]mid-night // [8]morning // [9]dawn]50L. from free-discourse, determining a-the correct-pattern, for forward-continuum-pressure-progression-matter-(s), by temporal-impasse, upon daily-parameter-objective-discourse-(s)...¶|||03m.:58s.78m.s.|||§

§...Count-[6 / Six]-preposition-of.. ..Article-Essay-[3 / Three].. // ..Book-Essay-[84 / Eighty-Four].. ..Lines-[51.76 / Fifty-One.1.seven-tenths-2.six-hundredths].. ..As revolutionary-circumference-pressure, revolve around a the solar-combustion-axis, not an individual conceive why solar-dating, is important in juxtaposition of economic-elemental-resourcing, for how much [silt / soil / dust / oxygen-air-vapors], incur throughout every inclination of above tectonic-crust, human-palezoid-thinking, Precambrian, Pennsylvanian culture, not understanding how distancing apply per individual whom exist, amidst an marines-population-cultivation, apose an army-ambition-population, working from under

25,000,000-genetic-population, having how those functions of schedule, apply along an commercial-effort-(s) of land-resourcing, apose human-objective-population-capacity-living, understanding how lifestyle, intertwine ambivalent cause, from how individual-perceptions, participate an role of schedule, per moment of wage-being-payment-refinement.¶|||45s.32m.s.|||§

¶85I prefer writing, so it is easy for me to write, by-when Time & Nature intercede, advertisement not guarantee impartial-trade-scope, perceptive an primarily-base interior, emitting in-to-out, while visualized inside a-the-mind, from whomever [see // hear // pulsate].163¶|||12s.88m.s.|||§ Literary-preference, upon legion-(s) per genre-(s), exist through non-fiction-aspect, intended to entertain a-the psyche, reading data, by-an persuasion for documenting, where time pass, reading interpretation, not by cause, wasting-purchase-(s), working within introverted-parameter-(s), elapsing visual-physical-limit-(s), restraint, in an continual-reset-process,| 334 |distanced from oneself, apposed other-individual-s' (superego-nature), as male-work-offset, affect feminine-discourse, for which idle-passing-(s), remain quaint comprehension per [book / movie / music weekly-compositional-pieces] accoutrement off-time-activity, interweave thought, ideal by belief, attempting through book-club, as grouping pertain minute-proportioning-(s) of data, upon perceptive-time-elapse-(s), temporally-maneuvered, by individual-self-being-composition-(s), at location-(s) for confidence, free-lancing factored data, collecting-material-(s) while alive in [customary / traditional / practice-(s) / ritual-(s) / trial-(s)] lease-data, in an passing of time, value-(d) by prerogatives per event-(s), to then incorporate an part, in interceding-impasse;166 while aging be part of those coordinated-systemized-operation-(s),| 335 |pertaining cardiovascular-simultaneous congenially interposed civilizational-act-(s), interval-(ed)-intermittent-amidst centripetal-force-perpetual-rotation-revolution-circumference-revolving-offset, when means' from motion-(s), barrier protect human-being-(s), from each spouse-relationship-segregation, not at-an off-time-cue, defining physics-appreciation, by those effort-(s) as physical-motion-(s'), stretch cue per date, unconsidered [condensation / evaporation / cloud-coverage / rain-charge-offset] deposited through land-soil-content, influctuate-present-observation-(s), of ecological-cycle-system-continuum, presently-irrelevant, matter-grandiose-impact, offset-attuned-collection-(s) matter-understanding that, if no activity, direct collecting those material-(s), for food-consumption, affecting [1vehicle / 2cellphone / 3utensil-(s) / 4appliance-(s) / 5furniture / 6tool-(s) / 7portrait-(s) / 8excess-material-(s)] then debt-credit inspire idle-furnishing, partaking passing-recollection-(s), upon simple-discourse, as elaboration by;167

|0Calendar-Present-Interval-(s)-Juxtaposition-(s)|

- *Gregorian-Year =equate=Present-Two-Thousand-&-Seventeen-Season-(s)*

  Week=equate=Modern-Hour-Day-Count-One-Hundred-&-Sixty-Eight-Hour-(s)
- *Elapse = Deviation-Rate:*
  - *Month = Six-Hundred-Eighty-Hour-(s)*
  - *Season = Two-Thousand-Forty-Hour-(s)*
- *Offset = Day-Night-Rotation-Twenty-Four-Hour-(s)-Elapse-Cycling·Secondths-tic-Metronome-Fluxuation-Simultaneous-Frequency-Rate*
- *Juxtapose*
  - *Minutes'-Numerative*
    - *Secondths-Impasse-Count-(s)-Fluxuation-(s)*
  - *Hour-(s)-Denominative*

- *Motion-(s)-Secondths-Count-(s)-Fluxuation-(s)*

|^0 Exterior-Compartmentalized-Date-Aging-Categorical-Decimal-Enumerated-Value-(s) Juxtaposition Article-Column-Offset-Order-Classification-(s)|

- *| | | | | | | | | | | | | | | |0w. |.10T. |.21H. |.32Th. |.43TTh. |.54HT. |.65M. |.76TM. |.87HM. |.98B |.109TB. |.120HB. |131Tri. |.142TTri. |.153HTri. | | | | | | | | | | | | | | | | |*

- *{.10T.} = Day-Hour-(s) - Rate-Mass-Weight-Deviation*
  - *{.21H.} = Day-Hour-(s) - Seven-Week-Day-(s)-Spacing-(s)-Offset*
  - *{.32Th.} = Day-Hour-(s) - Thirty-(one)-(twenty-eight-(Nine)) Month-Day-(s)-Substantial-Dry-Wet-Offset-Proportion-(ing)*
    - *{.43TTh.} = Day-Hour-(s)-Ninety-(one)-(two)-Seasonal-day-(s)*
        - *Solid-Density-Comparative*
      - *Horticulture | Agriculture | Mariculture | Apiculture | Floriculture | Terraculture | Plant-Crop-Algae-Cell-Expanse-Taxonomic-Classification-(s)*
        - *Earth-Crust-Topography-Surface-Measure*
        - *Quarterly-Earning-(s)-Reporting*
        - *Federal-Reserve-Quarterly-Reporting*

{Article:} {April}

- {.54Ht.} = Day-Hour-(s) - Solstice-day-night-hour-reset-(s)
  - {Summer-A 07:15A.M. - 08:15P.M.} {08:16P.M. - 07:14A.M.}
  - {Winter-C 05:45A.M—06:45P.M.} {06:46P.M. - 05:44A.M.}
- {.65M.} = Day-Hour-(s)-Equinox-transition-hour-offset-(s)
  - {Spring D 03:00A.M. - 09:00A.M.}
  - {Autumn B 03:00P.M. - 09:00P.M.}
  - {.76TM.} = Daily-Hour-(s) - 114.4783-Year-denominative-one- millionth-posit
    - density-volume-distancing-median-deviation-hundredth-thousandth-integer-offset-(s)
    - (for year, tangent-articling is not over Three-hundred-Sixty-Count)
    - {.87HM.} = Two-Year-denominative-Interval-(s)-resets
      - {House-Representative-(s)}
      - {Senate-Member-(s)}
      - {.98B.} = Four-Year-Presidential-Progression--Sixtieth-Election-Cycle
  - {.109TB.} = University-Two-{Associate-(s)-Four-{Bachelor-(s')}-six-{Graduate-(s')}-eight-{Master-(s')}-Ten-{Philosophy-year-Study-(ies)-incre-

ment-(s) per decade-per-year-aging whole-decimal-tenth millionth-mass-deviation

- {.120HB..} = Decade-Census-Mass-Population-Offset
- {.131.Tri.} = Century-Ten-Millionths-continuum-birth-death-population-mass- offset
- {.142.T.Tri.} = Millennium-Nation-Federal-Reserve-Debt
- {.153H.Tri.} = Treasury-Territorial-Objective-Denominative-Year-Currency-Count

- |°Quarterths-Rate-Twenty-Five-Percent(ile)-numerative-fluxuation(s) |
  - |In 15/Fifteenth of multiple-(s)-rate-interval-(s), (Percentile) |
  - 30 : 45 : 60 : 75 : 90 : 105 : 120 : 135 : 150 : 165 : 180 : 195 : 210 : 225 : 240 : 255 : 270 : 285 : 300 : 315 : 330 : 345 : 360 : 375;[168]
  - Upon

◊ |In 25/Twenty-fifth of multiple-(s)-rate-interval-(s), (Quarterths) | 50 : 75 : 100 : 125 : 150 : 175 : 200 : 225 : 250 : 275 : 300 : 325 : 350 : 375 : 400 : 425 : 450 : 475 : 500 : 525 : 550 : 575 : 600 : 625 : 650 : 675 : 700 : 725 : 750 : 775 : 800 : 825 : 850 : 875 : 900 : 925 ;[169]

- Upon

◊ |In 90/Ninetieth of multiple-(s)-rate-interval-(s), (degree-(s))-(right angle) |

◊ 180 : 270 : 360 : 450 : 540 : 630 : 720 : 810 : 900 : 990 : 1,080 : 1,170 : 1,260 : 1, 350 : 1,440 : 1,530 : 1,620 : 1,710 : 1,800 : 1,890 : 1,980 : 2,070 : 2,160 : 2,250 : 2,340 : 2,430 : 2,520 : 2,610 : 2,700 : 2,790 : 2,880 : 2,970 : 3,060 : 3,150 : 3,240 : 3,330 : 3,420 : 3,510;[170]

◊ *Upon*

◊ |*In 60/Sixtieth per multiple-(s)-rate-interval-(s), (minute-(s))-(secondths)* |

◊ 120 : 180 : 240 : 300 : 360 : 420 : 480 : 540 : 600 : 660 : 720 : 780 : 840 : 900 : 960 : 1,020 : 1,080 : 1,140 : 1,200 : 1,260 : 1,320 : 1,380 : 1,440 : 1,500 : 1,560 : 1,620 : 1,680 : 1,740 : 1,800 : 1,860 : 1,920 : 1,980 : 2,040 : 2,100 : 2,160 : 2,220 : 2,280 : 2,340 : 2,300 : 2,360 : 2,420 : 2,480 : 2,540 : 2,600 : 2,660 : 2,720 : 2,780 : 2,840 : 2,900 : 2,960 : 3,020 : 3,080 : 3,140 : 3,200 : √ 3,600;[171]

◊ *Upon*

- ◊ |In 24/Twenty-Fourth per multiple-(s)-rate-interval-(s), (hours) |
  - ◊ **48 : 72 : 96 : 120 : 144 : 168 : 192 : 216 : 240 : 264 : 288 : 312 : 336 : 360 : 384 : 408 : 432 : 456 : 480 : 504 : 528 : 552 : 576 : 600 : 624 : 648 : 672 : 696 : 720 : 744 : 768 : 792 : 816 : 840 : 864 : 888 : 912 : 936 : 960 : 984 : 1,008 : 1,032 : 1,056 : 1,080 : 1,104 : 1128 : 1,152 : 1,176 : 1,200 : 1,224 : 1,248 : 1,272 : 1,296 : 1,320 : 1,344 : 1,368 : 1,392 : 1,416 : 1,440 : 1,464 : 1,488 : 1,512 : 1,536 : 1,560 : 1,584 : 1,598;**[172]

|º Family-(ies)-Aging of Original-Biological-Specimen-(s) |

- |Seventeen-Sixteen through Sixteen-Fifty-Six=Ultimate-Past-Initial-Period of Physiological-Perceptive-Pulsed-Being |
  - |Sixty-Year-(s)-Growth-Death-Birth-Period |
- | Seventeen-Seventy-Six=First-Present-Past-Period|
  - |Sixty-Year-(s)-Growth-Death-Birth-Period|

{Article:} {April}

- |Eighteen-Thirty-Six=Second-Present-Past-Period|
    - |Sixty-Year-(s)-Growth-Death-Birth-Period |
- |Eighteen-Ninety-Six=Third-Present-Historical-Period|
    - |Sixty-Year-(s)-Growth-Death-Birth-Period|
- |Nineteen-Fifty-Six=Fourth-Present-Historical-Period|
    - |Sixty-Year-(s)-Growth-Death-Birth-Period|
- |Two-Thousand-Sixteen=Fifth-Present-Continuum|

|°Offset-Aging-growth-decay-cycling|

1. |Insemination-Pre-Growth-Trimester-Period-(s)=0-Nine-Month(s)|
2. |Child-Birth=Zero year-(s) to One-Half-Year|
3. |Toddler=One-Half-Five-Year-(s) to Five-Year-(s)|
4. |Child=Five-Year-(s) to Nine-Year-(s)}|
5. |Pre-Teen=Nine-Year-(s) to Thirteen-Year-(s)|
6. |**Teen**=Thirteen-Year-(s) to Eighteen-Year-(s)|
7. |Young Adult=Eighteen-Year-(s) to Thirty-Year-(s)|
8. |Adult=Thirty-Year-(s) to Forty-Five-Year-(s)|

9. |Aging-Adult=Forty-Five-Year-(s) to Sixty-Two-Year-(s)|
10. |RETIREMENT=Range-death-Aging from Sixty-Two to Eighty-Year-Average|...¶08m.:29s •72m.s.|||§145L.

§...Count-[9 / Nine]-preposition-of....Article-Essay-[4 / Four].. // ..Book-Essay-[85 / Eighty-Five].. ..Lines-[145 / One-Hundred-Fourty-Five].. ..For how this essays deviation of numerical-grammatical-decimal-ordering-scales, juxtapose how sedimentary-material settle, below ones average of resourcing of land, per ration of residential-housing-deviation, comparing genetic-species, signifies an literary-astute-practice, that is proportioned commercial-governing by [copyright // serial-documenting-deviation-standards], for an dynamic of free-at-will-voluntary-beings, offset [residential / commercial transit-location-(s)], comparing how land is centralized, from rural-resourcing-sites, for proportion-(s) of pounds into tonnes, for how grams and ounces circulate from human-cardiovascular-respirtory-reflexive-compositional-capacites.¶|||33s.54m.s.|||§

¶86.Past tense, historical-year-requisite, resets how one Identifies through season, characteristic (s) at focus, (secondths-continuum-interval-(ed) fluxing frequency-(ies) per hour, for no reference of time, measuring respirational-blood-pulse-beat-(s), (as an) offset per cent, conjuring matter from physiological-survey-(ing).164¶|||12s.72m.s.|||§ Referencing [1.Act-(ion-(s)) / 2.Maneuver-(s) / 3.Transit-(s) / 4.Task-(s) / 5.Agenda / 6.Survey / 7.Subject-Period / 8.Passing / 9.Laying / 10.Founding / 11.Standing / 12.Upholding / 13.Convincing / 14.Writing / 15.Conversing / 16.Studying / 17.Reading / 18.Seeing / 19.Hearing / 20.Listening / 21.Vibrational-gauging / 22.Observing from sense-conjoin-effort / 23.Enact-ing / 24.Course-Tiering / 25.Objecting / 26.Sermoning / 27.Receiving / 28.Functioning / 29.Engineering / 30.Placing / 31.medicating / 32.prescribing / 33.transaction / 34.Parish /

{Article:} {April}

35.Graving / 36.Raking / 37.Contorting / 38.Debating / 39.Counting / 40.Monitoring / 41.Washing / 42.Moving / 43.Baking / 44.Packaging / 45.Eating / 46.Digesting / 47.Expelling / 48.Clothing / 49.Running / 50.Hammering / 51.Noting / 52.Redacting / 53.Critiquing / 54.Lifting / 55.Excavating / 56.Collecting / 57.E-Numerating / 58.Pressing / 59.Affirming / 60.Dissecting / 61.Discoursing / 62.Pawning / 63.Sporting / 64.Driving / 65.Watering / 66.Personal-Cooking / 67.Purchasing / 68.Adorning / 69.Factoring / 70.Concepting / 71.Serving / 72.Saving / 73.Filing / 74.Piloting / 75.Conducting / 76.Currenting / 77.Plumbing / 78.Architecting / 79.Contracting / 80.Ordering / 81.Scheduling / 82.Nursing / 83.Incision / 84.Alerting / 85.Terming / 86.Disseminating / 87.Editing / 88.Stamping / 89.Sorting / 90.Stocking / 91.Affirming / 92.Sequencing / 93.common-law / 94.Judiciary-Presiding / 95.Arresting / 96.Automating / 97.Performing / 98.Measuring / 99.Distancing / 100.Warranting / 101.Form-Documenting / 102.Cashiering / 103.Modifying / 104.Mounting / 105.Painting / 106.Picturing / 107.Medically-Dictating / 108.Theatre-Performing / 109.Shoveling / 110.Planting / 111.Regenerating / 112.Memorializing / 113.Matriculating / 114.Teaching // 114.1Learning / 114.2Understanding / 115.Musical-Noting / 115.1Theorizing // 115.2Concepting / 115.3Arranging / 116.Plotting / 117.Planning / 118.Budgeting / 119.Federal-Reserve-Filing / 120.Butchering / 121.Operating / 122.Handling / 123.Currenting / 124.Constructing / 125.Envisioning / 126.Practicing / 127.Balancing / 128.Scaling / 129.Adjusting / 130.Advertising / 131.Broadcasting / 132.Prioritizing / 133.Shipping / 134.Train-Operating / 135.Printing / 136.Copyrighting / 137.(Carbon)-Dating / 138.Romancing / 139.Socializing / 140.Marketing / 141.Inventory-Accounting / 142.Accounting / 143.Playing / 144.Tinkering / 145.Embargoing / 146.Treaty-Cycling / 147.Treating / 148.Customizing / 149.Abridging / 150.Idling / 151.Contemplating / 152.Concealing / 153.Constituting / 154.Probing / 155.Talking / 156.Dialoguing / 157.Preparing / 158.Slaughtering / 159.Chopping / 160.Calligraphing / 161.Refrigerating / 162.Freezing / 163.Scrubbing / 164.Detailing / 165.Sewing / 166.Hemming / 167.Setting / 168.Informing / 169.Documenting / 170.Deliberating / 171.Extinguishing / 172.Recovering / 173.Saving / 174.Executing / 175.Legislating / 176.Adjoining / 177.Component / 178.Directing / 179.Ascertaining / 180.Usurpation / 181.Sensing / 182.Tense-Isolating / 183.Stretching / 184.Walking / 185.

Amending / 186. Billing / 187. Managing / 188. Breeding / 189. Grooming / 190. Reproducing / 191. Rendering / 192. Tendering / 193. Affirming / 194. Referencing / 195. Difference-Meticulating / 196. Scanning / 197. Stalking / 198. Escalating / 199. Clouding / 200. Qualifying / 201. Verifying / 202. Variating / 203. Variablizing / 204. Labeling / 205. Diagnosing / 206. Lessoning / 207. Lecturing / 208. Curdling / 209. Curling / 210. Ironing / 211. Stabilizing / 212. Motions' / 213. Showing / 214. Telling / 215. Projecting / 216. Rocketing / 217. Inclining / 218. Finding / 219. Examining / 220. Purporting / 221. Conveying / 222. Shooting / 223. Racketeering / 224. Plaguing / 225. Migrating / 226. Integrating / 227. Voting / 228. Collaborating / 229. Injecting / 230. Exterminating / 231. Fumigating / 232. Inventing / 233. Paving / 234. Sodding / 235. Soldering / 236. Welding / 237. Charging / 238. Returning / 239. Sculpting / 240. Carving / 241. Chipping / 242. Deteriorating / 243. Welcoming / 244. Hosting / 245. Habituating / 246. Testing / 247. Touching / 248. Smelling / 249. Tasting / 250. Sipping / 251. Drinking / 252. Counseling / 253. Therapizing / 254. Integrating / 255. Globalizing / 256. Aforementioned / 257. Crying / 258. Joking / 259. Rolling / 260. Rubric / 261. Benchmarking / 262. Determining / 263. Typing / 264. Texting / 265. Opening / 266. Closing / 267. Instrumenting / 268. Harmonizing / 269. Intonating / 270. Tempo / 271. Tracking / 272. Tacking / 273. Turning / 274. Sweating / 275. Pulsing / 276. Reaming / 277. Questioning / 278. Inquiry / 279. Proportioning / 280. Professing / 281. Rating / 282. Appraising / 283. Grazing / 284. Milking / 285. Cultivating / 286. Matriculating], motion-(s) for reasoning commercial-unit-factor-(s), practicing civilization-cultivation-(s)-effect-(s), timed simultaneous article-dating, reflex-(ing) motion-(s)-impact, asserted by tabling action-(s), in part for conjuring-data, for objective-tangible-inferencing, by constraint of space, through period-(s) intermittent hour-(s), for interval-(s) classing-momentum, along an kinetic-solid-guide,| 336 |transitional point-stall-out-(s), throughout various impasse parameter-(s) amongst spacing-(s), because addressing reason, not be capable for human-average-simultaneous-reflection, as pulse-defining, at tangent-topic-issue-(ing)-place-(s), serving case-in-point-(s), from action-(s), coordinating movements-upon-physics...56L.¶|||03m.: 36s.00m.s.|||§

§...Count-[2 / two]-preposition-of.. ..Article-Essay-[5 / Five].. // ..Book-Essay-[86 / Eighty- Six].. ..Lines- [56.06 / Fifty.1.zero.tenths-2.

six-hundredths].. ..cases of judiciary-history, remain filed in dormant-repetition, for how scheduling is not arranged, per individual, whom volunteer, directed-effort-(s), when land is not fully developed, from requirement for ecological-site-compositional-resourcing, amid an centripetal-force, proportioning, each fact of existence, for how population remain the circulational-piece-balance, amongst anticipated retirement biological-species, decaying amid lifestyle-living, separate from each individual-participation, due to facts of entity-objective-task-arrangements-(s), requiring periodical-revision, for directing those motions of entity, amid an offset of county-sociological-resourcing.¶|||32s.00m.s.|||§

¶87..Inhabitant-(s) remain far too affluent ignoring practicing, ethical-impasse-progression-(s).165¶|||04s.94m.s.|||§ This stir up interval-notion-(s) per day, intended [chivalry / guidance / direction] amidst adversity, upon presence, quaint from necessary yearly-effort-(s), resetting variable-(s) know from following dating-impasse, regimented-default-order-(s), where commercial-debt, sustain objective-product-faith-based-action-trade-(s), cycling order-(s) maintained commercial-debt, affecting belief-(s), from tradition, blind by wife, for implementing romantic-interaction-(s), by-an serried daily-proclamation-swearing, dialogued through hour-progression-(s), awhile romantic-procession, during duties per labor-task-(s), confirm yearly-regeneration for-of child, because concepting scheduled-agenda, direct chore-(s), by family-social-aptitude, particular familiar-proximity, interacted gross-exchange, denominating process-(es), at-an physical-emotional-intuitive-reaction-(s), from life-partner-(s'), upon tectonic-transit,| 337 |when dating-impasse, affirm secular-population-cultivation-development.166¶|||53s.50m.s.|||§ Children have yet developed an pre-programmed-routine, defining cognitive nature from self-id, defining adult-action-(s), elucidating daily-visual-impasse-focus, at topical-discussion-(s), variablizing [1motion-(s) // 2object-(s) // 3substantial-(s)) // 4article-(s) // 5tension-(s) // 6collected-belonging-(s) // 7word-(s) // 8article-(s) // 9et cetera...] Picking-Up & Resetting commercial-recipe, for Each & Every-Motion, programmed daily-diligence, throughout invigoration-(s), intermittent schedule-reasoning-(s), measuring objective-subjective-article-note-(s), acquiring inquiry data, standard upon place-locational-capable-impasse-(s), while article-(s)-sub-influence action-(s), when during an impasse for daily-voluntary-effort-(s).167¶|||01m.:26s.91m.s.|||§ Whole noted slurred-tie, indicate [movement-(s) / act-(s) / debate-(s) / dialogue-(s) / sight-(s) / hearing-(s) / blood-pulse-sensation-(s)] unaligned by [Governmental-Constitution / University-(ies) / Percentile-Mass-(es) / Psychological-Document-(ing) / subjective-matter-(s) / act-(s) of monetary-proportional-parameter- (izing)], reasoning payment-(s), from human-labor-act-(s), by-an superfluous-anomaly, unrationed material-deviation, objecting space-surface-area-cultivation-impact,| 338 | commonly-matrixed, governmental-planning, [metropolitan /// rural-terrain-mass-acre-area] influencing effect-(s) amidst daily-passing-interval-(s)-point-(s), conjuring human-fatigue-gauge-scaling, factoring celestial-revolutionary-rotation-impasse, apose observation-(s) from civil-ization-juxtaposition-(s), systemizing parameter-(s), apose fifteen-minute-interval-(s), where (un) / (sub)-conscious-motion-(s),| 339 |point-reference by Thirty-Five-Thousand, One-Hundred & Fourty-Lapse-(s), per-year,28L. impassing minute-(s), interposing [latitude // longitude centripetal-force-degree-offsets'];173 awhile miniscule-(s) per Being-(s) & Blood-Pulse-Life-Form-(s), serried acre-spacing-(s), designating purpose, per generated-life...¶|||02 m.:23s.23m.s.|||§

§...Count-[2 / Two]-preposition-of.. ..Article-Essay-[6 / Six].. // ..Book-Essay-[87 / Eighty- Seven].. ..Lines-[29.85 / Twenty-Nine.1.eight-tenths-2.five-thousandths].. ..an simple observation of human-ethnic-groups, can go an long way. Those individuals that function without an genetic-origin-species, do not think of an bastardized population, growing, not documenting yourself or family lineage, so how to understand Congressional-government-filing, demonstrating an consensus of family-educational-effort, unreasonable to how each individuals' (family) house, offset from conditions of commercial-education, not realizing those minuté subtleties, would not reflect the same outcome of temporal-community-housing-residential-perception, so retirement settle south of the Missouri-River, blank for understanding what signify lifestyle, confluence reasoning, for proportioning each purchase, with housing retentive-sectors, from an historical-fault at non-sequitur-parameters, from an youth-studies-educational-teaching-studies-grade-school-straining, demonstrated well in adulthood, by those gross amounts of commercialism, with waste, not directed formal-proportioning of human-population-weight-count, in proportion of [1.scales : 2.measures : 3.weights : 4.distances : 5.proportions], offset comparing how visual-harmony, gamma with refined-commercial-product-objective-unit-(s), split at market, from entity, still unlearned for how to proportion where each lot, remain barren, from unformal cities, for longevity of ones lineage, as Montana-Geological-Boundary, remain Barren not vacant, from industrial-commands, scheduling from human-mass-bias, not invigorating, from an genetic-offset-pursuit, due to family-governmental-adequacy.¶|||01m.:27s.16m.s.|||§

¶88Colloquial-Tongue, fashion from-an monetary-motion-(s)-reflex-(es), affluent inhabitant-behavior, as word-aptitude, not participle-grammatical-punctuation-numerated-rate-increment, participle subconscious interval-interposed-mentality-reading-(s), affirming-data, where [1friend / 2self / 3community / 4co-worker-(s) / 5church-member-(s) / 6alcoholics-anonymous-member-(s) / 7racial-population / 8ethnic-population / 9government-official-(s) / 10legal-practice / 11medical-practice / 12employee-Christian-balancing], affirm perspective, where-when composite-mass-reduction-(s) per material-based-trade-data, affirm blue-collar articlized, impasse-ground-level, plane perimeter-distancing-(s), taxed-continually, incrementing-focus, where perspective, weigh mass-pound, per secondths, surrounding spacing-secondths-elapse-(s), by objective-schedule-agenda-directive, fraction-scaling population, during-when deviation-(s), remain held-humanistic-accountable, in lieu per receipt-date-compound-product-tracking, expensed awhile day interpose,|340|cycle-location-focus, rested trade, without edible-origin-(s)-processing-packaging-producing-purchasing-preparing-cooking-tabling, understanding due-(s)-process-perspective, concurrent routine, creating an direction, from motion-(s)-impasse-(s), surrounding-terrestrial-development, by-an phonetic-harmony per letter, conceiving participle-glottal-stop-consonant-shift-notion-(s), vowel-letter-centralized-notion-(s)-(ed);[174] modifying-voice, elucidated [conjure / conceived perceptive-impasse] when perspective, pertain article-matter-(s), where exterior-value-(s), confluence [[1]presence / [2]matter-(s) / [3]mass-(es) / [4]scale-(s) / [5]measure-(s) / [6]weight-(s) / [7]engineering-tectonic-infrastructure] from-by [sounding / sighting] citing-physiological-tense-(s), on paper-ink-subjective-periodicalizing, equivalent notion-synapse-(s), sentenced-(s) paragraph-(s) per elapse-(s);[175] rather comma-vocal-expression-(s'), median-interval-article-deviation-data, formally comprehending, [[1.]act / [2.]mechanism / [3.]statute / [4.]fixture / [5.]placing / [6.]structuring / [7.]labeling / [8.]fashioning / [9.]emblazing-matter-(s)] their by-an perceivable-pertinence, verifying natural-affair-(s), progressing discourse, interweaved kinetic-energy, numeratively-mechanical, result-(s), placing potential-energy-effort-(s), in brief-prior-location, defining diagram-act-(s), Referencing & Discoursing-Incandescent-Observation-(s), upon secondths-impasse-increment-retrograde-article-measuring [apose // juxtaposed] article-note-(s), per letter-participle-word-article-term-definition-sentencing-tone-quartered-vibration-(s), inflicting reflex-(s) per muscular-activity, defining reflective-subconscious-interpretation-(s), using letter-participle-word-syntax-deviation-(s)| 341 |amid vivid-depth-perceptive-article-visual-tonal-interpretation-(s), depicting defined written-character-(s), instrumentally-inclined from self, reapplicated at measuring tone-(s), pragmatic-analysis, reasoning tone-(s) per instrument-(s), intonationally-phrased, syncopating-inter-action, per moment-(s) of extraspective-purpose, adrift physics, involving dating, book-account-(ed), focus where staffing-scheduled-inhabitants', define how incremental-thinking, continuum physiological-sensation-(s), instating sight-subconscious-sound-phrasing-(s), per word-sense, sounding reverberated-interaction, by method of life, undaunted suffering, from [Western // Eastern],34L. conductive-ethics...¶|||02m.:43s.96m.s.|||§

§...Count-[2 / two]-preposition-of.. ..Essay-[7 / Seven].. // ..Essay-[88 / Eighty-Eight].. ..Lines-[34.17 / Thirty-Four.1.one-tenths-2.seven-hundredths].. ..taxes have an insurmountable debt, to balance and proportion, yet noting really happens to stop the debt, because how each population-mass-inhabitants, are offset to forget, why trade is occurred, from an barren land, where each action, is not directed for assuring each entity-operations, suffice an material-goal, through [entrance // exit], timed-scheduled-wage-circulated-commercial-cycling, per individual voluntary at an offset genetic-prerogative.¶|||26s.13m.s.|||§

¶89Languished-Tense, naturally move pass physics-territorial-ambient, without conceiving inquisition (per of) [stated-responsibility / genetic-referencing / monetary-denominating / planned-act-(s) of family-practice by neighbor-scheduling / act-(s) of tasked-schedule] along those line-(s) remaining reasoned, physiological-genetic-territorial-claim, militarily-progressed free-sover-eignty, bordering nation-(s), from Being & Physics-Impasse.168¶|||20s.72m.s.|||§ [Day // Night Elapse-(s)] yet be subtly verified, during daily-routine.169¶|||25s.19m.s.|||§ So when are we affluently-coherent, from other-(s)-individuals', natural-progressions', when right-(s)-discourse, tense in being, residual location per inhabitant-act-(s), skewing nominative-self, by [work-talk / free-time-sensation-(s)] repetitive-role-(s), as [day-kinetic-method /// night-potential-method] from role-(s) of performance.170¶|||40s.91m.s.|||§ I presume mechanical-advantage, be energized day upon night, as effect per sum-potential-placing-(s), upon data-usage, view of [tool-(s) / hardware / lumber / cement-article-(s)] as variable-data-point-(s), upon spacing-aspectual-view, tangent by independent-individualism,| 342 |pursuing [booking / musical-orchestration], timed awhile moment-elapse-discourse, progressed median-article-passing, affirming conjecture of vocal-vernacular-data,| 343 |as centripetal-force, revolve-simultaneous, human-impasse-disposition.171¶|||01m.:07s.69m.s.|||§ For sleep evoke, I tense by missionary work, paying interactions', upon spacial-shifting's', from long-term-planning, directed state-climate-territory-perpetual-focus, pertaining daily-dynamic, territorially-inclined by physical-limit-(s), acted particular-dimension-concept-proposing, linearly-diagraming, [limit-(s) / composition-(s) / concept-(s)], per [product / architectural measure] measuring weight-(s), resourced from tectonic-composition-matter, determining placing-(s), based-soly-interval, self-lifestyle-miniscule-mass-impasse,20L. pertaining matter, from self, upon municipal-limit-(s), exceeding order of humanity by perceptive-tectonic-boundary...¶|||01m.:40s.43m.s.|||§

§...Count-[7 / Seven]-preposition-of.. ..Article-Essay-[8 / Eight].. // ..Book-Essay-[89 / Eighty- Nine].. ..Lines-[20.59 / Twenty.1.five-tenths-2.nine-hundredths].. ..I can not direct how to develop terrain, without an through search for an genetic-population-mass, whom work consistently on population-scheduled-entity-task-scheduled-agenda-operations, from commercial-factory-market-production-(s), per situated-settlement, that has not been directed an form of proportional-lifestyle-living, from an surrounding community-socio-logical-trade-populus-survey.¶|||25s. 69m.s.|||§

¶90As of now, human-(s) age per year, incrementing those [subterior day-(s) / week-(s) / month-(s) / season-(s)] by weekly-religious-commercial-reset-(s) amongst yearly-act-(s), never perceiving state-limit-(s), for youth-education-stagnation-practice, by individual-population-study, each post-year-(s)-non-fiction-age-seasonal-analytical-review, [trend / style-(s) / fashion / tasted] perceivable tangible-mark-(s), validifying-focus, at complex-intrication-(s), intermittent centripetal-force-impasse-miles-distance-mark-(s).172¶|||24s.78m.s.|||§ Year is denominative, revolution-ary-complex, thinking from who notice, all age-(s), impasse self-individual-purpose-(s), when median-place-purpose-(s), are [weighted minute-(s) / table-(s) of secondths] exponent input-reasoning, never of-an output-end, reasoning tangible-elemental-solid-(s), directing person-being,| 344 |objective-reason-use-frequency-trade-count-responsibility-(ies), adjecture observation-(s), in formed-sequences, affirming [type / class / form] attempting extract-focus, from inhabitant-(s), whom agree conscious-arrangement, for mass invigoration at state-reproductive-development-(s'), pushed astranged-state, volatile, era-modernization, defining-generation-(s) of developmental-focus, aging pass hourly-time...¶|||01m.:00s.34m.s.|||§

§...Count-[4 / Four]-preposition-of.. ..Article-Essay-[9 / Nine].. // ..Book-Essay-[90 / Ninety].. ..Lines-[12.75 / Twelve.1.seven-tenths-2.five-hundredths].. ..era of individuals, means not an decade of formal-commercial-educational-thinking, from how entity-development, offset an scheduled belief, from introverted-resourcing-cities, by how festivities, interpose work-effort, when developing ecological-terrain, by an geological-tectonic-political-boundary-offset-(s), per individualistic-mega-class, falsely interpreting how the origin-characteristics-population-effort, interpose ecological-metropolitan-effort-(s), from industrial-ingenuity, that not move far from Foundry-Commissioned-bills, economically-circulating transited-lands, that not interact from sociological-perspective, due to an inadequacy of attention-span, for [listening / speaking // interpreting], an class of human-citizens, whom comprehend from 14-cu.ft.-perspective, an motions-cycling upon terrestrial-physics, by those effort-(s) of ecology, interacting upon conscious-human-interaction-(s), per continuum-impasse-days, per [week /// Month /// Season]; intervalizing points of human-scheduled-ecological-entities-territory-impasse, amid centripetal-force-revolving-revolutionary-483-years-pressures-scales-forces, defining etymological-exist-ence.¶|||01m.:05s.59m.s.|||§

¶91At what familiarity, are heterosexual-being-(s) supposed to interact, when blank-minded-congenial-impasse-(s), affirm self-life-total-care, rating [Blood-Pulse & Word-Labeling / Inferencing] underthought-(ed)-notion-(s), during Passing-Time & Perimeter-Spacial-Area, [basically-talking / activity-interaction], intermittent Physical-Life & Mathematical-Reference, variating element-(s), interceded envelopment awhile moment, unravel stringent-awareness, sensitizing physical-being, for natural-basing, cardiovascular-respitorial-breath, adrift Centripetal-Force, resetting calibre per act, affirming [superior / inferior related-affair-(s)] spaced cause at developmental-purpose, for ascertaining scientific-documenting, tensed upon guidance intermittent moment-(um), experiencing an lapse-infinitive-unequitable-progress, from self-(ves), individually affirmed tectonic-matter-(s), leaning factual developments' (as an) part, for every-inhabitant, rating astute-select-general-effort-(s), acted in subservient-focus of leadership-role-(s), continuing-progression of daily-hour-dynamic, when intent-analysis, verify place, from person-one, interacting with object-(s) for product-purchase-influence, intricating-use of general-parameter-(s),13L. impassing property-spacing-(s), because general-congenial-thinking, rule greater-than, physiological-effort-(s)...¶|||01m.:05s.72m.s.|||§

§...Count-[3 / Three]-preposition-of.. ..Article-Essay-[10 / Ten].. // ..Book-Essay-[91 / Ninety-One].. ..Lines-[13.47 / Thirteen.1four-tenths-2.seven-hundredths].. ..men will either understand, or not comprehend, that with women, repopulation, does not serve an subjective-ethics, for repopulating from women yearly, to circulate life, by an interior-entity-process, by developing-terrain, from how tectonic-plates-physics works, when developing an population to accumulate tectonic-plate-billionths-square-meters-distanc-ing, have an developing-curb, that can only be comprehended, if an genetic-mass-population-survey-tectonic-characteristics, for trust of species, in developing an central-ized-genetic-territory, or else remain confused amidst ecological-county-metro-politan-city-tribunal-areas, for how each individualistic-inhabitant, affect surrounding impasse-territory-area, per mode of impasse-parameters, awhile offset-factoring, affect all that be constructed, for interior-space-timings, by task-moding under entity-premises. ¶|||48s.00m.s.|||§

¶92I write without direct-analyzing, physiological-perspective, affirming mean-(s) of tangible-instance-(s), unraveling-matter-(s), from inquiry at-of bill-(s), survey-(ed) through citizenship, considering-data, in role per data-article-(s), informing cause per daily-effort-(s), when fact-casing, instance-(s), impass for metropolitan-offset, stabilizing-mass, because programmed-progression-(s), affect notion-ordering, identified place-co-worker-(s)-confirmation-interval-(s), pertaining data, upon parameter-guideline-(s), put by-an effective-ordering-(s) of data, upon an general-abstract-free-will-purchase-deviation, directing sign-(s), (as an) habit curbing, from terrestrial-boundary-limits', 7L.[undefined / undeclared / undetermined] a-the way of Will & Territory...¶|||39s.28m.s.|||§

§...Count [4 / Four]-preposition-of.. ...Article-Essay-[11 / Eleven].. // ..Book-Essay-[92 / Ninety-Two].. ..Lines-[7.62 / Seven.1.six-tenths-2.two-hundredths].. ..Human-neglect is an unconscious-family-issue, that affect-(s), each individual, from how each piece of developing-matter, increment those ties of incremental-focus, around other introverted-individuals, when each location is designated by an entity-schedule, when each moment impasse, requiring the sustaining of human-organism-matter, visually-perceiving exist-ence, for an vision on daily-affairs.¶|||24s.03m.s.|||§

¶93Generation-(s) of inhabitant-(s), are pre-ordered by [age / effort-(s)] where efforted work-place, reason treasury-metre, verifying homework State-Study-(ies), from those [1condition-(s) / 2parameter-(s) / 3Temperature-(s) / 4Genetic-Inhabitant-(s) / 5Objective-(s) / 6Infrastructure / 7House-Planning / 8Social-dynamic / 9Daily-Theatrical-perspective] in obligation of religious-effort-(s), affirm-defining, each composite-product-(s), [concepted // diagramed] by-an colloquial-group-effort-(s'), requesting [resource-(s) / federal-aid / material-(s) / loan-(s)] upstarting upkept-business-(es), whom plead, civil-obedience, defining national-effort-deviation-(s), from notion-(s) of-those variable-(s) per [comestible-(s) / Clothing / Housing] upon lifestyle-application-impasse.173¶|||35s.97m.s.|||§ Since Eighteen-Eighty, has Commercial-Industrialism, been an part of a-the American-(Dream)-Lifestyle, fabricating-mass-effort-(s), involving [1Clothing / 2Construction / 3Farming / 4Transcendentalism / 5Communication-(s) / 6Technology / 7Government / 8Work-Place / 9Astronomy / 10Road-Construction]11L. enveloping metallurgical-development, enhancing metropolitan-lifestyle-(s')...¶|||51s.50m.s.|||§

§...Count-[4 / Four]-preposition-of.. ..Article-Essay-[12 / Twelve].. // ..Book-Essay-[93 / Ninety-Three].. ..Lines-[11.62 / Eleven.1.six-tenths-2.two-hundredths].. ..following is significant from how each individual demonstrates an reason of individual-focus, while cooperating on an extraspective-square-meters-distancing-balancing, from when determining celestial-boundary-(ies), amid an discourse of surrounding-centripetal-force.¶|||17s.47m.s.|||§

¶94Decade-Interval-(s) confluence century-(ies)-denominative-Gregorian-death-toll-rate-(s), [at-where // during-when], [day // night, hour-interval-four-fifteen-minute-(s)-period-(s)-cycling-(s)-dating-offset-(s)] impasse present-posture-timbre, defining census-omitted-past-present-reasoning-(s), aged through gradual-increments', per general-era-activity-(ies), socio-commercial-elapsing-(s), from birth-age-generational-road-conjuncting, due-to inhabitant-being-(s), whom share off-time-relations', quickly successioning sighted-act-(s), observed requisite at discourse, offset objective-usage-(s)-timing-(s), hardware-tangible-existence, at realm approxi-mate-surface-area-impasse,|345|comprehending meter-(s) through lifestyle, inquiring population, whom learn from-an intuitive-belief, where way of reliant-reaction-(s), conjure factifying, arithmetic-awareness, not instilled from natural-choice, indeed certain, assert method per daily-life-pulse, affording price-(s), held-at stocked-market-data, conceptive-unpurchased-discourse, thinking how lifestyle-dynamic, trail-along political-agenda-(s-(')), Basic-Human-Rights' & Retirement-Agenda, at impasse-checkpoints of individualism-distancing, as citizens abide law, enveloping attention per community, while existence, perceive human-behavior, (from an) tally-track, persuading-discourse, obstacle observable-perspective, forcing-effort-(s), practice-(ing) space-location-interval-presence, elemental, objective-documenting, conceded visual perceptive dating-article-(izing), congesting-street-(s), by housing-structure-(ing), interval year-(s) per-dating-of 16L.[study / style / spacing-passing-(s) / infrastructure / effort-(s) per hour of day / night-impasse-elapsing-(s).]¶|||01m.:26s.81m.s.|||§

§...Count-[4 / Four]-preposition-of.. ..Article-Essay-[13 / Thirteen].. // ..Book-Essay-[94 / Ninety-four].. ..Lines-[16.96 / Sixteen.1.nine-tenths-2.six-hundredths].. ..Anticipating decades is difficult to determine for where to schedule each individual of an socialism, practicing from male-transit-cycling, rather focusing on how to balance an cultivating of socialism, from when order of affairs, require an understanding of programmed-scheduled-tasks, when developing each sequestrated offset task, for conducting political-wage-affairs on entity-land-developments. ¶|||25s.65m.s.|||§

¶95Nineteen-Fifteen.174¶01s.34m.s.|||§ By calendar-dating, serve time from citizen-role-(s), budgeted-activity-(ies), remaining contested, in-an internal-services-(C.I.A.-{President George Herbert Walker

Bush}-infiltrated-war full of [destruction / terror / death / battle-(s) / conflict] entertained presence, (from an) idle-history, during earlier-timeline, factoring past-(history), when-during present-day-offset, serie-(s)-cycle-(s), [synapsing / elapsing // syntaxing] throughout individual-work-effort-hour-offset-discourse, offsetting particular-presence-active-location-(s) upon hour-objective-directive-(s), inlaying maximum-application, formal acting awhile affair-(s) imply [[1]effort-(s) / [2]luxury-(ies) / [3]place-(s) / [4]event-(s) / [5]occurrence-(s) / [6]object-(s) / [7]development-structuring / [8]eatery-(ies)] persuade-influence, for-those [[1]physiological-hyper-tense-(s) / [2]mind-(s) / [3]personality-(ies) / [4]place-(s) / [5]act-(s) / [6]cycle-(s)] where human-transcendentalism-resourcing, exude-from, human-still-act-(ion-(s)), for every lifestyle, embodying excuse-(s) for extraspection, affecting nationalism-reformation, by human-bias-psyche.175¶|||57s. 19m.s.|||§ Natural-focus, Expanse & Resource, (from of a the), [World War I // II reoccurrence-(s)] reconciled origin tiering, impassed event-(s)-article-(s),| 346 |generating production-(s), abided Religious-Governmental-Economic-Trancendental-Industrial-Commer-cial-Path-(s), undiscussed, from physiological-ethical-practice, at state-expansion, breeding [personal-individualism / family-orientation-activity-(ies) / working material-activity-(ies)] direct-contort, elemental-resources, payment-method-cycling, interposed conceived commercial-conceptional-cause, naïve-nature, inhibited mass, pieced at-an locale, furthering trial-(s), aspired driven-discourse, conjunct mass-exalt, cordial by-of perspective, impassing-interaction, juxtaposed territory, maintaining domestic-tranquility, stanced as reliance-work-payment-method, mediating ambient-expanse-territory, remaining unclaimed, as natural force from male-political-city-territory-tectonic-physiological-perspective-impasse, require-land, delving women-still-motion-article-(s)-impasse, developing cordial-conducted-interactive-affair-(s), where each [inclination-(s) // indication-(s)] of syntax, affirm perceptive-observation-(s).176¶|||01m.:58s.63m.s.|||§ Was alterior-align-(ing), coordinated from nation-(s), where modern-day-past-cold-war-allegiance,

affect day-elapse-(s), as [line / wait / battle / death-toll] recoursing
human-population, in-an account per developmental surrounding-di-
mension-(s), not [aware // prevalent] unless stated from [vernacu-
lar-vocal-tone-pitch-key // time-signature-perceptive-article-affirm-
ing-(s')] verify-ing how progression of article-event-(s), proceed each
dating of living-impasse;[176] I also inquiry, World War I as occurred,
demonstrating treasury-pathology, due-to American-Lifestyle, sci-
entifically-principalized, through rigor, incumbent free-will-mel-
ancholia, having psycho-logical-effect-(s), direct human-discourse,
amounting individual-prerogative, intermittent obstacle-discours-
ing, planned depth-perspective, where [height / altitude] vertical-off-
set-(s), per measure by-of individuals' distance-(s), upon-an [Length
/ Width-discourse-(s')] ambiguous auxiliary-device-(s), transited
along government-public-road-development-(s), influencing medi-
an-maximum-potential, repetitive construction-method-effort-(s),|
347 |influencing constructive-discourse, upon per, instrumental in-
terposed-impassing-(s), continuing unresolved-research, at param-
eter-(s), cause blissed-ignorance, conceive purpose, living through
elapse [objective-(s) // subjective] factual-historical-presence;[177] per-
suading oneselfs'-life, acted upon impulse-direction, during lesson-(s)
per mass-existent, influencing outcome by [race // ethnic] physio-
logical-effort-(s), accustomed perpetual-tradition, amounting faith-
confidence-based-trade, at guidance, by stock-life, per condition-(s)
upon land-territory-parcel-(s), amassing common-survey-inquiry, for
paying land-development, as objective-material-(s) device duty-(ies),
from-of-an existence, pertaining each value of individuals', upon
an [two-(conceptive) // three-(littoral-concept-developing-dimen-
sional-parameter-(s)] simultaneous all eye-plasma-spacing-recep-
tion-synapse-ligament-muscle-skin-blood-pulse-tension(s)-reflex-
(es).[177]¶|||03m.: 41s. 91m.s.|||§ As year-(s)-impasse from nineteen-fifteen,
through nineteen-eighteen, World War, invaded an calendar-period,
of industrial-progressive-development, distracted by internation-
al-trading-method, trusted deterring morale-(s) of mass, as new en-

terprise-(s), were relying on European-resourcing, created by American-Ingenuity, juxtaposing inhabitant-(s), left relatively unscarred, for interval period-(s) of engineering-industrial-boom, hoarded resolve from conflict-(s), alluding poor-common-mind-(s), controlling population, by person-blank-uncontributed-effort-(s), counted for-an clean-modern-present-crescendo-development, when-where, metropolitan Trade & Population, were too far unaware of religious-evaluation, as how to extraspectively-interact, as for after the war-period, their remain an [despair // triumph] focusing [sustained // objective development] instated metropolitan cleaning up, those debris put upon, plane-voluntary-existence, pertaining industrial-purpose, intent cultural-development-(s-(')), styling visual ambient-grandeur, supposed lavish exaggeration, living (un) [reclaimed // paying] for war-cost-(s), an debt, not educationally-focused, in-an routine through action-(s), that periodicalize-human-mass, for prerogative of matter-(s), pertain-upon an [2017 // 483-year-(s) of celestial-existence].178¶|||04m.:50s.13m.s.|||§ Musical-Influence enter-(s) in-to perspective, by-an genre-(s) [blue-(s) / swing / big band-influence] (un)-philharmonic city-off-time-preference, as off-time-activity-(ies), instill night-rendezvous, apose work-place-relief, those stressors from industrialism, individualizing focus, [where // when / why] Culture & Sophistication, remain unthought by those cause-(s) from era-nightly-homework, affecting humanity, by-an tensed-ease, classifying [birth-(s) / obituary-seniority / mid-age-commercial-production / method for economic-growth].179¶|||05m.:18s.03m.s.|||§ America would continue to rely on an Federal-Wall-Street-Treasury-analyzation-process, cordially-confirming, paper-confidence-trading, from Federal-European-International-mass-material-good-(s)-process-(es), put in to perspective, from habitual-cause, affecting era, per mass-affair-(s), while ignoring mutual [age / genetic quality-(ies) of mass-(es)] determining group-inhabitant-(s)-function-(s), liberally delving discussion with current-decade-discourse, per [affair-(s) / issue-(s) / trading-method-(s) / property-investing] moving formal-concert, by physiological-population-effort-(s), by-an mean-(s') for duty, by decade-prestige,69L. when glory, not reproduce offspring, present-requisite, how human-conduct prefer to perform thought, from ones' physical-perceptive-composition...¶|||05m.:59s.40m.s.||||§

§...Count-[16 / Sixteen]-preposition-of.. ..Article-Essay-[14 / Fourteen].. // ..Book-Essay-[95 / Ninety-Five].. ..Lines-[69.91 / Sixty-Nine.1.nine-tenths-2.one-hundredths].. ..Wages expense is such an difficult payment, to figure out how to proportion, per individual, when tasked at an designated-scheduled-entity-(ies), for circulated-objective-usage, from an application of directed-orders.¶|||14s.97m.s.|||§

¶96Dress & Show, define era by-an fashionable timing, differentiated as, 17-62-years'-old-retirement-aging, anticipate through-an present-continuum-view, timed, amidst an [future-location-impasse-anticipation-(s') // past-requisite-impassed-community-relation-(s')] abstract-relief, not focused by personality, or redacted from-an written-accent, variating city-position-(s), where state-constitutional-budget-proportion-(s), affect how individual-interpretation-(s), perceive abound surrounding territorial-religious-covered-tectonic.180¶|||29s.47m.s.|||§ This reveals an modern-fact, that human-nature, is from-an, feeling-sensational-placing-(s'), unfocused on [genetic-reproduction // repopulation / state / artistic-creativity] creatively-active, through daily-development-(s), perceivable abound barrier-(s), where-of physics-offset, deviate mutual-camaraderie, executing tasked-labor-(s), enforced upon male-driven-effort-(s), in force when cause, derive inhabitant-(s)-focused-nature, from-an, intercoordinated-resourcing-trading-method, working linear-directive-(s), from educational-background-(s'), pre-instated, career-mentality, sufficing traded-means' by litoral-tangible-dry-wet-substantial-(s)-precipitation-respiration-cardiovascular-eye-plasma-spacing-reception-synapse-ligament-muscle-skin-blood-pulse-tension-(s)-reflex-(s), digesting, through tract-process, chore by basic-means', interposing formal-pattern, yet by-an genitive-perceivable-moment-(s), acquiring tectonic-experience, resourcing objective-trade-pattern, constitutionally-revisable, ordering (from an) archetype-year-dating-impasse-scheduling, passing day-(s), lock-in an mass from such unawareness from study-(s), never conceiving how amounts' of numerical-prose, influence count-(s), circumstantial happening impasse-(s), per matter of daily-volatility, because requisite-prior-dating, either routine, pre-destine, unscheduled-matter-(s), or attempt through day-prose, isolation-(s) for [objective // subjective // physiological data] reworking ideal-(s) by retirement, fundraising adjunct time, rural-relocating, still remaining up in the air, due-to mass-survey-prerogative, uninquisitive concluded-effort-(s), from-of State & Physics, determining-discourse, deterred surrounding climate, affecting how daily-parameter-(s), impasse article-retrieval, sustaining substantial-influctuated-existence.[181] ¶|||02m.:03s.[34m.s.|||]§ Why does plainness, derive compulsive-behavior, idle through natural-way of living? It is beneath us to effort blue-collar-(dirty) action-(s), or does observation of [tectonic-plate-(s) / physiological-mass] omitting objective-data-collections', clearly from population-reclusive-thought, preferring non interactions', intend to keep an focus of governmental-transit-parameter-(s);[178] enclosed idle-belief, because discoursed-routine, refer location-(s), where [repopulated // covered rate-(s) of humanity] preferring idle-paved-surface-impasse-(s), as how whereabouts, juxtapose amongst transit-street-thoroughfare, select from intention, permitting perceptive-ground-(s), as-an formal-impassing of as-tranged-passing-(s), not coordinated extraspective-appreciation, for matter-(s), from narcissistic-mode, verifying mentality, proudly-stubborn, indulging through an international-Europe-American-Chinese-Japanese-trade-affair-(s) producing without philosophical-inclination, as trade matter where product is produced and developed, pertain elemental-disposition, upon surroundings', to emphasize an particular-local-engineering-construction-method-society, upon articles' of [local // national] development, because how dating apply upon present-juxtaposition-(s'), apose posit-tangible-being-(s').182¶|||03m.:02s.49m.s.|||§ This allude me fact by child-labor-law-(s), enforced by, condition-(s) of infrastructure and payment, succession education, over-inflated human-payment-method, for those dynamic-perspective-(s'), insisted instituted-cause by physiological-subjective-period-perspect-ive, determining notion-(s) per [payment / ordered-operation-(s) / parcel-space-lease-rent / taxes / date-deviation-year-sales / inventory-control] maintaining accurate-operation-(s), in light of blood-pulse-perspective, emitting view, in distance from mentality-feeling-effect-(s), not personable, conjecture by [sight / sounding, distanced-object-(s)] substantially-inclined, voluntarily-effort-(s), setting juxtaposition mass-literary-level-(s), affecting influenced mundane discoursed-prerogative-(s), set on development, per [cement /// metal // aluminum // iron-pound] yet conceived from fluent inhabitant-citizen daily-discourse-(s),46L. by original-resourcing-territorial-isolated-practiced-development...¶|||03m.:59s.22m.s.|||§

§...Count-[12 / Twelve]-preposition-of.. ..Article-Essay-[15 / Fifteen].. // ..Book-Essay-[96 / Ninety-Six].. ..Lines-[46.334 / Fourty-Six.1.three-tenths-2.three-hundredths-3.four-thousand-ths].. ..rebar is not supposed to be infrastructuring cement, yet the government never cese to notice, why such flaw exist, omitting engineering, by planned-construction, converting rebar better for geological sewer-systems, preferring architectural-developmental-flaws, rather pure-cement-residences, handling more than house or mansion, but in an fashion of family-hallways and room-(s), per indigenous-being-(s)-family-offset-population-classification-(s), from historical-solar-483-year-error.¶|||26s. 69m.s.|||§

¶97...Influence upon inhabitant-(s), has no capable truth for factoring, being of-an place of view, thus amounting non-fiction-literature, faulted figurative-reality, not in an realty-influence, perceiving mathematical-developmental-concept-mold-part-(s), affecting those conjecture-(s) by activity, shorted by day-payment-entity-federal-routine, in no other trade-data-method, sequencing from [rhetoric / historical-account-biographical-discourse / obituary-physiological-discourse / psychological-taunt] unobservant current [year / season analysis] from mass-modern-simultaneous-schedule-acting, based from political-agenda, structure-(d), from international-assurance, default aware-presence, [uncreative // unconceptive], literature-artistic-outlet, by [English-predicate-modern-civilization /// African-Ethnic-tarring /// West-Holy-Roman-Empire-Mexican-Italian-Archaic-Civilization-order /// West-Holy-Roman-Empire-Spanish-South-American-Archaic-Civilization-order /// American-modern-historical-protestant-Civilization /// Holy-Roman-Empire-French-English-modern-historical-predicate-Civili-zation-(s)] with an interrogative-converse, examined by our fatherly-lord, "whom art in Heaven" {Bavaria}, in an analysis for weather an nation can be built, from English-predicate-constitutional-method (further regalia-documenting {Magna Carta}) still drifting by physics, under scrutiny, for Sitting & Reading, when common preference, be more inclined to purchase-unit-data, without realizing genetic-population-mass-common-elemental-extraspection-extraction-practice, from formal-extension by-of term-definition-data, remained in an judgement, from contemporary-governance, for not psychologically-practicing an method to develop those genre-(s) from non-fiction-direction, upon present-continuum-maxim, pertaining shelf-life-impasse, upon movement-(s) from mass;179 forgetting sequential-operation-(s), in rule (for of) human-demand, because discoursing-perspective, occur far too individualistic, upon throughout subterior-effort-(s), when natural-physiological-direction-(s), remain incapably-immovable, actually-visualizing-footage, from tangible-dry-wet-substantial-precipitation-cardio-vascular-eye-plasma-spacing-reception-synapse-ligament-muscle-skin-blood-pulse-tension-reflex-(s), by whom serve duty, fulfilling demand-(s), because requisite incremental-interpretation-(s'), have ever only sustem momentary-output.183¶|||01m.:56s.86m.s.|||§ Temporal-at-will-employment, define longevity, when 401K-retirement-planning, yet feature reclusive-prerogative, in case form, Effort & Individual-(ism) incapable familiar-interpretation-(s), never meeting on grounds for mutual-interaction-(s), intelligibly-affirming, from-where article-affirmation-(s), confirm individual-extraspection, through physiological-parameter-development-(s'), interceding goals', by ones motions', for invigorating thought, apose idle-reclusive-interaction-(s), (un) confident because of an lack of fear for survival.$_{184}$¶|||02m.: 19s. $_{65m.s.|||}$§ What extravagant setting may humanity assess, amongst reset-(s) by-of physics, as-when executive-legislational-judiciary-fact, exist-by local-enabling, staying-locked, by city-living, unexperimented from nature, applying grandiose-ambivalent-mass-motions', condensate interpretations', because materialistic-objective-possession, affirm temperate-gauging, by perspective, in past-neutral-requisite-present, inform-(ed) dating-pacing-(s'), routine adjusted-repopulation-civilization-(s), common of an mass reproduction-cycle-process, still by women, intolerable from children, by [racial // ethnic front-(s)] sustaining enumerated-freedom-(s), requiring an mute-illiterate-mass-conjunction-social-efforts'-economic-extraction, casted out surveyed-opinion, from volume-vocal-tone-vibration-(al)-word-note-audible-term-syntax-frequencies, defining data-paragraph-tangent-(s), in-of-an consortium-reference, in due of reaffiliating piece-(s) of syntax, by formal-invigoration, offsetting presence, during mass by governing [domestic-trade-impact / Library-Congressional-Literature-Inquiry] primarily expecting mystery-fiction-romance-tangent-abstract-live-focus, in an parody of power, but boasted from what there is of, in infinitive-constitutional-reasoning, defining those barrier-boundary-(ies) of city-limit-development, when passing age of moved-effort-(s);$_{180}$ (un)-conceptualized surrounding-setting-physical-walking-impasse-time-dating-construing, to invigo-rate-tense-(s), live active-reflexes', weighing from mute-natural-ignorance, apose extraspective-interaction-(s'), those ideal-(s), personally-seen, to matter as data, from how history is instilled from basic-human-rights' required-implement-in g.[185]¶|||03m.:40s.[59m.s.|||]§ The good of such fashion of existence, pertains' reasoning by customary adage, Greatly-Advantaged & Fair-Practiced work sustainability, Reading & Lounging (rather remaining kinetic, while awake);[181] reclusive by self-(ves) literary-purpose, from how those invigorations' have been, because how interaction, is a-the illiterate-fact, that suffer-(s)-humanity, patent-copyright-fantasizing-{Artistical-Handwrite} register-trademark-objective-trading-{Technical} rather rating, each and every, piece of data, tense-applying, how our blood-pulse, affect hour our muscle-ligament-ventricle-skin-plasma-synapse-tension-(s'), motion, from our dry-posit-(s)' of bone, (due to an) matter-of-fact, defined simply so, with not an analysis, but reminding how such [fact // truth exist];[182] furthering such reality, by constant-inquiry, rather idle-entertained-bemusement.[186]¶|||04m.:24s.[66m.s.|||]§ I am an individual as such, for starting [project-(s) // interaction-(s) // motive-(s') // motion-(s') // (legislational)-act-(s) // et cetera] inhabitants' do not see how other-(s) comprehend what activity, I partake-suggestion, perplexing relevant payment-method-practice, commercially-religious-obedient, (never) deviating from pre-disposed-disposition, inter-cooberated awhile tense-(s') be in an educational-quality, visual natural-bliss, grandeur consistency, for-the still-requirement (s) of living, disturb movement, only from-an introverted-individualistic-freedom limit-(s), sequencing from requisite-date-hour-data-denominative-impasse-period-(s),| 348 |[subterior-incremental-deductive-reading // reordering-thinking // motioning-(s')] for whom have an way of reception, enjoyed by ticket-purchase-(s), (as an) result from short-minded-nature, venturing venue-(s) at [[1]basketball / [2]football / [3]Olympic-(s) / [4]Hockey / [5]Circus / [6]Nascar / [7]Club / [8]Festival-(s) / [9]Carnival-(s) / [10]Theme Park-(s) / [11]Soccer / [12]Tennis / [13]Formula I / [14]N.A.S.C.A.R. / [15]Movie Theatre / [16]Amphitheatre / [17]Theatre / [18]Hall / [19]Greyhound-Transit / [20]Airplane-Transit / [21]Baseball / [22]Concert / [23]High-school-event-offset-reset / [24]College-University-event-offset-reset-facility-site-(s)] preferred time-stretch, by daily-discourse [stop-(s) // pause-(s)] intermittent realty, ignorant newspaper-subscription-(s), determining political-discourse from international-illiterate-agenda, without my plause of littoral-basic-human-rights' sufficing development from-an present ten-millionths-intelligent-reproductive-population-reform (my fiction-utopic-ideal for literary-prose) juxtaposing, billionths-population-rating, that require an strict-emphasis, from when mass-citizen-(s) remain quite reclusive of political-social-security-openness, passing time, until away, obituary-aging, as do most naturally discourse life, continually unsegmented by the daily-prerogative-(s), participating in civil-extraspective-forum-(s);[183] instead-of individual-compulsive-consumerist, yet read consist literary-purpose, by focus of literature in two-non-fiction-fashions', one is writer-new-print-context, two book-editor-constant-editing-purchasing;[184] selecting a-the common-reader, by my editor-per-book-unconscious-extempo-rizing, apposed my initial-new-context-upkeep, attracting new readers', from my extenuations', [juxta-posed editor // myself-effort-(s)] to guide an crowd, by such book-(s)-nature, along-with governmental-cabinet-(s)-agenda-tier, naturally conceive requisite-hedonry, by-the ignorant-dismissal of literary-technique, as tool for subjective-data, impassing effort-(s), from youth-bias, through an aging of mid-age-retirement-anticipation-planning, showing an default year-by-year-conversion-(s'), without stress-fatigue, per product-purchase-parity-agenda, parallel those adjacent-discourse-(s) of tectonic-mass upon momentum, as awhile each day is in an unfortunate dismay, per schedule-(s)-natural-fatigue, for political-motive-(s), for like celebrity, is house-representative-(s')-thirty-thousand-being-mass-discourse-appropriation-discourse, supplement, a-the extraspective-requirement-(s'), for handling Paper & Ink that is relied historically-physiologically (continuum) by men, on an trust of Federal-trade, counted from university-Federal-loan, linked by-the state-system-(s), that remain in-an unilateral-juxtaposition of soveirgn-nation-(s), not identifiable, from partisan-action-(s), from Federal-Discourse & Concurrent-Circulation-Outlet-(s), reforming each method, upon those constraint-(s) by free-trade-policy, affecting legislative-outcome, by social-parameter-cause-(s), comprehend-ing, we all require extraspective-interaction-(s), (as an) method for interacting with one-another, because how [extracting // developing // resourcing // collecting // et cetera] require an helping-hand, in-of-ever objecting to-an national-order,95L. by elongated-grace, rather our present-offset-disposition-(s'), by introverted-individualistic-independent-idea-inclinated-indicated-interpre-tation-(s')-inhabitant-(s') ...¶|||07m.:30s.85m.s.|||§

§...Count-[32 / Thirty-two]-preposition-of.. ..Essay-[16 / Sixteen].. // ..Essay-[97 / Ninety-Seven].. ..Lines-[96.33 / Ninety-Six.1.three-tenths-2.three-hundredths].. ..I don't know why with educational-interpretation, individuals partake in ticket-traded-affairs, beyond retirement of nation. Maybe family-individualism, not partake with neighbor for of genetic-ethnicity-interpretation-(s), because how aging forget to align-American-English-grammar-algebra-educational-studies, with commercial-book-market-vending-sales, proportioning [Cum Laude /// Magna Cum Laude /// Summa Cum Laude]; educational-tier-empirical-astute-individual-(s), observing in my opinion from commercial-market, how none partake in passing an astute-educational-lineage, from university-studies, slumbering into centripetal-metropolitan-oblivion, by each passing human-composition-al-lineage, reproducing under ten-children per family, not thinking from literature, how word of mouth, would not slip from demographic-offset-bias, due to those population-effort-(s), thinking how to develop an order of personal-library-(ies), that concentrate the most specific-content, by an need of an preferred-genetic-citizens, referencing from an common-requirement for an bookshelf of 1,200-books per century, by an civilization-class, of 5-7-million-family-individuals, whom understand the transgression of literature, by aging of an book-solidly-collected, per month of living, as this book would suggest for an century, by an estimation of white-kidnapping-ambition, through New-York-Book-Surveying, and blonde-unconscious-laziness, primarily-bayou-European-Latin-American, gauging from my context, international-poverty-illiteracy, examining how the more stupid life remain, the less sales occur, and more calm human-illiteracy remain on my tear of one-hundred-non-fiction-books.¶|||01m.:23s.00m.s.|||§

¶98Who has ever discussed each Fact & Opinion, differentiating situation-(s), set-by variation-(s) of circumstance-(s), when self, post-age-birth-visualizing, tense-pressure, evoking existence, never in-an cuneiform-requirement, pre gerundive-interaction, motioning impasse lifestyle, not open to, off-time-scheduling, detailing psychological-designated-parameters', amidst an gamma-erroneous-abstraction, pertinent high-school-governmental-crestenza, tonal-visual, virtualistic-relative-success, pertaining present-movements', [inclinated // indicated] written-conjecture, for formal-scaling by balance-(d)-product-(s), accounting per mass-population-locational-practice-extent, an fully Federal, commercial-production, from commercial-market-shelf-stock-live-(s), premise-(s)-inducing, an Totalitarian-regard, pre-requisite-order,| 349 |from American-Historical-initiative-egalitarianism, requiring every [human // mammal-racial-ethnic-group-compositional-posit-(s)] as-an denominative-unconscious-mass-means', uninquisitive, (the) cause from Subjective-Feudalism-Rule, by debate, neutral bi (tenth year)-(bi) partisan-politic-(s), emptily deviating, succession-cause-(s), for reason-(ing) scarce instrumental, W-2-form false-filing-(s), judge legislation from house-analyzation-(s'), where senate-commercial-belief-responsibility-(ies), sacrament examination-retrograde, by those fault-(s) pertaining I.R.S, from-an 76-years-impassing-history-predicate, uninquisitive, human-population-judgement-(s'), idle belief, [where // when // why] whom Legislate living-purpose upon commercial-budget-(s), individuated rescaling vehicle-(s), when mass house-representative-title-order, overseeing reordering, where county-classification-(s), influence governmental-oversee, yet understood by common-mass, for millennium-judgements';[185] awhile men, ration [race / ethnicity] because educational-(re) probing, sustain mass-present-examination-(s), scaling denominative-season-(s), by protocol, for reclaiming punitive-cost-(s), existing through market-trade-(ing), defining human-prerogative, never capable in light of day, [to // at] stay up nightly-savings'-interaction-(s'),| 350 |observing an revolving-moon, offset proportion, solar-continuum-kinetic-friction-combustion-plasma-celestial-central-energy-force, as perceptive-competent-individuals', monitor celestial-metric-gauging, unmeasured of approximate-distance-(s), awhile each citizen, is out of-an natural-physiological-perspectives', defining tangible-causes', out of-an property-complacency, without an reason for event-(ing), human-individual-mass,| 351 |discoursing, daily-surrounding-parameter-(s), reasoning [emotional-intimate // colloquial-impasse-purpose] through our commonly-compulsive-cycle-(ing), that partake upon dating of receipt, because, being-(s) are in right-(s) of reproduction, by the freedom of civil-liberties, not invadable through path-(s) of roadway, covering an balance of population-mass, instilling national-inhabitant-(s), (not citizen-(s)), as citizen-count-(s) per 15,600,000-citizen-(s)-(family-(ies)), covered by Federal-government-discourse, (as my father be an title-owner{1452 N.E. 135 St.;[186] 33161;[187] North Miami, Florida});[188] measuring Five-Hundred & Twenty-Federal-House-Representative-(s), from-an default,| 352 |of literature-fiction-mystery-auto-biography-book-shelfing, where in lieu of historical-requisite, not have an common, book-shelf-library-article-reference, (for highlighting data, pertinent upon an, surrounding parameter-developments', influential, by [walking // bicycling impasses'] that persuade physiological-sensation-(s')), untensed by-an juxtaposition of, [vision / hearing / taste / smell] awhile blood-pulse pump pulmonary-gland-lung-respiration-transitions', through ventricle-intrication from skin, as central-means', during respitorial-response, assent intelligible-perceptive-inhabitants', by-an rated-article-interpretations', requisite celestial-universe-history.187¶|||03m.:14s.08m..s.|||§ Statistical-reset, applied while in day-passing-(s) having an offset per circumstance-(s), around dynamic-surrounding, metropolitan-city-(ies), pertaining issues' of individualism, upon locational-city, identifying developed land, from resource-(s), technologically applied, from when tectonic-plate-(s), offset-juxtaposition human-transit-impasse, awhile interpretations', not matter because of human-interval-extraspection.188¶|||03m.:32s.21m.s.|||§ Why not engineer an literature-prose-purpose, from local wood-work-(s), pressurizing discourse, by-when "tangible-going-(s)", participate view, by-of light, surrounding plasma-force-distance-mile-(s)-counting, per day, when tectonic-drift, affect mass-distancing, upon three-hundred-sixty-five-day-cycling-impasse-repetitions', continuum a-the surrounding-globe, by-an horizon-east-to-west-force-drift, when tectonic-posit-impasse, solar-conjecture, offset, continuum solar-finite-spacing-(s), where respiration-observations', are-from solar-fusion-combustion-friction-infinitive-plasma, never ceasing continuum-offset-pressure upon existence, from earth-posit, invested-conjure from belief, enveloping date, because claused-language, influence which article-(s) apply, religious-fervor, from idle nature from humanity, inclined-fact, for discourse from common-memory, believe mass-nature, not be from-an million-(ths)-inclination, during legislative-reform, as trust of individual-(s'), claim importance of data, by surface-area, measured from ground-contracting, dimension-measure-(s), when locational-birth, regard domestic-development, when state-movements', are as retirement-like-living, having Cubic-Limit, affirm [Height-(clearance) // Width / Length] in regard where observation-(s), construct product-dynamic, as trade be in awe of transaction-(s), because muscular-skin-ligament-bone-cartilage-organ-perceptive-blood-pulse-centri-petal-respirational-pressure-press-along-interior-ventricle-lung-pulmonary-mouth-throat-esophagus-stomach-intestine-primary-bend-digestive-tract-tension-(ing-(s)), case an point from self, to live by-an extraspective-group-formation-(s), in thought for living, while prose of human-behavior, constantly-adjust, revolving-surrounding-parameter-(s) per lifestyle, amounting from-when, respiratory-muscular-notion-(ing-(s')), volume-circulate, ventricle-tension-posit-(s), that coursework, connotational-colloquial-converse, influencing human-effort-(s), because dating be in order from personal due-(s), settling an home-abode, not affected by individual [sustaining / maintaining / obtaining] for developing-matter-(s), because independent-living, time numerical-appropriating, where-when, 66L.[boundary / limit / conception / tangible-quality-(ies) / perceptive-independent-individual-(s) / millibar-pressure-rate-(s)] upon natural-secondths;189 scale elapse-(s), upon mass-perspective, egalitarian human-posit-perspective...¶|||05m.:12s.19m.s.|||§

§...Count-[20 / Twenty]-preposition-of.. ..Article-Essay-[17 / Seventeen].. // ..Book-Essay-[98 / Ninety-Eight].. ..Line-[68.33 / Sixty-Seven.1.three-tenths-2.three-hundredths].. ..pending on human-perception, an sway of illiterate-hoc-bias-rhetoric, can happen, but not understand that with copyright-ethics, an serial-examination of human-mass-population-thinking, is observable of physics in tectonic-fashion, because how developing celestial-force, occur without an function-directive, (just yet), as when thinking is for an limited-population, circulating-books, by an pursuit of genetic-finding, thinking either to join along an group of individuals, whom understand why Communications & Books, are required in developing subservience, for thinking how to comprehend each developing fact, amid an turn of traded-affairs, for understanding the family-lineage-genetic-pursuit, of having an blonde-genetic-cooperation, by retirement-vacationing, to [Montana, U.S.A. // Manitoba, Canada] etcetera, creating commercial-university-educational-book-entity-(ies), for circulating the formal-objective-products, for developing architectural-lands, for arranging an commercial-developing, before planning governmental-constitutional-reformation, because how order works by male-developing for fine-voluptuous-(hopefully-blonde)-women, for serving menial-effort-(s), by each activity, because of territory-command.¶|||01m.:02s.34m.s.|||§

¶99Abstract-Physics remain uncalculatable, from those tension-(s) by-of physiological-physique-blood-pulse-plasma-tensions', affirming motion-(s)-deviation-(s), at-an human-intangible-interior-perspective, from exterior-sky-view-limits',| 353 |where celestial-secondths, are deductive-subterior-referenced, intermittent hour-interpretation-impasses', upon an {3 / 3 / 4}three-third-fourths-motions-elapse-cycling, balancing compositional-weight-(s), subterior, tectonic-centripetal-force-drift-impact, when ground-parallel, intermittent calendar-distancing, defining why [existence / weight] from-of biological-beings', remaining how tense-relay-(ing) exist, conjunct-sensational-reactions', when dating denominate-interpretations', interpreting observations', from physiological-eye-plasma-ventricle-blood-pulse (not [memory // thinking // theoretical] belief), continuum-day-reset-presence, as how hour-time-intervals, deviate from-an exterior-spherical-posit-offset-(s), far too difficult, for understanding, when we merely impasse as such, and sleep, by our nature of fatigue.189¶|||55s.52m.s.|||§ Today is ancillary-juxtaposition, by-when revolutionary-progression,| 354 |drift in perspective day-night, affect celestial-abstract-mile-metric-(s)-count-(ing), perceiving ancestry, passing-memory, influencing an still-human-posit-perceptive-memory.190¶|||01m.:12s.11m.s.|||§ I dearly enjoy non-fiction, for neglect of [rate-(s) / ration / statistic-(s) / calculation-(s) / Numerative-counting / denominative-counting] percentile twentieth-five-percentage-deviation-(s) juxtaposed whole integral-factor-(ing-(s)) proportioning Time & Calendar, for sum-equation-operation-(s), as how superlative-default-reasoning, not be deductive-reasoned, for refining article-focus-scaling, (as an) means' for learning, rather than, [period // block-schedule-subjecting], to gander inhabitant-human-population-balancing, from [issue-(s) // affair-(s)] unresolved, unrecognized-placing-(s), awhile religious-genetic-trust, affirm familiar-inconsistency-(ies), offsetted by debt, where operation-(s) profligate (un)-useful-self-adjunct-objective-application, having shelf-life-exponent, oversuperlative, from individualism, purchasing by volatility, when Supply & Demand, purchase-parity, from-of-an erratic-deviations', affecting human-scheduled-interactions', by religious-community-trusting, sustained hence far, by introverted-auxiliary-action-(s), offset banking age-(s)-tier-education-decade-interposed-year-month-cycle-(s), free-will-individuation-offset when year-to-date-interactions', partake as feature-biased-motion-(s), temporal developmental-fatigue, unstressed, for mass-(es) of inhabitant-(s), not physiologically, appropriate empty-terrain, from free-prerogative, directing an lifestyle, without congestion, yet not in-an totalitarian-educational-fashion, coming from electrical-effect-(s), awhile freedom exist per-at-of-reset moments', spaced-off order-(ing)-market-demand-volatility-article-(s), comprehending boundary-limit-(s), awhile subjective-feudalism, remain unperceivable, by metropolitan-developmental-population-offsets', [servicing / serving-due-(s)] by [military // commercial, corps-obedience's] when men not ruse [literary-pre-conceptive-architectural // political-currency-denominative-method // action-(s) of terrain // method of influencing morale for such ecological-boundary-(ies)] inspired from national-youth, aging from elder-statesman-posit, enticing [lotto / bingo // entertainment-activities'], idle primary day-(s)-thinking, by monetary-constraint;[190] not thinking how to balance currency, cycling [objective-posit-(s) / subjective-posit-directive-(s) // subjective-objective-article-(s)-ordering-directive-(s)] because pastor-worship-effect, persuade community, purpose of investments', because industrial-pattern,| 355 |rely ingenuity of American-North-Eastern-Centralized-Governing-Trade, requiring duty per thought, | 356 |neglected compulsive-consumeristic-purchasing of food, saving-(s) data, for self-embolizing, past locale-lifetime, either when obituary, part-way-(s), where memorially-parted-way-(s),| 357 |factify mass-human-disposition, apose national-international-agenda-objective-progression.[191]¶|||03m.:47s.[02m.s.|||]§ I attempt to analyze this notion, from only factual-observation, when lease-loan, affirm [post-year-purchasing-(s) / live-year-accrediting] not successioning union, for proper article-ambiguation, from when constitutional-legislating, intervalize subject-periodical-article-tier-(ing), prosing article-(s),| 358 |fatiguing thinking, balancing physical-characteristic-tense-pulse-(s), elapsing daily-discourse, affecting critical-thinking, by when case of present-budget-schedule-disposition, draw inhabitant-(s) from short-minded-thinking, furthering extent per lineage, making life pertinent by-an extraspective-limits', [conscious-output // input] upon communal-district-dynamic, materializing national-trade-observation-(s), that impact audible-interaction-(s), from competency per personal-limit-(s), surveyed while an hold of national-governmental-guidance, direct currency-confidence, affixed upon live-dating, from memory-transition, because conceptive-valuable-(s), pre-determine, hold of individual-personal-gain, instrumented Federal-resource-material-(s), ensuring lifestyle, by enumerated-living, uncalibrated from Semper-Fi-Nine, future-requisite-common-family-lineage-offset-war-morality-transition...¶|||04m.:20s.33m.s.|||§

§...Count-[15 / Fifteen]-preposition-of.. ..Article-Essay-[18 / Eighteen].. // ..Book-Essay-[99 / Ninety-Nine].. ..Lines-[53.6 / Fifty-Two.1.six.tenths].. ..how the globe sees politics, has an differentiation of sounds, that can not be interpreted, due to past-dating not formally correct, when dating revolve immensely swift, beyond an recognition of visual-surrounding-elemental-entity-factory-objective-developments, for layering motions upon an active-entity-(ies), circulating how developing premises, are suppose to fulfill an obligation of objective-upkeep, new-objective-circulating, pending how individuals are Creative & Standard, by an focus on continual-national-boundary-upkeep.¶|||33s.94m.s.|||§

¶100.Method of credit-equipment, remain bemused from each passing-weekend-(s), yet accustoming family-dynamic, thinking focused by-of objective-usage, through aging pass auxiliary-vehicle, by gasoline-deposit-(s), presided by [gear-shift-intake / odometer-speed-rate-intake / mile-(s) of total-part-(s)] elapse public-passing-discourse, cycling volatility, measuring human-metropolitan-continuum-impasse, from how trust of national-capability, can be truthfully unit-processed, when dating, juxtapose time, for [output-effort-(s) // location-unit-demand-expectancy] observing from how ["dime-outs" // expiry] affect each of those season-(s), for product-unit-development, per dating-impasse-period-(s).192¶|||33s.31m.s.|||§ To feature bind-lease-consent, an measuring-paper-subjective-demand-timbre, is required, proportioning routine-impasse-recollection, acting upon an sequential-destination-(s)-operation-(s), [working / housing / marketing / storage-center] sufficing basic-human-rights', formally conducting, scheduled-operations', as way of life, perceived in view, for constant-upkeep by colloquial-default-human-being-European-language-letter-word-article-term-definition-extent-usage-(s), not modified by literary-non-fiction-directive, from Grimm's'-Grammatical-Law, because fatigue affect vowel-vocal-mutation, interacting by requisite separations', because spouse-romanticism-off-time-upkeep, Gregariously-calendar-periodicalize, [[1]past-[2]human-activity-(ies) / [3]event-(s) / [4]happening-(s) / [5]occurrence-(s) / [6]forum-(s) / [7]discussion-(s) / [8]talk-(s) / [9]action-(s)] tepid volatility-outlook, when technological-dimension-(s), resource from government, where mutual-bi-partisan-unilateral-politic-(s), form land-(s) of law, equipping characteristics (from an) blood-pulse-interrogative, reacting by communal-virtue, [where // when] developmental-conjecture, persuade effort-(s) (not internally evoked, from introverted-individual-(s'), concerted upon an objective-ordering, spacial-dimension-(s)-cycling-process, affirming time in guideline from temporal-tense-(s), by observation at exterior-cubic-area-elapse-point-impasse-interpretation-(s), from title-dynamic, defining political-act-(s), while language-bill-documenting, not pre-ulterior, because of how conclusive-deductive-ordering, enhance fact-(s), from convey-(ed)-data, interior from self-need-(s), juxtaposed exterior-dimension-count-(s), required by [objective / cubic-spacing-(s) / activity-program-agenda / mass-continuum-circulation] admitting capable-effort-(s)-extent-(s), compartmentalized generalized-comparison, requisite-ordering-attempt-ing, [resourcing / citizen / dimensional-modification / product-object-dating-decay-perspective-motion(s)-schedule-impasse-year-season-month-week-person-regiment-impasse-ordering], verifying account-(s) from individual-(s), by person-bill-documenting, affiliated-association-(s'), for an amplitude of mass, similar by frequent-interpretational-sound-(s), overlayered [elapse-(s) // synapse // syntax-timing-(s)], upon present-day-dating-influx, requisite future-conceptive-accounting where Mass-Population, remain introverted from historical-interpretation, by ignorant-objection to object, by subjective-interpretation-(s).193¶|||02m.:32s.09m.s.|||§ Workweek can only time confirmed-referendum, continually-offset-(ed) per day-morning, on-an merit, based-of trusted-good-(s), informally-mass-appropriated, budgeted-market-demand-volatility-expenditure-(s), held under regulated guideline-(s'), confirming Subjective-Feudalism, when inhabitants' naively-follow justice, [uninquisitive / investigative] item-(s)-proportion-ration-per-mass-population, objective-tangibility, as-when testing-material-(s), attempt Enhancing & Invigorating formal-physiological-tense-(s), where moment-(s)-(um)-posit-(s), are created upon, exterior-ongoing-continuum, from [Men / Women Reproductive-Discourse] affirming affair-(s), technological of creativity, from Post-Industrial-Order-(ing), yet to be considered by an mass-aging, simultaneous-impasse, verifying-data, from those constraint-(s) of accomplishment, producing object-(s), from dimensional-concepting, curbing resourced-element-(s), by-an Universal-Degree, affirming matter-(s), persuading motion-(s) of ordering, from [self // extraspective-managerial-influence] differing religious-guidance, tiering decade-denominative-aging, upon live-pulse-existence, still unclear, fostered-day-effort-(s') upon year, when calendar-interval-(s), construe directive-schedule, as one desire(s'), from-requisite matter-(s) per dating-impasse, those dating-parameter-(s), where governed-constitutional-merit, continue an average-influence-impact, by citizen subjective-effort-comprehension, per individual-input,49L. mattering where active-article-confirmation-(s), are made at an impasse of competent-existing...¶|||03m.:51s.19m.s.|||§

§...Count-[13 / Thirteen]-preposition-of.. ...Article-Essay-[19 / Nineteen].. // ..Book-Essay-[100 / One-Hundred].. ..Lines-[49.28 / Fourty-Nine.1.two-tenths-2.eight-hundredths].. ..Balancing industrial-factory-goods, remain significant, from when location-(s) offset from human-physiological-cardiovascular-effort-(s), impassing circulated-entity-premis-es, for either retaining of [wage / product-(s)], when having residence, have an maxim-capacity, that influence how trade is conjured, by market-shelf-(ves), when impassing age, deviate how one formulates an perspective upon tectonic-plate-statehood-county-city-district-political-circumstances, which display how citizens-wage-circulation, determines currency-exchange-levels, from observation-(s) of elemental-compositional-development, by commercial-market-shelf-unit-price-inflation, to recover those expenses of utilities-operation-(s).¶|||41s.00m.s.|||§

¶101.How many eye-(s) have an chance to see my take on existence? American-Legislational-order, form-an commercial-impact, denoting government-resource-agenda, denominative governmental-technical-reliance, when vehicle-youth-spouse-lifestyle, not create from reproductive-method, but by-an offspring-production, [raising / breeding / lineage extenuating-focus of debt] enveloping purpose, by off-time-activity-(ies), serving delectable dialect deserved-action-(s), from motion-(s) by Physiological-Effort & Subjective-Action-(s).194¶|||27s.72m.s.|||§ I see those desire-(d)-(less)-interchange-(s), intermittent continuum-aging-article-impasse-(s'), wearing teenager-adult-impasse, before capable-inter-dialoguing, present dimension-(s)-input-measure, by mathematical-cause, relaxing-ideal-(s'), inhibited from historical-process, circulation-interval-motivated, from defining prerogatives' upon literary-cause, when off-time-preference, prefer [concert-(s) / movie-(s) / game-(s)] in regard by periodical-article-study-lesson-(ing-(s)), rendering physiological-mind, apose present-observational-perceptive, abstract, various material-(s), at hand, for allotting insight of merit-(s), [when / where], [day / night-offset-(s)], sustem from perceptive-observation of centripetal-force,| 359 |affecting cubical-aging-reflexive-interpretation-(s), consistently considered, from inanimate-product-purpose-(s), when amongst physiological-tangible-internal-ventricle-respirational-blood-pulse-liquid-contracting-vibration-(al)-(s)-eye-plasma-posit-iretina-spacing-receiving-ear-canal-ligament-reverberation-(al)-(s)-synapse-intangible-perspective-(s')-kinetic-tensions', persuading [mutual-co-worker-(s) // (colleagues)], by Bachelorized-Commercial-Retirement-Study-(ies)-practice, specimen-(s)-obituary-retirement-cycling, because fatigue-(s) by-of Tectonic-Plate-Revolutionary-impasse, deteriorate awhile metropolitan-glimpse, sustem from-when urban-perspective, not [stock merchandise / product-trade-finite-cause-(s)] when-where middle-age-trading-power, defining introverted-parameter-(s), when-where, personal-idea-(s'), over-superlative perpetual-opulent-surroundings', still fragmented-activity-(ies), saving volume-(s) per [open receipt / W-2-form / Check / application / Company Sales-(2900-form)] from those mental-limit-(s), corps-dynamic per work-environment, tendered-tradition, thematic among, temporal-climate-locations', listed in existence, passing-moment-(s), as spherical-circular-centripetal-force-second-(ths)-minute-(s)-measuring, distance-measuring-(s'), respirational-breathed, through centralized-areas'-ambient, timing [elapse-(s) // synapse-(s) // syntax-(s)] distanced-unconsciously, due-to, the idle-(mind / body / soul)-requisites', intermittent, physiological-tangible-blood-pulse-respirational-ear-canal-reverberational-eye-plasma-posit-iretina-intangible-perspectives'-(al)-exterior-kinetic-tension-focus.$_{195}$¶|||02m.:30s.$_{08m.s.|||}$§ Disseminated press-information, partially denominate-(s') human mass, as data routine, reams' of context, by literary-prose, sustained (sub) consciously, considering constitutional-merit, when placing write, out of an common-usage, from natural-routine, yet lesson-(ed), formal-practice, because outlet-(s) of usage, conjure guided mass-gestalt-behavior, cognitive by, psychoanalytic-(s), because valedictorian-enabling, can yet be affirmative fashionable-prose, unison by mathematical-interposed-adjecture, priority of matter-(s), continuing confirmations', from principle-mass-deviation-(s), per motivated-act-(s), upon individual-(ism)-mentality,| 360 |cumulative extraspective-participant-(s)-focus, from whom interpose mutual visible-area, developing an pre-architecture-constructive, during work-order, because human-payment-inflation, not verify all those payment-consistency-(ies), where spacial-governmental-influence, affect dynamic-interactions', from Human-Wage-Inflation & Market-Purchase-Parity, optically-pass, before cumulative-engrossing, per net-purchase, controlled by [corporation // government // military-(default)-inquiry] cased default-aptitude, intertwined by Dimension-(s) & Order-(ing) banked upon Area-(s) & Elapse-(s), crediting interaction-(s), per dialogue-agreement-confirmation-(s) interposing data, while interposing-existence-impasse, affecting how we posit-impasse, by local-affirmations', of statehoodship, because of inordinate-existence-subjecting-interpretation-understanding.$_{196}$¶|||03m.:43s.$_{59m.s.|||}$§ Local-fundamental-(s), are instituted from birth, enforcing differentiated-aging, upon station-(s)-impassing, centralized Metropolitan-foundation-(ing-(s)), meticulously-intricated from independent-variable-wage-being-(s), tentative statehood-constitutional-dynamic, incurred-reluctantly, dependent-variable-act-(ion-(s)), intermittent denominative-calendar-dating,| 361 |inferenced past day-week apposed month-interval-denominative-reference-(ing), juxtaposed those passing-(s'), while location-(s) or spacing-(s'), affirm denominative-effect-(s'), from ecological-planned-city-district-(s), developing an limited pattern, interested apprehensible-physical-self, not instating boundary-(ies) per terrain, but conjuring impassionate-belief, narrow-sight-(ed) numerative-reasoning, spacing objective-tabling, where-when Human-Fatigue & Passing-Momentum, consider tectonic-force-centripetal-drift, upon offset-impassing-revolutionary-system-(s), timing political-position-offset-(s), as data-conjure-(ing), influence each article-focus, because how information apply upon interpretation-(s'), from youth-studied-interpretation-focus.197¶|||04m.:35s.74m.s.|||§ Infinitive-Fluxuation, is-of-an agrarian-ignorance,| 362 |spacing-effort-(s), for decade-transitional-consideration-(s'), during modern-contemporary-circa-era, [elapsing-(s) // visualizing // synapsing // syntaxing-introverted-bias] intermittent constant-momentum, revoking revolving-tension-(s'), from [self / predicament] by-of [surrounding-area / intermittent-happening-(s) / Occurrence-(s) / cognitive-imagery] amounting presumed memory-(ies), (un)-sensed-tense-(s'), offspring-offsetted, relay continuing-precedence, of-an civilizational-merit-(s'), by-when-where [land / space / territory] cycle hour-(s) by-of entity-default-handling, manning per space-(s'), commercial-center-(s) upon-of [horti-cultural / metallurgical-constructive-extraspective-effort-(s)] moved by when centripetal-force, subterior perceived inhabitant-(s), by-an quasi-component-(s), unarticulated possessive-objective-existence, pertained article-application, per hour, upon daily-impasse.198¶|||05m.:18s.99m.s.|||§ Age (has an) heirloom for jading, self relied, from parental-discourse, not in-an active-present-(ce), culpable daily-routine, merchanting oneself in-an mass-visual-perceptive-effort-(s), uncoordinated cycle-schedule-routine-(ing), believing millennium-limit-impasse, omitting daily-labor-effort-(s), (thinking from consideration-(less))), per mode of consumption, interrogative self-impassing-(s'), defining Employment & Calendar-Dating-Impasse, aging deducing-evidence,| 363 |rather off-time-inferring, because rule-(s)-measure-competency, that legally-define State-boundary-(ies), not educationally-affirmed, from-an process of duty-(ies), set in past-incrementation, upon present-day-reset-(ing-(s)), from materials', conjunct weekly-impasse-interpretation-(s), from when day permit hour-pertinence, thinking of retirement (senile) / pre-adult (pre-mature)-aging-process, timing product-(s), from path aging, decaying [our // my // your general-tense-(s)] while perspective offset judgement-(s), errored where executive-location-(s'), continually disregarded, legislational-dating, Collecting & Sorting, parameter-value-(s), by presence, defining extent, when intelligible-article-objective-use-(age),| 364 |pass-ambient-abstract-disposition, requiring [Lease / Loan / Property-Right-(s') / Plan-(s) / Bond]| 365 |proceeding competent-interpretation-understanding, intermittent centripetal-force-existing, from general-directives', when freedom abstract upon individualism, making family difficult for conjoining, belief by individual-discourse, not listening-frequently, from instituted-directive-perspectives', as artistic-interacting, define language-literature-study-(ies), aware contemporary-fashion, empirically daily-discourse, determining white-collar-effort-(s), unconsciously-settled, from human-effort-(s), not in-an dirty-fashion for resourcing, comfortably-reliant, on trading production-product-(s),| 366 |forming territorial-centralized-saturated-ethnic-racial-conjunction-(s'), irreparable, fatigued generalized-pedestrian-cycling-continuum-impasse-(s)...¶|||06m.: 58s.03m.s.|||§

§...Count-[19 / Nineteen]-preposition-of.. ..Article-Essay-[20 / Twenty].. // ..Book-Essay-[101 / One-Hundred-One].. ..Lines-[86.15 / Eighty-Six.1.one-tenths-2.five-hundredths].. hardly counting any redact-paragraphs at this point, focusing on how human-error, can be adjusted, for making individuals focus on genetic-cooperative-communication-(s), inter-minglinging with other individual-(s), whom affect each outcome of monetary-wage... ¶|||16s.85m.s.|||§

¶102. For barely over an century, has education been an part of a the average-ambiguous impassionate-act-(s), intended to discover how human-behavior could conduct;191 relevant to those gross-masses, that have been influenced to conduct [commercial // militarial operations'] awhile each date, consecutive-succession-impasse, from tendon-(s)-motion-(s)-reflex-(es), conceiving religious-duty, as idle-worshipping, not be mathematical, as-of physiological-movement-(s), blandly-abstracting reflective-memory, by when an call of formed-act-(s), intercede motion-(s), dire comestible-substantial-needs, not preconceived [when // where] average-efforts', median-average-means', studied by educational-subject-(s), remaining disregarded beyond past diploma-requisite-interactions', commercially sustemming, from commercial [bank-(s) // Federal-Reserve // Treasury-trade-(ing)] upheld by local governmental-denominative-trading, populating-existence, by each individual-family, where effort has not mattered, because religious-impact, define how cronyism has been ethically traded, historically processing dated-objective-materials', passing interval-(s) from hour-time-(ing)-tense-interacting-(s), cycling (sleep), quantified Intricated-Population & Genetic-Mass, chastised-social-interaction-(s), merit-serving, civilian-duty-(ies), by-when living sustem-from grade-point-average-commercial-competency-output;[192] expected from-of tax-(s), occurring as cause per governmental-mass-denominative-stock-objective-trade, pertaining capable commercial-comparison-(s'),| 367 |not developing an alternative-juxtaposed-government-referencing, for defining cursive-(mass)-mutual-colleague-competency, in observation of passing-moment-(s) in thousandths-extent-interpretation-confirmation-(s);[193] omitting circulation by-of each point of view,| 368 |juxtaposed terrestrial-conversion, because denominative-governmental-commercial-access-point-(s), matter, while upon governmental [[1]patent / [2]trademark // [3]copyright / [4]registration // [5]article / [6]license // [7]I.D. {Identification} / [8]denominative-pertained-article-objective / [9]subjective-defined-motion-filing] encompassing, [[1]commercial-trade-mark-concepting / [2]interactive-fundraising / [3]planning / [4]dimensioning / [5]communicating / [6](em)-(sym)-bolizing / [7]demarcating / [8]purchasing / [9]aligning] personal-constructive-contracted-aging-location-optimized-objective-article-restataing, through self-reflexive-motion-(s)-activity-(ies), during time-schedule-consecutive-series-sequencing-(s) per season, during substantial-perceptive-effect, pertain-upon human-mass-requirement-(s)-offset-(s), afloat celestial-ambient-physics, from how we [1interpret // 2perceive // 3conceive // 4conduct // 5purport // 6interact /// 7behave] is all relevant upon interactive-agreement, from-an Lord-Historical-purpose, sustemming (from a the)...¶|||01m.:25s.96m.s.|||§

..[Holy-Roman-Empire: East-Byzantine // West-San-Marino // Rome-Trade-control-impact]...

§...Count-[7 / Seven]-preposition-of.. ..Article-Essay-[21 / Twenty-One].. // ..Book-Essay-[102 / One-Hundred-Two].. ..Lines-[28.52 / Twenty-Eight.1.five-tenths-2.two-hundredths].. ..Why is conduct such an important character-trait, for living, when [Women / Men / Children] are in different incremental-offset-aging's, per simultaneous-secondths-incremental-visual-evocation, accumulating experience, through various forms of physique, allocating perception, from how tectonic-plate-land-space, is barren, from an lack of common-genetic-intelligence, due to centralized-book-referencing.¶|||22s.41m.s.|||§

¶103...There are an numerous-amount of [tale-(s) // fiction-(s) // story-(ies) / building-(s) // fake // improper-thought-(s'), told] amounting each lie-(s), by methodical-impact of [rubric // benchmark // syntax // grammar // literary-technique // punctuation], altering-perspective, from blood-pulse-eye-plasma-ear-canal-synapse-ligament-muscle-skin-reflexive-tension-(s'), (un) emotional subtle-motion-(s)-timbre, walking during intermittent period-(s), from self-posit-reflexive-observation-(s), upon perceptive-planning, where [dimension // measure] define space-(ing)-(s), rationing parcel-distance-block-acre-(s), from how collection-(s) of crop-(s), augment intervisionary-territory, from amounting cycle-(s), tabling Taxonomic-Classification-(s), for defining property-(ies)-(approximate), through premise-(s)-routine, rotationally-resettled-tradition-accustoming, while revolutionary-influctuation, premise city-population-count-(s), by numerous-voluntary-assortment-(s), of introverted-personality-(ies), not fully capable of seeing stadium-grandiose, as-an common-usage, as cottage-interaction-(s), proportion-relay, compared-room-converse-(ation)-(s), conscious-interaction-(s), progress impasse-(d)-chain-(s)-cycling; 194 awhile current-event-(s) focus temporal-subjective-awareness, from local-surrounding-common-legal-history-denotative-confirmation-(s), that afflict existence temporal-outlook, as required, to be read, per agenda-tier-rate, pertaining-itinerant, from shape-decision-(ing)-(s), familiarized by planning religious-effort-(s), abnormal national-allegiance, by-an common-vocal-timbre, topically-elaborating, balance-(s) of word-syntax-subterior-article-(s)-deductive-interpretation-(s),| 369 |denominating-data, [inclinated // indicated] intermittent time-passing-(s)-interval-(s), [rating / accumulating / counting] cultivated-resource-(s), endeavor-(ed)-(ing), by-an reasonable-configuring, from-those age-(s) per month-day-birth-obituary-deviation-(s), articlizing tensable-perspective, because how dynamic of physics, confluent domesticated-development-impasse-existing, construed governmental-architecture, centralizing-planned-entity-(ies)-development, from governmental-passing-competency-percentile-deviation, laude-Vale-Dictorian, inescapable weekend-frenzy-sleep-awake-effort-physiological-impasse-(s), during workweek-impasse-reflex-ive-cycling-(s) where present year, substain interval-(s) of dating, resetting dating, from periodical-observation-(s)|370|per human-subjective-interpretative-awareness.199¶|||01m.:40s.95m.s.|||§ Indiscriminate momentum, pertain observation-(al)-(s), by place-label-(ed)-extraspective-physics-efforts, barricaded weekend-abode, uncharacterized conscious-(ness), for entity-acting, by intelligent-citizen-(s)-(ry), adorned objective residual-incandescent-peripheral-sight-(s’), glimp-sed indicated-quality-(ies), by-when present-discourse, practicing-realty, from self, commonly misperceived through warehouse-community-investing impasse, supplementing off-time-impasse-effort-(s), invigorating (entertainment-function) interval-program-period-(s), variated from common-blank-subjective-blank-plain-(ness), per inhabitant-conduct-impasse-(s'), behaved by-when effort-(s) pertain uninquisitive academic-purpose, apose congenial-reminder, why reproduction-impact, be restricted by federal-guidance, balancing boundary-(ies) where commercial-reliance, define human-behavior, enjoying simplicity by demonstration-(s') of human-conduct,| 371 |Working & Retiring-idle-fervor, as for-how, Great & Abstract,36L. tectonic-plate-(s) remain among,| 372 |Grandiose-Abstract-Dimensionless-Clearance-Plane-Existence...¶|||02m.: 26s.52m.s.|||§

§...Count-[9 / Nine]-preposition-of.. ..Article-Essay-[22 / Twenty-Two].. // ..Book-Essay-[103 / One-Hundred-Three].. ..Lines-[36.08 / Thirty-Six.1zero-tenths-2.eight-hundredths].. ..How to proportion Today, amid tense, per individual-perpetual-momentum, affect where each fact incur beyond elemental-table, because how scales are to gamma, numerical-elemental-condition-continuum-pressure-offset-density, from an buoncy continuum recep-tion, lanthanide¶|||14s.77m.s.|||§

¶104 Abundant-superlative-means', have their short-comings', where physiological-inhabitants', derive practical-methodical-reason-(ing-(s')), per daily-virtue, awhile-tense-(s), [indicate // inclinated] singular-reactive-operating-situational-place-(ing-(s)), obliged property-maintaining, because affixed-affirmation, occur as, consecutive-succession-motion-(s)-impasse, pertaining reflexive-(hyper)-tense-(s)-subjective-article-(s)-referencing, directing how person-(s)-think, per [visual / vibration-(s)-tempo-tone-reverberated-organ-tense-(s)], verifying referenced-act-(ion-(s)), pertaining place-distance-location-setting-(s), in realization of resource by-at [location / elapse / synapse // syntax-mass-reasoning-(s)], differing disclosure [from // of] kinetic-effort-(s), among conferred elect-citizen-(s), perceiving obligated topical-parameter-(s), amongst territorial-direction-(s), relevant list-(less-(ing-(s))), per Interaction, maneuvering present-age-cottage-class-(ification-(s)), comforting to those default-instilled-parameter-(s), sustaining Constitutional-Limit-(s), as how one motion-(s) by [push / pull / binge / press / empirical-guide-line-history];[195] distracted awhile voluntary-impasse-discourse, perceive-by (un) conceptive-liturgical-future-specifying, conducting extraspective-affair-(s), remained unacademically-periodical (ized), from youth-reclusive-introversion,| 373 |referencing intermittent-distance-parameter (s), by-of [commercial / governmental discourse-(s)] affecting how fraternity-dynamic, forget opposite-sex-visual-tonal-physique-arousal, for persuading intervoluntary-motion-(s), as men are before their subjective-passing (s) of diploma, conjunct congruent-aftereffect-(s'), pre-annotation-redacted, from physical-impasse-limit (s), where subject-(s) are unproportionable, interval-article-reference, as in restating conjure of [essay / lesson / paragraph / sentence-data] by means of objective-locational-preserving, investigating weather relocating substantial-material-(s), by spliced-information, better practice terrain [horticulture // agriculture cultivation] effectively using transit-impasse-goal-(s), where tangible-means', progress during hour-(s) under-in-of empirical-entity, drain-(ed) lit-living, from subordinate-inadequacy, dulled mutual-genetic-interaction-(s), [lustfully / vainly] impassing moment-(um), physiologically inept screenplay by-of centripetal-force, measuring parcel-(s) of common-impasse-spacing-(s), from grandiose-distance-(s), identifying class-local-denominative-limit-(s), by surrounding-boundary-constitu-tional-ordering-(s), defining inhabitant-perceptive-class, ruling-in scale per mass, from spouse-offset-arousal-frequency, pertaining offspring-generation-(s), percentile retrospective-interaction-(s), when religious-timing-(s), interpose intimate-discussion-(s), identifying reason, apose conjure-conjecting, article-data-agenda, referenced formal confirmation mass-perspective-order, upon an parameter-state-constitutional-spacing-(s), per house-representative-discourse, conversed population-rationing, as aging-discourse (or not), cottage intercede moment-(s) per day, dating hour, per minute-(s)-location-effort-(s), spacing mass, per individual-action-(s)-intrication, sufficing Property & Spouse, where objective-need-(s), pertain perceivable-population-reproduction, parallel surrounding-spacing-(s), in-an denominator-motion-(s)-impasse-count-(s), positing-upon [[1]numerative mass / [2]individual-(s) / [3]productive-object-(s) // [4]placing-(s) / [5]documented-ordering / [6]cubic-surface-area-dimension-(ing)-(s)] role-model-(ed), motion-apparatus, directly-contributing, expository-discourse, deducting minute-(s)-secondths-operat-ions'-(time), from temporal-response, as human-interpretation-impasse,| 374 |continue confluencing-physiological-indication-interpretation-(s).$_{200}$¶|||02m.:44s.$_{41m.s.|||}$§ Blending impasse-moment-(s), facilitate interval-conscious-awake-motion-(s) fatigue, better-operated, if ever mutual-unison-trust, can be appropriated, constitutional-mass, founding father-mass-operation-population-state-(s), denouncing city-county-metropolitan-revision aspiration-(s'), incompetent, from (un) inquiry rule-data, measuring centripetal-force-impasse-velocity-drift, by-an basis of ground-(s), blended deductive-tier-progression-(s), tiering motion-notion-(s)-data-discourse-retrieval, litigated for-of Trust & Act.$_{201}$¶|||03m.:09s.$_{28m.s.|||}$§ Forever is impossible, by-an variation-(s) per consecutive-sequential-ordered-denominative-outcomes', fostering constructed-effort-(s), | 375 |warping civilian-elemental-article-extraction, from tectonic-offset, not thought of an scheduled-continuum-system, for population-article-(s)-usage,| 376 |gained (as an) meander of total-mass-action-(s)-effect.$_{202}$¶|||03m.:25s.$_{56m.s.|||}$§ Pre-conceptive-documenting, has yet been ordered, per paragraph, for documenting subjective-interpretation, conceiving why we presently, can not maneuver every piece of matter-(s), by discourse of individual-family-perspective-(s'), adequate tax-(s)-expenditure, for while financing objective-article-means', perceptively-progress, American-population, blank, pre-conceiving an future-settlement-point, pre-settled by birth, because moving place to place, affirm district-regional-location-(s)-posit-(ing), predicate product-anomaly, for how, objective-referencing proportion Gregorian-Roman-Catholic-Calendar-Year-Reset-(s), living upon dated-conjecture, (Pre)-Pre-Control-Number-(ed)-Government-issuing, articulating article-emphasis, enticed collecting [plate-(s) / bucket-(s) / barrow-(s) / bag-(s) / trailer-(s) / box-(es) / plastic-packaging] requiring denominative-financial-material-(s), conserving independent-private-residence-(s), hidden by-of governmental-production, historically by [militarial-guidance / contracted-labor-enterprise] by-an retrieval of compositional-elemental-data, elucidated creative-intention, of market-volatility, individual-istically-inclined, idle-attempt, during national-constitutional-referendum, intricating [nation / state / count / city / principality] constitutional-reference-align-ordering-(s).203¶|||04m.:18s.93m.s.|||§ Subordinate-effort, is subjective-feudalism, caught on the axis of motion-(s) of physical-preference [impasse // action-(s)] of objective-article-designated-usage, grandeur venture-(s), from parameter-limit-(s), parallel individual-saving-(s), self-directed article-contribution,| 377 | fatiguing relevant-velocity, fusing visual-impasse-moment-(s), consecutively-sequentially-ordered, per tangible-elemental-patent-commercial-conceptive-denominative, constructing-competency, inked ((by of)) documentation, daily monitoring, hour-denominative-maxim-impasse-(s), as various effort-(s), sustemming cardiovascular-gauging, minute-(s)-second-(ths)-discourse, aware by perspective-affirmed-data, registering letter-word-ink-subjective-article-paper-count-(s), yet unfurled by possession, (for of) tectonic-ecological-national-agenda-direction, denominating nature, not understood by private-commercial-enterprise-purpose, giving enumerated-quality-(ies), tasked produced by mass-quantity-land-territory-frequency-data, indigenous individual-purchase-measure, offset per municipal-code, timbre through formed-assertion-(s'), per effort-(s), conjuring place-(ing-(s)), synthesized static-continuum-production-reset-(s), confirmed intermittent, peripheral-quantity-dynamic forging how men pertain lifestyle, because lifestyle-effort-(s).204¶|||05m.:16s.31m.s.|||§ Man-(en) is supposed to work for woman-(en), I not yet, by mutual-effort-(s) discoursing an modernization independent-interaction-impact, from how cause be specified, conceptive-discoursing, [when // where] pinkish-white-women-(an), not practice an preference for my tanned-presence, as I intend to bleach my skin at some impasse of living, elapsing ones'-memory, by normal-idle-passing-existence, from child-father-mother-education-work-monetary-activity-(ies)-dynamic,| 378 |aging apposed passed introverted-romanticism, ambiguous presence-formulation, requisite [high-school / collegiate-fraternity-discourse] encompassing an unconsciousness, per two-child-offspring-lineage-default-desire-choice, reactive upon mid-term-impasse-aging, reflexive article-formation-(s), requiring dialogue-period-iteration-(s'), twice an day, as program pertain upon intimate-relation-(s'), present discoursed community tectonic-approximate-surface-area, as article-usage, affirm-data, remaining unaccounted majority of common-manuscript-discourse, yet fully in "love",| 379 |by rated-activity-(ies), mutually-referred, when [**extra-spective // introverted**] article-(s), pertain article-objective-usage, per act-(s) proportioned tectonic-spacing, modeling posit-placing-(s), in light of centripetal-force, symbolizing night-spacing-(s), of surrounding-surface-area, balancing physiological-energy-(ies), superseded voluntary-genetic-dynamic, amassing population-monitoring, looked down from leadership-responsibility, as house-representative-outlook, pertain each [impasse // object // article // entity-(s) // thoroughfare // land-property-spacing // element // product // et cetera] (as an) predecessor to common-thinking, as-an unsubordinated-nature of superego.205¶||06m.:25s.52m.s.|||§ Will wilt away, for not listening (to of an upon), other [inhabitant-(s) / person-(s)] for advice, interposing-conjecture, by mathematical-principles, ordering consecutive-sequential-event-(s), filing-data, upon mass-common-consensus, reasoning in lieu of perceptive-space, instructed where discourse pertain parental-out-view, teaching-reference per [1.mathematical / 2.language-art-(s) / 3.Historical / 4.Biological / 5.Anatomical / 6.Physiological subjective-means'] maturing subjective-focus, upon written-inquiry, discoursing time-impact, adjacent base-perceptive-awareness, vocal default-means', from-an standpoint of weight-registration, denoting-mass-commonality,| 380 |yet projecting psychological-documenting, scheduling mutual-inhabitant-discourse, awhile constructive-implementation, pertain mass-population, upon circumstantial-matter-(s'), sufficing freedom, felt so good, as individual-independence, domesticate developmental-curb of action-(s), ordering ambivalent-mode-(s), inconsistent upon centripetal-force-passing-hour-(s), harmonizing, minute-(s) [action-(s) // motion-(s)] per hour-(s), upon tectonic-plate-drift, affirming an national-county-city-spacing-(s), positing Celestial-Hour & Centripetal-Force-Drift-Momentum-posit-(s), with regard to local-existence, juxtaposed the massive-quandary, persisting due to, method-documenting, existing from [individual-(s) / constitutional-numerative-denominative-deducting] remaining inconsistent voluntary-subjective-application, because daily-objective-parameter-(s), are of-an yearly-ephemeral-passing-(s), not in contention, aligned-coordination for whom can handle, what type of [elemental-composition-resourcing / dimension-structure-(ing), forming-of [brick / plate / cement-cylindrical / et cetera...], bended of elongated-plane-contour, supplemented societal-direction-(s), from-an central-constitutional-method, per [1Women / 2Men / 3Children / 4Genetic-Biological-Discourse-History / 5Territorial-Claim / 6Scheduled-Objective / 7Entity-mission-statement / 8transient-road-development-documenting / 9resourcing-amendment-mass-proportioning / 10Age-Comprehensive-Static-Manual-Labor-Ordering-(s)] each formal-physiological-requirement-(s), | 381 |aligned filing-tier-progression-circumstantial-discourse, needing University-House-Representative-Valedictorian-Year-(s)-extraspective-deviaion-(s), (not considered) in an Theatrical-comprehension, psychological from total-population, in attempt to confide, mass article-schedule-agenda, uninfluential for-of free-will, rating-reasoning, as subjunct-literary-origin, not independent per literary-data, because unconscious-impact from |African / Indian-Genetic-Mass-(es)| cycling, mass-physiological-existence-direction, intersected-dating, unsubjective those bastardized-fusion-form-(s) of humanity, not in an recollection of origin-(s), from a-the solar-impact, presenting continual-momentum, exhibiting those factor-(s) of common-mass-development, surrounding ordered-prerogative-(s) of humanity, by-when off-time-effort, compare an work-thinking (only-moment-(s) post education),| 382 |of Major-Focus-volume-mass-product-(ion)-development...122L. ¶|||08m.: 38s.15m.s.|||§

§...Preposition-Count-[41 / Fourty]-of.. ...Article-Essay-[23 / Twenty].. // ..Book-Essay-[104 / One-Hundred-Four].... Lines-[122.07 / Thirty-Two.1.zero-tenths-2.seven-hundredths].. ..education was suppose to discipline each child, under the circumstance of resourcing, awhile centripetal-force revolve, each year, since ||||1,534-{Origin Gregorian Solar Year, In American Spokesman English Tone {*Manuel Pagan Copyright Point}}!.||||¶16s.85m.s.|||§

¶..105·Pragmatic-misinterpretation-(s) are made, for how age-elapse-date, without an formal-referencing, per word-data-mode rate-median-date-count-(s), by daily-impasse-cycle-reasoning-(s), juxtaposing mass relative-outcome, when formal-referencing, not direct any mode-cause, as-during live-physical-presence-effect, controlled juxtaposed,| 383 |{I.S.B.N.}-dating, infinitive-continuum-dating, [1.cycling / 2.ordering / 3.periodicalizing / 4.placing / 5.dating / 6.sequencing / 7.routing / 8.directing // 9.positding // 10.forming // 11.bending // 12.welding // 13.periodical] sort-ordering-sequence-focus-program-verifying-denominative-tiering;196 awhile blood-pulse-sensation, in-tran-sitive-reaction, Centripetal-tectonic-time-divide-continuum, abstract-revolutionary-discourse, where second-metropolitan-contort, prelate population-discourse, variable-date-median-gauge-perspective-intercontort-denominative-time-numerative-dating-impasse;197 affirming daily-present, routinely featuring-conjure per Show & Tell-aspect-function-(s')| 384 |enticing superego through elder-age-post-sixty-two-year-retirement-focus (American-Will-Expectation-diminuendo), [withering / flourishing] relevant (the) perspective (an) sight-depth-tangible-reverberation-impasse-motion-visual-confirmation-sound-particle;[198] existing as-an pertained-focus, where placing-(s') of data, continuum-relay, past-requisite fact-(s);[199] pertain-defining, work-order-load, when aging-individualism, not be probed into, (as an) [extraspective-genetic-listening-vocal-cultivation-mass [100 // 1,000 /// 10,000 //// 100,000] quarterly-operation-(s')-mass-future-scheduling-effort-(s)], intertwining-requisite-effort-(s), comprehending why such [property /// device // article / self-composition-perceptive-visual-tonal-focus-tense-focus-(es)], weather constitutional-discourse proglamate, perspective [through // per // at // as],[19L.] not of per-of, inhabitant-free-voluntary-will-discourse.¶|||01m.:40s.[78m.s.|||]§

§...Preposition-Count-[3 / Three]-of.. ..Article-Essay-[24 / Twenty-Four].. // ..Book-Essay-[105 / One-Hundred-Five].. ..Lines-[19.25 / Nineteen.[1.two.]tenths-[2.five.]hundredths].. ..Common imbecile nature, has yet to perceive-proportion, from how density juxtapose matter of buoncy, when air-vapor-millibar-pressure, gauge apose pounds to tonnes reset-proportions of density, awhile [grams / drams -(dry weight) / ounces], vary how celestial-distancings, require an proportioning per interval set-product-objective, pertaining an subjective-book-entity-itinerant, per obliged-individual, whom determine to comprehend an purpose for lifestyle in retirement, when youth-mid-age-work, delve through each [quandary / conundrum /speculation / indeterminant-variable], without firm inquisitioning for how centripetal-force denominate an land-ecological-terrain-resourcing, per offset of designated-pronoun-wage-competency-limitation-(s), per [county / city / state], short-sighted contextual-reasoning, because of naïve-blank-mindedness, inhibit instinctual-subconscious-jargon, not rationale upon mutual-written-intellectual-comprehension, for when brief-decisions, remain dormant, impassing-by present-circumstance, and not living for future-savings-preparation-(s), awhile maintaining basic human needs, in competition of other illiterate-ethnicities, not capable of purchase, if ever formally directed, and followed through formally-proper, in proportion of calendar-reasoning, from wage and resourcing, where entity-operations, circulate manuscript-manual-operations, yet functioning from cursive ethical-mathematical-proportioning of paper and pages, due to how motivation is suppose to reduce how hoarded-savings, depend upon the confidence of an particular-economic-classification-(s), per dating thinking how mode of weeks, operate in segments, through dating, as discourse, eventually [impasse / occur / signify], why tribune limit proportions, because of an technical-physics-incapacities, judged by how performance from training, affect routine of affiliated-entity-operation-(s), awhile common-men, (*White and all in America), remain basic from wage-function, due to an international-county-state-city-county-book-geological-ecological-documenting, flawed by human-population-(s), impassing entity-(ies), awhile each group of thought, is suppose to remember how genetic-appreciation, (blondes for example), would have certain limited-functions, relying on opposite genetic-objective-trade-wage-values, defining why individuals serve an purpose by family-housing, not yet conceive from [clay / rock / composite], where to position in commordire-juxtaposition, an particular-ambivalent-trust, competing with family-(ies), for survival, where I would presume if blonde-word-of-mouth- illiteracy exist, inquiring an intelligent-white-pinkish-intellectual-group, to focus investing in to Montana-Ecological-Commercial-Corporation-(s), calculating physics, in minuté-ecological-fashion, from an books-continuum table-non-fiction-genre, (self) evidently observing how all the facts of existence, circulate through an bit of aesthetic-culture, forgetting how to interpose-commercial-intelligence, through post-educational-present-continuum-commercial-governmental-militarial-legal-medical-work-practice-methods, moding an work-wage-stress-eyesight-physiological-limitation-(s'), per period-ical-functions, amid daily-micro-86,400-secondths gamma 1,440-minutes moding, incrementing thinking from an secondths-milliseconds-molten-miles-dynamic, measuring above tectonic-plate-surface-area-crust, each fact of ecological-layers, intermittent from human-genetic-visual-perception-bias-hoc-offset, in juxtapositions of secular-17-mega-national-agglomerate-tectonic-land-regions, offset celestial-ambient, recognizing why duration of effort, circular-denominate, in an timed-revolutionary-rotation-fashion, from either {Origin Gregorian Solar Year-1,534} or {483-years}, as of September-2017. ¶|||03m.:06s.84m.s.|||§

¶[106]Today {January Fifteenth;[200] Two-Thousand-Sixteen;[201] Two-Thirty-Eight-Ante-Meridian-(m);[202] North Miami, Florida} since Nineteen-Ninety-One, *(con) current-total per [Nine-Thousand &Twenty-Five-Day-(s) // Twenty-One-Hour-(s) /// Thirty-Nine-Minute-(s) //// Thirty-Six-second(ths)]-{47,593-word(s)-influc(x)tuation-sum-(S)} thriving upon lifestyle, until objective-article-(s)-dating, influctuation-(s') {January Fifteenth, Two-Thousand, Sixteen},| 385 | rate per daily Day & Age, date-adjecture-adjacent-aging-requisite-sequence-(s), yet filtering perspective through [Momentum // Blood-Pulse /// Space-Meter-(s) // Feet-Distancing-(s) //// Mile-(s)-Transit ///// Secondths ////// Beat-Note-(s)] upon word-tone-vibration-pitch-range-note-equivalent-document-inferencing-compartmentalized-bit-(s), requisite-from historical-disposit-ion-perspective, dating-posit-(s),| 386 |interval-data-procession-spanning, measure-equivalent-(s) pink-pig-(let), self-data-comprehension, (un) configured (deviation) at modern-rate-(s), classifying-interposed-object-(s) (yet-complete), from Subjective-Exterior-Value-(s)-Classificat-ion-Age-(ing)-standard-article-comparitive(less(ness)), Column-Order-(ing), into-interior subter-ior-exterior-visual-tonal-audible-decibel-cycling, (intermittent) juxtaposed, calendar-denomin-ative-interval-(s), by-of-an historical-requisite-present-tense Time & Date, verifying case point-(s), incremental-article-particulate-focus, during facilitated-motion-(s)-gamma-(s), where inhabi-tant-(s) discourse an, at-will-scheduling, invigorated [physical-sensation //

untensed] reference documenting, confirm article-piece-(s) per-data-bit-syntax, pertaining hour, by-of-an-per, minute-(s)-secondths-mass-equivalents'-act-(s), sequencing secondths, where at an interpose-interval-(s)-continuum-requirement-values', Grandiose-abstracted, incremental-interval-interposed-focus, counting tangible-variable-(s), timing human-effort, upon article-usage, according [through // by perceptive-boundary-(ies) / land-(s) // territory-(ies) /// limit-(s)], (claustrophobic) where [dozen-(s) / hundredths/ thousandths-secondths-post-illustrate], [literate // illustrative // medical /// legal /// default-position-(s) of order] after (fanatic)· fundamental-(s)-tense, from birth-Twenty-Five-Aging, deriving formal-written-documented-inquiry-(ing), [location-perceptive-subterior-spher-ical-limits'-impasse-(ten)mile-direct-numerative // denominative-x360° × y360° = z129,600°D] identifying tectonic-plate-abstract-methodical-process, perceived subterior fundamental-independence-individualism, encapsulating ritually-repetitive-practice-objective-national-state-duties', focusing-material, where-during, district-independent-housing-saving-(s') per Dollar, attempting cultivation-inhabited-open-field-territory-parcel-(s)-species-$112,386-housing-pro-visional-self-family-means, self-functioning independent-apose-others', whom mutually-comprehend cohesive-interactive-effort-(s'), in-light-of, producing-resource-(s), receipt-development, data (from an) communal-detection, snared share load-(s') present-day-prevalent-commercial-maintaining, servicing duty-(ies), requisite-present-past, in-fluctuation future (modern // contemporary)-accurate-percentage-dynamic-(physical-continuum), pre past-day-confidence, for physiological-sensation, weighted-proportioning, by-of physical-life,| 387 | for an untested-continuum-sequencing-(s'), hoc-bias-child-continuum-elder, not clearly an extraspective-university-collegiate-post-examination-(s), where commercial-enterprise, supply Term-(s) & Condition-(s), protected by government-development, physiological-calculator-study-continuum-regimenting, formal-ambervision, effort-(s) where land, state-constitutional-space-(ing-(s)), control-(ed)

{Article:} {April}

Beck centralized-confidence, verifying layers by age-(s) at work-cycle-(s), mentality-mute-command-voiced-interaction-moments', dynamic-deviating monetary-policy,| 388 | systemized by At-Will-Freedom & Cultural-Schedule-Operations'-Standards-Foreshadow, (Talt's'-ancestry, calculate) present-junction-awareness, day upon year-hour-moment-subjunct-ion-method, currenting pig-(let)-sheep-oxen-Athens'-horse-animal-control (un-conceptive-collecting) century-aging, unconstructive throughout decade-(s)-ambivalent-(im) passing, how to schedule from free-will-municipal-order-individualism;[203] constant by year to date-continuum-primodial-630-Genetic-Human-Specie-(s), upon self-view, for comprehensible-fastidious-act, existing by the default-effort, in lieu of perceptive-means',| 389 |without an understanding of [Mammal-primary-species // Mammal-secondary-foul-species /organism-offset/ insect / mite-tension-reflexive-physical-tangible-offset], (un) juxtaposed physiological-human-interaction-(s) (unMedida-E-9-five-nine-comprehension), for now I would enjoy stressing day-life-count-(s), summing total-article-exothermic-purpose-function-deviation-(s), expressing discern-(ed)-frustration-(s), prior-day-failure-volume-value-(s), formulated-cultivation, where written-timbre-balancing, crop-variable [[1.]horticulture-variable-(s) / [2.]vocal-physiological-genetic-written-documenting / [3.]routine / [4.]schedule-(d)-activity-(ies) / [5.]light / [6.]space / [7.]citizen] natural-posit-grandiose-source-product-fragmented-factor-(s),|390|through observation-developing, arranged-surrounding-space-(s), visual-vein-plasma-blood-pulse-live-reflexive-repetition-usage-impasse, relative-motive, [personal // genetic-focus] for latter-day-aging, where at an permanent-post-mortem-memory, departing-after the age of one-hundred-year-(s), of passing-age, (or so), by ingestion-survival. [206]¶|||04m.:22s.[06m.s.|||]§ Stilled by dues repetitive-assignment-(s), [collection // spacing] have yet factored-land, corrugated through human-physiological-tension-act-(s), limited because of a-(the) cardiovascular-effect, neurological-deficient, land before modern-tectonic-plate-centripetal-force, [revolutionary

/ rotational] latitude-abstract-passing, vertical pressure-distance, perceived-horizontal, while subterior-manner, align by physiological-human-species-offset-racial-ethnic-pool-offset-(s), spectacular per ration-(ing-(s')), curve-spherical-property, convulsive-conceding, passing-effort, remained by personal-objection, (not totally property), inclined requisite-presence-offset, through curve-spherical-property, convulsive-conceding, passing-effort, remain personal-objection, (not totally property), inclined requisite-presence-offset, through-of mass, relative efforting, demands' of government, directly valued, workplace-account-logging, registered-perceptive-fundamental-basic-human-right-(s') at tendered-act-(s) accompany-(ing) from-self, volatility-cohesion, procrastinated-subservient-effort, due to act-(s)-completion, proportion [labor-pool-physiological-effort-(s) /// book-subjective-effort-(s) /// extraspective-events-voluntary-impasse-continuum-planned-calendar-mode-operating-decade-dating-(s)], required when mass, (further) function-(s) upon composite-tectonic-plate-matter-(s), supplementing article-existence, denote self-numerative, vocal-morale-function, before our very [perspective-secondths-stance // minutes-tangent-repassing-rereading-thinking-process] unfurnished land-tectonic-rural-space, for restriction, affecting fluidity,74L. through self-centered-compositional-conjunct-reflexive-body-speicies', during [perspective / mass-subtle-nuisance-matter-posit]...¶|||05m.:36s.93m.s.|||§

§...Preposition-Count-[12 / Twelve]-of.. ..Article-Essay-[25 / Twenty-Five].. // ..Book-Essay-[106 / One-Hundred-Six].. ..Lines-[74.875 / Seventy-Four.1.eight-tenths-2.seven-hundred-ths-3.five-thousandths].. ..Self is never safe, by family or neighbors soly, an tentative-respects is in dire need, to be subconsciously-considered, away from entity-(ies), amid offset rates per population-non-genetic-interaction-distancing-(s).¶|||13s.68m.s.|||§

¶107As for whom, can obituary-(ied)-perceptive-aging, when auxiliary-device, attain-(ed) attribute-(s), adjecture-actual-association-(s), amongst aforementioned lined-impasse, for furthering dimension-(s) [rectangular // spherical /// triangular-prism-(s)] inlayed, articulating surrounding-adult-total-age-era-reformation, that practice of reasoning, throughout direct-focus of matter-(s), forget defining-observation-(s), from actual-non-fiction, where as aorta-ventricle-present-(ing), of past-arrangement-mode-hour-anticipation-(s), variating an state-tectonic-momentary-control-government-impasse, juxtaposition [elapse-(s)-four-minute-(s) / hour // four-hour-offset /// eight-hour-offset //// date-(ing)] contra requisite aspect-hour-(un) pertaining-(ed), [[1]day // [2]week /// [3]bi-weekly-payment-period //// [4]month ///// [5]season ////// [6]year] juxtaposed [year /// decade //// century], millennium-season-present-year-continuum-offset-legal-trade-ordering-psychological-economic-trade-impasse, remaining false per date, consigned denominative-purpose, for control, (I prefer control (Library Congressional-Congress-Number confluenced with International-Standard-Business-Number sales) where subject-variation, shift adult-genre, interaction-capacity, [where // when /// who] will balance those primary-harmony-events', affecting whoms'-place station-(ed), [point // letter /// participle //// word ///// term ////// condition /////// Climate //////// Terrain ///////// physiological-posit-(s) ////////// motion-(s) /////////// eye-blood-spacing-interior-exterior-shape-color-dimensions-depth-syntax-synthesis-expository-reasoning-method //////////// shovel-stroke-exterior // broom-sweep-interior-method-aspect-(s) of living] pertained tangible-debt-monitoring-(objective)-article-(s)] only effective, upon arrangement where mutual-interactive-discourse, have [discussion / debate / dialogue / interaction-(s)] pamphlet-bible-progressive-church-sequence, rather an church-hour-hundred-written-desk-articlizing-parameter-(s), by artistic-architecture-amusement, when accustomed anticipated-antithesis, [perceive-syntax-(is) // pulsate-synapse-(s)] present of literary-advise, ancillary-(future) apose mode-hour-fifteenth-variation, as train-destination-state, commercial-scheduling, because cultivated mid-age-action-requisite, individualize-presume, default-requisite (national-aging), inclined spherical-rotation-revolution-combustion-spherical-aspect, ordering in juxtaposition-article-scale-referencing, when common-retirement-examination, (fail) to define each generation, by minute-particle-progression-schedule-procession-grade-point-subterior-redactive-extem-porizing, due to becoming only an family-individual-primordial-government-default-body of context, antithesisless, individual-self-explanatory-prose, prior requisite at fact-(s), time-usage-function-factory-impasse-discourse, graded from repetitive-intrication-grading-half-tone-referencing, accustomed by -discourse, graded from repetitive-intrication-grading-half-tone-referencing, accustomed by antiquity perceptive-physical modern-reset-live-production-trade,| 391 |temporal visual-sensational-common-mentality (I do not know if Congressional-Library exist of such literary-inclination), affirming seasonal-quality-(ies) where tangible-posit, viral visual-listening-placing, listening (from an) governmental-factionalism, significant by independent-subject-maximum-limit-collection, bubbled-in mathematical-empirical-article-tier-fashion,| 392 |apose literary-grammatical-increment-article-continuum-reference-impassing-punctuation-redacting, perplexed observation-(s), tiering subjective-data-double-primary-calculating, from mark-(s)-offset-indication-(s),| 393 |evident [commercial-mutual-compact-holding-agreement-(s)-extraspection // residential-independent-living-introversion], displaying discourse (per) of matter-(s), persuaded discourse, that daily-dynamic, place upon awareness of matter-(s), [Earth-Core-Rotating // Solar-Core-Revolutionizing] apose tectonic-seasonal-dynamic, as I am quite subjective-planning-physiological-season-competitive, extraspective common [motive // focus, requisite] [where // when] perimeter pertain an concept of motion-objective-time-frame-three-dimensional-planning, [psychological // architectural // written // labored-physiological // theatrical-ambient-progression], spacing independent-effort-(s), visual by physical-age, renewing lagged present-century-perfect-nation-(state)-impasse-territory-control, attributing birth-syncopated-characteristic-(s), particular in-through-of view by spherical-ambient, unangular skin-tension-oblong-thinking comprehension;204 stressed sweatless at work in-transit, enumerated-inference,| 394 |focusing denominating article-matter-(s), align-currenting, crusted-sedentary-engineering, directing our capability from understanding, from whom verify each [1notion // 2inclination // 3indication // 4motion // 5sighting // 6fact // 7tension-(s)] precision-articulated-attached, [to / from] per self, upon attained-subjective-objective-article-(s)-motions' of tectonic, pertinent-observation, adjusted for emphasis on an live-time-analysis, calendarizing modern-live-contem-porary-purporting, (from an) desire for comprehended influctuation-impasse-observation, primarily-extraspective, procession of construction, as for attain, remain in an offset of duties, meant relevant from how spacing of capable-motion-sweat-effort-(s), can sustem from [Male // Female motion-(s)] centralizing-focus, per construct, defining ground-(s) at approximated-development-(s), tubular cycle-ordering, intangible inferable-tangent-reasoning, per state, as culpable-focus, for youth-national-religious-metronome-age-ordering, scheduled-(40) by, involuntary-independent-impassing-(s), momentum-rotating-centripetal-force, minute-physiological-tense-(s') perceptive, when general-adequacy, lifestyle [seasonally // work seasonally // sleep-seasonally // subject-seasonally] imbuing nation-reform, when educational-process, can not identify living under 2000's-years-cycling, inspected mid-age-passing, mundane yearly-offset-effort-(s), in an present deficit of truths-judging-(s'), apose legislational-defining-(s), four score-(season-(s)) & late-eves'-ago, sleeping in dating of city-county-route-routine-cycling, not balancing those laudes of tangible-physical-subjecting, instrumenting-aging-effort-(s), in an congruent-motions'-apprehension, affluent-astute-present,| 395 |prepared operation-(s) by basic-conception, per matter-(s), by where mass-creativity-requisite, institute, firm-congressional-legal-grounds', yet understood of rural-territory-present-law.207¶|||04m. :18s.85m.s.|||§ Tensed-perspective, cardiovascular-siture, slip-pass, blood-pulse-beat-ration-(ing), perceiving limit-(s), from minute [muscle-(s) / tension-(s) / ligament / skin-tension-(s)] extraspective-mass-offset-ted, by probably 630-genetic-species-bone-posit-(s), defined from observation-(s) per [[1]ear // [2]nose // [3]lip // [4]teeth // [5]eyebrow // [6]jawline // [7]head-shape // [8]cheek-(s') // [9]hip-waist-bust-knee-contour // [10]hair-splice-characteristic-(s)-subjunction] comparative by only [Human-waist-bust-knee-contour // hair-splice-characteristic-(s)- subjunction], comparative adjusted (only) Human-Being-(s), or Mammal-(s), elapsing-by secondths-impasse-tension-scenario, through an (never) concerting respects', formally redacting-date-in-two-effect-(s) by Time & Calendar (for) refining-result-(s) of misinterpretation-(s), intended to comprehend by [subjective-write // subconscious-desire // vocal-vernacular-inclination-(s)], [morally-expressed // evoked-through thought] per millionths-mass-offset,| 396 |thousandths-driven-vicinity-approximation-introversion, not literary in cooperation with political-person-(s), ([person-(s) formal-utopian-tenths-millionths-collection-reformation // fourty-year-tranquil-transgression-repopulating-reproduction-community-formal-Montana-national-state-set]), per year populating, from feminine-reproductive-intuition, conduct territory-parameter-motions-method-impasse, operating order-(s), where mutual-entity, can not be illegitimate legal-trade, cause hiring-budgeting-scenario, not yet table copyright-register-trademark-patent-articulate;[205] act-(s) for elapse-sequencing, by calendar-incremental-dating, by present-reset-requisite, (as an) accord of housing-off-time, that define one-(s') genetic-discourse, where at an tectonic-stabilizing, be permanent of neighbor, for idle-interaction, as-by when co-worker-mediate, influctuation-demand, denominative at an later-impasse-motion-(s)-moment-(s), monetary-default-policy-(ing), from survival, evocating (rudely), by subversive-nature, not fully-cooperative of formal-interaction-(s), (I am more extraspective than the prior, just in an remark of aging-deviation), because individual-tension-belief-(s), cloud mentality, with person-thinking effective hour-dating-implementation-(s'), onsite control, self-editorial-registered-remark-(s), by awe of exterior-observation, upon surrounding-posit-dimension-article-(s), tangible-density-infinitive-weight-(s), from-of crust-composite-tectonic-(in)-tangible-grandiose-offset-(s).¶|||06m.: 22s.41m.s.|||§

§...Preposition-Count-[26 / Twenty-Six]-of.. ..Article-Essay-[26 / Twenty-Six].. // ..Book-Essay-[107 / One-Hundred-Seven].. ..Lines-[90.5 / Ninety.1.five-tenths].. ..we are suppose to docu ment physics, from [land / territory] subterior spherical-circumference-molten-density-(ies), by an estimate linear miles (no kilometers), observation of how revolution-{365-days}, intertwines rotation-{24-hours}, from an equator-circumference 35,928-miles, as tectonic-plates-oceanic-water-bodies, equivalent 12,934,080-spherical-miles-{deviations of cubic-two-dimensional-measures-5,280-ft. // 1,760-m.}, for how clearance vary by human-compositional-form, per inclination from physiological-spacing. ¶|||40s. 81m.s.|||§

¶108No detailed synthesis-agenda, has ever been in [session / affirmation / order / code / agenda] for how mentality react by control, without standards, for mutual [age / place / ethnic-referral] giving genetic-bastarization, away from syntax, conjuring those primary-mammal-human-hyper-tense-characteristic-(s), arranged incremental-tangible-confirmation-(s), unilateral, from [issue-(s) // matter-(s)], interval-article-relied, present congressional-legislation-period-(s), as how territorial-resourcing, rate Federal-Rate-(s), as Holy-Roman, remain in mystique of query, factual-account-(s), religiously-simultaneous-tangent-(s), interceded-dissolve,| 397 |debiting substantial-textile-under-$100-transaction-(s).208¶|||33s.35m.s.|||§ Observable-distance-territory, measure-(s) are by how constitutional-amendment, merit order, rightfully-proceeded-procession of operation-(s), juxtaposing [livestock // Terraculture /// agriculture //// horticulture ///// mariculture ////// floriculture /////// mammal-taxonomic-classification-(s)] envisioning affairs, evident (from an) organ-reproductive-communication-(s), pronoun-adjective-verb-transitive-dynamic, articulating mode of relationship, when [last // first] name-(s), identify point of territorial-distance-denominative-impasse-documenting, upon an consistent-hours-succession of language-liturgical-focus, noting, insight-(s) of objective-collection-(s'), in juxtaposed-responsibility-(ies), per personal-account-(s)-interactive-referring, at [local-place / communion-impasse] focusing developmental-progressive-interactive-interval-(s),| 398 |inconsistent from [interposed-referenced-interpretation-youth-mid-age-content-extenuating // using // directing] year-to-date, decade-ecological-routine-(s), sufficing survival, where commercial-scheduling, limit reproductive-creativity from offspring, as nature from place, think present-requisite future, article-purpose, awhile-routine, impasse an driven-transitive-county-university-transit-hall, not emphasized on the way, to agenda-discourse, rather sufficing-resourcing, proportional, parts of syntax-lengths-part-(s) of syntax-length-reset-(s)-identify-verifying, human-living-data, before objective-function, an mundane-retensing-(s'), by governmental-denomination, evident (from an) career-statehood-effort-impasse, pertaining environmental-ambient, apose perimeter-acreage-coverage, enumerative-denominating, elaborating in an relevance from federal-subjunction, meticulously-divided, far too incapable of extraspective-genetic-religious-refining, by-an [blue-eye // blonde-hair // pink-skin] regeneration-repopulation-reproducing-mass, intent [awake-invigoration-(s) // act-(ion-(s)) // motion-(s) // subjecting-(s)] reasoning from substantial-time, date-solid-interval-(s)-conceptive-article-posit-(s), placed at calendar-operation-(s), emphatic security, where air-space, is appropriated, conscientious conception-(s), in an thousandth-tens-thousandths-patent-article-purpose-impasse-defining, why product-usage, matter where posit-(s)-emplace, when impasse-motion, respirate instruction-(s) of fact, for what may be pertinent by the location-(al)-capacity, in rule of whom sequence (from an) centralized-denominative-effort-(s) of nation, rather than standard-business-class,| 399 |[revolution-momentum-interval-solar-composition-dating // rotation-momentum-interval-tectonic-timing] unconscious for accumulating article-(s) in an juxtaposition of community-preference, because how motivation from data, is preferred an idle-sensational-solace, sweat-vaporless, vapor-fission-pressure, subdued impenetrable-force, sustem from character-position, placed at order-role-(s) of deed-(s), conducted validification, during intermittent modern-day-pulse-tension-(s)-impasse, ruled-historical-outcome, commonly embodied, through transition-(s), intermittent matter-posit-mundane-emphasis-purpose, having [manual // instruction // pamphlet // military-blue-book] from century-abstract-relegated-respect-(s), without an firm modern-decade-year-perceiving, why accord of matter, act (as an) grounds for territory, from manual-labor, proposed-stability, remaining proficient, at operating function-(s), for which, verify matter-(s), pass an limit per receipt, not registered-arrangement-(s), from checked-aspect, continual basis, influencing imbalanced arranged perspective-(s'), wherever note-(s) apply, through trading of dynamic-market-product-(s), informing while plan-city, set trusted order-chain-operation-(s), civil state-federal-balancing, [waged / (ing) upon mass-(es)-subjective-extraspective-reliance-process] as individual-(ism-(s')), cater-(ed) factoring-moment-(s), inquiring what cent of time, one has way to agenda-discourse, rather sufficing-resourcing, proportional, effect, persuaded each and every passing moment, coming in contact by an self-capabilities'-repertoire, play, where at an moment of time, can develop incremental, guideline of [solid-density / Liquid-pressure / Gas-vapor-pressure], discoursing resulting of interval-(s), (by an) scientific-calibre, order-visual-impassing-data, A Book subjective, tier serie-(s) of essay-punctuation-comma-increment-article-impasse-tier-ordering, effectively defining interval-procession-outcome-(s), initially hypothesized, supplementary envisioned-tension-(s'), before one individual, ever define capability of logic, formulating those ground-(s) upon [short / mid / long] term-planning-(s) of data, attempting to influence reader, out of [fanatic //

loyal-objector // listless-rebuttal-citizen-inhabitant] socializing periodical-ordering, attempt constant fundraise article-furthering, an request-continuum-data-posit-bit, particular-property-designation-(s), for resourcing mass-continuum-agenda-inquire, as standard-business, has not thought to raise those stake-(s), near quality of homage-tradition, losing those archaic-customs', introverted, human-conduct-offset-Infinitive, interactionless (as an) part of our disposition of sociological-experiment, from those [state-(s) // county-(ies) // city-(ies) // nation-(s) // tectonic-territory-(ies)-offset-a-the-630-human-genetic-physiological-hyper-tense-form-(s)] (as in youth aging), quandary perplex why we require proportion-(s') upon-present-day-requisite-effort-(s), intercontort, each compositional-process, as individual-(ism-(s')), cater-(ed) factoring-moment-(s), inquiring what cent of time, one has upon an location, as well dollar for objective-influctuation-(s), [standing // sitting // laying-awake] for thinking how to enclose tectonic-anterior-roof-infrastructure Place & Location-(s), factoring [in // of] juxtaposition of legitimate-setting, harmonizing letter-word-sentence-paragraph-essay-chapter-syntax-deviation-(s), apose land-(s) of territory-designation-distancing sociological-repopulating-desire, confluence-influencing, gain-(s) of effect, impassing-moment, tensed juxtaposed time-elapse-(ing-(s))-(short-term), juxtaposition calendar-dating (mid-term), working-physical-objective-transit-offspring-populating, mutual human inhabitant-citizen-interaction-(s) (or so (renaming simultaneous moded-action-commonality-(hypothesis))), defining appreciation along long-term-posit-(s)-definite-trust-means', mattering how, [influence / persuasion] juxtapose confluence indetermine perspective-(s), from whom [Women / Men / Children] recluse [dialogue // subconscious, thinking] when sensational of theatrical-medium, supplementary Gregorian-year-reset-(s), tense identified by religious-disposition, evident seen, why all follow pre-requisite-impasse, conjuring prior conceiving, but as how position-(s) of Salary, are not employee-waged, for finding an middle-grounds' per notion of -perspective, because way of will, not be evident [covering // impassing] all humankinds-territorial-grounds', without natural supplementary juxtaposition-(s') of Government-Agenda-Empirical-Ecological-Effort-(s)-Position-(s)-Tier, collecting natural-compositional-weighted-resources, malformed commercial-patent-purpose-formulation-affirming, as discourse (per) of object-(ive)-(s), vague (by) tensing-live-perspective, (from-an) mass-individualism, (not) agreeable on extraspective-affair-(s), offset, representing repression-(s) personal-intimacy-(cies), harboring daily-living, (by an) creative-curb at infrastructure, efforted by extraspective-persuasion (rate) per fact-(s), from data-denominative-bit-(s), apposing denominative-note, upon those [patent // copyright // registered article-(s)] existing by compound-resource-dimension-(s), plane-depth-surface-evoked, three-dimensional-tension-(s)-articulation-(s), heightened by those weight-(ed)-matter, component-compartmentalized, developing, deviating, desire-ideal-foundation-(s), referencing concept-perspective, where designated-address, house-limit-juxtapose, relevant extraspective-objective-subjective-article-competency-(ies), dimension-structuring, barren-space-(s) relevant objective-commercial-production, agreeing on outcomes of [interior / exterior] calendar-frame-(ing), requiring impasse, (for) defining hypertension-evocation, eventual-passing at [stance (still) // sat (still) /// walk (constant-motion) //// indication-(s) (still-motion-(s)), where daily-discourse] remain considered, upon surrounding capable-impasse-infrastructure, individual-possessive routine, not understood by troop-(s) of district-(s), how to develop an community subterior-year, conversive by longevity, awhile development-impasse embrace continued-matter-(s), unsegmented by interaction-(s), preferred an basis of influctuated-human-composition-introversion-hobbies, (not) article-scaling, when protruding-reverberational-force, upon tectonic-plate-matter-infinitive, retain book-article-(less), constitutional-article-purpose-investigation offset [vernacular / vocal / written / read / dated construes]...¶|||07m.: 12s.58m.s.|||§

§...Preposition-Count-[41 / Fourty-One]-of.. ..Article-Essay-[27 / Twenty-Seven].. // ..Book-Essay-[108 / One-Hundred-Eight].. ..Lines-[96.9275 / Ninety-Six.1,nine-tenths-2,two-hundredths-3,seven-thousandths-4,five-tenths-thousandths].. ..vernacular-voice, is suppose to interpret-interpose, visual-tonal-vocal-physiological-inhabitant-offset, when dating, is left in ambiguous for of %§86,400-secondths, as how moded-deviations of numerical-timing, inductively-recollect, from etymological-finding-(s), unless commercial-educated-interpretation-(s), are coordinated, in an fair-operational-fashion, due to scheduled-timings, for objective-function, incre menting how individual-perception-offset, estimate-time, per hour varied from minutes-function, as secondths-operation-(s), continue, in laggard-preference, sat in calm-living, for barren-lifestyle-existence, making how dating is anticipated, with an interval-timing, from if human- competency-exist?¶|||40s.04m.s.|||§

¶109Then and when;206 are (analysis) momentary occur-(ing), for data-interval-increment-(s), still unsettled for the observational-interposing of matter-(s), for common-attention-span, peer-pressure, through live-past-requisite-present, influencing past-requisite, simply-secondths-letter-aspect, but with still-surrounding hour-minute-(s), of developed-spacing, dissolved discoursing, upon calendar-day, having limit-maxims, [syntax-interaction / blood-pulse-cardiovascular-repetition / centripetal-force-constant-moment-second-(s)-distancing-yearly-reset];207 proportion absolute-truths', define defense-future-outcome-(s), at merit-methodical-intuition, supplementary-interposed-act-(s), stanced awhile feet-approximate-meter-surface-area-surrounding-mile-(s)-centripetal-movement, adjust vernacular-verbal-vocal-interaction-(s), affecting setting of character-role-discourse, latitude-horizontal-parallel-posit-degree, integral intangible-pressure, upon ventricle-cardiovascular-respiratory-system-(s'), mobius [transit-tract blood-pulse / oxygen / carbon-dioxide-exhale-sedentary-posit-(s')] without stature-auxiliary, belief impassing-momentum, east to west-vapor-pressure-oxygen-respiration-air-drift, west to east tectonic-centripetal-force-drift, remained [[1]in // [2]of // [3]at // [4]an // [5]by // [6]as // [7]perspective]— [transit // housing // work-place // social-subjective-interaction-(s)], still of ground-revolving-mode, timed underneath ones' inferior-elated by (centripetal) force-prevalence, intermittent juxtaposed state-(s) of matter, premise-pertained-upon (a) the conception per matter, composite-retrieval, view-eye-confirmation, as depth-optological-perspective, [Etymological-Letter-Linguistic-Language-Participle-tone-word-tonal-two-double-particle-offset-consonant // vowel-interposed-article-denominative-scale-gauging-syntax-subjective-inanimate-tensable-matter];[208] discoursing activity, where moment-belief, not supplement-time, dimension-identifying, perceptive-take, person-live-part-continuum, disregarding Infinitive-Presence, (p)-articulating from-exterior-value-(s), extra-spectively perceived, due to regard of [density // buoyancy // vapor //// inanimate-solid-tundra /// soluble-motion-mobius-composition /// spacing-vapor-chill-respiration-oxygenation-tangent-blood-space-flucutating] agronomical-goods';[209] consistency-schedule-school-youth-mid-age-career-national-infrastructure-impasse, memory deceased-being-(s')-life-style-repopulation-breed-ing, [[1]constructive // [2]engineering // [3]storage-reproduction-technique-skill-interpose-master // [4]apprentice // [5]listener // [6]director / [7]dictator / [8]persuader, by-of live-perspective] inculpable, translation transition transiting, perceptive-live-visual-tonal-tense-conjunctive-subject-tier-art-icle-deviation-(s),| 400 |that combine [calendar-posit-hour-dating-central-time-minutes'-perspective-requisite-being // present-objective-denotation-impasse-effort-extent] visualizing, discoursed demand, urgent where life, relates' mutual-capacity, where such time, is to blood-pulse, an various-subject-paragraph-definition-articles, casting those momentum-sequence-(s), inter-mittent past-date-calendar-dating, apose future-time-presence-impasse-hour-location-interval-(s), expected influence where motion-(s), foresee mass-interaction-(s'), frame-adjusted, An-(s')-mind-mentality, utmost impossible, reference-relevance, prerogative-(s) per dating, pictured [electromagnetically // digitally // film /// photo-visual-reflective-gloss-stone-palm-plastic, ambient elongated spacings] inclusive daily-fervor by 2016 spacing-velocity-street-city-linear-routine,| 401 |amongst redact-quality-(ies), such as [Metropolitan, Urban /// Suburban /// rural-land-spacings'] secondths-minutes-tense-article-verifying, tested [mammal // human specimen-(s)] at university-county-distancing-(s)-worth, covering-ground there-of, shorted Bachelor-(s')-commonality-repopulation-planning, preferring commercial-idling sweatless lifestyle, limited-bounds', [theatrical-tone-volume-tempo-time // key-signature-staff-coordinating // time-signature // beat-(s) per minute-intonating-reverberational-resonance-vibration-(s)] Line-(s)-(pressurizing) & Space-(s)-(abstract) intonation-subjective-purpose (inanimate-abstract-infinitive-surrounding-black-silhouette-non-for, as how-per-se, pertained Ground-(s) of territory, offset perceptive-objective-gamma deviation-(s) minuté of distanced-trajectory-(ies)), improbable, [yearly-feminine-repopulation ///// seasonal-male-product-patent-reproduction-article-(s)] reproductive-quality-constant-thought-(s)-focusing, where women are amongst-variation-(s) past-historical, Holy-Roman;[210] East;[211] //{Yugoslavic-Boundary} // West {Primary-Culture-(s)}-tive-quality-constant-thought-(s)-focusing, where women are amongst-variation-(s) of Holy-Roman;[212] East // {Yugoslavic-Boundary} // West //-{Primary-Culture-(s)}, [Italian // French // Dutch // Belgic // English // Spanish // Portuguese-west-influx Germanic-(Swiss)-Schweich-Turkish-Israeli-Origin-Catalyst-pattern] remain impertinent idle-concern, relevant-(military-elemental-auxiliary-article-objective-reproduction-factoring-(process-(es))] fundamental consideration, gently-advised, pre-basis coverage impasse-genetic-trust-tier-territory, commonly unfamiliar subjective-study, not apprehended at youth-subjective-bachelors'-youth-adult-inquisition;[213] for comfort verify pertinent daily-discourse, through mark-article-currency-intervalizing, exchanged-experimented, past 1600's-Pronoun-Regal, transitions' 1700's-philosophy-articlizing, transitioning 1800's civil-industrialism, existence, [[1]where / [2]when // [3]why /// [4]what //// [5]how ///// [6]who] an experiment matter for juxtaposition of document-(ing), in an supplementing of data, by the constitutional-intention, to amend the article-(s) of form, for proportioning those mass-(es), to an psychological-educational-conscious-exponent-monthly-interaction-succession-(s),| 402 |year-(s)-century-discoursing, millennium-decade-discourse-month-minute-perspective-interval-point-articlizing, interacting competent citizen-(s)-{#$%-individuals'} whom partake intrigue [dust // cement // mortar-quikrite-composite-water // refined-concrete-dry-mineralizing] [dry-solid // dry-substantial-article-reference-deviation-(s)];[214] enlightening adapting tangible-patent-feature-(s), discoursing dissemination, by quality of saving-data, though perceptive-spiritual-experience, continuing adrift solar-spherical-infinitive-fission, unconscious mammal-in-Twelfths'-origin-celestial-impact, from omnipotent-human-civilization-(s)-offset-tectonic-origin-juxtaposition-(s), in through, physiological-human-being-posit, by characteristic-Eighty-Nine, apose 630-Genetic-speicie-(s)-Taxonomic-Classification-Tectonic-disperse-arrangement-adjustment-(s'), appose-(d) visual physiological-physique-conjunct, viewing-surrounding-interior-space-limit-(s), setting an quality, through our adjective-noun-intransitive-verbal-communicative-presence (not perceiving), why method of mass, not fully proportion, totalitarian-development-extraspective-means', mute-communicative, comprehensive-language-skills, upon-apose paper & pen-quality discourse,| 403 |elaborating (en) acting-storage-elemental-population-article-century-lineage-anticipation-article-limit-(s'), deriving product-space-usage, purchasing entertainment passing-moment-(um), common-study-(ies)-impact, suggest mass-socializing, far too gossiped, comprehending why politic-(s) remain mysterious, calculated-repression-(s') per [[1]matter / [2]resource-(s) / [3]political-boundary-(ies) / [4]infrastructure / [5]event-(s) / [6]population // [7]birth-rate ///

[8]death-rate / [9]state-boundary-(ies) / [10]commercial-enterprise-economic-demand-central-advises'] sharing an cumulative-distribution-functioning, factored trade posit-hour-minute-(s)-impasse-mode-purpose, emitting blood-pulse-tendon-tension-(s'), for living by [due-(s) / responsibility-(ies)] per day, at-will-basis, succession of prerogative-(s') serving-physical-interaction-tense-article-subjection-felt-sensation-(s), balancing human [candor // conduct // behavior // responsibility-(ies)] sat in an continuum, per-of comfort-idle-silence, contingent-elated investigation-(s), from youth-lesson-guideline-(s'), adult-parent-referencing;[215] yet central an state-government, [devising /// finding motion-time-article-(s) /{dating}/ still-calendar-article-(s), routine activity-(ies)] per proportion perceptive-offset-origin-date, compositional-physiological-self, intervoluntary defining [[1]government / [2]commercialism // [3]university /// [4]economic //// [5]ecological ///// [6]militarial ////// [7]residential /////// [8]Terrain-Topography aspect-(s) of element-(s)] tangent-(ed)-apose physiological-psychological-dating-impasse, when interaction occur, from during-when instated-interaction-action-effect-(s'), exposing interposed-individualism, independent elapse-{120-secondths — 300-secondths-deviation-motion-enumerated-effect-denominative-integer-rate-article-usage-categor-izing-article-bit-(s)}, observing humanity per [road-way / inhabitant] by [housing-structure // citizen-classification-offset] of [public-decorum /// church-member] at local-interacting through functional-auxiliary-mode-(s), per-of population-discourse-(s), renounced week, subjunct-occupant-(s), measuring interior-parcel-(s), when-(where whom) cubic-deviation-(s),| 404 | visualize-wall-frame-perceptive-elemental-patent-locational-usage-impasse-discourse-square-feet-grounding, [ration-grammatical-term-word-definition-(s)-deviation-(s) // ratio-mathematical-symbol-exponent-mark-articulation-defining-(s)] as living-room-length-twenty-five-feet, by width-twenty-feet, equivalent five-hundred-square-feet, centralizing perimeter-surface-area, erected [height-clearance-maximum / length / width] impasse walked spacing-(s), fulfilling dai-

ly-obligation-(s), from self, upon an impasse mutual-empirical-discourse, involving cubic-measure, elapse-(d) when surrounding meter-(s), pertain tense in cause of time, by measure, having an rate of elapses, day-Fourty-seven-thousand-six-hundred-second-(ths), per cubic-feet-measuring-indication-notion-(s'), validifying [month-proportion-subjective // objective-article-(s')] by when daily-activity-(ies), intersect prerogative-(s) per [family / self-individual-(s)-member audible] their-of momentary-activity-(ies), outlooking analytical-discourse, ancestral-influence, elapsing-(s) by [possible // probable] activity-(ies)-capable-impasse-motion-(s), placed where-at objectives, appeased entertained-mentality, being observationally-congruent, requisite tangible-impasse-hour-territory-article-data, routine focusing, daily-passing-(s), prior-equipment-article-posit-(s), (re)-furbished value-(s), generally-conglomerated, synthesizing-recounted-posit-(s), resettled daily parameter calendar planned article volume developmental feature spherical-density-offset-define-tiring,| 405 |relative-tangent-focus-groups', agreeing on relative-discourse-(s), when momentum-infinitive, have life-time-counting, circulate reach, at threshold, for commercial-cooperation, instructed paragraph-(s)-tangent-(s), arouse sequence-(s) per-of-an 110L.[Sentence & Terming // Word / Letter counts].¶|||08m.:56s.81m.s.|||§

§...Preposition-Count-[26 / Twenty-Six]-of.. ..Article-Essay-[28 / Twenty-Eight].. // ..Book-Essay-[109 / One-Hundred-Nine].. ..Lines-[110.445 / One-Hundred-Ten.1.four-tenths-2.four-hundredths-3.five-thousandths].. ..when inhabitants can not interpose an vocal-existing, no etymological-proof-exist, from mute-physiological-inhabitants, apose physics-elemental-objective- silence.¶|||16s.78m.s.|||§

¶110.Simply put, spring is abound [Rainstorm-(s) // clear day // surrounding-mineralizing] an cue for cause, growing natural-force-(s), reasoning invasive-impasse-covering, from local-area, fully-covered where atmospheric-matter, precipitate ground-level-vapor-saturation-(s), [when // where] [sea-(s) / lake-(s) / river-(s) / ocean-(s)] evaporate their of, lunar-revolution, upon solar-impact-drift-continuum, drift-tangible-level-means', forced [up / down] (con)-current-wind-heat-front-temperate-influctuation, naturally-cool, per water-body-(ies), where Columbus-Cloud-Formation-(s), naturally-posit an electrostatic-friction-temperature-tension, affirmed where jet-current-fluctuation, vapor-pressure amongst mid-air-current-stream-(s), [swipping // swilling // seeing-simultaneous-symbolized-effect-(s)] per-juxtaposed-of spring-depth-procession-infinitive, temperate-fashion reasoning, central-transit-condensing, droplet-(s) of water, falling down an immense-reaction, sufficient-living-repopulation-status, overall-perimeter-circumference-spher-ical-rectangular-prism-saturation-interior-tension-level-(s), directly-apposed latter-day minute-(s)-hour-day-ground-level-saturation-(s), preparing relevant from those anticipations' of elemental-tangible-article-trade-objective-article-(s).$_{209}$ ¶|||01m.: 02s.$_{65m.s.|||}$§ Why impasse routine of crop-(s), outside, anticipated winter-volume-impasse, verifying edifice-cover, façade, per arrangement at Individual-Crop-Discourse & Proportional-Rationing of mass-deviation-impasse-requisite-motion-(s), apose lighting, fulfilled obligation-(s), per year-populus-mass, physio-logically-weight-count-(s), per inhabitant, measure where [crop-weight-(s) / territorial-spacial-rationing [(per thousandths-occupant) / recipe-fraction-(ing)] market-product-(s)-transition, for organized-family-interaction-(s), tensing [smell / taste / vision] where method-(s) per confirmation, attempt inhabitant, operative perceptive-capacity, working commercial-socialism-thought, collect-(ing) those genetic-inhabitant-(s), realizing inhabitant-population where constraints of [mass / measure] per [feet-cubic-distance-(s) / activity-elapse-purpose-(s)] perceive less than an numerous other-(s)-roadway-transit-(s), when work-week pertain work-related beneficial-profit-(s'), characterizing-worker-purpose, following [rule-(s) / regulation-(s) / guideline-(s)] involved in cycling by repetitive-motion-(s), persuaded perspective-matter-(s), that has how existence, complicate comprehension of national-reform, amidst parameter-(s) per time, perceiving issue-(s), whenever proportional, term-data-competency-elapse-(s), synthesized impasse conceptive-analysis, by-an subterior-essay-extent, interposing transit, as how we receive our distance-coverage-credit, to claim an means' of sociological-resource-trade-purpose... ¶|||02m.:13s.93m.s.|||§

§...Preposition-Count-[9 / Nine]-of.. ..Article-Essay-[29 / Twenty-Nine].. // ..Book-Essay-[110 / One-Hundred-Ten].. ..Lines-[28.4975 / Twenty-Eight.1.four-tenths-2.nine-hun-dredths-3.sev-en-thousandths-4.five-tens-thousandths].. ..not many other redacts from here on out. |||10s.75m.s.|||§

{January-(31)-February-(28 / 9)-March-(31)-April-(30)-Month-(s)-Day-Hour-Minute-(s)-Secondths-Count}-{10,368,000 D.H.M.S. = 60S. × 60M. × 24H. × D.}- {2,880 H.}-{172,800 M.}-{120 D.}

[Article:] [May]

¶111.I personally find an difficulty of sexual-interaction-(s'), for personal-block, affect tangible-act-impasse-(s), yet discussed in-an comfortable-environment, deserving mutual-respect, for independent-specification-(s'), when self, interact amongst genital-specie-(s), honoring compulsive-act-(s), conjuring an clearly more effective style of hour-(s)-denominative-impasse, for-identifying, conjured-residence, adjunct extraspective-cycle-(ing), via monetary-means', circulating cost-(s), perceiving through an influctuation per variated [objective // substantial means'] (un) incorporated by territory, denominating [tangible // intangible] article-deviation-(s), as culpable-act, contemporarily-mission, destination-(s) of multiplicity-simultaneous-objective-article-(s)-tonal-visual-interpretation-functioning, sequenced upon rate-receipt-dating, occasion-ally time-variable-(izing), intercontact-(s') of [Mass-Infinitive-First-Person-posit-juxtaposition // second-person-posit-enabled-position] unable to comprehend why elapse amongst day-interval-impasse, revolve under physics, by term-scaling tier-article-subjective-issue, pertaining data, upon word-subjunction, as-how property station focus on homework, as [objective // subjective] task-(s), enveloping-intermittent-incremental-letter-word-syntax-referencing-data, by conversed-acting, enumerated adjourning individual-help, placing-incremental-envision-(ing), where sight confirm variable-article-subject-(s), instructing information, isolating self, as-an warm-volume-compositional-form, tempering data-bit-(s),| 406 |by distance-spacings', written-as typed-documenting-data-interconjecture, heightening plane-cartographic-surface-area, spherical awhile centripetal-force, clearance height, to actually-interpret, by [Length × Width] rectangular-parallelogram-squared-off-property, defining population-influx-position-(ing), from where subjective-study-(ies), pertain perspective, by tectonic-plate-(s), drifting west to east, from-of-an pressure-vapor-force, winded by east-to-west-velocity, furthering (in)-claritive-effort-(s), where earth-plasma-molten-pressure, move-kinetic-motion-(s), celestially-parallel, solar-fission-impact, affecting moon-lunar-solar-impasse, from tectonic-mass-dry-formations', simultaneously-aligned, in-an view upon axis-angle-dimensional-objection, when movements'-remain, incandescenary participle-interjunction, intersect-(ed)-discourse, generalizing centripetal-drift-elapse-(s), when use-of human-reclusive-introversion, continue forward-progression, by forward-depth-perspective-continuum-impasse, rated by public-road-interval-(ization)-(s), serving duty-(ies'), where latter-day-belief, celebrate-in-age-impassing, reminding daily-offspring-reproduction, litigated hospital-expenses, involuntary monetary-interposing, motioning by inductive, inhabitant-mass-cause-(s), defining how way-impasse, pertain live-existing-inhabitant-(s') influenced theatre, while being in-an repetitive-year-day-factoring, [simultaneous revolution // rotation-motion-(s')-progression] consecutive-simultaneous pre-molten-pressure-offset, not in-an comprehension article-extent, formulating human-perspective-retrograding-human-method, testing blood-pulse-nature, upon-an elemental-form, per requisite of article-matter-(s), baptized [how // why] data be an part per notion-(ing)-(s)-synapse-interval-(s), through [1elapse-(s) // 2syntax-(s) // 3synapse-(s) // 4referencing // 5reading // 6motion-(s) /// 7action-(s) // 8et cetera…] where self-independent [1inhabitant // 2citizen // 3person // 4being // 5deziden // 6homeless-societal-enabled-worthless-being] commonly-impasse daily-dating, defining blank-nature, at off-time-article-impasse, awhile commercial-production, test governmental-development, apose natural-space-(ing-(s)), comprehending affirmed individual-competency-aptitude, interposing-data, as-an prospering perspective, when observation derive-directive, received from perspective, as capability define invigoration, valuing-data, by past-comprehension-requisite, present-offset-(ting), crescendo [word-synapse // sentence-comma-elapse /// time-interval-summing], by [letter-notion / secondths-notion-(s)] for-an reasonable interposing per [rate / percent / integer-(s)] reasoning syntax-extent, every [day / week / month]-reasoning-(s), aware parameter-(s) per space, defining vocal-tone, by syntax-rating, affirming per data-progression,| 407

|furthering perceptive-focus, while in-an [reconstruction / construction / deconstruction period-(s)], surrounding territorial-advise, consoling present-day, upon age-self-littoral-day, when year-cycling, adjunct, year-to-month-operations'-reset-(s), by-of present-day-impasse, littoral-adject-ure, similar unread-adjunct-abstraction-adjecture, aging upon place, for tangible-circumference-dimension-(s), comprehensible-through mutual-respect, subsequent interaction-(s), defining dynamic-competency, rendering formal-discourse, of progressive ancillary-notion-(s), present article-consideration-(s),| 408 |when-where [place-hour / physique-cardiovascular-respiratory-minute-pulse-(s) / subjected-rate-grammar-reasoning]$_{51L.}$ mundanely-elapse-(s)-secondths through cycling of ambient-abstract-parallel, civilization-offset-distancing-(s'), impassing planear-level-height-perspective, motion-(s')-impasse-measuring-(s'), per civilian-interaction-motion-(s')-impasse, upon centripetal-force, visual-tense-impasse-effect... ¶|||04m.: 37s.$_{69m.s.|||}$§

§...Preposition-Count-[10 / Ten]-of.. ..Article-Essay-[1/ One].. // ..Book-Essay-[111/ One-Hundred-Eleven].. Lines-[53.65 / Fifty-Three.$_{1.six-}$tenths-$_{2.five-}$hundredths].. ..yet another chapter, I've not fully redacted for of complexity of type.

¶112Sort-Ordering, categorize-decimalize, through-an means', per an tens-hundredths-rating-dating-denominative-method, no tout-of minute-(s)-order-reset-deviation-(s), pertain hour or three, by-an timed-requisitve-relapses'-common-capable-periodical-subjective-documenting-method, filing visual-observation-(s) by-an physical-specimen-operations-data-year-dating-collecting, internal-revenue-services' pertained-infinitive-past, primary [fundamental-perceptive-comprehension-pieced-objective // subjective-article-data-posit-objective // physical // subjective-school-primary-rubric /// benchmark /// predicate // commercial-governmental-(militarial) // transit // constructive-architectural-(introverted not be, for of physic / tectonic-plate-aspect-(s)), upon an constitutional(-)

current(-)succession focused-presence-impasse-discourse, where mutual-objective-placings', access // location-perceive-(sedentary)-data] retained-(ing) at formal-article-conjure-progression-impasse-(s)-tense-motioning (historical-read-Dewey-decimal-(letter-word / participle-library-location-center-periodical-ordering-(s))), alerted process-continuum-operation-(s), timed as is dated, from-of deadline, characterizing each step-timbre, from-while-at-upon| 409 |[objective / experimental / commercial-production-examination] per word-increment-article-data-bit, embolizing syntax-period-ordering, designated-square-surface-area-feet, per developmental-reasoning, methodical still millionth-(s')-affirmation-impasse-conjunction, during an juxtaposed perceptive millennium-discourse, intervalizing-data, per a(n) article [time // calendar] dating-article-(s)-rate-interval-(s), [where // when] applicable-denominative-interval-(s), storage-refer, data, impassing several-moment-(s) while tectonic-accounting, form from-of-an Gregorian-Year-objective-housing-article-storage-impasse-calendar;[206] pertaining why Jeffersonian-syntax-elapse-cycling, {Seven-Billion, Five-Hundred-Sixty-Eight-Million, Six-Hundred-Forty-Thousand-Second-(ths)-{July 4, 2016}}, as-how human-driving, propel introverted-being-focus, (which is not basic human rights, for, from of extraspective-article-objective-timbre), in abundant-abstraction, per reputable-(introverted)-service-(s'), yet using library-(ies), (as an) extraspective-genetic-territorial-subterior-presence-(s), ordering at times', archived-material-(s), unverified word-article-intrication-(s), parallel surrounding present-second-(ths)-continuum, intermittent, [revolutionary-centripetal-force-calendar-articlizing // rotational-centripetal-force-time-denomin-ating]| 410 |subcommittee-literature-prose-congressional-executive-legislational-observat-ion-(political-aging-total-dating-presence-L.C.C.N.-ordering, apose an I.S.B.N-voluntary-entity-continuum-constructive-methods, practicing cursive-discipline) by-whence secondths-grandiose-currenting-omission, contemporary-synapse century-historical-compartmentalize-secondths, understanding how secondths-count-(s), signify confirmed-rotational-data;[207] from awhile comprehension schedule-work-interposed-effort-(s), conceptively-reason, deliberate purpose, retrograding physiological-article-objective-act-(ion-(s)), adjusting inhabitant-article-schedule-prerogative, consistently-corrugated, common-core-applicable-genetic-curriculum-crestenza-(s);[208] preferred continuum-subjuncting into, English-Predicate-reasoning-focus, rating-article-deviation-(s), referable by an transit-schedule-hour-conceptive-location-constructive-preferred-population-topic-genre-displace-align-(ment)-sort-article-interactive-formation-(s)-impasse,|411| from [tensable-colleague-{city-county-college} / university-{believed-city-county-state}-decade-habitant-renaissance-method of saturated-populus-statement-living] account-(ed) by realigning, territorial-decorum, administering date-documenting, vernacularized-by, vocalized-objective-article-repetitive-revolutionary-date-order-ing, forgetting material-(s) offset instructional-discourse, intermittent mile-(s) [upon-departure / to-arrival] post-educational-lapse-complexity-(ies), featured in-an developmental-distancing, affirming timing-(s) per minute-(s), per first-second-person-word-experience, by-an bias-hoc-reset, influencing general-material-market-(s), passing along inset-population, from-an plausible-mass-population-traffic-(ing), sheltered mass-population-silence, (un) affirming rotational-rate-(s), denominative-of Conscious-Tangented-Simultaneous-Cursive-Word-Participation-Coordination-inter-action-level-(s),| 412 |thinking [from // by] Stephen-King-Type-Manuscript-article-Write-Curb, non-inquisitive, fiction-self-analysis, upon authorship, without the-of-superlative, not reduce-exponent-term-definition-grading, cross-affirming data-referencing-(s), advising (un) syncopated-interceded-context, from formal-align, by Bias-Genetic-Virtue, fiction-story-boarding, grammar, by an adult-adjunct of month to date day-genetic-characteristic-(s), aligned of particular-feature-(s), of [vocal-manner / physical-movement-(s)] as for how currency is conceded by enumerated-enterprise, at point-(s) of establishment, which are denominating-trade, by the purpose of information, where at an impasse of deliberation, so data be vouched for, by saving-(s) in cause of [place // territory // land-pre-architectural-(individual-planning)] interposed extraspective-land-pre-concepting-impassed-structural-ecological-civil-engineering-constructive-geological-engineering-development-(s')) at an perimeter-base, contacting bank, from [labor // book] dynamic, protection of asset-(s), individually-count-(ed), (as for of a the) Census-Decade-Denominating, meriting passed paper-trade, counting account-(s), when general-cognition, determine an discourse-of [listening / interdialouging / personal-written-interposing] as fundamental-rubric-subject-tiering, of place-(s) upon habitant-Citizen-Student-(s), from prose of [past-fiction-decimating / historical-recurrence / tangent-(s) of Individual-(ism)] remain undefined exponent-reproductive-opposite-sex-(feminine)-[nine // seven] month-rate, for-of, perk-breast, not sag, while sag-breast, not press, blood, through an subjective-discourse, referring by, unconscious-limit-(s), comprehension-requisite, present-offset-(ting), crescendo [word-synapse // sentence-comma-elapse /// time-interval-summing], by [letter-notion / secondths-notion-(s)] for-an reasonable interposing per [rate / percent / integer-(s)] reasoning syntax-extent, every [day / week / month]-reasoning-(s), aware parameter-(s) per space, defining vocal-tone, by syntax-rating, affirming per data-progression, furthering perceptive-focus, while in-an [reconstruction / construction / deconstruction period-(s)], surrounding territorial-advise, consoling present-day, upon age-self-littoral-day, when year-cycling, adjunct, year-to-month-operations'-reset-(s), by-of present-day-impasse, littoral-adjecture, similar unread-adjunct-abstraction-adjecture, aging upon place, for tangible-circumference-dimension-(s), comprehensible-through mutual-respect, subsequent interaction-(s), defining dynamic-competency, rendering formal-discourse, of progressive ancillary-notion-(s), present article-consideration-(s), when-where [place-hour / physique-cardiovascular-respiratory-minuté-pulse-(s) / subjected-rate-grammar-reasoning] mundanely-elapse-(s)-secondths through cycling of ambient-abstract-parallel, civilization-offset-distancing-(s'), impassing planear-level-height-perspective, motion-(s')-impasse-measuring-(s'), per civilian-interaction-motion-(s')-impasse,| 413 |upon centripetal-force, visual-tense-impasse-effect...¶|||06m.:11s. 41m.s.|||§

§...Preposition-Count-[25 / Twenty-Five]-of.. ..Article-Essay-[2/ Two].. // ..Book-Essay-[112 / One-Hundred-Twelve].. ..Lines-[77.0435 / Seventy-Seven.1.zero-tenths-2.four-hundredths-3.three-thousandths-4.five-tenths-thousandths] ..Where are we amid centripetal-force, transiting along an grandiose-tectonic-plates, unmassed resourcing, per dating of calendar-dating, when human-physiological-function, still remain in conjunct with interior-anatomical-hyper-tenses, for evoking how displays of physics are aligned by different values of subterior-human-physiological-weight-micro-compositional-proportioning. ¶|||23s.10m.s.|||§

¶112Sort-Ordering, categorize-decimalize, through-an means', per an tens-hundredths-rating-dating-denominative-method, no tout-of minute-(s)-order-reset-deviation-(s), pertained hour or three, by-an timed-requisitve-relapses'-common-capable-periodical-subjective-documenting-method, filing visual-observation-(s) by-an physical-specimen-operations-data-year-dating-collecting, internal-revenue-services' pertained-infinitive-past, primary [fundamental-perceptive-comprehension-pieced-objective // subjective-article-data-posit-objective // physical // subjective-school-primary-rubric /// benchmark /// predicate // commercial-governmental-(militarial) // transit // constructive-architectural-(introverted not be, for of physic / tectonic-plate-aspect-(s)), upon an constitutional(-)current(-)succession focused-presence-impasse-discourse, where mutual-objective-placings', access // location-perceive-(sedentary)-data]| 414 |retained-(ing) at formal-article-conjure-progression-impasse-(s)-tense-motioning (historical-read-Dewey-decimal-(letter-word / participle-library-location-center-periodical-ordering-(s))), alerted process-continuum-operation-(s), timed as is dated, from-of deadline, characterizing each step-timbre, from-while-at-upon [objective / experimental / commercial-production-examination] per word-increment-article-data-bit, embolizing syntax-period-ordering, designated-square-surface-area-feet, per develop-mental-reasoning, methodical still millionth-(s')-affirmation-impasse-conjunction, during an juxtaposed perceptive millennium-discourse, intervalizing-data, per a (n) article [time // calendar] dating-article-(s)-rate-interval-(s),| 415 |[where // when] applicable-denominative-interval-(s), storage-refer, data,| 416 |impassing several-moment-(s), while tectonic-accounting, form from-of-an Gregorian-Year-objective-housing-article-storage-impasse-calendar;[209] pertaining why Jeffersonian-syntax-elapse-cycling, {Seven-Billion, Five-Hundred-Sixty-Eight-Million, Six-Hundred-Forty-Thousand-Second-(ths)-{July 4, 2016}}, as-how human-driving, propel introverted-being-focus, (which is not basic human rights, for, from of extraspective-article-objective-timbre), in abundant-abstraction, per reputable-(introverted)-service-(s'), yet using library-(ies), (as an) extraspective-genetic-territorial-subterior-presence-(s), ordering at times', archived-material-(s), unverified word-article-intrication-(s),| 417 |parallel surrounding present-second-(ths)-continuum, intermittent, [revolutionary-centripetal-force-calendar-articliz-ing // rotational-centripetal-force-time-denominating] subcommittee-literature-prose-congress-ional-executive-legislational-observation-(political-aging-total-dating-presence-L.C.C.N.-ordering, apose an I.S.B.N-voluntary-entity-continuum-constructive-method, practicing cursive-discipline) by-whence secondths-grandiose-currenting-omission, contemporary-synapse century-historical-compartmentalized-secondths, understanding how secondths-count-(s), signify confirm-ed-rotational-data, from awhile comprehension schedule-work-interposed-effort-(s), concept-ively-reason, deliberate purpose, retrograding physiological-article-objective-act-(ion-(s)), adjusting inhabitant-article-schedule-prerogative, consistently-corrugated, common-core-applicable-genetic-curriculum-crestenza-(s);[210] preferred continuum-subjuncting into, English-Predicate-reasoning-focus, rating-article-deviation-(s), referable by an transit-schedule-hour-conceptive-location-constructive-preferred-population-topic-genre-displace-align-(ment)-sort-article-interactive-formation-(s)-impasse, from| 418 |[tensable-colleague-{city-county-collage} / university-{believed-city-county-state}-decade-habitant-renaissance-method of saturated-populus-statement-living] account-(ed) by realigning, territorial-decorum,| 419 |administering date-documenting, vernacularized-by, vocalized-objective-article-repetitive-revolutionary-date-ordering, forgetting material-(s) offset instructional-discourse, intermittent mile-(s) [upon-departure / to-arrival] post-educational-lapse-complexity-(ies), featured in-an developmental-distancing, affirming timing-(s) per minute-(s), per first-second-person-word-experience, by-an bias-hoc-reset, influencing general-material-market-(s), passing along inset-population, from-an plausible-mass-population-traffic-(ing), sheltered mass-population-silence, (un) affirming rotational-rate-(s),| 420 |denominative-of Conscious-Tangented-Simultaneous-Cursive-Word-Participation-Coordination-interaction-level-(s), thinking [from // by] Stephen-King-Type-Manuscript-article-Write-Curb, non-inquisitive, fiction-self-analysis, upon authorship, without the-of-superlative, not reduce-exponent-term-definition-grading, cross-affirming data-referenc-ing-(s), advising (un) syncopated-interceded-context, from formal-align, by Bias-Genetic-Virtue, fiction-story-boarding, grammar, by an adult-adjunct of month to date day-genetic-characteristic-(s), aligned of particular-feature-(s), of [vocal-manner / physical-movement-(s)] as for how currency is conceded by enumerated-enterprise, at point-(s) of establishment, which are denominating-trade, by the purpose of information, where at an impasse of deliberation, so data be vouched for, by saving-(s) in cause of [place // territory // land-pre-architectural-(individual-planning] interposed extraspective-land-pre-concepting-impassed-structural-ecological-engineer-ing-constructive-engineering-development-(s')) at an perimeter-base, contacting bank, from [labor // book] dynamic, protection of asset-(s), individually-count-(ed), as for of a the Census-Decade-Denominating, merit passed paper-trade, counting account-(s), when general-cognition, determine an discourse-of [listening / inter-dialouging / personal-written-interposing] as fundamental-rubric-subject-tiering, of place-(s) upon habitant-Citizen-Student-(s), from prose of [past-fiction-decimating / historical-recurrence / tangent-(s) of Individual-(ism)] remain undefined exponent-reproductive-opposite-sex-(feminine)-[nine // seven] month-rate, for-of, perk-breast, not sag, while sag-breast, not press, blood, through an subjective-discourse, referring by, unconscious-limit-(s), human-capacity-visual-enumerating-counts-limits, during hand-held belief of Control-(Library-Congressional-Control-Number), as how personal-identity, shift mass-objection, [past // conceptive-future-presence-interposed-articulations'], verbal religious-purpose, intermittent clasp-clause-cardiovascular-limit-condition-(s), voluntary metropolitan-production-(s), pre-corollary-complex, aorta-coronary-organ-gland-aspect, visualize surface-area-approximate-height-limit-depth-competency-repetitive-conjecture, when mass not outweigh, crust-infinitive-surface-density-distancing-expanse, never point-plot-birth-retirement-covered, fully of individual-physiological-efforts'-impasse, going unscaled formal Count & Measure-(s),| 421 |intermittent individual abstract demand-selection-subterior-impasse-hoc-forms'-decisions', sectionalized divided conjugation-(s), abstract consensus-biblical-manual-literary-aspect, personally-divided, where self sleep-mode-interval, pertain-upon ecological-setting, respirating pre-read accelerator-reader-tense-(s), counting data, per notion-offset-extent, instilled per-of genre reference-relevant historical-truth-(s'), economic-industrial-boom, as legal-documenting-discourse verify literary-mass-compilation-constructive-aspect-denominative-efforts',| 422 |as when auto-biological-reference, majority-non-fiction-reality, from where distance-offsets, blank mass-literary-intrigue, from nature of per entitled-entertainment;[211] as some what equivalent from attention-span-(s) those matters' of limits', pertained per state, juxtapose abiding municipal-code, attain of jurisdiction, when local-city-jurisdictions', incline Federal-Law, article-merit-credit-debit-access-acceding, present sustainable-practice, by progression-(s) tier-per data, to be aligned, from ambiguous count-documenting, insisted sustaining;[212] [resource-(s) / edible-(s) / weapon-(s) / land-space upon territory / Terraculture / et cetera...] general-annexing, orthodox pride per tense-(s), monitored surrounding watt-(s), spacial directive avenue, enticed Man & Woman-individual-extraspective-opposite-sexual-interactive-work-perspective-offset-effort-(s) (marriage-relationship-effort-(s');[213] present-past-thought-unscheduled-interactive-extraspective-planning, as-when off-time placing-(s), merit substantial-matter-(s), uncredited dense-credit-resource-subterior-patent-dimension-compartmentalizing-article-posit-(s)-tiering, indicated-present-purpose, simply-appreciated, mundane-arousal-repression, where other-(s) around, affirm presence under entity, reset-(ed) by [identification / title / deed] trusting received reference-auxiliary-objective-deviation-article-(s'), [Greater & Lesser / denominative], [numerative-[1]inclinations /// [2]indications /// [3]motions /// [4]manning's /// [5]controlling-referenceable-Dewey-decimal-book-chapter-article-state-finding-(s) / [6]differencable-debate-data], interceding sleep, comparative-conjure, shelf-life-variable-value-(s), archiving, contemporary-ancient-progression-processions, parallel intersecting phase-(s) of year-season-month-impasse-aging, remained (un)-original,| 423 |age-(s) per Biological-Historical-Reference,| 424 |present secondths-calculus-configuration-article-impasse-function-usage-variablizing;[214] trigonomic-parcel-interior-spacing-(s), akin elapse-tangent-dating into time-juxta-position, averse [elapse-secondths-inclination-time /// minutes-impasse-syntheses /// hourly-dating-interval-competing] compiling [cent-(s)-(milli) meter-(s) // secondths] foreshadow dollar-cornmercial-cutting, ergonomic formulated interaction-(s);[215] conscious-discourse, by

[Article:] [May]

[1population / 2letter / 3second / 4written-tangent / 5form / 6count / 7word / 8sentence / 9term / 10paragraph / 11lesson / 12Question & Response / 13ancellary-response / 14debate / 15minute-(s)-numerative-operating of integer-tense-(s)] summer-day-temporal, denominative-future-time-deviation-(s)-Fluxuation-articulating, celestially-abstracted, age-drift-mode-offset-characteristic-(s), meriting Mendel's-Law-genetic-complex, enigmatically, commercial-unaligned-discourse, awhile present-schedule-perfection, [itinerary / agenda / schedule confirm-efforts], perspective (as an) character-act, upon physics [alluding / deriving] [to / from] incalculable (e)-numerative-population-notions-quotients, | 425 |peripheral spacial-requirement-reasons, stood from, prerogative of [leased / loaned Federal-Debt-(s)] reconciled reconfirmation, at operation-(s) per-of mass-basic-civilian-survival awhile-of existence-celestial-centripetal-force-tectonic-aquatic-vapor-evaporation-condensation-blood-pulse-compositional-perceptive-impasse-visual-retinal-corona-half-oval-solar-bend-imagery... ¶|||09m.:16s.24m.s.|||§

...Preposition-Count-[33 / Thirty-Three]-of.. ..Article-Essay-[3/ Three].. // ..Book-Essay-[113 / One-Hundred-Thirteen].. ..Lines-[99.46 / Ninety-Nine.1.four-tenths-2.six-hundredths].. ..Balancing suggests that human-population, is suppose to function with objective-elemental-resourced-refined-items, for functioning under entity, or land, for tilling those grounds, where conceptive-patents, help space those causes for commercial-trading, when individual-family-interpretative-dynamic, is how common-beings believe thought, to convey where extraspective-interaction-(s), determine from juxtaposition-offset, [whom / what] is suppose to proportion mode of territory, transiting [residential / commercial / governmental / militia means].¶|||30s.60m.s.|||§

¶114Statistic-(s) requisite prior, registered percentile-denominative-article-deviation-(s), corrugated sat-dating-placings, blood-pulse-tense-tangible-character-occupant-(s)-referred, under-at an height-tension-limit, revolving-distance,| 426 |developmental-constructive formed-deviation-(s), from [residence / commercial / religious-mode-(s)] secular-free-will-impasse-position-(s), evading time-posit-action-historical-precense-statistical-article-actions'-motions'-notionths-offset-(s), for tranquil-living, precipitate-(Thesis);216 awhile off-time-interior-requisite-ideal-position, held-stretch-conducting, article-objective-subjective-time-intermittent-affair-(s),|427|where human-(s) elaborate-tentative-quality-(ies), [passing-moment-secondths-century-millisecondths-millennium-milliliter // meter-(s)-drift-off] continuum taxonomic-sedentary-cycle-(ing-(s')), passing live-existence, aluft velocity-intrigue, microcosm perpendicular-oblong-impasse-location-pertain-article-(s)-usage, documented by an empirical-fourths-denominative-fashion [Artistic-Conceptive-Two-Dimensional-Envisioning-(rate-time-dating-deviation-histor-ical-marks // Cursive-Article-Paragraph-Deductive-Interpretation-Rate-(s) /// Manuscript-Biblical-Article-Trusting (belief in faint-illusive-transparent-invisible-vapor-evocerating-hoc-read-talk //// Typed-Quotient-Mathematical-Symbolism-Majority-Present-Spanning ///// Illustra-tion-exampling after dating-offset, physics-formal-483/4-year-boundary-present-limit-facts, apose presence-2018-selling-year ////// Tier-Diagram-physical-component-contra-juxtaposition-parcel-imagery].210¶|||01m.:15s.31m.s.|||§ ><><<><>><><<><>What operation-(s) have produced enumerated-enterprise, by Independent-Genetic-Article-Offset-(s), sectionalized district-(s) per-of humanity-belief-trade-complex? What interval of breath-cardiovascular-deviation, upon secondths per [week / month // season] pertain of hour-offset-conjunction-production, prevalent discard of Mathematical-Prevalence & Syntax, prior-arrival, pre-requisite-ordering, too far grabbed of (the) thing-(s), elapse-living, without littoral-prevalent-pertinence, from prerogative-preference-partake-tier-articlizing, [perspective > memorial-millennium] time-date-calendar-physiological-birth-death-Gregorian-Year-Median-interval-blood-black-iretina-eye-color-plasma-vibration-temperature-depth-degree-scale-gauge-comprehending,| 428 |variable-factoring, when date-present-(s), hour interposed-moment-(s), directive past-secondths, suggested present-future-independent-introverted-effort-(s), complied tomorrow-anticipated, orchestrated-article-arrangements', intervoluntary calendar-deductive-accounting-process-auctioning, fatigued when dating-information, impasse deviation-factoring,| 429 |reflective-interior-subconscious-anterior-alterior-reference-pace-day-elapse-syntax-synthesizing, Marxist-five-points-Sixtieth-second-(ths)-hour-five-minutes'-articles-tier-ordering, eighty-six-thousand-five-hundred-tic-(s) per hour, of season-2048 (72)-hours per reset-frequency-(ies)-maxim, juxtaposed present-century-historical-reasoning, without an future-conceptive month-date-temporal reset-article-hour-verbal-physical-action-method-rate-proportional, day-hour-decade-reasoning, primarily in county-sedentary-universal-weekly-belief, ending-week-recharging, exhaust inset, [objective // subjective] referencing, where product-utilization, Rorschach-Objective-Tangible-Factors', under [duty // perjury] of action, clearly defining calculation-(s) of-present [secondths-major-inclination // indication // action // motion // notion-exemplifying // affirming // et cetera...] minutes-compost-interpretation-elaboration-impasse-conjuring, breath-(s)-major-juxtaposing, climate-terrain-topo-graphy-effort-{Cartographic}, from [1dig // 2excavate // 3collect // 4fracture // 5tunnel // 6channel // 7bay // 8plain // 9forest // 10encamp // 11et cetera...] Eastern-Porkaholics, saturated-addicted present-objective-requisite-demand, believing in shelf-life-demand-processing, word-major-deviation-(s), requiring 41L.[written-{Cursive (Calligraphy)} // (Manuscript)-article-(s) typed parallel data // invoice-message-aspect] not delving aging-inquiry, expository motion-directive-discipline, expert-filed, by Academic-Constructive-Conduct & Residence, per comparative-(less) empirical-construction-denote-usage-factoring, when time-date-aging involve retirement-anticipation, confluctuating common-effort-(s), during modern-state-nationalism-status-articulation of pro-ceeded-impasse...¶|||03m.:37s.13m.s.|||§

§...Preposition-Count-[9 / Nine]-of.. ...Article-Essay-[4/ Four].. // ..Book-Essay-[114 / One-Hundred-Fourteen].. ..Lines-[44.965 / Fourty-Four.1.nine-tenths-2.six-hundredths-3.five-thousandths].. ..Continue to learn, is the premise of this book..¶|||05s.46m.s.|||§

¶115.What subjuncts' surrounding Genetic-Context, use auxiliary-objective-accentuating, when during Louisiana-Purchase-Territory-(ies), State Settled-Constant-Work-Residence-Shifts', per Daily-Voluntary-Act-(s), Developmental-Map-Article-Academic-Conduct-Debate-Effort-(s)-Convey-Tier-Display, [lecture-(s) < sermon-(s)] navigating Pre-Age-Interval-(s), with Limited-motion-present, causing Voluntary-Pursuit, during Fiat-objective-regard, as Woman-interaction, jewel-(ed) introverted-interactions', cause vocal-uncreative-uncertainty, form introverted-self-(ves), brought-together from childhood-obedient-(obesity), consumerizing to get know-how, fascism plague-method at recidivism-reasoning, that be through subterior-tier, placed-sensations', timed-tense-(s), from organ-posit-(s), as bone, stress cardiovascular-tendon-tension-stress;217 where juxtaposition per characteristic-(s), formulate whom [pertain-at // transit-through // work-upon // sleep-at-of-an] place-posit-interval-(s), simultaneously-offset, extemporaneous-fact-(or-(ing-(s))),| 430 |conceived through realty-economic-factor-conceptive-cooperative-processing, but preparation, manipulate order, while [sequential-article-rate-(s) / measure-(s) / petroleum-hertz-frequency-(ies) / watt-current-frequency-(ies) / cubic-meter] hierarchy interval-deviation-(s), where human-acts, define way-(s) by auditorium-focus, impassing inductive thinking, during [day / night, adjecture] politically-introverted,| 431 |developed-historical-municipal-legislational-halls', where common-commercial-governmental-spacing-(s), juxtapose [enumer-ated-product-object-trading // Federal-Agenda-Denominating, those metropolitan-Federal-Reserve-Offset-(s)] where officers' monetary-activity,| 432 |affect civil-state-federal-county-city-development reliant [Philadelphia // Denver Currency-Denominative-Production-Offset] from documentary-denominative-Subjunct-Analysis,| 433 |calling [constitutional-consonant-verbal-ordering-(s) // Dewey-Decimal-Observation-(s) /// Exterior-Compartmentalized-Value-(s)-Classification-(s)-Age-(ing)-Denominative-Juxtapositional-Article-Ordering], [two-dimensional-blueprint-(s) // three-dimensional-tangible-offset-impasse-variables'];[218] confirming composition-matters, resourcing acquisition-denominating, expanse article matter-(s), when live-perceptive-focus pertained comprehension at dimension-(s), ordering [Interior-Spacing // Objective-(red)-brick-dimensional-spacing-(s)-(wall)-offset-(s)],| 434 |where BBC-Macmillan-Sheep, fuse formed prior-blue-print-mechanical-facility-contraption-dues', needing order at stable-stationary-obligat-ion, conjured metropolitan-offset-(s), mattering-collection, by per data-article-bit-(s), survey-affirming, characteristic-(s) interval society interposed-motion-(s), Defending-position, by brawn-sworn-daily-understanding, how yearly-temporal-observation-(s) not definite-set, as objective-parameter-filing, as method per data-documenting-filing, formal-sequential-ordering-(s), confirm-ing-information, as interposed pre-requisite-prerogative-(s), defining entity-lavish-development-article-motion-impasse-schedule-progression-perspective, considering, interior-comfortable-factor-(s)-posit-mid-term-placing-documenting-present-process-impasse-articlizing, [forward / additional depth-article-usage-piece-(s)] as matters, explicitly define elapse-(s)-cycle-document-ing-(s), purpose-(s), verifying [daily / weekly] repetitive-maxim-individual-impasse-motion-(s-(')) juxtaposed entity-dynamic-dir-ection-order-finance-intention, personalized, reclusive-physical-dust-effort, hyposensitized, confirming how beams-composite-weight-distancings', yet have been ordered, for per se, use throughout off-time-constructive-interactive-cooperation-activities;[219] adjusting difficulty per individuals', suffice common-man-denying-sleep, or physiological-sweat-effort-(s), affecting through communal-outcome-order, where suggestion interpose daily-impassing-(s), (introverted-independent-sense) determined formal-impartial-observation-meticulating, natural-matters', impassed as infetesimal-recognition, throughout-microcosm-daily-impasse,| 435 |three-dimensional-surface-area [interior // exterior] parameter-(s)-deteriorating daily-use, forward-depth-auxiliary-west-rotational-origin-reactive-reflexes', validifying progress, statistically-perceptive, grasping-conjure item, using, article-matter-(s), where timing implement streamlining, empirical-development, Tier of militarial-engineering, not impact-national-commercial-governmental-construction, using effort-(s), prerogative-focus-action-(s), documenting aged guideline-(s), aspiring physical-space, climate-temperature-topographic-condition;220 instrue-(s'), purpose as insighting an [3 / 6] hour-period-daily-act-(s)-offset, laude-motion-transition-timed, ((un) pre-bound-weapon-age-usage)| 436 |simultaneous pick-up [percentile-category-date-manning // night-shipping // et cetera (per se)]| 437 | commercial-effect-discourse, quasi-Deutschland-Holy-Economy-Gregorian-Continuum, dollar in Thousandths-Hour, for working on Day-to-Lifetime-Expectancy-Hour-Minute-Impasse-Gamma, subterior-secondths-past-historical-requisite in-of-an Associates'-Effort-Temporal-Seasonal-Primary-Objective-Modifications' upon territorial-land-space-measuring-condition-(s'), not computated, comprehending how entity affect perspective of terrain, by vehicle-city-district-local, apose [stadium / arena purpose] while arcade governmental-constructive-frugality, rather simply-suffice, Retirement-Free-Will-Land-Savings'-conversion,58L. concise-repopulation-purpose-exist ence... ¶|||04m.:03s.23m.s..|||§

§...Preposition-Count-[4 / Four]-of.. ..Article-Essay-[5 / Five].. ..Book-Essay-[115 / One-Hundred-Fifteen].. ..Lines-[58.12 / Fifty-Eight.1.one-tenths-2.two-hundredths].. ..How else does question go into tier, by order of differentiated word-syntax-conjunctions-tier-deviating. ¶|||08s.12m.s.|||§

¶116{E Pluribus Unum}, is our Senate-Forum-Infrastructure.211¶|||02s.:88m.s.|||§ {In God We Trust}, [House-Representative-(s)-Forum-Infrastructure // Senate-Legislational-Written-Manu-script-Carptunnel-Caligraphy-Cursive, {Capitol Hill}] As (the) General-Infrastructure, holding prior under Grand-Legislational-(Carptunnel)-Observation-(s)-congressional-updating, State-Population-Census-Count-(s), that fundamentally implement, but divert-subcommittee juxta-position resource-concept-product-objective-substantial-balancing-means', where distance-(s) through basic-human-right-(s) capable-effort-(s), examine where University-Dynamic, confirm extents, per Government-Course-work-information, observationally-examined, as place remain important per data-ordering-(s), cycling daily-dormant-capacity, while living-blood-pulse-airpocket-pulse-impassing-(s), reflexive-motions', ever validating data, awhile House-Repre-sentative(s')-ethnic-title-error, elucidate collection ethnic-culture-federal-document-(ing-(s)), Federally-relied, yet unexpected (by NEW!!-Complex) stabilized from mass-enabling, giving up from youth-graduate-inquiry, how to collect data, from impasse-information, inquiry-(ed) during off-time-interaction-(s), self-religious as action I believe sustem [physical-year-aspect // subjective-season-Major-Six-Hour-Focus] (ill)-literate-comprehension, suggestion, for other inhabitant-(s), attain [(un)-vernacular-long-tone-extensions' // short-voiced-transaction-talk] juxtaposing an symposium of individualism, blind when Millionths-Mass-Effort-Impasse, individualize-swear, and the rich not direct or dictate, while-education-aptitude, not conform by term-per-perceptive-particulated-objective-count, from juxtaposing storage, awhile cause affirm national-boundary-discourse, confidently economical,| 438 |because impassed-secular proper-formal-interacting, giving [donation-(s) // fund-(s)] per [entity / organization] developmental discourse-(ing), creative process-due-(s), either pre-inclined present-purpose-reasoning, or form per receipt-allocation-documenting, constraint-(s) of trade, permitting government-trade-denominative-motion-effort-(s), primarily-gauged when mass melancholia, for love of eating, interval radio-transmitted-discussion, yet find a-the median for conveying, [mental-brainwashing // literary-staged-competency] intervalizing time-posit-(s), difficulty through dimension, to measure-recollect-avidly-view-dimensions, simultaneous-mathematics, {calculus}, pertaining {Statistical} observation-fraction-discourse, balancing numerated-integer-(s), sequencing whole-column-offset-(s), arbor-down, continuum-tectonic-posit-(s)-offset.$_{212}$¶|||01m.:54s.$_{04m.s.|||}$§ Compre-hension of discourse, distance where successive-article-competency, synthesis through an reading thousandths-words-ten-minute-sheet-rate, denominative of [2,048 / (72)-season-hours-impasse-juxapositions'] liturgically-worked, for transferring constitution-prose, [abridging // amending // practicing] method of common-written-articling-cursive-enumerated-denominated-reasoning;$_{221}$ stemmed coronary-synapse-conscious-blood-pulse-reality-tangible-audible-discourse,| 439 | metronome-counting, affirming per secondths-notion-inclination-exponent-aspect, proportion other calendar-dating-scales apose time-date-hour-tier-location-objective-posit-gauging as (such my Room at my Father's-House, Contain Four-Bicycle-(s)-(two-crusiers // one-mountain-bicycles / one-700c-bicycle ///// an closet of Clothing & Knick-Knacks ///// two laundry-basket-(s) ///// an bed (form // cushion-mattress-(es)) ///// an short-table (for my laptop) ///// an refrigerator ///// an lamp ///// an three-cabinet-drawers // wooden-book-ottoman-three-shelf-(ves) = Eleven-Feet-N.W.—S.E. /// thirteen-feet-N.E—S.W /// Height-Eight-Feet-Coordinate-28°N // 80.975°W = cubic-sum-1,144-Feet-Room-impasse-space)] pending how one would choose to recollect cubic-parcel-matter-injunction impasse, juxtaposition-(s) of person-(s)-(Federal-Hierarchical-Position-(s) // Being-(s)-(Commercial-Citizen), due default, diploma-antithesizing percentile-aging-twentieth-monetary-junction, ruling discourse of human-thought, parallel-rotation, gamma-data, that mathematically-reset, yet constitutionally-amended, per limit-(s) housing [receiving // acquiring] from historical-planning, tiering-constitutional-article-redactive-prose-topic-point-delve-amend-ing, mass-requisite-count-(ing), by formal-requisite-confidence, continuum data-order-(ing-(s)), proglamated-shelf-life, intermittent, [numeral // alpha // Roman // Greek // PoNNRR-(rōsy) // Germanic primary-tier-article-factor-(ing-(s))] daily-conjecture, conjuring motive, as elemental-technical-creative-quality-tier-order-impassing, visual-tonal-vocalizing, for verbal-motions-extemporizing, scaling distance purpose, from marriage-self-posit-repopulation-focus, granted bias-reclusive-extraspective-limit-capacity, inclined boundary-condition-spacing-(s), defining cause while each focus per inhabitant-blood-pulse-rate, timbre at product-objective-(s)-glad-mode-emotion, accomplished repopulating standards, at two-child-capacity-curb, claused by men whom formulate an Military-Marine-Five-Nine-Agenda, three-bit, when formally-practiced, not current-pre-eminent-discourse, basing sub-numerical-factor(ing), for enumerative-concurrent-responsibilities, past-idle-present-acting, when surface-area-spaces, are missing from analytical-dynamic, practicing, finding those median-central-numerical-subjective-subterior-tier-glossary-terms-sums, operating upon method of calculation-directive, [ventricle-blood-pulse-tension // cardiovascular-stress-tension-analyzed], for purpose by process-(es), due to consent of managing, impassing employee-discouraging, (which is fine), I only intend to proceed by merit of book-work-net-annuity-learning, measuring where massed-exponent-(ed)-conjugation, apply processes of governmental-restrains, book-shelf-data-ordering-(s), entity objective-property-routine-schedule-usage, applying mass-output-redeveloping, relevant visual-tonal-perceptive-raison d' e-tat, verifying constitutional-merit, (from an) literary-pre-cause-basis, impartial schedule-agenda-surveying, mass-motions-impasse-function-(s), riveted rated-cursive-mode-measure-redacting, synthesizing syntax-subterior-Article-Essay-deviation-(s), syntaxing-theoretical-continuum-schedule, yearly-resetting, upon alive-continuum-relaying, overlooked space-specification, particulating-impasse, through thought on data, considering daily-invigoration, finding oneself natural-timbre, amongst surrounding-elemental-topographic-climate-condition-impasse-perceptive-alive-

documenting, from how mass-physiological-preference, would phys-
iologically-impasse upon all of existence, when [away // at] industri-
al-production, when book-(s), should be bibliographized, from
book-shelf-genre-referencing, inform Literate-Competent & Ridicule
of inhabitants, as moral-juxtaposition, serving competent study,
awhile "Citizen" is subjunct per house-developing-order, default-dai-
ly-work-existence, which instill motive for focusing visual-tonal-per-
spective, alive-surrounding, blood-pulse-being-inhabitant-(s'), [in /
out] practice of daily-trade, [finding / supplementing] wage of
check-earning, from government-taxes-help, structuring 401k-retire-
ment-confidence, trusting those years of subjectiveless-working, as
aging-youth-differing, Death-Age-Expectancy, found by the duty to
nation, executive military-purpose, motioning pass, much of a the
unwanted-resourcing-elements-repopulating-opposite-sex-reproduc-
ing-substantials [[1]objects // [2]tools // [3]auxiliary-device // [4]toiletries //
[5]plants // [6]bricks // [7]windows // [8]interior-furnishing // [9]paint // [10]pole-(s)
// [11]sign-(s) // [12]et cetera...] for National-Theatre-Requisite, because
each passing-increment-time-minutes-twelfth-synapse-momen-
tum-influctuated-synthesis-cognition, complacent, not formally documenting, dislike of language-art-(s)-redacting, juxtaposed Social-
Study-(ies)-(youth-voiced-slick-studies-extraspective-enabled-cheat-
ing, providing from retentive-perceptive-data, uncreative article-ref-
erence-data-present-extemporizing-(formal-thinking), then furthering expectancy-(ies), from international-trade-market, extraspec-
tive-reclusive-reducing, legal-excerpt-extent-(s), rating under-of light
of day, pre-emanative-30,000-interactive-population-partici-
pant-maxim-lifetime-expectancy-interactions', reading-rated-re-
dact-reset-(ting-(s)), where written-emphasis, up to now, [understand
/ synthesis];[222] by (from) religious-forum, to dynamic-conscious-in-
teraction-(s), when vocal-vibration-secondths-beats-rate-propor-
tion-tone-inclinate, [Key // Time] signature-timbre, pres-
ent-day-(s)-volatility, through daily-communion, servicing sur-
face-area-spaced-locations, building upon interaction-(s), from [ac-

quaintance // co-worker // religious-members] minisculizing friend-(s), apposing how linear-parallel-distance-patent-objective-act-(s), pertain plane-two-dimensional-surface-area-measure, upon [objective-pricing-dating-weight-function // purpose-impasse-(s)], scaling interior-measure, where [cubic-feet // surface-area-(s)] require [voiced-audible // written /// typed /// printed-context-conjecture] [maximum // minimum] vapor-pressure-tensions-simultaneous-synapse-(s), unseen from notion-quantity-limit-(s), visible-eye-physical-motions, limited from impasse, upon boundary-transit-exhaustion-effort-(s), extenuating force-(s), defining elemental-comprehension, commonly-succeeded, undelved past-requisite-present-legislative-congressional-hall-(s)-{Capital Hill /// House // Senate} representing exponent [compositional-documenting / article-documenting / dimensional-documenting / mass-survey-signature-documenting / file-cabinet // laptop-server-documenting-(s) / entity-per objective-discourse-documenting / objective-time-table-documenting / human-act-wage-effort-documenting] mass-unconscious-orchestration, survey-collecting-analyzing, agenda-objective-(s), per practice by-notion-of rule, pre-book-affirming-article-data, discerning notion-(s) of article-(s), under Empirical-Subjective-Subterior-Elapse-Period-(s),|440|remain-prevalent-{Dry}-{Wet}-Substantial-(s)-offset-posit-(s), complicit land-territorial-ground-(s)-solid-deposit-developmental-ordered-sequences, tectonic-split desert-molten-composite-denominative-kinetic-frequency-matter-(s), reduced caress-cradling-offspring, emotive-dating-hour-denominative-aspect, apose, when similar perspective-only, impasse incendiary comprehended-function-(s), per [time-minute-secondths-posit-action // motion-subjunction] civilized, auroramatic elemental-daily-interval-hour-minute-(s)-secondths-recoursing, calendarized-year-denominative-time-posit-(s), juxta-posed long-term-aging-elapse-(s) [behavior // conduct] deducing purpose, from verified-action-article-data, present periodicalized requisite compulsive-natural-behavior, formally-impassing, aging-parameter-handle,| 441 |product-objective-(s)-articles-usage, request impasse-temporal-ordering;$_{223}$ surrounding eminent purpose, logging-distance-dialogue-discourse-(s), adjunct formulating terrestrial-interaction-(s), partaking impasse place, when-where daily-passing state-boundary-(ies), remain out of common-control-competency, from yearly University-High-School-Bachelor-(s')-Limit-(s), interceding data-article-extent-timed-applicating, per sworn-managed-institution, diploma-commercial-confidence, theatrical-prose, at [commercial-product storage // transit // store-shelf-life] pertaining trading-(s) at [common-shelf-open-hours-objective-cooridors // subjective-parameters // substantial-articles] fundraising by religious interaction, instructed, tier-deductive-isolation-(s)-operations, per elemental-physics-data-pertain-article-bit, then upon time-incremental-deviation-(s), where [Major // Minor // Subterior articles-posit-(s)] scoping data-usage, influctuated where setting, per article-place-(ing-(s)), shelf-demand, comprehension, while deductive-fashion, bid per se, this book-(s), as always lead through, an active-documenting, where group-(s)-familiar-activity-(ies), commonly-increment-article-(s)-data, deductive subjective-posit-(s), defining data-word-sentence-comma-paragraph-syntax-function-(ing-(s)), time-impasse-affirming, interlude-subject, better relied on physiological-action-(s), pre-affirmation, conceptive (lease)-discourse, existing, for furthering oneself, by monetary-constraints, calcu-latable calendar-aging, compulsive natural, but good-rule, awhile outcome per-of experiment-(s),$_{126L.|}$ 442 |or discourse, cause mundane-(un)-comprehension-individualism, throughout-mass-disposition, attempting daily incremental-deductive-subjective-practice, conceded where interior-articles-bias, build upon abridged-development, with mysticism of mammal-human-organism-effort-(s)...¶|||09m.:15s.$_{35m.s.|||}$§

§...Preposition-Count-[25 / Twenty-Five]-of.. ..Article-Essay-[6 / Six].. // ..Book-Essay-[116 / One-Hundred-Sixteen].. ..Lines-[128.63 / One-Hundred.$_{1.six-}$tenths-$_{2.three-}$hundredths].. ..continue following each essay break, awhile their be no interpretation. *Predicate fact dynam ic¶||||11s.41m.s.|||§

¶117Conceded by Grandiose, human-perspective, limited where gauged-act-(ion-(s)), will pertain one individual amongst store, from when data not be fully agreeable, departing literary-loop-hole, from contention-(less)-effort-(s), inertia-combustion, centripetal-latitude-force, informal-conjecture, as believed by gravity-effect, defining mathematical-scaling, measure-(d), distanced;224 wherever grouping-(s) of surrounding-space-parameter-(s), cause daily-impassing-(s'), competent human-being-blood-pulsed-individuals', whom can only be factor-(ed), from when date-supersede-(d)-comprehension, annotated volume-maxim-handing-budget-process, comer-cially-intricated, where those factor-(s) per spacing, define cultivated-land-means', per mass-term-requisite, rendered incapable placed-order, when myself, be where mass not conjugate, house-survey-orchestrating, individual-confidence, where restriction-(s) of [age / genetic / individual-prerogative / limit-(s), per article-religious-sect-distance-acts'] prerogative pre-requisites', remain persuaded amongst lifestyles of volatility, per surrounding-parameter-(s), as when surface-area, confluence-linear-articles-matter-(s), because of-an natural-effect, where momentum adjunct human-perspective, an quality for bias-hoc-output, under male-command of militia, where objective-purpose, require physiological-active-objective-article-usage, from when massive-effort-(s), conclude surrounding-article-compost-(s), as dimensional-objective-matter-(s), because [1cut-(s) / 2mold-place-(ing-(s)) / 3bend-(s) / 4weld-(s) / 5twisting / 6setting;225 metal / cement / lumber / rubber / cotton // et cetera...] modify elemental-offset-position-(s), fixated where life, inaudible individual-production-piece-(ing-(s)), where elapse pertain, [Private-residential-timing-(s) // public-com-mercial-timing-(s)] yet comparable, common-denominator, article-purpose-(s), from governmental-cabinet-agenda-scheduling, having impartial-ordering-(s), pertain human-physiological-effort-(s), remaining white-collar, comprehending syntax-data, balance [action-(s) // motion-(s)] spacing perimeter-territory, by paid lease-dynamic, retributing land-space-objective-property, intermittent cursive-documenting, developing an consensus, per survey as when acquaintance-(s), pertain daily-interposed [effort-(s) // effect-(s)'] not appreciated per perspective, unless physical-effort-(s), understand how individualism-dynamic, constraint militarial-historical-five-nine-line-ordering per-of men, because engineering tectonic-plate, scale working-nature of men, whom do not work for free, apose labor-wage-inflation, as why economic prestigious-discourse, formulate without perspective, as motive of trade-(ing), elucidated purpose, inputted upon terrain, while national-federal-state-governmental-constitution, having [county // city enumerative-value-(s)], accounting citizen-(s), where local-state of constitutional-examination, remained unconscious, of reclaimed-present-tense-wage-confidence, placed where discourse of daily-trading, input an Federal-examination, not defining city-governmental-constitutional-(other document-(s))-documenting, for examination of formal [inferencing / inducting], demonstrating hourly-time, by-an calendar-table-(ing), (in)-succinct mass-deductive-blood-pulse-deviation-(s'), comparative merit-confirmation-(s), where liberal-art-(s), default-conduct, intermittent affair-(s), not intensively-intricating [objective // subjective-data] by allure of merit, defining increment-data-contrasting, as simultaneous-mass-act-thought-intrication-deviation-(s), pertained daily-impasse, juxtaposed tectonic-calendar-repopulation-dynamic, as generational-era-(s), do not repopulate by per spouse (friend-interaction), yearly affirming trimester-reproductive-repopulate-discourse, where conceptive-planning, enabling tangible-documenting-succession-ordering-(s), | 443 |considering contractor-exponent-docu-menting-discourse, expanding objective-possessive-property, relative denominative-documenting-ethic-case, from when active-lesson-comprehending-interpreting, instruct ruling thought, where realm of time, proceed expanse-elapse-(ing-(s)), where desired-conveyed-obligated-observation-(s) isolate-individual-(s)-contemplated-self-(ves)-political-matters', where formal-interposed-convey of data, allude when mass-survey juxtaposition mass-requisite-purpose-(s), for what should account for adult-extraspective-dynamic, mathematically-grammatically-inclined, social-requisite, not interposed, redact-article-comparison-data-referencing, superlative subjective-intuition, [when / where / why] tangible-conceptive-compositional-interpretation-(s), are bragged at subjective-period-(s)-consideration-(s), juxtaposed residential-period-(s), articulating daily-impassing-(s), as community-limit, determine how life-span-maxim-{Thirty-Six-Thousand-Five-Hundred-Twenty-Five-Day-(s)} relate one-hundred-year-aging-presumption, census defining, statistical concept-planning, not corrugated commercial-forum, or national-reform-article-referendum, [ordering // following] an serie-(s) of pre-requisite [restriction-(s) / territory-(ies) / residence-(s)], procuring genetic-origin by-how, [metallurgical / brick-construction-(ing)], influence purpose of, surrounding-constructive-data, validifying verified-tangible-hard-data, hand-written-dating-documenting, order-(s), focusing denominative-calendar-passing-(s), juxtaposed, [time-day-secondths-reset-(s) // time-subjective-tier-deductive-elapse-period-increment-(ing-(s))] when moments can be pre-comprehended, from articulation-cause, reasoning spacing-(s) per [infrastructure / hand-written-denominative-documenting / objective-form-composition-(s) / state-national-boundary-dynamic] not discovering, surrounding-boundary-affair-(s), from pre-requisite-(s) of our founding-father-(s), whom are relied on depicted, live-article-usage-means', biologically, grave-anatomical-documenting, by pre-subjective-construing [law-(legal-grounds) / medical-field] as merit-based progressive-motion-(s), by splendor per subjective [Deducting // Incrementing]61L. {Enumerated-Denomin-ating}...¶|||04m.:40s.96m.s.|||§

§...Preposition-Count-[18 / Eighteen]-of.. ..Article-Essay-[7 / Seven].. /444/ ..Book-Essay-[117 / One-Hundred-Seventeen].. ..Lines-[61.275 / Sixty-Seven.1.two-tenths-2.seven-hundredths-3.five-thousandths].. ..Hundredths of Essay-(s), make difficult redact work, per tenths, of individual-(s)-interpretation-moding, omitting physics, for non-interest of circulated-documented-text, post-intermittent-educational aging; without an voluntary-retirement-planning, under substructural-governmental-architecture, amid transit-marine-barrack-voluntary-complex-(es), intermittent, blind-mentality-modes.¶|||21s.72m.s.|||§

¶118What reservoir of water, generalize [cubic-interior-surface-area-two // three-dimensional-feet] collecting an proportional-private-accumulation-(s), as foresight from self, attain volume-weight-(s), remaining intermittent populus, affirming mass-population-dynamic-exertion-practice, reset-(ed) where offset-distancing-(s), injunction diameter-circumference-two-dimensional-reasoning-(s), fashioned conduct, when-at an [fact-(ing) // passing of blood-pulse-impasse] upon pre-meter-(s)-conjure-application, verifying spacial-distance-(s), for what can pertain article-time-supplementary-focus, amongst year-lesson-aging, not exactly-socialable, but incorporable, interactive-commercial-focus, legible by means', understanding engineer-(ing), (as an) extraspective-work-order-(ing-(s))-aspect, where pricing-outlet-(s), verify raw-resource-(s)-effort-(s'),| 445 |interlude access-control, where human-interlunation, circumference cubic-metric-mile-(s)-offset, for core-article-distance-(ing)-primary-offset-impasse-(s), not celestially-documented,| 446 |awhile semiologist ration secondths per cartographic-centimeter-inch-(es)-degree-(s)-scaling,| 447 |those mundane attentions', weighted repopulation-reproduction-(s), maintaining-placings,| 448 |free from, naïve-humanism-duty, upon where trade-accommodation-(s), riddle refrigerated-voluntary-will, motivating forecasted individual-posit-matter-(s), when acquirable contractive-request-(ing(s')), prepare self-passing-posit-(s'), agrarian massive-capacity-perspective-action-usage, defining limit-(s) per effect, upon physiological-visual-tonal-perspective, intermittent-passing-intercontort, where human-present-historical-elapse-(s), illicitly reading, from-an serialized-picturizing, comprehensive-article-lesson-series-sequence-(s), confirming labor-act-rate-(s), (like N.F.L. Picture-Playback) of data, in multitudinous-medium-(s), when theatrical-production-(s), affirm article-(s)-curb-(s), per posit-(s)-count-(s);[226] remaining of will-directions, clarifying trade, by perimeter-momentary-control-distance-(ing-(s)), cubically-proportion-(ed), rendering reading under entity, after contextual-inclination-(s), efforting articles-conjure-presence-purpose, distancing limit-(s), from synapse-(s) of mirrored-reflected-matter-(s), affirming perimeter-act-(s)-agglomeration-(s), pertaining cubic-surface-area-space-(s), from legion-(s) of article-(s)-data, influencing fastidious, categorical-discourse, excavating minute-corridors of elemental-retrieval, disproportion Nation-Federal-denominating, apose county-metropolitan-civilization-(s), where inclined educational-prose, serve memory-lineage-passing-(s), inherent product-inclinated-application, not effectively offspring-(ing), [subjective-major // minor-dictating] by lineage-minor juxtaposition offset-(s), from antithesis of offspring-raising, along all stage-(s) of perceptive-existence, from extraspective-timing-(s), where state, require individual-perceptive-conceptive-focus, pertaining object-(ive)-(s) by production per commercialism-city-community-(ies), defining capacity-(ies), that of, comprehensive-schedule-ordering-(s), refashioning agenda, by survey-literature-etiquette, yet interactively-confirmable, awhile classism of entity, chastise layer-(s) of secular-reclusive-level-(s), from surrounding approximate-sight-juxtaposition-(s), dimensions-walked-impassing-(s) an routine of infetesimal-physiological-being, (as an) part of extraspective-interactive-existence.[213] ¶|||02m.:36s.[15m.s.|||]§ Stance upon area be physical, such is not cardiovascular-intensive, due to observational-tangible-cycle-prose, candidly display-(ing) truth-(s), remorse blank-sensation-(s);[227] affirmed characterization-(less)-realm-(s) of deductive-reading, juxtaposed denominative-date-time-interval-(s)-inducting-count-(s), posit-denominating, date-potential-effort-(s), scaling mass, for weather Globalization-dynamic, is not fully conceivable, for derivative-discourse-method-(ing-(s)), centripetal-impasse, individual [genetalia

/ genetic-identification-(s) / Torso-Head-Arms-Legs-Feet-Hands-composite-matter] not-redacted, as respects', by personal elaborated-point-(s), interval-offset-(s), cement-constructive-lock-limit-(s), loading stock, for posterior-live-(s'), as religious-fervor for family-individual-(ism)-(s'), consumed by present-distance-(ing-(s))-county-impasse-(s) {Geological-Timing}, from mute-behavior-habit-conduct-effort-(s), pertaining ideal-(s), unverified documented-declarative-defining-(s), while living-blood-pulse, exist where state-(ly)-pertained, climate-mean-(s'), fixate permanent-fixture-offset, by adult, middle-school-distancing from county-impasse-perspective, not true for universal-county-work-schedule-cycle-(ing-(s)), marginal managerial-third-person-power-monitor-(ing)-(s'), solving motion-(s), by-tense-(s) per-of [sight / sound / smell / taste] as sustain-(s) from data-articlizing, substantial-market-recipe-tax-goods, claused because their being an nature of word-(s), corollary, design paramount remains, Fixture & Physics, [showing // displaying mass-article-(s)-mean-(s')] directing those way-(s') of thousandths-year-belief-cycle-inhabitants', revolutionary-passing, Sixteen-Hundred-Year-Estimate-Existence, that present where-upon day-study-(ies), cause an rule of calendar, candor impulse-existence, believing self-matters', by various offset-(s) of distanced-impasse-effort-(s),54L. affirming exterior-crossing-(s), at interaction of binded-contract-agreement-(s), suave of design, colloquially affirming language, in educational-social-disinterest, mass-subjective-nature...¶|||04m.:11s.99m.s.|||§

§...Preposition-Count-[22 / Twenty-Two]-of.. ..Article-Essay-[8 / Eight].. // ..Book-Essay-[118 / One-Hundred-Eighteen].. ..Lines-[55.6125 / Fifty-Five.1.six-tenths-2.one-hundredths-3.two-thousandths-4.five-tenths-thousandths].. ..Recidivism is negative from an standpoint per perspective-honesty.¶|||11s.81m.s.|||§

¶119Whom define Core-Common-Reflexive-Contemporary-Consummeristic-Capitalistic-Liberal-Conduct-Calculating-notion-(s), among presence referenced, from [auto-biographical-community //

developmental-motion-(s)] date-twenty-four-hour-centralizing, Day & Age, consulting concept-(s) of effervescent-spacing-(s), by extraspective-discourse, of human-quality-(ies), summarizing how conceptive-interaction, remain introvert-reclusive, mathematical-simultaneous-general-common-population-effort-(s).214¶||||22s.26m.s.|||§ [Populus / Existence / Commercialism / Advertisement / Broadcasting-dynamic-juxtaposition] signify language-clause-limit-usage, comparable by conjure-facet, offset-(ed) by mass-nature-agreement, not intracoronary per dat-(ing), forgetting qualified presence, when objective-form-confirmation-(s), brief an stand-point, superlative-interest, having an national-enabling, not list-drop-off, [listening // seeing interacting] for an logical-prose, defining elongated-word-confidence, through historical-day-composition-(s), abrupted periodical-artistic-intrigue, where sport has not fashioned the mind, for body-composition;228 as torso-components-posit, ulterior physique, sight-sound-perceptive, formed-blood-pulse-breath-posit-(s)-offset-(s),| 449 |as count-(s)-deviation-(s), similar as secondths, juxtaposing article-minute-(s)-syntax-synthesis, without tangible-entity-grandiose-volume-(less)-façade, will passing interior-constraint-(s'), examinable where exterior-ambient-environment, surround rural-topographic-discourse, only (as an) approximate-distance-(ing-(s)), leveled limit-(s), implemented by density-(ies)-tension-(s), winter variated incremental-impasse-observation-(s), conclusive scaled-cause, reasoning surface-area-range of decay, by-when dating, permit solid-impasse-(s), where reclined-driving-motion-(s), affirm dating, time-limited, prerogative impasse place-(ing-(s));229 disusing prevalent-focus, to respond by Movement & Cent, where matter-conception, instill resource-originating-tectonic-means', supplementary, inter-mediary-fact-(s), pertaining self-moment-(s), around collective-reference-surrounding, partaking in local-district-city-county-state-district-regional-library-observation-(s), because children pre-tension-(s), are inclined, adult-aforemention, present-requisite, affirming mass-conglomeration-(s), where [0sect-denominal / 1Latter-

Day-Saints / [2]Pentecostal / [3]Episcopal / [4]Mennonite / [5]Presbyterian / [6](Ana) [7]Baptist / [8]Evangelical / [9]Protestant / [10]Christianity / [11](Roman)-[12]Catholic / [13]Judaism / [14]Muslim / [15]Hindu /// [16+]et cetera…-religious-sect-tectonic-impasse-denominative-(s)] skilled by regional-body-(ies) of faith, Demographic-Biological-Division-Pattern-(ed), funda-mental physics lessoning, from-when home-effects, centralize-mite-tectonic-plate-(s)-posit-(s), impassing an threshold of presence, defining tectonic-fashion, where [state-nation / tectonic-nation-calendar-date] afflicted politic-(s), domestically-trading, faith-based-beliefs' of-an scramble of ideal-(s), not interposed-inferenced, insured nature, of listening-visual-vernacular-skill-(s), intangible, [vocal-range-static-current-tone-time // key-signature-count-(s)] intertwined per work-harmony, amplifying parcel-space-acquisition-usage, ambient-fashion, interval-lecture, debate, (continually-aloof) prose-presence-nature, intent on directing distance-(s), per child-offspring-anticipation-offset, where year-meandering, methodicalize document-(ing), insufficient setting mode, per-of letter-(s), scoping archetype-figure-(s), for tier-tabling, accurate-account-(s), when act-in-secondths-data-filing, interval minute-(s) of presence, accumulating median-living, by-an curb of century, tension-directed, millennium-aforemention-impasse-(s), where physics are inclined, perceptive Day & Age, as complexity-(ies) of grandeur, suffice financed trade-agenda-issue-(s), free from mass-freedom, salvaged by hemisphere-social-security, securing those tiding-(s) of census-reasoning, default-curb-article-affirming, population-offset-visual-tectonic-surface-area-posit-offset-(s), where grading our energy-level, determine [date / year] not conclusive decade-impasse-historical-fashion, intent depth-elapse-(ing-(s)), measuring weight-(s), requiring verification-(s), from supplemental-demand, upon individual-(s') (ism), unclear asserting operation-cooperation, as technique of vocational-interaction, when objective-distress, filter vestige per practice, because universal-tabling, international-fragment-conjure, post-associate-(s)-class-fundraising-work-method, sufficed historical-requisite bachelor-(s')-cause-(s), colleting interest of product-objective, serving default-analyzation-ecological-work-mode-process-(es), which suggest from-an historical-calendar-observation-(s)-impasse, not enough supplemental-data, for total-mass-position-population-work-effort-realign-(ing) (for which I think Labor-Pool is instilled for, as pertaining Nationalism-Aspiring), non-fiction-subjective-article-tier-interest-(s), in schedule proportioning, afloat [heat-condition // cool-condition / warm-transitional-condition] scheduled by supplementing an natural-impasse-interval-space, cool-density amongst heat-density-plasma-core-universe-offset, while warm-continuum, condition passing time-secondths-comparative-metronome-beat-(s)-measure, tempo-gauging reverberated-kinetic-solar-radiation-wave-vibration-(s)-frequency-(ies), synapse-elapsing, intangible-perspective-comprehension-dynamic, retrograde-redact-bookshelf-referring, intangible-quality-(ies), per matter-(s), grading-(s) [procession // progression-interval-impasse-project-annotation-point-(s)], by Architect & Contractor,| 450 |where city-planner, arranging-affair-(s), when community-development-(s), that have an macro-denominative-compartmentalization-(ing), per character-affair-(s), articlizing from posit-(s) of [tectonic-state /// population-repopulation-schedule-order-accounting-dynamic /// motion-(s)] fatigued by [[1]surface-area-physiological-weight-(s) // [2]air-measure-impasse /// [3]transit-practice-(s) /// [4]substantial-Agricultural // [5]Horticultural-deductive-detail-article-input-impasse-tiering /// [6]Housing-Plan-(s)-Proportioning /// [7]Defense-Mechanism /// [8]Education-Aging-Order /// [9]Currency-Value-Appraising-Method /// [10]Engineering Technique-(s) // [11]Objective-Article-(s) /// [12]Death-Count-Continuum-Decade-Revise-(s) /// An Clause of constant deductive-article-update-(s)] pursuing perceptive-explanation, defining each minuté-nuance, from blood-pulse-hyper-tense-(s)-perspective;[230] enigma, pertain mundane-offset, when society-mass-capability, is deterred by perspiration-theatrical-nature, sitting reclusive under elemental-condition-(s),instilling variance-deviation-(s)-impasse-mode-extension-(s), moment-ary-planning how present-historical-view,| 451 |unraveling present-historical-progression-(s), in plain view, by depth-perspective-east-west-grandiose-untangible-ambient, peripheral carto-graphic-physics-perspective, never covered by an self-individual-effort-(s), remembering religious-nomadic-shifting, (Miami-Helena-Outskirts-concerted-movement-(s)), as should partake where understanding vacation-(ing), dissolve ones' capable-effort-(s), upon community-formal-practice-reformation, where live-mass-(es)-offset, categorizing,| 452 |hemisphere "tangible-trade-quality-(ies)", given an substantial-voluntary-reproductive-impasse, as how international-limit-(s), tolerate territory-impasse, passing under solar-galaxy-passage, nationally-conquered (as I have ambitioned of), or vacation-data-collect-(s)-(ing), remembering how time of space, is pertinent for usage, by clause of time, abided by myself upon other-(s), evaluating notion-(s) of perceptive-conduct;[231] swear-claiming transaction-(s), identifying self, from lineage-vague-remembering-(s), before (suppose to be after), product-factory-packaging-production-(s), idling celebrity-notoriety, drawing comparison-greater-than, due to the diligence upon introverted-physical-preference, existing concernless of passing-perspective, as said-property-production-(s);[232] residentially responded-upon, posit-(s) of *personal-property-morale-matter, where sensational-historical-reference, discourse how realty-board, pass-day-interaction-(s), not be understood genetic-religious-fundraising, by impasse-time-article-(s)-constraint-(s), upon date-pass-article-posit-finding-aging, simultaneous present year, to be delved [estimated-hour-(s) // minute-(s) // secondths] for dating-elemental-product-object-article-(s)-subjective-pertain-history-(ies), from origin placing-minute-(s)-inferencing, apose retirement-aging, slow from youth-commercial-ambition, simply balancing existing prerogative-(s), when persuasion, alter moment-(s) of data, yet synapsing grammatical-elapse-process-(es), an month per liturgical-piece, grading by hundredth-page-syntax, apose Five-minute-(s)-twelfth-period-hour-syntax-listening-thinking-writing-conjuring-vernacularizing,| 453 |per [day ///// week ///// month ///// season ///// year-denominative-calendar-impasse offset per data-inductive-increment] isolating point-(s) per specified-subjective-synthesis-key-extent, refine-(ing) Alphabet-data, examined by survey-motive-extent, affirm-information, while showing pre-requisite-bulletin-notion-(s), in discourse of dissemination, affected from illiterate-outcome-(s), of proglamated-governmental-commercial-discourse,| 454 |because dimensions-pertained, [sort-collecting / solid-composition-(s) / territorial-constituting / mass-schedule-agenda-supplementing] by architect-fashion per person, recidivistically-following proportional-formal-development-conjuring of data, rating-time, persuading amicable-orchestration, from conscious-day-focusing, in accumulation of [ounce-(s) / pound-(s)-trade-data], in depreciation per ounce-(s) in to gram-(s);[233] reclaiming confidence where perspective exist upon practice-continuum-impasse, mutely convincing blood-pulse-count, when self, deviational mass-exponent-collection-(s)-offset, imbue terrestrial-claim, resorting centralized-planning, as spiritual-movement-(s),| 455 |find barren-terrestrial-state, scheduling-book-article-(s)-mass-schedule-concepting, confirming method of payment-(s), discoursing mass-note-interchange, currenting debt-based trade, not mutating Grimm's-Law, for vernacular [{Consonant} / {Vowel}] word-interpretations [retrograde / syntax-retrograde], oriented scaled grammar, vocal-timbre-inconsistent-interpretation-(s) from human-recognition, elapsing one-on-one, in respect-(s') of redacted-articlizing-data-information, ruling grade-(s), where belief of worth, not denominate mass-discourse, per human-district-city-civilization, from complete-total-tectonic-plate-impasse, because mass-perspective-proportional-offset-(s), have how each and every notion, simultaneous-elapses;[234] idle by nature, not to synthetically formulate formal-mass-documenting, from signature-approval, when religious-contract not exist, because pastor-directing of mass, can not settle on an singular-vernacular-long-term-basis of national-residence, engineering-quality-(ies),| 456 |where educational commercial-direction, not guide governmental-denominating, from gross-domestic-product, pertaining 112L.[real estate / university / empirical-subject-probing] passing homework-competency, for sociological-classroom-purpose, interior-spaced, balance-article-(s)-offset, physical-percept-ive, basic-human-right-(s),114L. from sensational-point of view... ¶|||08m.: 41s.21m.s.|||§

§...Preposition-Count-[33 / Thirty-Three]-of.. ..Article-Essay-[9 / Nine].. // ..Book-Essay-[119 / One-Hundred-Nineteen].. ..Lines-[114.13 / One-Hundred-Fourteen.1.one-tenths-2.three-hundredths].. ..Article-(s) are complex, amid each essay intrication-(s).¶|||10s. 68m.s.||||§

¶120.I ponder which physique-type women prefer, for birthing babies;235 without personal-gain, what [neutral-government-trade-discourse // say] interjection of wage-budget-monitoring, where one will ever supersede such (two) (three)-dimensional-denominating, perceiving ambient, impassing physical-observational-conjecture, enumerated-denominating, infetesimal-posit-(s) of human-blood-pulse-quality, where those rate-(s) for mass-product-conceptive-creative-curb, merely sense purchase, not documented-redact-interposed-data-verifying, by the perspective, aging-discourse by [hyper // hypo] neurological-impasse-effort-(s), day-hyper,| 457 |night-hypo, tense-volume-glide-formation, aligned [evening / morning-fatigue-(s)] mute-routine, signaled government time, mundane educational-curb, satiated fatigue, of soil-based-energy-trade-(s), depleting juxtaposed self-ideal-(s'), from curb of comfort-(s'), invigorating formal-interacting, when placed at seasonal-notion-(s)-inclination-(s);236 juxtaposing day apose night, sleeping prior work-scheduling, simultaneous article-(s)-schedule-deviation, annotating mass-inclination-(s), by male-debt-discourse, when human-mute-behavior, breed by [possibility // probability] of matter-(s) for thought.215¶|||53s.06m.s.|||§ Daily-Biological-Deviation, decease daily from obituary live-date-handling, accredited debt-holding-(s'), when impasse of observation-(s), rate-reasoning-(s), responsible by case-point, estimating approximate-physical-activity-impasse-objective-product-usage, sequenced syntax-article-(s)-rate-deviation, accumulating through focus by self-visual-tonal-perspective, limited surrounding offset-effort-(s), in path for mathematical-progression, variablizing product-objective-usage, dimensioning cubic-measure, upon method of hour-(s)-passing-(s) of secondths-per-cubic-foot-momentary-purpose, as weather feet-grand-aspect, distance simultaneous, metabolic-discourse, intermittent [meter-(s) // mile-(s)] per mental-cognition, fatiguing through happening-(s'), containing commercial-presence, defining-dollar-(s), from physical-interaction-(s')-deviation, intricating perspective, by interval-notion-(s), upon daily-living, where impasse, listen of vocal-directive-(s), tense-(d) from awareness, (as an) future anticipation-view, at limit-(s) of distance-recourse, conjuring walk-passing-effort-(s), juxtaposed electrical-cognitive-component-(s), relying force of plane-surface-area, for relevant physical-subjection, articlizing focused-tentative-review (reading-(s')), per data-instruction-(s), inaudible individual-personal-directive-inference-(ing), spacing, ethical-survey-constituted-forum, for holding conventional-article-(s)-session-(s), in lieu of defining-perspective, upon civilian-constituent-(s), sworn by government, overseeing legal-reciprocal-discourse, interpreting chapter-(s)-lesson-(s)-filed-data-comprehension, ream through formal-financial-recoursing-(s), affecting monetary-passing-(s), as substantial-requirement-(s), proportion-verify daily-trade, affecting resourcing-effort-(s) per state, turning the corner of trade-current-(ing), syntax-equivalent-count-deviation, mass-physical-motion-(s)-perspective, recoiling those matter-(s) where nation, state-responsibility-(ies), from taxed-law, liberally represent citizen-(s), by a (n) rate of notion-(s), influctuated by bill-continuum-input-orchestration, from where university-facility-(ies), pertain county-operations, upon revolving-parallel-axis-centripetal-force-momentum, [when // where / who] intertwine-syntax-notion-(s), entering a-the realm of previous-tangent-discourse, while understanding mass-purpose, remain an introverted-challenge, which I find perfect, where I am aspiring for an 25,000,000-discipline-reading-repopulating-culture,| 458 |to develop an rate-deviation-interval-(s)-aging-logic, [fraction // percentage / percentile-pronoun-data] as obituary-present-documenting-fashion, from Four-Hundred-Eighty-Three-Year-Reality;[237] dating by observation of effort-(s), examining (not judgement), mass-human-prerogative-(s), bolstering collection-(s) of objective-(s), sharing communal-space-(ing-(s)), unkined, for understanding how appreciation by-of independent-effort-(s), remain impartially, not [bias // hoc] apart for what discoursed-remark-(s'), increment our understanding of aging, in-an comparison of data-filing-(s), confirming analytical-notion-(s), juxtaposed temporal-blood-pulse-reaction-(s), by celestial-revolution-rotation-default-offset, when centripetal-force, remain omnipotent, destiny-parcel-(s), as exterior-fundamental [color-(s) / shape-(s) / texture-(s) / Compound-Floor-Ground-Road-(Un)-rural-Beneath / Dimensional-objecting / elapse-syntax-comprehension] ranging property block-fashion, defining those merit-(s), as City-Planning-discourse, follow an governmental-property-denominative-responsibility, recoiling wave-reverberation-(s), kinetic-influctuated, infinitive-matter, where religious-rite, offset personal-perspective, because how mass-influx characteristic-(s)-observation-(s), not by-an syntax-exponent-quotient-reset, per exponent-decimal-incrementation, formally comprehend an One-Hundred-Year-living-process, defining objective-(s)-article-(s)-interval-(s) from-an {Four-Hundred-Season-(s)}, apose {One-Thousand-Two-Hundred-Month-(s)}, apose {Five-Thousand-Two-Hundred-Seventeen-Weeks-Seven-Days}, apose {Thirty-Six-Thousand-Five-Hundred-Twenty-Five-Day-(s)}, juxtaposed [time-continuum-second-(s)-elapse /// deductive-calendar-tier-juxtaposition-reading] by interval-(s) of calendar for {Eight-Hundred-Seventy-Six-thousand-Six-Hundred-Hour-(s))}, apose {Fifty-Two-Million-Five-Hundred-Ninety-Six-

Thousand-Minute-(th-(s))}, apose {Three-Billion-One-Hundred-Fifty-Five-Million-Seven-Hundred-Sixty-Thousand-Second-(ths)} as tangible-interval-conjunction-(s);[238] forgetting to anticipate an calendar-dating, (Thursday;[239] August 21, 2025), as intermediary-checkpoint, surveying [expiry // longevity-age-gauging] filament-(s) where depreciated-mass-object-(s),| 459 |concept refined-data-articlizing, when valedictorian-median-age-comprehensive-reasoning, suffice formal-recoursing-data, by when agenda, surmise under commercial-developmental-constraint, guiding physiological-human-tension-(s), by article-attempt, when afternoon-night-homework, balancing day-conscious-commercial-governmental-common-offset-(s), pertaining perceptive-intrigue, to have median-impasse-interval-(s) when Time & Calendar, date simultaneous-subjective-offset-(s), [when / where // why] information arise (from an) mathematical-measure, incomplete-comprehending, elapsed general-fatigue, per twelfth-hour-time-posit-(s), as denominative-count-(ing), juxtapose enumerated-product-(s), while government resource-composite-ground-terrain, [Soil / Clay / Rock / Coal / Oil] simultaneous elemental-offset-(s), too great for communal-cooperation, from territorial-requisite-(s) of development, where matter is waged in to conception, and then retired from serviced-effort-(s), defining generation-offset [yearly-aging / ethnic-background / commercial-comfort] where impassed-aging, process off-time-post-wage-individual-prerogatives', uncentralized state-national-reformation, where linear-view from self, sustain pre-requisite humanity, partaking sport, like N.A.S.C.A.R., when driving randomly mobius-deviation-(s), heavy-density-commercial-compositional-matter-(s) incapable genetic-reidentifying, as religious-intrigue;[240] formulating perceptive-focus-(es), at an subjective-discourse, understanding why motion is significant, by oneself, relevant juxtaposed parameter-(s)-fixturing, [sight / sound] juxtaposition-(s) of data, conjuring how life upon surrounding-limit-(s) amass-matter-(s), where personal-legacy, can primarily define self, secondary upon family, by when thirtiary survey-endeavor, influctuate weather further intrigue of Individualism & Trade, can be recognized common-citizen-impasse-interpretation-(s).216¶|||06m.:06s.91m.s|||§ This is why I explain how offset-lineage is so important upon living, when-where parameter-(s) of discourse, display an similar blood-pulse-interior-composition-effect, reacting awhile matter-(s), precipitate-perceive invigoration, enveloping subtle-nuance-effect, focusing exterior-being, upon housing-interior-setting, verifying duty-discourse, desiring prerogative-(s), by spacing-(s) of common-mutual-activity-(ies), routing residential-way, for identifying commercial-material-(s), significant where work on intricating document-(s), allude technical-means', affirming artistic-directive-(s), stated by relative-prevalence, because how parcel-dimensional-cubic-feet-per-second-timing-(s), associate unintricated cubic-meter-(s), height-impasse, upon centripetal-force-clearance, by-of Vehicle-height, upon meter-(s)-mile-(s)-distance-(s), offset by calendar-dating, affirming timed-fact-(or-(s)), by personal-mathe-matical-timbre, passing-presence-perspective, as how any of us attain visual-tonal-impasse-recollection...¶|||06m.:55s.72m.s.|||§

§...Preposition-Count-[28 / Twenty-Eight]-of.. ..Article-Essay-[10 / Ten].. ..Book-Essay-[120 / One-Hundred-Twenty].. ..Lines-[91.31 / Ninety-One.1.three-tenths-2.one-hundredths].. ..How to formally convey each essay, amidst each syntax of period and semi-colon-numerical-indication.¶|||12s.43m.s.|||§

¶121How does one reset dating upon day-hour-interval-(s), and than expand-elaborated time-posit-(s), [elapse-(s) // synapse-(s) /// syntax-(s)] while life proceed fifty-forty-nine-percent-balance, through Attire & Decorum, in depth-perspective, per movement-(s), attaining matter, by-when compositional-appreciation, devise hand-in-hand, actable-reset-space-(ing-(s)), constant [revolution-year-one-count // rotation-day-one-count] article-(s)-comprehending, extraspection-objective-ration, by when variable-mass-parcel-count-(s), dimension-tension-space, apose inanimate-intangible-dimensional-spacing-(s), precluding perceptive-vision-boundary, abstained article-(s) conjecture, while autonomous conscious-competency,| 460 |affirm article-interposed-status, wherever possible, interval-secondths-denominative-elapse-cycling-(s), persuaded subjective-lessoning, simultaneous particle-matter-offset-(s), in elucidated celestial-third-person-offset, as how enforcement per action, pertain an awareness of location-offset-juxtaposition-(s), because direct-response, affirm day-year-reset-offset, through perceptive-matter-(s), articulating accounted, focus-space (ing-(s)), predicted why ventricle-tense-response-(s), react where-at stress, from several timbre-act-elapse-(s), affluent by indicative-cause, provided core-efforted-genetic-longevity, when instilled-intention, reasonably merit-thought, because without such formal interposed effort-(s), lifted basin-notion-(s), merit-data, by when self-posit-juxtaposition, apose other inhabitant inquiry-quality-(ies), where technical-spacing-(s), define engineered-production, requiring manual-handling, because vocational-purpose, serve duty per each and every inhabitant that maintain an perspective, upon presence of existence...¶|||01m.:20s.07m.s.|||§

§..Preposition-Count-[2 / Two]-of.. ..Article-Essay-[11 / Eleven].. // ..Book-Essay-[121 / One-Hundred-Twenty-One].. ..Lines-[17.71 / Seventeen.1.seven-tenths-2.one-hundredths].. ..still with the article-count-(s), but less and less of a the syntax-redact.¶|||08s.97m.s.|||§

¶122Bias & Hoc affect our method of construe-(ing)-data, instruction-informing location-objective-pre-designation-(s), sustaining objective-hold, Reading & Ratifying-(-right-(s')) extending relay of compilation-(s), for contextual-prerogative, proceeded an difficulty of degree;241 for when perspective elapse upon centripetal-force-present, fatigue-inquiry-(ied), extent of outstanding-motion-(s), remaining on an historical-mechanical-technical-production-basis, because article-application remain measured per space, calling-to-action, enumerated-powers, for right-(s')-reasoning, judiciary-deviation-(s), per interval-(s) variablized amidst those called-ordering-(s), when commercial-output-production, plan prior objective-article-posit-(s), quantifying-count-(ing), [where /// when // who // what data] persist Term & Syntax an amount of contiguous-article-(s)-ordering from those fashion-(s), that exist predicate persistent-deviation-impasse-focus, where cardiovascular-breath, interval live-blood-pulse, on opposite-requisite-register, sustemming simultaneous-mobius-pressure, transition-(ed) interior-posit-(s), with an exterior-requisite, being posit-(ed) (as an) retrieval of exterior-parallel-article-form-documenting-(s), sustaining momentum-deviation-minimum-impasse, for affirming third-quality-denominative-maximum, where our perspective exist, in mundane-form, rated by velocity-frequency-(ies), per general-outcome, per prevalent compositional-outcome, different by each objective, for where distance range Time & Space, selecting objective-experience, (as an) modification of composite-matter-(s), during commercial-conception, inspiring those interval-level-cause-(s), from dimension-tangible-(ity), deriving outcome-(s) per-of relevant-area-impasse-coverage, aging constraints of perspective, relevant invigorated tier, upon live-aging-retrograde-reflection-(s), pertaining self-density-perceptive-surface-area, acted out when condition-(s) from formulated-particle-(s), identify timing-(s) as simultaneous [numerical-deviation // posit-(s) /// sum-(s)-factoring] rather superfluous infinitive-abstraction, not upon an competition of mark-(s), for comparative juxtaposition-(s), qualify-(ing)-notion-(s), as when ground isolate our peripheral-attention, conveying proper conduct, relatively-probable, as where dating pertain life, nominal reproduction-(s) of matter-(s)24L. genetically-excluded, physical-perspective of individualism, throughout-upon existence...¶|||01m.:56s.80m.s.|||§

§...Preposition-Count-[12 / Twelve]-of.. ..Article-Essay-[12 / Twelve].. // ..Book-Essay-[122 / One-Hundred-Twenty-Two].. ..Lines-[24.57 / Twenty-Four.1.five-tenths-2.seven-hundredths].. ..Consideration is primarily memory, as we impasse hourly-time-posit-(s).¶|||10s.06m.s.|||§

¶1.2.3.For oneself, living can become reclusive of ambition, not tense-stressed, enigmatically pursued tensed-invigoration, conditioning self-composite-matter, because dating (has been) humanistically-sold, perceptive-being, premise mass-possessive-data, in enabling desire, when living, for nature being more flat per space-(ing)-(s), intensive when fact of transient-backlog, attempt article-(s)-collection--passage, routine momentum-rotation-revolution-organism-age-cycling, apose factor-(s) per rate-(s), scaling congruent, elemental-staging-(s), isolated by rule for subjective-isolation-focus, rationing time-offset-posit-accumulating-interpretation-(s), defining how [Major / minor article-(s)-dynamic] pertain incremental-(s) of either word-(s), upon paging, or objective-concept-dimension-(s);242 ancillary grandiose-surrounding space-(ing)-limit-(s) when extra-spective-interaction, remain dormant, by natural human-introversion, because partial-limit-(s), bias men, when counted remain for, at capacity comprehension, interluding discourse, knowing why your sitting at an place, or, why your sitting in an transit-vehicle, or, why we stance inside our residence, because each passing moment, pertain exterior-visual-regard, forsaken exterior listening, fulfilling repopulating-requirement-(s), not by exponent-population-community-offset-reasoning, particular pastor-appreciation, enabling (not brainwashing), inhabitant-perceptive-focus, due to preliminary-article-trial-(s) of trade affecting blood-pulse, in regard where present-citizen-mode, virtue valued interaction-(s), deviating introverted-beings';243 while influctuated-article-syntax-interpretation-(s), maintain mass-motion-(s)-function per date, upon centripetal-force-physics-impasse-deviation.217¶|||01m.:25s.15m.s.|||§ Day is present.218¶|||01m.:26s.21m.s.|||§ Year is modern.219¶|||01m.:27s.37m.s.|||§ Hour is requisite.220¶|||01m.:28s.40m.s.|||§ Yesterday is past.221 ¶|||01m.: 29s.25m.s.|||§ Decade anticipate future-contemporary.222¶|||01m.:33s.18m.s.|||§ [Stance / Sitting / Push / Pull / Clasp / Grammatical-interpretation / Enumerated-denominative-elapsing / vocal-articulation / jogging / running / bicycling / Walking] as from physiological motor-sense-(s), upon [lever / pulley / axle-mechanical-function-(s)] where effort-(s), remain by entity-place-(ing-(s)), concerting-focus, for applying-effort-(s), from inhabitant-competent-being-(s'), whom understand guideline of matter-development, when infinitive-structure-constructive-development, pertain surface-area, per time-limit impasse-restriction-(s), considering component-aspect-(s), posit-(ed) religious-residence, for public-government-order-(ing-(s)), transition movement-(s) of men, denoting acting-(s) when municipal-territory, place inhabitant-person-populant-subjunctive-correlation, from youth-objection, transitional aged-era-(s), that have an impact of historical-continuum-effort-(s),29L. for how existence has been lived thus far...¶|||02m.:09s. 77m.s.|||§

¶...Preposition-Count-[7 / Seven]-of.. ..Article-Essay-[13 / Thirteen].. // ..Book-Essay-[123 / One-Hundred-Twenty-Three].. ..Lines-[29.0375 / Twenty-Nine.1.zero-tenths-2.three-hundredths-3.seven-thousandths-4.five-tenths-thousandths].. ..Timing upon Dating, makes what we consider memory.¶|||10s.37m.s.|||§

¶123,,,Interactive mode, is by individual-prerogative, designated past-requisite, by present-future-natural-reoccurring, on middle-ground-(s), for defining citizen-social-state-mutual-agenda-activity-(ies), touched up on checklist-(s) of fabricated-intrication, sequenced from voluntary-procession, per rationalized solidification of tensable-reference, crescendo developmental-boundary-(ies), autonomy interior-dimension-(s), Context & Tense-(ing)-effort-(s) when cause of subterior-deviational-limit-(s), place calendar-past-day-deviation-(s), referenced upon present-day, to an standard-deviation-reference, from subjective-isolated-incremental-reading-reference-study-(ies), when live-blood-pulse-existence, supply an numerative-interval-denominating, of human-existence, cycling through Gregorian-year-precept-reset-longevity-anomaly, by an subscripted-tense-visual-tense-(ing-(s)),| 461 |for subversive-constructive, per data-bit, referencing article-(s), by continually updating count-document-statistical-impasse-unit-(s)-collected, apposing pertinence of reference, validifying how [space-(ing-(s)) matter // cease-of-existing] coordinated by activity-(ies) per natural-mass-matter, at-of impasse-discourse... ¶|||52s.22m.s.|||§

§...Preposition-Count-[7 / Seven]-of.. ..Article-Essay-[14 / Fourteen].. // ..Book-Essay-[124 / One-Hundred-Twenty-Four].. ..Lines-[12.65 / Twelve.1.six-tenths-2.five-hundredths].. ..What convinces us to even think from celestial-physics-nature.¶|||09s.41m.s.|||§

¶123...Strict Education! Remain not an interval of physiological-activity-(ies), for comprehending what [subjective-article-(izing)-(s)] be from-an act of preference, during free-will voluntary-effect, factoring industrial-production-(s),| 462 |of individual-product-mass-composition-(s).223¶|||13s.12m.s.|||§ [Romance / Freedom / Interaction (I type / write)] does not mean that mass-competency, be capable of information in usage of data, proportion momentum-passing of centripetal-force, [revolution-calendar-dating // rotation-time-posit-dating] in explaining friction-pressure-density-infinitive-expansion, while individuals'-adding, influctuate uncentered mass-tectonic-state-infinitive-ground-coverage-article-(s), defining national-production, as state-isolation notion per Brook-(s)-style-institution-(s), haunt data, that matter, not by shifted act-(ion-(s)), in an fluent-formation-interpose-(ing), from [high-school / University-data-dissemination-enabling] pertinent per action, when each notion can be off of-an pertinence in movement-(s), having climate-temperature-modern-date-distance-hour-offset-(s), pertain environment temper-ature-scaling, per mile impasse, as how stress affect usage of objective-fact-(s), from when live-order-sequential-formation-act-(ing)-motion-(s), sequence environment-tectonic-topography-off-set-(s), affirming purchase of matter-(s), when fact be in an tectonic-subjunction, affecting interactive-social-prerogative, sustaining matter-(s), because involuntary-movement-succession, crescendo article-(s)-interpretation-(s), not fluently interpreted by mass-mutually-studied-lessoned-compositional-beings', for which observation-(s) pass present-observation-(s), affecting physiological-present-motive-(s'), remaining dormant interactive-conjecture, for in school-study-(ies), would each inhabitant, partake upon voluntary populant-limit-(s), that are in effect-(s) for locational-matter-influence, developing-dimension-(s), for-of Context & Tense perceiving letter-word-syntax-deviation-offset-(s), tiered-simultaneous-comparative article-(s)-location-purpose subscripted-text, where word-emphasis, not instill an thousandths-mass-mark-maxim, by standard-deviation-(s), from syntax-article-theatrical-documenting;[244] awhile attention-span, not induct from present-(ce)-date-(ing), because each moment of dating, articlize humanity waited-behavior, upon reproduction-cycling, ordering by dormant-disposition, alluded off-time-fifty-(ies)-dynamic, totalitarian-pre-emptive-juxtaposition-observation-(s'), that draw human-mass-matter-sensation-(s), not preferred blue-collar-natural-effort-(s), [wanting // alluring] synthetic-automatic-wrench-effort-(s) being entertained, work-model-mode, reading-context-elapse-(s), yet comprehended, for how dormant-century-(ies), are upon an set of ideal-(s), male-concept-inventor-production-creative-curb-affirm-(ed), influenced motion-(s), explaining medium-(s) of data, not from where perspective impasse [surface-area // cubic-interior-dimension-parcel-spacing-(s)] remain an modern-focus, for too much unnecessary explaining, remain unlistened, for naturally-redacting-context;[245] [where // when] data-referral operating populus-deviation-(s), as when, height-cubic-feet-measuring-space, by where sitting in to millimeter-visual-recollected-distancing, affirm per day, an interval-simultaneous-unconjectured, age-offset-(s), proportioning fact-(s) of compositional-anatomical-mass, relating physiological-unnoted-motion-(s), (as) for redacted-documenting, comprehending physiological-observation-(s), (as an) part of where mass-deviational-effect-distance-(ing-(s)), become an city-perimeter-surface-area-articlizing, when exterior-visual-default, not tone recognition, interior-residence-(s);[246] awhile commercial-government-public-offset-(s), have legislational-ground-(s), to be notion-(s)-(ed), when passing involuntary-existence, not understanding how written-space-(ing-(s)) fact-(ify)-information, premise juxtaposition-offset, per [rural / suburban / urban territorial-claim-(s)], remained-an part of police-(ing), where roadway-passage, define interactive-voluntary-fatigue per prerogative, from physiological-exhaustion-(s), motion-impassed per day, not at an active-physical-developmental-part, for how tectonic-plate-(s), pre-arrangement, define cubic-distance-(ing)-tangible-offset-(s), an [conceptive / hypothesis / theoretical-focus of individual-effort-(s)] from conversive-extraspective-offset, controlling-objective-substantial-subjective-affair-(s), by monitored natural-human-behavior, because mass-context, attain philosophical-article-(s)-count-(s), inclined by rhetoric, for defining theoretical-focused-notion-(s), because as perceptive-mentality assume perceptive-space-(ing-(s)), because self-individual-(s), can be free from will-living, unobservant surrounding-existence, but not intuitively-identified, religious-message-(s), extraspective because friend-count-(s), equate those amount-(s) of off-time-commercial-prerogative-(s), which can be schedule-(d), for influence per each of those fact-(or-(s)), that contain time-mathematical-injunction, not understanding calendar-Gregorian-induct-(ing) as [time-timbre-tempo-blood-pulse-respiration-inhale // exhale-reoccurring-impasse-sustaining], affirming secondths-beat-(s)-measure-frequency-(ies), measuring by numerical-median-(s), calculus-statistical-fraction-reciprocal-acting-function,| 463 |by longevity of long-term-documenting, determinant natural past, indelible by mark-(s), from legacy-output, deriving initiative consenting from act-(s);[247] as each reproductive-intuition, becomes' an method of currency, not fact-(or-(ing)) article-(s)-information, [when / where] space-(ing)-(s), pertain how part of a the cubical-feet-notion-extent, pertain secondths-per-elapse-denominative-count-(s), comparative-conjuring sat voluntary-posit-(s) of space, measuring [feet / meter / mile] as feet-denominative-comprehension, when three-dimensional-epilepticus-individual-pos-it-matter, only by sensational-trial, without incorporating an consis-tent-reading-(s'), per informative-matter-(s), remaining tempo-(less), for leader-configuration-process, when conscious-interaction-inter-val-deviational-follow-through, direct car passed enabled free-will-requisite-(s);[248] advantage endeavor, those fate-(s) of free-dom, in rule of matter, not assertively-active of extraspection, wan-ing-fatigue, of decrescendo-progression, article-(s) of celestial-uni-verse-county-existence, alleviating substantial-burden of effort-(s), because entertainment of matter-(s), pass aging by self-independence, simultaneous, observational-progression-matters', astonished perim-eter-surrounding-development, for internal-resourcing, material-(s) per trade, by [white / blue] collar-dynamic-(s),| 464 |aforemen-tion-(ing) [age / ventricle-muscular-cardiovascular-respitori-al-tense-(s)-response-(s)] when Earth-core-continuum, demonstrate elapse-(s) by karma-reciprocal-effect-(s);[249] entrenched where-when aging in progression, entrust offset lineage, for measuring mass-off-spring-reliance, as license-in-requisite-forward-depth-infini-tive-space-resourcing-formation-(s), deriving independent-individu-al-perspective, kinetic objective-living, hearse-(d) when mute-collo-quial-tounge, verify mutual-friend-(s), incumbent childhood-interac-tion-(s), not instilled cursive-tier-redact-progress-ion-(s), surfacing natural-disposition, by [[1]read / [2]write / [3]act / [4]work / [5]order / [6]save / [7]collect / [8]store / [9]intonate / [10]picture / [11]envision-(ing)] those formal effect-(s), where written-rigor, construct fundamental-concep-tive-space-(ing-(s)), by interaction-(s) of other-inhabitant-sched-ule-(ing)-impasse-period-(s), callisthenic tense-limit-physical-im-passe-perspective, ambiguated by bounded grandiose peripheral-dis-tance-(s), circulating enumerated-discourse, an retirement-aging-op-position, quantifying-space-(s), by extraspective-perspective, when millennium-judgement, qualify how routine-(s), inhibit inhabi-tant-self-age-dating, by when offspring be conclusive for no subjec-tive-furthering, due to resource-technicality-(ies), reluctant of creative-artistic-effort-(s), for exemplifying expertise-difficulty, because product-factors, remain patent-conceptive by-of refinery-factory-market-directive-dynamic, socially-trading for merchandise, antiqued as furniture, because rates' of routine, can not educate independent-individual-(s'), while I prose-fathom how letter-(s)-article-(s)-data, neither signify paragraph-essay-article-(s)-conjuring from organic-matter-(s) stapled tangible-particle-fragment-piece-(ing-(s)), binded by surrounding [[1]space-(ing-(s)) / [2]product-object-(s) / [3]mass-(in)-dependent-individual-inhabitant-citizen-(s') / [4]Public-roadway-(s) / [5]Public-Central-Planning-Entity-(ies) / [6]Private-Commercial-Operation-Entity-(ies) / [7]Private-Residential-Entity-(ies)] compiling perspective-(s) at will, due to mute-physical-document-(ing)-(s), that exist upon mass-magnitude-notion-(s), sufficing impasse-extent per [day / night] historical-deviational-hour-requisite, present-requisite-elapse-syntax-article-identifying, future exploring, conduct-impasse-grading-(s), when commercial-effort-(s), past tense-(s), past year-reset-(s)-sequence-(s), identifying seasonal-impasse, through focus [at // intransitive] perceptive-mass-temporal-matter-circulating;250| 465 |introverted-mute-meek-peer-pressured secondths-interval-simultaneous-continuum-stream-sequence-(ing-(s)), inducting by [[1]rested-act / [2]maneuvering / [3]transiting / [4]tasking / [5]passing / [6]arriving / [7]departing / [8]entering / [9]exiting / [10]coming / [11]leaving / [12]studying / [13]endeavoring / [14]injecting / [15]extracting / [16]standing / [17]sitting / [18]riding / [19]reading / [20]writing / [21]tasking / [22]seeing / [23]hearing / [24]observing / [25]enacting / [26]course-tiering / [27]objecting / [28]Sermoning / [29]receiving / [30]raking / [31]Baking / [32]Confectioning / [33]Subjecting / [34]entertaining / [35]believing / [36]contorting / [37]consorting / [38]dialoguing / [39]discussing / [40]debating / [41]lecturing / [42]lessoning / [43]redacting / [44]interpose-inducting / [45]chaptering / [46]vocal-align-harmonizing / [47]running / [48]clothing / [49]eating / [50]washing / [51]moving / [52]hammering / [53]lifting / [54]reasoning / [55]affirming / [56]dissecting / [57]discoursing / [58]paving / [59]sporting / [60]ruling / [61]judging / [62]executing / [63]legislating / [64]filing / [65]diagnosing / [66]ordering / [67]burning / [68]incinerating / [69]wasting

/ [70]trashing / [71]filtering / [72]measuring / [73]balancing / [74]scaling / [75]deriving / [76]collecting / [77]placing / [78]teaching / [79]monitoring / [80]babysitting / [81]watering / [82]driving / [83]showering / [84]cooking / [85]deeming / [86]packaging / [87]purchasing / [88]budgeting / [89]adorning / [90]factoring / [91]concepting / [92]serving / [93]saving / [94]checking / [95]banking / [96]piloting / [97]conducting / [98]currenting / [99]marketing / [100]advertising / [101]broadcasting / [102]disseminating / [103]liberating / [104]informing / [105]instructing / [106]directing / [107]agenda-orchestrating / [108]orchestrating / [109]stamping / [110]filing / [111]sorting / [112]labeling / [113]terming / [114]alerting / [115]nursing / [116]envisioning / [117]scheduling / [118]faceting / [119]screwing / [120]fixing / [121]tapering / [122]carrying / [123]performing / [124]automating / [125]marking / [126]denoting / [127]enumerating / [128]disturbing / [129]observing / [130]warranting / [131]distancing / [132]distorting / [133]modifying / [134]painting / [135]picturing / [136]medicating / [137]offsetting / [138]ranging / [139]allocating / [140]periodicalizing / [141]elapsing / [142]synapsing / [143]pastoring / [144]typing / [145]tiering / [146]instating / [147]habituating / [148]peaking / [149]sleeping / [150]waking / [151]allotting / [152]graduating / [153]dimensioning / [154]compounding / [155]compensating / [156]cutting / [157]rendering / [158]meticulating / [159]adding / [160]subtracting / [161]multiplying / [162]dividing / [163]fractioning / [164]variablizing / [165]integerizing / [166]geometrical-aligning / [167]comparing / [168]contrasting / [169]faceting / [170]fitting / [171]cleaning / [172]wiping / [173]scrubbing / [174]washing / [175]plugging / [176]pressing / [177]pushing / [178]pulling / [179]parameterizing / [180]scribing / [181]scanning / [182]faxing / [183]copying / [184](miter)-sawing / [185]tuning / [186]calibrating / [187]melodizing / [188]methodizing / [189]hitting / [190]smashing / [191]lifting / [192]positing / [193]documenting / [194]receipting / [195]trading / [196]clicking / [197]touching / [198]grabbing / [199]smearing / [200]telephoning / [201]texting / [202]heating / [203]welding / [204]bending / [205]breaking / [206]molding / [207]flattening / [208]noting / [209]denoting / [210]paragraphing / [211]biting / [212]chewing / [213]birthing / [214]inseminating / [215]raising / [216]jumping / [217]climbing / [218]falling / [219]perceiving / [220]retorting / [221]spectating / [222]dulling / [223]arising / [224]tangenting / [225]paralleling / [226]conceiving / [227]mattering / [228]clutching / [229]clasping / [230]arranging / [231]arraigning / [232]arresting / [233]convicting / [234]felony-matter-(s) / [235]misdemeanor-matter-(s) / [236]seriating / [237]serializing / [238]brushing / [239]snapping / [240]picturing

[Article:] [May]

/ [241]seeding / [242]germinating / [243]concealing / [244]firing / [245]pointing / [246]shooting / [247]snaring / [248]blaring / [249]attaching / [250]matching / [251]terminating / [252]growing / [253]stopping / [254]gaming / [255]pertaining / [256]telling / [257]honing / [258]honoring / [259]bothering / [260]coddling / [261]penmanship / [262]roasting / [263]fermenting / [264]charging / [265]basting / [266]basing / [267]ascertaining / [268]acclimating / [269]educating / [270]ruining / [271]rummaging / [272]dropping / [273]lofting / [274]lobbing / [275]lathering / [276]massaging / [277]asking / [278]interacting / [279]fathoming / [280]deeming / [281]appearing / [282]cornering / [283]clothing / [284]bathing / [285]sheltering / [286]housing / [287]batching / [288]verifying / [289]exonerating / [290]expelling / [291]exerting / [292]furnishing / [293]burnishing / [294]banking / [295]capping / [296]sealing / [297]expunging / [298]levering / [299]succeeding / [300]reaming / [301]serrating / [302]blooming / [303]slaughtering / [304]lining / [305]preparing / [306]proportioning / [307]freezing / [308]refrigerating / [309]paging / [310]sheeting / [311]surfacing / [312]counting / [313]compositioning / [314]dating / [315]referencing / [316]mating / [317]hating / [318]loving / [319]domesticating / [320]snowing / [321]raining / [322]blowing / [323]breezing / [324]projecting / [325]laying / [326]dreaming / [327]scrolling / [328]tolling / [329]stalling / [330]crying / [331]sighting / [332]timing / [333]nailing / [334]rolling / [335]weaving / [336]texturing / [337]tiling / [338]plating / [339]forming / [340]soldering / [341]wording / [342]connecting / [343]stirring / [344]mixing / [345]pouring / [346]touring / [347]scoring / [348]scorning / [349]pathing / [350]discoursing / [351]tempo-frequencing / [352]rowing / [353]mowing / [354]whacking / [355]detailing / [356]barreling / [357]tinkering / [358]pulsing / [359]beating / [360]reaping / [361]weeping / [362]teeming / [363]toweling / [364]draping / [365]caulking / [366]extinguishing / [367]pealing / [368]stealing / [369]eating / [370]feeling / [371]kneeling / [372]tearing / [373]wearing / [374]bearing / [375]staring / [376]layering / [377]thinking / [378]encompassing / [379]damning / [380]ditching / [381]stitching / [382]savoring / [383]smelling / [384]seeing / [385]mazing / [386]tazing / [387]routing / [388]rowing / [389]legalizing / [390]towing / [391]shipping / [392]handling / [393]fanning / [394]slighting / [395]slamming / [396]checkpointing / [397]spraying];[251] (in)-transitive-act-subjunction, variablize verbatim-action-(s), apart of preliminary-process-act-(s), following| 466 | succession of chain-order-past-historical-dating-incremental-forward-progressive-motion-(s)-operation-(s), acting by sequence-programmed-parameter-(s), for post-instructional-ordering-(s), defining caliber comprehension, exuding formal-focus, as trained upon mechanism-posit-(s)-tectonic-development, rating mechanism-composite-function, from directive-agenda-tectonic-activity-(ies), refining action-(s), [when / where] date have an [third // quarter] sleep-reset-process-cycle-(ing)-(s), considered when scheduling-matter-(s), merit pre-existence, awhile hour-(s)-impassing, are where, what indicated item-(s), pertain whom possess (prior), objective-storage-property, when pertinence affirm why action-(s) elucidated, formation-(s) of dated-timed-matter-(s), interpreting [symbol-(s) / sign-(s) / image-(s)] perceived upon limit of boundary-perceptive-matter-(s'), recoil-recoursing, reciprocal act-(s), requiring age, to realign sleep, by suffix -article-scheduling, meticulating impasse-matter-(s), memorizing pattern-(s) from breath-historical-population-impasse, morbid each matter-(s) of lineage, as live-presence-effect, verify mute-blank-numb-presence-memory, influencing technical-moded-way, without an natural-creativity, at Natural-Territory & Mass-Schedule-Arrangement$_{160L.}$ of human-disposition...¶|||13m.09s.$_{94m.s.|||}$§

§..Preposition-Count-[40 / Fourty]-of.. ..Article-Essay-[15 / Fifteen].. // ..Book-Essay-[125 / One-Hundred-Twenty-Five].. ..Lines-[160.11 / One-Hundred-Sixty.$_{1.one-}$tenths-$_{2.one-}$hundredths].. ..Thinking is important for any mass population survival past 100-years of age.¶|||12s.$_{29m.s.|||}$§

¶126How has construction post-humorous intertwined Federal-denoting, when storied-fore-closure, formulate upon personality, not making citizen actively-ascertain, dimensional-residence-infrastructure-design, by contractor-planning, surface-area-surrounding-limit-(s), controlled by, centralized work-routine-(s), when daily-practice define parameter-cycle-progression, unscheduled by extraspective-governmental-limit-(s), partaking in off-time-religious-activity-(ies), distancing perspective, while account-(ing), an preferred timbre-focus, instated temporal-position, enamored by aptitude, an approximate-compound-unilateral-symmetrical-allegiance, maintaining crust-under-ground-impact-control, resourcing from territory, as awhile dating impasse intervoluntary-purpose, upon condition-(s) of tectonic-plate-centripetal-force-deviation-divide.$_{224}$¶|||40s.$_{28m.s.|||}$§ This generally remain omitted by [youth-education // adult-commercial-working-competency-extent /// elder-retirement-senile-grandeur] illiterate from photographs, philosophical-tense per [[1]lineage / [2]territory / [3]resource-(ing) / [4]affair-(s) / [5]education / [6]agriculture / [7]horticulture / [8]terraculture / [9]apiculture / [10]mariculture / [11]floriculture / [12]taxonomic classification / [13]processing-plant-(s) / [14]mechanical-factory-development / [15]political-debate of tangible-mass-matter-(s)-denominative-ordering] thinking from hour-twelfth-minute-denominative-deviation, day to day effect-(s), from calendarized-impact, where motion-(s)-made, verify mass-family-individualism, for which religious-impact, incur through an series of transit-cycling-(s), defining [workplace / market / gas station / roadway-discourse / extemporaneous-entity / housing] afforded from tasked parameter-(s) of Character-(s), deviating spaces of statistical-population-state-territory, when forward-depth-continuum-perspective, reset from mile-(s)-distance-reset-(s), as day rotates' past birth-housing-offset-(s), simultaneous mass-equivalent juxtaposition, historical-timbre-sensation-(s), fatigued by personal-perspective-requisite, without tension, in perceptive contrast of past-present-progressive-reset-(s)-offset, (rather infinitive-matter (reset-offset-(s))) presented conscious-prevalence, interacted where momentary-influence, not alleviate incurred dating-affair-(s), simultaneous subjective-time-table-(s)-aligning, per [letter / word / sentence / syntax / paragraph / essay-(s)] from empirical-subjective-notion-(s), pertained upon, natural-voluntary-free-will-discourse, pertaining human-behavior, awhile impasse per inhabitant, visually-peripherally-offset, toneless because mass-volume-circulation, practice conjure-observation-discipline, yet to be academically-aligned, conducted relative-national-state, assigned agglomerated-past-compositional-matter-(s), acceding practice per-of Schedule & Concept-Resourcing, deriving mass-obligation, upon request of action, syncopated [note-(s) // rest-(s)-(ed)] interposing schedule-cycling, by-an psychological-mass-referencing, cross-referencing Dewey-Decimal-Referencing,| 467 |pertinent-arithmetic-geological-topographic-data-continuum-articlizing (abstractly-bounded).225¶|||02m.:33s.50m.s.|||§ Whom will allot checking alone, for comprehending dynamic-(s) of perspective, arranging prelate median-impact, generally-passed, from repetition-(s)-impasse-(s), per-of date-rotation (revolution), affecting memorial-cause, through self-pre-lineage (non history)-dynamic, anticipated-retiring, tense-(s) of perspective-comprehension, [interposing-relatable-faded-bias // hoc] inferring (by of) a the form of personal-action-(s), coordinating extraspective-activity-(ies), (un)-reference-(ed) [written-schedule / form-referencing / novel / biography / book] upon an perceivable-comprehension, person-monitored, from whom assess [1perimeter / 2cubic-area / 3objective-impact-discourse / 4wage / 5trade / 6stock / 7inventory / 8reading-material / 9redact-composition-(s)-dynamic-(s)] of population-function, tangible compositional-crust-tectonic-plate-state-boundary-matter-(s), simultaneous an mega-effort-(s), thousandths-percentage-decimal-posit-scaling, influctuating numerative-action-(s)-objective-(s)-location-(s)-interaction-(s)-subject-(ive(s))-completion, per-act, reasoning of article-(s)-data mechanical-denominative-trading, conceptive-purchase-curb, that order mass-social-inclination-(s), by allure, for still-luxury, not mathematical-calculator-computating, remained bored by Fifties-Social-requisite-education-dynamic, interacting from physical-mutual-inhabitant (citizen) of national-state-county-city-affair-(s), districted by mute-vernacular-tense-(s), inflated-pay for interpreting, subjective-objective-tasking-(s), with no off-time-subjective-article-(s)-requisite, where personal-consideration, posit upon an intimate-comfortable-extraspection by other-local-citizen-(s')...¶|||03m.: 49s.51m.s.|||§

§...Preposition-Count-[16 / Sixteen]-of.. ..Article-Essay-[16 / Sixteen].. // ..Book-Essay-[126 / One-Hundred-Twenty-Six]..

[Article:] [May]

..Lines-[48.07 / Fourty-Eight.1.zero.tenths-2.seven.hundredths] ..Dating is for longevity, while timing resets upon each day for calendar-contorting of past day notion-(s), not capable of intricating amid daily-physics-impasse; we can only intricate an day, not calendar-scheduled-cycling.¶|||17s.53m.s.|||§

¶127Youth-converse, slid pass superfluous-book-lesson-(s)-conjuring (not article-(s)), yet still-class-identified, from educational-point-(s) of (ticket) game-(s), instilling blank-idle-aging-memory of a the English-language-subjunction-(s),| 468 |subversive word-intrication-interpretation-inclination-indication-(s), defining point of view, [alone // interactive // communal] influencing domestication, where realty-contracting, bond memory-lineage, have linked upon approximate-distance-cycling-impasse, from governmental-state-(s), relying currency by Federal-inhabitant-(s), whom not take an simultaneous-active-part, of pre-reading-conjuring, for lessoning those article-(s), for notion-(s)-revise, apparent simultaneous-event-ordering, juxtaposed [district /// city /// county /// state /// Federal-nation-counseling-prevalence-(s)] reordering state-operation-(s), by commonality of act, upon proportional-mass-legislating, each of a the prior-impact-schedule-resolution-(s) of matter-(s), continuing an secondths-tics-simultaneous-age-meandering, offset by time-deductive-standard-deviation, [literary-objective-date-referencing // date-anticipating] those locational-requisite-(s), from individual-self, allotting an reasonable-routine, intermittent product-mass-count-(s)-cycling from denominative-trading, instated of product-conceptive-creative-purchase-parity;252 not tested per date, counting each product-(s) from social-extraspective-requisite-interaction-(s), filing-data, through lineage-trust-confidence, as pertinent-data, convene communal-interposing, upon those requisite-(s) per interaction-(s), devised relative-boundary, where [1population / 2territory / 3objective / 4physiological-timbre / 5boundary-limit-(s) / 6psychological-scheduling / 7university-subjective-article-redact-review-centralizing / 8territor-ial-tier / 9reproductive-cycling / 10territorial-cultivation-practice-(s) / [11]note-(s)-denominating // [12]currency /// [13]written-conjure] refining gamma-(s) of developmental-goal-secondths-denomin-ative-adjustment, upon [minute-(s)-scheduling // hour-(s)-location-ordering].226¶|||01m.:37s.62m.s.|||§ From child-birth, I forget how to calculate alone, past denominative-deviational-account-(ing), interactive compulsive-existence, for whom not write note-(s), served historical-lineage-presence, petulant dues from Jeffersonian-Founding-Fathers'-Article-(s)-secondths-continuum-impact-reliance, aligned Vatican-Gregorian-Calendarizing, upon present-day-interval-(s), timed interval-secondths, from a-the Seven-Billion, Five-Hundred-Sixty-Three-Million, One-Hundred-Forty-Thousand-Secondths-continuum as of {January Twentieth, Two-Thousand-Sixteen} {January 20, 2016}}-denominative-exponential-reset, of [ten-whole / one-hundred-whole / one-thousand-whole / Ten-thousand-whole-categories] simultaneously-proportioning-word-article-(s), upon [dating // time in hour-(s)] (from an) historical-time-line-background of information, by individual-perceivable-activity-(ies), regulated-federal-commercial-dollar-(s)-debt, resetting cent-(s)-deviation-one-hundredth-standard-deviation-denominative, through all matter-(s) pertaining mass-perspective, rationing maxim-denominative-numerative-impasse-reasoning, not handled equivalent Time & Date-account-decimal-ordering of mass-impasse-wage-matter-(s), formal by conscious-extraspective-interaction-(s), surmised object-(s) possessive-hold, subjective-(less) of article-mass-literary-pertinent-inquiry.227¶|||02m.:39s.18m.s.|||§ The average-inhabitant, does not reason affirmation-(s) for conscious-opinion-interactive-inquiry, so how can Person & Being-(s) become aware subconscious-inclination-(s), from disenchanting-possession, contra-opposite, personal-bigotry, from stated-volume-blood-pulse-beat-beings', maintaining objective-count-(s), Simply for obey-(ing) decade-(s)-census-elapse-(ing-(s)), fading from the subsequent-present-requisite, having intersection-(s), be lived through, turning subservient-individual-mass-perspective-(s'), under an historical-archival produced-individual-major-focus-(es), corrugated mutual-communication, excerpt extemporaneous-offset, for Listening & Reading-data from influence-(s) of [peer / colleague / co-worker] phasing [seasonal / monthly / weekly] stated-entity-article-(s) reasoning-(s), because circumstance-(s) obstacle-discourse, awhile mass-public-intersection-impasse, influence-(s) blank inductive-interest-(s), affirming Notion-(s)-Quotient-Extent-Inquiry & Article-(s)-Affirmation-Arrangement, by natural-selection of human-physiological-psycho-logical-subjective-article-survey-common-voluntary-impasse, appropriate-genetic-evaluation-(s)...¶|||03m.:32s.65m.s.|||§

§...Preposition-Count-[16 / Sixteen]-of.. ..Article-Essay-[17 / Seventeen].. // ..Book-Essay-[127 / One-Hundred-Twenty-Seven].. ..Lines-[46.167 / Fourty-Six.1.one-tenths-2.six-hundredths-3.seven-thousandths].. ..If we take each moment seriously, we receive the perspective of everlast ing enjoyment of time, amid each daily-reset of centripetal-force, amid revolutionary-pattern. ¶|||17s.43m.s.|||§

¶128Inaudible-discussion has been too commonly disregarded perspective-visual-deviation, from compartmentalized-sensation-(s), untensed by conjecture, contesting purpose of [locational-place / substance-(s) / product-objective-(s)] from experience through natural Trial & Error, from a-the mute-reflex-(es), state-national, common reliance, for commercial-basic-product-(s), simply-presumed, by city-citizen-(s), where offset national-grading-(s), conjugate cabinet-mark-(s), sustained by prevalent-articlized-person-(s), whom represent an body of faith-based-merit-trade-note-(s)-currenting, incorporating affair-(s),| 468 |where filed-data, subpoena upon requisite-(s) of executive-blood-pulse-live-anatomical-being-(s')-impasse, considering age, (as an) ultimate-biological-death-date-passing-continuum-infinitive, reincarnating offspring, from those value-(s) of appreciation-consideration, attuned by exterior-vibrational-sound-(s), tone-minuté-instrumental-dimensional-vibration-(s)-exponent-quality-verify-conjuring, worth valuing mea-sure-(s) of [beat-(s) //

note-(s) // cent-(s)] proportion-evaluated per use, by volumized abstract-concertion, synapse-rating, elapse-rate-median-measure-numerative-(denominating)-fragment-ing-data, referencing past-historical-data, requisite contorting interactive-notion-(s), thinking how future (past-requisite have no mathematical-referencing) individual-contorting, balanced tangible-matter-(s) from subversive-effect-(s), [mathematically-calculating // liturgically-referencing] moded-time-posit-(s), per mutual-perceptive-comprehending fact-(s) of perpetual-public-common-existence, (de)-crescendo perceivable-place-posit-(s), revolve-rotating moment-(s), where public-passing, reside cycle-homogenizing, tenses-elapsing, rated simultaneous-offset-reset-juxtapositional-environment-climate-temperature-setting-inclinated-article-(s)-volume-(s), enticed by [option-(s) / decision-(s) / random-choosing-(s)] by mediocre-patience, aspiring peer-persuasion, while [question // response-collecting] affirm practical-article-(s)-usage-method-(s), validifying various-private-venture-(s), from storage-material-(s), interluded vernacular-presence, where value through live-act-(s), delusion melancholia, inset on perspective-prelate-requisite-mode-cycle-status, denominating-thought, by trade-purchase-parity, that design aligned affirmed-observation-(s)-confirmation-(s), tensing-peripheral-parameter-surface-area-space-(ing)-(s), daily-conjure-abstraction-offset, common-article-(s)-usage, per [secondths-denominative elapse-(s) // synapse-(s)-cycling] identifying schedule-impasse-time-posit-(s) simultaneous-offset, juxtaposed time-zone-(s), quadrants thousandth-exponent-reset, per {86,400-notion}={$_{10,000}$86.4-notion-reset} comprehending notion-(s) by comparison per [secondths / note / beat / rest / synapse / elapse / period-(s)-subterior-mark-(s)] issued day-impasse, annotating from day-posit-act-requisite-ordering-process, in to an, year-secondths-millionths-exponent-reset-(s), of {31,536,000-notion}={$_{10,000,000}$31.536-notion-reset}, comprehending how mass-scaling, percentile-sur-rounding-arrangement-(s), denoting mass-denominative-goal-(s), from offset-(s) of interaction-(s), conferred awhile, each fact is by an perceptive-requisite, affirming article-perceptive-requisite-data, [where // when // why // what // whom, mass-population] act as liturgical-competent-(s), maintaining standard-mode, upon period-(s) of activity-(ies)-discourse, reading denoting-data, pre-affirming surrounding-matter-(s), with respects to government, (no genocide-mechanical-mass-usage), conceding article-(s), (from an) lazy letter-word-syntax-usage, not diligently following, an process of mutual-subservient-means', programmed simultaneous-serie-(s)-re-set-(s), contemporary-transiting, an numerative-physiological-conjuring, where threshold of debt, value physical-skill-(s), by merit-(s) of act, impending an waged-budget-balance, of financial-debt-conceptive-product-usage-trading, unstressed by parental-enabling, to pass adequate dynamic-(s) of time, remaining under an control of tangible-product-object-(s), Planning & Scheduling-lineage-requisite-agenda-scheduling-impasse-(s), of tectonic-mass-denominative-verification-(s),| 469 |handling interactive-article-(s)-affair-(s), by technical-acquisition-(s), amassing each of those piece-(s) of pos-it-set-objective-data;[253] juxtaposed Earth-Centripetal-force-drift-molten-plasma-substantial,| 470 |adjacent grandiose-untangible-peripheral-obser-vation, not tallied fact-(s), but sorted an equivalent of space-(ing) (s), as data-understanding, verify temporal-impasse-effect,| 471 |underling inhibited-action-(s), ((un)-subversive) naturally-naïve, pre-requisite, live-presence-offset-distancing-(s') guiding each purpose-(s) of life, Envisioned & Sounded for posit-(s) of perceivable-desire-effort-(s), as how [length × width × height] not ingrain-insignia, into natural-perspective, denoting how mutual-impasse-effort-(s), can be acted for every note, not in an article-(s)-arrangement per formal-fact-(s), rationalizing information, where each passing moment, affirm territorial-space-(ing-(s)), by cursive-subjecting, for validifying those confirmed-verification-(s), of subjective-article-fact-(s), making each day impasse per enumerated-denominating, continuing circulated-currency, as how those fact-(s) by deviational-supply-(ies), relate mass-exponent-offset-article-(s)-deviation, accrue by-an perceptive-influctuation, of stock-shelf-life, not inquired at religious-investment-(s), mass-direct-purchase-(ing), centralized-perimeter-terrain-space-(ing-(s)), (as an) activity-diagraming, of planned-surface-area-square-perimeter-space-(ing-(s)), exterior-prioritizing, cubic-dimensional-surface-area, where interior-prioritizing, identify mass-individual-(s),58L. personal-prerogat-ive...¶|||04.m.:45s.93m.s.|||§

§...Preposition-Count-[24 / Twenty-Four]-of.. ..Article-Essay-[18 / Eighteen].. // ..Book-Essay-[128 / One-Hundred-Twenty-Eight].. ..Lines-[58.176 / Fifty-Eight.1.one-tenths-2.seven-hundredths-3.six-thousandths].. ..Timing is superior Dating, Dating is superior subjective-discourse. ¶|||09s.00m.s.|||§

¶129How do any of us formulate an consensus of mass, by signature-oblige, when matter-(s) of surrounding-limit-(s), interact affair-(s), genetically persuading reformation, by an means' of genetic-state-referendum-mass-population-agenda-scheduling? Geological-timing remain simultaneous mass-population-matter-affair-(s),| 472 |when the person-(s)-elected, [maintain // sustaining-sovereignty-responsibility-(ies)] for inhabitant-discourse, where exterior-impact-effect-(s), pace passed general-observation-(s), when individual-(s), remain district-approximate, as natural-expectancy, making independent-impasse, upon locally managed township-(s), where off-time-living, act (as an) social-right, prose indicated, inanimate-action-(s), when substantial-requisite have an difficulty of comprehensive-patience, comprehended by Speech & Dialect, when by vocal-vernacular-tone-pitch-vibration-measure-frequency-ranging, intend interaction from visual imagery-vibrational-recollective-thinking, as while [conversing / reading / writing] formulating data, cognitive reference-data, in of an confluence per individual-influence, yet incrementing formally, theatrical-succession-(s), where citizen-(s)-cultivating-tangible-matter-(s)-capability-(ies), remain evident at territory-impasse, incurring through indelible fact-(s), as independent-individual-(s), offset-great-action-(s), simply from differing-bias-hoc-opinion, when subjective-prevalence per date, fatigue hour-(s)-impassing, when date continue, (as an) singular-injunction, ordering-responsibility-(ies), common communal-ambient, because locational-effort-(s) define physics, when remaining an bit too difficult to physiologically-maneuvering, demonstrating why repopulation is important,| 473 |by an genetic-mass-reproductive-offset, showing why dirtiness of [perspiration /// exterior-filth] embattle individual-(s'), emitting idle spouse-notion-(s), for offset-denominative-system-(s)-installation-interaction-(s), by mechanism over natural-exponent-effort-(s), defining those boundary-(ies) of territorial-limit-(s), by an pre-tense, to perceivable-living, simply-there-of, time-voluntary-impasse-discourse-(s).$_{228}$¶|||01m.: 42s.$_{60m.s.|||}$§ Ability is simultaneous of what other-(s') perceive, (as an) expectancy distance trajectory of territory, (un)-excavatable, for not an common cultivating independent-property, from agriculture-transition-development, when territorial-spacing-(s), cooperate with article-(s) place-(ing-(s)), pre-set at an surface-area, defining human-perspective, cognate-memory-(ies), influencing distance-(ing) persuading youth-adolesance-sustaining, by an auxiliary-device-purpose, where comfort per drive, believe in inherent persuasion, as class-effort-(s), contain The United States of America, as those independent-quality-(ies), not signified singular-total-industrial-development, by each state-space-(ing), (as an) preliminary-obstruction per space, interceding expressive-pursuit of physical-act-(s), when object-(s), remain tension-(ed), where prerogative of compositional-inquiry, affirm when individual mass-impact, can not move mass-inhabitant-(s), from inclination of dating, routine each notion where linear-de-facto, requisite perceptive-conjure, by those perspective-tense-(s), type-manuscript-reading, attention-span, at grandiose-abstract-focus, interceding-adjecture of opinion-(s), with natural-silence, define perspective-respiration-blood-pulse, remaining unconvinced those difficult-action-(s), for impassing voluntary action-(s), rendering utopia-mutual-effort-(s), incapable from birth-offset-juxtaposition-(s), remaining inclined to adequacy, spurred from self-decision-inclination-(s), convened where conventional-practice, is approximate public-effort-(s)-aligning, human-introverted-prerogative, by an relay of synapse-(s), that contort, compact-elemental-piece-(ing-(s)) of matter-(s), believing denoting state, fluently-adjust-perceptive-mark-(s), by offset-territory;254 spaced-off, because daily-parameter-(s), cause each inhabitant an embarking on an continuum-journey, gauging-effort-(s), form those caliber-(s) of act-(ion-(s)), (un)-characterizing-surface-area, by when, instant presentation, not move beyond mark-note-inferencing, because acceded-journey, pre-destiny permeant-federal-influence, as where solid-object-(s), are instructed handling maneuver-(ed)-material-(s), accented through Subduction-Zone-atmospheric-air-vapor-wind-pressure, never digging under those property-(ies) of state, not understood unless conceptive-concerted-effort, remain placed in an concertion of action-(s), as no individual, can notion as mass, through election propitiating, recidivistic perspective upon objective-schedule-succession-action-(s), annotating how progression-(s) of basic-human-need-(s), charter (un)-orchestrated-present, defining physical-parameter-(s)-limit-(s), of boundary-terrain, when perceptive live-tense-(s), impede contracting-development, (due to an) lack of where-with-all, by developing-standards'.229¶|||03m.:51s •19m.s.|||§| 474 |We not work from Subduction-Zone-atmosphere-pressure, in to crust-posit-position-(s), thus mantle-depth, confine from comprehension of reality, so much that national-effort-(s), remain claused for improper university-in-depth-inquiry, as those school-(s) remain without an impact of trade, which merge an magnificent grandiose-territory, remaining barren by mass-individualism-faith, for idle-mentality pass rotation-interval-(s), when fact be simultaneous-deviated [hour / minute / secondths-(s)] of ambiguous-view, through conscious-interior-metropolitan-accounting, balancing hoc-adjustment-(s), as how present be built up, by past-requisite, for maintaining common oversight, by committee-translation-(s), affirming why action is motion-present, where signature-conduct, curb effort-(s)-made, to accede accelerated-data, as serie-(s) of (im)-passing-(s), not remaining capable of physical-method-output, upon circumstantial-lifestyle-(s), reset-(ed) from forward-approach-simultaneous, for balancing-moral-regard, when prerogative of self, position regard at area, (as an) applicating greater direct-value, where limit-(s) of physical-perspective-cardiovascular-respitorial-tense-pulse, are due-upon default-ambition,62L. to live-free, by impractical-flaunt, elaborating self-impasse-motion-(s), where involuntary-limit-(s), pertain lifestyle of exhausted-living...¶|||04m.:51s.47m.s.|||§

§...Preposition-Count-[32 / Thirty-Two]-of.. ..Article-Essay-[19 / Nineteen].. // ..Book-Essay-[129 / One-Hundred-Twenty-Nine].. ..Lines-[63.275 / Sixty-Three.1.two-tenths-2.seven-hundredths-3.five-thousandths].. ..Competent-Verbal-Consideration is superior timing. ¶|||08s. 62m.s.|||§

¶130. An millennium impression century, when century impression decade;255 decade impression months-year;256 as year impression season;257 an season impression month;258 and month impression an week;259 week impression day;260 from day as ruled by time, calendar-date-positing sequential-orders, by an influctuation of data-parameter-limit-(s), availing-focus of organ-tense-(s)-reflex-(es), upon written-subjection, per observation-(s), surrounding terrestrial-ambient, handling-space-(ing-(s));261 by when limit-(s) of family-perspective, intricate surrounding-environment, to partake in modern-review, an group-aligning, for generalizing-article-(s)-function-(s), when letter-word-deviational-reset-(s), define [paragraph-elapse / essay-chapter-period-elapse-(s) / book-tangent-date-reading-period-(s)-elapse-(s)-counting] meant as account-time-interval-period-(s), denominating-secondths-adjusting-(s), for gaining an control-contortion for syntax-predicate, amid mutual-setting-order-time-table-schedule-aligning, fervor as cause hold an regard per [act / movement / applicable-exertion-(s)] simultaneous [verse / converse / inversing-vocal-vernacular-extent-practice] in off-time-requisite, approaching entity of core-common-interest-(s), coinciding consideration, for how calendar-maintaince, approach living, sustem requisite mass-obedience, projecting as when the (re)-lessoning-(s) of mass, remain far too temporal, for literature-alone, persuaded motion-(s)-impact-article-(s), while interceded-conjuring, factor state, by (in)-tangible cite-(ing), dormant library-referencing, virtuing mathematics considered throughout literary-revise, relaying-register greater than syntax, due-to followed-brainwashing, voluntary meaning information, relay-(ed) upon mass-impasse-injunction, apposing mass-prerogative-inducting, from a-the nature of perspective, where mathematical-guide, direct notion-exponent-extent-(s), where examined independent-individual-(s)-prerogative, gather-together, encountering act-(s), defining material-longevity, from dispel of tier-group-redact-interposed-align-inferencing-(s'), successive-relayed, day-monitor-ordering-(s), by when general-progression-(s), sustem from subject-conjure, for objective-documenting, where entity-space-(ing-(s)), offset order-succession-comprehending, rotating day-routine, interposed hyper-tense-(s), for relay of passing-reaction-(s), conjure-(ing)-data,24L. by analysis of acting-use, where each piece of objecting, remain from location-acquisition, as each concept of elemental-resource-(ing), pertain space-tier-date-time-table-order-defining, at mass-interposed-inclination-(s) upon existence...¶|||02m.:13s.40m.s.|||§

§...Preposition-Count-[11 / Eleven]-of.. ..Article-Essay-[20 / Twenty].. // ..Book-Essay-[130 / One-Hundred-Thirty].. ..Lines-[26.32 / Twenty-Six.1.three-tenths-2.two-hundredths].. ..Thinking is ink on paper, or type after the fact of ink on paper. as this book be written, in cursive, initially, and then manuscript-typed for publication.¶|||15s.40m.s.|||§

¶131Calendar-Word-Per-Day-Effect, sustem from county-city-mundane-nature, upon, physiological-major-perspective-vision-hearing-macro-exterior-view-(s), sounding tensable-fatigue-rate-(s), by agreed-conjecture, virtue-(ing) faith-trade-based-debt, informal,

[Article:] [May]

written-inquiry, when individualism perceive human-affair-(s) (from an) non-interactive-preference, not considering why those school-subject-(s) matter, for significant posthumous extraspection, by [[1]arithmetic / [2]algebraic / [3]geometric / [4]trigonomic / [5]statistical / [6]grammatical / [7]benchmark / [8]rubric / [9]syntax / [10]chemistry / [11]biological / [12]anatomical / [13]earth-space-science-(s) / [14]orchestra / [15]dance / [16]theatre / [17]film / [18]picture-(s) on an median-canvas;[262] depicting-data from [read / write / rate / document / integer-(s)-inductive-influctuation-(ing)]] comprehending language, incumbent literary-documenting, attempting to persuade community-effort-recourse, referencing article-book-paragraph-data, by means' of tensable-impasse-perspective, relating with other-individual-(s), an approximate-discourse, to [[1]refer / [2]infer / [3]place / [4]current / [5]observe / [6]document / [7]verify / [8]confer / [9]debate] numerical-operation-reoccurring, simultaneous-sum-(s)-factoring, in variation-(s), per-of offset-distance-(s), perceiving still-pulse, where (hyper)-tense-(s) weight compositional-mass-matter-limit-(s'), [subterior celestial-grandiose-facet-(s) / tectonic-plate-(s)-offset / ocean-water-body-(ies) / Political-State-National-boundary-(ies) / County-City-Boundary-(ies)] secondths-time-line-deviated, (hundredths)-millisecond-comprehensive-truth-(s), where positional-offset-(s), can not verify sum-(s) of article-data, regarding vocal-vowel-consonant-word-article-mutation-(s), offsetted by matter-(s), yet intercontorting syntax-redacting, synthesizing written formal-tangent-(s), by individual-perceptive-hyper-tense-(s)-posit, meriting quality-(ies) of day, by hour-subterior-moment-(s), making impasse of thought, be how pronoun-discourse, approximate-location-impasse-motion-celestial-hour-deviation-effort-offset, because purpose upon secondths-continuum-effect, not interpret an independent-inherent-article-(s)-interpreting, by birth-to-present, presence in transit, comforting house-residence, by pre-requisite-idling, not actively using what resource-(s) pertain upon existence, because secular-matter-(s) affirm delineated article-(s), not actively comprehending, mystified belief in biblical-Chaplin-appreciation;[263] not wrong or right, but awaiting weather article-(s) can be individually-referenced for comprehension of subversive-comprehension, conscious-requisite present, from when surface-area-offset, virtue setting, interacting human-candor, upon lifestyle-pattern, where inhabitant-(s) tend to have an disinterest of interacting-impasse-vernacular-dialogue, for when complacent determinant-bi-nominal-numerative-denominative-fraction-standard-deviation-{calculus}-oscillation-kinetic-wave-frequency-influctuation, occurring without measure of letter-word-sentence-paragraph-essay-article-(s)-note-(s)-synthesis;[264] determined count-(s) advise visual-color-depth-distance-perspective-texture-shape-(s)-imagery-scenery, juxtaposed centralized-place-hearing-posit-(s), permitting wave-sound-reverberational-vibration-(s)-resonance-sound-frequency-level-(s), juxtaposed seeing-hearing-tense-posit-(s);[265] impassing notion-mass-unconscious-responsibility-(ies), expected amongst settled interior-lifestyle-(s), claustrophobic lifestyle-location-(s), not inquiring expanse-vacant-spacing-(s), pre-conceptive-constructive, vacant-lot-territory-observation-(s), determining mass-development, by fixated-posit-(s), because lease-mentality, sustain an derived perspective of mass, not aware of listening for an ultimate-live-reference-reasoning, because control of data-usage, concur independent-(ce)-(s'), ignoring exterior-international-trade, not realizing realty, where space-governmental-housing-title-offset-(s), suffice landlord-title-proprietorship, arised instated pre-requisite-effort-(s), living by tense-(s) of calendar-existence, silver-spoon-(ed), official-duty, existing upon vocal-decibel-vernacular-timbre, rating beat-(s)-per-minute-tempo-elapse-(ing-(s)), from prose Through & Gamma-Differentiated, tone-quality, per pitch-four-note-(s)-per-measure, as quarter-note receives-one-beat{4/4}timing-per-measure, measuring, (in error of timing to comprehend fact-(s) by state-developmental-space-(ing-(s)), incurring-debt-history, counted-of, post-extemporaneous evaluat-ion, when character-role-posit-(s), [[1]father / [2]mother / [3]child / [4]peer-(s') / [5]neighbor-(s') / [6]self / church-member / colleague-(s)] serve grand-requisite-interpose-challenge-(s), inhibiting blood-pulse respiration-impasse-effort-(s), by when an nature of silence, inform individual-(s') by [attention / caution / warning / alert / advisory] to [remember / remind / recollect] hardware as receipt-data, periodicalizing commercial-physical-objective-creative-(ity), for influencing piece-(ing-(s)) of article-issue-(s), comprehending [schedule / timed-act-interval-count-(s)] juxtaposed theatrical-ambient, due to [physic-(s) / earth-space-science-(s)-tectonic-Plate-(s)] when their remain an expanse that be, because [solar-total-mass-exponent-algebraic-weight // Earth-solar-core-impact-total-mass-exponent-algebraic-weight /// Tectonic-Plate-(s)-offset-political-map-deviation-(s) of perceptive-posit-(s)] are weight-(s)-offset, an denominative-value-(ing), rank-ordering, managerial-decision-(s), in an commercial-determining, from capacity of human-perceptive-act-(s), remaining interposed, fragment-increment-data, when synthesis-comprehension sustem-from subjective-discourse-deviation (cum laude limited off), by order-(ed)-factor-(ing-(s));[266] followed by governmental basic-human-right-(s'), influencing mass, from communal-prerogative-(s'), affirming individualism, for grandeur of interpreting, not total per centripetal-force-impasse-(s), variating data-confirmation-(s), because time-posit-(s)-elapse, without synapse-pick-up-verified-data, occurring by article-comprehension, tabling mass-observation-(s), currenting energy-effort-(s), labeled upon [[1]crop-(s) / [2]tree-(s) / [3]soil-deposit-(s) / [4]government-pavement / [5]plant-(s) / [6]product-objective] amassing conceived off-time-requisite-(im)-passing-(s), central rotation-year-(s)-existence, because perspective is (from an) false-aesthetic-by-self, restraining physical-effort, upon existence, when objective-(s), have an engineering-extraspective-balancing (per) of hardware, pertaining responsibility-(ies), not (yet) pertaining data, upon conjuring of individualism-effort-(s), as mass-(es) of ethnic-retardation, affect communal discourse of all perceivable-aspect-(s), where genetic-linguistic-(s)-individualism-fault-(s), continue portrait per medium-canvas-conceptive-language-skill-(s), remaining greater than preference-of, physiological-inquiry, when well-accustom-living, remain prevalent present-vision, wave-vibrational-intonating-thought, as how future be of sounded-data, vocal-verbal-vernacular-tonal-vibrational-wave-intonation-exponent-offset, an pitch-quality, zealous physical-fatigue, unconscious scaled independent-mathematical-grammatical-article-incrementing,| 475 |affirming presence at live-existence, seeped from aging-impassing-(s), due to matter-(s) of state-(s), pre-requisite-live-vision, collecting information, pass exterior-self, in eve where genetic-extraspective-prerogative-(s), define genetic-disposition of state, affecting extenuating-factor-(s), that remain collaboratively-collectively-pro-portionally-simultaneously-syncopated, from origin-(s'), questioning proportional-mass, not individuating particular-difference-(ing-(s)), for how mass-enabling, follow substantial-guideline-(s), altering collective-fault-(s), (as how a the) curb of tectonic-plate-state-development, may be for payment-(s) (per) of (offset) reproductive-impasse-method, lazing religious-effort-(s), by white-mass-commercial-eye-color-ethnic-deviation-(s), deriving national-offset-(s), per pre-requisite of timbre-discourse, regarding human-preference, understanding austerior-posit-(s'), halo ambient-ambiguous-surrounding-interval-(s), Subduction-Zone-anterior, crust-nth-miniscule-effect-(s), perceiving interior-self-surrounding-limit-(s), present our natural-effort-(s), conceiving life, by trade-(ing) human-effort-(s), due to [calendar-period-interval-(s) / decade-cycling / generation / era / century / et cetera….] contrary Limit & Fatigue partaking intermittent daily-development-(s)-impasse, verifying parameter-dynamic, from day-to-day-primordial-limit-(s), aged-tectonic-impasse-reference, discoursing century-millennium-progression, by century-living-reproductions, not exponent-(ing) women reproductive-offspring-cycling-(s), (due to) controversial-fatigue while trading, bellied unilateral-bi-partisan-politic-(s), slating state-operation-(s)-impact, affected from foreman-perfect-tier-(ing), of populant-(s')-dynamic-impact, where wage-trade-orientation-(s), continuum-existence,90L. by alternative-mass-(es)-juxtaposition-trading, during daily political-orientation, due to daily-work-labor-purpose-(s)...¶|||07m.:11s.00m.s.|||§

§...Preposition-Count-[31 / Thirty-One]-of.. ..Article-Essay-[21 / Twenty-One].. // ..Book-Essay-[131 / One-Hundred-Thirty-One].. ..Lines-[90.71 / Ninety.1.seven-tenths-2.one-hundredths].. ..Thinking requires extraspective-individual-input upon political-topical-executive-cabinet-matter-(s).¶|||13s.28m.s.|||§

¶132How does Solar-dynamic affect physics, interposing competency per [day-affair-(s) / tomorrow / yesterday-living-mutual-suffice-life-style], where reliant-independent-borrowed-matter-(s), sustem-from Federal-Guideline-(s)-inner-working-(s), personal-House-Development-Congestion, erroneously reasoning, unthought impasse interval-existence, modernly-sufficed, community-article-(s)-means', as still-mark-(s), Americanized by enabled-offspring-lineage (which is great) due to routine of historical-impasse-recognition-mark-(s), tiering apprehension-(s) [tangible-hardware / crop-(s) / cultural-piece-(s) / territorial-parcel-(s) / objective-schedule-(d)-routine-(ing) / Off-time-interactive-dynamic] as how mode pertain parameter-(s) per article-(s)-place-(ing-(s)), uncoordinated perceptive-individuation-deviation-(s), from [fan / member / athlete / crowd-dynamic-(s)] entertained greater-than inhabitant-lecture-listening-interaction-(ing-(s)), resonating vocal-vernacular-verbal-significance per Day & Age when self muscle-compart-mentalization-(s), celebrate celestial-subduction-zone-pressure-inlay-perspective-(s'), ecology-ically shadowed by mere-effort, as when recuperated fatigue of terrestrial-ambient-transition, document millennium-aging, juxtaposed each organism apose celestial-parameter-(s), because how development, of surrounding-(s) impasse an [millionths-block // billionths-city-district county-dynamic-developing] not enticed an genetic-reduction of acre-effort-(s), apose idle-payment-contingent-sustaining, from independent-individual-effort-(s)-house-grounded-impass-ing [[1]long-sleeve-t-shirt-(s) /

[2]sock-(s) / [3]underwear / [4]bra-(s) / [5]toaster-(s) / [6]paint-brush-(s) / [7]paint-(s) / [8]table-(s) / [9]vehicle-(s) / [10]train-railway-system-(s) / [11]helicopter-(s) / [12]can-(s) / [13]glass-(es) / [14]window-(s) / [15]textile-(s) / [16]toilet-(s) / [17]bathtub-(s) / [18]tile-(s) / [19]brick-(s) / [20]cement-refining / [21]musical-instrument-(s) / [22]commercial-truck-(s) / [23]packaging / [24]truck-trailer-(s) / [25+]etcetera...] by elemental-exterior-value-(s), naturally-resourced and collected, for production of tangible-object-(ive)-(s)-product-mass-common-means', verifying conceptive-schedule-(d)-request-(s), confirm-ed by reasonable-article-(s)-schedule-usage-(s), when objective-analytical-discourse-recoursing-(s), fervor affluent independent-citizen-(s'), as why State & Lifestyle, remark acting-participation of terrain, when relied independent-act-(s), verify ambiguous-developmental-commercial-value-(s), continued extraspective tangible-hand-led-entity-spacing-(s),| 476 | involuntary proportional-data-document-verification-(s), dated from past-ordered-information, present upon daily-entity-tensable-tangible-development-(s), cycling mass-effort-affair-(s), maintaining invigorated purpose from receipt-data-collection-(s), envisioning object-(s), at posit-(s) where weighted-volume, verify natural-cause, because human-extraspection, enable male-off-time-duty-aging, unconceptive patent-production, amongst present individualism, as why post-humorous-understanding, depict present debt-impassing-(s)-operation-(s), affirming territorial-handing-difficulty, when-where linear-major-study-method, influence conscious-interaction-(s), as how perspective pertain living-juxtaposition-(s), for scaling mutual-mass-offset-common-weight-(s), balanced-compositional-offset-individual-posit-reason-(s), comprehending commercial-production, from knee-jerk-reaction-method of wage, substantially inclined, purchase-parity, because elemental-objective-reliance on trade, sustain an informal-clause, conditional temperate-climate-topography-offset-(s), not measuring distance-(s) of depth-perspective, spherically-concepting, each desire of corresponded-disseminated-data conducting-present-requisite-article-subjective-constitutional-affair-(s) unemotionally-intertwined, due to blank-post-subjective-thinking,

where monetary-currency-denominating, affirm date-largo-fluctuation, juxtaposed time-minute-(s)-hour-moderato-fluctuation, tempo-timbre-pacing, placed-posit-requisites', of spacing-reset-resource-objective-(s), when movement-(s) of impasse-exaggeration, feature feet-measure per secondths-deviations', clear per time-posit-(s), where pertained journey, story-board, human first-floor-limit-(s), creative by American-Treasury-debt-comfortability, effort-exhibited, breve-inclination-(s) from-of Solar-impact calendar-timing-(s), when each date is imbued past-requisite-fixation, greater-than, past-future-article-dating, pertaining word-term-article-(s)-data, from word-contortion-present-year-decade-interval-(s), pertinent, live-data, affirming continuum future-week-(s)-cycle-act-anticipation-(s),| 477 |verifying physical-observational-written-documenting-requisite, where religious-word-structuring, validify various-impasse-outlet-(s) of interval-pick-up, including daily attempt-(s) at extraspective-honor, resource-(ing)-(s) collect-(ed) mass-individual-wage-scheduling-elemental-article-(s), ordering how state-affair-(s), crop-propitiation title-residence, faltering greatly, agriculture-fertile-soil, by when those relative-residence-(s), sit idle in surreal-infamy, dull-idle-communion,52L.| 478 |enabling how belief in timing-(s) of date, pertain collective-surrounding-information, per article-(s) of independent-individual... ¶|||04m.: 28s.48m.s.|||§

§...Preposition-Count-[15 / Fifteen]-of.. ..Article-Essay-[22 / Twenty-One].. // ..Book-Essay-[132 / One-Hundred-Thirty-Two].. ..Lines-[52.55 / Fifty-Two.1.five-tenths-2.five-hundredths].. ..Consideration is difficult to expect out of introverted-individual-(s).¶|||10s.31m.s.|||§

¶133...Why not Gardner perfection through physiological-activity-(ies), Balance-(ing) & Verify-(ing), [visual // tonal-vernacular-observation-(s)], affecting objective-creative-production-(s), from book-subjective-article-interaction-(s), [gesturing // elaborating elapse-(s) // syntax-(s)] where psyche-impasse-(s) revolve throughout existence, scheduled intervoluntary affirmed state-mattered-affair-(s), because national-responsibility-(ies), articlize pertinent-matter-(s), mass-impassed-accounted, from-of-a-the Muslim-African-Latin-pseudo-European-cent-state-(s), affecting each European-American-Australian-Chinese-Indian-Russian-State-Dollared-Mass-Assumption-(s), visa-exchanging-(s), for how Human-Idle-Behavior & Jailing-Imprisonment, affirm dollar-cent-international-trade-person-article-constitutional-transaction-(s) not giving an historical-reference, of a-the Holy-Roman-Empire-Method of Currency-Exchange, going through an moded-cycling-(s), of monitored-trade, where past-substantial-impact affect present-objective-credit-trading, as fabricated by conceptive-understanding, fatigued from daily-event-(s), affecting sleep-impasse-unconscious-pattern-(s), where pieced-matter-(s), impasse elemental-compos-itional-developmental-subjective-article-(s)-fact-(or)-(ing), as prior-day-effect-impasse, variate sleep, when life-time-cycling, influctuate [major-visual // tonal-deviation-(s)], by sixty-year-genetic-offspring-repopulation-calendar-cycling-(s), apose minor-decade-reproductive-calendar-cycling-(s), inclined unanimous-bi-partisan-decision-(s), applying impasse-setting-objective-article-(s)-focus, upon pre-setting-calibre, instilled comprehensive-instructive-usage of hardware, by-an technical-stance, apose creative-individual-interpretation-juxtaposition-aspect-(s), contin-uum-defining title-proprietorship, entrenched aged-nature, not aware from denominative currency-amount-(s), how celestial-orbit-year-(s), impasse [week-(s) of month-(s) /// quarterly /// bi-annual /// seasonal-past-prior-future-year-present-extent-requisite-pattern-(s)],| 479 |by Four-Hundred-Thirty-Eight-Year-(s) of Two-Thousand-Sixteen-Written-Present-{Spring}-future-publishing-deviating, because how habitual-patterns, designated lot-proportion-(ing)-(s),| 480 |as posit-(s) remain one-by-one-offset-independent-extraspective-unilateral-matter-(s), undelved interval-(s) from [task / act / chore / communication / movement / text / reading / placing / period / day-sequencing-(s) / et-cetera],| 481 |tensing impasse-movement-(s) when-where subterior-fact-(s), (constitutional-data) underlie constitutional-common-fact-(s), as perspective (major-articling of constitutional-data) define-affirm individual-state-accounting-impact-discourse, offsetting climate condition-(s) during discourse, where at an stance of recited-contorting, in part for how, reciting data, is through impasse-occurrence where identity confirm date-verification-(s), exceling road-mile-(s)-meter-(s)-visual-reset-county-impasse-meander-virtue, affirmed where at daily-practice-repetition-conjecture,| 482 |interpose-incremental-deducting, outdated mass-observation, because how monetary-effect-(s), influence ruling-(s) of data, where dynamic of existence, pertain human-voluntary-impasse-effort-(s), (for an)33L. off-time-extraspective-mass-schedule-impasse-comprehension-exchanging...¶|||02m.:44s.44m.s.|||§

§...Preposition-Count-[11 / Eleven]-of.. ..Article-Essay-[23 / Twenty-Three].. // ..Book-Essay-[133 / One-Hundred & Thirty-Three].. ..Lines-[33.465 / Thirty-Three.1.four-tenths-2.six-hundredths-3.five-thousandths].. ..Timing refers to [Hours / Minutes // Second-ths].¶|||08s.84m.s.|||§

¶134Entitlement without proprietorship is blatantly-flagrantly retarded, as juxtaposition per [superior / inferior-aspect-(s)] understand guidance, requisite mass directive of article-(s)-data, affirmed subjective-inclination-(s), apposing timbre-metre-movement-(s), upon distance-measure-(ing)-(s) of timed-impasse-physical-subjective-movement-(s), coinciding data-logging-(s), by incumbent-impact-(s) of mass, where construction conjuring, continue effort-(s) of payment, when sleeping-fatigue, wear on a-the unconscious-impact of human-mass-existence, not comprehending active-invigoration-(s) of mathematical-scientific-analyzing, each post-study-(ied)-(ies), by commercial-work-place-impasse-merit-(s), inconclusive incarnate-imagery, through real-estate-property-scheduling, extraspective-perspective, inquiring no means element-composition-(s) pertaining schedule-conceptive-discourse, as agenda pertain upon mass-individual-(s') (ism), swearing on trade, for default debt-trade-transaction-(ing) (s'), at public-commercial-lease-access-retention-(s'), when month temporal-sale-(s), identify discourse from day-article-reset-(s), per quarterly-mass-conjunct-impasse-update-(s), denominating year-to-date-business-(es), (un)-incorporated by national-requisite-prose, where [New York / Washington D.C. {District of Colombia}] amount sociological-mass-swear-debt-purchase-(s), detailing deadline-(s) of repopulation-objective-article-(s)-usage-(s');[246] scaling parity by an consensus of interactive-affair-(s), when carbon-dating rely, active-article-(s)-appraisal, for according-semi-articulated-truth-(s), judging from [extraspective-envoy-(s) // introverted-basis-perspective-(s')], forced-vessel-interceding-(s), disliked by peer-(s), at time-(s) when longevity affect dating-mentality, through natural-fatigue, awake blood-pulsed-perspective upon centripetal-force, defining mind by dissolution of dating, timed absolute-retirement, by an barren-requisite of [crop-(s) // mammal-horticulture] from farmer-agriculture-method-(s'), agrarian by common-regard, sufficing annual interposed subjective-article-(s)-objective-data-filing-(s), by Internal-Revenue-Services, confirm-ing time-dating-(s), preinstituted, mass objective-purpose-(s), articulated social-class-talking-requisite, present-economic-effort-(s), not naturally inclined, mutual-common-impasse-article-(s)-effort-(s), pertaining off-time [reading // redacting] intermittent neighbor-(s)-intrigue, syntaxing educationally-applicable population-observation-(s'), uncited present-dating-fatigue, by contin-uum-past-requisite, furthering future-day, by present-day-impasse, referencing from past-dating-act-(ion)-(s), present-day-method-effort-(s), incapable by-of Grande-territory, defining extraspective-impasse-limit-(s), voluntary from introverted-independent-individual-(s'), subver-sive sleeping in contraposition of awake-active-motion-timbre, never prevalent by tense per motion-time-impasse, counting in coincidence, where each routine cycle amidst, centripetal-force-abstract-latitude-longitude-revolving-revolutionary-momentum-drift, conceiving [day night-grandiose-effect-(s)] inductive by [sound-(s) / sight-(s)] exterior where boundary-limit-(s), levitating molten-gaseous-plasma-density-liquid-pressures', [under // above // central] our natural-comprehension (phenomena), by interactive offspring-commercial-impacts', invigorating vocal-tense-interaction-(s), awhile work-place-forty-hour-fatigue, transit upon an week-hour-(s)-denominative-impasse, one-hundred-sixty-eight-hours, sleep-cycle-fatiguing, in an opposite forty-hour-(s)-recuperation, numerative-valuing, juxtaposed weekly-denominative-subjective-article-interpretation-comprehension-limit-focus-(es), having how week remain with eighty-eight-hour-(s) or so (transit-impasse), of [market-place / idle-entertaining] (un)-pertain-(ed) dormant-lifestyle-activity-(ies), not repopulating an mass conscious-living-requisite, admiring daily-impasse-(s), of Trade-(ing) & Saving-(s), offsetting those depressing-matter-(s) of international-state-county-city-district-monetary-article-objective-trading-(s), bemused transit-impasse, when still of physiological-effort-(s).230¶|||03m.:33s.03m.s.|||§ How does any human-mass, extra-spectively-evolve from modern-work-schedule-routine, when those production-(s) of factory-(ies), not align naïve-nature, from constitutional-basic-human-right-(s')-backgrounds', separately defining [male // female organ-subjective-requisite-(s')], assented through vehicle-impasse-clearance-(s), not contacting, surrounding neighbor-(s), annexed (sub)-urban incremental-work-payment-trade-sustain-(ing)-(s), following, resolute-interaction, managed by temporal-guide-directive-(s), locally-set, upon centripetal-force-drift, obscuring physiological-capability-(ies), when dating remain without an interval-scaled-conjuring, per conducted affair-(s), alleviating burdens of daily-duty-(ies), discoursing [1territorial-space-coverage-claim / 2objective-production / 3agricultural-crop-(s) / 4horticulture-mammal-(s) / 5clothing / 6Housing / 7etcetera perceptive-dated-time-motion-(s)-cycling] amongst parameter-documenting-(s), confided human-physical-perspiration-capability-(ies), creative as language apply from literature, upon an median [canvas / parchment blank-measured-neutral-color-subjective-article-premise], finite-tuning, those [read-written-inter-pretation-(s) of [1]letter-(s) // [2]word-(s) /// [3]syntax-synthesis //// [4]paragraph-article-subterior-decimal-matter-focus-sort-concentrating ///// [5]essay ////// [6]chapter /////// [7]book-pertained data] post-humorous from perspective at location, motioning subjective-articlizing, thus acting objective-posit-(s)-function-purpose.231¶|||04m.:42s.94m.s.|||§ This enable understanding, for defining present-date-to-year-impact, influencing traded-act-(ion-(s)), continuum aged-capabilities, where tectonic-decade-scheduled-planning-(s), elicit centralized population-schedule-agenda-cycle-ordering-(s), not covering an tectonic-plate, per decade, (from an) mass-genetic-schedule-agenda-perspective, contexting production, from county-population-tier, concerted interval-article-objective-subjected-data, interposing self along succession articlizing, litigating-extraspection, intriguing interaction at common-basis-human-needs-talking, discussing [matter-(s) // affair-(s)] when serviced extraspective-interaction-(s), not value individual-(s'), featured-action-pertinence... ¶|||05m.: 19s.88m.s.|||§

§...Preposition-Count-[20 / Twenty]-of.. ..Article-Essay-[24 / Twenty-Four].. // ..Book-Essay-[134 / One-Hundred-Thirty-Four].. ..Lines-[64.07 / Sixty-Four.1.zero-tenths-2.seven-hundredths].. ..How to plan on an entity-offset-ecological-scheduling, of human-hourly-wage, is an governmental evaluation by balance from word-observation-competency, evaluating weather an individual can or can not handle those circulated-objectives, for maintaining an entity-(ies) ob jective-shelf-production-trading.¶|||19s.90m.s.|||§

¶135An cul-de-proxy, interpose [city // county library-reference] apose Congressional-National-Library-Consortium-article-(s)-reference, thinning data purpose, by citizen-impasse-interposed-influence, ascertaining extraspective-interaction-(s) of human-relation-(s), in development for state-article-(s)-affair-(s), affirm-(ing) longevity per mutual-community-effect-(s), verifying vindicated-article-(s)-interpreted-continuum, regardless, mutual-inhabitant-cooper-ation,| 483 |because [introverted-bias // hoc-effort-belief-(s')]| 484 |surmise centripetal-force-revolving-revolutionary-tectonic-plate-state-nation-trade-article-impasse-(s), in recognition of human-recognition-effort-behavior.232¶|||32s.40m.s.|||§ Repetition of work-place-developmental-effort-(s), allot self-experience, upon perspective article-(s)-action-conjuring, from-by muscle-blood-pulse-tension-cognition-reaction-(s), aligning physiological surrounding parameter-(s);247 for feature-(s) of data, when dating present-continuum-sleep-reset-(s)-cycling-(s), of human-mass-offset-(s), fatigue indelible-movement-(s), where impassed-belief, interpose data-contortion-elapse-fatigue, time-elapse-(ing) (s), [reverberated // wave-vibrational-individual-conception-(s)] subjective-objective-article-(s)-extending, notion-interval-extent, identifying human-character-(istic)-(s) per mass, as when executed action-(s) of terrain, limit human-effort-(s), (un)-characterized monetary-trade, where individual-article-(s)-impasse, place locational-settlement-(s), for discoursing tempo [beat // rest-(s)] measuring secondths-interval-progressive-impasse-elapse-(s), experiencing simultaneous-mass-requisite-interactive-article-(s)-reset-(s), character-izing subjective-article-(s)-word-data, preserved-repetition, aligning subterior-mass-tier-identity, verifying schedule-verification-(s), resulting default-parameter-(s), offspringing by compulsive-behavior-awareness, children of nominal-transit-retained-volume-means', of state-propagated-past-article-(s)-reference-assurance.233 ¶|||01m.:35s.63m.s.|||§| 485 |Vocal-vernacular-vibrational-tempo-sustained-tone-inclination-indication-intonation-count-(s), virally vitalize empirical-subjective-juxtaposition-(s), resonated cubic-surface-area-setting-(s),| 486 |for figurative-comprehension-fixation, by [individual-posit // commercial-extraspection] factoring consistency of aging, from spouse-momentary-extraspective, in lieu of lifestyle limit-(s),| 487 | ordered by [base-A / adjacent-B // Hypotenuse-C-lined-demarcation-(s)] of triangular-enclosed-cycling-parameter-(s), forward-depth-line-aligning, discoursed interior-angle-(s) of solar-physics, when upon (un)-tangible-impact of natural-rotational-drift, revolutionizing mundane-impasse-article-(s)-form-(s), intercorolary valuing introversion, as primary-characteristic-(s) of humanity, tied in to, respited-behavior, conduct formal-impasse-amicable-affair-(s), elicited human-wage-function-(s), serving responsibility-(s), when metric-transiting, interact vernacular metric-momentum, [repeated // exaggerated mass-motion-(s)-directive-elemental-extraction-purpose-(s)] gamma-inquiring, human-developmental-effort-(s), examining metropolitan-climate, per scale of data-distancing, covering gas-station reset-offset-(s), where inhabitant-space-(ing-(s)), locate-intelligence, displaying ordering-(s) of time-date-calendar-posit-(s)-interval-(s), affecting date-aging-cycling-(s), pertaining minute-gamma-posit-numerative-denominative-integer-deviated-approximation-(s), by mass-relative-comparison-(s), impassing live-activity-(ies), [where // when] this book {written {December 2015 – April 2016} // typed {April 2016 – July 2016} // edited {September 2016 — June 2017}} deduce exterior-value-(s), upon relative-interior-perceptive-offset-(s), from mass-individual-(s), whom tense existence, through an workman's'-anxiety, for referring-prior, aging-prevalence, by birth-offset-posit-(s), myself-{April 25, 1991}, apose {Age-25.12328767 to 26.65205479}, adjusted by dating, whole year-posit-(s), apposing decimal-article-(s)-adjusted-numerative-fraction-calculating, [integer // whole (roman)-numerical-standard-deviation-(s) per division-(s) at-incorporated constructive-entity-rate-deviation-(s)] age-perspective-juxtaposing, age-month-action-rate-interval-(s), offset influential fluent-inter-pre-tation-understanding-(s), following instructed logistical-reading-manual-bible-instructions-muni-cipal-sign-aspect-conjunction-(s'),| 488 |competency deviated, percentile-Mass-Briggs-Myers'-self-monotheistic-redacting-letter-word-term-article-interpretation-English-data-tangent-(s), (hoc) where paragraph-count-(s), age intangible-mysticism, motioning subjective-lesson-quality-deviation-(s), under period-(s) of hour, supplemented communal-voluntary-interaction-perspect-ives, calendarizing place-posit-(s)-setting, comprehending why rule of reasoning-(s), recon-aissance objective, in lieu of collecting-tangible-data, aware of aging-continuum-crescendo, not decrescendo-(ing), efforted-activity-(ies), balancing physical-energy-effort-capability-(ies), when conforming congressional-activity-(ies), rate-apose, mass-civilizational-action-article-effort-offset-(s),52L. defining data-period-(s), in rule of order-(ing)-(s)...¶|||04m.:31s.98m.s.|||§

§...Preposition-Count-[21 / Twenty-One]-of.. ...Article-Essay-[25 / Twenty-Five].. ..Book-Essay-[135 / One-Hundred-Thirty-Five].. ..Lines-[52.19 / Fifty-Two.1.one-tenths-2.nine-hundredths].. ..Leaf-Clams casino.¶|||08s.13m.s.|||§

{January-(31)-February-(28 / 9)-March-(31)-April-(30)-May (31)-Month-(s)-Day-Hour-Minute-(s)-Secondths-Count}-{13,046,400 D.H.M.S. = 60S. × 60M. × 24H. × D.}-{3,624 H.}-{217,440 M.}-{151 D.}

{Article:} [June]

¶136§¶¢Moment-(s)-perplex when other peripheral opposite-space-visual-effect-(s), tense signify cause by mass-population-impasse reproducing-ethnic-article-(s)-offset-(s), that affect each deed-title-place-cubic-space-(ing(-s)), during day-perimeter-fact-(or-(ing)),| 489 |while day-elapse-(s), serve as {3,600S. × 24H. = 86,400-Secondths}, where hour-(s) remain [plug-(ged) / draw-(n) / place-(d)] by an place-posit-setting-article-(s), conjuring how daily-affair-(s), work by seldom, impasse-clearance of height, offset-(ed) centripetal-force, as temporal day work-clause-order-(ing)-(s), (are) due to origin-notion-development-(s), convinced where cause of fact-(or-(s)), have an impact second-person-character-error, interacted when default effort-(s), note notion-extent-(s), having an historical-account-(ing-(s)), tense-(d) where engagement by Bias & Hoc, exhibit blood-pulse-kinetic-active-perspective-effect-(s), gauging valid-physical effort-(s), because natural-prevalence, remain upon work-load-activity-(ies), from treaty-accord-(s), suiting posit-(s) of tectonic-matter-(s), instilled where building-block-(s), are articlized, per [objective // subjective] tangible-piece-(s), juxtaposed free-will-inhabitant-perspective-(s'), for when article-fact-(s) sustem type-(s) of time-(d)-schedule-(d)-effect-(s), aligning rigor upon each work-week, elapse-conjure-(ing)-intermission-(s), when [intermittent / amidst / during / fatigued / upon / while momentum-aforemention] beyond year-day-requisite-(s), as when [era / legion-(s)] of men pertain mode of extraspective-denominate-matter-(s), from abstract-natural-posit-(s), (as an) synthetic-perplex-(s),| 490 |of (un)-reviewed-free-will-population-breeding, tensing physical-objective, from ascertaining where article-objective-fact-(s)-origin, reoccurring, temporal concealed-parallel-depth-base-passing-(s), subducted crust-posit-placing-(s), abound scaled weight-(s), acknowledging [1.account / 2.factor / 3.managing / 4.accounting / 5.deliberating / 6.transit-operating] effect-(s) of effort-(s), in coverage per measured-surface-area-land-space, individually-possessed, when future-person-responsibility-(ies), elapse-(s) duty, servicing distributed-tense-(s), worn those verification-(s) trade, supposed national-article-(s)-transition, sustaining-an method of technical-science-rate-(ing-(s)), devised from survey of relative-reflection-(s), adjecture declarative-space-(ing-(s)), ratified reset-(s) of activity-(ies), offset placed-cubic-space-(ing-(s)), | 491 |that have manuscripted-commercial-method-interaction, stop informed-practice, "by the book", at Effort & Place, uninvigorated by literature, for mass-American-English-Language-usage, sustemming default-education, instituted-imposed, by symbolism of velocity, not involved by lecture-place-location-(s), upon transit manifest-destiny, as for thinking how our reproductive-process is applied, in an series of base-primary-focus-(es), requisite-future pre-indoctrinate-acceptance, from [age-(s) / subject-contort / cognitive-memory] that influence illiterate cycling-(s) of data, without-an vernacular-human-right's-limit-(s), for conscious-physiological-tense-(s) mass-effort-(s)-proportion-defining.232¶|||02m.:28s.59m.s.|||§ So why continue to read? While all of our living existence, prefer soft-physical-act-(s), greater-than [reading / writing / redacting / syntaxing / genetic-dialogue-interposing] registering religious-pastor-presence,| 492 |from-an intercorolary-adjunction, comprehensive individual-impact-impasse-purpose-aging, when cardboard-box-in pattern-progressive-sequencing-(s), affirm loyalty through vocal-intelligence, as story-(ies) compile, confided by bemused-effort-observation-(s), maneuvered subconscious-tense-(s), conjuring date-(ing) from past-numerical-operation-(s), pertaining roulette-date-week-interval-(s),| 493 |per day-count-(s), present-infinitive personal-purpose, adjunct population-matter-(s), because as each factor of industrialism, remain incredibly difficult for personal-retention, no mass has made a-the reasoning to collect objective-article-(s), when governmental-political-focus attend mutual-interposing of past-date-article-data, written-conjured, for documenting each inhabitant, apose [person / inhabitant] lifestyle-dynamic, insisted physical-enabling, for adjoining judiciary-revise, when credit of denominative-currency, pass revolution-millennium-impasse (not correct), as considered by those simultaneous secondths-tier-deviations, sorting numerical-ordering-(s), by literary-book-shelf-comparison, apose present-Gregorian-year, juxtaposing numerical-tense-article-progressive-ordering-(s), defining space-terrestrial-usage-(s), in regard of off-time-stress, from calculus-analytics, far too in-depth, basic-conjuring, when continuum of article-(s)-data-conjecture, factor issue-(s), from premise of constitutional-observation-(s), articlized uncharacterized-ordering-(s), by "lah-sey-faire"-flaunt, incapable perceiving how claused-trade, define article-perspective, at impasse-space-(ing)-(s), denoting major-focus per terrestrial-space, [minor-subterior / major-subjective-subjunct] when ambiguous-fact-(s), remain impartial by trade-confidence, discoursing article-(s)-fact-(s), by disseminated article-(s) balancing, verifying how leadership, count common-role-subterior-relay, when human-interaction, not wage-payment-solely by objective-count, intermittent act-(s) for land-territorial-cause, because factored-article-(s)-discourse, figurative-conjuring, date-impasse-ordering-(s), fictional present-past-minded-society-value-(s);248|494 |when native-liberal-inbred-perspective, unfold extraspective-power-usage, conceiving how directive of inhabitant-order-discourse, affect each of those fact-(s) of passing-interval-(s), as day-elapse-(s)-reset-positioning, retain an variation of [week-(s) / day / month-(s) / season-(s) day-interval-cycling-(s)] affecting why each thought follow order of person-ordering-(s), not being participated, where inhabitant-mass-count-(s), sustem from influential-effect-(s) of matter-(s), affecting how inanimate-affair-(s) of state, moving posit-(s) for maneuver by general-context, per personal-affair-(s), rescinding no extraspective-yearly-movement-(s), when time continuum document apose calendar-dating, for of an rationale per interaction-(s),62L. premised upon extent of interaction-(s)...¶|||04m.:47s.54m.s.|||§

§..Preposition-Count-[30 / Thirty]-of.. ..Article-Essay-[1 / One].. ..Book-Essay-[136 / One-Hundred-Thirty-Six].. ..Lines-[62.43 / Sixty-Two.1.four-tenths-2.three-hundredths].. ..Extraspective-interaction-(s), are suppose to happen, by when each individual, is offset by an physiological-composition, interacting, reasoning why ones' place, is for circu-lating product-composition-(s), per titled-entity-offset, from [residential / commercial courtesy-(ies)], made by when factory-product-ingenuity, affect how one perceives title-land-infrastructure, comparing barren land-(s), for extracting resourcing.¶|||25s.25m.s.|||§

¶137At 17:30{5:30-P.M.}-{12:30-P.M.-Miami} on 9 September 2015, Queen Elizabeth II had reigned for 23,226 days, 16 hours and approximately 30 minutes;249 surpassing the reign of her great-great-grandmother Queen Victoria.233¶|||15s.84m.s.|||§ This milestone ration upon present-live-memory, defining modern-dating, from those perspective-(s) of letter-(s)-word-(s)-term-definition-(s)-article-(s)- grandeur-data, that Movement & Discourse, age-simultaneous-direct, decade-aging-marking-(s), milestone-marking-(s), interval-article-interpretation-focus, in percept-ive juxtaposition of regular-inhabitant-effort-(s)-offset-major-increment-method.234¶|||34s.93m.s.|||§ Modern-reasoning-error, confluence between Mysticism of Mass & Repopulation-Count-(s), moving past-requisite, citizen-human-inhabitant-(s), whom determine moment-(s) upon lifestyle-hour-impact, not concerting an direct-effort, by means' of consideration (for thinking), while daily-impasse continue affecting how our conjured-interpretation-(s) pertain self by religious-syntax, extending under one-million-word-(s)-extent, supposed to be rewritten, when chalice-discourse, exemplify oneself, as dated product-(s) per [inedible / edible] purchase-deviation-(s), refrain collected-rate-conjuring, where subjective-article-(s)-interval-(s), devise centripetal-force-momentum, ascertained grandiose-momentum-passing, enveloped molten-plasma [inner // outer] core-mantle-lithosphere-asthenosphere-kinetic-solar-energy, dispositional-offset-discourse-(s), viewing perspective-matter-(s), (from an) person-mass-wage-activity-(ies)-count-(s), when state-measure-tax-bill-proposal-(s), interact default-population-trust-guidance, for inhabitant-(s) do not infer, by claused-proportion-property-(ies), per [title / lease] living each relative-moment-(s), when synthesized from duty-(ies) of worship, meandering static-dissolve of metropolitan-recourse-(s), forgetting why cursive-simultaneous-grammar, is instituted learning-interpretation-(s), from resolve practicing workplace intercorolary common-repetition-(s) (per) of dating, timed by data-hour-moment-(s), remaining intertwined physiological-pulse-tension-(s), pulsed interval-occurrence-(s), due-to an motive of living, based from independent-family-individual-(s'), whom interact when formal-usage for understanding how information follows succession order-(s), continue because impassing centripetal-force, hover above tectonic-plate-(s)-calculation-(s),| 495 |considering active-interactive-tangible-article-(s)-rationing, from proportion-(s) of surface-area, ascertained pre-date-aging-requisite, upon collected [data / hardware / software / article-(s)] that have an conscious-social-creative-injunction, where [natural-talked // discussed-interaction-(s)]| 496 |influence periodical-article-retention-area-(s), because Physical-perceptive-observation-(s), instill offspring-cause, relevant upon tectonic-national-directive-guideline-(s), asserting how physiological-impasse, define requisite-present-motion-(s), intervalizing, tectonic-plate-national-individual-schedule-agenda-focus...¶|||02m.:32s.52m.s.|||§

§...Preposition-Count-[10 / Ten]-of.. ..Article-Essay-[2 / Two].. ..Book-Essay-[137 / One-Hundred-Thirty-Seven].. ..Lines-[30.7375 / Thirty.1.seven-tenths-2.three-hundredths-3.seven-thousand-ths-4.five-tenths-thousandths].. ..interposing along an continuum-secondths-ticking, would have an method of etymological-reasoning, which pertain how developing title-land-grounds, prefer to decide weather to [1.construct / 2.irrigate / 3.extract / 4.transit / 5.commercially-operate / 6.station / 7.library / 8.residence].¶|||19s.59m.s.|||§

¶138Mass-(es) not formally-group, tangent-impasse-intercourse-dialogue, deviating circumstance-(s) of hour-time-frame-interval-(s), conjuring why purpose of extraspection, sustain debt-boundary-limit-(s'), incalculable vast existence, when metropolitan-public-domain, discourse voluntary-inhabitant-(s), because pre-requisite-intrigue, affect article-lesson-subjecting, influenc-ing syntax-intrication-(s), as subterior-subject-(s)-mark-(s), core common-living-observation-(s), pertinent by [observation / count / elaboration-(s)] for continuing an common-perspective, capable pre-historic-accounting-(s'),| 497 |un-rated per statistically-applicable-live-pulse, learning from past-requisite-interpretation-(s), identifying leisure-lifestyle-pattern-(s), abound self-limit-capability-(ies), when mass not comprehend subversive-tangent-liturgical-lessoning, intermittent constitutional-article-(izing), where primary-university-high-school-subject-lessoning-(s), limit intended social-reveal, cohesive-comprehensive-document-(ing) per semester, when social-reclusive-commercial-interactive-prose, intertwine entity-lessoning-(s) from physical-observation-(s), developing structural-note-(s), for making how effort is enumerated (numerated)-denominated, by debt-trade-effect-continuum, conditioned ambiguous-judgement-(s), at purchase-point-(s), per inhabitant, executively-monitored, while conducting general-affair-(s), (as an) default of reasoning-(s), monetarily-align-(ed), when monitoring dating of human-act-(s), as representative-initiative-(s'), implicate physical-cubic-area, limiting cardiovascular-respiratory-mode-moment-(s), [where // when] data is comprehended by physical-impasse-place-(ing-(s)), sustained [state-funded // city-planned /// county planned-surface-area-place-(s)] that interval family-off-time-individualism, affecting mechanical-continuum-articlizing, due to those difficulty-(ies) of extraspective-monetary-agreement-(s), differentiate by each individual-prose, by limit-(s) of physical-act-(s), exerting cardiovascular-respitorial-muscle-blood-pulse-aorta-offset-effect-(s), tensing skin-exterior-offset-(s), from exterior-vibrational-wave-(s)-value-(s), significant of conducted-article-affair-(s), because misgivings' of self-(ves), vestige pursuit of basic-human-need-(s'), as alliteration-difficulty, not succinct each impassing-moment-east-to-west-Geological-twenty-four-hour-timing-offset-(s), ambivalent-juxtaposed subductive-location-(ing)-(s), validating product-(s), remaining scarcely purchased (from individual-wage-payment-cycle-(ing-(s))), because motive-(s) pertain velocity, propelling clearance per metropolitan-approximate-offset-(s), verifying latitude-degree-(s), at impasse-area-injunction-(s), when differentiated-article-(s)-application-approach, capture simple-vestige-(s) of self-idling, happily-content to lie of product-objective-collection-(s), not invigorated by communal-verbal-interaction-(s), coherent in form, by centripetal-drift along roadways, meandering live-purpose-(s), apose entity-subjective-stability, defining how wage is unsubjective mathematical-superior-quality-(ies), from applicable-activity-(ies),| 498 |used upon common constraints' of payment-survival, in conceptive-curb-interpretation-(s), individualizing parcel-planned-offset-(s), where individual-effort-(s) sustain allocation-(s) for word-subject-conjure-post-application-effect-(s),| 499 |ruling non-fiction-philosophical-constitutional-impact-(s), (for) juxtaposing pre-constitutional-obser-vation-(s), considering (as how) tier-observational-issue-(s), diligently-elucidate absolute-mass-population-documenting-means', remaining complacent by impasse constraints per timed-dating, where period-(s) of day, are salvaged from monetary-effort-(s), due-to clearing of monetary-objective-(s), as when mere fact-(s), not reinvest family-religious-saving-(s'), for assimilating mutual-adult-aged-peers-crestenza-purchase-(s), asserting territorial-space by hundredths-account-aspect-(s) (religious-examination), as how population build by minute-increment-(s) of principality-district-design, greater-than conscious-attempt-(s), for orchestrating human-(s), revere an similar instrument-(s), refurbished fabricated-existence-impasse, considering why those elemental-observation-(s), define method of post-subjective-purpose-(s), hastily habitual, disregarded idle-off-time-effort-(s), because extraspective-article-(s)-issue-(s), remain questioned by such ambiguous-behavior, during placed-center-(s) of political-body-ordering-(s).235¶|||03m.: 28s.41m.s.|||§ What come before [where // when] colloquial-conjure, compulsively object, human-impasse-concern, from idle-word-prosing, verifying sensational-reaction-(s), that are inhibited by blank-minded, perceptive-nature, atoned (from an) wave-vibration-(s)-percussion-offset-effect-(s), rather than vocal-verbal-woodwind-pitch-tonal-vocalizing-(s) of word-(s), when [chorus // verse] define vocalized-vernacular-mark-(s)-sound-(s), congruent listening, interposing, date data-verse-conjecture, limited timed-motion-(s)-interposed-data-juxtaposing, demarking day-denominating where juxtaposition, offset dated-night-cycle-denominating, timed scaled limit-(s), where observation per terrestrial-expanse, clearly distance, subductive unscheduled collective-unscheduled-input, affirmed topical-article-(s)-matter-(s), because mass-territorial-examination, strengthen attempt to kindly enable terrestrial-constructive-effort-(s), from when other-individual-(s') of state, offset national-restructuring of territory, barred barren-roadway-offset-development-(s), in lieu of American-Claused-Production-Objective-(s), (yet to be) post-subjective-considered, when non-fiction-prose, pertain past-mortem-memory-lineage-age-offset-(s), differentiated inhabitant-mutual-developmental-grandiose-concepting, for thought (by of) fundraising construct-ion 58L. [1roadway-(s) / 2entity-(ies) / 3residence / 4university-subjective-construction-planning-tier-(ing) / 5government-border-resolving // 6engineering-factory-conveyer-belt-(s)-product-concepting / 7mega-ton-truck-(s) / 8shovel-(s) / 9camera-(s) / 10wiring / 11window-pane-(s) / 12soldering / 13welder-(s')-butane-sword / 14stove-(s) / 15refrigerator-(s) / 16pipe-(s) / 17computer-(s) / 18printer-(s) / 19sink-(s) / 20tool-(s) / 21pan-(s) / 22pot-(s) / 23utensil-(s) / 24pen-(s) / 25paper / 26toilet-paper / 27paper-towel-(s) / 28jean-(s) / 29+et cetera...]...¶|||04m.:57s.53m.s.|||§

§...Preposition-Count-[20 / Twenty]-of.. ..Article-Essay-[3 / Three].. ..Book-Essay-[138 / One-Hundred-Thirty-Eight].. ..Lines-[62.535 / Sixty-Two.1.five-tenths-2.three-hundredths-3.five-thousandths].. ..rooms are suppose to isolate cubic area, for an concertion of particular-hygiene-values, that sustain an variation of variable-moded-usages, on an count deviation per 86,400, at 3,600-one-hour-timings, with intermittent titled-spacings, verifying common-decorum, security-act, verifying each individual-being, with an duty to physics, or else decease vacant of objective-lineage-collections, amid territory-religious-genetic-retirement-communion, for how individual-bias, pertain amid physics, for of each daily parameter from culcable-acts, deciphering why every dating remain complete by an focus from weekly-repetitive-acts, per location-hoard-practice, comparing where stationary-remain, practiced by log-interpretation-registering's, by composite-shelf-life-trading. ¶|||45s.75m.s.|||§

¶|139As a-the Sixth Month past-present-requisite-motion-(s)-impasse-reset-occurrence-(s), notion animate-being-(s), abiding [day // night-impasse-elapse // syntax-centripetal-force-progression-(s)] directing inanimate variable-volume-accumulation, superlative proportional-conscious-article-(s)-relay-(ing), reasoning market-factor-data, by constraint-(s) at interpretative-impasse-mode-(s), when citizen-(s') affirm, subjective-conjecture, through colloquial-language-predicate, referencing data, by when, visually-tone-letter-participle-word-term-definition read-after-thought, defined term-definition-word-notion-observation-(s), conjecturing where inter-action-(s), verify geometrical parameter-space-(ing-(s)), having surrounding atmosphere, confirming respiration-blood-pulse-beat-(s)-organ-(s)-tension-(s), apposed key-time-signature-(s)-tempo-reverberated-wave-vibration-(s)-deviation-(s),| 500 |manual-motioning acceded act-article-(s), upon interactive-to-do-basis, per time-constraint-cardiovascular-tense-blood-pulse-compartmentalize-(d)-motion-basis,| 501 |experiencing depth-perspective-dimension-parameter-offset-(s), defining [space-dimension-pressure-interior-ambient-measure // solid-dimension-exterior-measure // entity-cubic-subjective-feet-per-secondths] moding [timbre-blood-pulse // respiratory-inhale-exhale-breath] accentuating simultaneous-resonance, for stanza-measuring frequency-extent, indicating measuring beat-(s)-tone, pitch-vibrating frequency-motion-impasse, as notion-extent, decibel-one-point-per-{0.000}thousandth-inhabitant-contrast,|502|volume-sounding, constant-statistical-mass-wage-production-introversion-exterior-trade-family-male-responsibility-offset-(s), not traveling lineage, nomadic sleep-impact-cycle-(ing), dating parameter-self-posit-(s), by an total-population-age-deviation-interval-month-day-year-offset-(s), individualizing past-climate-offset-posit-(s), characterizing condition-(s) of humanity, undocu-mented extent of barren-grandeur, off-time-provisioned, work-instructive-objective-variable-(s), intermittent behavior of citizen-(s), conducting simultaneous-mass-effort-(s), piecing a-the micro-ecological-biological-matter, wilted by aging, repetitive strenuous-muscle-tension-(s)-fatigue-(s), sat as commission of city-geometropolitan-block-(s), story-(ied) historical-ambiguous-tradition-(s'), prelate by day, constant of week, contrasted among month-(s), ingrain-(ed) during season, reset-(ed) aforementioned work-grind, past-requisite, live-conjecture, reverberated in an atmosphere, interior from stack-posit-(s), isolated from day, sweatless from centripetal-force-revolutionary-velocity-pressure-impact, by-of-an kinetic-tension, vaporized as water-surface-vapor-plane-contour-canyon-valley-impact, variablize locational-place-(ing-(s)), distanced by grade point average-(s) of [thousandths-word-rate-period-(s) // block-(s)-hour-reset-testing] {3,600 words per hour, median-standard}, lagged by peer-to-peer-co-worker-interaction-(s), dating by-through, modern international-federal-government-physical-impassing-momentum, aged at offset-(s), where exterior-surface-area-climate-condition-(s)-constant-continuum,| 503 |subversive object-(s) of purchase-value-variable-depreciation, excerpted as obituary congenial [morning // night stillness] auto-biographically trade-(ing) by glory of grandeur-tranquility, glazed of marble-surface-plane-effort-(s), stilly operating production-(s), at metropolitan-public-domain-(s), possessed by property, rather balancing extraspective daily-living, while methodicalizing moded private-terrain, objecting from genetic-morale, limit-(s), by those purchase-parameter-(s), collecting elemental-composite-matter, from individual proportion effort-(s), constantly measuring existence, from extraspection of mass isolated fatigue-tense-(s), while petroleum is limited, from the dynamic of [[1]entity-(ies) // [2]distance-(ing-(s)) // [3]transit // [4]substantial-recipe-comestible-(s) // [5]raw-resource-(s) // [6]subjective-extraspective-interaction-schedule-method-inquiry // [7]elemental-refining /// [8]objective-factory-patent-product-mass-production // [9]physiological-fatigue // [10]opposite-sex-arousal-attraction /// [11]repopulating-conversive-agenda-schedule-planning] for an numerical-ten-reset-thousandths-notion-scaling, in receipt of food, edible simultaneous degree-level-competency-application, present-continuum, foretelling retirement, as each candid depiction of grammar, not yet hyphen hyper-focus-tense-(ing), by-with formal-median-rate-deviation-(s), of [where / when / what / why] question-deposit-(s), pertaining deductive-mass-production, (as an) free-will-challenge, over-producing, imbalanced-purchase-expectancy-(ies), influctuating shelf-life, through (that of a the) limit of mold-collecting-(s), handling surface-area-development-(s), for denouncing word-reference-(ing-(s)), as each season Seven-million-Seven-Hundred-Eighty-Five-Thousandths-secondths-continuum-reset-sequencing -(s), remain [simultaneous / acted / slept] of secondths-hour-blood-pulse-sensation-(s), of an exponent = thousandths-count-notion-reset-offset-operating, ranging from refined-resource-(s), [[1]beef-(s) // [2]cow-(s) // [3]paper // [4]pen-(s) // [5]computer-(s) // [6]mouse-(s) /// [7]superlative-et cetera…] simply for, offset-matter, sustained by natural-prerogative, instilled, from reading-preference, over [Writing // Reading // Redacting // Revising] co-aligning period-(s)-extent, adjusting affirmation-(s), from vocal-vernacular-volume-decibel-tone-contrast-visual-posit-(s)-perspective, conjuring from market-warehouse, environment-(s) of ethnic-immigrant-naturalized-unconstitutional-preset-history, by white-silhouette-brightness-contrast, tectonic-plate-(s) [distance-(s) // excavated-terrain // cultivated-terrain // development-terrain // constructive-terrain] sovereign by [House-{E Pluribus Unum} // Senate-{In God We Trust} // Presidential-{Oval-Office} // Supreme-Justice-(s')-{Federal-Superior-Court}] as legal-logistical-municipal-jurisdiction-(s), proportioned parcel-repayment-parity, from those property-ownership-practice-(s), by free-manner, accessing interior-dimension-space-(ing-(s)), adjusted by-of work-living-method-(s), cycling experience, by effort-characteristics, only significant, if conveyed by an extraspective-competent-population-collective, perceiving fact-(s) by formal-sort-order-adjustment-(s), deliberating how interaction, should affect national-economy, in adjustment of basic-human-right's, present of memory by visual-contact-brainwashing, ascertaining cent, by no deviation of consistent-off-time-extraspection, affirming introvert-(ed)-conception-(s), where reception is pertain-(ed) per purchase-brainwashing, while our house-mass-proportiated-count, is at an natural-past-requisite-low, from total-auto-biographical-history, counting Five-Hundred-Twenty-year-(s), as Two-Thousand-Sixteen-Two-Thousand-Twenty-year-(s), calendar-approximate-scaling, literate-competency, not measured along literal-distance-(s), federally [restricted // tracked] of cartography-hour-day-geological-degree-measure, bordering perimeter-boundary-(ies), conject-ured of terrain, (from an) inconsistent-relevance, when past-requisite, remain as present-day, apose past-to-present-day-week-count-(s), for example, I am at Twenty-Five-Year-(s)-Age, [9,131-Day-(s)-expired // 1,304-Week-(s)] while at an [writing // typing // editing // publishing, effort-(s)-aging of mode-9011-day-(s)-{December 25, 2015} through an estimate of March commencing publication dating, 9,458-day-(s)-{March 17, 2017}] deviate-date-aging-{447-day-(s)}, where constitution not amend age-population-decade-interval-census-reasoning-(s), definite universal-past-finding-(s), from commercial-literature, continually not pertaining non-fiction-genre, for observational-scaling self-age, juxtaposed nuance year-month-date-mass-count-(s), simultaneous alive-awake [timing-period-count-(s) // hour-count-(s)] complementing mathematical-posit-(s), upon distanced spherical-flat-plane-surface-area-mile-extent-(s), planning centralized-metro=-politan-directive-(s), as unconceptive of rural-acre-parcel-ecological-maintaining, [cardio-vascular-blood-pulse-respitorial-inhale /504/ exhale-musculatory-reflexive-tension-skin-liga-ment-joint-bone-Deposit-(s), {c.0.b.p.r.i.//e.1.m.r.t.2.s.3.1.j.4.b.5.-D.}-interior-perspective-pre-mise] to conjure through depth-parameter-offset, where mode-(d)-moment-(s), translucently-perceive, by respect-(s') of nature, that without blood [pulse // inhale /// exhale] tensions may not elongate singular-exterior-posit-(s), per three-hundred-sixty-degree-circumference-clearance, where currency-note-(s), premise interpretive-observation-(s), of worth when time-elapsing, succincts' an [substantial // objective // auxiliary-value-(s)] recoursing from governmental-regulated-currency-pricing-guidance, deteriorating from birth, through expiry-matter-(s), clogged of documentation, not writing, but weight-discoursing, measure-(s) (per) of impasse, scaling out those (offset) fact-(s) awhile dating viscount per [present-date / calendar age / time-interval-(s) / centripetal-force], [when // where] count-interval-(s), pertain calendar-deviation-extenuation-(s), not yet by inhabitant-competent-interval-interpose-(d)-count-cycle-(ing-(s)), forward-mention what future-parameter-(s), can set, instilled-constant-practice-(s), (from an) mass-introverted-requisite, (un)-invested or concerted for how extent living to retirement, not jolt an charge of economic-literary-planning, by prior-conceptive-word-term-syntax-notion-deviation-(s),| 505 | bridled-by solid-objecting-(s), yet thinking how [vernacular-interaction // vain-attraction] are superior-greater, objective-usage, which is in an shared aspect of acre-land, proportioning auxiliary-distance-(s), requiring vernacular-vocal-tone-volume-decibel-audibility, appreciating existence, from when no total-rotation, simultaneous-mode-module-method, assimilate why distance-(s) remain far off, Academic-grade-point-average-median-standard, simultaneous-exponent-mass-thousandths-words-per-hour-inquiry, from letter-note, upon G.P.A.-deviations', as secondths-elapse range-mean-{mathematical} past the Eighty-Six-Thousand-Secondths, resetting daily,| 506 |[calendar-anticipation-(s) // act-(ion-(s))] from individual-requisite-(extraspective-interactive-attempt), where-of Tectonic-plate-(s), posit as Galaxy, having [political-state // nation-universe-(s)] upon planetary-County-City-Government-(s), where mid-day-effect-(s), reset aging-parameter-dynamic, sat under primary-infancy, fringe still-perspective, as when tectonic-stability, not tense kinetic-momentum,| 507 |inanimate, human-ethnic-racial-physiological-component-being-(s'), deductive-inductive-observation-(s), from exterior-observation, from our eye-plasma-cornea-iris-pupil-fovea-sclera-interior-requisite-reception, when pupil-tension-perceive exterior-wave-vibration-(s), [substantial // intangible solid-density-(ies)] awhile bickered-trade, not posture-tense-perceive, why commercial-effort-(s), genetic-religiously-matter, as how our way of pavement, is relied on bastardized-human-effort-(s), uncleansed those weight-(s) of density-genetic-extraspective-impact, governmentally-homogenized, by patent-intelligence, historically cooperated, upon terrain-temperature-latitude-degree-separation-(s), exceeding county-terrain, without an literary-reproductive-cycling, arthritis athletic-physical-retardation, forgetting time, and how to date-coordinate, concert of transit, for then [constructing // legislating]115L. (humanity is (from an) loose interpretation of (executive-nature // judiciary-monitoring), morally inclined by duty-requisite, as why payment-method, following an order (per) of congressional-regulated-dollar-cent-commercial-patent-registered-trade-mark-brand-currency-conjecture-morale... ¶|||09m.:59s.25m.s.|||§

§...Preposition-Count-[41 / Fourty-One]-of.. ..Article-Essay-[4 / Four].. ..Book-Essay-[139 / One-Hundred-Thirty-Nine].. ..Lines-[117.4815 / One-Hundred-Seventeen.1.four-tenths-2.eight-hundredths-3.one-thousandths-4.five-tenths-hundredths].. ..education is difficult for any individual to conceive, without comprehending those deductive-empirical-entity-(ies), developing those offset territory-boundaries which demand an capacity-population, when dating elemental-tectonic-terrain, range from Minutes and Secondths, directing an command of parameter-residencing, per inhabitant from unfamiliar-family-ties, upon an block of family-rooms-construction, seeing how masonry, not remain common by American-Family-Retirement-Values.¶||||28s.56m.s.|||§

¶140. Motion-(s) & Manner-(s), Semi-Colon & (Con)-Current-Action-((s) - derive from Person-Historical-Recognition-Observation-(s), relaying-obituary Ethnic-Racial-Impartial-New-Faded-Men-Elder-Retirement-Holdings'-Aspect, maintained New-York-Protestant & Yorkshire-Offset-(s') offset while compositions' (per) of June-month-solar-transit-twelfth-thirty-day-offset,| 508 |six days longer [rotation // revolution] after-effect-(s), encompass [Spring-end // Summer-commencing] Tectonic-Observation (perspective), upon Standard & Poor-Senile Aged-Suicide-Progression-(s), (Death of an Sales-Man (Anticipated)-Mentality), per [Feminine // Male] June-Twenty-First-Dating-Reset-interpose-recollection, ellipses-continuum-offset, detest average-rate-grade-point-average, per cent anticipated under one-thousandths-words-thought per hour, listless of [verbal-word-rate-(s)-{being} /// written-word-rate-(s) /// typed-word-rate-(s) /// Vernacular-Verbal-Word-Rate-(s)-{person}];250 Dating maybe continuum-forgotten-dating-straight-historical-past /or/ Time-Hour-Continuum /or/ Time-Minute-(s)-Define (Century-Expectancy-Circa-Era-(introverted-trade-reliance)),| 509 |inclination-(ing) solar-juxtaposition, in an {720°}-scaling apose Lunar-dry-Solid-Tangible-(s) unsure ration /-/ proportion-Solar-Ten-Thirty-North-West-Southern-Hemisphere-Summer-Four-Thirty-South-East-Northern-Hemisphere-Summer-Revolutionary-Offset-Fission-Aspect,|510|horizontal-climate-temporal-massive-tectonic-densities-offset-grandiose-depth-passing-(s), circumference-distancing-(s), where self-family-impasse-age-(s), have-an reproductive-trauma-impact post-past-historical-moment-(s)-count-deviation-denominative [1sub-ject-(s) / 2scale-(s) / 3periodical-elemental-weight-(s) / 4Ingestible-Market-Weight-(s) / 5glossary-word-redact-(s) / 6Windows'-Pane-Perspective / 7Watch-(Stop)-time-alarm-compass-secondths-count-cycling-(s)] where each subversive-rate-(s), come (from an) frequency of product-shelf-life, intermittent peripheral-sight-physical-perspective, anterior-recollected, by miniscule-inferior-micro-millisecond-mass-weight-offset, premised from self, non-influential-matter-(s) impassing-(s) affecting-blood-pulse-ventricle-muscle-ligament-skin-organ-blood-plasma-sedentary-water-posit-(s)-bone-tense-reflex-(es), mitigated temperat-ure-compositional-timbre, where stance per-of fathom-(less)-{Six-Feet (oo)}-feature, formulate individual-pulse-sight-sound-listening-perspective, apose tense-momentary-posit, yet acted by formal-vernacular-conjecture, by-when-date mathematical-synthetic-synthesis, regard religious-grammatical-historical-lessoning-nature, apose an (opposite) expectancy of womb, two-child-maxim-repopulating, Father & Mother & Child-Developmental-Cycle-Phase-Stage-(s), under an formal letter-word-impasse-rate-deviational-subjective-retentive-purpose, defining extraspective-objective per measured-layer-series-schedule-structure-sequenced-length-width-two-dimension-al-conceptive-compariative-three-dimensional-confirmation-pre-action-quantifyable-mass-pur-chase-article-objective-action-motion-notion-(s), upon measuring-(s) of distanced-mile-(s)-acreage-family [maintain // contain // sustain] land, handle-(d) skill-(s), as examined through force-pressure-clearance-alignment-purpose-(s), exquisite by superfluous [bias // hoc] intermittent decade passing day-hour-rate-interval-(s), per [parcel / perimeter-planned-area] engineering / federal-mass-county-classification-(s)] characterizing fact-(s) by daily-impasse-land-(s), fortuned from only one-title-owner, upon property, when under an legal-banking-method-(s), for practicing how to develop tectonic-plate-territory-(ies), dating per age-(d) trusted-tangible-resource-(s),|511| fostering-an farmer-substantial-commercial-origin-split-(s), spearheaded handle-(d)-crop-count-(s), apposed comestible-edible-(s) of horticulture,| 512 |existing as [mentality-live-digestive / personality-auto-biographical-lineage-trace-undocumented-origin-reset-(s)-posit-(s)], without an lineage-legislational-historical-influence, fathered-off, founded-means', willfully collecting, citizen-adorned-object-(s), perceptive upon visual-sight of physiological-restriction-(s), [when // where] each vision spherically-parallel, crust-exterior-surface-area-solar-hour-clearance-offset-(s),| 513 |path an journey by input /(in)-transitive/ output-cognitive-memorized-data informing depth-clearance-effect-(s), perceptive distanced, tangible-objective refined-productive-tangible-(s)-space-(ing-(s)), per moded-independent-citizen, where effort is tended to instill various characteristic-(s) of ecological-ambient-theatre, numericalizing eye-contact-perspective-offset-(s'), as word-grammatical-copyright-juxtapose-register-documenting, from inventor-patent-invention-production-development-(s);[251] intentional international-standard-business-number-posit-(s), where rated-market-volume-value-(s), not deviate [person / subterior work-routine-effort-(s) / being-dynamic-(s)] continued where individual off-time-silence, effort an juxtaposition per manual-transit-drive-(er)-(s'))-posit, apose working-place-instructional-schedule-reading-(s)-article-fact-defining, force impasse-effort-(s), as under largo-tempo-perspective, per measured-currency-note-(s), apose article-(s)-notion-noting-frequency-sustain-(ing-(s')), celestially-ambient, rotational-revolution-tectonic-plate-offset-(s)-spherical-(galaxy) centripetal-drift-(s), not meaning anything for appreciation from student-subjective-study, hence dormant subservient-effort-(s), idle-independent-confidence, continuum velocity-transit-tranquil-domestic-cycling-continuum-impasse-uncreative cardiovas-cular [distance-picture-covering // vehicle-transit-(im)-passing-(s)] realm-experience-past-requisite, each being personal-possession, as trade-(d)-object-(s), know none other discourse, from historical-dating, remain perceived by clearance-cubic-feet-per-second-moment-(s), [where / when data], pertain an simultaneous by-an [cardiovascular-blood-pulse-respitorial-breath dimensional-objective-offset-(s), tier-documenting] when hour relation per simultaneous-moment-(s), for how concept-(s) of [vision // hearing] confirm how momentum elapse away of molten-matter-conjunction-density-(ies), land-deposit-accumulating, tectonic-plate-tangible-quality, juxtaposed-offset offspring-perspective-appreciation-(s), enamor-ed from distance-parcel-depth-perspective-posit-(s), when day-night-temperate-rotation-influx-deviation, affect physiological-individual-temperature, apose topography-posit-(s), at-an celestial-season-year-impact-effect-(s), conjuring place-(ing-(s)) of miniscule-micro-millisecondths-personal-perspective, subterior-align, revolutionary-count-maxim, adjacent-spherical-rotation, oblong-aging-reset-(s), when commercial-experience, presence-aging retirement-prerogative, grandeur whenever acceptably processed from solar-heat-kinetic-constant-friction-molten-pulse-wave-infinitive, juxtaposed, lunar-solid-freezing-point-maxim, apposing infinitive-space-(ing-(s)), by Earth-Tectonic-Climate-Topography-Celestial-Temperature-Posit-(s), perceptive matter-civilization-(s)-offset-(s), empirically-denominative, objective-subjective-article-(s)-trading-bind-ing, live-present-continuum, (un)-thought from when mass-extraspective-wilting, impasse each present-date-daily-aging, where at an mode of premised accustom physical-effort-(s)|514|sufficing limit-(s) per perceptive-distance-impasse-coverage,75L. for how [constructive // engineering development-(s), are mattered by an motion-action-(s) // elemental-refined-factory-dimensional-posit-(s)-tangible-impact]...¶|||06m.:42s.70m.s.|||§

§...Preposition-Count-[14 / Fourteen]-of.. ..Article-Essay-[5 / Five].. ..Book-Essay-[140 / One-Hundred-Fourty].. ..Lines-[76.58 / Seventy-Six.1five-tenths-2.eight-hundredths].. ..library-referencing is held by an conjunction of literary-shorts, not reasoning how ink and paper modify perceptive surrounding existence, when each dating conjunct various circulated-product-variables, which are to be scheduled for modification to premises, when ordering from factory- production, variablized weekly-daily-box-units-packaging, for [backstock / sales floor shelf-life] by an customer-arrangement-purchasing, by timed-hourly-circulated-generalized-unit-(s), which order developmental approximate values, for thought on how to retain solid-compositional-objects, by decisions per common-housing-residence-retention, of primary-voluntary-collective-objects.¶|||41s.81m.s.|||§

¶141 Approaching daily-interaction-(s), rely influence of political-congressional-regulated-bill-personality, prior-season-(s)-medium-climate-clause-effect when summer completely plasma-molten-melt, intermittent-temperate-soft-elements, awhile independent-perspective-offset, apose momentum upon centripetal-force-revolving-rotation-impact, influencing [elapse // syntax // synapse-method-(s)] synthesizing-literary-context, for interacting where data pertain upon each individual of prevalent-awareness, focusing tandem-(s), from population-decade-year-reset-(s), at tangent by national-mundane-basic-human-right-(s')-article-(s)-count-(s)-requisite-past-obser-vation-(s), where mass-population, space away-apart from each and every moment, simultaneous written-principle, gauging blank-mass-notion-(s), by product-conceptive-creative-(ity), as-when fact can be of an translucent-interval-piecing-(s) of matter-(s), per independent-perspective, governmental for no genetic-house-representative-mass-population-collecting, of each type of genetic-background, where [ethnicity // race] is of an conjunct-effort-(s)-tandem, omitting mathematical-predicate-timbre, not simply thought, by case-point-(s), influenced by effort-(s) at work-place, convinced monetary-check-note-cash-debit-trade-disparity, dictating why recourse of impasse-journey, affair article-(s)-mass-directive-offset-(s), pertaining cultural methodical-mode-(s), by-of Transit-Distance & Trade-regulated-currency-note-(s) where citizens' whom origin by men, believe their be an truth, not needing to think! Inspecting interposed-mass-inducting by-an ruse of listening-practice, [sight // sound vibration-(s)-twentieth-gamma-article-note-miniscule-observed] reverberating from centripetal-force, human-fathom-height-perspective, drifting peripheral, from [involuntary-elemental-solar // lunar-perspective upon distancing-maxim-offset] aging awhile parity-disposition, not fully [discuss // write-tangent-articlize-intricate] term-syntax-contort-(s) of data, affirming notion-(s) by-an genetic-identifying, as tense-(s) are by-an vocal-larynx-pressure-note-vibrational-reverberation-(s), listen-visualizing, letter-character-word-tone-sentence-paragraph-syntax-(s), defining depiction-(s) of image-listening-mark-(s)-syntax-inter-pretation-perspective, awhile subjective-tense-(s), become elaborated from sequence-(ing)-(s), when psyche not concept-morale, for producing have-an off-work-time-week-end-offset-economic-extraspective-judgement, unused relevant-elongation per tangent-article-(s)-interposed-interpretation-(s), significant state-location-place-(ing-(s)), clearance-maxim by Height & Interior-Ceiling-Constructive, pre-subjective-inhabitant-habit-process, default-limit-boundary-parameter-(s), [adding / passing-observational-tense-(s)] affected by family-lineage-inter-pretation-(s), present-(ce) modern-effort-(s), ordering operation-(s), juxtaposed appreciated sight-visualization-(s), adjusting [elapse-(s) // syntax] period-(s), incoherent talking-elapse-(s) (vocal-focus), for listening-elapse simultaneous blank-idle-perspective-period-(s)-offset-(ed), present-impasse-reference per moment-(s), deposited individual-topographic-posit-(s)-juxtaposition, from where, subjective-lesson-question-(s)-contexting, apose rated-article-(s)-redact-data-period, when language initiate our journey of imagery-silhouette-visual-image-(s), while date, impasse parameter endeavor-(s)-cognition, task-(ing)-(s) individualism, uninspired by wealth, for superceding [death-rate-eighty-one-women // seventy-nine-men-year-aging-expectancy-from-birth] verifying continuity at community-effort-(s), unstressed written-data, preferring federal-inquiry, rather congressional-library-legislational-thinking, pending notion-extent of exponential-production-fund-(s), validifying data-observation-(s), abstracted by personal-mass-possessive-collection-(s) per space-(ing-(s)), not affirmed by-the constitutional-title-deeding, acting upon bank-transaction-account-(s), active of fundraising, proportional by framing of living-impassing-perimeter-territory-parameter-(s), secluded-present, natural-karma, pertaining calendar-infinitive-belief, not formally-succession, incendiary-remark-(s), as subconscious-individualism-offset-(s), discourse an monetary-extraspection, because each fact is deviate-(d) from those grandiose-elemental-extraction-weight-(s),| 515 |comforted governmental-historical-past-planned-space-(ing-(s)), without an architectural-impasse, upon congested-claustrophobic-land-surface-area-space-(ing), labor-(ed) through mid-aging-work-effort-(s), not subjecting alive-awake-living-intrication-(s), at methodical-practice-existing-perspective-matter-(s), apose-juxtaposed dynamic-(s) by anticipated point-(s) of obituary-death, complacent-intermittent, life-cycling-(s), impassing secondths-movement-offset, without an cardiovascular-physiological-common-mass-presence, for affirming block-period-(s)-deviation-(s), awhile interacting, by lifestyle-affair-(s), not (subconscious) thinking-attempt, tectonic-lifestyle-discourse, preferring inhabitant-(s)-citizen-method of life, meandering centralized-Federal-Reserve-National-Sixty-Two-City-(ies)-Requisite-continuum-development-(s), requiring an specification from subduction-zone-median-offset-perspective, comprehending interaction-(s), as a-the primary-influence of living, impacted elemental-object-(s) upon commercial-factory-patent-focus-factor-(ing-(s)), databased, intellig-ent-interposed-interaction-(s), sustemming [individual-ism // genetic] subjective-objective-article-(s)-inclination-(s), diatribe-collecting, decisive depicting denotative-matter-(s), defining self-perspective, mid-age-work-effort, modernly-commonly-attempted, work impasse county-vehicle-traveling per surrounding-location-(s), documenting setting-(s) where tectonic-terrain, prefer retirement-senile-aging,| 516 |apose purposeful-traveling, demanding commercial-participation, by those payment-(s) of hour-timing, incapable formal-determining, where one would enjoy retiring, by locational-effort-(s)-constraint, maintaining acreage of terrain, (as an) goal for independent-living, cycled by substantial-proprietary-intelligence, inquiring weather their are other mutual-individual-(s'), whom conceive why interaction prelate such effort-(s), by important-aging-impasse, for-an crescendo of tense-(s)-(ed)-(s')-motion-(s), outlet ergonomic-development, when tectonic-plate-drift, affirm reproductive-ingestible-substantial-spacing-examination-(s), retirement-living, by resource-recipe-(s), not self-developed, juxtaposed those commercial-recipe-enabling-(s), trading by state-parameter-dynamic-(s), per action-(s), affirmed by voluntary impasse-motion-(s)-timbre, pertaining action-(s) upon an site of developmental-purpose, [where // when dating];252 past-historical-count-(s), present-continuum-requisite-deviation, by when hour-(s)-elapse-effect, prompt action-(s) of article-(ized)-mass-tectonic-elemental-dimension-compositional-refined-matter-(s),| 517 |determining-effort-(s) where national-Federal-state-control-order-development, sequestrate physiological-tasked-cycling-(s), organically purposed, due to the nature of sight, where perspective cognate congested-claustrophobic-reasoning, as how inhabitant-(s), deteriorate from (sub)-urban-living, not abiding by rural-mutual-(s')-family-(ies)-vernacular-interaction-(s'), conceding lackadaisical-effort-(s), by residential-spacing-offset-(s), rural-county-commercial-center-(ing (s)), enriched lifestyle, per object-interposing, upon individual-vocal-subjective-interposing, requiring person-vernacular-article-(s)-regulated-swearing, progressive inductive-interpretation-intrication-(s), pertaining-(ed) [task-(s) / territory / time / schedule / tier] per genetic-bias-lineage, focusing extemporaneous-activity-(ies), by designated-rule-(s)-state-formation-(s),| 518 |nationally-denominating, tectonic-inclined post-developmental-purpose, similar to youth-class-subjecting, remaining undefined common discourse-characteristic-(s), at support for argument per [introverted-personality // extraspective-interaction-(s)]81L. rather preferred living by seasonal-year-decade-century-millennium-perspective-(s')-offset, when intuitive-religious-act-(s), past-continuum-requisite, routine-(s) of lifestyle...¶|||07m.:00s.26m.s.|||§

§..Preposition-Count-[26 / Twenty-Six]-of.. ..Article-Essay-[6 / Six].. ..Book-Essay-[141 / One-Hundred-Fourty-One].. ..Lines-[82.42 / Eighty-Two.1.four-tenths-2.two-hundredths].. ..underutilization.¶|||07s.60m.s.|||§

¶142Thirteen-Million, Sixty-One-Thousand-Five-Hundred-Secondths;253 extenuate an estimate of Fifteen-million, Nine Hundred-Fifty-Six-Thousand, Five-Hundred-Secondths-reset-elapse-impasse, by June through July-Interval-(s), as Periodical-Day-Secondths, impasse-motion momentum upon elapse-perspective-(s), when day be conscious awake-fatigue-factoring, balancing (ob)-conjunct-tandem-(s), by population-(s) per [year // season-(s)] divided where written-principle, simultaneous form-documenting, justifying inductive-non-fiction-adept-tan-gent-largo, by single-story-auto-biography-reference-practice, astute subjective-tense-(s), by no congressional-citizen-(s)-timbre-reading-(s), parallel parcel-space-area-(s), when elaborated-tense-(s), particulate focus of paper-(un)-documented-tense-(s), tangibly influencing human-Fashion, from what thing-(s) stabilize introverted-being-(s);254 subjective form-documenting-(s), as-when lifestyle, be viewed-common-article-(s)-impassing-(s), relying on Governmental-literary-control,| 519 |for defining Scientific-Development-(s), default programmed-developed, maximum-accommodated-spacing-parameter-(s), pronounced by work-occupant, parameterized day-(s)-physiological-posit-impasse, related as, relied-assurance, for enveloping familiarity of tasked-labor-structure-(ing)-(s), when pre-set-developmental-block-structure-(ing-(s)), emanate other-inhabitant-natural-articulation-(s) of belief, lacking subservient-obedience, upon shelf-factor-(s) of governmental-locational-development-(s),| 520 |(un)-construct-(ed) genetic-default-offset, balance-(d) through timed-twelfth-hour-five-minute-(s)-testing-(s), for progressive land-appraisal, at detectable-surface-area-measure, not value-(ing) training, by-an Reasoning of Topical-Matter-(s), generating economy, for when trade state-organism-compositional-apprec-iated-value-(s), pertain where order-(ing) volume-series-sequenced object-(s), form an Act, for having how, Task-deductive-act-empirical-age-ordered-repetition-process-(es) by-of actual-variable-(ization-(s)), rate-(d) ethnic-racial-perceptive-default-(s);255 for only physical-measured-rate-(s), could demonstrate human-fatigue, upon those Right-(s'), perpendicular-intersect, individual-aisle-purchasing-parity, as cost-(s) from commercial-interior-trade-timings', regard industrialism, as religious-effort-(s)-junction-reset-continuum of elemental-patent-product-developmental-Day, [smooth / Season method-order-(ing-(s))] orchestrated-tangible-article-(s)-objective-directive-sorting-(s), by-of Place & Period, working labor, by limit-(s) of fatigable-being-(s)-effort-(s), acted from-an mental-capacity-(s), ignorant physiological-human-elapse-(s), educationally-self-accrediting an belief per Virtue involved at check-payment-method-(s), observed from terrestrial-objective-subjective-article-(s)-act-(s), with requisite of [Relationship / Time / Calendar schedule-order-(s)] to-be proportioned by disobedient-inhabitant-limit-(s), fatiguing from those transgression-genetic-fact-(or-(s), free from (un)-orchestrate-(d)-matter-(s), per political-Territorial-Space-(ing-(s)), verifying state-order-mass-impasse-bill-article-(s)-constitutional-preamble-filing-(s),32L. adjunct formal-common-reason-(ing)...¶|||02m.:46s.47m.s.|||§

§...Preposition-Count-[11 / Eleven]-of.. ..Article-Essay-[7 / Seven].. ..Book-Essay-[142 / One-Hundred-Fourty-Two].. ..Lines-[32.285 / Thirty-Two.1.two-tenths-2.eight-hundredths-3.five-thousandths].. ..still continuing those redact-sentence-(s).¶|||09s.22m.s.|||§

¶143Dating remain limited, for impassing-effort-(s), (from an) humanistic-stand-point of individualism, Transcendentalism & Industrialism, foresee those factor-(s) by auxiliary-complex-(s), naturally-selected, when Commercial-Educational-Forum, debate weather to study, dateable-agenda, skewing encounter-(ed) [Romantic-Relationship-(s) / College-planned-partner-(s)] default-fatigueable-cause, amid from [state / county place-(ing-(s))] (per) of [locale-activity-(ies) / Genetic to-self-effort-(s) of longevity-discourse], defining rigor of practiced-Genetic-Secondths-aligning by voluntary-bastardized-census-effort-(s), fundraised planned-governmental-developmental-Value-(ing-(s)), perceiving awake-observation-(s), by topical-material, while factoring, individuals', whom ecologically premise natural-state-place-(ing-(s)), for operating default currency-trading, willing to pass perspective-affair-(s), while order-(s) of daily-operation-(s), prefer a-the inconsistency by-an infetesimal-mode, sharing America, when place-(s) count, by Day-Developmental-dimension-(s), through [surface-(s) // object-(s) // panel-(s) // window-(s) // et cetera…] viewing (from an) lackluster attention-span-limit-(s), instating blank-idle-feature-programming-effort-(s), attempting to cut-corner-(s) of work-effort-(s), incoherent, cognitive-elapsing-(s), when actual-moment-(s), individualize affirm-(ed)-article-(s)-interpretational-fact-(s), re-setting sighting-(s), paced by parcel-parameter-(s), guided by limit-(s) per outlined-budgeted-agenda, itinerary-auxiliary-transiting, basic-act-(ed)-event-(s), commanding expert-directive-(s'), when elaboration-(s) "de facto", Expanse & Terrain, when rescinded unexplainable-objective-collection-(s), require substantial-matter-(s),18L| 521 |perceiving exterior-subduction-zone-atmosphere-centripetal-force-water-vapor-pressure, or inanimate-motion-(s), pertain how existence be...¶|||01m.:30s.10m.s.|||§

§...Preposition-Count-[6 / Six]-of.. ..Article-Essay-[8 / Eight].. ..Book-Essay-[143 / One-Hundred-Fourty-Three].. ..Lines-[18.77 / Eighteen.1.seven-tenths-2.seven-hundredths].. ..What is the territory-developmental-rate, from human-physiological-competency, per situated-territory-settlement.¶|||11s.97m.s.|||§

¶144Devout male-citizen-(s'), remain unfamiliarzed by religious-family-inhabitant-(s')-inter-action-(s), locally positing lifestyle, because national-enfranchisement, prefer county-city-limit-(s), particulating spacial-area, by scheduled-routine, because particular-regimen, fatigue conglomerated-unit-cycling, when verified-specialty-detail-(s), are for enhancing order-(ed)-word-sequence-(s), by consumer-demand-arrangement, accustoming (un)-orchestrated-manner, enabling idle-nature of Free-Will & Default-Labor-Effort-(s), basing Free-Enterprise work-routine, from industrial-solution-(s), as-when Governmental-monitor,| 522 |require Product-(s) & Resource-(s), from daily-inventory-shelf-life-compilation-(s), for each numerative-variable-integer-count-(s), maintain physical-limit-(s), qualifying still-mentality-mode-exertion-(s'), defaulted characterizing, maneuver-(ing-(s)) through pre-expert output-inductive-diminuendo-objective-article-(s)-incremental-principle-impact, affecting classification-(s)-offset, apose those Mile-(s) & Hour-(s), sorted by Meter-(s) & Minute-(s)-relevant-proportioning, distance-transit, because of option-(s) while adrift mile-(s), extraspective-technique, near-an approximate-distance-(ing)-measure-(s), viewing from centripetal-meter-(s)-(ing), each passing-individual, while road-development, stall at [maxim / continuum] physical-limit-(s) of parallel-offset-circulation, gas-pump-petroleum-limited, from Federal-Regulation-(s), timbre by-of pulsing-age-(s), whom not cursive-direct, an sociological-genetic-civilization, through legislational-congressional-library-mode-(s), affecting [live-conscious-blood-pulse-breath-inhale // exhale-fatigue] intermittent local-congregated-sect-(s), figuring out how pseudo-limit-interaction-(s), can cycle away from formal-ethics, ugly from those bastardized-genetic-comparison-(s) of map-cartography-degree-centripetal-force-drift-peripheral-impact, where sight-sound-scale-comparison-(s), are suppose to invigorate-tense-(s), working from spaced-observation-(s), as university-study-(ies), order term-tier-logic, from juxtaposition filling of [state / commercial-enterprise / national-Federal-Denominative-Currency] apose secondths-elapse-(s), past vast vision of creative-balancing, formulating literary-congressional-reason-(s) of practice (for [racial // ethnic-(less)] cooperation, to change from genetic-study, as from how an total-society-(ies)-physical-superlative-activity-(ies), demonstrate [congressionally // regulate-economically] how illiterate-trade, is not an formal premise, for basic-human-right-(s)), affecting whom remain prevalent, from reiterating letter-(s)-word-(s)-syntax-article-(s)-data-posit-(s) by signature, as not all would purchase per se, this book, (for) featuring an new method of trade-(ing), recalling data upon the locational-offset-reparametrizing, proportioning term-cubic-feet-distance-(ing (s)), per room basis, amidst constant-measure-(s) of those parameter-(s), count-repetitive-ordering, exponent room-(s) of ensemble, where discourse per interaction, reoccur, by vocal-larynx-verbal-vernacular-word-syntax-limit-tone-pitch-volume, affirming an timbre of denominating, numerative-note-circulation-(s), meandering [tangible-graspable // ton-(s)-mechanized weight-(s)-{1pound-(s) // 2liter-(s) // 3quart-(s) // 4pint-(s) // 5dram-(s) // 6grain-(s) // 7ounce-(s) // 8gram-(s)} / 9measure-(s)} {1Feet // 2Meter-(s) /// 3Yard-(s) ///// 4mile-(s) / 5centimeter-(s) / 6millimeter-(s)} / 7composite-(s) of elemental-compound-materials}] numerically counting those local-domestic-market-(s), by when shelf-stock-ordering-routine, yearly-reset-birth-aging, not realize international-impact, from European-Trading, in part due to Default-Existence, evaluating Daily-hour-schedule-reset-count-(s), from Literate-career-tier-conjecture, influencing an common-fluent-social-competent-interaction-(s), syncopating extraspective-interposed-data-affirmation-(s), rather trivia per miscellaneous-instructive-information, when Ways & Means', have founding-father-premised an Two-Hundred & Forty-Year-Lineage-Impasse-History, mass-census-population-count-de-facto, [1Use / 2Apply / 3Develop / 4Commit / 5Conjure / 6Debate / 7Resource / 8Market / 9Ship / 10Handle / 11Concept / 12Deliver / 13Produce / 14Develop] as functional-schedule-task-(s), [blood-pulse-beat-respirational-breath-inhale // exhale-organic-being] under imperial-impression, fatiguing an worn independent-individual-perspective-(s'), aging by an physiological-phase-(s)-{1Birth // 2toddler // child // 3pre-teen // 4teen // 5young adult // 6career-adult / 7pre-retirement-adult / 8retired-elder}, perceiving as one grows old, by an past-historical-dating-to-year-lessoning-(s);[256] stubbed method-(s) of Remembering, never inquisitioning [world // global-affair-(s)] fatigued-tensable-limit, [blood-pulse-(d) / breath-(ed)-respirating-inhale // exhale-circulation-count-(s)] conscious by-of-an Unconscious-mass-continuum-registering, denominating by scaling objective-weight-(s), balancing regulated-currency, through-an routine accustomed by [1dry / 1.1wet-substantial-(s) // 2metal-object-(s) / [3]plastic-object-(s) / [4]rubber-object-(s) / [5]wood-object-(s) / [6]recipe // [7]technique // [8]skill] dimensional-observation-intrication-analysis, precision [[1]cut / [2]rivet / [3]sanding / [4]mechanical-manufacturing / [5]saw-stroke / [6]position-(s) / [7]fitting / [8]painting / [9]architectural-posit-(s)], through "memory"-(memory I believe incur through "obituary-death-live-human-observation",| 523 |as I prefer [tense-visual-tonal-syntax-synthesis-observation // written / manuscript-documentation // alluded-present-day-past-recollected-term-definition-glossary-word-(s)-instruction-blue-print-patent-copyright-registering-intelligence] as term for such quality) in day-requisite-continuum, apose calendar-age-dating-denominative-cycling-(s)-offset, verify-ing from lifestyle-juxtaposition, infetesimal-incremental-crescendo-miniscule-document-ation of population-opposite-sex-affair-(s), deducting from exterior-clue-collecting, simultaneous process-conjuring, on scaled-property-(ies), referring by live-blood-pulsing-inhabitant, alluded intellig-ence, (Fathom)-height, having weekly-conjure-period-(s), thinking from past-data-collection-(s), about future currency-note-(s)-data-expenditure-(s), referencing passing-word-notion-(s), where empirical-subversive-{Subterior}-articlizing, derive duty by underlying meaning, not conveyable, as how tectonic-plate-(s), remain an part of Federal-Boundary, (as an) example to humanity, that development of territory, require Denominative-Elemental-Resource-(s)-count-(s)-documenting, objecting limit-(s), when independently-responsibility, offset-location-impasse-limit-(s), not visible-interior-room-human-limit-aspect, where entity-structuring, yet flaunt intelligible-trust, as neutral-unbiased-opinion, pertaining workplace, in maintaining intimate-affair-(s), because during extraspective-unreligious-socialism, genetics selethisize by American-governmental-swearing, ordered-objective-motion-(s), arranging progression, of article-(s)-bill-count-(s) in process of work-week{5-Day(s)} // weekend{2-Day(s)}, that have an {Article-Count per [Time // Calendar-posit-(s)-reasoning}, relative-relevant individual-being-(s), upon mass genetic-requisite-surveying, awhile capable agreement-(s) establish planned-living-arrangement-(s), slept and passed, faster than any one of us can be applicable for identifying those arrangement-(s) of schedule, synapsed of sight-identifying [1color / 2shape / 3distance-depth-perspective / 4texture / 5density-(s) / 6liquid-(s)] (from) adjunct sound-vibrational-resonance-volume-time-key-signature-pitch-registering (per of) [Minute-(s) / elapse-(s) / Tense-(s) / synapse-(s) / hour-(s)] evocating subjective-thinking, by an season-year-perspective, dubious cubic-area-per-cubic-foot-proport-ional-realizing, time by minute-(s)-synapse-syntaxing, per hour, aging by hour-day-count-(s), by height-clearance-parallel-centripetal-force-gravity-pull-subduction-zone-perspective-maxim, gauged from step-(s) of physical-movement-(s'), adjusted from lot-tower-zoning-construction, or digging underground upon high-altitude-terrain, where coastline undercurrent, depression deductive-prose apose grandiose-tectonic-expanse;257 continuing work, from city-county-ambient-development, retiring by-an old-age, contracting oneself to an house of open-limit-(s), offset expansion, though yearly-aging, merely breathing to synapse, for direct-individual-physical-limit-(s), noting exterior perceivable-value-(s), through an instance-(s) of independent-existence, as grandiose-mass-peripheral-untangible-celestial-tectonic-offset, not proportion oneself, by locational-developmental-surface-area-space-(ing-(s))-impact, motioning-past solar-unconscious, infetesimal-miniscule-fragmentizing singular-physiological-posit-(s), supposedly insighting humanity, that effort is never completed, while physical-extraspective-government-corporate-militarial-effort-(s), are required,90L. apose family-neighbor-rural-space-(ing-(s)), maintaining territorial-claim, to construct along an [Tectonic-Impact / Water-Pressure-Impact / Molten-Plasma-Fission-Combustion-Drift-Impact-Infinitive-Continuum!!!!]¶|||07m.:41s.56m.s.|||§

§...Preposition-Count-[26 / Twenty-Six]-of.. ...Article-Essay-[9 / Nine].. ..Book-Essay-[144 / One-Hundred-Fourty-Four].. ..Lines-[91.815 / Ninety-One.1eight-tenths-2.one-hundredths-3.five-thousandths].. ..Fundamentally government pays for human expenses, when commercial-wage, remain new in those visions of human-disposition, per individual-wage-worker, leaving how retirement is intended to refine those progressive-aging's of humanity, from an disposition of circulated-elemental-resources, overspending commercial-money on trivial-present-product-gifts, when daily-centripetal-force, would evidently display an cause for mathematical-elemental-proportional-prose of each [[1]location / [2]place / [3]lot / [4]entity / [5]infrastructure / [6]designation].¶|||32s.31m.s.|||§

¶145Where are those personality-(ies) amongst city-district-limit-(s), awhile inhabitant-citizen-(s) remain judged by capable tensed-motion-(s), typically remaining dormant, possible elapsing through hour-(s), as assess conjuring, merit-avid-scholar-independent-citizen-(s), for whom affect exterior-surrounding-existence, attempting to verify, article-(s)-arithmetic-grammatical-punctuation-decimal-exterior-compartmentalized-Classification-(s)-age-(s)-denominative-value-(s)-juxtapositional-Article-(s)-column-row-offset-(s)-Ordering;[258] prosing constitutional-purpose, inquiry-surveying, weather human-retirement-aging-passing-(s'), never have had high-school-education, soly resolve mass [issue-(s) // topic-(s) // affair-(s) // bill-(s) // resolution // et cetera…];259 omitting article-amendment-functional-purpose, by psychological-conscious-interaction, forcing physiological-formal-interaction, from time-rate-count-(ing-(s)), apose terrestrially spacing, entity-(ies)-development, when reverberated by mass-requisite, for-of blood-pulse-breath-perspective, incredulous citizen-perceptive-influence, from arid-continuum-effect, working in-to-an solid-retirement, by rural-space-(ing (s)), amid mid-age, wet-sexual-substantial-youth-studies, furthered-transition, dry-sexual-substantial-career, introverted-aging, commercial-blue-collar-work-muscle-reflex-(es), simultaneously-repeated, again and again, for no common-average-citizen-inquiry, verifying land-space-(ing (s))-rate-impasse-impact-(s), interpose-interpreting, substantial-ingestion, intermittent posit-(ing-(s)) of tangible-solid-object-(s), in distance of family-neighbor-house-representative-dynamic, (formal-trust-examination), supposed to be naturally investigated, as where existence pre-mature procure, ambient-peripheral-untangible-(s), celestially cycled, apose person [reason / trade / currency-balance-measure / state-territorial-responsibility-(ies)-(ing-(s))] predicate conjunct, trading-standards, sufficed by mass-introverted-continuum-requisite, gossip-talking, without an mathematical-prose-cause, under-taking grammatical-punctuation-literary-prose, defining article-(s)-data-confirmation-(s), intricat-ing why an pertinent-wage, should [sustain // maintain // object-(ive) // subject // contain], inhabitant-pertinence [planning / scheduling / tasking / labor-ordering], from organization-(s) of labor-worked-action-(s), constructive-managed-directive-(s), impassing-results, attempting to avoid free-will-bi-partisan-requisite, comprehending politic-(s), when adverse-effect-(s), inclinate independent-introversion, limited by blood-pulse, constraint by tectonic-currency-trading, unaware of-a-the reproductive-intelligent-genetic-repopulation-objective-article-ordering-impact, through trade-(s) being independently-attain-(ed), without denominating an mass-proportional-extraspective-interactive-usage, defining why war gradually incur time to time, reiterating why year require perceptive-affirming, in an constant-continuum Three-Hundred-Sixty-Day-(s)-cycling, apose day-denominating, by Twenty-Four-Hour-(s), scheduling other-(s), (from an) abiding to-an entity-location-agenda-directive-(s), expanding-extraspection, focusing on avoiding fatigue, balancing an Ambervision of [Blue / White Collar-populant-(s)-manual-written-effort-(s)] by how articlizing define new leader-(s), from survey-inquiry of mass-participation-effort-(s), tasking territory, by an agenda, relevant literary-syntax-capability-(ies), furthering physiological-limit-(s) (or remain unaware) of great-expanse, by repopulation-methodical-offspring-reproduction-agenda-planning, not feeder-patterned from decade-effort-(s)-maxim-offset, when universal-systematic-organization-(s), not formally interact trade, as those offset-(s) of celestial-miles-distancing-(s), cluster upon [Federally-Centralized-State-(s) // County-(s) // City-(ies)] parameter-mass-population-legislating, Treasury-Federal-Reserve-Commercial-Labor-Wage-Operation-(s), trading when Federal-Reserve-Count-(ing) (s), [Bond // Grant // Scholarship-Bachelorizing] introverted-individual-(s'), remaining an part of insoluble resolution, denominative-accounting for incorporated-city-planning-county-state-taxes, for how inter-national-Federal-trade, is relied upon, governmental-ingenuity, elemental-resource-constructive-planning, natural [customs / traditions / responsibility-(ies) / duty / service / retirement] sedentary by individual-preference-lifestyle, dated when by monetary-affair-(s), circulating (from) lineage-memory-existence, biblically-referring, by local-governmental-library-dormant-reference-(ing (s)), prelate commercial-fiction-economic-historical-posit-(s) from auto-biographical-littoral-observation-context, not infusing arithmetic of conjured-observational-order, regimenting schedule-agenda-day-secondths-week-minute-(s)-month-hours-cycling of data,| 524 |upon {Sight}-{Sound / Listening}-{Taste}-{Smell}-perspective-complex, comparing tense-(s) to vibration-(s);[260] assessing how synapse-(s), impasse elapsed-moment-(s), understanding why scheduling-complex, intertwine [work / off-time-requisite-mode-(s)], intervalizing momentum-inducting juxtaposed by documented [constant-denominative-counting / room-count-(s) / bathroom-count-(s) / living-room-count-(s) / hallway-count-(s) / kitchen-count-(s)] separate-of [[1]utensil-count-(s) / [2]lamp-count-(s) / [3]chair-table-count-(s) / [4]stove-count-(s) / [5]refrigerator-count-(s) / [6]furniture-count-(s) / [7]mattress-count-(s) / [8]toilet-paper-serrated-sheet-count-(s) / [9]cooking-materials-count-(s) / [10]door-count-(s) / [11]lock-knob-key-count-(s) / [12]tool-count-(s) // [13]et cetera-superfluous...] articulating accumulated-objective-data, by exterior-posit-(s)-aspect, for humanity intricate mass-object-production, by those timings of objective-collect-(ing (s)), requiring mass-population-inhabitant-product-usage, when dating alive-perspective, proglamate introverted-opposite-sex-affair-(s), not thinking how women's-historical-wage-effort-(s), matter intermittent self-male-opposite-sex-literary-effort-(s), demonstrating an gay-undertone, from marital-repopulation, incapable of dozens-hundredths-thousands-exponent-repopulating, for succinct university-bill-note-documenting, denominate-note-currency-trading, by cubic-parcel-schedule-progression-ordering-(s), determining data, upon conceptive-conjecture-requisite, simultaneous independent-character-acting, accounting interval-pre-siding-expenditure-(s), for proceeding familiarity of formal-city-county-communal-impact, by the ecological-consideration, surrounding surface-area-development, when barren-rural-terrain, remain dormant, in ever considering why their is such an disparity of documented-action-(s), from when America was never intended to serve every bastardized-governmental [state // county // city], waged by Constitutional-Surface-Area-Space-(ing (s)), reaffirmed Federal-soft-hold-helm, congressionally-controlling, Senate-Bill-Resolution-Passing-(s), apose an soly-strict House-Representative-Article-(s)-count-(s), using literature for improvising amendment-(s), to constitutional-merit, civil where-at city-decorum, as how each [city-district // county-district-(s)], should have had an house representative, per district, managing duty-(ies) of social-prose, for understanding the disparity of human-behavior, upon trade, by survey-examination, per state, understanding why [education // literature] have not fluently-undertaken, an revolutionary-impact, awhile act-(s) should be intricately-literary-expositoried, referencing instance-(s) for [new-article-impasse // amendment-developing] fortifying wage-commerce, rather bill-regulation-(s), business-marketing why human-effort-(s)-requisite, should attempt to persuade an genetic-mass-formal-repopulating-element-patent-documenting-product-reproduction-effort-(s), practicing an legit modification of-an future national-present-requisite, when wedged-extraction of conducted-effort-(s), have not yet evolved in-to an mutual-genetic-conducted-timbre, amongst an all...¶|||05m.:50s.$_{78m.s.|||}$§

{Article:} [June]

§...Preposition-Count-[21 / Twenty-One]-of.. ..Article-Essay-[10 / Ten].. ..Book-Essay-..[145 / One-Hundred-Fourty-Five].. ..Lines-[80.13 / Eighty.1.one-tenths-2.three-hundredths].. ..Literacy is an commercial-evaluation, observing how human-ethnic-inhabitants, populate per generation, not extracting a-the fruition-cause of feminine-repopulating, amid male-activity, devising an basis for living for offspring, by thought of years per monthly-week, of daily-sleeping-cooridors, affecting how each of those facts, pertain each variation of perception, when dating is suppose to scale exterior meter-mile-(s)-distancings, apose an density-compositional-structure, balancing commercial-product-capacities, per composit-ional-respitorial-cardiovascular-physiological-anatomical-beings, per clearance-capacity-centers, of circulated-target- population-monetary-expectancy-(ies).¶|||36s.50m.s.|||§

§146 Exponents: {Dime-{1933-1945-Era} / Penny-{1860-1864-Era} / Nickel-{1776-1801-Era} / Quarter-{1776-Reset-Present-Executive-President-median-era-interval-(s)},| 525 |culminated from Woodrow-Wilson {1914(3)} — Dwight D. Eisenhower{1957}, moding our way of Good-(s) & Service-(s)-mean-(s'),| 526 |trading from-an clearance-surface-area-perimeter-planned-territory-measuring-method, convening current-(ed)-government-aligning-process-(es), as free-will-belief, not stress from physics-economic-natural-disposition, post-dating-relying, those commercial-objective-collection-(s), using mass-effort-(s) by Age-(s) of Decade-Bachelor-Era-Reset-(s), by when [youth / adult] careers, how to Federally amend legislation through the House, is an fancy of mine, by identifying how exponent-mass-offset-(s), [pertain // contain // maintain // sustain // object // subject], State-Territory, articlizing to discuss those condition-(s) of other state-territory-(ies), by when living-house, abide by senate-ruled-congressional-legislation-(s), affirming bill-adjusted-tier-legislating, for accounting an general-mass-population-needs, vague by-of analytical-survey, detailing mass direct-recognition, where position-(s) of illiterate-congressional-non-fiction-(legislational-cursive-context)-concensus,| 527 |not influence an Victorian-Henry-Magna-Carta-conducting, because idle daily output, balance an senate-commercial-adjusted-control, from-an Governmental-commonality-basic-human-right-(s')-premise, without an due-(s') of [physical // subjective // social-interactive-living] intricating from birth, impassed routine, while common-exterior-perspective, not sustain fluent-succession of terrain-space-proportional-adjustment-(s), because human-subconscious-lazy-preference, omit why effort-(s) by state-constitution, matter for handling those matter-(s),| 528 |by self-opposite-sex-repopulating, an mass-quantity-exponent-wage-proportioning, reveiwing why creating bills, are according to State-House-Operation-(s), handling sexual-affair-(s) by ordering processing-action-(s) per mass, motioning upon extraspective-interaction-(s'), past-requisite, present-term-impasse, advising bill-collateral-debt-(s), for not amendment incrementing, Singular-Person-elemental-constructive-engineering-factory-work-order-pattern-(s) offset per inductive-definable-physics-impasse-formal-developmental-career-work-trait-(s'), measuring masses at per thirty-thousand-population-resolution-present-legislational-handling, incrementing a-the house-representative-group-ordering-(s), from a-the abysmal 431-(81)-Federal-House-Representative-(s), consisting State-Representative-sustained-offset-(s), where each state process those duty-(ies), in coordination of Federal-House-Representative-(s), whom mutually-maintain, coherent-conceived-Bill-(s), regulate-balancing, tier-weekly-session-communication-(s), intervalizing scheduled-consulting, for survey of mass-population, by economic-Treasury-Federal-Reserve-Corporate-trade-transaction-(s), seasonal, by attempt to apply conscious-principle of mass-human-entity-interaction-(s), when interaction-period-(s), have [self-(ves) / political-party-instructor-(s) / State-member-(s) / County-member-(s) / City-Member-(s) / community-designated-elder-teen-volunteer-(s)], prestigious upon an serie-(s) of regular-meeting-member-(s), whom are supposed to attempt, rectifying a-the formal human-$100(.00c.)D./$1(.00c.) //

{Article:} [June]

r.D.:$1(.00c.)r.D./$100(.00c.)D.}-Note-Notion-Inclination-(s);$_{261}$ verifying Enumerative-quality-(s)-quantity-(s), by-of-an {0.10c.}{0.01c.}{0.05c.}{0.25c.}-decimal-integer-Deductive-Ration-(s), upon Whole-One-Hundred-Cent-(s) juxtaposed Dollar-Note-Enumerated-Value-(s), pertaining Reciprocal-Cent-Exponent-Denominative-(s), from effort-(s) of unconscious-mass-purchase-(s)-present-day-impasse-retention-rate-(s), furthering mass-quantity-(ies)-deviation-(s) of Federal-Commercial-Logic-(s), by Interval-Numerative-Governmental-Fifty-State-Spread, as [[1]Transcendental / [2]Industrial / [3]Revolutionary / [4]Commercial / [5]Federal / [6]Executive / [7]Legislational / [8]Judiciary-act-(s) of production] surface upon Residential-Lot-Parcel-(s) where [Legal / Medical-pre-present-governmental-recourse-(s)] Sevenths-Federal-Impact apose [National-Constitution // Fifty-State-National-Constitution-(s')] practicing how political-power, gauge-handle-maintain, topographic-terrain, by Educational-Grammatical-Arithmetic-Article-Documenting-(s), apposed tangible-definite-matter-(s), Tiering mass-population, Time-Syntax-Elapse-Act-(s)-Period-(s),$_{48L.}$ upon [Article-(s) // Amendment-(s)], [Mass-Impasse-Observation-(s) // Interior-Entity-cubic-feet-per-secondths-moment-(s)-time-posit-(s)-deviation-(s)] apose [Calendar-Date-Period-Order-Planning-(s), as Exterior-Compartmentalized-Age-(ing)-Classification-(s)-Denominative-Value-(s)-Juxtaposed-Article-Column-Order-(ing)-(s)...¶|||04m.:00s.$_{29m.s.|||}$§

§...Preposition-Count-[16 / Sixteen]-of.. ..Article-Essay-[11 / Eleven].. ..Book-Essay-[146 / One-Hundred-Fourty-Six].. ..Lines-[51.24 / Fifty-One.$_{1.two-}$tenths-$_{2.four-}$hundredths].. ..I wonder where automatic-manual-transmission is going? amid each past historical-numerative-monetary-exchange-piece, for product-circulated-shelf-production-(s), when population forget from [high-school / University / Commercial-Literary-Geological-Genetic-Table-Non-Fiction-Books] how to define each decade-aging, by an century of 1,200-formal-commercial-non-fiction-books, for one-hundred-million-population-mass, thinking how to event book-article-(s), at stage-stadiums, for understanding how to proportion those transit-resources, by entity-population-commercial-scheduled-capacities; documenting societal-wage-scheduling's, per empirical-commercial-tasks, order in succession most important-elemental-tasks, which matter to circulating the global-tectonic-plates, by metal-extraction-process, practicing when petroleum is restricted from elemental-crust-deposits, observing where entity-developments, are to be controlled by literary-documenting, to progression-proceedings of schedule, for maintaining wage of regular-living-quarters.¶|||50s.70m.s.|||§

¶147.Precision for paid-action-(s)-effort-(s), yet be documented formally, by Commercial-Bachelor-(s')-Era-Politician-(s), for rural-terrain, suffice an commonly underuse of (hyper)-tense-(s), surrounded by entity-(ies)-past-historical-present-usage-requisite-(s'), (from an) navigational-directive-scope, determining individual-populant-mode-(s), mute of political-mass-Congress-ional-Library-evaluation-prose, annotating word-(s),| 529 |by initial-combustion of magma-molten-solar-earth-core-plasma, peripherally upon our dilated-pupil-receptive-visual-hyper-tense-sight-(ing-(s)), (as an) Grande-ambient-mega-kinetic-fission-mass-continuum-infinitive, constant-ly of vapor-dense-friction, by an default-phenomena, unexplainable by human quality-(ies), requiring observation from deductive-thinking, for estimating-calendar-dating, from-our Four-Hundred-Thirty-Eight-Year-(s)-Past-History, as prior what Gregorian-Two-Thousand-Year-(s)-Historical-Present-Requisite;262 remaining in an contention of biblical-disguise, (for of a the) nature of mass, disrespectful from common-qualified-legal-legitimate-person-(s), in religious-attend-ance-counting, where mass not qualify an singular-citizen-loyal-objective-director, of lord-appreciation, due to the individual-prose, dormant by nature, and satisfied by-of mere basic-effort, for stressing tense-(s), when perspective can not be fully ordain-(ed) individual-preference-listening, [where // when observation-(s)] of individualism, contain those limit-(s) of attenable-focus, as movement-(s) of clearance-height-(hyper)-tense-perspective, aurora by physical-tangible-component-composite-composition-(s)| 530 |conjunct [interior-ventricle-blood-pulse-breath-respiration-inhale // exhale-muscle-(s)-bone-(s)-ligament-(s)-cartilage-organ-(s)-function-ing(s)-exterior-skin-hair-follicle] offset by-of [1head / 2neck / 3torso / 4arm-(s) / 5leg-(s) / 6hand-(s) / 7feet / 8finger-(s)] where those joint-(s)-{1knuckle-(s) / 2wrist-(s) / 3elbow-(s) / 4shoulder-(s) / 5hip-(s) / 6neck / 7knee-(s) / 8ankle-(s)} are point-(s) of movement-(s'), for which human-(s) have not understood by [sea-level / altitude] (at) of [tectonic-plate-Mass / clearance of subduction-zone-atmosphere-vapor-pressure-respiration-effect] intermittent centripetal-force-tectonic-plate-(s)-momentum, indicated where incandesant-peripheral-after-effect-(s) of daily-rotation, exemplify our limit-(s) from Sleep & Fatigue, when tectonic-plate-(s), appear like galaxy-(ies), separated by political-boundary-(ies), remaining fatigued by independent-maneuver-(ed)-movement-(s);263 | 531 |which are maintain-(ed) by commercial-taxed-governmental-operation-(s), from state-person-representative-(s')-interaction-(s),27L. where boundary-surface-area-offset-(s), pertain cardiovascular-respitorial-musculature-breath-ingestion-cycling, from those location-(s) of birth-impasse, apose singular-incapable-movement-(s), upon the physics impasse, remained at present -hand...¶|||02m.:11s.44m.s.|||§

§...Preposition-Count-[18 / Eighteen]-of.. ..Article-Essay-[12 / Twelve].. ..Book-Essay- [147 / One-Hundred-Fourty-Seven].. ..Lines-[29.465 / Twenty-Nine.1.four-tenths-2.six-hundredths-3.five-thousandths].. ..reflexes of human-motions, have yet been documented subterior, commercial-wage-action-(s), counting how many motions occur per task, then counting each motions secondths-rate, for understanding how to process-input, refined-scheduled-operation-(s), per entity-capacity-product-objective-operation-(s), developing each entity, offset for elaborating why motions are pertained per individual, when mass-population, have an individualism-error, per major-grouping, apose common-govern-mental-individualism, sufficed by air conditioning climate, for proceeding each actions of commercial-government-guidance, per offset city-population-living-cooridors-competen-cy.¶|||32s.56m.s.|||§

¶148June is an perplexing month, as how work-continue, children enter summer-vacation, commercial-workplace-environment, adjust to those droves of children with off-time and daily activity-(ies), when individual-(s) whom have children, may have their offspring, bear in-to existence, having educational-summer-vacation, idle summer-season off-time, where those lessoning of children, stop and forget, remaining dormant, adult-mentor-(s), whom are dissuaded by lessoning, for how subjective-interaction, is intended to practice, article-(s)-application, from-an locational-observational-vantage-point-(s), amidst common-living-standards, due to, limit-(s) of mind, courageous from lineage-offspring-monetary-passing, where each citizen, perceive from their continuum-existence, an commercial-guidance, freely-impassed, discussing natural [governmental / religious / social / economic-freedom-(s)];264 premised manual interaction-(s), by those various ground-(s) of terrain, for social-effort-(s), by coordinating time-posit-article-(s), intended for administering citizen-peer-(s), when planning is to be an yearly-foreshadowing, awhile presence of dating, remain because of boundary-limit-(s), implemented due to human-mass-incapability-(ies), envisioning applicated sound-barrier-limit-(s), where physical-mass-clearance-tangible-area, not have an formal-physique-tangible-coverage, of surrounding-grounds', for comprehending developmental-purpose.236¶|||01m.:04s.40m.s.|||§ Holiday is different from vacation-cycle-(s), for their impact, not have taken focus of religious-scheduling, upon the commercial-workplace, to effectively apply [fund-(s) / saving-(s) / investment-(s)] from the house-representative-legislational-reformation, due to the factoring of data, upon the requisites of dating, for appraising those exterior-variable-(s), of mass-production, to apply upon locked-interior-warehouse-cubic-feet-entity-(ies), by a-the dynamic of objective-progression, in succession of local-development-(s), approximate a-the living-location, pertaining scale-(d)-topical-article-(s), by independent-inhabitant-(s), whom roam freely by off-time-requisite-(s'), when of where place-position, pre-pertain discourse of mass-action-(s) in-an timely-fashion of event-(s), hyper-tensing primary-year-calendar-offset-(s);[265] juxtaposing secondary-{Week}-{Month}-{Season}-{Year}-count-interval-(s)-offset;[266] following an succession of progressive-operation-(s), through aging by individual-(s') capable of convincing, reexamined national-requisite, where those nation-(s) set forth from birth an favor of religious-persecution, seeing weather-if, any inhabitant-(s'), have formal where-with-all, reconjuring national-requisite, upon-an industrial-mechanical-technical-engineering-developmental-observation-(s),| 532 |ordering-patent-commercial-part-(s), by formal-process, through House-Representative-(s), abiding Congressional-Legislating, for continuing motion-action-effort-(s), interacting-by mutual-association, informing each of those adult-peer-(s'), why natural-digestive-cycling, interval holiday-week-input-(s) of {Day-order} [1.Sunday // 2.Monday / 3.Tuesday // 4.Wednesday /// 5.Thursday //// 6.Friday // 7.Saturday-week-day-reset-order];[267] as those reset-(s) of [rotation // revolution];[268] vary juxtaposition by tier of [activity-(ies) / neighbor-interaction-(s) / work-place-interaction-(s) / church-interaction-(s) / commercial-market-interaction-(s)];[269] defining an major-hour-(s)-cycling-perplex of dating, [calendar-date-inclinate-recollect-referencing // time-hour-minute-(s)-motion-(s)-deductive-elapsing-(s)] how we perceive methodical-motion-(s), amongst exterior-physiological-value-(s), concluding interval-interpretational-point-(s), avidly attentive focuses where limit of development, should be documented of periodical-Dewey-decimal-chronological-order, for [National // County // City compiling-(s) of data], interposing [time-(d)-interval-(s)-observation-(s) // calendar-mass-action-(s)-dating-observation-(s)] that matter upon presiding-act-(s), scheduled location-juxtaposition-objective-interposed-designation-(s), reconstructing those formal-free-intrication-(s) of [inhabit-ant-(s)-bias // hoc-behavior-preference], believing in cooperation of activity-(ies), when their be-an realm of objective-use, for [objective // substantial comprehension], allotting tangible-article-(s), from when formal-request, merit such virtue-(s) of recognized-objective // substantial-functioning].237¶|||03m.:16s.35m.s.|||§ Think of [A.M. // F.M. Frequency-(ies)-Transmission-(s)], as-an broadcast-mass-population-dissemination-deviation-estimate per hour, updating human-mass-free-will-distance-auxiliary-objective-user-(s), how county-tradition-(s), past-historical-requisite, present date, impassing roadway-(s), determining average-(s) of meter-(s)-mile-(s)-transit-elapsing-(s), measuring distance, by-an mundane-coverage, reducing those independent-individual-work-cycle-(s),| 533 |to have no vocal-tonal-vernacular-visual-letter-(s)-word-(s)-syntax-synthesis-interaction, from self-independent-preference, as receiver-reception-cognitive-memory-method, receptive-inclinating, statistical-ranged-distance-data,| 534 |odometer by mass station-signal-transmission-commercial-program-deductive-day-order-(ing), comparative those mass-product-volume-(s), not listening for how to measure those commercial-program-(s), similar to program-(s) by [county-city-(ies) // commercial-city-district-(s)]| 535 |as article-(s)-order-adjustment-(s), use hertz-kilowatts-frequency-adjustment-(s), as-the comparative-observation-(s) for Ways & Means {senate-congressional-judiciary-committee}, pertaining contained-maintained commercial-product-(ion)-shelf-life-means, comparative state-constitutional-county-house-representative-handling, of mass-quantity-product-patent-developmental-count-(s), for devising residential-wage-comparison-(s), where those means are proportional, those participating market-(s), purchase-deviate-cycling, yearly-impasse, not constitutionally-adjusted, for firmly affirming each exponent-notion-quotient-extent, meticulating particular-market-(s), under statehood-legislation, legally-progressing national-incorporated-filing-(s), at division-(s) of state-responsibility, handling unilateral national-tectonic-duty-(ies), from when United-States-govern-ment-holding-(s), matter by an mass-consensus, not written of past-decade-tier-government-aging-order-progression-(s), as modern-date-elapse, listless millennium-moding, hyper-tensing product-factor-(s), [subjunct // conjunct] by market not in order of (hyper)-tense-(s)-motion-(s)-act-progression-(s), thinking how to legislate formal-action-(s), where tense-(s) pertain by timed-ration-surface-area-clearance-spacing-(s), [dry-substantial-solid-(s) / wet-substantial-liquid-(s) / dense-solid-(s), as where elemental-composite-resourcing] clear as to how, purpose of patent-commercial-conceptive-objective-article-(s)-use, premise [independent-resourcing // production-factory-(ies)] by those mass-quotient-factor-(s), banked commercial-production, directing advertised intention, for examining humanity, from those pre-requisite-(s), of mass-ethnic-culturing-(s), never surveyed from those state-dispertion-(s), in an [bias // hoc space-(s)-handling], Louisiana-Purchase-Territory, examining Constitutional-measure-conducting, by those person-(s)-physical-character-observation-(s), literary-congressional-(less), upon mass-act-time-rate-(s), reviewing zoning-(s) of formal-constitutional-documenting, by county-legislating, as for why war occurs, due to those inhabitant-religious-dismissal-(s), over-swearing, daily-effort-(s), in adjunction to political-city-district-convention-(s)-(center-(s)), [vehicle // vessel // ship // spacecraft] ignorant weekly-meeting-(s), where political-boundary-(ies), not inspire an genetic-literary-repopulation-reproduction-method, from commercial-worker-investment-(s), attaining objective, when offspring matter,| 536 |(from an) religious-segregated-genetic-physiological-mass-population-developing, shouldering weekly-mobius-inclination-(s), by how city-district-timing-(s), interpose upon date-restraint, planning scheduled-transiting, church-attendance, suppose daily, for ever being able to gather formally mass-requisite, upon those ordering-(s) of mass, before an warring of tectonic-commercial-free-market-purchase-method, not enticed, factory-direct-mass-dated-accumulated-investment-purchasing-(s), expensive-item-(s), for grad-ually adjusting community to state-order-(s)-denominative-function, segregating Federal-value-(s), when mass still would not even respect say this book, by timely-fashion, to attribute an purpose for communal-persuasion, efforting off-time-weekend-work-week-church-requisite-present, referring-literary-data, by congenial-means', for commercial-purchase-inquiry, elongated formal-year-(s)-purchase, thinking how many repetition-(s) of singular-objective per dollar into cent-category-(ies), would matter per juxtaposition at church-objective-interactive-centralizing, how objective-purchase-communal-usage-(s), pertain greater-than congressional-library-number-purchasing, as real-time-dating-fashion, exceed personal-objective-article-(s)-usage, for enjoying introverted-action-(s)-existence, simultaneously passing by, with no daily-reading of literature, (this An book formally is two-week-(s) to an month, of reading, pending work-load-time-constraints), where work intercede an greater period-(s) of elapsed-mode-(s), ignoring an fluent-religious-interactive-reading-conjuring, by communication-(s) of inhabitant-(s), calmly appreciat-ing, independent-requisite-lifestyle, leading fate by common-means from elder-purpose, because if handling objective-(s), continue to hoard product-matter-(s), no thought can commence to think how each objective is overstored, without-an proportioning of density-objective-item-(s), juxtaposed-an surrounding aurora of [formal-objective-action-(s) // posit-(s) // sculpture-(s)] pertaining given-state-operation-(s), for containing tool-(s), centralized community-storage-center-(s), flowing interdialogue of land-terrain-act-(s), preferred by communicating through [literature // city-count-government-interaction-(s)], [locational-latitude /537/ longitude-mile-(s)-meter-(s)-surrounding-topographic-climate-calendar-timed-action-(s)-scheduled-agenda-volantray-participated-directive-(s)] delving from population-literary-prose-signature-confirm-ation-(s), signifying an formal process for interacting spaced-off Genetic-Lineage-existence, sorting through culture-(s) of conjunct-identified-individual-(s), whom personal-belief may prefer through [literary // commercial-internet // industrial // engineering // constructive-action-(s)], initiating effort-(s) of commercial-interactive-operation-(s), in grandeur of tectonic-plate-national-commercial-demand,| 538 |presently not demanded, from-of direct-commercial-religious-orchestrating, accustomed deficit-perspective, using [independent auxiliary // tool-(s)] understanding how proportions from physiological-motion-(s)-exertion, acted-experience, paper-documented, when impassing daily-requirement-demand-(s), listing those action-(s) required, per [tool // auxiliary object-(s)] upon-an [time-frame // calendar-weekly // monthly // seasonal-count-usage-(s)] resuming mass-cognition, contrary century-mid-term-interval-pre-emanation, inconclu-sive independent-introverted-lifestyle.$_{238}$¶||||08m.:11s.$_{67m.s.|||}$§ Secondths-tic-continuum apose day-(s) of [week-(s) / day-(s) of month-(s) // day-(s) of Season-(s)] so why not count hour-(s) of month, and apply an mathematical-calculated-legislative-commercial-labor-action-(s), arrange-order-task-(s), per [[1]constructive // [2]industrial // [3]engineer // [4]factory // [5]medical // [6]legal-(proceeding-(s)) // [7]militarial // [8]congressional // [9]judiciary // [10]executive // [11]agricultural // [12]labor // [13]clothing // [14]supermarket // [15]shipping & handling-plant-(s) // [16]warehouse-(s) // [17+]et cetera...], [economic-empirical-field // entity-(ies)] devising how interval-range-(s), mode-set, gamma-perimeter-map-scaled-territory-space-(s), cell-millimeter (or less), by mapping of third-person-juxtaposition, from self-interior-first-person-singular-seeing of matter-(s), [article-(s) // amendment-document-cycling] premising [[1]physical-tension-(s) // [2]locational-statehood-offset // [3]auxiliary-transit-device-(s) // [4]agriculture-yield-(s) // [5]horticulture-yield-(s) // [6]archetype-blueprint-basic-human-right-(s)-living-corridors /// [7]commercial-market-(s) /// [8]outlet-(s) /// [9]mall-(s) /// [10]warehouse-(s) // [11]fact-ory-(ies)-constructed-entity-(ies) // [12]objective-tool-(s) / [13+]et cetera...] for defining per [cent-(s) // dollar-(s)], those process-cycle-(s), from time-offset-rationing, why each dollar-per hour, is made from twelve-five-minute-(s)-period-(s) per hour ascertaining formal denominative-political-Federal-Bank-National-State-County-City-Note-Legislating, (much to do with why) surrounding-terrain, remain barren by American-standard-(s), omitting common-documentation-cursive-data, expecting expenditure-(s), adjusted from enabled (im)-partial-second-person-legislating, cooperating as always been an difficulty from human-behavior, not involving an Genetic-Political-Referendum, when Federal-Individual-enabling, bemuse mass-effort-(s), by university-trade-school-skill-(s), where pertained-data, not consider literary-congressional-review, sequencing-seasonal-reading-(s), apose monthly-new-book-count-(s), trying to balance the [physical // subjective-focus-level-(s)], incremented from deductive-tier-specification-(s), per age-experience, where their has been an lack of Live-Gregorian-Sociological-Psychological-Political-Scientific-Non-Fiction-Observation-(s), posing natural-genre, for Constitutional-attempt, where their is an depletion of observational-tier-contortion, when data, is not mass-perceived, requisite that of bias afloat upon, introverted-stubbornness, (when an) total of Federal-National-Governmental-Reform, not meet Federal-State-Genetic-Inquiry, purifying international-trade, cultivated [regenerating-crop-(s) / edible-mammal-(s) // constant-total-tier-legislating // develop-ing-new-engineering-technique-(s) // developing-rural-acreage-constructive-classification-(s) (not exceeding 10 story-(ies) of entity-development) / et cetera...] prosing weather it is possible, to affirm such decision-(s) of state, when inquiry-(less-(ness)) of common-daily-work-off-time-interaction-(s), historically-continuum, human-behavior-introversion, not morally persuaded away from free-will, because how Block-Routine-Act-(s), center-an comfortable-method of space-(ing-(s)), by rule per city-advise, uninterested in common-post-subjective-examination, focusing major-commercial-career-work-place, by exponent integer-objective-activity-(ies), cycling author-labor-(er-(s)), whom censor personal-introversion, listening by preference genetic-interaction-(s), generating-an common-hundredths-thousandths-dollars-per-year-common-twenty-million-population-circle-(s), from familiarity of internet-book-forum, by monthly-reading-(s) {25,000-Reader-(s) × $4-$6 per book = estimated $100,000 per year by Author (reader 12 book reading-(s) per year) = 89,400 Author-(s)-Group-(s)-Classification-(s) per year} when [labor-pool // genetic-enabling] could foresee purchasing circulating-chain, by working (from an) standpoint of [author-$100,000 per year // 22,400-reader-(s)] per Author-month-year, whom would work author-reader-workplace-outreach-congressional-persuasion, enhancing pronoun-recognition those supple-mentary-individual-(s), by non-fiction-forum-subjective-technique, making Author & Reader, attempt to repopulate yearly, from such economic-curb (presumption)}, making literature mission, housing-expansion, by congressinal-literature, influencing Federal-Venue-(s), Major-Metro-politan-Twelve-Cities-Treasury-Abundant-superfluous-Monetary-Saving-(s) {Estimated 321, 000,000, 000,000 Trillion-Dollars-stockpile-(continuum)}, forbidden by modern illiteracy to protest-propagate, participating by those various other career-field-(s), required to learn how to generate product-(s), when consumer is an primary component in trading, expanding by formal-denominative-denoting-impasse-(s), figuring how mere-trading, by basis of basic-human-right-(s), can be congressionally-interceded, from internationally-instated-trade, when-an millionths-intelligent-mass, would require constant-month-yearly-updated-pronoun-observation-(s), detail-ing projection-(s) of lifestyle-aspect-(s) {when in all honesty, I know I want [1]monetary-security // [2]daily-feminine-intersexual-intimacy // [3]stable-U.S.D.A.-food-cleanliness-standard-(s) // [4]housing-expansion // [5]an subjective-light-labor-work-routine // [6]constant nightly festive-(i-ties) // [7]an light vehicle-usage-routine / [8+]not including my nationalism-pursuit} without compromising identification of collected-object-(s), by personal-individualism, futile state-territory, fatiguing oneself, through interaction-(s), as required by subjective-physical-extraspective-effort-(s'), but when lazy-nature of men, not appreciate woman-repopulation-yearly-nature, wilting when men are overfatigued by eight-hour-work-method, [ignoring // not enjoying children], not similar to children, in fact of behavior, but dominating by age, passing religious-communion, because where reading is suppose-(d) taboo, by subjective-education-guide, youth-peer-pressure-socially-induced, ignoring why literature is fundamental an Congressional-freedom, meriting our very existence, when-an balance of other-(s), remain deliberately, common, preferring independent-superego-mentality, basis for introverted-work-reclusion, sustemming male-off-time-superlative-intoxication (why prohibition happened, {1920 to 1933}), believing blue-collar-work-method win the day-(s')-work, from whom individually-ascertain, adult-youth-transition-peer-enabling-attitude, without-an book-club-initiative, conceiving why off-time-interaction-(s), remain incum-bent from living upon tectonic-plate-centripetal-force-celestial-impact, offset linear-literary-punctuation-redacting, proportioning space-(s) parcel-plot-establishment-(s), by [book-context // objective-function-maintaining /// mass-population-scheduling //// calendar-dating ///// timed-act-(s)-physical-motion-(s)-elapse-sequence-(s)-major-detail-documenting-agenda-ecological-routine-effort-(s) ////// Seasonal-Month-Weekly-inspection-(s) revising [operation-(s) /////// dimension-observation /// cubic-meter-height-clearance-impasse]-{motion-(s)-impasse-discour-se}-{Mile-(s) // Meter-(s)} /// objective-posit-dimensional-observation / function / aurora-distance-spacing-offset, {sedentary-posit-distance-proportion-observational-determining}-{Meter-(s) // Feet // inch-(es) // centimeter-(s) // millimeter-(s)}, implicit of constitutional-religious-direction, because coherent-purpose, bog White-American-ethnicity-(ies)-support, from how partial-inadequacy, stain from North-Civil-War-Mentality-Method, by southern-Northern-Hemisphere-bigotry, drinking alcohol, more than defining literary-grammar-punctuation-book-Congressional-observational-existence-denominative-impasse-producing, forgetting to balance children-repopulation, with drinking, or lavish living, while space-(s) remain cort, showing display-(s) of congested 1,240-cubic-square-feet-general-bedroom-(two)-one-person-living-cooridors'-(including closet and a / c-unit)-{my room at my Fathers'-house}, when each independent-individual is off-mark by extraspection,| 539 |disoriented by Federal-trade-history.$_{239}$ ¶|||14m.:07s.$_{50m.s.|||}$§ American-(s') believe in oversight is over-waging-individualism (especially in this war {Afghanistan // Iraq},| 540 |without an inclination for denominative-conceptive-envisioning, [diagramming // sketching // literary-dimensional-detailing] proportion-adjustment-(s) of off-time-requisite-effort-(s), expenditure party-material-(s), not conjuring, state-commercial-reform directing national-change, because those difficult-effort-(s), passing-momentum, in preference of dues, "calling it an day", as why America has had an influx of outsourcing, that is not reviewed of outstanding-debt, balancing product-trade, from physiological-timbre, where human-population-classification-(s), have long been forgotten, because bewildering mass-millionths-confidence, where I suggest an nation in-an Federal-non-treasury-comparison, require Twenty-Millionth-Inhabitant-(s'), whom have those capability-(ies), comprehending, schedule-extraspective-ordering-purpose, per daily-impasse (as weekend example interaction-(s) of inhabitant-(s) abstract requisite-work-modes for working-together, on those civilization-component-(s), from Federal-House-representative-inquiry, along with Commercial-factory-direct-investment-fundraising-saving-inquiry, by genetic-aligning, (that is not from) say, an singular-edition, affecting outcome of act-(s), as tectonic-plate-offset-centripetal-force-hour-deviation-(s), affect the manner, for which we believe in, detesting discourse, continued along by now, powered from those [Executive / Judiciary / Legislation-Branch-(es)] vehicle-impact, ignoring totalitarian-effect, by basic-human-right-(s'), protest-influence concern, long directed, unpleasant-inbred-(s), whom prefer introverted-sole-physical-bigotry, paying those [objective // substantial-nutritional-debt-(s)] apose-an extraspective-interacting-communal-centripetal-force-deviation, not mutually-assessing, the nine-hundred-word-conscious-interposed-rate-deviation-sustain per four-group-note-taking per hour,215L. thinking in contrast of redacted hour-(s)...¶|||15m.:35s.34m.s.|||§

§...Preposition-Count-[83 / Eighty]-of.. ..Article-Essay-[14 / Fourteen].. ..Book-Essay-[148 / One-Hundred-Fourty-Nine].. ..Lines-[215.135 / Two-Hundred-Fifteen.1.one-tenths-2.three-hundredths-3.five-thousandths].. ..Executive-tiering have an offset-wage-historical, scheduling unspecific, per moment individual-(s), impasse where market-strips, centralize-petroleum, and abandoned-terrain, not be discussed, for of climate of terrain, never consider why developments of tectonic-plate-(s), Continental-offset, commercial-wage-thought, yet serial-redact-copyrighted, exist post literature, verifying how each note, collect those sources of fact-(s), for then tiering layers, of products at traded-zones, validating each population-impasse, for an token of appreciation, without an vocal-cue from moral-faith, unliterary, common-centralized-purchase-cause.¶|||33s.12m.s.|||§

¶149Referencing, (from an) honest standpoint, has yet ever been examine-(d), by congressional-non-fiction, not implored, mutual-respect-(s) by-when personal-possessive-national-identity, define religious reliance on monetary-account-(s), where Commonwealth & Community, vary extraspective dynamic, from those calendar-cycle-factored-effect-(s), [1sleep / 2transit / 3eating / 4bathing / 5trading / 6working] ordered schedule-revision-(s), while psycho-logical-inquiry, order objective-(s), from shelf-demand-cycling, invariable-time-(ing-(s)), per place-(s), per territorial-boundary-extent, from where orchestrated-effort-(s), predict capable inhabitant-(s), defining effort-(s) of terrain, cycling, processed-schedule, defining-an devel-op-mental-curb, per energy-level-(s), moding common-activity-(ies), required by populus, through survey-inquiry, when interrogative-directive, follow mundane-presence, pre-empt, elapse-(s), (for how) historical-pertinence, have an variation, ignorant-an adequate-formation, per daily-effort-(s), not planned, calendar-dating, to cycle-act-(s) by time-requisite, to interval-mass-voluntary-place-effort-(s), where responsibility of territory, incur momentum-limit-(s), never perceivable by the naked eye, only an camera-posit-documenting, to adjust understanding of phenomena, where directive of study-focus, is contingent by concerted parcel-space-(s), when weight-distribution, juxtapose terrestrial-affair-(s), furnished territory, [exterior / interior place-(ing-(s))] for having impasse-(s) of routine, be upon [hour / minute-(s) / secondths] per blood-pulse-elapse-(s), counting breath-(s), [inhale // exhale respiration-count-(s)] gauging how tension affect-(s) act-secondths-interval-(s), per formal-documenting, by psychological-schedule-routine-(ing), per daily-prerogative, reset-(ed) by common-physical-cause, upon perceptive-tense-(s), where exterior-value-(ing), is interpreted (from an) observational-subjective-calculated-grammatical-punctua-tional-purpose... ¶||||01m.:19s.48m.s.|||§

§...Preposition-Count-[5 / Five]-of.. ..Article-Essay-[15 / Fifteen].. ..Book-Essay-[149 / One-Hundred- Fifty-Two].. ..Lines-[20.125 / Twenty.1.one-tenths-2.two-hundredths-3.five-thousandths].. ..I reduce the first book redact-article-(s), for intent on monetary-trading, adjusting how extensive each redact can be, apose those fact-(s) of historical-literature, working on later-book-edition-redact-(s), perfecting my model of success, when I had not previously thought of such an inclination, prior [writing / typing] my first book.¶||||18s.94m.s.|||§

¶150[1Aspect / 2blood-pulse / 3article-point-(s) / 4plane-(s) / 5parallel / 6height-limit / 7depth-perspective-limit-reset-(s)] from transit of Consciousness by hand-eye-coordination, interior linear-parallel-interior-boundary, when physical-limit-(s), restrict boundary-perimeter-limit-(s), pertain where [County / City-limit-(s)] have personal-possession, individualistically-inhabit, human-being-routine, reclusive, housing-sector, [electric-current // water-meter-pressure-gauging] monthly-payment-(s), enveloped privatized-agency-inquiry, where reclusion remain conversational, conventional motion-(s)-motive, as daily-dynamic, insist inset-(s) of reset-(s), by calendar-schedule-(s), arranged by ordered-trade, processing community-mass-article-(s)-common-objective-development-(s), default miniscule-ordering, live-act-operation-(s), using [place-(d)-objective / auxiliary-transit-parameter-(s)], producing composite-labor-resourcing, continue market-purchase-parity-volatility, refreshed by shipping private-company-purchase-(s), for market-outlet-(s), company-gross-domestic-product-sales-production-ordering, daily-impasse-(s), by free-will-inhabitant-impasse, when night-sequential-reflexive-infinitive, invocational [median-hardware / specimen-article-filing / commercial-company-2900-(50)-Federal-Tax-Fil-ing-(s)] handling tangible-data, by-an means of common-trade-operation-(s), soly inclined parcel-remains', in formal-sort-order, of legal-format-property-(ies), using auxiliary-effort-(s), conduct-ing elemental-operation-(s), of Federal-work-state-demand, where Schedule & Labor, define an Ways & Means legislating, by Territory & State, product-variable-(s), defining dynamic of common-inference, where present-historical-progression-placing-(s), time-elapse in lieu of crop-edible-production-(s), apose horticulture-offset-mass-present-day-eating, affecting how those Goods & Services, become socially-accustomed, through memory-lineage-tradition, not aware purpose for retirement, thinking constantly, how throughout each aging per human-behavior, adjust rural-terrain, initiated where State-Condition & Population, persuade wage-labor, from offset-(s) American-national-religious-examination, by where each individual, remain inclined dating parcel-impasse-(s), by free-nature, moving from home-position, in-to road-control, of commercial-resource-(s), affecting outcome-(s) of Federal-Wall-Street-Treasury-monetary-production-(s), by percentile-(s) of {Cum-Laude-{Top 15%}-{(myself)}} / {Magna-Cum-Laude-{Top 10%}} // {Summa-Cum-Laude-{Top 5%}}-dichotomies-of-power, explaining how introversion-95%, is why equal-poverty-rate, remain against position-(s) of extraspection, study-analyzing, compared commercial-median-order-refining, until

50% of the mass, can be order-(ed) by those means of political-totalitarian-natural-purpose-(s), that linger the unfortunate-truth, no action can be pertain-(ed), if legislational-cooperation, (not) be acted through judiciary-executive-impact-juxtaposition, because the balance that order now, clout adequate congressional-library-literary-reference, viewing those Legislator-(s) & Person-(s) (of) power, in scope by, neutral-balancing, instated by North-Federal-Centralized-Development-(s), confirmed from timed-dated-calendar-history, compared decennial-census, to affect the direction of free-will, by international-talk-(s), intended to compare compiled-fragmented-fact-(s), by continuation of Executive-Branch-Conduct, where power, pertain legislational-competency, of [Enumerated / Reserved / Concurrent-Power-(s)] not extenuating balance by basic-order, maintaining morale of ethnicity-(ies), by educational-evaluation, because gauging historical-percentile-scaling, is evident an non-sequitur-Congressional-Library, not comparing [ethnicity // race // career-class // educational-standard-(s) // Library-content // aging-contribution-(s) // Commercial-Private-Book-Purchase-Rate-(s)-Deviation-(s), (per) those educational-lessoning-(s), Grammy // Emmy // Pulitzer // Nobel // Academy-Award-(s)} exposed, guide-(d) from inhabitant-collective-peer-pressure-herd-effort-(s), steered-by mass-collection-contrast, at individual-objective-storage-collected-trinket-(s), internalizing an physiological-envy, uninquisitive by collective-adjustment-(s), that would point out subjective-isolation, [contrast // compare] Federal-Instituted-education, by G.P.A.-hour-word-rate-deviation-(s), per thousandths-secondths-period-(s), referencing-subconscious-retained-data, affecting how mode of perspective, conjure from confidence of past-requisite-experience-(s), conducted from subject-material-comparison, by glossary-lesson-deviation-(s), elapsing-period-(s) objective, [where // when] simultaneous-matter documenting, apose lesson-reading, intervalizing fifteen-minute-(s)-period-(s) in [ten-minutes-introverted-reading // five-minutes-extraverted-interposed-elaboration-opposite-lesson-insighting];[270] awhile hour-exponent-lesson-maxim, would require contained, Fourteen-thousand, four-hundred-word-thinking, per working-eight-hour-impasse, extraspective-four-person-work-group-exponent, as-an efficient-method of documenting, where instated-power-(s), remain judiciary-blocked, by congressional-intuition, mandating Federal-Historical-requisite, constantly collecting, collateral-debt, when nationalism would better fact I think, an Constitution, under Congressional-affirmation {I.e. To raise and support Armies, but no Appropriation of Money to that Use shall be for a longer Term than two Years}-{Constitution;[271] Article-I;[272] Section VIII;[273] Twelfth-Enumerated-Power}, yet serve executive-impact by commercial-intention, for national-shifting, thinking how you would read, an book per month would matter as your filling up an book-shelf, decoring your personal-room, when meander of those political-power-(s), influence, how city-governmental-planning, extend an personal-housing-property, along with an theme of literary-prose, enriching oneself by [exterior-architecture-permitting // interior-décor-lavishing] an extraspective-influence per self-bias-preference, recollecting fact-(s) of engineering-(auxiliary)-object-(s)-usage, affecting each motion influenced (from an) historical-shift by navy-army-requisite, command-ordering form-(s) of military, governmentally-regulated, from becoming too powerful, in comparison focus-(es) of state-territory, where engineering, is not made an Ten-Million-(s)-Population-Reform, persuading competent-intelligence, understanding [elemental-resourcing-arithmetic-population-request // demand-expectancy-calculating] by pre commercial-move-(s) in addition to-an two-year-Federal-regulated-period, by match-contact, where territorial-claim-(s), affect condition-(s) for development-(s), required for mutual-contact, where territorial-claim, affect condition-(s) per development-(s), supposed by investing participated-act-(s), claiming where national-fundraising, focus applied various means, mapping out territory-(ies), from population-settlement-claim-dispute-(s), migrating from formal-state-reformation-boundary-perimeter-space-(s), ordained when constitutional-dating, model American Government, comparative Genetic-National-State-Reform, where each fact of mass, concert responsible-duty-(ies), where constitutional-direction-(s), place weather formal-objective-enabling, power mass-requisite, from planning an decade transition {2035-2045}, transposing [investments / fundraising saving-(s)], at Federal-Contracting, [data-transfer-turnover // conjuring] (re-written-copying-{Formal-Letter-Cursive}-(I can carry out such fact-(s), as this book be Cursively-Written-& Manuscript-Typed), where each piece of fact, not directly-count, from physiological-offset-visual-tonal-perspective, offset displayed earlier "I.e." where population remain free until competent-intelligent-subjective-individual-(s'), collectively choose to enact such right-(s), (un)-clear through legislational-enacting, from army-disposition, per territorial-engineering, constantly-efforting, reformation-(s) of state, (or not so), by how means by power, intertwine rebuilt territory-(ies), by onsite-mapper-territory-impasse-conjuring, how two-year-Federal-curb, correspond avid-inquisitive-assistance, relative to celestial-impact, by-when communal-voluntary-interposed-effort-(s), commonly not attempt reformation, per offspring-reproductive-repopulating, where capable-extraspective-objection, effort Federal given-concerted-purpose-(s), restraining self-limit-(s), by collective-mass-complacency-(ies), forever in congruent-fashion, by territorial-claim-(s), abiding Gregorian-Calendar-Year-wage-effort-(s), monitored for greed, affecting how planned-statehoodship, interact where each piece of matter, should have an multiple-effect, placing generated-effort-(s), from default monetary-survival-National-Allegiance, impacting daily-planning, amid conductive-effort-means-(s), where those control-(s) of trade, not effectively-balance, national-budget-genetic-prerogative, for human-repopulation, not by instinctual, yet blank human-average-subconscious-urspuration-(s'), when outsourced-elemental-resource-(s), remain brought in (from an) cheaper-waged-nation-(s), for American-(s') reeling on waste-inflation, from independent-introversion, without thought for whom, legislationally-enact army-disposition, per territorial-engineering, constantly-efforting, reform by state, (or not so from how) means per power, intertwine rebuilt-territory, by onsite mapped-territory-impasse-conjuring, (from how) two-year-Federal-fiscal-curb, avidly inquires' assistance, relative to impact by communal-effort-(s) trading-wisely, upon an concerted-direction-(s), when mass-extraspective-effort-(s), commonly not attempt reform-offspring-reproductive-repopulating, where capable extraspective-objection, effort Federal given-concerted-purpose, self-limited, collective-mass-complacency, forever in congruent-fashion, by territorial-claim, abiding Gregorian-Calendar-Year-effort-(s), interacting (as an) part per planned statehoodship, where each piece of matter, should have an multiplicit-effect, placing generated-effort-(s), from default monetary-survival-National-impact, by outsourced-elemental-resource-(s), per trade-(s), not effectively balancing national budget-(s) from nationalism, amid international-impact, per outsourced-elemental-resource-(s), brought in cheaper-waged-Nation-(s), for American-(s'), reling on inflated-wage, from independent-introvert-(s'), without thought for money, as how international-goods', affect those labor-wage-(s), by national-bill-regulation-(s), affirming how statehood-development-(s), affirm what notion-(s) are supposed to affect each individual-object, mattered by cycle-physiological-usage, amid physics-centripetal-force-momentum-impact, affecting basic-right-(s)-observation-articlizing, when balancing order-(s) of mass, by governmental-religious-order, required conscious-interaction, by approximate-location-(s), succinct-objective-filing, by bill decimal-tier-agenda-directive-ordering, entity-(ies) of representative visitation, approving why bill-(s), suffice requisite-interaction-(s), not forced by free-will, but required when foundational goods', are serviced when daily-Federal-trade, remain bill-regulated loopholed where commercial-trade, senate-approve (from an) past-congressional-history of sequestrated-measure-(s), not formally-accorded, asserted Federal-House-Representative-(s), (from a the) entire-conscious-past-histor-ical-requisite, of listless regard, not marking, those failure-(s) of legislational-execution, as daily-weekly-session-fundraising, is intended for accounting an mass-(es) of commercial-wage-payment-(s), through fact-(s) at interactive-communal-programming (for possible segregation), where impasse-(s) have yet been achieved, honest-expression-(s'), from internal-subconscious-exterior-visual-observation-(s), apose dated-fatigue of mass-documenting, sworn upon legal-standing-representative-(s), ancient-referencing, dishonest-squabbling-interaction-(s), scared of forthright-interactive-regard, (from when) discourse of action-(s), follow-an meander of actual-tense-(s), not corollary-grasped, awhile footheld, still be basic state act-(s), for operational-process-(es), requiring desire by interval-exponent-objective-article-(s)-tabling, an total-statehood-data-accounting-collection-(s), when political-person-(s) lack formal-regulation-(s), by-an blatant lack of congressional-imperial-imposed-strict-povertizing-article-(s), suffering humanity, for-of illiterate-mass-population-mentality-disparity, [unloyal // ignorant // undeserving // lackluster // disobedient behavior-(s)] demonstrated why (when) vehicle-mass-error, by basic-human-right-(s), political-person-preference from fear of World War, erroneously regard intelligent-observational-inquiry, common-wealth-appreciated, when work of formal-locational-representing, require-an understanding that credit an [congressional // national deviation-credit-clause] tracking objective-dating, when Three-Hundred-Sixty-day-cycling, halt on weekend-(s)-{104-day-(s) per year}, accumulating conservative-commercial-credit-collateral-debt-spending, not corrected-ordering-(s) by Federal-Economic-Regulate-Bill-State-County-City-Population-denominative-ordering, believing the right to educate-enable, all those mass-(es) of degenerate-understudy, when in fact, Education & Labor, would demonstrate an mode for which extraspective-cooperation, examine weather mass care live or not, and further an Totalitarian-approach America, rather an Egalitarianism-approach, maintaining politics, variating twentieth-five-whole-posit-(s)-population-percentile-ordering, mass-national-genetic-invention-patent-contribution, upon commercial-mundane-wage-balancing, analyzing how education has brought an globe of inhabitant-(s), for evaluation, under Swiss-Psychological-Principles, not (yet) understood, as for why, abroad-economy, has premised our livelihood, but when avid-continuum-effort-(s), are required from those mass-(es) interactive-discourse-(s), because referral-process, not pertain non-fiction-literature (examination of other genre-uselessness by {Live-Gregorian-Socio-Psycho-political-scientific-observational-non-fiction} self-proclaimed) to affect perspective, by natural-reading-timbre-offset-(s), meriting writing-offset, by manuscript-confirmation, compre-hending exponent-notion-deviational-rate-(s) as unread-illiteracy, median-focus through common-social-hoc-bias-youth-study-affecting-adult-aging, not enhancing lifestyle-timbre-presence, when not settled-down, calm study-(ies)-approach, to then intercontort literary-subjunction, clearance-sphere, tensable-perspective, using hyper-tense-(s) of perspective, understanding how (sub)-conscious-effort-(s), are interposed humanity-impasse-(s'), with night-time-retrograde-religious-appreciating, affirming pertinence of meander-observation-(s), while [rereading // referencing // reoccurring // rewriting], sum-(s) per composition-calculator-calculating oneself individual-compositional-posit-(s), as when simultaneous-dimensional-weight-interior-measure-premise, [scale // gauge // measure], exterior containable-tangible-(s) juxtaposed [hour-minute-(s)-elaspse-(s) / synapse-(s) // syntax-(es)-time-posit-(s)] constantly working on how physics denominate [interacting // impassing with other-(s)], worshipping neighbor-religious-family-interchange-(s) factify objective-resolve, by extraspective-interactive-claim-(s), rather individual-possession-belief-(s'), admonished ecological-chain-order-(s) of affair-(s), requiring an encyclopediatic-dictionary-literary-contortion, chronologically depicting unilateral-space-(ing-(s)), when dating present-pertain offset-(s) of mass-documentation, defining prestige of Federal-Responsibility, not formally-passed to contend against state-(s)-deliberation-(s'), not formally interposing cooperation, mark-measuring-(checks) each individual, through incendiary-remark-(s), or accumulating mass-observation-(s), defining sequestrated-congressional-budgeting-(balance-(s)), | 541 |creating article-scale-(s), pending how physiological-observation, incur through genetic-birth-age-transgression-(s), evaluating national-congressional-library, those curtail-(s) of Ignorance & Free-Bliss, where commercially-inclinated-ideals, found why thinking, origin by human-organism, comestibles-substantial-mass-matter-means, [regulation // referendum revision-(s)] upon [Dating // Calendar // Time], [where // when // why...] national protest-(s), have propagated fifty-percent (of a the) mass-population-regulating, (by an) discourse of mass, motion-(s)-efforting, trade where national-constitutional-amendment-objection, is how principality is suppose to Count & Track, all piece-(s) of weight-(s), from location-withdraw of composite-material-(s), refining-plant-(s), upon factory-pane-plate-rectangularized-government-refined-corporation-material-(s), for conceptive-constructing, from mass-patent-product-(s), under an Federal-Government-spiritualism, where capacity of means, requisite order, (from an) history of auto-biographical-congressional-fiction-analysis, of Constitution & Inhibited-Observational-Inhabitant-Existence, upon parcel-median-proportioning, from psychological-vernacular-meeting-(s), affirming regulated-bill-scheduling, apose [constitutional-article-(s) // amendment-(s)-merit-(s)] per government-count-control, on goods, cycling [objective // subjective matter-(s)] fluidly continuing elemental-resourcing-recycle-means, as mass-consumption, adapt Federal-Bank-object-Bond-(s)-giving, apose commercial-market-empirical-hierarchy-politician-balanc-ing, an juxtaposition-of ethnic-mass, examining education, by those means of trade, practicing loyal-objection, to help space currency of state, trading upon market-(s) of Federal-control, by when goods of American-requisite, keep only best [fashions / food-market-(s) / tool-(s) / auto-shop-(s) / et cetera...] not delving in-to how entity-(ies)-historically-offset, remaining without-an formal distance-measure-space-(ing-(s)), each objective-storage-requirement-(s), debating how ecological-system-ordering, has been subjectively-studied, by educational-want, warranted dormant literary-prose, comparing redacted educational-subject-(s), furthering entertainment-moment-(s) when [day // night] remain in comparison of work-mode, furthering off-time-schedule-cycling-(s), not though-of, concepting formal-model, national-reformation, continuum-extenuating, expedited-trade-method-(s), when-of common-commerce, warring intermediate-period-(s) of state, verifying how mode by which common effectivity, partake upon [inhabitant-(s') bias // hoc-perspective-(s')] living an means of statehoodship, present-article-(s), national-requisite, bettering mutual-timed-trade-(ing-s)), enhancing resourced-pattern-matter-(s), when method of inhabitant [blood-pulse / breath-respirational-inhale // exhale] count as point-(s) of natural-discourse, for making cent-per-object-count-(s), matter by national-documentation-concern, denominating [objective // subjective matter-(s)] in comparative-adjunction, by altitude-prevalence, comparative sea-level-disposition, extracting those [solid-(s) / wet-measure-(s) / dry-measure-weight-(s)] territorially-offset, phonetically remodeled Nation-State-County-City-development-(s) by common-means of mass-matter-weight-usage-count-(s), formally-equivalent, mutual-interactive-affair-(s), having matter-usage, affirm [where / when / why / what-matter-(s)] define whom, assess how [issue-(s) // topic-(s) // matter-(s)] conduct an balanced-enhanced-efficient-interactive-effort-(s), rather an system of unclear-enabling, not orchestrated-understanding (of) national-increment-method-(s), referencing from inhabitant-(s)-personal-work-experience, yearly-spouse-repopulating, by national-revision-(s), from such mass-inhabitant-(s), for whom means of survival, conflict confident-balance-securing, (per of) territorial-entity-perimeter-named-development.$_{240}$¶||||14m.:16s.$_{91m.s.|||}$§ Responsibility, accrediting payment-(s) by parental-enabling, affirm how living-space, adjust surrounding-community, when those rate-(s) of elapse-(s), are impassed present-effect-(s), of quick-time-elapse-period-(s), apose hour-day-dating-interval-(s), where Calendar & Age, hyper-tense slow, throughout day-effect-(s), living from omitted hour-impasse-elapse-(s) of dating, collecting [image-(s) / sound-(s)] by self-live-memory, developing circumstance-(s) of payment-(s), where community not learn (how to) communion, for living upon existence, by-an base of means, for gracing transition of self, minor-biographical-impact, Aging & Surviving,207L. centripetal-force-impasse-directive-impasse-notion-(s)-extent...¶|||1 4m.:47s.48m.s.|||§

§...Preposition-Count-[87 / Eighty-Seven]-of.. ..Article-Essay-[16 / Sixteen].. ..Book-Essay-[150 / One-Hundred-Fifty].. ..Lines-[207.385 / Two-Hundred-Seven.1.three-tenths-2.eight-hundredths-3.five-thousandths].. ..How to claim domesticated-land, is an challenge of utilities, upon physics, sourcing in concerted-saturated-miles, tonnes of cylindrical-cement-planning-infra-structure-pieces, proportioned from governmental-census-city-foundation-planning-mundane-survey-(s), appropriating each land, by tasks, offset particular-major-focuses, offset scheduling minor-off-time-activity-(ies), deducing from population-consensus, for how physics would always remain, evaluating how population-performs, by an criterion of standards, observing conjunction-(s) of entity-capacity-space, proportioning what makes each notion pertain daily-parameters, proportioning first balance of entity (residential (savings) / commercial (wage)) operations, per independent-perceptive-competent, whom comprehend what to trade and collect, and what tools affect an extraspective-entity-operations, for developing ones resting-time, in collaboration with other-scheduled-individuals, whom comprehend why developing from purchased-ranged-pot-lucks, would require an reasoning of where wage require an location to participate, with how context define an cause for an entity, yet convinced by government-ingenuity, still thinking how to survive pass the age 62, due to each stride not sustained by an singular-effort, when development of titled-land, not know what to object for developing territorial-expanse, in proportion of population.¶|||01m.:01s.75m.s.|||§

¶151Focus of [blood-pulse-breath-respiration-inhale // exhale-reflexive-muscular-ligament-skin-tension-(s)] contain bone-density-weight-posit-(s), standing at-an beginning of place, reaffirm existence, where daily-hour-impasse-interval-(s), 3,600-secondths-per-hour, impasse reference, through individual-self-determinant-observation-(s), through physiological-perspective-(s'), exterior those wind-pressure-(s) of solid-(s)-tangible-posit-(s), tension apose tectonic-plate-offset-continuum-impact, affecting blood-pulse-(s)-cycling, for currenting memory, by millimeter-interior-pressure-measure, not juxtaposed ventricle-aorta-central-pulmonary-mark-(s), formed by interior-organ-(s), measuring dry-wet-substantial-composition-(s), impassed centripetal-force, awhile initiate momentum, process existence, in careful balance of still-reflexive-movement-(s), when circulation-reset-(s), interpose Twelve-Cubic-feet, {6H. × 2L × 1W. = 12C.F.H.L.W.}, aligning tense-(s) counted by-an individual-focus response, from when perspective-visual-tonal-playback, contravene evocation of exterior-limit-(s), idle deciding call-(s), from pressure-(s) of aging, evolving by simultaneous-offset-dynamic, where tectonic-denominative-soil-composite-metal-weight-(s) of political-state-(s), observe base-notion-abstraction, pertaining those [sine / co-sine / tangent], ordered-progression-(s), remembering, after minute-(s)-elapse-line-align-(ing), (from an) abstraction to maintain perimeter-defining-spacing, following, angle-juxtaposition{180°}, denominating each notion-deviation-(s)-extent, pertained article-(s) matter-(s), identifying interval-progression-(s), per count-(s), giving label-identification of [context / content] simplifying matter-(s), upon intrication of subject-lessoning-(s), tabling-impasse-elapse-(s), in-an diagram-coordination of material-(s), when each fact scale [weight-(s) / taped-measure-(s)], per [distance-(s) / dimensional-production] where tangible-count-(s), not extrapolate, creative-readjustments for newly-identifying-material-(s), meticulating mathematical-prose, apose perspective based observation-(s), Articling & Elaborating dimensional-objective-ordering-(s), by where those approximate-commercial-steel-corporation-outlet-(s), weld-project-(s), in contingency for configuring product-concept-(s), by-an mass-purchasing, from communion-interaction-(s), intervoluntary elapse-sequence-gauging-over-lay, per calendar-interval-dating-(s'), aligning time-(d)-impasse-motion-(s)-effort-(s), per purpose of living-daily-impasse, [when // where // who...] collect commercial-product-object-(s), for cycling ordered proportional land-space-capable-effort-(s), by limit-(s) hardware-objective-data, injuncting property-parcel-surface-area, ascertaining [storing / use / activity], from hardware, where cycle-(s) through an debt-base-trade-method, suppose an question, [yearly-use / seasonal-use / monthly-use / weekly-use] developing storage-center-(s), for hardware, where daily-use, is clearly not per individual, and yet planned of commercial-discourse, as community-storage-center-scheduling, yet cycle, [auxiliary // tool-(s) hardware-data] upon those [residence-(s) / commercial-entity-(ies)] which required labor-wage-output, upon those condition-(s) of county-city-district-Federal-bank-development-(s), attempting to incorporate, planned-extension-(s) of family-owned-property-dimension-development-(s), when routine-(s) from those [tool-(s) / contraption-(s) / mechanism-(s)] by mass-quantity-intervoluntary-action-(s), not be though [purchase-parity-dating-interval-tangible-objective /// subjective /// interactive-dialogue-discussion-usage] pertain each solid-tangible-material, by juxtaposition of impassing centripetal-force-impact, understanding why date-counting of solid-object-(s), matter-(s) by-an fact-(s) of objective-usage-count-(s), per city-district-space-(ing)-limit-(s), fluently comprehended why object-(s)-matter, require addressing mass-production-usage, [where // when...] influence of vernacular-data, factor patent-product-information, instructing usage, when factual-commercial-output, sustem from Supply & Demand, governmentally-resourcing, corporation-mass-product-quantity-purchase-modifying, by direct-bulk-quantity-item-resource-(s), reducing those price-(s) of pre-manufacturing-mass-count-(s)-data, in those concept-piece-(s) of commercial-quantity-(ies) of data, where commercial-market-(s), offset independent-capable-use, apose interactive-formal-directive-usage, from component-(s) of decay-life, compared human-live-anatomical-aging, where an influence per [temperature / condition-(s) / climate / topography / timing of daily-activity-(ies) / communal-extraspection-interaction-(s)] influence letter-word-specific-subjective-topic-data-accumulating, [when // where // what…] remain pertinent;[274] by subjective-quality-usage, extenuating formal-repetition-product-use-(age-(s)), as hardware-data, influence outcome of manual-effort-(s), while simpler commercial-labor-effort-(s), forge those after-effect-(s), from any inhabitant, which would require circulation-understanding, subjective-prosing, required in neighbor-co-worker-religious-member-housing-surrounding-community-interaction-(s), in ever conceiving common-factor-(s) of product-data, intricating mathematical-prose,| 542 |from grammatical-punctuation-identifying, an ever enhancing capacity-(ies) of theatre,| 543 |{Drama}-{Ensemble}-{Orchestra}-{Chorus}-{Philharmonic}-{Painting}-{Sculpting}-{Ballet}-{Mural}-{Entity-Carving};[275] surrounding community-entity-(ies), defining an local impact, from developmental-timbre, as each fact, remain through step-perceptive-progression-(s), not developing an constructive-composition-(s), where daily-anatomical-schedule-designation-(s), are purposeful in developing, those surrounding area-(s), where life continue an significant relocating, from climate-temperate-environment-(s), [where // when…] temperature affect daily-dynamic, of constructive-development, enveloping cause-(s) of tally-(s), tabled modern-act-(s), at an juxtaposition of dating, rescheduling participating-inhabitant-(s), by-an routine of vernacular-read-data-affair-(s), storing subjective-article-(s), in-an graceful-transition of effort-(s), for proportioning visual-sight-(s) of surrounding-territory, enhancing [63L.] [visual-imagery // tonal-reverberated-vibrational-volume-pitch-time-key-signature-tempo-beats-per-minute-measures-sounds...] for defining what is around us at any given moment...¶|||05m.:11s.76m.s.|||§

§...Preposition-Count-[32 / Thirty-Two]-of.. ..Article-Essay-[17 / Seventeen].. ..Book-Essay-[151 / One-Hundred-Fifty-One].. ..Lines-[64.67 / Sixty-Four.1.six-tenths-2.seven-hundredths].. ..For an appreciation for learning have not ever exist, how could any individual compare what learning methods, affect those outcomes of living, when subconscious-sounds are instinctually impassing every single billionths [individual / insect / mechanical-sounds / mammals / mariculture / collision-impact] (* as I would presume from Los Angeles, my subconscious be sound-emitted, as the police interrupted my book sales in L.A. for some odd reason), because an lack of subservient-documenting-discipline, not efficiently-work on capable-acts, from an motion-(s) of morale, to comprehend how trading is understood, by each patent-product, circulating an infinitive unclear shelf-life-demand, yet structuring cities to be formally comprehended, to order each inhabitant, by an capacity-juxtaposition, each five-minutes-mode per day, at sequestrated-entity-(ies), serving an purpose for populant-product-retaining, for residential-housing-storing, determining when to trash those products that serve an extent of dating, by counts of day, per commercial-factory-production.¶|||01m.:00s.00m.s.|||§

¶152Indication remain out of question, for individual-tangible-bias, offset common-effort-(s), when no effective-production, count human-mass, proportional, objective-substantial-impasse-scheduled-goods-requirement-schedule-dating, pertaining Federal-duty, for developing tender by labor-capability of trading, when an mutual-unilateral-balance of metropolitan-international-tectonic-plate-(s)-dynamic-impact, affect [mass-tangible-breath-respiration-inhale // exhale-blood-pulse-musculature-contraction-(s)-tense-(s)] that inhibit perspective, (from an) exterior visual-sight-sound-perspective, ranging distance-(s) upon an peripheral-location, where our motion-(s), apose our hyper-tense-(s) [Seeing / Hearing / Tasting / Smelling tense-(s)] evoking tensed-contraction-(s) of movement-(s), further passing centripetal-force, in-an visual-apparent-repetitive-abstract-grandiose-ambient, as revolving-rotational-pressure remain vapor-tangible, seen aurora, when subducted from solar-earthly-centripetal-force-pressure-friction, comparable of electricity-iron-magma-soil-tectonic-rock-fire-water-impact, similar to those current-(s) of metal-wiring, from water-aqueduct-pressurizing, basing exterior shaped-form-(s), of elemental-impact, by-an dimensional-refining, uncreative of default-purpose, for than stabilizing, objective-handling, by feature-(s) of language-art-(s), as creative-medium for literary-non-fiction-prose, attempting to influence, voluntary-competent, while remaining in-an clear examination of offset-will, not to invade perceptive-effort-(s) when independent-citizen-(s'), think of remembrance after living upon existence, considering whom will observe oneself, by an non-concern from extraspection, numbing how a-the nature of humanity-work, not to consider after-life-living, existing in-an luxurious-reality of circumstance-(s), that while during living, should be pre-anticipated, for repopulating an lineage-neighbor-territory-aspect, that would require an indefinite-effort-(s), infinitively unconscious-future, by one-on-one-momentary-perceptive-interpretation-(s), askewed when present-anticipation-effort-(s), show how State of Florida, would exemplify retirement-sector-(s) of inhabitant-(s), whom consider the location of Florida, worthy of living, by hot-condition-(s), not in-an form of territorial-sector-(s), rural default-government-guidance, literary-tangent-expository-pronoun-observation-(s), denominating spouse-relationship-(s), from Author-Reader-interaction-(s);[276] moving away from civil-right-(s), with duty of nation, by latter-day-judgement-(s), to expanse-period-extent-(s), when planning-principle-judgement-(s), unscheduled for independent-commercial-governmental-medical-legal-grounds, which foreshadow those truths of bliss, imbalance army-effort-(s), by-an common-national-regulating, for appraising surface-area-dimension-personal-family-denominative-space-(s), incrementing terrain of "Louisiana-Purchase"-Tectonic-Coverage, inhibited from Biloxi-effect, rather an Mass-Genetic-Population-Nation-State-Reordering, from independent-effort-(s), reasoning why deed-(s), pertain where documenting is important from those aspect-(s) of longevity, concentrating-elemental-resource-(s), by-an unified data-agenda-directive-tasking, affirm those astute-reasoning-(s), when afterlife-anticipation, apose lifestyle-metre, understand why work-effort-(s) remain an part barren-terrain-development.$_{241}$¶|||02m.:49s.$_{66m.s.|||}$§ The General-Challenge, of existence, is premised from detail-(s), [[1]limit-(s) / [2]reasoning / [3]reference / [4]hemisphere-(s)-offset / [5]citizen-mass / [6]objective-possession / [7]property-claim-(s) / [8]substantial-resource-(s) / [9]reproductive timing-(s)] where withering of existence, is by-an aged-impact, from [mysticism / belief / superstition / bias / hoc] for how [visual-observation-(s) / vernacular-vocal-chord-ascertaining / sounding of place-surrounding] by [live-effort-(s) / offspring-development / stated-legacy] meticulate living-blood-pulse, by those deviation-(s)-count-(s), pertaining measured-enumerated-locational-parcel-space-(s), promenade because micro-fibromyalgia-impact, signify effort by-an difficulty of state, when infetesimal-incremental-interpretation-(s), abide Enforcement & Law, instilling purpose of parameter-parallel-prerogative-(s'), by common-reserve-trading, impassing age-(s) of adult-periodical-responsibility-(ies), yet by focus mass-(es), from possessive-introverted-individualism, from when date-time-interval-(s), elapse under period-(s) of hour, interpreting-task-(s) of property, currenting cycle-flow, when passing generation-(s) of era-(s), by development, stagnated from conscious-period-(s) of unscheduled-locational-off-time-requisite-(s), reorganizing, [[1]objective-act-(s) / [2]route-discourse / [3]trade / [4]custom-(s) / [5]reading-(s) / [6]instructing / [7]redacting / [8]modeling / [9]act-elapse-tabling / [10]activity set-up // [11]clean-up / [12]designation of parameter-(s)] thinking freely from our [cardiovascular-circulation-respitorial-breath-inhale // exhale-blood-pulse-musculature-tension-reaction-(s)] measured by [inch-(es) / centimeter-(s) / millimeter tension-scaling] remaining out of order, where we are supposed to-be inherent by family, while exceeding those expectation-(s) of perspective, from those recipe-granule-(s) of data, indulging our taste of inductive-reasoning, that smell no problem of hyper-tense-(s), for how each moment remain sensational, awhile purpose, without regard of effort-(s) elapse upon tangible-plane, of existence, imbued living, where each of those set-(s) of limit-(s), can only (be of an) deterioration of living-progression, where offspring is suppose to take up the living-progression of self-identity, by volatility of production-(s), balancing [budget-work-timing-(s) / sleep-timing-(s) / free-timing-(s) // Individuality /// extraspection //// off-time-work-effort-(s), for accumulating those piece-(ing-(s)) of data] for mutual-interaction-(s), marking up materials, when of-an creative-pursuance, by interception-(s) of bulk-quantity-trading, intertwining purpose of settled-area, where those parcel-(s) of individual-offset-(s), can affect historical-balance, by-an national-reidentifying, per inhabitant-interaction-(s), by live-actual-active-Gregorian-calendar-year-reset-offset, impassing tectonic-plate-dynamic, while mile-(s) of impact-distancing, span an life-pattern, in giving an default purpose to humanity, where wilted-lifestyle-(s'), procure natural-timbre, regulating standard-(s'), comfortably reflecting exterior-image-(s) of interior-sight-(s), for living-timbre-account, when moment-(s) elapse-continuum, by matter, constraint of living, by individual-day-to-day-labor, in regard greater-than that of future-schedule-planning, those factor-(s) of existence, because aging in approximate-distance-(ing-(s)), muscle-impasse reality, during regimented-work-order-load-(s), preferred of inhabitant-(s)-nature, introverted without extraspective-preference, when date exist by hyper-tense-(s)-fatigue provocative of periodical-obituary-growth, with-an compulsive-desire, generally inclined of physical-reconnaissance-placing-(s), by routine-dull-command, by what objective-collection, maintain morale of individual-introvert-(s), before the scaling of wilting, for how illiterate-nature, demonstrate those effort-(s) by-an rare-characteristic, in comparison of natural-behavior, due to technological-impact, by centered-hardware, developed to examine weather perspective is pertinent by individual-interaction-(s), from time-posit-(s)-act-(s), observing impassed tangible-development-(s), where interposed-character-trait-(s), require an furthering by in-depth-coverage, zoning zoomed-in macro-perceptive-sighting-(s), from micro-particle-form, celestially identifying population-dynamic, upon component-effect-(s), exceeding property, by rural-claim, in transition of daily-effort-(s), sorting through notion-(s) of neighbor-(s), to interact by discussion of existence, past-historical, requisite-observation-(s), apose future-impasse, that can only be, present-day-act-(s), through sleep-impact, upon cycling juxtapositions [solid-(s) / dry-substance-(s) / wet-substance-(s)] from anatomical-physical-cardiovascular-respitorial-muscul-atory-composite-posit-(s), hyper-tense-(s)-perceived, when formulated linguistics, origin collo-quial-communal, commercial-usage-effort-(s),83L. because mutual-inhabitant-mentality-impact... ¶||||06m.:50s. 76m.s.|||§

§...Preposition-Count-[63 / Sixty-Three]-of.. ..Article-Essay-[18 / Eighteen].. ..Book-Essay-[152 / One-Hundred-Fifty-Two].. ..Lines-[83.07 / Eighty-three.1.zero-tenths-2.seven-hundredths].. ..reading is about referencing from ones family-house-city-block-library, focusing on surrounding title-land-space-(s), for orchestrating an circulated-economy, focusing on proportioning each variable of geological-expanse-ecological-refining, proportioning those density-(ies), per interior-cubic-area-spacing judging from [resident-ial / commercial] grounds, how each space is suppose to expand by an progression of handling, maintaining daily, how each grounds are suppose to develop an purpose for alterior customs handling, having an genetic-mass, understand what timings are suppose to be regimented, for currenting those trades, by inhabitant perspective, by an genetic-allegiance to land-(s), for prospering each inclination of geological-nation, from state-ecological-territory-(ies).¶||||43s.28m.s.|||§

¶153..Present-retrograde-numerical-ordering-(s), are defined by posit-point-(s) of [elapse-(s) // syntax-(s) // synapse-(s)] from when secondths are through time-interval-(s), by interior-tension-(s), for labeling-activity-(ies), from place-(d)-entity-(ies), limit-(ed) surrounding competency-value-(s), [where // when..] {Academic / Effort / Conduct}, affect our method of purport, transcending entity impasse, by self-objective-relay-(ing), attaining tangible-product-means.242¶||22s66m.s.|||§ Maxim future-limit-(s), are not century-anticipated, while present serie-(s) of event-(s), [process / budget / schedule / labor / agenda-elapse-continuum-infinitive] by no dated-secondths-symposium, from how orchestrated-time-(s), monitor prerogative-(s), by when [1industrial-development / 2construction-development / 3university-subject-tier-constructing / 4objective-conjuring-debate of quality-(ies) / 5week-secondths-synapse-(s)-elapse-period-(ths) / 6season-(s)-secondths-synapse-elapse-period-tangent-(s) per 0.ayearly-reverberation-interval-(s)-daily-hour-place-transit] sequestrate blood-pulsed-fatigue, by ventricle-tension-existence, coherent voluntary-interpretation-(s), from calendar-denominative-time-variable-(s)-locational-obligation-(s), upon dated-hour-(s), at place of structural-manned-activity-(ies), where week-to-month-secondths-denominative-influctuation,| 544 |present-live-reasoning-(s), upon archived-documenting, when observational-data,| 545 |still be pertained without constitutional-existence-denominating, (in)-tangible, state-observational-referenced-citing-(s), by-an denominative-claritive-coherent-citizenship,16L. due to clairvoyant-prose, by read-subjective-tensing... ¶|||01m.: 19s.22m.s.|||§

§...Preposition-Count-[5 / Five]-of.. ..Article-Essay-[19 / Nineteen].. ..Book-Essay-[153 / One-Hundred-Fifty-Three].. ..Lines-[16.575 / Sixteen.1.five-tenths-2.seven-hundredths-3.five-thousand-ths].. ..Observing physiological-eyesight, is significant for understanding how each moment is portrayed amid other individual-introverted-perceptive-bias-hoc-preference-(s), adjusting by each elemental-compositional-pound-(s), human-lazy-population, whom not understand an formal-focus, in contributing each notion by governmental-monetary-purchase-historical-disparity, countering each populant-entity-capacity, apose an scheduled-pronoun-entity-population-operation-(s), for considering how each day is proportioned apose each 24^{th-}hour-(s), amid centripetal-force-dating, apose revolutionary-365-day-(s) yearly-process, resetting common population-observation-(s), by an 86,400-secondths-countdown, per day, proportioning millisecondths by product-purchase-objective, by elemental-compositional-development, compared as well with an 31,536,000-yearly-secondths, when molten-pressure, not be titled-tracked.¶|||55s.$_{51m.s.|||}$§

¶[154]Life on the Florida-peninsula-marsh-archipelago-state, (of my birthed-living), can be quite an drag, with scenic-view-(s), remained firmly entrenched, by aged-retirement, (for how) effort-(s) of individual-(s'), ignore physics-elaborating, as when way of transit, encompass focus, from-an superlative-clearance-abstracting, purporting entertainment-culture, rather fluent-literary-reading-(s'), focusing on book-shelf-house-life-style, after labor-week lifestyle-trading, retaining resource-objective-product-trade, which effect [mass / affect-individual-(ism)] intermittent mutual-literate-decimal-incremental-grading-(s), from [date-(ing) / (s) // hour-(s)] upon perceivable-blood-pulse-(ing)-component-literature, denoting [saving-(s) / reading / lecturing / fundraising / directing-activity-(ies)] upon mandated-scheduled-space-(ing-(s)), from territorial-claim-(s), surveyed-consensus, of private-industrial-effort-(s).$_{243}$¶|||40s.$_{15m.s.|||}$§ When here on such lovely-land, what direction of territory, [can / will] be used from introverted-individual-(s'), awhile commercial-access, of industrial-goods, intertwine self-idol-appreciation, of service-(d)-property-objective-(s), by mundane-account-international-ecological-chain-effect-effervesant-(ce), relying managerial-order, for thinking from Titled-operation-(s), collections of mandate-historical-task-(s), built by still-blank-elapse-(s)-period-(s), of infinitive-physiological-entertainment-effort-(s), that inhibit time by mass-population-(s), of county-census, upon central-genetic-state-reform-intuitive-civilizational-schedule-ordering-(s), embodying cultural-custom-(s), text-continuum-entity-restructuring, every personal-matter-(s), persistent astute-educated-demeanor, from syntax-intelligent-(s), conversing mutual-effort-(s), as day-denominative-time-interval-(s), continuum secondths-minute-(s)-hour-synapse-(s), arranging [city-block-(s) / acted-sequence-(s) / industrial-order-(s)] upon minute-(s)-decision-(s), as hour-{Tabling}-denominate-deductive-impasse, remain far too abstract, from elapse-(s) of blood-pulse-(s), to gauge-perceive, mile-(s)-terrestrial-expanse, diminuendo-decrescendo objective-dimensional-work-product-(s), idly-blank, from off-time-(s), as secondths remain how we perceive-interpret, influctuated from minute-(s)-impasse-interpretation, requiring an reference-point-(s), for illiterate-introverted-reclusive-individual-(s), not pertaining an genetic-desire, for quality of State & Order, flat default-momentary-moding, accumulated-crescendo populant-existence-(s), staffing word-(s), not range-(d), note-instrument-method-juxtaposition-intonation, vocal-timbre, apose vernacular-grammatical-punctuational-interpretation-(s), synapse-frequency-(ies), measured in numerical-synapse-(s) per [tenth-(s) / hundredths / thousandths-exponent-measure-(s)] syntaxes sentence-paragraph-expository-extent-(s),| 546 |isolating denominative-population-article-(s)-reference-retrograde, of predicate-interval-point-(s), directly-referenced [leader / expert / Writer / Politician / Professor-subjective-career-path-(s')] unilateral attention-span-focus, by population-(s) of [mass / objective / nutrition / Entity-(ies)] Governmental-Constitution-tiering, 32L.[developmental-condition-(s) / ambient-observation-(s)] by perceptive [elapse-(s) / iteration-(s)]...¶ |||02m.:26s.49m.s.|||§

§..Preposition-Count-[13 / Thirteen]-of.. ..Article-Essay-[20 / Twenty].. ..Book-Essay-[154 / One-Hundred-Fifty-Four].. ..Lines-[32.75 / Thirty-Two.1.seven-tenths-2.five-hundredths].. ..paper is scarce, remember not to waste any of it in your lifestyle-generating.¶|||09s.22m.s.|||§

¶|155 For me, each period, is like an day, with those hour-(s), similar as comma, crescendo-crescent-impression work-effort-(s)-impasse, while semi-colon, defining dollar-(s)-per-hour-payment-method-(s), claritive Zero-Reset-Presence-Commencing-Interpretation-(s) & Decimal-Order-(ing)-(s)-Increment-Synapse-Offset-Moment-Interpretation-(s), resetting date, upon posit-range-offset-(s), [1whole / 2integer / 3number / 4numerator / 5denominator / 6sum / 7equation / 8geometric-surface-measure-compartmentalized-observation-(s) // 9length // 10width // 11height /-exterior-reset/ circumference // 12hypotenuse // 13base // 14adjacent /-perimeter/ 15sine // 16co-sine // 17(co)-tangent // 18opposite // arc // 19section // 20extension // 21radiant // 22factorial // 23absolute // 24reciprocal // 25polygon // 26parallelogram // 27circle // 28rectangular // 29(tri)-angle-(gon) /-prism/ 30square // 31tetragon // 32pentagon // hexagon // 33heptagon // 34octagon // 35adding // 36subtracting // 37multiplying // 38standard-deviating / pertaining tangible-modifiable-solid-(s) / liquid-substance-(s) / dry-substance-(s) // vapor-pressure-(temperature)] through distance-(s) ungraspable, per three-quarterths-mile-expanse.244¶|||52s.47m.s.|||§ I stop time to time, stressing how history has interpreted such method of context, awhile extenuating much (if not all), celestial-observation-(s), untangible, [Physics & Pulse-Perceptive-Life-Form-(s) / Distance-Depth-Perspective / Solar-Kinetic-Continuum / Lunar-Tectonic]| 547 |apose physiological-perspective-compositional-structure, where approximate-relevant-movement-(s), are more significant, than intricate-elaborated-subjective-school-lessoning, for from the four-fundamental-aspect-(s) of existence, in-an split-(s), from Human-Perceptive-Invariable-Premise-Posit-(s') viewing,| 548 |[territorial-property-expanse // dry-liquid-substantial-compositional-nutrition /// composite-resource-(d)-(s)-commercial-concept-dimensional-product-(s) //// live-organism-culture-(s)] apose [grandiose physiological-anatomical-biological-mass / earth-tectonic-mass / oceanic-water-body-(ies)

].245¶|||01m.:35s. 94m.s.|||§ My view is that of intangible, as how individuals can not clasp abstract-grandiose, from pupil-vision, when humanity of introverted-infinitive-character-measure, not age vision, by covering territorial-ground-(s), and settle-(ing)-down, on such terrain, by rural-requisite of major-focus-limitation, for [comprehending those movement-(s) // motion-(s) upon place-(ing-(s)) of tangible-object-(s)] mass-genetic-repopulation is how to elucidate, defined parameter-boundary-dynamic-(s), in cubic-feet-clearance-rectangular-prism-depth-measure-rate-impassing-(s), when each variance-frequency, be perceivable, tangible-clasp-clearance-limit-(s), from sequential-progression-(s) of activity-(ies), upon rule-(d)-aspect-(s) of surrounding-surface-area, temporal-momentary, reset-(s) of [objective-weight-(s) / participant-activity-(ies) / entity-space-(ing-(s)) / monetary-method / extraspective-oversee / distance-timing-comparison], absolute syntax-elapse-(s), periodicalizing, mass-ordered, [handwritten // manuscript-tangent-(s), on Pen & Paper] for latter-present-requisite-date-affirmation-(s), per daily-tasking-(s), of routine-act-(s), crescent-crescendo numerical-subterior-current-(s), containing all those fact-(or-(s)), of tangible-vision, affirming truth from tensed-act-(ion-(s)), upon exterior-dimensional-meters-composite-requisite, having an height-ten-story-clearance-maxim-denominative-existence, from conscious-view of existence, do we have any cause for life, by means' of payment, as-an gauge of living, quaternary of seasons, [1cycling substantial-ingestion / 2clothing / 3housing // 4electricity //// 5water / 6transit-monthly-insurance / 7gasoline-miles-destination-(s) / 8Taxes / 9Company-Patent-Production] as for what purpose-(s) matter, living on-an short-term-basis-infinitive-reset-(s), when free-will-dynamic sustain operation-(s) of existence, intending to indulge those luxury-(ies) of retirement, not differentiated from work-age-lineage-trait-(s').246¶||||03m.:06s.63m.s.|||§ So why does this effect continue? It is an error of [questioning // inquisition] for belief of inquiry, not exist by an natural-means' of populant-interaction-(s).247¶|||03m.14s.97m.s.|||§ Inquiry is an method of investigative-interval-perceptive-observational-survey, focusing Government-empirical-mode, decrescent-decrescendo mass-quantities of deducing-reference, incrementing matter-(s), across an time-period-(s), affecting limit-(s) of year, for uncovering those fact-(s) of Pronoun-existence, not Congressionally-denominated, prevalently aware of impasse-sequence-(s), repeating decrescent-decrescendo article-(s)-notion-(s), crescent-crescendo individual-mass-characteristic-(s)-impasse-inclination-(s), defining property, where their are those requiring Treasury-Congressional-Literature, (as an) desire of exponent-maximizing-repopulation-existence, upon limit-(s) at Perceivable-Physiological-Tension-(s) & Boundary-(ies) of State-(s)-Territory, which inhibit existence, referred by-of payment-(s), merited by hour per [hour // salary-payment-primary-basis] not enough, as for what has been done, from the prose of intuitive-interposed-into-national-reference-(ing), enabling population-(s) of surveyed-populant-(s), by nature of cultural-development, directing statehood, as when religious-matter-(s')-offset, where continual-development, circulate live-perspective, observing long-term-longevity, maintaining reproduct-ive-existence, by situation-(s) of Territory & Regimen, scheduling inhabitant-(s) per designated-task-(s), by day to week, territorial-coverage-period-(s), inaudibly-impassed, maximum-capable-documenting-covering, per expanse, in reminder of life-style-quality-existence, by mode of direct-simplicity, by lifestyle-focus, through Genetic-Inquiry-Resourcing, formally conjuring-data, upon directive instructed-schematic-blue-print-data, cycling human-perceptive-organism-work-lifestyle...¶|||04m.:31s.87m.s.|||§

§...Preposition-Count-[34 / Thirty-Four]-of.. ..Article-Essay-[21 / Twenty-One].. ..Book-Essay-[155 / One-Hundred-Fifty-Five].. ..Lines-[59.43 / Fifty-Nine.1.four-tenths-2.three-hundredths].. ..Where would you enjoy retirement at, amid an commercial-continuum, product-purchase-rates, trading from commercial-markets, shelf-life-products, for maintaining leased-utilities-entities, where circulated-physiological-beings, convey an purpose for residential-retained-product-objective-(s).¶|||19s.43m.s.|||§

¶156.For how massive is The Celestial-Ambient, through commune-(s) lacking pertinent-focus, of daily-regimen, directing desire-(s), from extrapolated-effort-(s) by extraspective-voluntary-domestication, per city-district-state, directing how house-(s) are suppose to be incorporated in to an block-singular-family-residence-(s), developing unincorporated-county-territory, by intervoluntary-interposed-raw-terrain-land-domestication, appraising expanse, when first-hypothesis-measure, use tool-objective-count-(s), [cultivating // constructing // founding-territorial-distance-(s)] by fashion of [stancing / walking / statistical-notion-(s)-ordering] complementary compartmentalized-citizen-(s'), crescendo-crescent interactive-stylistic-directive-developing, an effective-mass-literary-intuition, diligently ordering-practice-(s), per pertinent-continuum-activity-(ies), ruse cultural-enhancement, when if none of those human-ethnic-mass-(es), can be aroused by tectonic-plate-covering, calendar-aging, feminine-opposite-sex-repopulation, with millennium-children, not to achieve yearly-children, in an transition of century-children, whom understand how practicing sociological-psychological-"common"-literate-reproductive-constructive-developing-government-order-(ing-(s)), by grammatical-punctuation-mathematical-musical-cursive-written-daily-documenting, an genetic-state-national-county-state-(s)-city-county-(s)-city-district-city-(ies)-block-principalities-constitutional-family-(ies)-classifi-cation-(s)-lifestyle-(s'), yet pursued by attention to minute-tasking-(s), handling how our movement-(s) of time, matter by tentative-use of counted-objective-ordering-(s), of statehood-function-(s), documenting [motion-(s) // blood-pulse-(s)-breath-respiration-inhale /// exhale-(s)] by minute-(s) into secondths-documenting-method, enamoring reference, by 19L.[Author // Reader-intimate-Pronoun-family-populating-order-count-cycle-schedule-forming]...¶|||01m.:31s.31m.s.|||§

{Article:} [June]

§...Preposition-Count-[7 / Seven]-of.. ..Article-Essay-[22 / Twenty-Two].. ..Book-Essay-[156 / One-Hundred-Fifty-Six].. ..Lines-[19.545 / Nineteen.1.five-tenths-2.four-hundredths-3.five-thousandths].. ..community has never been an Western-Hemisphere-practice, due to those arrangements of policing, separating community-members, from citizen-discussion-(s), per moment of moded -development.¶||||14s.69m.s.|||§

¶157As the Solstice offset, summer-tilt, of Northern-Hemisphere-Perspective, [Men / Women] remain interior, developmental-locale, product-shelf-purchasing, (from) extenuating, governmental-reliance, awhile human-(s') remain lazy, per conscious-interaction, approximate surrounding-property-perimeter-terrain, dynamic from Bias & Hoc clouding mind-(s), function-(ing-(s)), inhabitant-human-(s), by-an unliterary-balance, ulterior [time / date / motion-(s) / effort-(s) / mile-walking-impasse-(s)], calendarizing fashion-(ed) nature-living, for how reclusive-individual-(s'), become adept at settling down, on-an lifestyle of belief, religiously abiding American-National-enabling, vague characteristic-(s), tie-sustain religious-trait-(s), by sect-denomination-(s), not alluring an genetic-purity-{Blonde Hair /// Blue Eye-(s) /// Pink-Skin}, extraspective character-trait-(s), when statehood-trust, abide made-goal-(s), Twenties-yearly-retirement-repopulating-reproduction, offspring, handling opposite-sex-affair-(s) per state, interacting Vernacular-Book-Median-Perspective & Physiological-Tense-(s), fact-(or)-(s) (per) labor, enabled wage where intrication (per) activity-(ies), are due (to) those mental-natures', undirectable independent-perspective-(s'), (by) free-liberation, (of) those said discourse-(d)-action-(s), (not) backed daily-written-typed-documenting, reparted-an vacation-mode, by those requisite-(s) of inhabitant-behavior, for how interaction-(s) are formulate-(d), through an continuum-reset-requisite, of those fact-(or-(s)), place-(d) in an contingent Western-Civilizational-Existence, subjective inept common-purpose, in youth as generally disregarded, where perspective requires reference, or remain dull of perspective-quality-(ies), due to fact-(s) of existence, [hoc // bias, inhabitant-daily-effort-(s)] because how dating has not been adjusted by the juxtaposition-(s) per day, of present week, in comparison past requisite-inductive-data, intertwining week-(s) per passing-imagery-recollection-(s), (memory), decrescendo-decrescent, deducing grade-(d)-rate-(s), from intervoluntary interval-(s), listless book-(s), per citizen, as how grade-(s)-crescendo, have not found an common-auto-biographical-documenting, congressional by individual-(per)-legislating, defining why bill-resolution, sustain how denominative-documentation, exist from how week-interval-offset, lesson-(s) classification-(s), alive (from) physiological-imagery-tier-recollection-(s), remaining domestic;[277] American-Government-Quality per state, pertaining day-count-(s), decrescent hour-impasse-(s), decrescendo five-minute-(s)-synapse-posit-(s), defining an premise denominating count-(s), formal (where) interactive-referencing, discourse daily-effort-(s), kindred pertained-data, upon property-surface-area-handing, requisite-continuum dating, intended to centre an focus per-of-prerogative [objective // subjective posit-(s)-data] furthering organism-pattern-(s), by year-(s)-present-hour-secondths-physiological-perspective-effect-(s'), as inani-mate-organic-matter-(s) apose anatomical-biological-physiological-blood-pulse-composite-perceptive-hyper-tense-(s)-being-(s'), pertaining daily-personal-independent-prerogative-(s), when fashion per (of) act-(s), denounce-data, mutual (by) physical-effort-(s) per of statehoodship, date-denominate-(ing), (all of) those impasse-motion-(s), offset [pronoun-territorial // pronoun-independent-individual-(s') // pronoun-animal-(s) // pronoun-tool-(s)] jargon-arranged, instruct-ive-manual-pamphlet, (not) intricated by book-lessoning-(s), (for) bettering construes (per) commercial-production-(s), (by-when) effort-(s) per musical-timbre, enhance living by where-when-depth-perspective, invigorate tense-(s), inhibited daily-effort-(s), (from) prerogative-preference-sensation-(s), apose perceptive-hyper-tense-(s)-muscular-ventricle-blood-pulse-(s), simultaneous secondths-percentage-(s), when off-time-requisite-(s'), not fashion formal-enabling, due to (an) requirement for total-life-live-active-thinking, through time-(ing-(s)) per dating-(s), that imply those effort-(s) where daily-requisite-enhancing, [matter of issue-(s) // living-requisite] juxtaposing perceptive-article-(s)-matter-(s), where state-observation-(s), convey coverage of territory, not by tectonic-national-requisite, continuing an abashing, due to negligence of constitutional-order-(s), by {Article-I;[278] Section VIII;[279] Twelfth-Enumerated-Power}-{I.E. To raise and support Armies, but no Appropriation of Money to that Use, shall be for a longer Term than two Years}, where each fact of daily-life, matter from individualism, or monotheism, not extraspection, polytheism-in-thirdths, for where polytheism base (from) introverted-individualism, through-an scope of personality-disorder, not in-an formal-order-orchestration-(s), by method of interactive human-mass-extraspective-impasse-(s'), where introversion rotten-(s')-perspective, pertain fact-(s) by human-timbre, far more significant, than other-organism-(s'), for how perspective is affected by volume-(s) of introverted perspective, affecting outcome-(s) of behavior, by when individual-live-example-(ing-(s)), rather-than, mutual-perspective-(s), perceive, exterior-value-(s), where each posit-(s), are of-an matter per relative-purpose-(s), appointed leader-(s) of designated-order, by-an rule of totalitarian-command-(s'), carried out by physical-labor, upon perceptive-observational-requisite-(s'), tentative-grandeur, where each inhabitant, pertaining perceptive-ruse, fundamental, core common-action-(s) of presence, by-when present-past-requisite-effects, interact human-behavior, crescent-crescendo, presence by perspective-requisite, posit-(ed) at an impact of distance, waking-effort-(s), cardiovascular intermittent inhabitant-behavior, affecting those outcome-(s), where receptive-behavior, is-an natural-affect, where each person is to duty upon mass-(es) impasse-requisite, where the mass choose to be free of inhibition-behavior, without an reasoning of cause, when count-(s) of individual-(s'), pertain product-factor-(s), by natural-freedom, inhibiting natural-perspective, by vivid-visual-variable-(s), defining mundane religious-examination, for how each individual is never capable of commercial-act-(s), refrigerated-perishable-dating, effecting dynamic of locational-prose, for inhabitant-(s) do not consider [high school / University inhabitant-(s)], as citizen-(s') transgression-(s), classifying sociological-specimen-(s), interacting upon campus, for transferring worded-term-definition-article-(s), valuing life, [where // when perspective] remain tended upon, interposed tangible-technical-value-(s), as how modification-(s') of composited-material, not be in-an useable-purpose, unless Genetic-extraspective-inhabitant-(s'), remain capable of action-(s), from statehood-constitutional-act-(s), offspringing by-an army-reproduction of inductive-increment-(s), valuing tangible-grasp-(s), upon each visual-area of statehood-tectonic-offset-posit-(s), affecting outcome-(s) of off-time-requisite-(s'), from natural-value-(s);280 observable property-territory, per action-(s) of subjective-documentation, affirming how condition-(s) influence reasonable-state-population-(s), co-signed by interaction-(s), manipulating ground-composite-extraction-(s), when centripetal-force-requisite, enhance living of off-time-requisite-(s'), by-an arithmetic-reasoning, adjusting terrain, where [wake // sleep] remain an part of conscious-prevalence, passing those corridor-(s) of surface-area, increment-crescendo, ordering-impact-trade, from tangible [composition-(s) // patent-product-object-(s) / intangible-perspective] affecting surrounding area development, from inhabitant-effort-(s), intervalizing momentum, where tangible-mass-posit-(s), miniscule article-(s)-interpretation-(s), for how each method of purchase, progress beyond presence-requisite, past-pertained, present-day-continuum, where life can only be perceived of present impasse-event-(s), cuing existence, adrift from mechanized action-(s), per inhabitant-(s')-dues', warning how tool-objective-usage, without an pertinence of an constitutional-economic-balance-(ing), guiding direction-(s) where each [tendency // currency] is not of-an extraspective-purpose, from historical-presence-requisite, naturally of human-mass-behavior, [inclinated // inclined // indicated // innuendo // interpreted // understood // comprehended human-quality-(ies)-effort-(s)]...¶|||06m.:44s.93m.s.|||§

§...Preposition-Count-[46 / Fourty-Six]-of.. ..Article-Essay-[23 / Twenty-Three].. ..Book-Essay-[157 / One-Hundred-Fifty-Seven].. ..Lines-[63.425 / Sixty-three.1.four-tenths-2.two-hundredths-3.five-thousandths] ..end?

{January-(31)-February-(28 / 9)-March-(31)-April-(30)-May(31)-June(30)-Month-(s)-Day-Hour-Minute-(s)-Secondths-Count}-{15,384,400 D.H.M.S. = 60S.× 60M. × 24H. × D.}-{4,344 H.}-{260,640 M.}-{181 D.}

[Article:] {July}

158

Month-Numerative-100%-Percentage-Denominative-Gregorian-Year-Aging-Intervals-Progress

0 = 0%
1 = 8.33%
2 = 16.664%
3 = 25%
4 = 33.33%
5 = 41.664%

——
——

6 = 50%

=

7 = 58.33%

———

8 = 66.664%
9 = 75%
10 = 83.33%
11 = 91.664%
12 = 100%

At what time is [First / Second Person-(s)] crescent-visual-perspective, by Ways & Means';[281] regulating congressional-senate-Majority-Leader-Judiciary-Review, by patent-production-factor-(s), variable-(izing)-saying-(s'), stated-proportional-visual-posit-(s), defining surface-area-limit-(s) {H. × W. × L. = C.S.A.}, per climate-terrain-latitude-longitude-temperature-offset, formulating meteorological-interval-influctuation-(s), per Day & Night, perceptive-visual-effect-(s), soluble-gauging blue-centripetal-vapor-pressure, apose Solar-day-Kinetic-Vapor-excess-discharge, eminent perspective-(s) by-of Friction-Kinetic-Explosion-Expansive-Pulsat-ing-Vibrational-Frequency-(ies), through Radioactive-Reverberational-Resonated, Fixed-Global-Rotational-Revolutionary-Physics-Untangible-Matter-(s), annotating figurative-exponent-(s)-Expanse, per abstract-celestial-drift-motion-(s'), affirming perimeter-surface-area-pressure, where earth-solar-split-infinitive-plasma-extreme-force, constant mobius-fluctuation, live-figurative-untangible-intangible-force-(s), peripheral visual-tense-(s), as July, commence commemoration-(s) of Summer-weather-{Orlando-Solar-Centripetal-Drift}, symmetrical-month-(s)-Northern-Hemisphere-aligning,| 549 |of surface-area-tectonic-tilt, pertaining area-approximate-temperate-reflex-(es), by consistent-solar-influctuation-(s), persisting molten-pressure-centripetal-force-drift, latitude-rotational-hours-deviation-four-minute-degree-(s), per non comprehended, state-square-mile-(s)-extent, apposing natural-expanse, centralized-regulated, mundane-reason-(ing-(s)), from-of superfluous-mile-(s)-distance-(ing-(s))-juxtaposition-(s), gauging perspective, from distanced subduction-zone-perspective-visualization-(s), above crust-posit-continuum-discourse, cooled from-of, molten-central-universe-juxtaposition-white-solar-halide-pressure, lifting each series-of [[1]form-(s) / [2]request-(s) / [3]order-(s) / [4]file-(s) / [5]docent-(s) / [6]book-(s) / [7]novel-(s)];[282] while aptitude yet confluence comprehension, from when revising [format // sequestrate-(d)-muscle-reflex-(es)] alluded-direct invigorated-intransitive-human-effort-(s), free conservative contort concert-concentration-(s), at-an point of colloquial-legal-guidance, intransitive-pulsed, from-an superfluous-right-(s), per five-Nine-Five-order-(ing)-method-(s), placing how Supply & Demand, work-place-environment;[283] pertain human-passing-age, for soly introverting friend-(s), as family-cause, define Workermens'-extraspection, confluence intro-verted-perceptive-characteristic-(s), length-crescendo inalienable-separation-(s), tensing transit-visual-imagery, believing influx of Temperate-characteristic-(s), per state-national-perspective-(s'), supporting common-interaction-(s), suspended boundary-extent, per limit-(s), because in fact-(or-(ing-(s))) by approximate-time-at-spacing-(s), distancing-(s), synapse-elapse-(s)-subterior-period-(s), from when tense-succession-order-(ing-(s)), divide fluent sequential-coherent-operation-(s), for acquaint-space-(ing-(s)), as day-(s) upon week-(s)-year-(s), require visualized-tonal-write-(ing-(s)), for maneuvering present-infinitive-visual-imagery, surmounting an method of payment-(s), post-received, surface-area-trajectory-discourse, identifying type-print-exchange-(s'), from reliance of individual-character-(istic-(s)), deviating individual-claimed-objective-property-(ies), for-an intersecting-passage-(s), by normal-human-behavior;284 as how form of habit-(s'), subconsciously-sustain through self-independent-prerogative-activity-(ies), identifying astute-awareness, from-by direct-activity-(ies), leading reclusive-work-impasse-pattern-(s), housed by inhabitant-(s), for-an maxim-limit-(s), by mental-physiological-adeptness, figuring language-skill-(s),53L. by-an [commercial-development / Governmental-regulating] how [populus / parcel / resource / objective-(s)] sustain product-(s), unsubjective entity-city-county-parcel-dynamic-(s), for-an daily-cultivation-effort-purpose, staking state-territory, for attempt at file-ordered-data, observing-measure-(s) due to political-handling, as populus-impasse-affair-(s), considering those mass-requisite-living-method-(s'), when developing an Genetic-Observed-territory, by effect-(s) of commercial-product-conception, by the basis of, human-need-(s), synchronized by mass-motion-(s), awhile daily-impasse, affect direction of prerogative-(s), for formal-daily-living.¶|||04m.:31s.00m.s.|||§

§...Preposition-Count-[14 / Fourteen]-of.. ...Article-Essay-[1 / One].. ..Book-Essay-[158 / One-Hundred-Fifty-Eight].. ..Lines-[59 / Fifty-Nine].. ..cycling sequences of human-physio-logical-effort-(s), requires an examination of which motion-(s) matter most, in proceeding each daily-voluntary-ecological-act-(s), serving entity-(ies), which conjure how developing surface-area is developed, by an intent to circulate an waged-human-population-capacity.¶|||17s.03m.s.|||§

¶..159..Thomas Jefferson;285 for whom drafted The Declaration of Independence, predecessor {Constitution of a The United States of America}, enabled a-the primary-ancestry, existing upon

[Article:] {July}

Western-Hemisphere-States, from religious-rite-(s), based on-an ancient-prose, from when initial-push of sailed-voyage, established the primary denominal of modern-trade, foreseeing tectonic-plate-expanse, where organization-(s) from free-will-means, had been alluded fundamental-economic-discourse, when population-directive-(s'), interact-(ed) by state-offset-(s) of terrain, accumulating mass-territorial-space-(ing-(s)), from construction of eastern-shores-territory, conjoining cognitive [judgement-(s) / execution-(s) / legislating of matter-(s) // issue-(s) // topic-(s)] from the unconscious-breath-blood-pulse-requisite-(s') of muscular-motion-(s)-tension-(s), (un)-document-(ed) for formal-intrication-(s), by-when daily-interaction-(s), serve where an series of natural minute-(s)-motion-(s)-tense-(s), perceiving by impulsive-conduct, how digestive-cycling, remain temporal of natural-effort-(s), uncultivated by barren-terrain, growing crop-(s) by acreage-planned-spacing-(s), for-an comparative superlative, of topical-article-(s)-matter-(s),| 550 |tied by commercial-reproduction-(s) proportional national-census-inhabitant-(s), for whom [were / have / are] rationed by state-matter-(s);[286] growing awhile present-day-cycling-(s'), not exponentiated by mass, for how farmer-technique, skill developmental-production of cultivated land-parcel-(s),| 551 |by where [water / fertile-soil / organism-focused-cultivation / nutrient-supplementing / heat from solar (lamp-technology) / massive-acreage-surface-area] define how chronicalizing observation-(s), should have had an stagnant-flow of confirmation-(s), from how cursive be disregarded for over three-hundred-year-(s), one-hundred-plus-year-(s), by educational-international-judiciary-judgement-(s), offset by when mass-inhabitant-(s), continue without an common-subjective-interpretation-(s), that state those fact-(s) of personal-possession, in-of-an physical-reflex-(es), not considered for respitorial-cardiovascular-movement-(s), when their are an [serie-(s) / sequence-(s) / count-(s)] not specified from pertained developed-territory, apose counting each stroke-(s) of movement-(s), presently-incrementing formal-conjure of territorial-development, when their are those fact-(or-(s)), for trade-usage, upon those planned-projection-(s), from ulterior-purpose, sustained from Thomas Jefferson;[287] logical-impasse-documenting-historical-reference.[248]¶||||01m.:55s.[35m.s.||||]§ {Charles Theodore} // {Maximilian I}, are historical-past-initial-requisite-(s), from of a the, Holy Roman Empire;[288] chaliced where American-States, never saw an transition of currency, via England, for how American-Independence, be from Western-Holy-Roman-Empire, labored an difference of theatre, held by-an cultural-ploy-society, not by an pure ethnic-culturing, due to those mentality-(ies) of inhabitant-freedom-(s'), feeling their way to Royal-Family-(ies), not by an impact of true-fact!, due to the nature of hedonry, existing as-an independent-individual-(s'), not having an formal-thinking-method-process of language, where their has been an deficiency in religious-pursuit, by undertaking-action-(s), languished commercial-enterprise, for nominal national-pursuance, when aging fatigue, by-an blank-period-cycling-(s'), of sensational-impasse, not consider through enacted-observation-(s), how data-material-(s), remain without act-(s) of succession, documented from hedon-(s')-nature, reclusive of extraspective-interaction-(s), where communication is by-an enabling of dialogue, exampling information, greater-than conscious-vocal-vernacular-affirming, by when volume-key-tone-pitch-vibration-(s)-frequency-(ies), staff language, where inhabitant use-(s) serve possessive-use, by fact-(s) of item, not in an constant-arranged-dialogue, from how interaction-(s) of dating, are by-an present-juxtaposition, from family lineage, that can not be accede-(d) by natural-hedonism, because how organized-arrangement-(s), not orchestrate-reference, mathematical-data, not by an usage of data-information, using natural-means', for how populant-(s) not arrange scheduled-activity-(ies), by communal-religious-building-edifice-(s), sustemmed from dimensional-planned-property-spacing-(s), efforting engineered object-(ive-(s)), by-an collection-(s) of [aluminum-can-(s) / plastic-bottle-(s) / glass-bottle-(s)] from-an corrugated-recycled-fashion;[289] when dating be incumbent, participating at academic-affair-(s), for terrestrial-space-(ing-(s), conducting G.P.A.-word-point-deviation-(s) of hour-class-study-(ies), having exponent-thinking, demonstrate effort-(s) per inhabitant-impact-(s), where state-territory, consider how lifestyle serve an national-duty, not scheduling signature-(d)-mass-written-cursive-article-(s)-interpretation-confirmation-(s), upon-an agenda of activity-(ies), for loyal-effort-(s), referring an conjunction of, daily-impasse, apose [1construct / 2cultivate / 3develop / 4concept / 5create / 6produce / 7schedule / 8arrange / 9time / 10count / 11interact / 12reflection-(s)] where (re)-claimed-territory, remind an effective remoding of living, [when / where] inhabitant-(s') can only perceive natural-physical-space-(ing-(s)), when each fact not meander an ordering an successive-formation-(s), as each inhabitant (has an) natural-bias, limit-(ed), by independent-effort-(s), arranged by-an living, not conceded consideration of-an roulette-constant-syncopated-interval-effort-(s) where night remain dormant per action-(s), for-how behavior is not formally-successive, in responding for litigating reference, as how reference not recollect-(ed), an natural-motion-(s), from how individual-belief, remain monotheistic, per settling of Christian-Faith, believing Governmental-Act-(s), resolved instructive-effort-(s), commercially-wage-(d), [when // where] fact-(s) of data' not further construe, for how limit is lock-(ed) by behavior, because data injunction-(s), not televise off-time-working-(s), comparing family-(ies)-theatrical-intrigue, as when, [television // social-media] insight important-issue-(s), evaluating mass-objective-article-(s)-usage-(s), of productive-matter-(s), evaluating humans daily-inclinated [exhaustion // tired-behavior // natural-effort-(s)] because the general-belief in (American) Standard-(s), sit above superior-belief, because attempt-(s) of greater individual-(s), differ when plane-axis-centripetal-force-spherical-impasse-(s), affect each individual-(s), by routine of behavior, that can not be in-an dialogue by interaction-(s), where government-constitution, depict truth-(s) of behavior, by "free-will", live in-an fogged-presence, by-of present-day-reset-continuum, never inclined to cooperate with other-(s), by off-time-commercial-home-work-interaction-(s), because when the "daily-grind", forsaken dating, not in-an emphasis of time-posit-(s), date-impasse-(s), what would be the duty we are held to, by commercial-mass-market-production-(s), regulated from free-will,71L. when relative to purchase-parity, serves' a requirement of Supply & Demand-process-(es)...¶||||05m.:20s.42m.s.|||§

§...Preposition-Count-[47 / Fourty-Seven]-of.. ..Article-Essay-[2 / Two].. ..Book-Essay-[159 / One-Hundred-Fifty-Nine].. ..Lines-[71.5 / Seventy-One.1.five.tenths].. ..Why is it from those aging's of our Founding Fathers, no group of individuals, come together, for understand ing how to interact by those parameters of titled-land, for communicating what objective-(s), to offset-collect, and implement, for using upon ecological-geological-terrain.¶||||19s.86m.s.|||§

¶160Time is sustained (by-an) [letter-word-font / numerical-measure] interpreting Earth-Celestial-Tectonic-Plate-(s), perceptive-juxtaposed, pending objective-usage-act-(s), instated by article-(s)-defining-extraction-refined-purpose-(s), by when dating human-being-(s'), remain capable of impassing summer, by-an heat-index, indicating, what clothing we would put on our skin, [food-(s) // drink-(s)] are consumed, how movement-(s) of labor-effort-(s), contention extraspection, where introverted-consumeristic-social-labor-interaction-mentality, concert from only oneself-quality-(ies), when conception of self, remain impossible, for no contemplation by self-compositional-posit-(s), enact straight-line-point-(s), verifying factorial-determinant-(s), from shape-(s) of base-adjacent-hypotenuse-sine-cosine-opposite-tangent-rectangular-epilipisoid-prism-(ing), layering two-dimensional-figure-(s), observation-accentuating, three-dimensional-shape-(s), as-a-the observation of sight, not to [have // be] of an exterior-prose, identifying composite-(s) of celestial-ambient, from relevant interactive-discourse, of intervene-(d)-action-(s), affirmed-by exterior-skill-define-(ing-(s)), from when those [state-territory-(ies) / county-territory-(ies) // city-territory-(ies) /// county-district-(s) //// city-district-(s)] requisite [mayor-(s) // commissioner-(s) /// councilman] whom pertain observation-(s) of tangible-matter-(s), (from an) mundane-effect, breathing by blood-pulse-[inhale // exhale tension-(s)] denominative day-year-period-(s), (ac)-counting for as many act-(ion-(s)), remain apart of Federal-Sixty-Two-City-(s)-Offset-(s), having cookie-cutter-Civilization-(s), resource by the Bastardized-Society-(ies)-grandeur, when governmental-subterior-distance-(ing-(s))-(High-School-parcel-(s), enable-crescent, inhabitant-(s)-mass-illiteracy-rate-(s)-interval-(s), skated pass classroom-subjecting, from-an dull subconscious-intrigue, regarding commercial-objective-(s), confident by national-trading, not educationally-regulated, for handling paper-executive-judiciary-legislation-action-(s), filed into an reserve-note-database-(s), for how timid executive-effort-(s), remain in duty of passing-freedom-(s'), incapable of those subjective-ordering-(s) of data, for-an succession hour-periodical-daily-impasse-intervention, where genetic-tangent-dialogue, yet understand, how inhabitant-(s') serve an mammal-consumption-impact, reflexive human-effort-(s)-energy, where ambiguous-agricultural-goods, often relay digestive-balancing, per internal-organ-systemizing, calculated sight-sound-tension-confirmation-(s), through local-distance-(s) that are limit-(ed) due-to natural-effort, examining appearance-(s'), of each waking-hour, through human-physical-maneuver, by-an natural-prose that would naturally-superlative physical-tension-(s), when each fact of natural-disposition, serve an none consideration upon continuum-false-belief, flawed through dynamic-(s) of state-territory-(ies), upon living-method-(s), inaugurated intersection-injunction-(s) of [ethnicity-(ies) // Race-(s)] unprecedented retirement, at mountain-time-national-referendum-conjugation-(s), from how commercial-economic-impasse-(s), serve upon an government-examination-(s), where each individual-(s'), disperse an metropolitan-Federal-reserve-deviational-timed-distance-(s)-offset-(s), where calendar-increment-count-(s), per {Week-(s)}-{Month-(s)}-{Season-(s)}-{Year-(s)}, retain daily conception of [anatomical-physiological-motion-(s)-tension-(s) // Earth-Celestial-Grandiose-Ambient] through each maneuver of tangible-offset-perspective-(s), not be an part of natural-equivalent, invigorated-matter-(s), pertinent [adept / coherent / intelligent / competent / civil-matter-(s)] where daily-impasse-(s), yet be confluently-understood, per night-scheduling, exonerated exhausted-fatigue-(ing-(s)), by civilian-impasse, where routine of city-residence-county-district-work-mode-cycle-(s), extend birth-to-present natural-view-(s), when inhabitant-prose, is proseless from school-government-subjective-construe, when literature has not been generationally-transitioned from physiological-specimen-(s)-to-specimen-(s), of lineage-generational-succession-(s), military-ulterior, exterior-commercial-work-cycle-impact, upon residential-exterior-post-set-impact-(s), where continuum national-government, remain the work of extraspective-interaction-(s), if ever to consider an nature of belief, due to the historical-requisite-(s'), upon present [[1]militia / [2]congressional-{E Pluribus Unum}-{In God We Trust} / [3]white-house / [4]supreme-court / [5]county-city-hall-(s) / [6]Federal-reserve-(s) / [7]treasury / [8]commercial-corporation-(s) / [9]Wall Street-New-York-Stock-Exchange / [10]State-Supreme-Court-(s) / [11]University-(ies) / [12]residence-(s) / [13]Monticello-impact-(s)] insured of trade, crescendo collected-goods', by local-national-market-(s), when commodity-good-(s), affirm community condition-(s) of latitude-longitude-distance-offset;[290] apose [temperature / topography / climate / weather / time / date / activity-(ies) / interaction-(s)-religious-clause-(s)] oblivious of such fact, for not read-studied-confidence, affirming claim-(s), by present-free-effect, with no direct-examination-(s), by-an natural-human-timbre, for how those behavior-pattern-(s'), conduct academic-affair-(s'), (when / where) an presence of tense-(s), serve upon cubic-clearance-space-(ing-(s)), incendiary cubic-feet-per-secondths, affecting formal-impasse-moment-(s), interacted-offset, from discourse of consumer-ingested-digestive-gesture-(s), for how humanity can be solvent natural-posit-(s), from when integration of variable-mobius, are in-an influctuation of factoring-(s), for modify-(ing)-tense-(s), where in-an usage of tense-(s) by action-(s), (due to an) default of tense-(s), observing natural-tension-(s), from breathe-pulse-tension-(s);[291] as when during momentum, influctuate interval-tense-(s), evaluating objective-matter-(s), from tangible-means', because each article-(s) of objective-posit-(s), factor an crescendo requisite, applicable-visibility, pertained throughout article-information, from when each impassing-(s) elapse, from spherical-tense-(s), decrescendo decent, visual-distance-relation-(ing-(s)), by depth-continuum-matter-(s), ascertaining parcel-depth-infinitive-surface-area-spacing-(s), because how each fact-(s) by-an mobius observation-(s), remain where-at-an usage of particular-objective-variable-(s), double-constant from [physiological-being // entity-structure] offset differentiated of tectonic-plate-(s), by state-topography-offset, affecting each outcome of reference-(ing-(s)), variable each independent-inhabitant-(s), ecologically (from an) use of article-(s)-factor-(ing-(s)), from how each posit-point-(s), pertain-(ed) by accentuated-usage, for affirming each moment, by independent-objective-article-(s)-usage-(s), as when each piece of matter, pertain retained data, volume-count-(s)-mass-requisite-future-offset, from each piece of article-(s)-issue, differentiating each scenario of intermission-discussion, formulating formal motion-(s)-act-(s), upon an formal-issue, of perceptive-tense-(s), because of a-the momentum-purpose, where factor-(s) of state, matter by the perpetual-interpose-momentum, pertinent of tectonic-plate-(s)-rotational-revolutionary-centripetal-force-drift-impact, where political-economic-accord, offset handling those dated-affair-(s) of interval-state, because each objective-matter, remain upon pre-requisite-factoring, present awhile motion-tense-(s), invigorate handling, of developmental-matter, per space-(ing-(s)), maintaining duty of space, by factoring routine of action-(s), rather schedule-(d)-interval-purpose-(s), maintain-sustaining, article-(s)-requisite-affair-(s), physically grounded from-of momentum, affecting peer-interaction-(s), where state-place-(ing-(s)), confirm district-inhabitant-(s), whom handle those place-(ing-(s)), by sedentary-perspective-(s), as how each factor, be relevant, by human-interactive-article-(s)-usage-(s), during an pre-requisite-holding-(s) of matter, by-when state-dynamic, influence individualism, self-pertain-(ed), interaction-(s) of handle-(d)-material-(s), weighted each piece of matter, when due-(s) of material-possession, not be from requisite-(s) of non-resource-(d)-confirmation-(ing-(s)), shelf-life-product-handled, allotted state-affair-(s), as politic-(s) meander demonstrated material-(s), where usage-continuum, sustem from tense-(s) per observation-(ing-(s)), from where each factor, crescent value-(s), by-an perspective of sight-sound-tense-wave-vibrational-observation-(ing-(s)), because the nature of movement-(s), commonly-reflexing, inhibited, from where each fact of humanity, observe an exterior-vision-(s), by vocal-vernacularizing an purpose of monitored-control, due to those daily-fact-(s), non-sequitur, independent-control, by unilateral-constitutional-article-(s)-(auxiliary)-objective-usage-(s'), from where resource-(s), deplete daily-night-effect-(s), from human-voluntary-effort-(s), affecting the duration of event-(s), impassed to handle common-accounting of interactive-affair-(s), where article-ordering-(s), temporally variate visual-processing-effect-(s), when movement-(s) per ordering-(s), process human-undertaking-(s), for how each inhabitant is supposed to fluently-affect, interval-intercession-(ing-(s)), that matter where integral-purpose, support daily-itinerant, because individual-tangible-mass-responsibility-(ies), suffice fluid social-security-process-(es), trade-(d)-product-(s), amassed by secular-limit-(s) of distance-(s)-offset, approximate where local-matter-(s), maintain moment-(s), pertinent by daily-impasse-interceding-(s),96L. decrescent-crescendo, limit-(s) of distance-volume, human-behavior-(s')...¶|||07m.:4 6s.51.m.s.|||§

§...Preposition-Count-[63 / Sixty-Three]-of.. ...Article-Essay-[3 / Three].. ..Book-Essay-[160 / One-Hundred-Sixty].. ..Lines-[96.834 / Ninety-Six.1.eight-tenths-2.three-hundredths-3.four-thousandths].. ..Where do we resource metal from, (lakes / rivers / ocean shores), while proportioning compositional-clay-silt-land for title-lineage-ancestry-developing, of four-story-family-residential-entity-(ies), when daily prerogatives occur pass waged-commerc-ial-purpose, from inhabitant-communication-cooperation-(s), by intervoluntary-thought-(s), by what is required for developing the tectonic-plate we stand on.¶|||25s.09m.s.|||§

¶|161Cronyism;292 the appointment of Friends & Associates;293 to positions' of authority, without proper regard to their qualifications.249¶|||09s.74m.s.|||§ Human-(s) current civil-liberal-discourse, as how enable-(d)-physical-effort-(s), ratify family-primary-religious-ancestry, by-an constitutional-theatrical-prose of human-effort-(s), yet reasoning, formal rigorous-effort-(s), as Non-fiction-legislational-prose, subjective [narrative / expository] paper-(ed)-documenting-confirmation-(s);294 secured where their pertain an temporal-live-elapse-(s), upon an simultaneous calendar-date-(ing-(s), periodical deductive-informational-interval-(s), (by-an order of) [financial-historical-calendar-reference // generation-filing-expectancy] present-impasse-thinking, simultan-eous each dynamic, for how physical-affair-(s'), annotate perceivable-existence, applying citizen-(s') undaunted mutual-trust, by our perplexive-perceivable-verntricle-blood-pulse-air-space-interior-foot-torso-hand-eye-vision-ear-sounding-motion-reflexive-tension-coordination-(s), offset approximate-exterior-value-(s), from read-vernacular-nature, label-(ed) per association-(s) of observable-tangible-(s), intermittent physical-value-(s), for when an confluence of act-(s), are envoy-(ed), through an familiarizing-oneself, by routine-schedule of [elapse-(s) // motion-(s)] settled by acquainted place-(ing-(s)) of enabler-(s'), finding objective-memory-reflection-(s), during-intermission-(s) extraspective-response-reasoning-(s')), for how cause of variable-(d)-effort-(s), remain uncounted through an sequence-(s) of action-(s), affirm-(ed) through evocation-(ing-(s)) of relate-(d)-reference-(d)-site-objective-effort-(s), form-documenting assertion-(s), envisioning [task-(s) / labor-(s) / chore-(s) / concert-(ed)-effort-(s)] where-by direct-(ed)-mean-(s'), appraisal for calendar-impact-(s), denote denominative-handled-objective-use-(age-(s)) of tangible-routine-(ing-(s)), where each focus of introvert-(ed)-personal-self, carry-over an perpetual-effect-(s), of convince-(d)-cause-(s'), interact-(ing-(s)) inhabitant-tense-(s), reflect-(s) mode-movement-(s), when welfare of tangible-tensable-form-(s), are of an convey-(ed) purported-observational-contortion, corrupt elicited-counts, through each fatigue counted, by an inference-(ing-(s)) of sequential-algorhythmn-discourse, due to tangible-technical-mean-(s'), analytically-account-(ed), for an way of function-(ing) by tangible-sight, where those default-(s) of physic-(s), enable an general-trust, conflicted by elapse-(s), maintained designate-(d) place-(ing-(s)), in an notion from human-effort-(s), upon exterior-dimensional-development-(s), endeavoring default-physics, configured as-an trusted-influence, by aware-notion-(s), consistently sustaining daily-practice-(s), as I generalize mutual-response-(s), for commencing analyzation of elapse-(ing-(s)) from men, designate-(d)-place-(d)-routine-(ing-(s)), on-an total-life-spamming-(s), solicited work character-role-(s), typical as is by men (mans')-reaction-(s), from-an childishness, through adult-youth-carry-over, hormonally-natured subject-reasoning, by off-time-extraspective-mean-(s'), for interval-converse-subject-lessoning-interpose-interval-pick-up, upon an nature of Physical-Developmental-Affair-(s') & Envision-(ed)-Affair-(s'), where temporal-adjustment-(s), from calendar-schedule-elapse-period-(s), premise an basis of variate-(d)-sequencing-(s), per populant-free-prerogative-(s'), enabling-hoc, for success of meander-(ed)-progress, typical product-(s)-procession-(s), through subterior-sector-(ed)-cultivation-(ing), of rural-terrain, balance-(ing)-(s), those method-(s) of payment, per state, by-an analytical-degree, from when past-year, be due to date, consequent consumption-(ing), where Supply & Demand-Method-(s), trade an free-introvert-(ed)-nature, as populant-existence, support an natural-tone of relatable-thinking, comprehending-matter-(s) of state-interaction-(s), where national-posit-offset-infinitive, continue an construe of developmental-impasse-(s), for visually affirming terrain, from how human-perspective[41L.] apply upon tectonic-plate-(s)-physics-plane-parallel-continuum-infinitive... subjective [narrative / expository] paper-(ed)-documenting-confirmation-(s);[294] secured where their pertain an temporal-live-elapse-(s), upon an simultaneous calendar-date-(ing-(s), periodical deductive-informational-interval-(s), (by-an order of) [financial-historical-calendar-reference // generation-filing-expect-ancy] present-impasse-thinking, simultaneous each dynamic, for how physical-affair-(s'), annotate perceivable-existence, applying citizen-(s') undaunted mutual-trust, by our perplexive-perceivable-verntricle-blood-pulse-air-space-interior-foot-torso-hand-eye-vision-ear-sounding-motion-reflexive-tension-coordination-(s), offset approximate-exterior-value-(s), from read-vernacular-nature, label-(ed) per association-(s) of observable-tangible-(s), intermittent physical-value-(s), for when an confluence of act-(s), are envoy-(ed), through an familiarizing-oneself, by routine-schedule of [elapse-(s) // motion-(s)] settled by acquainted place-(ing-(s)) of enabler-(s'), finding objective-memory-reflection-(s), during-intermission-(s) extraspective-response-reason-ing-(s')), for how cause of variable-(d)-effort-(s), remain uncounted through an sequence-(s) of action-(s), affirm-(ed) through evocation-(ing-(s)) of relate-(d)-reference-(d)-site-objective-effort-(s), form-documenting assertion-(s), envisioning [task-(s) / labor-(s) / chore-(s) / concert-(ed)-effort-(s)] where-by direct-(ed)-mean-(s'), appraisal for calendar-impact-(s), denote denominative-handled-objective-use-(age-(s)) of tangible-routine-(ing-(s)), where each focus of introvert-(ed)-personal-self, carry-over an perpetual-effect-(s), of convince-(d)-cause-(s'), interact-(ing-(s)) inhabitant-tense-(s), reflect-(s) mode-movement-(s), when welfare of tangible-tensable-form-(s), are of an convey-(ed) purported-observational-contortion, corrupt elicited-counts, through each fatigue counted, by an inference-(ing-(s)) of sequential-algorhythmn-discourse, due to tangible-technical-mean-(s'), analytically-account-(ed), for an way of function-(ing) by tangible-sight, where those default-(s) of physic-(s), enable an general-trust, conflicted by elapse-(s), maintained designate-(d) place-(ing-(s)), in an notion from human-effort-(s), upon exterior-dimensional-development-(s), endeavoring default-physics, configured as-an trusted-influence, by aware-notion-(s), consistently sustaining daily-practice-(s), as I generalize mutual-response-(s), for commencing analyzation of elapse-(ing-(s)) from men, designate-(d)-place-(d)-routine-(ing-(s)), on-an total-life-spamming-(s), solicited work character-role-(s), typical as is by men (mans')-reaction-(s), from-an childishness, through adult-youth-carry-over, hormonally-natured subject-reasoning, by off-time-extraspective-mean-(s'), for interval-converse-subject-lessoning-interpose-interval-pick-up, upon an nature of Physical-Developmental-Affair-(s') & Envision-(ed)-Affair-(s'), where temporal-adjustment-(s), from calendar-schedule-elapse-period-(s), premise an basis of variate-(d)-sequencing-(s), per populant-free-prerogative-(s'), enabling-hoc, for success of meander-(ed)-progress, typical product-(s)-procession-(s), through subterior-sector-(ed)-cultivation-(ing), of rural-terrain, balance-(ing)-(s), those method-(s) of payment, per state, by-an analytical-degree, from when past-year, be due to date, consequent consumption-(ing), where Supply & Demand-Method-(s), trade an free-introvert-(ed)-nature, as populant-existence, support an natural-tone of relatable-thinking, comprehending-matter-(s) of state-interaction-(s), where national-posit-offset-infinitive, continue an construe of developmental-impasse-(s), for visually affirming terrain, from how human-perspective~41L.|~ 552 |apply upon tectonic-plate-(s)-physics-plane-parallel-continuum-infinitive... ¶|||03m.:01s.~63m.s.|||~§

§...Preposition-Count-[29 / Twenty-Nine]-of.. ..Article-Essay-[4 / Four].. ..Book-Essay-[161 / One-Hundred-Sixty-One].. ..Lines-[41.125 / Fourty-One.~1.one-~tenths-~2.two-~hundredths-~3.five-~ thousandths].. ..life has an elemental-grandiose-basic-aspect, not thoroughly-explained, from how work-effort, determine those periods of hour-(s), for extracting those resources, from when each date is positioned upon an [city / county] ecological, [state / nation] geological position of tectonic-plate-(s), affecting how each timed-moment, is perceived, juxtaposed physiological-action-(s), for validating common-resourcing-effort-(s).¶|||23s.90m.s.|||§

¶162Appeal of appellate, serve given-effort-(s), latter from-an cautious-present-progression-fluctuation-impasse, for how introverted-independent-genetic-birth-offset, remain injuncted per live-active-presence, from interval-(s) of [state-taxed-goods / Federal-commercial-wage-operation-(s)-yearly-taxing-(s)] in an continuum-presence-offset-(s), for gauge-balance-scaling, [territory // human population-mass-(es)] upon instruct-(ed)-entity-(ies)-budget-balance-(ing)-relation-(s), by those commercial-corporation-(s')-directive-(s), secure by-of demand-(s), proportional upon a-the market of trading, from [parliament / international / congressional-daily impasse] not interceding government-road-way-impact, which require intervention of independent-inhabitant-(s'), posturing upon physical-presence,| 553 |an vernacular-redact-conversive-inquisitive-extemporaneous-miscellaneous-superlative-extent-(s), retaining tangible-article-(s)-matter-(s) by crust-deposit-offset, apose dynamic of physics, where tectonic-divide-(s), secularize posit-facet-(s), from original-compositional-patent-product-retrieval-(s),| 554 | dispersing kinetic-influctuation-magma-combustion-density-double-extent-matter-infinitive-pos-it-(s), continually undated, as for how, such lunar-tectonic-satellite-drift, (where // when) progressive opposite-motion-pull, contrast [lunar-tectonic-observation-east-west-drift // earth-kinetic-influctuation-magma-combustion-density-matter-infinitive-physics-posit-mobius] remain-untraceable, for detecting human-motion-(s), where spherical-combust-infinitive, apose prose of peripheral-perspective, when language for one reason or another, not be humanistically-conjured, proportional where place-(ing-(s)), live-breath-blood-pulse-muscular-ligament-organ-tension-(s), as how each reflex-(es), pertain an vague-primary-raw-composite-interpretation-(s), refining-along, those time-(s) of dating-(s), influencing each fact of matter-(s), short of empirical-individual-subject-intricating, where subjunction-count, harmonized-intonation-interpose-redact-cause-reasoning-(s), not clearly frame-(d), from standard-deviation {Division}, perplex-(ed) by cause-(s) per individualism-posit-(s);[295] when presence supersede ascertain-affirmation-(s), at-an word-notion-(ing), per account-(s) of check-(ed)-exchange, deposit-(ed)-affirmation-(s), when-where guarantee of article-(s), be interval-(ed) for-an purpose of tangible-(s), as [dry-substance-(s) / liquid-substance-(s)] not ensued resource-labor-effort-(s), during an syntax-exponent-secondths-offset-reset-counting-(s), when information, come instructed of data-posit-(s), inform-(ed) due to social-sensation-intuition, by-an aging-relative-fatigue, intermittent taxed-means', not vocalized of prudent-upkeep, (as I attempt to intonate subjunct-paragraph-data-context), for how freedom naturally limit-(s) oneself, as civilian-right-(s), define where common-mutual-mass-populant-redact-affirming-(s), not exist awhile each particular of [Declaration / Constitution / Bill / Resolution] indicate, legislational-revise-upkeep, by median-data-dating-documentation-(s), not rate construct-(s) of interaction-(s), for Rule & Order, intricate casing-(s) of inhabitant-(s), from view of appeal-(s), that are tested due to grandeur non development-(s), juxtapose-(d) inhabitant-purpose-religious-examination-(s),| 555 |non incrementable technical-boundary-(ies), foreshadowing [custom-(s) / tradition-(s) / ritual-(s)] ingrained by unconscious [via-dolorosa // De census Averno] (as an) Ways & Means, for physical-action-(s), motionless of senate-committee, for how process of social-Darwinism, methodically pay, each general-option-(s), pertain-(ed), an nature of physics, ruthless of those fatigue-(s), intermittent observational-behavior, when counting-(s) can be pre-empted, requisite those familiarity-(ies) of destiny, resettled by sight-(s), offsetted by sound-(s), where those reverberation-(s) of matter are seeable, and those resonance-wave-vibration-(s), sound in-of-upon tangible-presence, because how every interaction-(s), have an limit of timing-(s), when capable progression, verify objective-article-(s), undocumented per entity-dating-(s), for how independent-individual-(s'), use an basis of data, for affirming notion-(s) of trade, where usage is matter-(ed) from past-day-documenting, voluntary receipt-data-presence-confirmation-(s), not define parcel-subject-documenting, directing order-(s) of mass-article-(s)-redact-(ing)-{manuscript}, per person-(s) (not) writing {cursive}, by position-(s) of enterprise, which are maintain-(ed), for handling, shipped trailer-container-(s), from purchased market-product-goods, formulating fact-(s) of production, for instilling mass-(es) mode-(s) of purpose-(s), by-an forward-factifying throughout each daily-requisite, temporal from [pork / beef / chicken] in posit-(s) per type-decay-process, from when juxtaposition-(s) of tangible-dimensional-composition-(s), digest by-an product-temporal-fatigue, clothed by human-posture, from inset-(s) of populant-parameter-(s), [using / pertaining] thought of present-day, upon those fact-(s) of biological-requisite, upon anatomical-physical-present, awhile future remain constrained of present-continuum-day, perceived apose present-visual-anticipation-planning-(s), where destined-location-(ing-(s)), parameter managerial-maintained-tradition-(s), policed by those theatrical-piece-(s), of entity-article-information, for mass-requisite, retained upon those parameter-(s) of common-social-theatrical-order-(ing-(s)), pertained affair-(s) of state, to [sustain // maintain]56L. those quality-(ies) of continuum-offset-development...¶|||04m.:43s.05m.s.|||§

§...Preposition-Count-[42 / Fourty-Two]-of.. ..Article-Essay-[5 / Five].. ..Book-Essay-[162 / One-Hundred-Sixty-Two].. ..Lines-[56.33 / Fifty-Six.1.three-tenths-2.three-hundredths].. ..Presence is all we have in an lifetime, on an estimate average of 2,750,380,270-secondths, per individual, offset how population-mass, have different-tasks, to maintain, handling loads of pounds, incrementing how with mathematics, new subjects, have not been made, creating formulas for denominating each individual-work-place-traits, by counts of acted-modes, upon entity-capacity-limits, weighing out each [product / tool / page / quantity] for comprehending along an presence of hourly-time-frame, how each tangible-item, is suppose to be functioned, by an intent to circulate those market-values, in count-weight-proportion, per secondths-moded-act-(s).

¶|||39s.82m.s.|||§

¶163Character-act-(s).250¶|||84m.s.|||§ For while aging-exist, listening yet litter through an superfluous reasoning-(s), (for how) humanity subterior time by [hour / minute-(s) / secondths] (per of) [elapse // syntax-(s)] synapse-effect-(s), intermittent [rotation / revolution of-physics];296 from tectonic-position offset perceptive land-(s), maneuvering Self & Family, by physical-objective-mean-(s'), when daily-procrastination-(s), subjective-reasoning, incur each fact, per-an natural-factoring-(s) from respirational-breath, blood-pulse-tension-(s), by independent-perceivable-being, awhile territorial-offset, posit-(s) stance-(s), which signify why effort is pertain-(ed), from the dynamic of perspective, occurred through [school-house / university-ground-(s)] unexpanded from established-founding-(s), by when cycle-(s) of physical-perspective, have an effect from how individual-effort pertain indicated-ink-paper-mark-(s), (per) of year-season-month-week-cycling-(s), impassed at pertinent-interval-(s), where individuals' participate upon settled setting-land-space-(s), visual sounded-measured-reasoning-(s'), for succession through of-an progression by-of product-fact-(s), their of an usage for modifying each parameter, by-an effective pursuit of value-(s), in lieu of general-means', from muscle-memory-repetition, confirming daily-objective-(s) under entity-dynamic-(s), for value-(ing) cycled-motion-(s), relevant by effort-(s), for general-action-(s), acceded from past requisite-cause-(s), pertained from-by-of fact-(s) for whom {1.[George-Washington] / 2.[Thomas Jefferson] / 3.[Benjamin Franklin] / 4.[John Adams] / 5.[John Hancock] / 6.[Robert Lee] / 7.[Woodrow Wilson] / 8.[Theodore Roosevelt] / 9.[Martin Van Buren] / 10.[James Monroe] / 11.[John Tyler] / 12.[Abraham Lincoln] / 13.[Ulysses S. Grant] / 14.[Theodore Roosevelt] / 15.[Howard Taft] /

[Article:] {July}

16.[John D. Rockefeller] / 17.[Thomas Edison] / 18.[Eli Whitney] / 19.[Andrew Carnegie] / 20.[Calvin Coolidge] / 21.[Herbert Hoover] / 22.[Franklin Delano Roosevelt] / 23.[Harry S. Truman] / 24.[Dwight Eisenhower] / 25.[John F. Kennedy] / 26.[Lyndon B. Johnson] / 27.[Richard Nixon] / 28.[Gerald Ford] / 29.[Jimmy Carter] / 30.[Ronald Reagan] / 31.[George Herbert Walker Bush] / 32.[William Jefferson Clinton] / 33.[George Walker Bush] / 34.[Barack Hussain Obama] / 35.[Donald J. Trump] / 36.[Hillary Rodham Clinton] / 37.[Millard Fillmore] / 38.[John Pierpont Morgan] / 39.[Adolph Hitler] / 40.[Wilhelm the Second] / 41.[Joseph Stalin] / 42.[Winston Churchill] / 43.[Eric Cantor] / 44.[Dimitri Medvedev] / 45.[Vladimir Putin] / 46.[Angela Merkel] / 47.[Janet Louise Yellen] / 48.[Ben Bernanke] / 49.[Margaret Thatcher] / 50.[David Cameron] / 51.[Tony Blair] / 52.[John Boehner] / 53.[John Bercow] / 54.[Queen Elizabeth Alexandra Mary] / 55.[Queen Victoria] / 56.[Queen Christina] / 57.[Carl Gustaf Jung] / 58.[Sigmund Freud] / 59.[Hermann Rorschach] / 60.[John McCain] / 61.[Jerry Brown] / 62.[Rick Scott] / 63.[Nathan deal] / 64.[Robert J. Bentley] / 65.[Charlie Christ] / 66.[Jeb Bush] / 67.[Danial Macmillan] / 68.[Alexander Macmillan] / 69.[Brian Murray] / 70.[James Harper] / 71.[John Harper] / 72.[Richard Leo Simon] / 73.[Max Lincoln Schuster] / 74.[Susan K. Reidy] / 75.[Susan Muldrow] / 76.[Madeline McIntosh] / 77.[George Von Holtzbrinck] / 78.[Monika von Schoeller Holtzbrinck] / 79.[Stefan von Holtzbrinck]} as how a-the physio-psycho-anatomical-affect-(s), are by-an usage of objective-matter-(s), when abstract-terrain, exist under period-(s) of impasse-development-(s), that remain an part for each endeavor, approximated, while limited present-day effort-(s), succeed from census-population-count-(s), apose how each tangible-factor, sustem from-an article-(s)-documenting-method, pre-concepted from territorial-responsibility, when dating each matter, remain legislational upon our natural-offset, from human population will, because every undertaking of affair-(s) matter by-an perspective from imagery, while those factor-(s) of commercial-production, rely upon an called governmental-basis-human-rights, determined upon setting per configurated-planned-construction-(s),| 556 |directed handling-(s) of substantial-routine, by product-objective-article-(izing), verifying dues of data, in variation of substantial-matter-(s), in their of, daily-passing-(s), elapse-(s)-impassing, aged-day-interval-(s), for all we can perceive, awhile an mutual-territorial-genetic-development, not think from religious-action-(s), interacting ecological-habitat, for mutual-moral-survey, when daily impasse favor our tensable-objective-(s), by hand-eye-coordination, determining mean-(s'), for understanding why data is for use upon physiological-substantial-schedule-coordinating, where their pertain practice of action-(s), repetitiously examining, which effort-(s) are formally applicated upon cerebral, district-local-vision, pertinent of those tensable-objective-(s), ancillary extraspective-observation-(s), generally-maneuvered, by those act-(s) put upon daily-effort-(s), articulated concerted-direction, (as an) type-form of subjective-intrication, for-when date is not acted of continual-progressive-interaction-(s), adhered by common-focus, conducting intersexual-family-affair-(s), for-an genetic-regeneration, by-an future-state-requisite, planned scheduled-agenda-development-(s);[297]| 557 |retaining those reasoning-(s) per-of cubic-area-3-D-dimensioned-handling, by-an two-dimensional daily-impasses', checklist-affirming, communal-required-effort-(s) from-an ambig-uous agenda, when inhabitant-(s)-offset, live as pertained upon the major-focus-objective-interpose-(ing), due to place-(ing-(s)) by-of matter, upon tensable-developmental-prerogative, pertained by-our breath-blood-pressure, for making existence as we see such, where abounded by physics, that be for how parameter-(s) of impasse, crescendo-crescent, our tense-(s) of daily-impassing, where an impact of deviation, formulate terrestrial-offset-posit-(s), among our dating-routine, imported from politic-(s), which are collect-(ed) from-of commercial-trade, for how parameter-(s) become apart of-an purpose, for recalling data upon the relevant usage of extraspection, where those complacency-(ies) of physical-limit-(s), offer how those affair-(s) matter upon, compositional-adaptation for existence, from where we give our faith of belief, upon an systematic-deviation-(s), of worded term-definition-topic-(s), by our proportional-competency, applicating modified clearing-(s) at surface-area-space-(s), from cause-(s) from mass-repetitive-observation-(s), sworn to abide such principle-(s) of daily-wage-impasse-(s), making pursuit, purpose interaction-(s), by territorial-development-effort-dialogue, rather game-ruse, inclined sensation-(s) of tense-(s), for-an muscular-tension-(s), upon our impasse [seen / heard] option-(s) of inclination-(s),| 558 |evoking vision, from muscle-memory-tension-eye-sight-ear-organ-recollective-perspective, individually factor-(ing-(s)), where those limit-(s) of perspective, remain bounded by ancillary-effect-(s), for man-made-factory-development, [inputting // outputting composite-(s)] for an cause of state-territory, made by formidable-effort-(s), for extraspective-appreciating, is how perspective remain upon an celestial-ambient-grandiose-tectonic-existence, elongating reason, from-an temporal-present-time-maxim, of day-reset-(s), where matter is by-an mutual-genetic-upkeep, referencing topical-matter-(s) of state, articulating epitaphs of Episcopalian-development, as Lutheranable-seizeures, when Catholic invasion, is how American-dynamic, not understand evangelical-immigration, defined by those body-(ies) of Protestantism, where each identity of governance, widely watch, mundane maintaining of those effort-(s) of extraspective-trusted-reference-(ing), as how method-(s) of action-(s), should pertain upon an optimized-setting, for reworking wage-standard-(s) of inhabitant-disparity, upon an natural-land-concept-product-resourcing-(s), from Christian-Fervor, understanding why engineering, is dated by-an political-surveying, yet to be understood, at religious-attending, for interposing pastor per house-representative-(s), an communion-deviation by conscious-extraspective-interaction-(s), as shelf-turnover, apply yearly ambient-examination of human-behavior, viewing individualism greater-than extraspective-neighbor-community-belief, when their remain an usage of objective-collection-(s), fortified natural-affect-(s), ascending-descending blood-pulse, upon tectonic-condition-(s) an part of why dating-(s), apply an kinetic-pressure-fluctuation, throughout revolving-revolutionary-pattern, premising daily-action-(s), effecting each monetary-outcome of day, through elapse-period-(s) of tense-usage, for perceptive-comprehension, circumference daily-effect-(s), as how daily-intervention matters, constraint constant-factoring, parameter-perimeter, physical-tensable-realm, offset-claustrophobic-corridor-(s), hall-impassing our perspective, juxtaposed temporal-parameter-(s), elapsing synapse-count-(s)-effect-(s), or mass-offset, comparative perceptive-tense-(s) of [sight / sound / smell / taste] apose-juxtaposition of mass-simultaneous-tectonic-political-offset,| 559 |respirate-blood-pulse-per-minute-rate-count {145-B.p.B. /// 90-H.B. /// 50-70-B.P.M.}, exterior our frame of time in secondths, generalizing-fact-(s) for labeling how each day is pertained from inductive-analysis, when patriarch-objective-usage, matters from incur of matter-(s) due to those issue-(s) at hand, when meandering open field-(s), vast of those stretch-(es) throughout passing day-(s), for denominative-reasoning, in circum-spection per space-(s), by-an means of objective-article-(s), according articlizing, dimension-(s) surrounding developmental-purpose-(s), per daily-effort-(s), amassed transitioned daily-effect-(s), for impact of parameter-(s), understanding Limit & Mass-Population-Count-(s);[298] intermittent process-(es) of daily-cultural-development, [composite-collecting-(s) // article-collection-(s) // horticultural-collection-(s) // agricultural-collection-(s) // subjective-collection-(s)] information for inquiry of posit-(s), space-elaborating place-(ing-(s)), by-an developmental-discourse, enveloping an formal path of debate, when simultaneous-awareness, can be put to only an comprehensive-understanding of matter-(s), by mutual-interpretation, [where // when // why] whom intervention discussion, from subjective-post-study-(ies)-application (liturgical fervor), from how each independent-individual, is-of-an natural-piecing for living, by [relative // political] boundary-limit-(s), encamped by human action-(s), for representative-responsibility, their of an usage by general-mean-(s), for how attempt of population-balance, affect those tense-(s) of reflexive-invigoration, factorial-multiple-addition-usage, human-objective-subjective-article-(s)-effort-(s), at an impasse of daily-interaction-(s), from Men & Woman, having children be grow in to those role-(s) of ecological-community-matter-(s), when day-kinetic-photosynthesis, affect each piece-(s) of surrounding sight, in light to attempt, transitioning physiological-effort-(s) of commercial-tradition, from those lineage-(s), adhering following timbre, per schedule-refinement, where carried-memory,| 560 |verify-define,111L. impassing developmental-conjugated-human-interpretation-(s)... ¶|||08m.:22s.90m.s.|||§

§...Preposition-Count-[71 / Seventy-One]-of.. ..Article-Essay-[6 / Six].. ..Book-Essay-[163 / One-Hundred-Sixty-Three].. ..Lines-[111.08 / One-Hundred-Eleven.1zero-tenths-2.eight-hundredths].. ..Performance from each inhabitant is not expectable, when individual-preference, muddle how geological-ecological-terrain, is suppose to be developed, from when tectonic-plate-(s), is suppose to be developed from an ambition of blonde-talks, interacting by subjective-discipline, how each land-parcel, is supposed to be impasse upon, from daily-physics-revolution-revolving.¶|||25s.78m.s.|||§

¶164Ancillary-peripheral-view, undertakes how humanity observes natural-existence. 251¶|||03s.19m.s.|||§ Each being-(s') focus, is introverted from religious-prose, intent on remaining locked-in by national-metropolitan-review, per proglamated-area, amid relevant concerted-subjective-effort-(s), (un)-contorted formal-documenting, from meticulated calibrated-focus, count-(ing) article-(s), ambiguous solid-(s)-objective-proportional-decay, when date authenticat-ing, mundane-simultaneous-Gregorian-year-count-(s), centralize crescent-progressive, present-date-continuum-reset, referring numerative-reasoning, for observable-data, progressing formal-prerogative-step-(s), which are for interpreting accumulative [motion-(s) // action-(s) // transit // conversive-objective-product-(s)-use] from-an general-dynamic, by outstanding account-effort-(s), per topical-use, time-interval-(s), intent of long term date-passing, from finding-point-(s), where concurrent-purchase-product-usage, require contemplated communal, objective-instruct-ive-action-(s), gathering from tectonic-political-offset, decisive objective-article-(s) observing, natural-impasse-blood-pulse-posit-(s), perceivable, by genetic-mass, applicating tentative-use, where orchestrated person-(s), succinct inhabitant-population-mass-government-commercial-center-(s), from developmental-action-(s), allocating demand-(s), which are met, by-an serie-(s) of production-(s), partaking impasse of those perceivable-being-(s), whom react from individual-guidance, amidst an unconscious centripetal-force.[252]¶|||01m.:07s.[53m.s.|||]§ View will only confirm only ones' personal-existence, to an extent of one-hundred-percent, apose the fact of exponent-mass-comparison-count-(s), circumference of 360°-denominative-percentile-aspect, unfamiliar from human-fathoming-deviation, apose [solar / terrestrial / tectonic / political-boundary-rock-soil-mass / water-body-(ies) / entity-(ies) / industrial-object-(s) / vehicle-(s) / factory-system-(s) / etcetera...] masses greater than human-mass-comparison;[299] while those piece-(s) under human-mass, remain calculating decimal-fragmentation-(s),| 561 |per piece-(d)-data, from when article-(s) are not an incremental-understanding, but when object-deduct, upon an offset-usage from article-(s), requiring pockets of locational-expert-application-offset, at an comprehension of instructed-information, understanding universal-study, remaining limited, on an Concord of commercial-scale, where literature-finding-(s), have had an drought of non-fiction-liturgical-usage, due to the technical-religious-imagination-(s), introverted from one another individual-(s'), not understand-ing how to be literary-documented, in congressional-contingency,| 562 |for observing each and every denotative-human-being, (from an) literary-writer-dissemination-interaction-(s), controlling mathematical-prose, by-an proportional repetitious-fashion, to partake upon [geometric-surrounding-(s) // objective-dimension-(s) // interior-cubic-feet] candor amid balancing state-constitution-(s), under an state-federal-nationalized, handling county-state-constitutional-mass-strict-count-obedience, practicing subjective-principle, comprehending orchestrated trade, by [public // private // religious-dynamics]33L. adjusting Supply & Demand, where private-engineering has not existed...¶||||02m.:22s.92m.s.|||§

§...Preposition-Count-[10 / Ten]-of.. ..Article-Essay-[7 / Seven].. ..Book-Essay-[164 / One-Hundred-Sixty-Four].. ..Lines-[33.445 / Thirty-three.1.four-tenths-2.four-hundredths-3.five-thousandths].. ..Humans tend to emit an illiterate-pacifistic-nature, demonstrating how each individual is not calculating each observation of physiological-two-dimensional-physics, yet calibrating how proportions of land-parcel-(s), are Titled & Owned; by an cooperation of elemental-eco- logical-resourcing by governmental-inquiry.¶|||20s.75m.s.|||§

¶165[Hypothesis / Antithesis / Theory / concepting / Thesis] all have in common, present-anticipation of date-impasse, as for how hour-(s) juxtapose, circumstance-(s) of event-(s), by-an twenty-fourth-hour-measure-reset, per day, remaining greater-than calendar-year-reset-maxim, from miniscule-gauge-(ing),| 563 ||[synapse-(s) / elapse-(s) / period-(s)] from syntax-context-locational-application, molecular-pressure, is not defined, as is centripetal-force, so how abstract does tectonic-plates-offset-mass, remain upon solar-impact, where our nature of perspective, is limited formal-progression of action-(s), when their remain an setting of ambient-entity-construction, perimeter-boundary-in, by-of political-parameter-(s), guiding human-purpose, when prevalent presence, verify action-(s), which are centralized, due the human-interaction-(s), self-concerted, on habit-action-(s), for pertaining off-time-motion-(s);300 where cubical-surface-area-measure-(s), deviate weight-dating, upon commercial-presence, emanating-spherical-posture-aurora, apose plane-subduction-zone-offset, where referenced measure of human-thinking-timbre, come from past-history, apose present-requisite-(s), [supplementing // complementing], objective-usage of article-juxtaposed-hyper-tense-(s)-documenting, instructed upon managerial-command, for how resolution is altered by disposition of population-mass, from political-position, where power of international-relation-(s), pertaining inhabitants, for an cause of trading in America, from-an constant disparity of [Domestic // Foreign-Trade-(s)] proglamating inference-(ing) of citizens, due to, mundane-observation-(s) of Action-(s) & Thought, not formally-documented, {Holtzbrinck}//{Macmillan}//{Hachette} //{Harper Collins}, conjuring mass-locational-offset, by state-religious-fashion, developing an army based on ethnic-background, motivating tense-(s) of existence, because how duty of data, is concerted through territorial-claim, by mass-survey-requisite, before ever engineering any of those metal-object-(s), affirming action-(s), by daily-obligation-schedule-(ing), stressed by constant regulation-(s), where international-affair-(s), serve from national-requisite-(s), an following of international-trade-guidance, from human-inception, (un)-arbitrated by matriarchal-defined-effort-(s), providing daily-sustenance, where article-(s) of data, have yet counted with intention, comparison-(s) of routine week-{condition-(s) / temperature / climate / humidity / moon / sun / parameter-boundary-(ies)}, when month continue an alternating cycle of inference from ambiguity of property-parcel-(s), intermittent, (un)-referenced inhabitant-interaction-(s), urspurating daily-effort-(s), as animated-extraspective-prose, measuring from usage-count-(s), timed-note-(s)-payment-(s), identifying frequency-(ies) of tone, registering pitch from vocal-vernacular-(vocative)-interaction-(s), only if astute musical-comprehension, crescendo-crescent-observe, Perspective & Act, as daily-encountering-(s), from [motion-(s) // action-(s)] such (as an) requirement to eat, if ever understanding why [practice // method of trade] are absolutely-pertinent for delving in-to the catacombs of perspective-(s), where critique (not critical-(care)), has seldomly been comprehended through political-prose, where affected community subjective-quality-(ies), result-(s) political-leader-conscious-measure, from maxim-(s) of pre-started-territory-revision-(s), regulating schedule-pattern-(s), of living-cycle-(s), serving condition-(s) of dating, lived-by, and passing along, offspring-memorabilia, extenuating semantics of perspective-(s), subterior, ethnic-mass, for whom cycle-(s) of survey, have reclusive, upon mass requisite-(s), exemplary from belief;[301] cycled by vehicle-impassing-(s), but when intelligent-comprehension, not be preferred, once vehicle is [claimed // possessed] from the social-dynamic, learned from school-youth-anxious-anticipation, [cronyized-hoc // bias-vocal-monetary-discussed-trade-interaction-(s)] abusing [social-studies // history] for how specific-point-(s) of focused glossary-lesson-competency, variate difficulty of time-elapse-(s)-interim, balance-(ing) prerogative-(s) upon lessoning of dated-hour-(s), where-ever standard-(s) are implement-(ed) due to the fashion of development, requiring interaction upon subjective-comprehension, emitting glory of development, from commercial-labor-effort-(s), while literary-prose, is intended to enable concertion of commercial-sale-(s), influencing preferred outcome-(s) of decade-life-transition-(s), by extraspective-appreciation, tested by religious-subject-(s), from subjective-feudalism-rule-(ing-(s)), persisting amid commercial-objective-(s), governmentally [monitored // regulated] from social-individualism, continuing perceptive-bias, relevant auxiliary-objective-usage, covering perimeter-boundary-grounds', which can only pertain requests of application, processing commercial-mean-(s), enabling trade-method, from legislational-article-(s), mattered upon requirement-(s) of documented-referenced-information, in an fashion of calendar-millennium-ambient, hourly-interceded, paused-moment-(s), crescent-decrescendo, consideration-(s), for what motivate-(s) yourself, by either an [extraspective // objective // ambient // working // transit // entity-mean-(s)] thinking why it has not been written down, article-(s)-collaborating, amongst an communication-(s)-portal, on {www.ancestry.com} or {www.linkedin.com}, an formal-historical-prose, not referencing before the [1600's // 17th Century] and present day, what to make from demand of trading, by the faculty-(ies) of hardware-documented-information, inferring those process-(es) of instructed-operation-(s), following guidance of industrial-objective-(s), difficultly-assembled, from [factory-direction-(s) // instruction-(s)] that stem an historical-guidance, by facility-(ies) of inhabitant-enabled-conceptive-production-(s), not receiving university-inquiry, for how mechanism-component-(s), compartmentalize work on engineering political-boundary-(ies), from requisite-(s') of longevity, intending an mid-term-perimeter-parameter-(s), objective-articlizing-usage-(s), complying upon derived-composite-resource-(s)-ordering-(s), which have yet been commonly-survey-(ed), by agenda-literary-legislation-reform, for how religious-background, affect semantics of inquiry, per decade-interposed-scheduling, (from an) pre-genetic-surveying, harbored, due to American-Technological-precision, willing to share regulated-mandated-object-(s), and privately-meet-up, at retarded-event-(s), rather objecting to private-means, of (un)-incorporated-article-(s), serving an daily-presence, focusing [[0]calendar-dating // [0.1]hourly-timing each [1]object // [2]person-independent-leadership-work-wage-position-geological-metropolitan-offset /// [2]synapse-(s) /// [3]inhale-(s) /// [4]exhale-(s) /// [5]blood-pulse-(s) //// [6]aorta //// [7]wrist-(s) //// [8]ankle-(s) //// [9]bi-ceps / [10]tri-ceps // [11]secondths // [12]beats per staff-measure, for count-(s) per secondths, from [motion-(s) // action-(s) // objective-function-(s)] through minute-(s)-increment-notion-(s)-extent], defining point-(s) of [deterioration // usage // pronoun-inhabitant-impasse-timing-(s) // adjustment-(s), to exterior-value-(s)] for how each duty is tasked upon living mechanism-(s), due to parameter-action-(s), emitting prerogative-(s) of inhabitant-direction-(s);[302] practicing sustain per intangible-focus, indeterminant-variablized, fact-(s) per commercial-product-(s)-production, rate-requisite, of industrial-performance, sustained from political-bound-ary-constructive-limit-(s), upon maintained mass-city-function-(s), partaken by-an repetition of territory, not conceived of genetic-class-survey, upon the house-representative-inquiry-dynamic, recalibrating state-scheduled-territory, repopulating simultaneous of methodical-modern-commercial-trading, insisting instated requisite, obliged to Germanic-Religious-political-economical-social-impact, where their negate usage of article-(s), from generated work-production-(s), fleeted self-relevance, either forever, by political-stability, or by-an limit of national-transition, from how each inhabitant transmission, from natural-birth-existence, breath-blood-pulse-tension-(s), abiding basic-human-right-(s), until an class can make an political-transition, [[1]understanding // [2]comprehending // [3]instructing // [4]directing // [5]retaining // [6]maintain-ing // [7]sustaining // [8]pertaining] mechanized-action-(s), intervalizing time-posit-(s)-directive-(s), scheduling adjustment-(s) of future-year-dating, as how political-handle, remain upon mass-genetic-inhabitant-requisite-allegiance, tested by effort-(s), [when // where] their-has-been-(is)-an Germanic-Hierarchy, offset bounded, political-requisite-(s), for industrial-capability, developing how American-trade-watch, initiate ingenuity, as for why after Alaska & Hawaii, political-boundaries, now continue to be tested, merit-(s) of state-hood-battle-dynamic, examine mass-trade-(s), from industrial-merit-(s), referencing patent-blue-print-commercial-production-operat-ion-(s), testing post-educational-institutional-commercial-method, by religious-extraspective-communion, not practiced outside those demand-(s), from general-mass-(es) of church, through [city // county-weekly-planning-(s)] identifying [what // whom] activity-(ies);[303] matter from extraspective-subjecting of data, when periodical pertinence, inform proper handling, reproductive-object-(s), regenerating genetic-mass, by communication-(s) of statehoodship-development, connecting with one another, from offset of conscious-communication-(s), learning from leadership, relevant effort-(s) upon the trade-impact, per daily-interaction-(s), requiring count-(s), for affirming why subjectve-extraspective-emphasis, is an point for invigorating interaction-(s), when each day pertain upon religious-mechanism-(s), defined by physical-usage, from date-documenting, those political-parcel-(s), from cubic meter-(s), through feet, in-to inches-(s), until interpretation can minuté through an

Adriatic-memory, by-an [centimeter // millimeter comparison-(s)] practicing dating exponent-notion-(s), per day-(s), because amid the general-scheme of things, harbor intuition, premise independent-perspective, from Family & Self, relevant to daily-parameter-activity-(ies), covering pertinent act-(s)-aligning, of Goods & Services, measured from Ways & Means, Time & Date, where Supply & Demand, post-institutionalize commercial-effort-(s), relaying effect-(s) of action-(s), affecting how future-outcome by those objective-material-(s), remain (un)-proportional repopulated tectonic-plate-terrain, by an century-curb, of planned-balanced-article-(s)-reproduction, measured-acre-coverage-maintaining, per inhabitant, sustemmed from mass-family-lineage, juxtaposed daily-interaction-(s), factoring industrial-object-(s), upon commercial-market-dynamic, practicing human-extraspective-coordin-ated-limit-(s), from mass-survey-count-(s), remaining an part of muscular-limit-(s), dexterity-coordinating objective-action-(s), while conjuring fluent process-(es), for long-term-memory, of those emphasized action-(s), from when dating-(s) are applied upon time-interval-(s), per device-function-(s), amid time-constraints, following human-routine, flowing parameter-(s), registering product-object-(s), formulating future requisite-(s), synchronizing human-mass-existence, (from an) subjective-objective-article-(s)-amendment-regulating-production-bias, when value-(ing-(s)) offset-(s) of notion-(s), that matter from action-(s) of article-referencing-(s),106L. due-to mass-survey-subjective-reference, deviation, comparison-person-lecture-dictating, when each fact of observable-data, is mattered from perceptive-competency, measured from reference-mutual-inhabitant-interactive-applicating...¶|||08m.:53s.70m.s.|||§

§...Preposition-Count-[69 / Sixty-Nine]-of.. ..Article-Essay-[8 / Eight].. ..Book-Essay-[165 / One-Hundred-Sixty-Five].. ..Lines-[107.83 / One-Hundred-Seven.1.eight-tenths-2.three-hundredths].. ..is this the last redact?¶|||08s.97m.s.|||§

[Article:] {July}

[166]Wall of law-(s), proglamate our use of dialogue, for area in ventilation, affecting how we prose annual Ways & Means, requiring dating, inductive-referencing, refined comprehension, amid grandiose-abstract-ambient,| 564 |perplexing learning, from claustrophobic-location-(s), awhile going through intersection-(s), hyper-tensing, reflexive-subterior-tense-(s)-{Seeing //// Hearing //// Smelling //// Tasting}, which [tropic // frigid offset], demonstrate-how [law // legislation offset], from how law rely judiciary-review, for legislating congressional-law-(s), apose literary-review, intertwined population-letter-writer-population-documenting, multiple plus deviation-(s), by hyper-tense-(s)-action-(s)-specification-(s), pertaining senseless-written-apprehension-logic, reflexive of general-identity, for how each action-(s) affect an general-impact, (from an) developmental-territory, [interior // exterior] interval-space-(ing-(s)), where each individual, not comprehend why life matter, without literary-impact, apose an wall of law-(s), lawyer-defining how daily-interval-impasse-(s), affect independent-breath-circulational-blood-pulse-respirational-muscle-contraction-tense-(s)-joint-(s)-reflex-(s)-interior-composition-set-(s), apose-along sovereign-day, progressing along those means of axis;[304] tonally volumizing registering, through applied thoughts, where point-(s) for comprehending, value-(s), by those rate-(s) needed, pertinent visual-decibel-frequencies-hertz-amperes-present-progression-(s), from individual posit-(s), judication timbre, relieving duty (from an) rested group function-(ing-(s)) due to those factor-(s), from individualism upon person obedience pattern-(s), method of thinking, trained by application for functioning, discoursing conscious tension-(s) awareness, relaying each formal requisites of documented observation-(ing-(s)), by-an ordering of (common) law, maintaining those independent composures' of freedom-requisite-(s), Amid blood-pulse along an simultaneous of time, having purpose gather those belongings for self-relative purpose-(s), upon location-(s) of extraspective group morale, for-as why one mass, can not cause an mass [reasoning-cycle / time period / interval-(s)] of Date & Time, [where /

when / why] purpose pertain those tension-(s) of blood pulse fractioning, while organ identify-(ing), not be understood by those intangible point-(s), for which pertain functioning of rationale by genetics extraspection, assimilating article-(s) by grandiose-belief, for assuring of referencing fragmentation, by ordered progression-(s) through an process of behavior, conducting action-(s) usage, cycling information in-an allegiance by trust-(ed) -effort-(s), coordinated by-an mass, for whom influence-pleaded-effort-(s), continually periodically affirming, dues'-process-command-(s), acted from designation [registered / copyrighted / licensed / incorporated-planning-(s) per objective-noted-data] from elicit nature by human bias-personal-subconscious-individual-preferential-decisioning, pertaining solicitation from habit-(s), having tendency-(s) overlook present obstacle-(s), for how we interact by-an basis of individual character-(istic-(s)), based from-an Biological-cycling, of blood-family-line-(s), interconnected, by daily-procedure, operating ways' of mean-(s), enabling those resource-(s) of product-purchase-collection-(ing-(s)),| 565 |correlating abstract-parallel-(s) from birth-posit-aging-fatigue-deterioration-offset, where various method-(s), can count-scale, what is not evocated through post-academia prerogative-(s) because reasoning an effective case, for enumerating such fund-(s), from the dynamic of trade, upon relative-mass-action-(s), accompany how mode-(s) for commutating label-receipt-data, cause how information,| 566 |be deem-(ed)-worthy of usage, from astute-individual-comprehension-(s'), for while those means', of substantial-weight-(s), contrast fact-(or)-(s)) of [time / date / dollar-currency] influence from objective-refining, defining parameter-(s) of property-claim, by usage in frequency of article-paper-detail-coin-engraving-live-trade, validating information by-an subjective-astute-post-reading-comprehension, referencing prerogative-(s) of timed-placing-(s), deviating standard improvement-(s), of physiological-anatomical-functioning, upon an designated-boundary-distancing-(s), for improvement of those particular-parcel-(s), that quantify how development-(s), pertain upon general-daily-impasse-(s), that persist continuity of entity-infrastructure-(s), by-an common disregard for mass-general-schedule-agenda-action-(s),46L. amassing how impasse of existence, maintain an formal conjuring of prepared-entity-structured-activities...¶|||03m.:37s.03m.s.||§

§...Preposition-Count-[32 / Thirty-Two]-of.. ..Article-Essay-[9 / Nine].. ..Book-Essay-[166 / One-Hundred-Sixty-Six].. ..Lines-[46.74 / Fourty-Six.1.Seven-tenths-2.four-hundredths].. ..General common hearing is sense, having an less than common median-average, per attention-span offset, for how every act-(s) are pertain through an intermediate-focus, at stationary-sustained-interaction, directing how attention can recollect & Proportion;305 per individual-bias-hoc-perceptive-interpretation-(s), poor by human purchase-rate-(s), offset city-(ies) like [1.Minnesota // 2.New York / 3.Washington D.C. (Post) // 4.Los Angeles Times /// 5.Houston-Chronicle //// 6.Atlanta // 7.Denver Post /// 8.Chicago Tribune // 9.Montana-(what? could be)].¶|||27s.46m.s.|||§

¶167Amateur of reasoning practice, does municipal-count-(s), preserve cause of action-(s), upon an median-documenting-platform, for such data, as non-fiction would suggest, cycling through daily rig-a-ma-role, over-stressed common reasoning in juxtaposition of (cum) (magna-cum) (summa-cum)-Laude-honor-(s) by-an ulterior notion-(s), formally-interposed, for indigenous comprehension, by family-objective, merely sufficed in whole, to no emphasis of particular-purpose, gathering through an means' of significance, cursively-investigating tectonic-temperate-climate-(s), per season, upon where climate birth-origin-surroundings', determine an usage-frequency-charge-apparatus, uncertain for handling I-count-(s), mass-configurated, disciplining non practiced formally-formulated ominous-expectancy, firmly held by weapon-(s') of executive-power, remaining idle by crafted cursive-cause, not ending from visual-perspective, in focus by inputted article-(s), from–an self-reference-method, not listened by-an attention-span, fixated-concertion, due to posited-parameter-(s') abiding property-possession-guidance, as each momentary-interpretation-concert, elucidate, reclusive (super)-ego, persisting by the way of visual-sensational-idea, at dimension-(s)-count-(s) of tangible-concept-debt-possession-creation, concepting from envisioning-(s) at incremental-point-(s), within-an designated-zoning, by-an unilateral-goal-(s), sufficing familiarities, instated those supply-(ies), ordered from planned concepting-zone-(s), aligned by communication-(s), put from mutual-focus present-progressive, sequential-succession, human-right-(s'), ever believing per notion-(s)-confluences, as rationale being-from physiological-human-being-posit, suffice upon grandiose-surrounding-nature, for articlizing an remainder of facet-(s), not formally in conscious-verbal-mutual-reference-interpretation-(s), due to, [major-deductive-empirical-hardware /// minor-inference-objective-posit-(s)] modifying each specified-designated-property, according to an objective-collective-dynamic, for whom decide formal-progression-(s), by each inhabitant involved, because how verifying article-data, pertain upon repudiated director of operation-(s'), addressed incorporated governmental-collective-resourcing, by how each familiarity of interaction, subconsciously-judge, subsequent moments per daily-impasse-(s), where factoring informed-data, pertaining along mediums of Calendar & Date, intermittent faculty-hour-(s), examine minute-(s)-continuum-offset-impasse-(s), by-an contrast per individual-adult-independent, interposed modified conjunctive-predicate-referencing, valued when purposeful-effort-(s), affirm-data, to be learned by Calendar-Day-(s)-Limit & Twenty-Four-Hours-Reset-Denominative-posit-(s);306 adjusting how each inference vary one-being, apose one another.253¶|||01m.:41s.43m.s.|||§ I will never infer for you, or others, due to the method of action-(s), differentiated due to those factor-(s) from aging-posit, not as what others can expect-freely, from one another, but commonly required by registered-regulated-decision-(s), based on-an environment, recounting how dating is in an posit from visual-tonal-perspective, validifying article-subjecting, through sequential-period-(s), differentiated daily-offset-(s), as payment-method-(s), conjure reasoning-(s), by-an inflow of objective-(s), mattering for how each state, date cognitive-practice, per products-item-(s), placing cycling-order-(s) on moment-(s), where condition-(s) of construction-period, infrastructure, not revising entity-(s), defining growth of corporate-development-(s), where daily-elapse-period-(s) adjust rate-(s) of article-data, intervalizing-focus of item-(s), by-when direction-(s) of locational-work-order, require subjective-purpose, by usage of tense-(s) where frequency of time-frame-interval-(s), adjourning productivity, hold truth-(s) through momentary-purpose, focus tier-article-(s)-data, informing how held-possession, occur prior communal-orchestration, causing burdens of trade, for survival in discourse of ethical-interaction-(s), identifying itinerary, where personal-prerogatives', practice daily-impassing-(s) [substance-(s) / transit / comestible-(s) / clothing / work / lifestyle-(s)]...¶|||02m.:35s.21m.s.|||§

§...Preposition-Count-[23 / Twenty-Three]-of.. ..Article-Essay-[10 / Ten].. ..Book-Essay-[167 / One-Hundred-Sixty-Seven].. ..Lines-[42.525 / Fourty-Two.1.five-tenths-2.two-hundredths-3.five-thousandths].. ..tiering has yet been commonly-interposed, by vernacular-vocal-read-interpretations, from various purposes, for retaining why information is pertinent, by voluntary-value-(s), arranging how an genetic-scheduled-mass, can cooperate by an function at basis from an pre-youth-destination-(s), for lineage-offspring transition, from an historical-proglamation, by back-order-info, not corrected by commercial-interpretations, by illiterate-common-colloquial-tounge-talk-function.¶|||26s.63m.s.|||§

¶168When individual capability reference, population require reasoning from evident-tangible-variable-(s), cuing denominative-posit-(s), for calculating impasse-dimension-(s) upon Pen & Paper, from letter-word-participle-sentence-paragraph-syntax-essay-(s), deductive of each chapter-book-sequencing-(s), juxtaposed article-(s)-posit-(s);307 where conjuring of data, formulate ordering of logical-empirical-deductive-article-read-inferencing, by what pertain point-(s) upon focus of deductive-ordering-(s), adjoining reasoning-(s) for influencing, action-(s) of Physiological-anatomical-subjective-conjecture, practicing general-disposition, through thought, upon timbre-tempo-focus, in-an objective-usage, for application, where one routine-(s') posit-referencing, where circumstance of Time & Calendar, function impact-effort-(s), as secondths-conjunct-method, pertain moment-(s), for comprehending variable-(s), where proper-acting upon factor-(s) of dating, where either [1regiment / 2acting / 3presiding / 4listing / 5subjecting / 6articlizing / 7canvasing / 8moralizing, observation-(s) per-of general-way] as general-motion-(s), apply inhabitant-(s), as free-time-colloquial-interacting-(s);[308] dating major-community-documenting, at impasses' for factual need-(s), amongst period-repeat-interaction-(s), by communal function-(s), functioning common-appreciation, conducting-timbre, as method abide by regulated-resourcing, formally-conjuring objective-(s), using objective-(s) for parameter-(s) at Designation & Ordering factor-(s), pertaining article-comprehension, when thought interim median-means' where perceptive value-(s) identify mark-(s), understanding data, because de facto interpreting, verify purpose Constructing & Reading, article-(s) by glossary-diction-objective-locational-purpose-(s), upon standard-(s) of trade-respects', when motivation-(s) inform sequential sequestrated-period-(s), when awake-aware-conscious-blood-pulse-intestine-cardiovascular-respiration-digestive-organ-process, invigorate objective-posit-(s), prior subjective youth-perspective of circumstance-(s), {for me, [po-ke-mon – card-(s) / cartridge-game & Yu-gi-oh card-(s) & game-(s)]};[309] dissuading perspective purpose, as subjective-comprehension, defined where peer-independent-reading-(s), practice group-response-(ing-(s)), [affirming / defining] article-(s) that pertain studied redact-deducting-(s), as each daily-lesson, pattern each inhabitant for how each moment is intermediate-confluence of data, apprehending social-studies-interacting, upon an group similar by inhabitant-(s), for whom affirm article-subjecting, through pertinent-article-isolating-(s), in lieu of restructuring adulthood, where Non-Fiction, monetary-output, focus terrain upon claused-territory, where geo-metropolitan-population-(s)-subjunction, remain confided by constraint of relative-intelligence, for physiological-anatomical-functioning-incapability-(ies), not concert, subjective invigorating, as for why Means' of Trading & Retention, perpetually ignore oneself upon mass-group-inhabitant-(s), that have an purpose of nationalism, due to, requisite past-history, through present-progressive-lifestyle, asserting objective-refining-(s), by-an listed-agenda, in and amongst confluence-(s) of genetic-reformation-confluence, nuance characteristic-(s)-modificat-ion-(s), deliberately served piece-(s) of data, used by quantitive-piecing-(s), from-an usage-frequency, by each ordained-swearing,| 567 |swearing unison-allegiance, upon an designated-ordered-reformed-(treaty)-constitutional-territory, trusting dated-timed-effort-(s), ordering-(s), elemental-refined-product-object-ive-(s), (at present current date, over-produced in inaccurate-fashion, of aging-period-specification-(s),38L. as objective-article-effective-count-allotment-production-development-(s), define contingent-congruent-congenial-interaction-(s))...¶|||02m.: 42s.88m.s.|||§

§..Preposition-Count-[18 / Eighteen]-of.. ..Article-Essay-[11 / Eleven].. ..Book-Essay-[168 / One-Hundred-Sixty-Eight].. ..Lines-[38.6845 / Thirty-Eight.1.six-tenths-2.eight-hundredths-3.four-thousandths-4.five-tenths-thousandths].. ..congenial-distancing, is not purchasable, phycological-effort, can not yet be metered, so what count each individual-incremental-focus, when perpetual-molten-distance-proportion-pressure, not fulfill each act-(s), served duty, for an false interpretation-(s), per ecological-workers-cycle-(s), when dating is per product-shelf-life, as is objective, or titled-offset, when idiots are not person, and not legislational-rhyme exist, exerting how to conducted executive-action-affairs, from an offset of bias-hoc-illiterate-international-cultures, in various universes, not cohesive past retirement aging, per independent-notion-(s), blank-omitting, to document at an centralized-national-genetic-entity-location-purpose, from an calendar-continuum-Gre-gorian-2017-through-error, per perceptive-unpurchased-individual, whom not conform along my preference per literary-syntax-purchase-blonde-demand, as how subjective-action, is ceased from dormant-redact-religious-genetic-lineage-documenting, under National Library-Congressional-Control-Numbering, at national-grade-omission, by Federal-commercial-aging.¶|||54s.84m.s.|||§

¶||169Freedom from responsibility, define effort-(s) as each inhabitant-(s'), decline whom preference for adaptation, by logic, where lineage-genetic-dynamic, forego conclusions', drawn by each conjunction, where objective, not firmly appeal factor-(s) of genetic-communication-interactive-patience, when proglamation of conduct, remain unconcerned by sweat-effort-(s), through daily-development-(s), holding objective-fact-(s), apose celestial-physics', when thousands of year-(s) have cultivated-construction per inhabitant, default-characteristics', pertained upon Each & Every mass-objective-participant-(s'), for whom remain inclined dry-preference of work, as blue-white-collar-method, affecting environmental, metropolitan-distance-(s), repetitive when-where metropolitan, not equate Territory, sequencing natural-heat-post-radiation-(s), independently-defined, when dues' are expected by-from, natural-default-observation-(ing-(s));310 noticed (from an) constant influx of action-(s), determining formal impact-(s), upon such environmental-posit-(s), as whom periodicalize [period-(s) / block-(s)] where time-posit-(s), derive dues', by means', omitting individual-(s), for-as why ones' way, be pertinent when constant-motion-(s), crescent process through relevance perspective, only when subjective comprehension, pertain formal-mutual-common-studies, at-an conscious point-(s), where natural-cause, determine why reasoning [means' / signifies'] an proportional-effort-(s), when [individual-effort-(s) / collective-objective-elemental-compound-modificational-adjustment-(s)], elapsing pass cycling-(s) of Terrestrial-Physics.254¶|||01m.:08s.82m.s.|||§ While confided-comfortably, human-(s) remain naturally not mutual, individually concerned by purpose of an weak-work-ethic, for how daily living, object purpose of daily-motion-(s), elapsing so frequent by our timbre of aging, slept by routine in evocation of natural-circumstance-(s), considering idle-evocation, in replace of reference-read-thinking for how each independent-individual, conceive logic, by composite-organ-skin-blood-ventricle-system-ligament-bone-respirational-oxygenation-tense-perceiving, where commercial-purchase, verify data in usage of information, particular awareness from progression of process-(es), dating related item-(s), signifying why result-(s) of [operation-(s) // mission-(s) // action-(s)] by which an person can communicate task-(s), [through // to] those inhabitant-(s), whom conceive pre-requisite-reading-(s) upon present-interaction-(s), amid article interposed-referencing, for reviewing data, conjuring cause of action-(s), among those reasonable purpose-(s),| 568 |for national-county-state-populational-documenting.$_{255}$¶|||01m.:54s.$_{38m.s.|||}$§ From youth, we be friction-prone, upon our youthful-peer-(s), pressurizing how any of us consider commercial-trade, not [reformed // referendum] upon western-grandiose, national-social-reform, where state-territory, nationally refer, objective-usage of article-(s)-data, crescendo-crescent-peaking, past-day-impasse-posit-(s), when obituary-cycle-(s), product-object-(s), applied in an yearly-manner, influencing article-understanding, alluded monetary-objective-purchase-factoring, where modern-inflation, not suffocate inhabitant-(s'), of all those stocked-shelf-product-(s), for developing an objective, pertinent through genetic-population-objective-usage, upon an territorial-claim, using referenced-data, applying effort from those competent-being-(s), whom oblige subservient-allegiance, abiding superior-elder-(s), as decade-tasking, locational-rationalizing, confirm eforted-experience-(s), pertinent by class of comprehended balancing of subjectives, where purchase-parity, comprehend through discourse, past hyper-cycle-(s)-present-comprehen-sion-subjecting, for each of those element-(s) of conglomerate-mass, along with data, in-an effective view of distance, allotting-time-frame-attention-elapse-(s), by each piece of matter-data, revolving centripetal-force-parallel, when exterior-rotation, not determine an formal-cause, crescent interpreting, crescendo registering, upon legion-(s) of space-(ing-(s)), significant when subjective-day-hour-intonation-(ing), serve past-year-day-elapsing-data, from whom matter by oneself, intermittent juxtaposition-offset, variate inhabitant verified fact of logic, using article-reference-(s), from those mutual genetic-variable-(s), pertaining inhabit locational-proximity, relating data of listed-information, as posit-(s) through past-historical-date-requisite-(s(')), accomplishing an agenda-purpose, while opening further data, from referenced literary-syntax-(s), identifying proclamation-(s) of Physics', not ordered, for grandiose can not be defined, or fully identified, intertwining inhabitants required efforts, at local-level-(s), so they can determine such cognitive-practice, closed by those mutual-inhabitant-(s), when natural-birth-cycling-(s), limit natural-effort, from accomplishing parameter-(s)-goal-(s), per distance, influencing those variation-(s) of mutual-focus, where-when temporal-temperate-climate, have no clasp of [intangible-oxygenation / untangible-tectonic-plate-(s) / untangible-molten-fission-pressure-combustion-kinetic-energy-infinitive] close to those Terrestrial-forestral-aquatic-mountainous-terrain-spacing-(s), pertaining tangible-supplies, receiving consumeristic-revise, as colloquial-mass, verify notion-(s) per day, adjusting night rest, as how fission apply-(ies) variated circumstance-(s), applying an physics'-plane-parallel-revolving-continuum, subdued per [rock / sand / mountain-terrain] where dry-density-tension-(s), perceive density experience, subdued live-barricade, natural by comfort-(s), naturally-ambient, (from an) applicational-processing, through subjective-day-hour-intonation-mode-(s), pertained each factor of lived-existence, objectively-used, for pervading motion-(s), as whom matter upon typified-quality-(ies), impassing-elapse-(s), when motion-(s) of notion-(s), pertain variation of efforted-motion-(s), constructively-defining, settled adjustment-(s), by morning of week-

(end)-(s), enjoying miniscule-objective-hard-data, ensuring confidence of offspring-reproduction, limited as physiological-anatomical-coherent-interpreter-(s), discuss fatigue-(s), upon juxtaposition of century-(ies), sweating by inhibited territory-space-(s), individually-biased, procrastinating manual-labor, not rationalizing purpose of effort-(s), for functioning effective-tasking during period-(s) by work-place-invigoration-(s), achieving fundamental-article-(s)-objective-developmental-infrastructure-schedule-agenda-period-(s), first finding ones' genetic-identity, from sparse-spaced diocese-(s), of (inter)-national-dispersion, formulating conjugation, at an local-view from crescendo of task-(ed)-objective-(s), sequencing local operation-(s), from-an default of elemental-resourcing-(s), in thought of national-exchange-communication-(s), working apart, from agreed-state, deliberated-mutually awhile Pastor-(s) & Communion, concert fabrication-(s) of Modern-Present-Discourse & Future-Genetic-Appreciation-Sustaining,31L. post-elaborating those article-(s) intended for, objective-agenda-tasking-(s), upon an population-direct-intention-territory-purpose-(s)...¶05m.:05s.11m.s.|||§

§...Preposition-Count-[42 / Fourty-Two]-of.. ..Article-Essay-[12 / Twelve].. ..Book-Essay-[169 / One-Hundred-Sixty-Nine].. ..Lines-[31.71 / Thirty-Two.1.seven-tenths-2.one-hundredths].. ..millimeter-letter-rates, are constantly omitted by individual-interpretation-(s), when dating is not specified by an proportion of millimeter-measures, apose meter-mile-distance-impasse-(s), where progress of daily-function, is intended through interaction by product-thinking, inbetween-retirement, from active-duty-commercial-aging, preferenced at physical-effort-impassing, serving how duty is to find an schedule for centralizing an purpose for living, while waging for an objective-retentive-entity, from an mass common-communion, agreed upon objective-usage-scheduling, opposite idle nature.¶|||32s.97m.s.|||§

¶[170]How do we [generate // circulate] an monetary-currency, by-an means' of direct-genetic-population-literary-mass? When we are born into an community at birth, where our potential is limited from our capable-potential, for how notion-(s)-extent, completely-restrict, motion-(s)-requisite-(s), where effort-(s) executed by each inhabitant, require motion-(s), for defining density of matter-(s), not by substantial-tension-(s) from mutual-response, but pre-placed-effort-(s), which occur upon daily-impasse, as survival of means', concert capable output-(s), frictioning density upon our substantial-fatigue, while celestially-solar-tectonic-ocean-iron-metal-rock-sand-central-circulating-combustion-auxiliary-rotation-revolving-motion-(s)-matter, offset collective-object-ive-(s), haste output-preference, as our tense-(s) have meaning for (inter)-coordinating subjective-task-(s), at an point-(s) by group-communicated-designated-trade-point-(s), meandering daily-articulation-(s), as fine nuance-(s), at proximity from night-centripetal-force-celestial-physics, completely surrounding ourself-(ves), in any endeavor we partake in, as whatever will ever happen, that never been by German-warfare, provoking those points of an absolute-judgement! For which balance, payment-(s) not presently requiring [regulated // mandated // registered] read-evaluation of human behavior;[311] defining judgement-(s) of Academic-Effort-Conduct, mundane present-commercial-efforts, intonation national-exchange, never orchestrated by those particular bodies of humanity, for how individualism prevail in line of reasoning, without much extraspective-communion-experience, concerting fact-(s) of local-approximate-terrestrial-claim-ing, as inhabitant-(s) pertain mutual-appearance-genetic-class, without grandiose-ambiguous, of homogenized-characteristic-(s), ignoring elemental-resource-(ing-(s)), as how, perspective-(s), selethisize by callous-individual-bias, proceeded through ancestral-lineage-offspring-passing-(s), incoherent living, as-for [why / how] whom intercede collection-(s), by our effort-(s) to extenuate family-lineage-impasse-(s), transitioning the process of community, for identify-(ing) what can be Done & Accomplished, while alive, processing living, from solar-propel-logical-impasse-(s), affirming offspring-comprehension;[312] furthering ideal-(s), through fruition of fact-(s), affirming trusted-posit-(s), as [genetic / family / neighbor / statesman] require objective, by totalitarian-control-factor-(s), for maintaining each progression of patented-product-(s), where inventor-(s), circa-era, an American-patent-conceptive-developmental-generation-(s), prior [literary-write // type conceiving variable-(s)] that are mass-produced, for-of development of metropolitan-territorial-space, identifying title-proprietary-space-(ing-(s)), in lieu of those voluntary-result-(s) (politically-nationally-shrouded), extracted from workplace, for how individualism prefer live-reluctant-effort-(s), understanding how production-means', deduce understanding, relying global-effort-(s), for not having anything, differ monetary-trade, where defined capable-conception-(s), develop those means' of elemental-resourcing, informing inhabitant-(s'), parameter-(s) of preference, weighing out option-(s), from consumeristic-requisite-comprehension, continuing [government / commercialism / economic-trade / military-religious-historical-transgression-(s)] emphasizing variable-(s) of patent-factor-(s), articulating request-(s) for-an intelligent-coherent-creative-interpretative-society, verifying logic as, daily-living fixate bi-weekly-yearly-payment-method-scheduling, incapable of giving similar-genetic-impasse-(s), where mass-education-level-(s), suffice upon each factor of posit-trade-(s), where natural-focus from independent-trade, conceive why tense-(s), delivering result-(s), lackadaisically invigorate succeeded gamut peripheral-article-(s)-conversion-(s), not evaluating sweat-level-(s), remaining (un)-perspired, individual-effort-(s), by-an vain-jaded-paper-fiat-trade-belief, when-where, evocation of thought, remain dry, from those drought-(s) of condition-(s), richly shed through collective-excuse-(s), based on $, conducting-effort as intonation of other inhabitants'-perspective-(s), remaining ignorant, from knowledge of [trade / communication-(s) / off-time-social-interaction / work-place / mutual-neighbor-commodore] not fluently-coherent, interaction-(s), by our nature of organ-tense-(s), interact-(ing-(s')), individualized birth-mundane-posit-(s), from-an unclear defining, Gregorian-Calendar-Progressive-cycle-(s);313 inputting-data, abstract, randomized posit-(s)-effort-(s), as inhabitant-blood-pulse-living, worked on from restricted-entity-(ies), which control central-primary-post-(s), impacting motion-(s), from individual-focus, confluencing-factor-(s), affecting approximate-outcome-(s), remaining apart, for influencing common-pattern-(s), continued along, sequence-(ing-(s)) of person-being-(s)-interaction-(s), consuming for perspective, calendarized neglected daily-genetic-interaction-(s), where way of self contrasting practice on [[1]cursive-subjective / [2]objective-engineering / [3]tectonic-engineering / [4]elemental-resourcing / [5]terrestrial-collecting / [6]mega-ton-vehicle-terrestrial-ordering / [7]floricultural-collecting / [8]agricultural-eating-cycle-ordering / [9]horticultural-eating-cycle-ordering / [10]population-psychological-scheduling / entity-(ies)-territorial-space-ordering / [11]artistic-creative-claimed-territory-input], cultivating character, when momentum, direct focus, through an fatigue-influx of effort-(s), which persist in and upon notioning-motivation, by-an various median-(s) of article-(s)-organizing, input data, referencing, through momentary-posit-perspective, remaining completely useless, rendering literature pointless, from library-congressional-local-offset-impact-effect-(s), as each proportional common-hyper-tense-(s), remove blood-pulse-tension-(s), from each significant sensation-(s), pertained alliterated-educational-comprehension, when [sight / sounding / tasting / smelling] transition respiration-evocation, as perceptive-limit-(s), matter-self, as other-(s), invigorate formal-tension-(s), where locational-predicate, process blood-pulse-(s), intermittent Task & Functioning, when-whom, such time of revolutionary-dating, continue daily-impasse-develop-mental-barren-terrain-construction-(s) .256¶|||04m.:26s.06m.s.|||§ Presence-motivation-(s), while each moment of daily-motion-(s), require an defining-action-(s), to emit characteristic-(s) of crescent-crescendo-tense-motion-(s)-observation-(s),66L.|

569 |amid constant-revolving-revolutionary-centripetal-force-tectonic-plate-(s)-drift... ¶|||04m.:44s.50m.s.|||§

§...Preposition-Count-[33 / Thirty-Three]-of.. ..Article-Essay-[13 / Thirteen]....Book-Essay-[170 / One-Hundred-Seventy].. ..Lines-[66.125 / Sixty-Six.1.one-tenths-2.two-hundredths-3.five-thousandths].. ..dating is intermediate Hour & Week; as when timing is moded minutes, through secondths, subversive millisecondths, per moment as when offset circumstances, incur pass one another, due to morale at placings to circulate various forms of thought, by when dating is formulated by two-dimensional-planning-area-(s), title, by how Stories & Layers, have an [Marine / Army-Housing-Sleeping-dynamic] offset by an cause for living, from when dating is restricted by 86,400-secondths-(tics), per duplicit [geological-track // impasse-action-mode- (s)].¶|||33s.25m.s.|||§

¶171 Immigration & Domestication, for so long have been an instrumental-cycling of human-fate.257¶|||04s.82m.s.|||§ Morally-directing common-mass-inhabitant-(s'), whom along an basis of lineage, generation to generation, offspring capable citizen-basic-act-(s), [where // when], [eating / sleeping / transporting / clothing / working] an part of daily-impasse, kinetic-circumference-celestial-pulse-liquefied-untangible-molten-frictioning, light-photosynthesis, accelerate correlate-ed-factor-(s), imminent locational-posit-(s)-application, forhow none of those characteristics', remain an part by greater-entity-body-compositional-development-(s), without an [military // police, defense-(s)] from minute-incremental-piece-(ing-(s)), drawn by elemental-modification-(s), pre-conceiving dimensional-shaped-hard-data-figure-(s), as empirical-religious-hierarchical-commercial-production-(s), plant those modified-material-(s), for human-creativity-evaluation-(s), upon directives of instruction, [understanding // comprehending-article-(s)-usage] pertaining Dexterity & Optometry, coordinating seldom physiological-respirational-blood-pulse-act-(s), | 570 |for assisting common-being-(s'), in directing elemental-engineering-constructive-transit-trade-practice-(s), while [preserving // sustaining] common-civil-free-liberties, impassing congru-ent-function-(s), from pre-dated-concepted-article-(s), [cut / molded / bent / pressed / hammered / modified] from mechanical-leverage, intended on modifying [soft // hard data];[314] moving acted natural-configuration-(s), extenuating pre-requisite, lineage-obituary-existence, documenting from present-dating-impasse, [past-motion-(s) // action-(s)-incurred] awhile planning [future-action-(s) // motion-(s)] pertained from-an natural-morale-stand-point, discovering findings' made, by loose-interpretation-(s), deterring singular-article-directive-retrieval, premising government-data-motive-(s), while restricted-analysis, regulate common-effort-(s), upon impassed-material-(s), focused on superlative-idle-gluttony, (im)-pertinent by Language & Literature, reasoning daily influctuated extent-(s) of [past-action-(s)-document-accumulating // future-action-(s)-planning, calendar-dating] present grandiose timed-presence-momentary-evocation-(s), elapsing Short & Repetitive, offset-notion-(s) pre-layered Gregorian-historical-calendar-input-directive, through timed-article-(s), intended on referencing from book-(s), an literary-bibliographical-chrono-logical-present-syntax-synthesis-revisement-(s), impassing weekly-sequential-activity-(ies);[315] unilateral not intricated-ambiguous [season-(s) // month-(s)] distancing human-being-(s),| 571 | by elemental-objective-article-pertinent-usage, measure-evaluating, how physiological-male-elemental-tectonic-coverage, ration from literary-subjective-constitution-amending, posit-(s) of daily-prerogative, where purpose for practice-(ing) action-(s) of [objective / subjective] (physiological-subjective-objective-referencing), by measure-(s) of sight and dexterity-action-(s), modify schedule-(s), median numerical-count-(s), compiled by pertinent grammatical-punctuation-proportioned-action-(s) through state-activity-function-(s), safeguarding, modified-elemental-object-(s), aging from present-impasse, into future-present-expectation-(s), as past-(blank) / (Auto-Biographical)-historical-impasse-(s), affirming motion-(s), assisting present-comprehension, living by present-means', offspring-blooming, population-inhabitants', as subjective-objective-entity-order-construe-(ing), direct modification-(s), through year-week-acreage of daily-impasse-territory-(ies), examining seasonal-planning of comestible-(s), in proportion of constitutional-district-population-eating-table-count-(s), accounting for those water-pressure-count-(s), affecting [nutrition / dispelling / showering / supplementing / cleaning / washing / watering...] variable-(s) of ecological-development, acceding population-perspective-mass-function-(ing-(s)), where unilateral-politics, modern-sequence, each inhabitant, whom is prone to fail, by independent-effort-(s), for accomplishing group-task-(s), slowly-believing linear-prerogatives of concerted-action-(s), ecologically inclined, self-aging-developmental-preference, ignoring [neighbor-(s') / other-(s)] estranging oneself, upon spacing-limitation-(s) & inhabitant-(s')-activity-capacity-(ies)-prerogative-(s'), interposed those surroundings, enriching develop-mental processes of life, for getting more out of each evocation-(s), awhile present-lifestyle-(s'), plan ones' offspring, in-an genetic-allegiance-coordination, post-life-orchestration, from when immigration continues to this day with America, rather progressing existence [further // forward] by those national-neighbor-allies, an reset of ideology, influctuating monetary-focus of Neighbor-(s') & State-National-Territory;[316] stifling genetic-population-commercial-domestic-trade-modification-shift-(s), from when those reliances' of globalizational-trade, not be technically-required, but demonstrate an Congressional-Literary-Starving, (that be my primary-copyright observation), over-aiding those foreign-nation-(s), without an stern free balancing by our national-family-favoritism, not illegally constraining international-ethnic-racial-communities, but emphasizing an literary-genre, Pronoun-documenting, all individual-(s), whom matter by their primary-climate-territorial-intertwined-life-story-(ies'), documenting congressionally, verifiable-data, for contraposing an common-balanced-$36,795-$50,000-

upper-class, repopulating, to then apose incumbent-requirement-(s), demonstrating the conundrum of Poverty & Census-Population, deeming percentile-tenths-hundredths-thousandths-decimal-column-documenting-contraposition-(s), validating unworthy-civil-working-classification-cultural-population-(s), in confidence, if ever, for formally concerting, difficult judiciary-executive-impacts, upon an legislational-political-method;[317] congressionally-documents-referencing, commercial-non-fiction-literature, explaining those errands of [eight-hour-(s)-per-day // forty-hour-(s)-per-week-lifestyle-work-method-(s)] basing confidence from political-analysis, balancing work-mode-(s), from newer-corporation-(s), working an bias of handling, by required-task-(s), attempting to offset, blank-perspective-pre-requisite-present-set, uncreative Savings & Checking, from-an genetic-perceptive-comprehending why vernacular-physiological-subjective-perceptive-being-(s), are more important fundamentally, than product-generating-economy-(while still primordially significant), when labor is tasked from [living-corridors // personal-mansions // commercial-developments] where literary-economy, has yet infused with Labor & Perspective, an offset-nation focused by awareness of intertwined-funding-(s), worked upon through serie-(s) of conversation-(s), discussing how elemental-topographic-birth-place-origin, matter from where to extenuate an territory from a-the Louisiana-Purchase-Barren-State-Territory-(ies), based upon dialogue differing retirement-savings-purpose, from continuum-repopulating-intent;[318] ridiculously vivacious, for furthering customs of culture-(s), by traditional-reformation, from those old-instilled-set-patterns of living, guiding Interaction-(s) & Mass-Occasional-Planning-(s), of population-interactive-count-function-(s'), while electronic-device-(s) now have been considered acceptable, consolidating space, rather than expanding through literary-prose, governmental-land, for pre-planning measured-purchased-land, for then building though decade-(s)-continuum-effort-(s), new-installment-(s), of artistically-creative extension-(s), of family-repopulation-lifestyle-(s), thinking how [work-pattern // lifestyle] interpose, from when [male // female-nature-(s)] (have not been) concerted formally, having women stay home-primarily, and have women input literary-book-(s), for trade, having Library [Congressional Control Number-{L.C.C.N.}-governmental // International Standard Business Number-commercial-purchase-(s)] make up the money, for repopulating, and moving career-work-center-(s), from male-interactive-commercial-literary-conversation-(s), practicing yearly-repopulating, intermittent literary-rebirth-offspring-lifestyle-anticipation-genre, attempting to make an maternity-like-center-(s), for then generating an new method of offspring-currency, by an $10,000 per child, from children-five-(5), practicing yearly-repopulating to matter upon an thousandths-population-year-deviation-(s), interacting, by-an purpose of currency-generating, by literary-intellect, purporting male-transit-work-method, apose feminine-house-child-raising-method, trying to expand career-extent, from book-reading-rate-lifestyle-(s), evading the continually sub-par, reclusive nature, of independent-individual-identity, practicing an raising-method, to influence offspring to repopulate, through [read // write // study] to not overspend on [eating // clothing // interest-credit-commercial-product-(s)] not partaking amongst frivolous action-(s) of physiological-anatomical-presence;[319] insighting that, introversion, is truly our natural-basis of living, but entertaining oneself, from extraspective-interaction-(s), more than deeming valid commercial-product-(s), without an place to store those object-(s), after purchase, require an article-interval-comprehension, for reference-deducing-data, from format-lessoning, by ways of significance, verifying installment-(s) of data, incrementing functional-purpose of community, [visualizing // envisioning, ambient-environment] subterior ecological-space-(ing-(s)), seeing though sphere-surrounding-condition-(s), an conceptual-linear-dimensional-compos-ition-(s), researching discovered, pre-occupation merit worthy, extraspective-article-confirmation-(s), enveloping simultaneous-function-sequence-(s), furloughing for how momentum-offset, interpose perspective, by where exterior-posit-(s), require sustaining language, prior to subconsciously-sustaining dictionary-newspresse-magazine-mail-contract-lease-title-encycloped-ia-manual-plans-bill-advertisment-article-inductive-infinitive-documents-compiling... meriting various outlet-(s) of consideration, when progress-impasse, by-an sequestrated-grading, understanding [mutual-formal-lessoning-(s) // article-(s)] proportioning location-(s), by distance-(s), to be considered for scaling by [meter-(s) / acre-(s) / feet / mile-(s)] median natural-perceptive-sight, when concentration of sound-(s), transition air-pressure, apose tender-density-(ies) of [wet / plasma / dry-substantial-(s)] visualize-sounding, from letters-words-syntax-subjective-interpre-tation-(s), after post-immigration-domestic-purpose, be shipped from-an nature of action-(s), swayed [written / read-base-posit-juxtapositions] pertaining existing-efforted-order-(s), from tensed-contraction-reflexive-muscular-ligament-motion-(s), offset distance-limit-(s),109L. deviating [city-(ies) / count-(ies) / district-(s) / block-(s) or mansion-(s)-count-(s)]...¶|||09m.:07s.75m.s.|||§

§...Preposition-Count-[39 / Thirty-Nine]-of.. ..Article-Essay-[14 / Fourteen].. ..Book-Essay-[171 / One-Hundred-Seventy-One].. ..Lines-[109.75 / One-Hundred-Nine.1.seven-tenths-2.five-hundredths].. ..Literary-Data is made so that those ecological-observation-(s), increment interposing a-the geological-terrain around, from physiological-being-visual-tonal-observation-(s), that define how each word-posit-labeling, proceed through an progression of effort-(s), impassing at an property-rights-terrain, exampling how each fact is due to timing, enjoyed laboring the land, proportioning resourcing-land-(s), with residential living, from factory-production, military-guard, practicing written juxtaposition of population-character-traits, from labor-work-characteristic-aging, building those ages of mass-population, intending through vocal-intellectual-morale, how each basis of interaction, is suppose to offset at an focus of factory-objective-(s) concerting the goal of an residential-mass-family-hygiene-cleaning-supplies, for understanding where daily-impasse-intervention, is suppose to go out an attempt to circulate each position of terrain, when thriving on extraspective-interaction-(s), to refine ones titled-position on American-Trade, for directing an geological-national-terrain, to focus on an harmonious perfection of subjective-documenting of all terrain population-inhabitant-(s), as well as each piece of bounda ry-limit-(s), and those elemental-extracted-refined-product-resources, for practice-ing how to pass upon lineage, those transgressions of impasse, to increment each notion of pre-anticipated-genetic-lineage-commercial-governmental-scheduling, when dating rou-tine those timings of designated-objective-concerted-entity-(ies), for parameterizing how developing each notion of circulated-anticipated-product-objective-(s), specifying each secondths purpose per day for every individual-inhabitant, proportioning mass apose those 86,400-secondths, of daily-centripetal-force-revolving.¶|||01m.:38s.$_{13m.s.|||}$§

¶[172]By what basis does humanity understand interaction-(s')? [Race / ethnicity / family / colleagues-(s) / nation / youth-schooling / co-workers] are an few of a-the factor-(s), affecting how interacting, requisite historical-background-(s), only communicating for requirements, such as food, by-an nutritional-supplementing, of self-character, sufficing an natural-basis of living, merely evoking from our relative-birth topographic-condition-(s);$_{320}$ undetermined to sweat out those labor-(s) of living, implicit by self-comfort, in and amongst usage of transportation, supplementing those spacing-(s), pertained by mode of activity-(ies), when metropolitan-basic-human-need-(s), implore common-developed-facilities, partaking in group-organized-activity-(ies), (from an) cultured-identity, from-whom attend those designated-area-(s), premature natural-(s)-passage, where occupying cubic-dimensional-spacing-(s), suffice activities, (from an) simple-historical-direction-(s), affirming Length · Width · Height;$_{321}$ by-an formal-progression, topical-matter-(s), piecing data, in-an congruent-fashion, of item-assistance, configuring factor-(s) of locational-placing-posit-(s), concepted by merited-duty-(ies), imposed by-an general-common-good, influencing human-behavior, unexamined from the slow nature of movement-(s), by flawed extraspective-live-conscious-interaction-(s), upon segmented-work-place-method-(s), uncult-ivated from those origin-factor-(s), by territorial-location-elemental-resourcing, through an dynamic of [action-(s) // motion-(s)];[322] impendent, verifying motions' repetitive-daily-impasse-(s), upon-an territorial-claim, for from nature, nothing would be objectively-existent, for-how incapacitated not subjective-inhabitant-(s'), would idle-by, an pleather of non-preferred-interactions, worshipping cubicle-eight-hours-lifestyle, calm from [fan / air conditioner / large-leaf] circulating surrounding-location-(s), without an passage of obstacle-(s), from tectonic-resourcing, Findß & Establishβ;[323] territorial-claim-(s'), primarily from-an introverted-nature, observing hedon-population-inhabitants', not involved in-an religious-social-commercial-government-militarial-university-entity-(ies)-modification-(s), pertinent by mass-community-neighbor-commodore, enacting-functions, in ever balancing how respiration-blood-pulse-wave-vibrations'-cycling-(s), remain an part of our default-perceptive-existence, forming natural-development, partaking amongst active-role-(s) of community-effort-(s), broad [protection / intrigue / repopulating / eating / sheltering / et cetera...] affirming contravening-interim-median-period-(s), at locational-claim, of designated-structured-variable-value-(s).[258]¶|||02m.:13s.[48m.s.|||]§ So as for July, each day passage along an extent of degree-(s), consecutively hot by tension of day, notion of temperature upon climate-condition-(s),| 572 |at an Earth-tilt-30.7391°Lat., by an Solar-Rotation-North-horizontal-Baton-Rouge-continuum-momentum-influctuation, fission-continuum-infinitive, from solar-combustion-molten-plasma-continuum when an constant feature to existence, date during an pleather of mass, inverted from those origin-(s) of matter, by common-stance, placed amid limit-(s) of visual-reverberation-(s), for as we persperate, by-an nature of evocation, awhile wet-blood-pulse-tension-friction-response-relay, electro-thermally conducts our natural-perspective, friction microcosm of vein-muscular-ligamental-reflex-(s), between [skin / bone / excess-blood] gandering existence, requisite [posterior-stance / anterior-visual-respiration-al-smell-taste-charge / lateral-volume-reverberation-wave-decibel-(s)-sound-charge / respiration-al-organ-intestine-aorta-ventricle-blood-flow-friction-charge] friction offset, our temperate-physiological-system, per Ninety-Eight.Six-Fahrenheit-degrees, by-an general cubic-area, as my weight is one-hundred-Seventy-seven-pound-(s) by twelve-cubic-feet-sum-composition-posit, from height five-feet-seven-inch-(es), impassing an tectonic-plates-centripetal-force-clearance-territories, as how perspective remain limited from physiological-impasse-fatigue, without vehicle-requirements,| 573 |upon our natural-motion-(s), of basic-voluntary-movement-(s)… ¶|||03m.:35s.13m.s.|||§

§..Preposition-Count-[24 / Twenty-Four]-of.. ..Article-Essay-[15 / Fifteen].. ..Book-Essay-[172 / One-Hundred-Seventy-two].. ..Lines-[41.96 / Fourty-One.1.nine.tenths-2.six.hundredths].. ..margins of production, have an error for not formally being counted, from when each individual pertain how market unit count-(s), make how developing commercial-production, has an subversive issue with introverted-individual-family-two-child-bias.¶|||16s.75m.s.|||§

{January-(31)-February-(28 / 9)-March-(31)-April-(30)-May(31)-June(30)-July-(31)-Month-(s)-Day-Hour-Minute-(s)-Secondths-Count}-{18,316,800 D.H.M.S. = 60S.× 60M. × 24H. × D.}-{5,088 H.}-{305,280 M.}-{212 D.}

{Article:} {August}

¶173 Pending on your point-of-view, the globe have various basin-pos-it-position-(s), juxtaposed by grandiose-abstract-distancing-(s), culminated by-an independent-individual-offset, when [racial / Ethnic characteristic-(s)] identify relative-juxtapositions, for no militarial-religious-effort-(s), predicate act-(s) upon preferenced-effort-(s), perceived through commercial-social-economic-standpoint, in-an constant-influctuation of trade, juxtaposing objective-(s), which matter, for how each independent-natural-bias, effort an pending Leader-Subordinate-Dynamic, perplexed [where // when] leader-individual-detachment from employee, rather prefer community-commodore, in lieu of dating, for comprehending an vocal-vernacular-median, as logic, timbre upon symbolic-conjuring, upon participle-subjunction-(s), which word-syntax, conjunction-(s) of data, by-an usage of article-(s)-referencing, compartmentalizing (apose cognate) data in juxtaposition, district-dynamic-reading, balancing physical-movement-(s), understanding how we are [to / at], [plan / arrange-order-(s)] from subjective-data apose elemental-composite-piece-(s), particulating modified-data, as each fact of placing-(s), fit in to, or sit upon, an ambient-distancing-(s), proportional, from adult-stance-visual-tension-(s), for how an certain-schedule, designate-action-(s), when purpose for daily-impasse, factify consideration-(s), while developing, government-data, filling-in gap-(s) of information, as common-mutual-ignorance, refer library, through Federal-Reserve-Trade, from educational-peer-pressure-cheated-enabling;[324] all awhile, rule of thump, hush-hush subjective-interpretation-(s), idolizing physiological-preference-motion-(s), as standard-(s), factoring issue-(s), yet to be considered, resolve of independent-perspective-(s), obliging ignorance, as bliss, rather totalitarian-critique, appreciating free-American-belief, when order of juxtapositions, forego intrication of concerted-substantial-elemental-mass-product-objective-developed-variable-value-(s), when life pass by, elicited-manner, not enough for balancing Listened-Population & Talking-Orator, as socialism runs an transmission-belief, forging friendship, enabling an imbalance of idle-perspective-(s'), not critiquing-mechanism-(s), on guard by direct-attentive, militarial-focus, at all point-(s) of conscious-spacing-(s);[325] fabricated as volumized-valuing, procrastinate efficient-effort-(s), from beings' over-developing tower-(s), (as an) symbol of human-repression-outlet, for how each being ignores surrounding-peer-(s), deciding life, when focused on idle-independent-individualism, unaware those impacts' of church-(s), upon government-infrastructure-trade-reliance, concerted analysis by those elemental-product-concept-composition-(s), pertaining day upon timed-counter-moment-(s), dating-tense-(s), which remain in part for using an [hammer // wrench // screwdriver // crowbar // jackhammer // et cetera…] (as an) notion-extent, which persist when, grouping perceptive-characteristic-(s), defining objective-article-(s)-origin-(s),| 574 |from-an black-genetic-historical-origin-Jordan-background-offset, because those meanings' for which signify conscious, sustem from continuum-literary-offset, yet evoking commercial-product-(s)-rate-(s), for partaking in respect of such context, where disrespectful-nature of inhabitants', scarcely involve any physiological-act-(s), counting undetermined offspring-neighbor-community-interaction-(s), because each independent-individual-(s'), raise-silently, an stressed bred-offspring-living, incrementing by-an blank-nature of work-method, rather formulating mutual-intuitive-interposed-interactive-listening, opening venue-(s) of reasoning, for elicitation of dialogue, directing forum at focus of inhabitant-behavior-(s'), for whom conduct academic-subjective-glossary-reading-lessoning-inquisition-(s), at each point from reasonable [cause // purpose] revising [article-(s) // amendment-(s)-usage] specifying letter-participle-recognition, interposing other inhabitant-(s') social-action-(s), delving analysis per mentality, apose peer, thinking apose idle-success, lying quite unhealthy, as each independent-mass-objective-notion-(s), elapse drift Further & Further away, from fact-(s) of trade, inputting receipt-purchase-(s), in-an comparative-usage, by disparity of product-objective-article-(s)-collection-(s), accrued through factoring [objective / subjective, article-(s)] pertinent by mass-communal-execution, per motion-(s)-action-(s), practicing daily-involvement of action-(s), alluding timing of date-(s), when an singular-tense-subjective-composition, conjunct [1moment-(s) / 2city-(ies) / 3state-(s) / 4nation-(s) / 5globe / 6district-(s) / 7housing-(s') / 8work-place-entity-(ies) / 9ambient] because natural repopulation, can work placed existence, fatiguing by stress-tempered-evocation-(s), affecting those outcome-(s) of energy, when output affect payments of service-(s), in-an loose interpretation from each level of mass-weighted-extension-motion-(s), tied from still-tension-offset, variablizing objective-product-factor-(s), put in and upon an focus of individual-(s), signifying timbre of effort-(s), enabled at entity-structural-operation-(s), juxtaposed competent-person-(s), whom conceive [dry // wet-substantial // solid // vapor-pressure-formula-(s)], by-an integer-input upon collected data, inputting information, by-an configuration of article-(s), formulating worded-syntax, where each piece of information, remain an part of an integral-function, assessing vernacular-data, by educated-trusted-reference, culminating objective-article-(s)-usage,| 575 |from locational-width-length-height-distance-denominative-offset, pertaining physiological-anatomical-movement-(s), through-an strict-formal-purpose of activity-(ies), inputting developmental-dimensional-count-(s)-curb, where each piece of information formulate matter-(s) upon governmental-creativity, handling mass-(es) of population, unconfigured comprehension of trade-(s), judged by when long-term-executing-discourse, hoc considered purpose, writing (from an) tangible-animate-mutual-interactive-means', trading by-an nationalism-offset, because each factor-(s), being merely trusted in to, sustem from natural-birth-posit-(s), as elemental-resourcing, developing an dimensional-capability, furthering objective-article-(s), in-an chemical-orchestration, for reducing an curb of tangible-objective-usage, counting Wear & Tear, evaluating each objective-item-(s), at locational-hold, understanding how each piece of information, instruct directed parameter-(s), configuring an bigger-puzzle of construction-development-(s), relevant for instructing circumstantial-factor-(s), in line of culturing designated-perimeter-territory-(ies), understanding how [1color-(s) / 2shape-(s) / 3space-(ing-(s)) / 4count-(s) / 5article-(s) / 6elapse-(s) / 7clothing-8fashion

/ 9.comestible-(s) / 10.discourse of sequenced-event-(s)] order controlled-patent-registered-object-(s), agreed by mass-controlled-sequential-function-usage-discourse.259¶|||05m.:51s.24m.s.|||§ Introspective extension-(s), of family-repopulation-lifestyle-(s), thinking how [work-pattern // lifestyle] interpose, from when [male // female-nature-(s)] (have not been) concerted formally, having women stay home-primarily, and have women input literary-book-(s), for trade, having [Library Congressional Control Number-{L.C.C.N.}-governmental // International Standard Business Number-commercial purchase-(s)] making up the money, for repopulating, and moving career-work-center-(s), from male-interactive-commercial-literary-conversation-(s), practicing yearly-repopulating, intermittent literary-rebirth-offspring-lifestyle-anticipation-genre, attempting to make an maternity-like-center-(s), for then generating an new method of offspring-currency, (by an) $10,000.per.child.,per year offset.. child-five-(5)-years-mother raising reciprocal [interior / exterior],79L. practicing yearly-repopulating to matter upon an thousandths-population-year-deviation-(s)...¶|||06m.:36s.03m.s.|||§

§..Preposition-Count-[44 / Fourty-Four]-of.. ...Article-Essay-[1 / One].. ..Book-Essay-[173 / One-Hundred-Seventy-Three].. ..Lines-[79.57 / Ninety-Nine.1.five-tenths-2.seven-hundredths].. ..I am trying to [make / circulate] an new way of money through literary-ethics, comparting each literary-documenting piece apose Federal-Reserve-Control-Currency, denominating from how American Weights & Measures, [Scale / Measure / Weight / Distance / Proportion], ecological-terrain.¶|||17s.87m.s.|||§

¶174Human-(s) had never been patient enough, from those various detriments' of driving, superlatively using gasoline, in ambivalence of extraspection, without ever considering talking, by local community-member-(s), for natural-existence, vernacularizing vocal-dynamic, reverber-ating-pitch-tonal-vibration-(s), applicating self-presence, from conscious-interaction, as momentary dialogue, intend an numerous amount of community-modification-(s) from elemental-compound-(s), not by-an singular-compositional-form, unstructural, visual-development-(s), which would be without an influence of common-characteristic-(s), where physiological-anatomical-impact, isolate pertained circumstance-(s) of time, hourly involved, from-an pretense of development, because daily-impasse-(s), envelop proper conjuring of articlizing, conjuring evocation, as prevalence of issue-(s), matter upon hyper-tense-(s), rehabilitating neighboring-group-(s), sect-(ed) article-adjustment-(s), from elemental-matter-(s), at local-national-developments, where populational-mass, await awhile, present-requisite, variablize obituary-deceasing-progression, instrumental for driving, when developing an natural need-(s), voluntary incumbent-focus, where temperature remain incredibly hot, frictioning from circumference-three-hundred-sixty-degrees-axis, when molten-tension-friction-(s), fight nature, from our nature of effort-(s), due to each density-(ies) of water-rock-tectonic-plate-solar-kinetic-circulating-molten-friction-pressure, dense-gaseous, when-where, natural centripetal-force, revolve upon our exterior-placing-(s), impacting-fission, against, each motive, positing friction, (from an) nature of humanity, while northern-hemisphere, display natural-implicit-force, expanse-revolving, water-solublesce-tension-(s), as-of dry-wet-circulation, continuing exterior-pressure, remain explainable yet, for how an intricate-tier of letter-word-respects', pertain Germanic-appreciation, from-an lyrical-fashion, where vernacular-vocal-volume-timbre-interpretation-(s), confluence wave-length-ear-canal-tonal-vibration-(s), interpreting factor-(s) of commercial-mass-item-volume-load-(s), In & Amongst surrounding-population-notion-(s), emitting an buoyancy-density-tension-dry-wet-volume-pitch, respirating blood-pulse-evocated-(hyper)-tense-(s), impacting hemispher-ical-respirational-friction, upon our natural-timbre-ambiance, where celestial-ambient, centralize motioned-effort-(s), as we emit our tension-(s), from particular-hyper-fixation-(s), active, while live-conscious-evocation-(s), intricate natural-vibration-(s), pertained by daily-living, commonly ignoring, those message-(s) of extraspective-interaction, where abided tension-(s), remain frivolous by [product-substantial // objective // auxiliary-trading] disregarding moral-character, by-an blatant omission religious-direction, hyper-tension-(ing-(s)) our location-(s), awhile we age, by-an natural-degree of self, through elemental-resource-developing, not considering enough, the discourse of travel, as other-(s') forsake, personal-possessive-journey-(ies) of passage, amid natural-inertia-effort-(s), invoked by daily-impasse, because how, Collection & Giving basis impasse, where not an singular-component, emit evocation, during recoil-(s) of interpretational-competency, circulating means' of travel-territory-(ies), when mode of population-discourse, forcome daily-impasse, at regard of tectonic-plate-(s)-offset, from birth-origin-(s'), as our cycle of locational-visual-impasse, regarding item-(s) influencing motion-(s)-timbre, intermittent physiological-territorial-impasse-coverage-action-(s), evolving out, comprehension of [sight / sound / taste] by tensed-notion-(s), involving an daily-discourse, from how our favor of action-(s), require documenting any individual-(s'), when workers' [factor // register product-(s)], as how competency-limit-(s), determine whatever extraspection is exerted, remaining out of our limit for comprehending interposed-reasoning, upon involved rejection, where uninhabited involvement, not motivate documenting-visual-tonal-vernacular-article-(s) consistently interacting, by natural-progressive-involvement, per [issue-(s) / matter-(s)] of state, due to population, proportioning watched-notion-(s), minutély presuming participle-variable-influctuation-(s) of data, fathering time, (due to an) largo-moderato, preamble across constitution, not instating opposite-value-(s), aging from-an pre-lineage, identifying density of space-(ing-(s)), for how independent-nature, remain discoursed, by-an variable-(s) of literary-article-(s), not yet formally orchestrated from common-inhabitant-(s), processing action-(s), from-an ambiguous interpretation of moment-(s), resourcing elemental-composite-density-(ies), yet conceived how objective-usage, comprehend what motion-(s), affect impasse-action-(s)-defining, timing-impasse-rate-(s), as action-count-(s), interpose tension-(s) of sight, (by-an) vocal-vernacular-ear-canal-organ-(s), reverberating-conjunction, wave-vibrating, effects of blood-pulse-eye-(sperm)-plasma-vision, singular-double-extending, repetitive reverberated-vibrations, concerting article-(s)-usage, per letter-participle-word-modification-consistency-(ies), verified (by-an) multitude of entity-empirical-deductive-variable-means', as independent-introverted-applicability, verify conscious-read-reference, where those page-(s) influence an deeper [licensing /// registering /// patent-production-(s)], pertaining past-date-data, mattering for tangible-objective-article-(s)-usage, where incompetent-inhabitant-(s'), wilt-off, from presence-sight, uncited by direction of article-(s), through a-the library-dynamic, not congressionally-relied, commercial traded-book-(s), pertaining relevant impasse-time-posit-(s), overlooked again & again, due to idle-physiological-anatomical-nature, where inhabitant-neighbor-(s), perceive packaged-product-production-(s), applicating process-(es) (per-of) article [issue-(s) // topic-(s) // affair-(s)], balancing day-to-year, mass-notion-(s), when daily-impasse, either require [weekly-repetitional-frequency-(ies) / monthly-repetitional-frequency-(ies) / seasonal-repetitional-frequency-(ies) / holiday-repetitional-frequency-(ies)] gauging scaled-measured-weighted-factor-(s), existing from posture of individual-interposed-economic-exchange, in lieu by off-time-trading-(s), appraising objective-worth, along those formal-denotative-impasse-measuring-scaled-data-value-(s), interacting human-character, from when we are to evoke an designated-function-(s),| 576 |per celestial-physics-geological-cartography-metric-miles-spherical-rotational-east-west-continuum-perceptive-universe-offset,$_{67L.}$ (idle by copyright-offset, applied by state-legislational-federal-law)...¶|||05m .:24s.$_{43m.s.|||}$§

§..Preposition-Count-[34 / Thirty-Four]-of.. ..Article-Essay-[2 / Two].. ..Book-Essay-[174 / One-Hundred-Seventy-Four].. ..Lines-[67.52 / Sixty-Seven.$_{1.five-}$tenths-$_{2.two-}$hundredths]..

{Article:} {August}

..Book-Congressional-Legislational-Documenting-Military-Security. ¶|||09s. 00m.s.||| §

¶|175August (is in the) center of a-the temperate-range-standpoint, ambiguous from [North-American // European // North-Asia-tectonic-plate] centripetal-force, month-interval, (by of) initial-standpoint, amid combustion-fission, referencing from Gregorian-Calendar-Perspective, where planning crop-(s)-collecting, intuitively prepare amid winter-solstice-hibernating, (due to at those) field-(s) at crop-(s), from physiological-agriculture-cultivating, inclined centripetal-drift, distancing fission-impact, as whom numerous perspective-(s') gauge terrestrial-land-space, where fertile-soil-grounds, offset latitude by state, when community pinpoint, monthly Daily & Hourly-Minute-(s)-elapse-(d)-cyclings, of-an celestial-combustion-drift, friction-pulsating extremes of a-the globe, when [solstice / equinox] season-month-passing, from depth-(s) of longitude-latitude-calendar-day-demarking, interval-(s) at linear-geometropolitan-extremes,| 577 |where intuitive-harmony, encounter influctuation-(s) (weathered-of) temperature, adjunct body-temperature, experiencing tension-(s), (by an) competency of celestial-drift, degree-(s)-(ing), [Fahrenheit // Celsius condition-(s)] instilled factor-(ing-(s)) Population-Observation & Tectonic-Traditional-Customs-developments, not capable pertaining per individuals', independent-introverted-comprehending, celestial-impacts of physics, amid Geology & Commercial-Militarial-Resourcing, uncalibrated literary-prefixing-article-(s), aging resource-(s), created by-an govern-mental-control, allotted an timbre for self-focus, containing object-(s), where inhabitant-(s), not fully succinct citizen-(s'), awhile legal-parameters, affect affair-(s), because each action-(s) have an timed-aging, counting yearly-calendar-dating-article-(s), (not) hourly, intonating interposed notion-(s), understanding extent (per of) counts, revealing genetic-identity, flat by those mundane-action-(s), appreciating subjective-comprehension-verification-(s), placing oneself in-an position at location-(s), tuning or adapting, comfortable routine-(s'), exerting motion-(s), incrementing matter, at vantage-point-(s) of interaction, Invigorating & Observing, procrastinations at work place, as why cronyism is far to erratic, serving sub-par-efforts, through condition-(s) of dating, decrescendo tempo, beat-measure-metronome-tempo, subterior perspective-trait-(s'), defining no effort-(s), required enabling contortion of act-(s), not synthetically-computated, when beat-(s) of wave-(s)-tangible-vapor-pressure, not be commonly recognized, compartmentalizing, under elucidated empirical-subject-(s), formulating further hyper-tense-(s)-observational-reason-(s), article-control-counting, tangible-observational-matter-deviation-(s), upon tectonic-plate-impasse-control-analyzation, furthering formulated-subject-(s), when east-to-west-horizontal-rotation-centripetal-force, processing-order-(s), by demand-purchase-minute-rate-deviation, quasi of [elements / being-species / ingestibles / tectonic-plates / taxonomic-classification-(s)-(animal / plant-group-(s))] retaining by hyper-tense-(s)-visual-sounding-observation, for affirming data, through an routine of letters-words-glossary-term-definitions-syntax-referencing, from subjective-articles-lessons-tests-examinations-comprehension, ranging reading-rates, where those collection-(s) of referenced-data, influctuate, quality-natural-theatrical-population-denominating, during-an continuum-cycling-(s),| 578 |circulating spherical-circumference-dimensional-pressure, displaying centripetal-force, ascertaining visual-evident-confirmation-(s) conceiving action-(s), for how constantly revolving-rotating, axis pressure, amid molten-central-force-fission-pressure, inert by our tensable-hovering-stance upon tectonic-cooling, independent variable-(s) of, [tectonic-plate-(s)-formation / mountain-solidification-formation / temperature of summer-effect, seasonal-influctuation-(s)] as I personally presume, a-the Earth spontaneous-abstract-solidified-molten-combustion-impact into existence at February 6, 1536.$_{260}$¶|||03m.:08s.$_{81m.s.|||}$§ Gregorian-calendar, askew dated-measure of common-incompetence, where each formation-pattern, indicate (from an) deep-space-winter-effect, ballooned from the sun-solar-impact, evident by celestial-time-signature, (from an) axis-rotation-impact, revolving natural-default-fission-combustion-pressure-force, where posit-(s) upon tectonic-plate-(s), exist by-an posture, capable enough, to affirm physiological-observation of celestial-ambient,45L. not evoked, from common-mentality, disposition-perspective...¶||||03m.:30s.84m.s.|||§

§..Preposition-Count-[18 / Eighteen]-of.. ..Article-Essay-[3 / Three].. ..Book-Essay-[175 / One-Hundred-Seventy-Five].. ..Lines-[45.525 / Fourty-Five.1.five-tenths-2.two-hundredths-3.five-thousandths].. ..I have conjured this redact-second-person-insight-observation-query, in attempt to create an offset genetic ethical-Literary-Scamming, in approach to my sexual-desire, for various blonde-genetic-deviation-(s)-{79—1000 divorces (me or my offspring)}, of an population-mass, for an aging into literary-retirement, still proving women can repopulate through one-Hundred-years of aging, while developing an literary-investing-population-survey-fundraising, for establishing Canad-ian-Montana-Retirement-Barrack-Construction, for furthering an way to circulate careers past retirement-aging, instead of playing frivolous-gay-games with money, as men choose not to serve women-aging-stationary-repopulation, upon an entity-schedule-cyclings, from sociological-market-impasse-circulate-pre-ordained-wage-disparity.¶|||41s.09m.s.|||§

¶176Everyday, impasse-(s) ecological-microcosm-matter, unparticular from of our natural disposition of event-(s), influctuating longevity maintaining, from human-equality-basic-needs', surviving the fire-storm of temperature, wearing down, our existence, by-an evident nature of kinetic-energy, pressurizing all there is around us, etiquette daily-decorum, effecting our candor by mundane-discipline;326 so are we supposed to be-an active-part, of-an programmed-progressive-routine-(s) through daily-territorial-impasse-(s), inputting our capable-creativity, directing contingent-effort-(s) from-an visual-listening, pertaining elective-matters, from [product-factor-(s) /

person-(s) / beings' / objective-posit-place-(ing-(s))] subjecting our perceptive-qualities' of observation, displaying act-(s), which involve participation of Federal-central-empirical-deductive-note-(s)-subjective-constitutional-declar-ation-article-intensive-(s), establishing our modern-contextual-framework, incorporating our posterior of health, eminent of land-space-objective-action-(s), upon schedule-(d)-routine, processing-national-agenda, itinerant transit-purpose-(s), surviving from-an nutritional-supplementing, by our physiological-stature, posturing against natural-centripetal-force-effect-(s), as why we serve duty-(ies) of physics, for our mission while living, offset kinetic [Solar // Earth // Lunar // abstract-ambient-infinitive-spacing-celestial-impacts-dynamic] remained peripheral by, physiological-observation-method-(s'), while [facility-(ies) / factor-(s)], remain pertinent by each reaction motioned, due to blood-pulse, respirating our [supplemental-nutrition // complemental-physical-activity-(ies) (by of) perspective] intermittent moon-lunar-celestial-rotational-cycling, from how out bone is an cool-posit, affected through-of-an| 579 |solid-metal-rock-tectonic-aquatic-friction-kinetic-central-energy-heat-solar-fission-pressure-universe-originating, perceptive-respirational-ventrical-blood-pulse-visual-sounded-neuro-synapse-(s)-cycling, when muscle-plant-ingesting, react by, tangible-sustaining, hyper-tenses-offset-notion-(s), requiring labor-factor-(s), acting upon, crescent-crescendo, physical-capabilities, structuring from [bone / moon / tectonic-posit-(s)] substantial-tangible-compositional-offset-(s), without thought for how we densify and substantially-subtly-evoke, perspective, from our stance of existence, as I see it, physiological-impasse, remain difficult from our natural-reflexive-tenses, [ordered // ordained] for how Intelligence & Communication-(s'), have forgotten to identify from nature, an literary-term-etymological-article-(s)-origins, still-comforted when, linguistics-notion-(s), modify letters-words-interpretations, perceptive, not documented, from influence-(s) of [territory / self-(ves) / juxtapositions of community / efforts / prerogatives / action-(s) /

routine / family // et cetera...];327 for how national-dynamic, applies an requisite-(s) of populational-effort, continually reling upon method-(s) of payment-(s), rather than Trust & Confidence, when those unison unilateral-bipartisan-constitutional-objecting-(s), imply order, by mundane-society, merely having something to appreciate, from-an observational-viewpoint, affecting human-morale, in order to ever have an generated-circulation of product-objective-(s), by-an commercial-congressional-continuum-trade-effort-(s), noting registered-trade-motion-(s),34L. from internal-revenue-services, non-congressional-regulated-denominating, of all those offset masses per proportional-census-population-mass...¶|||02m.:42s.92m.s.|||§

§..Preposition-Count-[19 / Nineteen]-of.. ..Article-Essay-[4 / Four].. ..Book-Essay-[176 / One-Hundred-Seventy-Six].. ..Lines-[35.265 / Thirty-Five.1.two-tenths-2.six-hundredths-3.five-thousandths].. ..I **boldly** presume an Post-Educational-Commercial-Trading-Illiteracy, that has yet been explained by any author before, for how visual-tonal-vocal-American-Commercial-English-(*My New Language >|), outcome various purchased-circulation-(s), tiering individual-redactings, for my reader-purchase-blog-inquiry, of each inclinated-notion-(s), of dating-purchase-(s), depicting those minute-(s) of hour-(s), for making post-serial-books, documenting my literary-circulation-endeavor-(s), formulating how stock-market-survey, would commence, for circulating commercial-markets, for governmental-credit-dealing, ethically-surveying how pop ulation is balancing literary-redact-subject-ive-trading, by an aging fatigue for verifying those live paper-bound-documenting, factually-verifying how existence is coming into existence, from my [Etymological-Geo-logical-ecological-County-city-Districts / Ontological-Literary-observational proofings], developing an purpose for living through death, while existing in offset celestial-ambient, enjoying how my blonde-intent, intend to provide auto-biographical-stats, through my literature, for practicing match-making, in an pre-nuptial-agreements, for understanding my post-consistencies of raising children, while taking children to formulate formal geological-tectonic-boundary-development-(s), allowing such agreements to cite how match making pleasing my significant-other-(s'), for literary-auto-biography-subversive-entertainment, exposing subtly, how I need word of mouth from my readers, to circulate an limited-selling per 5,000,000-Y-readers, through those years of book-publishing-production, from those offsets of white genetic, in tier of [opaque / bland / tinge / Blush / Magenta skin variations], aim ing at 25,000,000-book-sales-per-year, after book-4, having purpose of survival, change through assertive residential-thinking, how saving and retirement, affect an outcome from Geological-metropolitan-county-city-(ies)-offset, variating how each family-individuation-(s), start consider from offset conditions, where to saturate in Canada-Temperate-Climate, an counter thought from Florida-California-Retirement-Mentality-South-Interstate-40-East-Interstate-95-West-Interstate-!7-contin-uum-conundrum of heat, commercially-literary-circulation-missioning, how daily-trading is significant posturing lean-fasting-eating-stance-(s), balancing how conducted-impasse, offset concerts better functioning from Industrial-Continuum-Production, from paper-ink-literary-subjective-enumerated-denominating-dynamic, teaching from an genetic-subject-ive-bias, an preference for [purchase-discipline / reader-offset-preference-redacting] learning upon Etymological-Existing & Ontological-Proofing, how formal Pronoun-Literary-Survey-Documenting, applies apose city-planning-dynamic, for otherwise residential-inquiry, thinking how women have yet been commonly-waged, from male-expenditures, influx family-population-offset-savings, retiring obituary-decrepit, due to an lack of commercial-literary-control-monetary-balance-subjective-prose-ethics, having how by my consumer-purchase-understanding, an Non-Fiction-Monetary-Sarcasm, con-trolled from those earlier-white-genetic-skin -purchase-deviation-(s), littoral-bias-non-fiction-literary-controlled-scamming, from those observations of my first book further, how population conceives of book-publishing-circulation, per 4000-books, initial, second-wave repeat, third-wave 6000-books, gauging from word-of-mouth and season-(s), how swift those book-sales, affect lifespan, from those readers whom engage in offset-home-book-collecting, determining why literary-directive, define how each aging by human-species, die in Age-62-Deterioration-Circumvention, due to an wrinkle-blank-instinctual-silence, from how each individual continues procrastinating physical-effort-(s), because an lack of youth-subjective-studies-discipline, extenuate-continuing through commerical-production-(s), because free-voluntary-unconcerted-effort-(s), never will define those survey- citizens, for scheduling post-labor, why to develop commercial-wage-ethical-disparity, affecting those international-illiterate-wages, apose an blonde-appreciation-hoard-offset-family-repopulating, by how women can technically repopulate yearly (every third trimester) from how literary-governmental-bias, practice around those impasse-(s), common-political-literary-neglect, apose House-Senate-Bill-Passing-(s), for cultivating those lands currently-hot lived upon, and furthering an retirement-Montana-Geological-Center-(s), for concerting pragmatic-Commerical-factory-production, when illiterate-common-political-individuals, not hear the subconscious, but consciously-vocalize, why genetic-offset, saddens poor-existence, practicing an religious-redacting-political-literary-congressional-Table-Non-Fiction-compiling-(s), when dating juxtapose timing, for sorting those etymological-shelf-factory-residential-objective-(s), for surveying human-illiterate-population, for religious-church-pastor-Stock-Market-County-Genetic-Retirement-Cen-ter-(s)-Developing, practicing an way to live over Age-62, in through Age-100, as do so many individual-(s),| 580 |[1.wilt-{950,028} / 2.die-{1,250,853} / 3.fatigue-{12,304,393,390} / 4.scoff-{7,202,398,293} / 5.fade-{6,839,327,038} / 6.assault-{259,308} / [7]cremate-{123} / [8]funeral-{360,028,296} / [9]mortuary-{36,500} / [10]Battle-(s)-{12,038,469} / [11]Ritual-Cannabal-Eating-(s)-{7,083,536} decay sectors of death], undocumented-lives, because how life is (*my-subconscious-living-inquiry-only)-[what-{109,850} / why-{46,902} / who-{36,938} / where-{67} / when-{24,334} conjure-inclination-(s)] ones' interior-perception, [Patent-{7,302,475} / copyright-{189,567} / register-{67,209} / Trademark-{24,595} ethical-etymological-ontological-documenting] (makes') of [such / it]. How many Incorporated-Markets,| 581 |juxtapose those [barbecue-shacks-{500} / hot-dog-stands-{750} / Burger-Houses-{20,000} / Gyro-Corners-{360} / Braut-Haven-{639}], apose Supermarkets, awhile malls have food-venues, and stripmalls, shacked of construction-directive-(s), not refining by supermarket-(s),|582|an Genetic-title-resident-ial-literary-population-survey-expository-(ies)-ontological-Commerical-publishing-cir-culating-defining, adjusting bookshelf-(ves), for common-illiteracy-(Genetic)-survey, His-torical-483-Years-American-Literary-English-Gregorian-etymological-2017-ontological-documenting, defining an past-death-cycling of international-population-circumstances, from the mode of Thirty-Year-Mass-Offset-Ancestry-Undocumented-Population-devia-tion of...

0.||1536-Population-Initial-Offset-Estimate||: 27,000,000 |583|

1.||1566-Population-First-Stage-Offset-Estimate||: 52,951,369

2.||1596-Population-Second-Stage-Offset-Estimate||: 104,795,864

3.||1626-Population-Third-Offset-Estimate||: 196,853,269

4.||1656-Population-Fourth-Offset-Estimate||: 359,166,175

5.||1686-Population-Fifth-Offset-Estimate||: 425,911,509

6.||1716-Population-Sixth-Offset-Estimate||: 542,860,369

7.||1736-Population-Seventh-Offset-Estimate||: 650,123,987

8.||1766-Population-Eighth-Offset-Estimate||: 910,640,610

9.||1796-Population-Ninth-Offset-Estimate||: 1,250,750,500

10.||1826-Population-Tenth-Offset-Estimate||: 1,512,149,193

11.||1856-Population-Eleventh-Offset-Estimate||: 1,878,304,573

12.||1886-Population-Twelvth-Offset-Estimate||: 2,362,590,751

13.||1916-Population-Thirteenth-Offset-Estimate||: 2,779,587,329

14.||1946-Population-Fourteenth-Offset-Estimate||: 3,279,543,190

15.||1976-Population-Fifteenth-Offset-Estimate||: 5,450,197,384

16.||2006-Population-Sixteenth-Continuum-Offset-Estimate||: 7,083,270,536

17.||2036-Population-Seventeenth-Anticipated-Estimate||: 7,387,892,910

18.||2066-Population-Eighteenth-Anticipated-Estimate||: 7,745,190,622

19.||2096-Population-Nineteenth-Anticipated-Estimate||: 9,100,000,000 |584|

¶|||08m.:04s.47m.s.|||§©

¶177Summer-season, is an celestial-transit, emitted from solar-molten-plasma-centripetal-force, awhile mundane-default-occurring-observation-(s), occur through presence-impasse-effect-(s), furthered from present-modern-documenting,| 585 |not containing Mass-Clothing-Production-Entities-Operations & Tailor-Entity-Local-Offset-Developing,| 586 |where commercial-methodical-motion-(s)-impasse-pertinence, relied national-default-fashion, formulated by, commercial-mechanical-dynamic-(s), existing prior from-of perspective, upon-an routine of [1.facilities / 2.banks / 3.government centers / 4.storage-center-(s) / 5.mom & pop-market-(s) / 6.cash checking center-(s) / 7.auto repair shops / 8.clothing-center-(s) / 9.thrift shops / 10.labor premises / 11.fast food restaurants / 12.city halls / 13.police stations / 14.cell phone premises / 15.outlet center-(s) / 16.fire fighter stations / 17.home improvement markets / 18.supermarkets / 19.office supplies markets / 20.attorney offices / 21.camps / 22.youth centers / 23.schools / 24.universities / 25.day care centers / 26.bicycle shops / 27.vehicle markets / 28.office buildings / 29.pharmacy-center-(s) / 30.doctors' offices / 31.factories / 32.warehouses / 33.shipment plants / 34.resource facilities / 35.Water & Power facilities / 36.park centers / 37.interstate rest areas / 38.company headquarters / 39.regional offices / 40.district offices / 41.single family houses / 42.apartment complexes / 43.duplexes / 44.mansions / 45.condominiums / 46.section 8 housing / 47.homeless centers / 48.ranges / 49.farms / 50.Federal Reserve & Treasury] accommodating human-effort-(s), meandering where their are duty-(ies) of premises-factor-ordering, designating an purpose, their at daily recurrences', when relevant-observation, by each individual, demonstrates how political-superego, ignores' religious morality, demonstrating an blank-congressional-non-fiction-Dewey-decimal-tier-system-ordering of books, pertaining an nationalism-fervor, premised on a-the genetic-qualities, of an small-flat-back-side-head{50 centimeter or down}-{I 53 centimeter-(s)}, weak-jaw-line-{Hittites-(Western) // Anatolia-(Eastern)-genetic-bastardizations} straight to no chin, mild squaring of a-the jaw-line, weak-thin-cheeks, blue-eyes, white-pink-crackle-skin, straight blonde hair, when pre-Egyptian (Ageptian)-no-jaw-genetic, never formally mixed with Yorkshire-similar-hog, for being far too light-brown, heavy-eye-(brow)-narcissistic, leading (to an) [vain // lazy // inept // et cetera...modern-black-African-statehood-civilization-(s)]| 587 |ignoring our Pre-Germanic-Historical-Biblical-Lord, pre-Hittite-Historical-Civilizations, not listening to their directives, for developing an bronze-age-era, apart of why we suffer under governmental-belief, following-an Gregorian-(housing)-thousandths-years-calendar, thinking from scientific-principle, {Mexican State population Demonym-(gentile)-(s), Jalia-(sci)-ense // (en)-d // (c)-irca (e)-ra = science}, through various national-splits, bastardizing from the 46-centimeter-average-head-size-no-jaw-line-genetic-offset, formulating youth, (by an) mixed-perspective-(s'), limited (due to) collegiate-routine, idle by conduct, setting an timbre of actions', not pertinent by repopulating, proportional to physiological-compositional-respiration-impasse-digging-motion-(s)-fatigue, measuring each distance, understanding that tectonic-plates-composition, (is of an) ambiguous superlative of pre-conceptive-resourcing, idiotically unwritten into billionths of pages, formally, differing from economic-stock-potlucks of exchange;[328] from (by those) church-house-(s), premised (by an) unison-genetic-order, imperially-interacting, yuppy-flagrant-ignorance of physiological-hyper-tense-perspective, and its' offset-impact, upon ethnic-state-demonym-culture-(s), locational-posit, (from an) fatigue-offset-impact, making those surrounding person-(s)-lifestyle-(s), matter due to those condition-(s), which impasse, because how our Bastardized-Pattern-Trade & Lifestyle, follow an unilateral-agenda, devising routines for inhabitants' Offspring-Generation-Lifestyle & Conscious-Interaction, in part for emitting-invigoration, (as an) pleasure, not [comprehending // understanding // subject-reference-literary-control-tier-progressing, executive-decision-(s)], in light of kinetic-pressure, while alive (primarily), anticipating an living transgression of [impasse-(s) // objective-retention-(s) // housing-(mansion)-expansion-(s) // extraspective-vernacular-(preferenced)-interaction-(s)] for after life offspring community lineage-appreciation-continuum-infinitive-pleather (secondary), alluring for homage, to appreciate-continue, [tradition-(s) / customs-(s) / habits'] where instated upon grandeur of governance, apply each living-physiological-perceptive-motion-(s)-fatigue of character, [imply-(ied) / direct-(ing)], sentimental-objective-value-(s), enhancing maintaining-order for every bite of comestible-substantial-(s), interposing consideration, from offset-juxtaposed-perceptive-opinion-(s), through-an fatigue-secular-scope, visualizing how, sequence-(s) of fact-(s), pertain retention, for grasping understanding of information, sweating interval, maturation visual-sonar-territory-(ies'), reverberated by distance, because how dimension-(s)-define those ambiguous-surrounding-parameter-(s), circulating centripetal-force-vapor-air-pressure, distancing independent-individual-(s'), acting where identification of notion-(s'), signifies' an purpose at location-(s) of tensable-moment-(s), cuing [physiological-motion-(s) // action-(s')] crescent-crescendo-measure formulating, sequential-order, by incremental-progressive-article-referencing, affirming at impasse, fact-(or-(s)) of minuté-particular-incremental-interposed-objective-article-(s), intermittent posit-article-(s)-directive-referencing, rue of opinion, outside of rule, sustaining regulated-mandated-bill-(s), from-an mass-intuition-appreciation, for product-matter-inferencing, variablizing each form of elemental-reliance, through an dimensional adjustment, patent-commercial-reconfigured, given abstract-grasp-untangible-(s), in lieu [from / to] location-(s) (per by of) municipal-designated-action-(s), relational-manner, interval-location-project-action-(s), upon [tectonic-plate / national-order / state-fashion / county-order / city-district-lifestyle] which affect why each [piece / item / object / article / thing / product means'], something [significant // important] by those objective-value-(s), not subjecting Independent-individuals', because freedom of will, never impasse upon reference, crescent-registering, unit-(ed)-product-(s), supplying means, for accounting population-item-(s)-denotation of fact-(or-(s)) intervened at, interval particular-point-(s) of locational-posit-(s), for how bias of homage, determine an continuum of successive-issue-(s), proglamated by [attitude / misconduct / sub-par-effort-(s) / focus of abstract idle inductive-thinking] unreferenced [locational-objective-(s) // article-(s)-posit-(s)-offset-impasse-progressive-motion-scheduling] practicing sequential-demand-impasse-factory-ordering-(s), those facet-(s) of barren-county-terrain-population-vernacular-interaction-development, [exhausted / tired / deteriorated] (by an) fashion from steadfast tradition-(s), yet embarking, (on an) ritual (for of) nationalism, conjuring means' of alleviated burden-(s'), by-per daily routine, cycling weekly-monthly-seasonal-routine-(s), which apply by their consensus of personal-observation, through parameter-(s) of (hyper)-tense-(s), deducting from blood-pulse-rate, [respiration-inhale // exhale-count-(s)] from joint-impasse-(s'), intaking by mouth-aorta-transit-pre-posit-(s), an singular-claritive-hyper-tense-(s), affirming data, balancing [finding-(s) / searching-(s) / referenced-cited-data / et cetera...] identifying an land-parcel-space, for conducting experiment-(s), isolating independent-variable-physiological-tense-action-progressions, (by an) double-dependent-offset-variable-(s), from self-independent-hyper-tense-(s), product-objective-modification-item-(s), amid an triple-constant, by [land-perimeter-altitude-depth-length-width-height -rectangular-prism-posit-(s) /// physiological-hyper-tense-(s)-perceptive-com-position-(s) /// mathematical-algebraic-geometrical-measure-(s)-weight-(s)-grammatical-punctuation-subject-ive-ratified-constitutional-article-(s)-(amendment-(s))-bill-(s)-section-(s)-senate-legislational-congressional-literary-denominative-creative-patent-blue-print-mass-schedule-agenda-itinerant-(s)-modification-(s)-directive-(s)] timing motion-(s) of lifestyle, along an serie-(s) of [secondths-timed // twenty-four-hour-constants] continuously sequencing (in)-dependent-variable-(s), where [visual // sounded-tense-retentive-observation-(s)], identify [territory / land / entity / object-(ive-(s)) / voluntary-person-(s) / scheduled-dating-purpose-(s)

/ hour // minute-(s)-interval-timing-(s) /// interval-action-(s)] subterior-tier-progressive-reference-factor-(s), considering specific-point-posit-impasse-(s), defining tradition of trade-custom-culture-(s), customizing Genetics & Lineage-(s'), from whom interact sociological-cycle-(ing-(s)) understand, grasping article-(s)-logic, by extraspective-interaction-(s), confluencing circumstance-(s), per [element-(s) / terraculture / agriculture / horticulture / engineering / calendar-dating-(s), hour-action-(s) /// motion-(s) // near-future-storage // medium-future-storage / construe of concepted-objective-(s) / social-lah-sey-faire-subjective-sustaining], in lieu, secondths'-continuum-objective-(s), subjectively-brainstorming,| 588 |constant-compiling, social-security, apose retirement, amongst nation-inhabitant-(s');[329] whom consider (present)-requisites', of live-day-progression-(s), with each [limit-(s) / boundary-(ies) / extreme-(s) of terrain] identifying perceptive-adept-inhabitant-(s'), whom Conceive & Elongate, lineage-(s) who custom develop, effort-(s)-(I not know how many generation-(s) per se, contradict one another, but as for High-School & Retirement), as generally clear university-diverging, not chance, literate-conception, through rationalizing-fatigue-(s), aging impasse (through an) focus on self-independent-individuals', rather-than family upon, religious-schedule-shift-(s), where activities, use an weekend-style, conducting commercial-factory-required-function-(s), intermittent elemental-location-positioning-(s), confirming community-communication-(s) (by mail) amongst other religious-protestant-sect-(s), apose Sunday-sport-customs' of savagery, unaware common-perceptive-subjective-interposed-interpretation-interaction-(s'), where perceptive-retaining, object-physical-motion-(s), in folly rather, peaceful-conduct, amongst death of tectonic-state-(s)-territory, transitioning calendar-event-(s), which matter per Scheduled-Agenda & Extraspection-Perspective-(s), where genetic-populant-(s'), serve-duty-(ies), by locational-settling or moving-terrain, resettling on new territory, by means' of pre-survey-analyzation, as genetic-religious-territory, requires subjective-article-(s)-reforming, (by an) measure-(s) (per of an) [liturgical-referencing / scientific-elemental-objective-sociological-experiment-referring / mathematical-point-plot-reference-inductive-deviating] those tangible-clearance-probed-inhabitant-(s'), whom volunteer adjusted range count-(s), of aged retirement-inhabitant-(s')-objective-usage, modifying yearly judgements, by-an singular-genetic-group (blonde-hair / blue-eyed / pink-skin-characteristics) for whom would pertain subjective-objective-physical-schedule-routine-planning, per perimeter, at territorial-topographic-climate-conditional-claim-population-survey-plan-probing-inquiry-method...¶|||08m.:45s.77m.s.|||§

§..Preposition-Count-[48 / Fourty-Eight]-of.. ..Article-Essay-[5 / Five].. ..Book-Essay-[177 / One-Hundred-Seventy-Seven].. ..Lines-[114.6125 / One-Hundred-Fourteen.1.six-tenths-2.one-hundredths-3.two-thousandths-4.five-tenths-thousandths].. ..Rule signifies [measure / legal-Federal-reference-parameters]..¶|||10s.06m.s.|||§

¶|178Provocative etymological-elucidation, would symbolize an inadequate-functioning, when method of life, continue an insufficient collection of human-effort-(s), resourcing patent-objective-form-(s), from-an philosophical-doctrine-cycling-(s), with American-White-Caucasian-Objective-inventor-concepting influenced through law,| 589 |from Treasury-Federal-Reserve-Holding-(s)) & Chinese-Outsourcing-Natural-Tectonic-Resources-Production-(s), inclined discourse, as truth of Pre-ßavarian-origins, monitoring those [component-(s) / mechanism-(s) / conveyer-(s) / shipping / handling / ordering-(s) / et cetera...] requiring an basis of sweat-labor-effort-(s), where Americans' are under an imperial-duty-(ies), for blue-collar-effort-(s), similar to white-collar-labor-(s), naturally identifying specimens of nature, under pseudonym bi-partisan constraints, ordering factory-commercial-reproductive-product-(s) from-an minimalist-mentality, preferring dating-mentality, idle from work-place-order-(s), fatigued through encountering-(s) at an impasse of volume-current-purchasing, conducting free-will-objective-purchasing, weekly occurring, under fifty-two-weekly-intervals-year, resolving elder-learning, from-an historical free-will-tranquility, affirming objective-(s), per ones'-physical-demeanor, serie-(s)-aptitude-(s), by-an vague-primary-basis, referencing observational-article-(s)-subjecting, yet not instructing an formal-merit-(s), diction by Vernacular-Perceptive-Observational-Embouchure & Term-(s)-Memory, identifying composite-modified-mass-patent-objective-developed-tangible-(s), from element-(s), designating objective-product-component-structure-(s), rather than, totalitarian-rule, counting rigorously, each census-populant-objective-(s), tiering patent-product-objective-notion-(s), subsequent subjective-merit-(s), while each passing moment, interval vapor-(s), from centripetal-force-impact, affecting how [hourly-timed-location-(s) // dating-temperate-effect-(s)] identify numerous-superfluous circumstance-(s), not factifying opinion-(s), for presence conceive physiological-perceptive-sight-sounding-(s), when independent-individual-ignorance, over classify, coherent-perceptive-mass-purpose-(s), comprehending outcome-(s), at locational-routine-(s), when-where-whom,| 590 |(be an) figure, pre-present-archaic-perceptive, common-impasse-intersected-metropolitan-routines, as-when commercial-success, guide-point-(s) of interest, matured through, introverted-perspective-offset;[330] uninclined [to / from] location-(s), using auxiliary-device, to over-velocity, hyper-active-autism, not coherently-still, for tensing-movement-(s), feeling over-exerted, due to physical-fatigue, not walking intermittent respirational-cardiovascular-rotational-pressure, synchronizing observation-(s) of momentum, according to hundredths-three-hundred-degrees, proportional by One-Second, as decimal-equivalent, of one-degree-four-minute-(s)-two-hundred-forty-secondths, & centripetal-force-degree-synapse, to the hundredth-decimal-whole-integer-count-posit, {1-S. = 0.0041667-Sy.De.-(four minute-(s)) = 0.0002778-H. /or/ 0.01667-M. /or/ 0.00001157407-D.(0.00000001822822-Y.)—-time-posit-(s)-apose-date-rotation-impasse-deviated-functional-tic-operation-denominator-integer-mode-notion-(s)-frequency-deviation-count-(s), or secondths-decimal-deviation-(s), into whole-count-(s)-impasse-(s), deviating from measure // weight of One-Whole-Count-(s), proportional by objective-mass, equivalent by-an variated-scaling of, Centripetal-Force-Synapse-Degree-timbre-operation-mode-integer-count-(s)-(counting motion-(s) by date)}-{1H. = 0.041667-D. /or/ 0.005952381-W. /or/ 0.0013889-.M. /or/ 0.000462963-S. (Q.) /or/ 0.0002314815-B.A. /or/ 0.0001141553-Y.-Dating-Denominator-decimal-percentage-fraction-count-(s), deviating celestial, degree-approximate-distance-(s) // (estimated)-solid-degree-measure-density-weight-(s)} of solar-kinetic-friction-(s), affirmed from select, perceptive-coherent-few, whom can respect desire, for comprehending subjective-purpose, among articlized market-serie-(s) of factory-product-order-(s), cycling capable commercial-market-dynamic-order-(s), by either managerial-superior-(s) or higher ranking-individual-(s'), rather than myself, whom [write // type] non-fiction-affirmative-article-(s), apose reader referencing, [where // when] reference, (be a the) primary-principle of all [language / literature / studies / off-time-inquisition] otherwise (from our) lords' biblical separation, (due to) ßavarian-rule, (by an) concertion of historical-ambiguity, (upon an) unconscious psychological-revelation, practicing an societal-repressive-minutè-catering-coverage, not liking an origin-linguistics-structure, stagnant through trade of article-(s)-object-(s), constitutional, by the body-(ies) of physiological-specimen-(s'), ratified elemental-commercial-product-modification-(s), not genetically-cooperated, reasoning without government-dollar into cent-(s), counting [from notion-(s) // per] traded-effort-(s), linear by each individual belief, not stressed-focusing, [mathematics / socialism / linguistics / genetic-mutual-loyal-mass / order of operation-(s) / reference of article-object-(s) / scientific-intelligent-experiment-referencing / ratified-constitution-(s) / literary-article-(s)-interposed-reference-citing-loyal-respect(s)] validifying data, when there are material-location-(s), in focus of perimeter-parameter-schedule-function-(s), continuing agenda, (through an) natural-inherency, falsely-efforting action-(s), (in an) superior-belief, cursive-(less), from generation-continuum, deliberately cronyized-illiterate-militarial-political-practice-(s) from youth, (through an) adult-physiological-comprehension-(s), informing lesson-matter-(s), (by an) article-(s)-usage, upon physical-objecting, not subjecting-physical-objective-(s), pertained where perimeter-parameter-(s), require extraspective-interaction-(s), by median-day-week-observing-(s), apose-upon range-season-year-observing, sequential per monthly-reflective-interval-(s), exacerbate-extenuating, particular-fundamental-article-(s)-data, when community-boast from issue-(s), rivalry-(ies), for retaining product-good-(s), grandeur relative-genetic-population-(s), periodicalizing-sub-serial-copyright-bordering-means', that offer an cultural-progressive offset action-(s), formally-concerting reference-(s), bibliography-book-retaining-data, for identifying perspective-distance-limit-maximum-physiological-impasse-extent-(s), concerting effort-(s), loyal by objective-interest-(s), designated, for modifying perimeter-purchased-land, when their (be an) reasoning of objective, for [1constructing // 1resourcing tectonic-composite-value-(s)] cycling each physical-motion-(s), awhile daily-hourly-impasse-(s), personally-have an deterioration apose momentary-minute-(s)-action-(s), affecting how perspective, develop, (upon an) series of means, based when [interior / exterior mode-(s)] continuum throughout political-county-boundary-(ies)-focus-(s), for assuring formal-Federal-Congressional-Senate-House-Legislational-Patent-Register-Operation-(s),71L. fluently impassing daily-article-(s)-sociological-developing...¶|||05m.:07s.49m.s.|||§

§..Preposition-Count-[22 / Twenty-Two]-of.. ..Article-Essay-[6 / Six].. ..Book-Essay-[178 / One-Hundred-Seventy-Eight].. ..Lines-[71.505 / Seventy-Five.1.five-tenths-2.zero-hundredths-3.thousandths] ..Voice is an argument, for of a the physiological-anatomical-compositional- beings-offset-(s).¶|||07s.54m.s.|||§

{Article:} {August}

¶179 Inherent-comprehension, embodies general-socialism-conduct, requiring an [leader // manager], to calculate daily-product-order-(s), not organized in anticipation of following-season-development, when free-trade, rely an international-effort-(s)-basis, to create commercial-market-trade-value-(s), through mass-notion-(s)-conjunction,| 591 |outsourcing-effort-(s)-compromising, by peer-interposed-comfortable-interaction-(s), simultaneous-calendar-offset-aging, not sweating per daily-discourse, through hours-minutes-secondths-dynamic, leaving an loophole in Legislation & Basic-Human-Right-(s), identifying those issue-(s) from individual-vehicle-cooperation-(s), (V.I.N.) leaving an judgement of blue-collar-sweat-effort-(s), (as an) evident error of managerial-trade, not [extracting // watching // judging] working employees-(s), where job-process-(s), require motion-(s) for trade, but when continuing the disregard of human-behavior be, using-objective-means, (would have an) congressional-legislational-literary-practice-(s), for intelligent-interposed-inquiry of dating, fluently affecting motion-(s), (by an) individual-primary-basis, when purpose of tense-(s)-evocation-(s), sharply contrast, skin-sweat-interior-fluid-(s), accumulating fat, interior ones' hyper-tense-(s), amid celestial-posit-(s), separate from kinetic-solar-pressure, affecting how our temperate-circulation-perspective, posture-stance in character, for perceiving [exterior-colors // shapes // qualities // et cetera..] continuing activities, for enabling subjective-objective-significance, upon ones' physique, enjoying the ambient, in fervor, when tense-(s) not be descriptive of those [requirement-(s) / want-(s) / collectable-(s)] for process-extenuating, product-material-(s), in lieu of article-function-(s), from tectonic-plate-posit-(s), retaining elemental-compositions, pertained specific-concept-developmental-function-(s), when visual-effect-(s), emit an median-means', for comprehension-functioning influenced from direct-objective-vision-(s), characterizing exterior-peripheral-surrounding-sight-(s), proportional [air-space / cement-rock-space], calcining in-an double-numerative-fashion, juxtaposing numerative-article-(s)-objective-(s)-value-(s), having an healthy amount of exertion, be imposed, for freshening a-the physiological-digestive-tract, for appreciating comestible-value-(s), pertaining an confluent-order-(s), finding invigorating-schedule-routine, intermittent work-order-applying, sequence-(s) of kinetic-photosynthetic-visual-sound-taste-smell-hypertenses-muscle-reflexive-day-night-cycle-repetitive-motions-posit-articlizing-(memory), enduring climate-condition-(s), as challenge-(s) of solar-kinetic-combustion-radiation-energy-infin-itive.261¶|||01m.:44s.75m.s.|||§ As summer remain an intensification of solar-heat-friction-plane, temperate-measuring-(s) of climate, where local-terrain, serve an basis of adapting each fragment of kinetic-vapor, directing photosynthesis of [<u>tree-(s) / plant-(s) / culturing-(s) of organic-matter</u>] from-an water-transfusion of vapor-pressure, upon those natural-element-(s), that would be-an part of an pace for conducting each trade-matters, formally using objective-article-(s), in part of an fluent-sequence of activity-(ies), progressing conducted-issue-(s), among an surrounding realm of environment, influencing human-effort-(s), from developing element-(s), marking genetic-disposition-(s), adapt concertion of locational-scouting, by American-view, transitioned statehood-reset-trading, asserting how group-(s) of inhabitants', have-an genetic-political-order, intent on formal-developmental-reformation, factoring product-commercial-composite-item-(s), from-an view of aspectual-design, reconjuring mechanical-concept-component-(s), worked on comprehending-data, as ethical-reproductive-time-frame, elapsing year-(s) of schedule-cycle-(s), constraint constant-schedule-rotation-impasse-(s), fragmenting dating, when their require application of object-(s),| 592 | requiring legislational-ordered-act-(s), per objective (construction-material-(s));331 count-secondths-elapsing, from numerical-alpha-time-count-interval-(s), perimeter-parameter viewing, an cardiovascular-exertion-limit-(s), balancing sequential-presence-impasse-defining, populant-action-(s), determining weather an applicant is value-(d) or disregarded, when directed through formal-fact-(s) of population, where relative-discourse, affirm affair-(s), when discourse of action-(s), validify-object-(s), from-an geoanaalgemathmatical-reasoning, individual-component-purpose, for thinking how to persuade concepting-sociological-developing, by commercial-daily-impasse-action-(s), whenever life demonstrate an needs' from subjective-inductive-input, ascertaining function-(s), for invigorating-intrigue, appreciating what thoughts of others mean to oneself, as when each fact matter, by article-(s)-directive-discourse, formulate-an means' of modern-impasse, creating feature-(s) for objective-article-(s)-function-(s), at land-location-(s), governmental-ly-document-ing, each notion-(s), of mass, from [physiological-perceptive-presence // objective-article-(s)-mass-matter-usage // perimeter-parameter-land-dynamic-cultivating]52L. when dating have an purpose for clarifying reference of notion-(s), from when hour has yet been incorporated in documenting, from progressive-impasse-sequ ence-(s)...¶|||03m.:34s.10m.s.|||§

§..Preposition-Count-[33 / Thirty-Three]-of.. ..Article-Essay-[7 / Seven].. ..Book-Essay-[179 / One-Hundred-Seventy-Nine].. ..Lines-[53.46 / Fifty-Three.1.four-tenths-2.six-hundredths].. ..I Verse each chord of vernacular-essay-context, for understanding how to arrange each notion of visual-tonal-vibration-(s)-tiering, so inbetween context, I allude moding of mathematical-offset-predicate, aligned time-stopwatch-counting, deviating how to label physics-geological-ecological-terrain-spacing-(s), not simply [extract (elemental-compositional-metals-soils-clay-rocks-tectonic-mesh) / cultivate (Irrigated crops / plants / trees) / Plan (human-mass-population-default-government-agenda) / Designate (pop-ulation residential-mass-offset-schedule-timing) / (Patents-Factory) Agenda] from an nature of individual-off-time-prerogative, not extraspective by familiar-intrigue from an genetic-population-outview.¶|||33s.78m.s.|||§

¶[180]During each dating-(s), how are hour-(s) influctuated by subterior denominative-impasse-currency-note-notion-(s), arranging cubic-surface-area-locational-land-space-(s), interval minute-(s)-comprehensive-motion-(s), where successive-action-(s), denominate from secondths-motion-(s)-redacted-progress, angling grades, retrograde motion-(s)-act-(s), constantly-offset-unconscious, apose subconscious-observational-deducting, referencing word-article-glossary-point-(s), as segment-(s) of location, increment-continuum, amid time, under day-week-hour-progressive-order-(s), unconscious by-an stress-fatigue-cycle per agricultural-equivalent-matter-comparative, when blood-pulse be Animalia-comparative, rather sight-sound-perceptive-intonated, at-impasses of observation, respirating-blood-pulse, fatigued-physically, apose subjective-article-reference-retrograde-comprehension, understanding how Dating & Hour-(s), deduct from [extent // notion-(s)] focusing concerted moment-(s) of particular-focus, for how fatigue applies, time-frame-interval-(s), for-an capable effect of notions-extent, intermittent centripetal-force-simultaneous-pressure-adjustment, when dating perspective-possessive-offset, for how community-(ies)-function, by-an standard of living, repetitive because how ungiving possessed-soul-remain, [understanding // comprehending], elemental-tectonic-plate-(s), thus affected from continuum-effect, without incurring with individuals', involving oneself, at group-activity-function-(s), motioning by merit, lesser-effort-(s), made by those subordinates to-an major superior, when environment-setting, place activity-(ies), functioning in order-operation-(s), through-an method of default-motion-(s), adapted-from commercial-trade-act-(s), due to governmental-currency-trust, in confidence at locational-posit-order-action-(s), affecting how purpose is meant from effort-(s), offset time-date-calendar-reset-day-night-rotation-cycle-(ing-(s)), meeting at executive-cabinet-agenda-basic-human-rights-legislation, default-requesting, fashion-(s) of [work / relaxing / interacting / cleaning / socializing et cetera...] juxtaposed characteristic-(s) complicit for non-sweat-expectation-affects, amounting-lackadaisical-perform-ed-task-(s), signifying-effort-(s), as individuals' > 23L.[community // family]... ¶|||01m.:25s.28m.s.|||§

{Article:} {August}

§..Preposition-Count-[7 / Seven]-of.. ..Article-Essay-[8 / Eight].. ..Book-Essay-[180 / One-Hundred-Eighty].. ..Lines-[23.115 / Twenty-Three.1.one-tenths-2.one-hundredths-3.five-thousand-ths].. ..Convincing is not in this book, after my excursion of Los Angeles, I realized from Fort-Benning, that without fine-referenceable-context, no mass can define an purpose from university-studies, for serializing lifestyle-mathematical-population-survey-context, thus I intend contexting in tier, various fashions of year-dating, layered by essay-redact-tier-subjective-function, expressing how human-mass-survey-(success / failure), affect outcomes for lineage, not in rich fashion, but poor from an silent-mentality for living.¶|||29s.37m.s.|||§

¶181Everyday is not fluently in an extenuation of article-(s), from-an possessive-clasp of article-(s) identifying longevity of genetic-integration-(s), where elemental-resource-product-(s), ship to factory-order-requested-entity-(ies), in an realm of purchase-(s), as judgements of community-integration-character-(s), have forgotten how each daily-sequencings, perform throughout [customary // traditional-action-(s)] embedded by those preferenced-opinion-(s), not contorted because how daily-hour-interval-(s), are suppose to implore an continuum-impact, when an dispersion of matter, not practice comprehensive-vernacular-fluent-commercial-operation-(s), because of how personal-perspective, retain when-whom-why, what pertained (as an) interest of perspective, believing from theology, mentality as perspective (illiterate-slight), from when sufficient-adequate-inference-(ing), understand managerial-order-(s), without employee-colleague-interaction-(s), where positioning objective-(s), are designated by an governmental-agenda, municipally-tracked, defining goal-(s) of retirement, by-an religious-social-judgement, serie-(s) subjecting, topical-article-(s), intended for [refining // specifying] paragraph-section-(s) about perceptive-hyper-tense-(s)-physiological-lifestyle-direction-(s)-observation-(s), where-when group-(s) of citizen-inhabitant-(s), remain in an mass-notions-extent-counting, as population-portion-control-specimen-(s), regulate executive-cabinet-agenda-legislation, by-an congressional-ambiguity, through post-auto-biographical-lifestyle-literary-interpreting, not find-ing new literary-interposed-procedure-(s), from congressional-house-representative-mass-executive-agenda-regulated-legislating, from when Executive-Branch & Judiciary-Branch, define the method, of Egalitarianism, when I practice that through an remodifying congressional-non-fiction-statistical-grammatical-punctuation-population-repopulating-control-reproductive-literary-documenting, from literate-citizen-analysis, sustained use of books, in House & Senate-Congressional-Hall-Session-(s), referencing from paragraph-section-(s)-article-(s), an attempt to convince various outlet-(s) of power, for how to create new [constitutional-(article)-(s) // amendment-(s)] for percentile-tiering (inter)-national-census-population-(s)-control-referencing, tiering each decade-census, apose each aging-tier of commercial-retirement-income, balancing from Youth-Education-Scores & Post-Military-Sixty-Two-Year-Enlistment-Aging-Retirement-Impact, an restraining of both aging-subjunct-era-(s), as well as commercial-mid-age-living, [annexing /// housing /// mansion-(s)] based on decimal-tenth-percentile-career-types-(s)-development-conceiving Totalitarianism, conducting-an formal-national-genetic-reformation, where at an unconscious-abiding, from international-politics, never literary-observing-documentation, through subjective-article-(s)-modification-(s), tiering loyalty of other-(s') from national-extent, to then draw up bill-(s), implementing annexing of national-individual-(s), for practicing national-resourcing, alleviating our [0.German-{Bundestag Republik Der Deutschland} /// 0.Japanese /// 0.Chinese-{Zhang Zhou}-outsourcing-reliance], from national-sixty-percent-F-rating-Class, tiering those other position-(s) of power, by [1.D-Class // 2.C-Class // 3.B-Class // 4.A-Class-Metropolitan-City-Offset-Civilization-(s)], centering major-City-County-Development, by F-Class-Citizen-(s), tiering their class by [5.G-Class // 6.H-Class // 7.I-Class // 8.J-Class // 9.E-Class] relevant to Federal-County-Development-(s), practicing an transgression of working annexation-(s), to reduce international-outsourcing, by-an national-restricting, using literary-congressional-control-documenting, to then, relocate-housing, by rural-exterior-county-(ies)-life-style-(s), for balancing how power is by transportation-coverage-factor-(s), influencing how international-classification-(s), would require an tier of [10. K-Class /// 11. L-Class /// 12. M-Class /// 13. N-Class /// 14. O-Class /// 15. P-Class /// 16. Q-Class /// 17. R-Class /// 18. S-Class /// 19. T-Class] as an suggestion to invert how outsourcing has been, because of an lack of [sweated-effort-(s). // congressional-library-development-(s)], variating why America simply give power away, ridiculously not controlling those value-(s),| 593 |per mass-factory-patent-register-product-production-(s).262¶|||02m.:53s.75m.s.|||§ In my life, a-the prior not occur, as I intend to formulate Five-nation-(s), of Five-Million-Citizens, throughout [Canadian-Montana-Geological-Fervor] working a-the white-intelligent-feminine-perfection-position, to then extract literary-effort-(s), from fine blonde-white-men-repopulating-technique, working on an prior societies, by-an common $50,000 per year- adult-individual-(s'), working an literary adult-repopulation-offspring, then influctuated-children-repopulating, using literature to inflate monetary-position-(s), of loyal-objective-followers, thus, trying to sarcastically-spite the repopulating-method impassed historically-set, expositorially-explaining self-centered-ignorant-nature, as how humanity remain martially-heterosexual-(flat), from how opposite-sex-intercourse, not matter to each independent-individual-(s), relying on international-outsourcing, blatantly explaining an neutral-prose, for how truth of natural-routine, continue by-an trust of trades, not interesting, due to common-compliance, making sure an superlative of statistical-fact-(s), make Commercial-Sales & Congressional-Sales, working on publishing-an book per year, after earning my first period of $25,000, complaining of blue-collar-labor, every way possible, apose sweating-vigorously, by an respect-(s') to [nature / peer-(s) / neighbor-(s) / co-worker-(s) / pupil-(s) / elemental-extraction / objective-concept-developmental-method] analyzing those factor-(s), for why national-resourcing is not possible, with primary-vehicle-machinery, because how [no jack-hammer // old-(primary)-fashion-shoveling // other-technique-(s) /// tool-(s)], intricate resourcing, rushed by dream-imagery-product-market-impasse-(s), when daily-evocation, numbly notion, serie-(s) of task-(s), idolizing executive-default-agenda-order, itinerant mundane-practice-(s), not article-(s)-modifying, where governmentally-birth-(ed)-religious-language, follow through of aging by commercialism, in droves of masses, focused by-an routine of effort-(s), acted on, for maintaining objective-(s), when product-trade, serve an time-frame of calendar-periodicalizing-(s), affirming those piece-(s) of matters, when middle-age requirements, have politics suggest how to have children, when limited of agreement-(s), unclear from how bias-partial-possessive-opinions, matter by-an Default-Means' & Return-Mean-(s), under [solar-magma-kinetic-friction-rapid-speed /594/ lunar-tectonic-rock-rotational-horizontal-celestial-impact // above-tectonic-plates-oceanic-earth-core-composition...] surrounded by constraint-(s) of, national-trade-dynamic, enabling an historical-Gregorian-Calendar-observation-(s), not learning from education, or pre-education-constitutional-article-(s)-documenting, idle-population-means', apparently focusing information, when acting on any-impulse, not capable of designating natural-posit-purpose-(s), combining-product-component-(s), sequencing intermittent [eating /// digesting /// dispelling] daily-locational-scheduled-task-(s), substantially, placing piece-(s) of [hardware // software data] intended to maintain our natural-posture, when usage of instructed-data, apply inhibition, rather diligence, by-an nature of humanity, unloyal to communion, from-an lackadaisical-efforts, comprehending-data, upon objective-product-effort-discourse, amassing the dynamic of religion, mattering through fixated-perimeter-surface-area, applicated in routine of successive-event-(s), interpreting from-an method of visual-tonal-perspective, interpreting subjective-comprehension, influencing commercial-trade, when at an site or interior-design constructive or designated-subjective-article-posit-(s), cursive-contexting, hyper-tense-observation-(s), by-an limited-physical-tense-perspective, [to / through] parameter-(s), of dimensional-article-subjective-purpose, defining observation, where conscious-function, constantly-peripherally-perceive, kinetic-centripetal-force-rotational-pressure-impacts, as ambient-impasse, be similar to youth (plant-like-growth), instilled by parental-nurturing, until capable-consideration, conduct thoroughly, ones' own notion-extent, by focus of-an numerous venue-(s) applicating data, through an interacting-experience, developing an relations between other inhabitants', conjuring thought, for juxtaposition by timed-one-hundredths-eightieth-secondths-interposed-interval-individual-conversing, influencing periodical-sequence-(s) of traded-product-value-(s),88L. enamoring hyper-tenses juxtaposed balance-denominative-awake-day-evening-night-scaling, invigorating moment-(s), invigorating pizazz, by live-physical-hyper-tenses, while-upon existence...¶|||06m.:11s.88m.s.|||§

§..Preposition-Count-[46 / Fourty-Six]-of.. ..Article-Essay-[9 / Nine].. ..Book-Essay-[181 / One-Hundred-Eighty-One].. ..Lines-[89.95 / Eighty-Nine.1.nine-tenths-2.five-hundred-ths].. ..Operation-(s), continue an meander of stress, I have encountered, wonder in context, weather formal-non-fiction-genres, every be developed, from population-mass-currency-trading, when motion-(s), of timing, juxtapose dating, from how human compositional-physique define an continuation per individualism-industrial-development-error-(s), not sourcing an copyright defining, why each practice matters upon objective-circulated-entity-(ies), for orchestrating sleep-patterns, from an mode of mathematical-scale-range-distance-factoring, from how title-restriction, is applied through genetic. ¶|||28s.34m.s.|||§

¶182Every county I presume (has an) [1local-dump / 2county-hall / 3police-force / 4firefighter-station-(s) / 5hospital-(s) / 6sewer-system

/ [7]Federal-street-thoroughfare-(s)-police-maintaining / [8]residential-commercial-entity-(ies)-planning / [9]interstate-commercial-entity-(ies)-trade] follow-ing residential-commercial-religious-development, where offset-capabilities, are acted upon national-interstate-trading-method-(s), climate-offset, implored for habitual-impasse of product-(s) advertised at particular-local-market-(s), without subject-discourse, for-an dynamic by [corporation-production / local-private-customary-venue-(s)] amounting up collection-(s) of product-(s), in ordering-(s) of commercial-national-factory-plant-(s), focusing from social-corps-dynamic, instructed-manual-booklet-referencing, without literary-referencing, those weight-(s), [elemental-tectonic-composite-resourcing // disposal] amounting those [intake-(s) / collection-(s) / storage-(s) / disposal-(s)] of elemental-commercial-fabricated-product-exchange-(s), cycling matter-(s) of population, proportional subjective-thought-write-discussion-interaction-method, placing objective-(s) upon an lend-land-surface-area-space-entity-surface-area,[13L.] in repetition of impassing-schedule-(s), admiring those pretense-(s) of count-interpretation-(s), passing year-(s), in reference of pronoun-blood-pulse-tense-invigoration-(s), harmonically-discorded, where which moment-(s) of living-impasse, influctuate discourse of [event-(s) / venue-(s) / office-(s) / et cetera...] while no biblical-central-topic-article-(s)-lesson-subjective-coordinate-(ing), perspect-ive-focus, rhythm-feeling interpretation-(s), mattering by focus upon [stance / formal, spaced-sitting-(s)] understanding how points of view, allot significance of [affair-(s) / relation-(s)] requiring point-(s) in surface-area-perimeter-parameter-(s), measuring materialized element-(s), from when commercial-governmental-dollar-social-dynamic, be at an posit-purpose-(s), for [[1.]function / [2.]observation / [3.]conjure / [4.]reference / [5.]idiom / [6.]vernacular / [7.]placing / [8.]modifying / [9.]forming / [10.]construing / [11.]balancing / [12.]scheduling / [14.]collecting / [15.]incorporating / [16.]ordering / [17.]offspringing / [18.]catering / [19.]commenting / et cetera...] from when modification-(s) of effort-(s), show an deductive-reference-process, for how momentum-sequence-(s), rotate by-when-of cool-tectonic-plate-(s)-composite-metal-rock-taxonomic-classifications-aside-oceanic-water-still-velocity-rotation-pressure-(s)-above-earth-lithosphere-asthen-isphere-core-magma-volcano-like-molten-central-velocity-revolution-density-pressure-(s''') = Earth;332 inverted-apose, solar-exosperical-molten-circulation-kinetic-hyper-velocity-circumference-revolving-densities-central-vapor-gaseous-core-spacing-pressure-(s''') = sun, revolving-apose, lunar-cool-molten-tectonic-rock-singular-revolving-revolutionary-composition-(having no metal or water, on the moon) = moon, upon segment-(s), calendar-day-ordering, juxtaposed time-hour-(s)-activity-function-(s), considering numerous amount of notion-(s), (from an) influctuation-repetitive-median-counts-extent-paces, affecting each moment by-an method of prose, of distance-measure-existence...¶|||02m.:22s.00m.s.|||§

§..Preposition-Count-[17 / Seventeen]-of.. ..Article-Essay-[10 / Ten].. ..Book-Essay-[182 / One-Hundred-Eighty-Two].. ..Lines-[31.95 / Thirty-One.1.nine-tenths-2.five-hundredths].. ..I hate how boring life exists, but understand from an lack of communication-(s), how each individual affects an circumventable-outcome, that if I can not influence those mathematical-plans, for convincing other individual-(s), then grade-point-average-thinking-deviation show an illiterate social-enabling, that remains unethical, due to those processes of common-interaction-(s).¶|||20s.50m.s.|||§

¶183Count & Distance, affect Grade-Point-Average & Distance-Measure-Extension, When & Where, elongated-infinite-effect, remain limited by-an transit-tectonic-plates-regard, except for genetic-distance-reformation, upon dispersed-populations where idle-idiomatic-sensation, primarily evoke those process-(es) of possession, abstracting distances, as how logic remain comprehended upon interaction of means', perceiving focused intermediate-interpretation-observation-(s) inclaritive-subjective-article-(s)-synthesis-examination, attempting to find out, weather effort-(s) without sweat, at work-place-environments, justify introspective-bias, opinionated by extraspection, from how socialism-perspective, pass time, enveloping experience from consistency of redundant-article-(s)-objecting, putting those effort-(s) of ambivalence, for ordering sequential-location-operations', rather than indifferent capacities of act, annotating the nature by-of blood-pulse-invigoration-(s), subjecting pattern-(s), from the need to survive, when impassing centripetal-force-momentum, by-an focus from mode-(s), periodicalizing article-topic-(s), invariabled, consistent offset hyper-tenses-interpretation-(s), for becoming affluent upon an mass of interactive-inhabitant-(s), working on one another-(s') literary-bias-best-interest-(s'), working from repetitive-confidence, amongst locational-posit-(s), pronoun-observation-literary-writer-defining, all encapsulating hyper-tense-(s)-reflex-(es), accommodating focus by-of blood-pulse-physical-hyper-tenses, emphasizing-impasse-effort-(s), prosing the sake of setting, in fashion of presence, attempting to first see weather an mass-group, can figure out physics, and then investigate, all the tectonic-plate-(s)-desert-(s)-forest-(s)-marsh-(es)-(swamps)-jungle-(s)-plain-(s)-mountain-terrain-(s)-tundra-prairie-(s)-shoreline-(s)-topography-condition-(s) offset ocean-sea-lake-river-canal-aquadect-water-bodies-latitude-longitude-miles-meters-feet-length-witdth-altitude-rectangular-prism-square-coverage-estimate-(s), for if we are not well to do, what activity-(ies) might an inhabitant participate in their lifestyle, while unreferenced-inferencing, reman by-an timbre of [mind / body / soul] because antedote-(s) of idle-glut, embellish those fact-(s) of interaction, when during nation-perspective-point-(s)-offsets, settle through view of commonality-repetition-(s), rather such homogenized-ethnic-claustrophobic-transit-annexing-municipal-metropolitan-conjunction, repressed-ignorance, retarded from push of subjective-emphasis, deducting from posit-(s) of objectives upon individual-(s'), lacking an totalitarian-reference, conjuring why commercial-means', not library-congressional-control-number-purchase-hoard, for how egalitarianism abide by executive-cabinet-duties-legislation, following descendent-(s) of lineage, which offspring by enabled superior-genetic-quality-(ies), comprehending-data, by-an relative-means', counting functions of notion-(s)-exponent, occurring to person-(s), article-(s)-paragraph-section-(s)-segment-arrangement,| 595 |for scheduling agenda-public-private-location-(s)-objective-(s), sequestrating cubic-surface-area-feet-spacing-(s), apose meter-mile-(s)-distancing-(s), using subjective-qualities, ranging direct-approximate-blood-pulse-feeling-sensation-tenses, where arrangement direct-arrangement-(s), progress pattern-(s) of elapse-(s), for marking checkpoint-(s), in part of either [addition / subtraction / pre-planned-factoring / pre-conceived-zoning-deducing / fraction-numerative-fragmented-unit-(s)-locational-observation // denominative-denotative-(10 / 100 // 1,000 /// 10,000 //// 100,000 ///// 1,000,000)-offset-unit-(s)-sum] analyzing from city-county-(ies)-objective-article-(s)-sums, in population-offset consideration, apose [whole-(s) / piece-(s) / decimal-fragment-(s) of commercial-article-(s)] for particulating amongst-along, calendar-date-hour-period-interval-count-(s), fragmented through secondths subterior time, considering estimate-(s) of motion-(s), through repetition-(s) of impasse-count-(s), psychologically-reflecting, period-(s) of [production // action-(s)-payment // community-interactive // documenting-mode-(s)-(et cetera...)] compre-hensive-editing, Individual-(s') & Reason of Mass, by elemental-commercial-product-exchanges, from revolutionary-survival, along-an basis of extraspective-mass-qualified-merit-(s), Showing & Telling, (from an) actual-basis of referencing of currency, through way of trade-(s), using article-(s), hence forth, where posit-(s) of application, merit-value-(s), from commercial-market-(s), allotting factor-(s) upon human-(s), premising an basis for judgement-(s), when community-character-(s), re-emphasize religious-center-(s)-purpose, interacting by opposite-sex-comfortable-opening-up-of, when repopulation-theatre, work intermittent those period-(s) of day-(s), where sequence-(s) remain limited, by independent-extraspection-reclusion, yet able to conjure through fluent-hoc-commercial-community-political-congressional-bill-regulating-mandate-(s), maintain-ed by an municipal-effect-(s'), directing free-work-load-(s), genetically offset, where an ultimate basis of effort-(s), innuendo-push, freedom by community-function-(s), rather currency-primordial-trade, instilling an unconscious patterns, infinitively-continued, by what merits', be conceived by-an longevity of community-offspring-generation-thirty-years-era-cycling-(s), understanding how worship of collective-member-(s), when notion-mass-population-extent-(s), suffice no more than 100,000-beings, per approximate capitia-surface-area-square-mile-(s), where their has not been an family-subterior-goal, developing from how city-street-sixteen-house-(s)-per-block-(s)-dynamic, would be better to reestablish an cultivating of [comestibles-practices // clothing-center-(s) // activities],60L. of [physical-land-terrain-coverage /// subjective-inquiry-establish-maintain-sustain-retaining // repopulation-cycle-(s)] transitioning method of lifestyle-(s), from metropolitan-work-living, harmonizing natural-effort, always considering in confluence of other families, the way of fatigue, and the impact of physical-posture, upon physics... ¶|||04m.: 47s.28m.s.|||§

§..Preposition-Count-[41 / Fourty-One]-of.. ..Article-Essay-[11 / Eleven].. ..Book-Essay- [183 / One-Hundred-Eighty-Three].. ..Lines-[62.6 / Sixty-Two.1.six.tenths].. ..Quantity & Proportion by food-groups, should be more focused on, developing each notion for daily-living, demonstrating how population is suppose to cooperate, by an means for living from self-voluntary-sufficiency, due to Parameter & Interaction-(s).¶|||17s.75m.s.|||§

¶[184]Timothy McVey & Columbine, remain an part of white propaganda, so entranced not concisely comprehending, from ignorance is bliss-perspective, because no guilt of common-living is understood by our requisite of executive-judiciary-independent-individualism-factoring, apose concept-factory-community-elemental-factoring, for how thought is not thought from politicians, population-percentile-classifications, sorting primary logics of mass, by Twenty-5%-groups {Jefferson-Cent ration Washington-Dollar};[333] working (from an) denominative-monetary-commercial-internal-revenue-services-Federal-Reserve-offset, of [supply /// goods /// demand /// services /// cultivation-technique /// literary-congressional-business-currency-denominating /// legislating /// tangible-effort-qualities-et-cetera… congressional-documents-tier-operations-balancing] following an time-frame, upon 10:32 p.m.-thinking, where work-stress-fatigue, is not recognized from such fact, amid [eating / housing / vehicle / phone / direct-family] where thoughts of other-(s'), remain conveyed, upon sequential-act-(s), conducting in an limited-city-county-ordinance-district, contained through community-populations, analyzing Extraspection & Faith, for whom worship-interaction-(s) among other-(s) presences, per-encounter-act-(s), intertwining motion-(s) per blood-pulse, from physical-hyper-tenses-evocation, intermittent-modes;[334] while motions of schedule, require an array of interposed activity-(ies), influencing perspective from minute-secondths-act-single-effort-elaboration-(s), superimposed upon calendar day-hour-dating, meandering function-(s) of mass, directing [festivity-(ies) / operation-(s) / affair-(s) / communion-(s) / study-(ies) / act-(s)] considering segment-(s) in [oral-oration / retort & (re)-inquisition] as article-(s) press-human-independent-individual-subjunct-(ing), by-an clause of currency, from condition-(s) of territory-(ies), Federally-monitored (inter)-national array-(ies) by [[1]week-issue-(s) / [2]monthly-issue-(s) / [3]seasonal-issue-(s) / [4]quarterly-issue-(s) / [5]yearly-issue-(s) / [6]bi-yearly-issue-(s) / [7]quaternary-yearly-issue-(s)] adventured from progressive-continuum-conjunction-observation-(s), from count-state-reasoning's, where each community remain an part of, uncoordinated political-default-coverage, against comparable handling of elemental-resourcing-(s), relying Federal-patent-commercial-mass-product-item-objective-(s)-concept-developing, remaining idle when impassing formal-recourse of interaction-(s), from Hellenistic-Hessian-behavior, remaining stubborn by judiciary-executive-individualism, thinking from family, not repopulation, how work-place upon church-communion-(s) of mass, find an discourse of objective-(s), lifestyle-state-reformation, upon collected-article-(s), subject-object-functioning, upon [entity-(ies) // territory-(ies)], through engineered-constructed-progressive [voluntary / funded-effort-(s)]| 594 |devising-an tectonic-transportation-elemental-patent-objective-commercial-concepted-trade, from unilateral-bi-partisan-politics, serving those travesties historical-from millennium-two-thousandths-year-calendar, affirming [popularity-(ies) / rivalry-(ies)-dichotomy-action-(s)], by-from-an synonym-individualism, coalesced where-at symbolism of Federal-International-Trading, feeling no pressure Perceived & Conveyed, from mass-population-requisite-history, an paradox-parallel-genetic-nation, by-when American-pre-Fifty-states-constitution-state-county-city-districts-constitution-analysis, rely on an Federal-Constitut-ion-International-Regulation-(s), tiering how secular-religious-sects, conduct, mass-resource-product-trade-factory-production-articlizing-dating-amendment-tiering, where-of [longitude-degree 110° / latitude-degree 51° (Montana)] not filter through [person-(s)-(personality) // being-(s)-(mentality)] an rural-terrain, for segregation-remodification, preparing how to fend-off, an mass-impasse-invasion, (from an) resourcing-method-(s), to comprehend how pre-retirement-commercial-communion-planning, assist conceiving vernacular-fluent-articulated-article-(s), for [review // redacting // (pre)-revise // dating-amending] tangent-syntax-interpretation-era-refining, adjusting when each era of humanity, concert from-a-the finest-religious-cursive-redacting, an centralized-saving-(s) by-an balance of Individualism & Genetic-Domesticated-Communion-(s), sort-ordering, an Federal-National-Constitutional-Subjective-Tiering of {Economic / Sociological // Militarial /// House-Representative-(s) //// Senate-Congressional-Literary-Legislational-Article-(s)-Stance-(s)], of [Psychological-Person-(s)-Work-Classification-(s) // Being-(s)-Work-Classification-(s)-(tiering from Primary through present)-(later offset omitting from genetic-review of Literary // Psychological // Medical-Documenting, apose legal /-/ judiciary-documenting-order-(s), denominating human-kind-bastardization-historical-territorial-resourcing-claim-aspect) / International-Religious-Obligated-Decade-Revise-Declaration-Document-(s)-{N.A.T.O. (North Atlantic Trade Organization) // U.N (United Nations)} ///// Ten-States-designated-Nation-Denominative-Territory-(ies) /// Tier-Land-Protectorate-Constitutional-Sector-(s), Denominative-Constitutional-Factor-Person-Property-Account-Documenting for every single-piece of Distanced-Territory / Population-Citizen-Being-(s) // Motion-(s) /// Action-(s) //// Work-Mode-Impasse-Count-(s)-Cycle-Documenting ///// Factory-Mass-Objective-Production, box-quantity-(ies)-Count-(s) // per-item-count-(s) /// measure-(s) / weight-(s), of each and every measurable-distance-impasse-territory-space-(s) // vapor-pressure-(s) // liquid-viscosity-(ies) // solid-density-(ies){5-(I)-Na.Con.Var. ///// 2-Psy.Var. / 1-p.Dec.-Na.U.N.Doc.Var. // 10-State-Constitution-(s)-Variable-(s) /// 100-Principalities-Constitution-(s)-Variable-(s) //// per national-family-lineage-cursive-constitutional-denominative-documenting}, following a-the fundamental-subjective-principles of denotative-paper-documenting,| 595 | [1.Grammatical-Punctuation-Literary-Prose / 2.Mathematical-Algebraic-Count-(s) / 3.Geomet-ric-Measuring / 4.Mathematical-Weighing / 5.Constitutional-Core-Article-(s)-Documenting / 6.Art-Currency-Note-Mode-Range-Documenting / 7.Retrograde-Reading, fundamental-point-(s)-Redacting;[335] Repeat of these prior Fundamental-Subjective-Principle-(s)] identifying mass-population, for monetary-documenting, upon territorial-objective before product-(s)-collection-identifying, product-(s)-collection-identifying before funding, funding before savings, savings after funding (funding = extraspective-probe-surveying / savings = introverted-confidence of reformation-calendar-year-(s)-anticipating), product-purchase-objective-storing before national-constitution-referendum and then an continuum of [[1]engineering / [2]constructive / [3]subjective / [4]conceptive / [5]objective / [6]communion / [7]social / [8]festive-(ities)] for-an territorial-psychological-mass-scheduling-agenda-calendar;336 put upon an self-genetic-mass-pressure, upon tectonic-plate-continuum-constant-continuum-reproductive-development-(s), anticipating-an centralized-terri-tory, reasonable for furthering an constituting of each title-land-surface-area-spacing-(s), by family-parental-cursive-comprehension, upon visual-vernacular-reverberational-sound-wave-key-signature-pitch-metronome-volume-perspective-observation-(s), per genetic-extraspective-mass-collective-effort-(s), surveying physics-theatre, per offset of genetic-mass-populus... ¶|||05m.:15s.00m.s.|||§

§..Preposition-Count-[31 / Thirty-One]-of.. ..Article-Essay-[12 / Twelve].. ..Book-Essay-[184 / One-Hundred-Eighty-Four].. ..Lines-[75.62 / Seventy-Five.1.six.tenths-2.two.hundredths].. ..When I was learning in school, I noticed children have an issue studying, when dating impasse so quick, and timing even faster, not learning from deductive-moding-inductive-scaling, how dating is applied apose celestial-physics, when rule of thumb signify predicate-lineage-subjective-interpretation-(s), from spacings of location, not yet formally-documented, tangent by how compositional-physiological-being-(s), move from an range of introverted-competency-extent, sufficing how daily-prerogative-(s), instrument each notion-(s) of daily-living.¶|||30s.25m.s.|||§

¶185Momentum from-of centripetal-force, revolve revolutionary, where-of-an natural-impact, when grandiose-celestial-abstract-universe, impasse upon existence, too great, to be felt by ones' natural-vapor-respirational-ventricle-blood-pulse, handling physiological-compositional-contained-vernac-ular-vocalized-sustained-visual-sounded-retained-subconscious-perceptive-conceptive-pertained-extra-spectively-substantially-cardiovascular-maintaining-motion-(s) / action-(s), retaining, perceptive-paper-bound-notion-(s), in line of municipal-county-bill-mandated-code-(s), for-an basis by free-will, between-abstract, celestial-untangible-(s), when posit-(s) of government-currency, count free-sociological-commercial-specimen-(s), by-an [Military-Police-Sociological-Executive-Branch //596// County-State-Federal-legal-Judiciary-Branch], watching the two internationally, for-of-an evident-bastardized-illiteracy, freely-examined by American-Ambassador-Leadership-Mundane-Observation-(s), as hedon-citizens, prefer (from an) primordial-historical-nature, to abide by order of national-free-enabled-judicial-complaint-notion-(s), basing perspective upon physics, rather partaking in an active-physical-perceptive-subjective-interpretation-(s), understanding how [Congressional-Literary-Federal // State-Senate-House-Legislational-Order-(s)] remain negated during those impasse-(s), denominative-parchment-paper-documented-action-(s), inputted from observational-decisions (not opinion), those independent-piece-(s), pertained-an confluence of contained-community-lived-motion-(s), not understanding Schweiz-Psychologie, upon legal-mass-populational-ink-cursive-paper-physical-motion-(s)-denominative-documenting, replacing barriers built from governmental-secrecy, when resources of [college-ruled // wide-ruled, loose-leaf // composition-binding-paper-(s)], have now in the millennium-shift, been over-disposed, from when historically, law initially carefully-documented all those cases of social-religious-illiterate-mythology, which makes me think, legal-judiciary-law, is over-done, because of-an neglect of paper-lifestyle-documenting, in course for metal-petroleum-jail-records-documenting, as when judiciary-legal-grounds, not wish to change, an leave medical-objects, to an military-states-presence-(s), because of how still-numb-compulsive-movement-(s), rely on [1.tradition // 2.heritage // 3.customs // 4.favor // 5.belief // 6.mysticism // 7.faith] mattering by-of default-monetary-impasse-motions-sequence-(s), relying on university-multiple-choice-thinking, rather then rigor of daily-development-(s), for how over-congestion, has not understood literary-commercial-copyright-repopulating, putting more efforts in to an pre-thousandths of dollars-youth-transitioning, swearing technical-application-(s) by parental-enabling, rather labor-effort-(s)-savings, book-(s) [copyright-documenting // serializing {$85}-{$85}];[337] imploring from blank-historical-congressional-non-fiction-(s)-genre-(s), an tier of factor-(s), evident by how judicial-newspresse-obituary-continuum-documenting-occur, unspecific of those issue-(s), to become an part of segmented-periodical-ordering-(s), when independent-individual-(s), affect driven-slight-impasse-(s), when compositional-matters, denominate, breaking-period-(s) of date, but when from child-to-adult-enabling, not [hundredth-(s) // (ten)-thousandths of intricated-statistical-grammatical-punctuation-sociological-psychological-military-engineering-mechanics-political-bran-ches /597/ powers-constitutional-literary-construction-tectonic-topography-tectonic-elemental-resourcing-economic-population-prose] topically delved, by physical-observation-(s), literary-subjectively-documented, because how pattern-(s) of augment, retain formation national-notion-(s)-article-(s)-sequence-(s), when rhythm of action-(s), construct entity-(ies), issuing an year-(s)-objective-discourse, of corporate-ordered-scheduled-activity-(ies), balancing wage as of now, from how extraspection upon population forget those elemental-physics-objective-demand-anticipation, waste-superlative-stagnate-circulating, remain uncalculated those extremes of exponent-notion-extent [grammatical-algebraic-population // item-observation-cycle-ordering], omitting from paper-sheets-analysis, the author-superlative-factor (edge), per [book-(s) / Article-(s) // Paragraph-Section-(s) /// sheet-(s) //// page-(s) ///// syntax-(comma)-prose ////// specific-word-usage-(s) /////// synthesis-notion-(s)-intrication-extent //////// total-word-(s)-per-page / paragraph-section-(s)] reaming into remark-(s), prosing interest-[($) % /// ?(x.x)) / desire-(s) // observation-(s) // tectonic-plate-(s)-sociological-historical-calendar-dating-impasse-(s) // superlative-continuum-et cetera...] from [perceptive-respirating-inhale // exhale-circulated-ventricle-live-blood-pulse-eye-pupil-ear-canal-wave-vibration-(s)-friction-spacing-(s)-physical-muscle-ligament-bone-inanimate-posit-(s)] configured from-an historical-observation-(s) of physiological-anatomical-hypothetical-structure-(s), not partaking in pathological or militarial-death-anatomical-decay-extrapolating-post-live-examining, tensed-components, insighting vocal-vernacular-conscious-interpretation-observation-(s), sequential-ordering, natural-elemental-required-operation-(s), where perimeter-parameter-(s), constitute an intelligent-consensus-mass, mute-reactive-consciously-aware, of calendar-dating-cycling-(s), per [season-(s) / month-(s) / week-(s)] from signature-inhabitant-(s), whom rewrite such constitution-(s), through continuum-denomination, of constant-new-impasse-(s) of secularized-terrain, by exponent-notion-quotient-constitutional-counting, from other intelligent-configuration-(s) of data, directing Territorial-Communion-Operation-(s), where elemental-patent-concept-objective-commercial-developmental-handling, from [Hard / Soft data] convince [Cause-(s) / Reason-(s) of Mass-Effort-Action-(s)], under Exomolten-kinetic-fission-pressure-energy, basically from what encompass-(es) a-the solar boundary, permitting any included visual-tonal-perspective-(s'), notioning aggressive-fact-(s), that expand existence, from-of-a-the-fact-(or-(ing-(s))), of [terrestrial / agricultural / horticultural / physiological-structure-perceptive-ingestion-posterior-stance] conducting by-an natural-behavior, pre-set on vehicle-impasse-(s), through secondths of momentum-inclination-(s), due to those cause-(s) of Physique & Reasoning of Mass, deducting ambivalent-grandiose, where sequence-(s) of means', differ per logic, upon tensable-coverage of locational-physiological-daily-walked-impasse-(s), verifying through hyper-tense-(s)-response-(s), those place-(s) for coverage, in light of conscious-tension-awareness, apose unconscious-ambient, sleep-offset-vapor-respirations, tempo-timbre intellectual-reader-physio-logical-compositional-respirational-cardiovascular-response-blood-pulse-fatigue-independent-being-posit-understood-comprehension, interpreting subjective-article-(s), notioning objective-schedule-directive-(s), because how genetic-mass-survey-mutual-appreciation, deviate sounded-requisite-visual-present-future-existence, from how agenda, thesis-population, without-an fluent-comprehension-understanding, of how those circumstance-(s) of mass, require less, an more imprisonment, from cabinet-labor-pool-judgement-(s) of human-efforts...¶|||05m.:15s.00m.s.|||§

§..Count-Preposition-[42 / Fourty-Two]-of.. ..Article-Essay-[13 / Thirteen].. ..Book-Essay-[185 / One-Hundred-Eighty-Five].. ..Lines-[73.905 / Seventy-Three.1.nine-tenths-2.zero-hundredths-3.five-thousandths].. ..We are birthed from an generation-continuum, that has passed down those objective-instruments, for using at an designated-entity-developing, each notion by developing standards, for how dating has elapsing-period-(s), affecting each outcome interval, for designating an tier of task-(s), for reset-entry-observation-(s), formally intended on documenting, by perspective-reading in secondths-through-minute-(s)-dynamic, for understanding daily- developmental-function.¶|||25s.91m.s.|||§

¶186.Theory is an subject, analyzing socialism in religious-slight, when transgressions of comparable-compositional-being-(s), serve by-an natural-reclusive-place-(ing-(s)), retarded at those suppression-(s) of action-(s), putting an order of fact-(s), because engineering-physics, remain governmentally, construed by entity-(ies), remaining in commercial-private-inclination-(s) of nationalism, apose Americas'-public-inclinations, of historic-traded-affair-(s), for no method of trading, exist without currency, because how Barbarism, insistent on basic-human-needs', apose an structural-psychological-schedule-genetic-mass, imploring cursive-lined-ink-parchment-documenting, from college-ruled-paper-written-ink-inquiry, of tectonic-plate-(s)-elemental-constant-resourcing, because how neglect of technological-creativity, parallel-developments, conjuring through physics-developmental-inclinations, upon territory-(ies) of surrounding-expanse, throughout physic-(s)-boundary-elapses, implied by no distances of measurable-features, prevalent by default tectonic-plate-coverage, serving age-(s) of existence, as-by-when simultaneous-autonomy conjunct through impassing-revolving-pressures, when micro-calendar-dating, upon day-secondths-repetitive-influctuation-(s), per subject-social-census-count, as educational-deductive-empirical-extent-(s), not inspire youth thought-process-brainstorm-critical-thinking-skill-(s), affirming notion-(s), amid ambivalent-confluence, of respirational-breath, exhaling-interpretation-(s), influctuating extents where hour-(s)-maximum-three-thousand-six-hundred-secondths, intercede moment-(s), through minute-(s)-momentary-impasse-(s), juxtaposing secondths, by notion-(s)-elapse-(s)- extents, influencing cause for Respect & Loyalty, offset balancing from hours, resourced-effort-(s), awhile segment-(s) of set-settled-referenced-visual-tonal-perceptive-observation-(s), affect those outcome-(s) of lifestyle-(s')-relation-(s), extenuating-genetic-modification-(s), preferring movement-(s), from bill-document-exchange-(s), at Federal-domestically-trusted-local-shelf-item-(s), from national-transited-efforts, serving-an deductive-position of fact-(s), grandeur unexplainable-celestial-ambient-impasses, learning by monetary-currency-note-method-of-trade, interpreting at secular-location-posit-(s), posterior-perceptive-physical-stance-(s'), aforemention from how those ways of human-structure, evidently rely patent-item-(s)-conceptive-objective-guidance, from neurological-perceiving-respiration-[inhale /598/ exhale]-circulating-ventricle-live-blood-pulse-eye-pupil-plasma-ear-canal-organ-wave-vibration-(s)-friction-vernacular-vocalized-sustained-visual-imagery-sound-key-signature-tone-(s)-retained-subconscious-perceptive-conceptive-pertained-gaps-(s) apose extraspective-substantial-cardio-vascular-maintained-physiological-muscle-(s)-ligament-(s)-bone-(s)-inani-mate-compositional-contained-posit-(s)-component-(s)-motion-(s) & action-(s), in light of repetitive-reflex-(es), as how timbre-tempo-interposed-impasse-perspective, respirate at-an call of Mythology & Language, from-when commercial-sociological-purchasing, influence how each [ethnicity // race // state // division-(s) of sect-perspective-human-qualities], not be an part of common-retention, of Congressional-Observation-(s), when Perceptive-Observation & Legislature, yet be conceived from those article-(s), under a-the body of Law, instilling per product-(s), an anomaly-usage of particular-destination-auxiliary-common-transit-object-(s'), protest-sworn, by common-population-envy, not controlled by congressional-revise, for dating those article-(s) from sociological-observation, when strict-congressional-disciplinary-fashion, superlative judiciary-fright, afraid to impose-percentile-reference-standard-(s), supposed to be learned by now, from Dewey-decimal-system-reading-(s), but when general-application, by social-merits', impasse beyond community-religious-genetic-offset-communion-pursuit, and recluse introverted-effort-(s), relying instructed-notion-(s), manual-transit-continuum, not to see their be an entity-(ies) imposing error, where annexing-facilities, would be more required by living-corridors, then requisite of effort-theory, having vehicles, be an joy ride for individual-(s), rather imposing bus-labor-pool-service-(s), for those lower-E-Class-(es), believing fervor of natural-idle-movements, but when Auto-Biographical-[hoc // bias], suffer along an lack of statistical-creativity, inanimate like bone, for swearing from White-Blue-Collar-Dynamic, on lower-class-sweat-labor, apose an [national-offset-independent-common-physiological-labor-three-(3)-hours-sweat-effort // subjective-three-(3)-hour-(s)-introverted-intensive-studing or reading // six-hours-(6)-(s)-including-labor-chore-(s)-by-ones'-land /// territorial-dues // three-(3)-hours-sociological-effort-(s) // one-(1)-hour-transiting // six-hours-sleeping // two-hour-(s)-extraspective-studying or reading // remaining eighth-day, by an influctuation of either work // transit // sleep // studying or reading // enjoyable-activities // sitting-four-six-day-work-offset] pertaining sweat from temperature-(s), by-an relayed-responses, inclining fatigue-count-(s), per hour-impasse-(s), upon daily-operation-(s), when non-level-response-preference-(s), validify merit-effort-(s) of trade, from educational-subjective-word-glossary-interpretation-graded-comprehension-(s), when vehicle-reliance, basis product-trade-(s), for trading article-(s) when commercial-reliance, maintain data, by-an production of statehood-offsets, where national-tectonic-need-(s), cause-objective-usage-(s), for physiological-practice-(s), operating through [temperate // condition-(s)-environment-(s)], conducting affair-(s), premising fake or theoretical-notions-observation-(s), by [1factories // 2markets // 3outlets // 4malls // 5centers // 6warehouses // 7barracks // 8courthouses // 9congressional-hall-(s) // 10capitol hill // 11state-capitol-house-(s) // 12libraries // 13city-county-hall-(s) // 14white-house, population-free-will-operations] not percentile-factoring, legislational-overhaul, conceiving tier-climate-location-(s)-procedures, by conditions of class-location, upon, [career // labor-effort-(s)], intended for perceptive-confirmation-(s), from individualism abound live-state reality, for national-border-(ing-(s)), that determine how conjuring of existence, formulate an blind-trusts, where objectives, safekeep packaging-affirmation, not exploring why communion remain an fundamental-observation of economy, because of [cubic // rectangular-prism-surface-area-measured-distanced-space-(s)] practicing how to retain mass-commercial-objective-items, where creative-posit-(s)-usage-(s), not recur from rates by-of book-(s), an causable-reasoning for expansion of perimeter-parameter-boundary-grounds, thinking from how repopulation, how an child per house on an block, {16 — 32 Consecutive-Family-Offspring}, count by territory, how developed-territory, expand, when unincorporated-territory, not have private-street-development-funding, when no blue-print-hypothetical-scaling, can verify how to scout, visit, proportion, evaluate, pre-plan, execute an determined-operation-(s), for growing an multitude of population, far more pleasing to me, then entertainment-idle-behavior, practicing documenting each piece of territorial-developmental-data, understanding the premise of location-(s), (as an) measure of territorial-distance-(s), apose human-perspective, revising repopulation-method-progression-(s), responding where-at article-referencing-(s), requiring further congressional-non-fiction-genre-(s)-article-(s)-document-refer-encing-data, when currently, it is evident from common city-county-library-(ies), no state-library-bus, reroute national-Congressional-Library-directive, as like in high-school, physical-achievement-state-activities, for-an adult national-prussing, reporting those observation-(s) thus far, from actions conducted during-an natural-routine of cycled-developments, entity-uncoordinated, by common-subjective-directive-interpretation-(s), since their never be an political-person, to emphasize an state-subjective-repopulating-error, leaving an national-hole, for blatantly-exposing, how dumb from the (inter)-national-Federal-dynamic, not ratifying newer constitutional-subjective-documents, because an common-compulsive-subjective-neglect-behavior, not conduct formal-vernacular-interaction-(s), when vocal-timbre, tone-out other persons-vernacular-term-definition-interpre-tation-(s), not debating legislation, from-an common-book-denotative-constitutional-observation-(s), sustaining bibliographical-reference-(s), for-then annotating-Federal-State-constitutional-extents, for then observing how their dynamic apply, (from an) ambiguous of human-independent-congressional-effort-(s), not understanding what to claim as premise of revise, when their require initially an context-reference-prose-inquiry-(ies), when I suggest an confluence of the prior, and literary-new-nation-(s)-revise-(s), conjuring from those article-(s) for constitutional-equivalent-tier-ordering-redact-amending, making other form-(s) of Magna-Carta-constitution-subject-(s), refined from [subject-(s) // literature], incrementing from-an constitutional-article-(s)-point-(s), balancing [intermediary-extenuating // refining], lengths of extents, implied, amid transgressions through term-(s), centering focus of thoughts, where perimeter-parameter-land-effects, impasse sight validification-(s), confidently tasking at site-confirmations, from-an way of logic, deriving normal-injunction where impasses of trade, bow human-(s) in attempt to avoid sweating as-an natural-way of thinking, without an creative-reproduction, ordering thought by recurring-location-motion-(s), succeeding nomadic-movement-(s), where natural-understanding of physics, upon human-fatigue, require present Thought & Extent-(s), affirming determination-(s) by-of effort-(s), referencing article-(s), supplementing [liberty / freedom / sovereignty] rather [physics / population / subjective-objective-ordering of article-(s)-data], intermittent posit-notion-placing-(s), interlaying purpose of effect-(s), for impassing-momentary-walking-settling observation-(s), | 599 |directing from off-time-extraspective-locational-scenic-ambient-observation-(s), how to conduct work-mode-resource-develop-ment-operation-(s), when now, labor works from those prices of resources, by-an awareness of theory, not origin-factoring-resourcing, apart of county-state-territory-patent-constitutional-developments, because how national-patent-effort-(s), rely on outsourcing-effort-(s), because how boredom settle-in, discoursing-unexplored-condition-(s), not wanting to move, (for some odd reason), when tectonic-spacing-(s), require paid-technological-standpoints, affecting how daily-decision-(s), sequentially-periodical-ize, seasons-months-weeks-dates-hours-simultaneous-dating, as yet been formulated in juxtaposition-(s), from individual-birth-posits, apose present-dating, intermittent an decade-birth-following-census-reset, juxta-posing differences of implicit-view of order, because how conduct is supposed from, intermittent-period-(s), at locations conducting-article-(s)-issue-(s), in an fluid-fashion, gathering [seeds // clay // rocks // refined-composite // metal // iron // gold // silver // slaughter-house-edible-mammal-(s) // plant-(s) // cultivation-techniques // et cetera…] performing by an natural-basis of conduct, supplying sequential-demands, for reiterating what supply-(ies) exist, when progressing reference-interpretation-(s), pertained at locational-outputs, reviewing progress, for enhancing-order, by-an notion of progress, encapsulating thought, by-an ruse of data, relating amongst peer-(s), methods of comprehension, when planned-direction-(s), require-impasse-directing, daily-impasse-method-(s)-dynamic-(s), understanding why we are [under // above // horizon circumference-centripetal-force-revolving],119L. amid an (un) // (in)-tangible-existence...¶|||09m.:02s.08m.s.|||§

§..Preposition-Count-[63 / Sixty-Three]-of.. ..Article-Essay-[14 / Fourteen].. ..Book-Essay-[186 / One-Hundred-Eighty-Six].. ..Lines-[119.36 / One-Hundred.1.three-tenths-2.six-hundredths].. ...I have layered those Article-Essay-(s), with redact-paragraphs, instructing how to reference in book-(s), (Intended vocal) Vernacular article-essay-numerical-tier-line-count-(s), so that those individual-(s), understand on an book-shelf, where to reference those thoughts, that order how formal-thinking is supposed to [1.write / 2.type / 3.print / 4.publish / 5.distribute / 6.circulate / 7.retain] those inclinated-thoughts, for reader-word-of-mouth-opinion, on how to formally-distribute those books, as I would like to be an Best Seller known from [1.New York Times / 2.Washington-Post / 3.Philidelphia Inquirer / 4.Minnesota Presse / 5.Atlanta Presse / 6.Dallas Star / 7.Los Angeles Times], to increment the norm of [first-production 4,000-books-circulation / second-production 4,000-books-circulation / third-production 6,000-books-circulation] leaving the rest up to those competencies of [human-being-bias / hoc-dynamic], upon vocalized-geological-ecolog-ical-temperate-terrain, while I'm hoping to circulate this book in first-year-fashion, to hopefully an monthly-mode, transition weekly-mode, by the fifth-month, hopefully for an decade, circulating an goal of 750-production-(s), (9 day-circulated-printing) per 6,000-books-printed, generating from estimate $45 per book an $20,250,000 potluck, for an Author-separation of currency, having about $3,375,000-dollars, author right-(s), thinking as an Author, how to inspire an industrial-compartmentalizing, of different required industries for developing an contingency, for understanding formal-religious-interaction-living, identifying ones' genetic-place, separate from those observation-(s) of daily-hourly-impasse-living.¶|||01m.:25s.07m.s.|||§

¶1187Evident in August, is youth-grade-school-education-enlisting-re-entrance, unfocused on those task-(s) per day, focusing on community-comradery, perceiving-circumstance-(s), when direct-setting, meander development-(s)-monitored by government, agenda-directing outcome-(s), for which subject-(s), grade from merited article-position-(s), conjoining mathematical-dimension-practice, learning primary-operation-(s), for-how government-elemental-process-(es), adjust-article-(s), [inputting // outputting genetic-relative-research] serving university-means', when during sequential-lifestyle-progression-(s), pertaining daily-hour-denominative-impasse-interval-checkpoint-(s), proportioning mode-minutes-momentary-impasse-(s), influencing word-interpre-tation-(s), influencing each outcome of demeanor, for how month-(s) prerogative-(s), deviate from extraspective-religious-communion-impact, free-will-retirement-off-time-effort-(s), rather than practicing critical-reflexive-referencing, for identifying enumerated-practices, by house-legislational-articlizing, genetic-mass, not strictly-ordering, inclined-deductive-reference-ordering-(s), for tentative-perspective, from free-will-bias, liberate independent-action-(s), so claused-precipitation, mode following uninvigorated-inhabitant-(s), outside an entity-basis, of progressive circumstantial-impasse-(s), implying objective-formal-usage-(s), (from an) historical-behavior, when of tectonic-national-product-conceptive-road-trade-route-system, elementally-resourced, at particular-educational-books-[Pearson // Houghton-Mifflin // Glencoe // Macmillan]-primary-glossary-word-interpretations-trade-system, implicit on neutral-standards, regulating weapon-(s), not per independent-citizenship, book-controlled, upon mechanical-technology-(ies), from-when human-behavior-(not conduct), remain monitored by those youth-social-effort-(s), ignoring youth-peer-(s)-student-(s), for later communication-(s) of vernacular-social-interaction-(s), because the lack of internal-self-discipline, extenuate-an normal-basis of coax-action-(s), without inclination of [[1]letters-reset-(s) / [2]words-compiling / [3]sentencing-(s) / [4]paragraph-section-fragment-focus-case-interpretation-(s) / [5]essay-article-extension-(s) / [6]expository-statistical-subject-annotating / [7]narrative-bias-hoc-interposed-story-board / [8]chapter-setting-occurrence-impasse / [9]term-definition-(s) / [10]glossary-term-(s)-definition-(s) / et cetera...], interpreting fundamental-principle-benchmark-subjective-article-(s)-usage, surrounding parameter-(s) of [Terrestrial-land-surface-area-space-coverage // usage], through-an clearance-height-limit-maxim,| 600 |respirating cardiovascular-compass-direction-length-width-side-forward-movement-(s) of matter, for if any one us, are not specific enough by-an aspect of referencing, what can be assured by affirmed-confirmation-(s) from article-(s), imploring designated-rhythm, by-when article-understanding, clench fist-(s), for tension of skin-vein-blood-pulse-muscle-ligament-reflexive-reaction-(s), central-positing-bone, by-an linguistics-vernacular-vocal-decibel-volume-frequency-pitch-reverberatioal-resonance, hearing-sounded-tense-wave-reverberational-range-vibration-(s), registering visual-traditional-note-(s), upon an market-ordered-shelf-product-(s), in array from letter-word-item-article-interpretational-registering-(s), referencing from glossary-dictionary-term-definition-article-observational-interpretation-(s), relaying secondths upon minute-(s) of time, identifying [[1]how // [2]whom // [3]where // [4]when // [5]why // [6]what] is [appreciated // signified // allured // pertinent // revered] amongst obituary-newspaper-commercial-daily-periodicalizing, affirmed notion-(s), thought by-an denominative-calendar-year-date-(s)-notion-(s), split at [hourly-minute-impasse-(s) / secondths-minute-(s)-elapse-incrementing-cycle-orchestration] yet of subjective-article-(s)-concertion, in spite of [Ink // Paper // Predicate-Syntax-Thinking], concerting-concentration-(s) by free-will, upon Natural-Condition-(s) & Interior-Entity-Development-(s), for which we [[1]sleep / [2]eat / [3]purchase / [4]walk / [5]sit / [6]stance / [7]learn / [8]interact / [9+]et cetera] denominating-instructed-data, by subjective-comprehension-focus, tiering from exterior-observation-(s), functioning apprehended-information, intermittent [elapse-(s) / syntax-(s)] vernacularizing implicitly-describing-data-article-(s), because trading-method, trade interbinded congressional-mass-(es), affirming shelf-life-product-(s) not formally-furthered, by homework-study-interaction, interposing paper-ink-fact-(s), by-an objective-environment-timed-usage-rate-(s), informally-colloquial, historical our unconscious-lord-(s'), linear-tradition, moving anyone along an impasse-(s'), by elemental-objection, progressing from apprehension, whenever location-function-(s), require elemental-solid-tangible-(s), advancing movement-(s), their-of-an locational-motion-(s), limited from interaction-(s) as genetic-mass-mutual-expanse-resetting, plan elemental-resourcing-article-(s), settling wherever visual-sounded-perspective-impasse-(s'), contain tradition-(s), creating an presence of morale, furthering how pre-cognitive-identification-(s), advance an premise-reset-boundary-means', claiming morality, by-an Judiciary-prose, coroner-practiced, because how massive Celestial-Ambient & Physical-Limits-Fatigue affect human-comprehension, upon requisite-vernacular-competent-limited-impasse-communication-(s'), when-where implicated objective-article-(s), [fortify // homogenize], our need for substantial-survival, retaining means' of perimeter-parameter-efforts', where as America & The Globe, remain offset celestial-ambient-abstract-effervescent, separate-form resourcing, because illiterate-inhabitant-perceptive-nature, learn because government instill educational-standards, callous toward-(s) physics, not understanding a-(n) cause for linguistics-vernacular-vocal-tonal-interrogative-documenting-observing, rather ignorantly-numb, to effort-level, repetitively-fought, sweating at last-line-stance, nine-five-nine-air-conditioning-setting,| 601 |adept to common-hospital-repopulation-infrastructure, upon our condition-(s) of Tectonic-Plate-(s) & Personal-Evocation-(s').263¶|||05m.:18s.15m.s.|||§ Academia, posture-slouching, in preference of American-Desire, when-their-of place-focus-recognition, from male-family-communion-reclusion, not refer [method-(s) /// practice-(s) /// trade-(s) /// action-(s) /// communion-(s)] whenever conjective-posit-(s) upon an jurisdiction, population-order-(s), by common-mass, resourcing-inept-understanding, not believing that commercial-book-illiteracy, affect how legislational-practice, not exist, in legal-bill-mandating, ordered from executive-primary-needs-superior, over-installing-cabinet-seat-(s), clausing how constitutional-free-inter-pretation, follow a-the pattern of self-learning, from how formal-extraspective-literary-action-(s), partake, upon, conduct of municipal-affair-(s), in having an survey-intent of presence-action-(s), while retaining an writer-personal-word-term-per-dollar-offset-juxtaposed-article-(s)-subjunct-economic-pur-pose-per-year, not lifting an mass-purchasing, unless trusting oneself-subconscious-effort-(s), from aesthetic-obligation, upon physics, will-then-offspring-communion-confirm or not, lifestyle intermittent commercial-production-(s), because of how objective-article-(s), overwhelm independent-individual-millionths-repopulating-mass-(s), (from an) origin-millionths-offset-century-(ies)-continuum-learning-method-(s), influence measured-distanced-mile-(s)-posit-(s), for enjoyment of metropolitan-velocity, circulating pass all that matter, (from an) family-extraspective-interactive-power-(s)-safety, affecting each (of a those) effect-(s) of visual-blood-pulse-tension-(s), which can be affected, or affecting payment-influence, relevant to the regard of how itinerant affects'-progression-(s) of circumstance-(s), because how data pertain an purpose for location-objective-collection-(s), currently either spaced or stored, issue-(s) of placement-impasse-purpose, requiring communion-impasse-(s), or not enough security will defend against entity-(ies)-objective-retention-data-holding-(s), when no citizen steals' subjective-document-(s), upon an mass-superlative-press-historical-observation-(s), exterior-value-(s), not evoking an desire for an perspective-mass-retention-usage-purpose-(s), when trading of article-(s), fatigue duration-impasse-(s), from celestial-ambient, (physics), mundanely-influencing pattern-progress-ive-parameter-(s) perimeter-locational-ruled-instructive-trusted-directive-(s), non-contextually-docu-mented, to persuade Elapse-(s) & Syntax-(es)-cycling-(s), when physical-quality-(ies) permit enclosed-parameter-(s), to be contacted, amid amidst family-city-perspective-(s') in confluence by routine-action-(s), deriving letter-word-posit-(s), for focus from interior-eyes-blood-pulse-hyper-tension, observation-method, apose ear-canal-organ-hyper-tense-pre-predicate-reference-infer-practicing, Visually-Positing & Vernacularly-Sounding, posit-(s)-intercontort-perspective-pur-pose-(s), only when upon physiological-rectangular-prism-cubic-clearence-surface-area-space-(ing-(s))-impasse, comprehensively-comprehending [mathematical-whole-integer-(s)-numerative /602/ denominative-fractions-percentage-(ile)-operations-proceedings], fragmenting decimal-article-(s)-topic-redact-adjustment-(s)-to-whole-number, integer-inputting, [supplement-ing // complementing-moded-moment-(s)], by [project // task-(s) // season // impasse-(s)-lived-purpose-(s)], pertaining moral-data impasse-writing, defined-interpretation-(s), along day-hours-act-(s), count-influctuating subjective juxtaposition referencing, contraposed article-(s)-deductive-aspect, pre-setting imagery, (ethnic / racial / sex),| 603 |for later sociological-location-community-non-fiction-setting-(s), to elaborate from present-situation-(s) of [comestible-(s) / housing-premise-(s) / clothing-trend-(s')], which denominate present-metropolitan-impasse-living, apose those move-(s) capable patent-product-interim-resourcing-goal-(s), extenuating how century can pass by, ranging decade-87,648-hours-mode-denominative-momentary-impasse-(s), when-where mass-pertain, none of those skill-(s) of education, while peer-(s)-attention-span, remain centered-focused by physiological-admiration, not rigid-strict-absolute-discipline, implementing focus based on syntax-lesson-orchestrating, practicing subjective-imploring, when future-planning of [entity-(ies) / resource-(s) / territory-constitutional-constructive-(s) / objective-article-patent-confirmation-developing-(s) / population-scheduling] remain utmost significant, avoiding misconstrues of free-will, as natural-fact, not invigorate basic-human-right-(s)-[life / liberty / pursuit of happiness], idle-entertaining, where we are required to live by an degree for workplace-schedule-routine, upon classification-ordering, from housing-annex-(ing-(s)), overwhelmed by-an basis of genetic-characteristic-(s)-identifying, extraspective-interactive-location-person-(s)-under-10-encounters-per-hour-goals, for-of territorial-impasse-(s), relatively-inclin-ated, under-standing lifestyle, apose comprehensive-communion-commercialism-product-objective-extra-spective-goals, balancing subjective-lesson-deductive-reference-peer-(s)-objective-article-contin-uum-interpose-plot-posit-inducting-(s), amid practiced writing, qualifying how degree-(s)-merit, pertain95L. upon covering direct-latitude-longitude-impasse-barren-terrain-metropolitan-surround-ing-(s). ¶|||09m.:47s.33m.s.|||§

§..Preposition-Count-[40 / Fourty]-of.. ..Article-Essay-[15 / Fifteen].. ..Book-Essay-[187 / One-Hundred-Eighty-Seven].. ..Lines-[95.125 / Ninety-Five.1.one-tenths-2.two-hundredths-3.three-thousandths].. ..Conditional-Cause, is not set from an individual-position, due to those factored-fatigues, affecting how outcomes per elapse-(s), are due to those factor-(s), placed from patent-intellectual-inquiry, apose population-genetic-survey, interposing how reading by close approximate-view, affect those factor-(s) per individual, not formally voluntary upon survey of context, those antidotes for considering how [male / female] work effort-(s) are affected by wage, balancing books denominating those survey-population-observation-(s), for labeling everything proportional offset-(s) from an introverted-human-population-family-mass-dynamic, affecting those outcomes per individual-bias-hoc-preference, not examining why ones' lineage should cooperate with an limited-group of individuals whom comprehend where formal literary-book-shelf-documenting, should affect each moment per rotation-revolutionary-impasse, clearance continuing inductive-impasse-human-offset-impasse-physiological-moment-(s), height-perceiving from vision, how transit-titling is not capable of escaping police-ecological-ethical-stationary-space. ¶||||50s.16m.s.|||§

¶188Condition-(s) per Time & Calendar-posit-(s)-interval-(s), date piece-(s) of matter, from storage-transgression, using lineage-aging-safe-keeping, securing local-community-shared-communion-appreciation-(s), when trinket-intrinsic-value-(s), serve common-citizen, relax-off-time-work-method, entertaining live-audience-live-fanatics', whom generate circulated-motivation, as mass repopulation, not be subconsciously-formulated, by-an grade-extents, where constitutional-exponent-deviation-(s), derive various genetic-deviation-(s), putting fashion by [head-shape / nose-size-form / ear-size-lobe-variation-(s) / jaw-line / cheeks' / eye-color / eyebrow-form / waist / hips' / skin-tone / hair-tone-texture-style] for protestant-genetic-identity, which remains subjunct, by means' of trade, first referencing literary-article-(s) when no origin-engineering, by American-means', sustem without Germanic-tradition, demonstrating an Federal-governmental-executive-congressional-deviations', linking [family / individualism / mass], by-an Calendar-History-Lineage & Hour-Time-Frame-Moment-(s), from-an National-Fashion, having how mysticism, rely conditions' of trade, inclinated from how conduct is portrayed through family-motivation, affirming self-means', because how truth be evident, from trade-handling-(s), amounting motion-effort, part of a the way of impasse, rather means' of possession, settled-down, by-an manipulation of mind, having trusted-confidence-life-expectancy, be how free-will, develop the lack of etiquette upon city-county-state-constitutional-reliance, ignoring yearly-Federal-constitution-reading-(s), conducting [residential / commercial currency-commerce-trade] physique-fatigued where at objective-rate-function-(s), apose physical-motion-(s)-rate-(s), rating posterior-entity-structures, where-when location-posit-focus, not envision, why newspresse-article-(s), are evaluating [cum-laude-graduation / magna-cum-laude-graduation / summa-cum-laude-graduation / valedictorian-graduation-effort-(s)] focusing commercial-objective-schedule-subjective-(s),| 604 |unethical by literary-aging-retirement, because of how each male-inhabitant, elect by-an historical-Federal-Treasury-currency-note-trade, stockpiling American-humane-effort-(s), relied psychologically theoretical-bill-mandated-hoc-conversive-executive-cabinet-legislational-senate-bill-(s), are auto-biographical, instilled from currency-economic-enabling, rather human-conjunct-morale, straying-off-focus, from attention-deficit-disorder, hypo-neurologically-moving, though vehicle-entities-observation-(s), without referencing, interior décor-lifestyle-impasse-documented-progression-(s), remaining short of everything, apart of family-individualism, inclined male-economic-electorate-obedience, impassing wherever physical-tension-(s), inclinate-ignorance,| 605 |remaining intentionally-reminded, future independent-male-electorate-white-(black)-men-(man), pseudo-progressive-competent-capabil-ities, remaining completely-ignorant, for how male-effort, remain by true-blue-collar-effort-(s), eliciting service-(s), by peer-enabling, not tensing elemental-condition-(s), when Tectonic-Plate-Birth-Origin-(s') & Genetic-Reproduction, not [sustain / obtain / maintain] literary-article-(s)-(not press-article-(s)), for devising an schedule using patent-objective-production-origin-(s), counting human-population, from-an complacent-theory, not evoked of Person-(s) & Human-(s), synthesiz-ing syntax-elapse-synapse-(s), irrelevant from possessive-bias, upon living from tenseable-posture-visual-reverberated-synapsed-blood-pulse-inhabitants', insistent on poverty, as male-work-place-effort-(s), conduct conjured-denominative-population-devising, because deliberate-ignorance, from self-bias-tense-(s'), require an truths that maintain cause-(s) from daily-developmental-living, awhile motion-(s) of mass, can not pertain cause when denoting population-inhabitant-existence, throughout those national-county-city-lifestyle-denomination-(s), for documenting an total-American-perspective, as when thus far, white-liberal-Americans', have shown with Obama, that their incompetency-shows, without much intrigue by literary-mark-(s), where I would want to sell my book to an 25,000,000 millionths book-(s) along an genetic blonde-hair-blue-eyed-feminine-civilization-(s),| 606 |whom consider by national-international-national-rebuilding, an yearly-repopulating-male-transition-method, for working in two to three-decades, 500,000,000 to 750,000,000 million-offspring, by mid-age-{20 - 50-year-old-(s)-{2016}}for then exceeding offspring-reproduction in thirty-year-(s), by 2,500,000,0000-3,750,000,000 billion-offspring-population, totaling-concert-literary-documenting-feminine-reproductive-effort-(s), by 4,500,000,000 billion-offspring-thirty-years-goal, between {2069 - 2099-commencing}, prac-tic-ing how to incorporate literary-legislational-barren-terrain, for repopulating-reproduction-(s), while maintaining an food-cycling, to first focus from family-(ies), how to sustain food-soluble-(s), extraspectively-invested in barren-western-unincorporated-land, and develop such territory, spacing off housing-(s), by each development, extending house by week-per-year-rate-(s), to then maintain-food-resourcing-regenerating-hyper-cultivation-tecniques, repopulating animal-(s), as well wheat-grain-(s), from establishing literary-cultivation-techniques, affecting male-female-food-reproduction-literary-education-American-product-international-trading-resourcing-transition-progress, when I would not forget other women, from my semantics by-of women-electorate-historical-suffrage, when-male-offspring-preference-head-size-reproduction, believe change so rapidly, as Americans sits-in, without an true Tresurigal-repopulating-desire, amid brotherly-eight-hour-wages, devising how to supplement-existence, by what matters most, without tampering extraspective-fine-looking-genetic-comradery, practicing those fundamentals of feminine-repopulation-worshipping, as a-the primary-practice, for tiering any other effort-(s) of living, understanding that invigorated-desire, remain the subversive-repopulating-millennium-analysis, by common-mass-lifestyle-methods, for how male-idle-thinking, (has an) continuum of such, while I would attempt to persuade an civilization of soly [blond-hair /// blue-eyed /// pink-skin-inhabitant-(s)], awhile [yellow /// brown /// black /// tan-civilization-(s)], not commonly congregate-literary-economic-trading, when living by-an means of two-children per lifetime-offspring-method, idled intermittent other mode-momentary-impasse-(s), circulating waste of product-shelfs, apose influencing horticulture-breeding-techniques, along with my book-chef-Boyardee-grain-primary-eating-habit, for considering how [grain-plant-(s) // tomato-plant-(s) // maple-tree-(s) // beef // pork // chicken] supplement our nutritional-balance, proportioning lifestyle-eating-habit-(s), complementing existence, whenever we are intermittent, daily-living-operation-(s), because how quick those motion-(s) for living, can affect outcome-(s) of long-term-future-development-(s), as well as motivate great-mathematical-quandaries,72L. when not of an repopulating-technique-aspect, from human-history-thus-far...¶|||05m.:39s.18m.s.|||§

§..Preposition-Count-[22 / Twenty-two]-of.. ..Article-Essay-[16 / Sixteen].. ..Book-Essay-[188 / One-Hundred-Eighty-Eight].. ..Lines-[72.45 / Seventy-Two.1.four-tenths-2.five-hundredths].. ..With subjecting, where do we reference from, to then affect artistic-common-population-constitutional-Documenting, setting predicate-constitutional-inclin-ation-(s), from human-population-moral-voluntary-survey-(s), pertaining where one will retain those objective-elemental-extraction-resourcing-(s), juxtaposed each individual-offset-populant, whom [bias / hoc], voluntary-position per [dating / timing]-denominating, point-(s) of view, that remain indetermined on physics, for an interest of tectonic-land-developing, those outcomes from compositional-perspective, as incremental-thought-(s), have not clear antithesis for communicating from context, what ethic-prerogative-(s) matter by common human-population-survey.¶|||36s. 63m.s.|||§

¶189Object-(s) & Saving-(s), for how Interaction & Vernacular-Subjective-Interacting, limit- individuals'-comprehension, ignore purpose of study-(ies), from when [mathematical // grammatical // punctuation defined-data], for-an method of reference, juxtaposing our invigorated-desire-(s), where white-women-repopulation, not ever be filtered from free-will, because the thought of nothing.264 ¶|||16s.78m.s.|||§ Ascertain from accredited-reference, is suppose to subject-criticism of major-professional-(s), by Opposite-Sexual-Desire & Lifestyle, when coincided directive-(s), allude an nature, unconfigured, by data, because male-physical-eating-clothing-retention-effort-(s), are lacked by-an method of territory, rural-county-developing, when apposing such voluntary-developmental-mean-(s), when natural-resource-(s), serve sociological-climate-environment-condition-(s) of time-posit-(s), yet specified per work-field, because those term-definition-aspect-(s), (have not been) literary-articlized, per notion-(s), proportioning cubic-measure-parameter-(s), from sociological-observation-(s'), claiming how independent-individu-al-ism-freedom,| 607 |remain-inept literary-non-fiction-continuum, of [1pronoun-extraspective-documenting // 2introverted-auto-bi-ographical-documenting // 3offspring-lifestyle-discourse-non-fic-tion // 4sociological-commercial-work-type-system-non-fiction // 5historical-work-impasse-(s)-non-fiction // 6mechanical-techni-cal-weight-dimensions-posits-non-fiction // 7horticulture-breed-ing-technique-(s)-non-fiction // 8agriculture-growth-technique-doc-umenting-cycle-(s) // 9seasonal-clothing-developmental-non-fiction // 10+et cetera...] uninclined objective-place-sharing, identifying item-article-(s), segment-(ed)-operational-routine-function-(s), de-vel-oping from product-(s), when proportional-count-(s), have not prior-superseded, formal-commercial-production-(s), apose shelf-life-decay, consistently-recluse trust of others', from-an difficulty of open-relation-(s), to partake in engagement-(s), when-at impasse-(s) of register-trading, as when those piece-(ing-(s)) of territory, not com-mence tight-compact-space-(ing-(s)), for serving national-effort-(s), paid for by tax-(es) from bill-(s)-instating, directing an confident trust, as hospital-birth-(ed)-inhabitant-(s), reproduce, by creative-sen-sation, for without women, how does man offspring? As well as for women, how does male inseminate? Will work-place, historically never be literary-superseded? Or has visual-physical-tone-vibra-tion-(s)-perspective, not thought how engineering-nation-state, would need to recycle-production-(s), rather mechanically-revised, (from an) method of consistent-practice, but by what method of direct-obser-vation-worship, when modern-procession, market-trust, such trad-ed-method-(s), when ethnic-racial-neutrality, posture by male-ef-fort-(s), when century-(ies) of women, have been trophy-prized, by release of male-possessiveness, when men have seldomly sweated, in an formal-comparison to women, where audible-decibel, is not considered upon thought-(s) of vocal-vernacular-syntax-subjunc-tion-review-extent, simultaneous Secondths & Minute-(s), denominative physical-visual-tonal-posit-(s)-item conjunction-time-frame, requiring an topic to deduct from, as like [book-shelf // drawers // closet // bedroom // refrigerator], for what can be [attained // retained] apose stored, by community-purpose, as-when inhabitant-(s)-cooperation, require communication of product-objective-ordering-(s), where allotted-count-(s), put forward, an confluential-superfluous of [project-(s) / planning-(s) / issue-(s) / agenda-(s) / patent-(s) / affair-(s)] because their remain an lack of creativity, implored by community-creative-development-(s), where day pass from depth-perspective, an night rotation-grandiose-impasse-(s), as each parameter, not incorporate-objective-literary-amendment-schedule-data, for revising subjective-group-interposed-referencing, when dating billow out to future-millennium-periodical-(s) of obituaries, displaying wilted-fatigueless-age-(s) of citizen-inhabitant-(s), whom never write, commonly-conveying, how design of Architecture & Documented-Extension, incorporate role-(s) of deed-(s), circulating from surveyed-mass, continued personal-effort-(s), considered by action-(s) by such group, which remain futile, through book-commercial-inquisition, incapable for understanding, why each inhabitant, passed through an itinerant-vehicle-notion-(s), self-devised, wherever communally capable, because how religious-protest-fervor, has cheated daily-impasse-written-discourse, for why no common-humanity-written-content, exist by means of Paper & Ink, because those dry-substance-(s), [(dry & wet) / solid-(s) / element-(s) / product-(s) / plant-(s) / tree-(s) / flower-(s) / vine-(s) / entity-(ies)] remain an part of our perceptive-characteristic-(s), applicating-order, instilled from hour-dating, understanding that natural-solar-lunar-progression-sequence, influctuate from personal-observation, not purchased-discussed, by an formal-recognition, for understanding how distance contradict personal-measure, when human-mass-introverted-reset-(s), can not function (from an) exterior-impact, but when intuition can be practiced, methodical-impasses can persuade, [collect / observe / appreciate / incorporate / motion / (s)-et cetera…] as composite-offset-notion-(s), remain without

{Article:} {August}

Written-Subject-Tangent-Extent & Currency-Objective-Product-Circulation, with an [weekly // monthly // seasonal // yearly-intent to live // premise-(s) at], so daily-impasse, pertain an millennium of confidence, at posited-entity-(ies), reset-dating, planned-governmental-terrain, no longer than human-death-remain-compost-(s), apose resemblance of human-subjunction, unparticular those extenuating surrounding-terrain-demarcation-(s), by-an consideration to Male & Female-Repopulating, always referring from population, those means' of reasoning, [1.suggest / 2.imply / 3.elicit / 4.innuendo / 5.denote / 6.insinuate / 7.direct / 8.incite / 9.tangent / 10.inclinate / 11.incur / 12.cause-(for-of) / reason-(from of a-the subsequent)] what century (has an) maxim-approximation of extinction of perspective from awake-conscious-fatigue-pulse-force-perspective, that is idle in place of rotating-centripetal-force-revolution-revolving-force-pressure, beyond full comprehension of tangible-coverage, when community-intricated-interior-common-façade-development-(s), comprehensively seeing [picture-(s) / context-(s) / form-(s) / currency-(ies) / objective-(s) / subjective-format-(s)], [1Macmillan / 2Glencoe / 3McGraw-Hill / 4Pearson primary-article-subject-(s) // 5Macmillan – 6Holtzbrinck / 7Doubleday / 8Simon & Shuster / 9Penguin-Random-House / 10Hachette commer-cial-contemporary-subject-article-(s))], adapting to technical-cultural-mundane-lifestyle-(s)), when masses-morale-currency-examination, seasonally incline from expert-(s), comprehending how referencing is supposed to direct, totalitarian-segregationist-separatist-order-extenuating-legislative-literary-thought-process, overcoming an lazy-idle-nature, (un)-interposed-mentality, affect-(ing)-(ed) [by / from] personality [politician-(s) / servicemen / Chief Executive Officer-(s) / manager-(s)] all whom retire and wilt, amid dawn of existence, comforted from safe-reclusive-passage-(s), inclinated by Parental-Existence & Currency, in sole-survival-role-(s), without an open acceptance to communal-communicational-saving-(s), informed by genetic-similarity, specifying objective-characteristic-(s), observing inhabitant-life-form-(s), when decade-check-point-(s), attempt no legislational-subject-article-(s)-referendum, by how territorial-develop-mental-communal-survey-compiling-inquiry-(ies),78L. require constant century-millennium-calen-dar-observation-reset-(s), for time, upon conscious-perspective, convey why we find an purpose for existence.¶|||05m.:45s.48m.s.|||§

§..Preposition-Count-[36 / Thirty-Six]-of.. ..Article-Essay-[17 / Seventeen].. ..Book-Essay-[189 / One-Hundred-Eighty-Nine].. ..Lines-[78.9675 / Seventy-Eight.1.nine-tenths-2.six-hundredths-3.seven-thousandths-4.tenths-thousandths].. ..I am intending to inspire residential-sales, for surveying my readers, for whom has an bookshelf per family-household, orchestrating rooms, by exo-circulated-physiological-work-wage-effort-(s), from the celestial-physics ambient apose our hand-eye-vocal-coordination-(s), affecting perception by outcomes for lineage-sustain-maintained-control, by block-family-lineage, genetic-sociological-women-lifestyle-male-work-wage-tectonic-development-(s), for understand-ing how formal proportions of matter exist, ecological-geological-continental-tectonic-global-dynamic, from an 483-years of physics-existence, apose an Gregorian-American-English-Contextual-2017-years-common-illiterate-belief, continuum daily-location-posit-decision-(s), for comprehending weather or not hourly-moment-(s) depict how ones'-genetic-stance, apposes those error-(s) of individualism-nature.¶|||45s.31m.s.|||§

¶190Period-(s) are hourly, so should occur under five-minute-(s), progressing-sequence-(s), while matter-(s) of mode, influctuate, from [1.program-(s) / 2.conversation-(s) / 3.work-place-task-(s) / 4.shopping / 5.demand-(s) / 6.task-(s) / 7.chore-(s) / 8.driving / 9.transporting / 10.free-will-monetary-designated-activities / et cetera...] where citizen-right-(s) are suppose to sustem from [religion // press // (conducted-vernacular)-free-speech-(s)], as how orator convey speech, or pastor sermon, when press-denote, as I try to follow an practice commencing from press, through [speech // religion-influctuation] pertaining observable-fact-(s) of human-disposition, inflowed from bias-partial-mentality-interpretation-(s), for how transportation expedite fluent-conjure of event-(s), when there are process of day-(s)-cycling, in-an trade-abstraction of cycled-hour-(s), producing major-isolation-(s)-process-production-(s), in usage of article-(s)-factoring, from action-(s), through elemental-composition-(s), insetting infrastructure, around surrounding-(s), because each notion is commercially-conducted, throughout familiarity-(ies) of unit-(ed)-process-progression-(s), ascertaining [truth-(s) // fact-(s)] from family-idea, rather than understanding-(s), intonating how desire by off-time-notion-(s), is supposed to be conveyed, for inclination-balance, at work-place, occurring through currency-Federal-governmental-developing, where-when daily human-work-relay, repress ideal-notion-(s), independent individually-obscured-elemental-grandiose, affecting-an greater range of inclination-(s), when dating is out of an denominative-day-to-year-documenting, from Day & Hour-(s), upon those restriction-(s) of momentum, because inclined-motion-(s), remain required retaining paper-ink-filed-data-year-reset-I.R.S.-information, using numerical-count-(s), culprit from free-will, as why legislational-documenting, (has an) error from House-Inquiry, from citizen-politician-survey-desire-influence-(s), remaining (un)-pertinent, awhile mass-population-(s), sit by bias, not voluntary-efforting, to repopulate-lineage-offspring and recultivate-food, for how historical-preference go, male-brother-peer-(s), not subject, from-an visual-denotative-fashion, an sounded-vocal-vernacular-reverberational-articulation-(s), for denominating-application-(s), of children, when working on subject-(ing), during those period-(s), for articulated article-tangent-extenuation, literary after initial-day-reading, for later day refining such article-(s)-(literary), for concise-clarity-data, intonating-read-article-(s), with an perceptive-period-(s), not circumstanced from vernacular-subject-lessoning, but instead an gossip-frivolous-discussion-conversation-(s), not comprehending how short calendar is, and tight time is, upon our off-time [condition-(s)

/ clause-(s) / term-(s) / period-(s)], as-when-of momentary-elapse-(s), retain emotional-timbre, coinciding with territorial-posit-development-ambient, recognizing trade, upon our way of human-conduct, affirming through requisite-(s)-continuum-visual-tone-sound-perspective, existing from lineage, through presence of appreciation, or wilted-recognition, methodically impassing, human-natural-behavior, ambivalent-grandiose greater-than an minutes, but when impending-stress, pass-voluntary-recourse, for-how during each parameter-impasse, their become an idiosyncratic-motion-(s), that not sustain duration, pertaining physiological-limits, from retained-imagery-perspective (as is myself), while dating contain product-part-(s), motion-(s)-occurred, limiting featured-vapor-pressure-parameter-(s), influencing outcome-(s) of develop-ment, designing each location, when during factoring ideal-(s) from White-Political-Racial-Demographic-Outlook & Black-Racial-Repressed-Hatred, affecting how Latin & Japanese, sit in on American-Treasury, all ambiguous to any universal-faltering-trade, because the handling of politic-(s) be far too difficult, for other-(s) to comprehend, trade upon literal-subjective-accreditation, traveling by an neo-nomadic-pattern, from nuance-items-intrication-(s), from individualism-family-trade, where inhabitant-(s), require referencing from historical-thought-process, understanding how procedure affect [re-cultivation /// population /// production] by those means required, for impassing duration of [[1]article-(s) / [2]topic-(s) / [3]product-(s) / [4]lesson-(s) / [5]chapter-(s) / [6]edition-(s) / [7]manual-(s) / [8]form-(s) / [9]application-(s) / [10]transaction-(s) / [11+]et cetera] interposed-progressive-visual-tone-sounded-interpreting, expediting timed-motion-(s)-location-impasse-documenting, from information upon deviation-(s), intended for applicating morale-mutual-mass-genetic-population-referencing, from [person's-(present) // being-(s')-(supposed to be confluenced)], not influencing each independent-perceptive-referencing-rate-(s), simply accelerating passed, while article-(s) remain congruent for proportion-(s) of dry-sweat-documenting,49L. not formally documented, by psychological-referencing, for verifying each dollar-cent-formally...¶|||03m.:39s.43m.s.|||§

§..Preposition-Count-[17 / Seventeen]-of.. ..Article-Essay-[18 / Eighteen].. ..Book-Essay-[190 / One-Hundred-Ninety].. ..Lines-[49.95 / Fourty-Nine.1.nine-tenths-2.five-hundredths].. ..Dating is passing, so what to retain, from subjective, to matter with objective, may pend upon opinion, affecting those natures of perceptive-feminine-masculine-offsets, determining away from residence, while trading lineages of shared-resource-ethnicities, averaging from illiterate-inconsistencies, not showing how offset-interposing, remain illiterate, from an tendency of learning, because of difficulty from (cubic-area)-space-(s), subversive alterior moded-range-impasse, minuté common-illiterate-congressional-commercial-purchase-survey-production-evaluation-(s), putting down those other customs of illiterate-space-(s), without each formal tan gent conjuring where to steadfast na longevity of titled-property, due to [Canadian-North-West- ern / Russian-North-Eastern]-geological-offsets, offset-barren-residential-industrial-ecological-developmental-pockets, determining how aging is deceasing south of most lands. ¶|||47s. 12m.s.|||§

¶191During my extemporaneous-tangent-paragraph-article-(s)-expository, human-behavior, I intertwine-intercontort, my intention of nationalism, from my curb of age twenty-five-year-(s')-old {26-years-old;338 April 25, 1991} exemplifying why I [write / type / re-edit / send for further editing / agent-contact-contract / publish for sale] my literary-piece-(s), which can only be opinionated upon pre-congressional-number, or completely-ignored, except for how statistically-explicit, the fact-(s) of such books-are, not in truth of diplomatic-justice, or fact-(or-(s)) of commercial-economy, throughout such outlandish belief, among western-one-hour-religious-custom-(s), not understand that extent from birth-educational-perceptive-social-offset-(s), should compile monetary-saving-(s), amid youth-physiological-growth, competing over territory, by physiological-objective-practice, unhumble from Free-Will & Observation-(s) of commercial-schedule, to insight mutual-vocal-change, upon those inhabitant-(s) whom perceive-approximate-genetic-communication-(s), where religious fervor, remain dull, upon an Weaker & Weaker educational-system, never sound from inception, broken up by vernacular-glossary-feature-(s), counting those motion-(s)-impasse-(s), for denominative-data, occurring upon mechanical-commercial-intelligence, unintuitive enough, for perceiving an creative-curb, formulating thought, understood by those limited process-manufacturing-duration-(s), extenuated among social-requisite-(s'), from independent-comradery, far gone from conventional-practice, pertaining educational-impasse-(s), out of work subjective-interposing, being completely-exhausted, from eight-hour-(s) an day fatigue, when their remain an payment-fundamental reasoning, blankly pertaining thought, from memo-docent-weekly-documenting, where monetary-trade, not stress exhausted inhabitant-(s), from place-monitoring-(s), in safe-bank-operation-(s), (due to an) form of objective-handling, where debt-(s) of free-will-purchase, are retained at apparent commercial-entity-(ies), alleviating-debt-(s), by-an comprehensive-understanding| 608 |of American-Patent-Commercial-Governmental-Study-(ies) & Religious-Genetic-Communion-Communication-(s), adhered by surveyed cultural-decision-(s), from-an leader-agenda, from substance-(s) [(dry) / (wet) territory-(ies)], of [climate-cultural-development / commercial-objective-purchase-parity of technological-instrument-(s) / practices from subjective into objective-fabrication-(s) genetic-branded-item-(s) / elemental-compositional-factory-output-population-count-developing], when awhile daily-impasse, advise inhabitant-(s), that counting-notion-(s), is incumbent, for validifying-data,| 609 |in clue of Dewey-decimal-private-reference-documenting, abridged daily-objective-substantial-need-(s), when through approximate-offset-location-(s)-impasse-ordering-(s), property remain land, as when possession, is objective, for both, Inhabitant-(s) & Politician-(s), not understand how religious-extraspection, is intended to apply, by each storage-facility, into an new-center, shifting objects from one-place, to another, when distances upon physics, inclinate prerogative of effort-(s), while impasse of daily-parameter-(s), not be opened from pastor-politician-officer-author-military-leadership-instructor-(s), remaining thoroughly inept, alluding why apprehension, is required of citizenship, continuing freedom for how no independent-individual, likes traveling, from local-offset-terrain-lifestyle, amid genetic-population-reformation, of hedonistic-nature, instinct-superego, inhabitant-disposition-(s), inclinated to sustain free-will, when dating occurred from place-activity-(ies)-count-(s)-impasse-(s), implicit upon demand of article-activity-(ies), during daily-cultural-comprehension, drifting pass momentum, not counting each [four-minute-(s) // two-hundred-forty-secondths], an tectonic-plate-(s)-drift, adjunct our sedentary-exogeotectonic-posit-offset-(s), mile-(s)-impasse-(s), when such drift have what miles per [second // minute] deviate upon natural-momentum {0.289-mile-per-second // 17.3502-miles-per-minute /// 69.408-miles-per-latitude-drift-degree} is inclinated through secondths-continuum-tic-process, proportional local-developmental-perceptive-impasse, but with how scientific-intervention, not exist, or rationalize American-Government, Weight-(s) & Measure-(s), intertwined by mathematical-prose, for thought occurring into cycled means', dating by argument, affirmed population-means, by free-will, legislated by executive-initiation-(s), incentive believing free-population enhance parameter-prerogative-(s), at ground of earth, restricted cultivation, for-of congressional-literary-count-(s), needing to proportion land-mass-possession, by-an self-documenting-method of survival, from how individual-(s') of mass, remain unsubjective, naturally, making diplomatic-impasse of article-(s), impossible by constitutional-trade-reference, influencing how implementation, extenuate from-an confluence-article-reliance of data, in term-(s) of English-language-reliance, serving custom-(s) of trade, that impasse daily-configuration-(s), when blood-pulse is evocated, but not thoughtfully perceived, because affirmed-physiological-impasse-(s), allude referenced particular-subjective-article-(s), from means' of commercial-lifetime-national-trust-(s), from post-war-enhancing, as war remind us, clear as day, that we are suppose to accede from United States of America-Union-(s), for an significantly more enhanced national-impasse-(s), amid dating-reference, from [century-day / month-day / year-day] concurrently intricating primary-(ies) of literary-article-(s)-interpretation-(s), for newer subjected-book-(s), far beyond Youth-High-School-Education & University-Degree-Median-Primordial-learning-tool-(s), when extent of article-dating-confluence-extent, forget War & Literature-Standards, by-an confluence of perceptive-morale, for what would I presume after this war, be The Final-World-War, against American-Tresurigal-Mentality, as would World War II contrast Korea-War, apose Vietnam-War, when not we not finish battling international-affair-(s), and the male-ego, {I enter // exit, Fort Benning, Georgia;339 March 12 –17, 2013;340 for six day-(s), with nationalism being evidently not in an militarial-capability} continuing-an brotherly-bias, not intending to repopulate-yearly, because writing-{Cursive} & Typing, is too [hard / difficult], for those love-ones', whom continue an mute-idle-youth-sexual-desire-(s), by thirty-years-two-children-generational-era-(s), as I intend to publish-nationally, for an push of such teetering from weather citizens are psychologically-locked-into-national-belief-(s), and I move out of here, or weather I am successful in erecting an nation,68L. by means, of billionths-thirty-year-(s)-national-attempt...¶|||05m.:17s.27m.s.|||§

§..Preposition-Count-[34 / Thirty-Four]-of.. ..Article-Essay-[19 / Nineteen].. ..Book-Essay-[191 / One-Hundred-Ninety-One].. ..Lines-[68.275 / Sixty-Eight.1.two-tenths-2.seven-hundredths-3.five-thousandths].. ..concertion by referencing, is only powerful if those intelligent-off-time-purchase-(s), are interchanged, trusting those development-(s) of tree-(s), at each dating-outcome, as Dating-Offset {1.1.Muscular-(s)-2.Bone-(s)-3.Organ-(s)-4.Interior-5.Skin-6.ligament-7.wall-8.genetalia-(s')-9.Aorta-10.Lung-11.Thigh-12.Calves-13.Forearm-14.Exo-15.Plasma-16.Skin-17.Being-18.Compositional-19.Respira-

{Article:} {August}

tional-[20.]inhale-[21.]exhale-[22.]hyper-[23.]sensory-[24.]brain-[25.]tension-s(s)-[26.]nose-[27.]oder-(s)-[28.]aerosol-[29.]sound-[30.]megaton-[31.]volume-[32.]vapor-[33.]static-[34.]sound-[35.]pressure-[36.]wave-[37.]vibration-(s)-[38.]taste-[40.]muscle-[41.]rice-[42.]grain-[43.]salt-[44.]enzyme-[45.]spice-([0.]flavor)-[46.]molasses-[47.]sweet-[48.]tangibility-[49.]sensory-[50.]perception (have an genetic-ethnic-tectonic-spacings-offset) /// [2.][1.]Terracultural-[2.]Tree-(s)-[3.]paper-[4.]Taxonomic-(*Population)-[6.]Circulation-[7.]Classification-(s)-[8.]Essay-(s)-[9.]Redact-[10.](First-Person / Second-Person)-[11.]Tier-[12.]documenting /// [3.][1.]Clay-[2.]Silt-[3.]Soil-[4.]Tectonic-[5.]Metal-[6.]Stream-[7.]Coal-(oil)-[8.]Solar-[9.]Combustion-[10.]Deposit-(s)-[11.]Offset– [14.]Supratangible-[13.]Matter /// [4.][1.]Metropolitan-[3.]Residential-[3.]Commercial-[4.]Entity-[5.]Lot-[6.]Cubic-[7.]Space-[8.]Parcel-[9.]Transit-[10.]Warehouse-[11.]Mall-(s)-[12.]Market-(s)-[13.]city-[14.]County-[15.]Planning-[16]Centralized-[17.]Sociolog-ical-[18.]ecology-(ies) ///[5.] [1.Universities-]([1.]Book-(s))-[2.]Entity-[3.]Classroom-[4.]lecture-[5.]Theatres /// [6.][1.]Barren-[2.]Vacant-[3.]Desolate-[4.]Uncultivated-[5.]Geological-[6.](Inter)-national-[7.]Preservation-(s)-[8.](if the air force ever thought to missioned to cultivate-daily-hourly-per-millisecondths every forestral-piece of preserve land, by Storm-Columbus-clouds-centripetal-force-air-vapor-mathematical-exponent-rain-drops-calculation-concentration-documenting-{i.e.10,302-meters × 3,294-pounds-per-approximation-acre × 1.75-feet-of- altitude-water-surge × 365,950-pounds-sedentary-organic-compositional-layer-matter = 0.10302-m. + 0.03294-p.p.a.a. + 0.000175-f.o.a.w.s. + 3.65950p.s.o.c.l.m.}}. ¶|||02m.:02s.[65m.s.|||]§

[192¶§a] Duty to those loved ones', should be-a-the first-part of effort-(s), implored through creating (an) community-interaction-(s), religiously-intertwined, from where statistical-numerical-grammatical-letter-deviation at-(equivalent), attempt modifying through Labor-Pool & Religious-Location-(s), populant-(s) whom share an genetic-identity, where statesmanship, whenever during focus to accredited-motion-(s), investigate commercial-product-(s), for religious-mass-trade-(ing-(s)), but tensability-intercede-interpose-vernacular-dialogue, yet how many inches with these words, can fully conceive, or slip past-conventional-mode-(s), for conducting-entity-operation-(s), having a-the duty, for, common-calendar-date-manuscript-documenting, (populational-offset), forever concede, beyond focus-age-era-mid-recognition.[265]¶|||38s.[35m.s.|||]§ If year-(s) by present-conjecture, have an ruse to forfeit, would mathematical-advantage, not christen-past-conception, or does hoc = rhetoric? Perceptively-conceiving-initial-era-rate-(s), reset-(s) by time-convention, has for as while interceded-motion-(s) simply compartmentalize-drift! part-(s) of-an sociological-experiment, specimen patriarch-mass-social-handling, not knowing how to pastor-fundraise, for commercial-founding, situated because how data is perceived, by ones' hyper-tense-(s)-{sight / hearing / taste / smelling}, observed direct-location-(s) [sight / hearing], when substance-(s) [taste / smelling], as for our blood-pulse, visually-sounded-vernacular, tone [observe / evaluate] physiologically moving-impasse, or blank-mode-hour-man, sit through drift-annotation-mathematics, those [[1]condition / [2]distance / [3]element / [4]density by object-(s) / [5]buoyancy / [6]substance-(s) or liquid-(s) / [7]millibar-pressure-(s) through air / temperature-entity—(ies) // [8]object-(s) // [9](comestible-(s)) substance-(s) // [10]plant-species // [11]tree-classification-(s) // [12]soil-level-(s) // [13]water (bodies) // [14]human-mentality // [15]metal-(s) // [16]cement // [17]roof-tops' // [18]empty-field-(s) // [19]road-way (s' // [20]vehicles' // [21]tools' // [22]pots' // / [23+]et cetera (s))];[341] objective when subjective-interpretation-(s), adapt schedule & hour, minute-(s)-perplex, millennium-capitalist-second-superego, past mass-hours-influctuation, without vocal-timbre, so I suggest, human-mute-vocal-aspect, mean all universe not fully perceive, for-of optometrist-[visual // Stethoscope-sound], at location-(s), partake mathematical-conversation-(s), conversive of [comestible-(s) / light-charge-water-pressure] upon battery-charge-drain [(lithium-ion-battery-(ies)) / housing distance space-(ing-(s) apose objective-posit-(s) / et cetera...] practicing mailing-habit-(s), by inhabitant-recipient-fanatic-spectator-third-person-observer, dichotomy each particular-position of government, not subjuncting-work-trait-labor-qualities, not creating an character-identified-objective-progression-(s), delving how to motion from pail-stroke, through sweep-stroke, for brick-lift, be primary-mid-major-payment, when inhabitants swear as citizens, and are rewarded, from executive-default-basic-human-living, so cars mainstream spectacularly, without father-listening, or passengers- the flotsam of misinformation, for creating an greater practicality of interaction-(s), as note fragment-compart-mentalize, nth-data-gamma {Megabytes}-{Gigabytes}, bytes-bandwidth-frequency-junction, reminding how deductive-composite-(s), count [trailer-box-unit-(s) // warehouse-(s)-box-unit-(s) // storage-facilities], apose how market-(s)-shelf-life, stock-coordinate economic-(s)-government-communities, when direct-setting, amongst all other human-being-(s), not pre-populate-coordinate, as why executive-cabinet-fifteen, senate-one-hundred-legislational, supreme-court-nine-judiciary, balancing Ink & Paper document-(s), by visual-toned-reference-(s), yet formally scheduling-trade, as when house-demographic-genetic-makeup, historically-pre-sign, what errors of common-legislation, auto-biographically-produce, so paper & ink remain as scarce, as [metal // steel // aluminum // wood // petroleum // oil // et cetera…] because time is an aspect, hour is median-timbre, minute is [rate // mode // impasse // interpretation // elapse // synapse], and second is [[1]tic // [2]cent // [3]unit // [4]count // [5]letters-{$_{\text{Aa-Zz}}$} // [6]numerals-{$_{0\text{-}9}$} // [7]whole-posit // [8]integer-mode-rate // [9]nth // [10]characteristic // [11]breath-{$_{2.76}$-second-sixty-seven-inches-one-hundred-sixty-seven-pounds} // [12]quarter-note-beat-timbre-sixtieth] offset {four-beats-per-measure // 4/4-timing // subjected-note-amount-of-beats-per-measure}-{$^{6}/_{2}$-timing = six-beats-(s)-per-measure / two-quarter-note-beats = rate of staccato-flute-trumpet-short-frequency-tone-range = treble-clef-classification}-{$^{3}/_{8}$-timing = three-beats-per-measure / eight-quarter-note-beats = Tuba– Sousaphone-Timpani-long-frequency-tone-range = bass-clef-classification}, tone-quality, for adjusting Mode & Setting, so from what quality to intricate, at what location, remains whom fund-operation-(s), from stock-provisions, cycling from [interest-(s) // intrigue] yet when Kroger-Los-Angeles-Count-(s), is not Publix-

Miami-Dade-Count-(s), from how market-(s),51L. do not solicit location-(s)-impassed, by-an article-usage, for [entertained / subjective] pleasing... ¶|||03m.: 56s.96m.s.|||§

§..Preposition-Count-[9 / Nine]-of.. ..Article-Essay-[20 / Twenty].. ..Book-Essay-[192 / One-Hundred-Eighty-Two].. ..Lines-[51.65 / Fifty-One.1.six-tenths-2.five-hundredths].. ..harmonizing existence, is not including those ugly-illiterate-ethnicities, that can not formally comprehend how to think, from when dating is shown pass an solstice-impact, affecting one mode of retentive-maintaining, apose those parameters of dating, sustaining from vocal-tonal-voice, an purpose for territory-genetic-commercial-work-wage-retire-ment-Montana-genetic-intelligent-experience, serving duties by how timing sorts [motion-(s) / stance-(s) / action-(s) / thinking-extent-(s) / vocal-interaction-(s) / task-(s) / order-(s) / reading-mode-(s) / et cetera...] concerting how scheduling is suppose to comprehend those subjective-focus-(es), understanding where ecological-title-deed-planning-centralized-perimeter-parameter-dating, coordinate-(s) an national-geological-communication-(s), with those ecological-stationary-parameters, defining how population stupid-intellectual-ignorance, waste [money / moment-(s) / effort-(s)], on trivial-introverted-bias-hoc-presumption-(s), determining how claim on objective-(s), are required for those readers to understand, that patent-product-(s), are served an right-(s), of unknown-literary-origins, due to those legal-error-(s), not denominating where perimeter-boundary-(ies), affect how tectonic-development, stall from human-mass-population-introverted-error-(s), illiterate of competency, and evident as I plan on exposing those faults of book-purchase-waste-(s), not recycle-circulating, each book to matter, building along those parameter-(s), for adjusting [objective / machinery / intricate-mechanism-elaborating] how dating maintains those timed-elapsed-retention-(s) of informed-data, past-dating-referencing, along an guidelines of present-purpose, making how books serve both [lifestyle-extraspective-purpose / workplace-perceptive-word-label-retentive-observation-(s)] while sleeping each day, demonstrating why reset-awake-designation-(s), matter by impasse-voluntary-effort-(s), when voice matter with an genetic-group, placing each item, at an year-dating-mid-term-plan for gauge of weather an item is deserved for being furthered through each [decade / century], as I intend to develop those syntax-essay-point-(s), for reader-community-interaction-intrigue, understanding why timing is steadfast around voluntary-timing-(s), referencing from books, how to perform [physical-chore-(s) / task-(s) / writing], practicing how physic-taxonomic-classification-cyclings, are limited from development-(s), by human-mass-population-comprehension, thinking which books are required for referencing, in asserting how purchase-rate-(s), affect book-publishing-production-(s), when using each fact of letter-word-participle-data, to exemplify how geological-terrain, remain barren, from an common-human-inhabitant-discipline-error, because where cause for human-morale, is suppose to interpose an nature of thinking inquiry, monitoring why moment-(s), are with common-behavior, not understanding why superlative-excess, is not tolerable, and needed formal-monitoring every piece of matter, denominating mass-population-documenting-survey-existence, by when millisecondths-secondths-timing, interpose dating awhile factor-(s) of market-production, have not been formally-used, in making corrective-judgement-(s), by entity-denominating-statistical-short-ethics, from how each worthy-individual, requires an living purpose due to extraspective-morale, by other individual-bias, not having strong men, affection with lovely-women, by their prerogative of dating, because rule of order, is always in an retarded-bind, by lack-luster, religious prevalence, serving each rhythm from celestial-regard, due to lineage > Ancestry > Genetic-state-population, not making an conscious-awareness, for routine, because sequential-order demonstrates where every human-being is not secured by an common-mass-intellectual-interpretation-(s), because [bias / hoc], is not formal by human-subconscious-preference, because displaces per date, not coordinate those subjective-topics, in subjective-deductive-category-(ies), for arranging with physiological-notion-(s), how to display appreciation for other inhabitants-existing, cooperating on mathematical-predicate, for subconscious-reasoning, by individuals, apose those literary-writer-ethics, examining how one would find their Genetic-Lineage, for enjoying cultivating geological-terrain, for planning those outcome-(s) as placed-impasse-scheduled-anticipation-(s), make work be from an dynamic for thinking space by [writing / typing / subconscious-thinking / vocal-conscious-direct-reference-thinking] why before we die, we should put our utmost effort-(s) to characterize our living-existence, when tectonic-plate-(s), having other civilization-(s) to compete with, not instill-motivation in other individual-(s), because those proglamation-(s), by moment-(s), are not fully sorted, because when handing-materials, require an formal-legal-documenting-dynamic, by book-trust, working away from [bible / manual / blank-store-compositional-book-shelves / wasted-classroom-paper] for under-standing how to proportion those boundary-(s) physiological-moded-impasse-(s), as each notion -(s) are not fully-documented, when verifying who can volunteer clear-participating-effort-(s), per moment at 75-meter-circumference-range, from peripheral-view, as action-(s) remain formed due to conjuring-perspective-error-(s), by still-motion-practice-(s), not motivating an total-economical-society, from formal-trust by how banks hold-money for saving-(s), while circulating from credit, an centralized-release per bill-passed, making evident how timing is elapsing amazingly-quick without talking much with others.¶|||04m.:37s.60m.s.|||§

193b§¶Legal-ink-paper-corrections-teeth-chatter, graduated, as mass-function-period-(s), post-studies-revert, to hobbies, after [customs // traditions // values] as earlier allusion of X-Class-(ification-(s)) perpetuate, [daily-celestial-impasse / tectonic-plate-(not plates)] forget 100%-deviation-technical-one-offset-aspect, from points of minute-(s), contra One-(an)-four-minutes-degree-momentum-impasse, apose-still-core-spherical-offset, translusion-solar-core-halide-momentum-impasse, counter-pose, non hour-(s)-day-(te)-map-degrees, scaling without planets of universe-subterior-county-handling-nation-state-mechanical-cherishing, apose translusion-solar-core-halide-momentum-set-impasse, juxtaposing centripetal-matter, mobius of tectonic-posits, beyond plant apose planets, by industrial-calibre, as present-medium-energy, not fixate Technometallurgy, defining (not define // de… facto), clear-economic-literary-prerogative-(s), existing by various-attainments, for all Federal-offsets, Treasury. \\low(ww)-tone//-Bored... //high-tone\\-Done.. ||An-tone-invariable||Finished. Gregorian-Season-Historical-Present-Calendar-American-English-Economy, by Brazil-aspect, a-verse, (A Book;342 An Series of Essays);343 for no predicate-coverage-exist, hence congressional-basic-human-rights, is like manifest-destiny, when destination go on having word-letter-participle-interpretation-(s), direct instruction-(manual)-daily-function-(s), pertinent by-self still-sleep-allure, not ethically-morally-historically-timed, by any formal-grandeur, for write-loop-hole-(s), so time can tell with literature, arc by cosign-deficiency, not to bold-(face), after work-eight-hour-extraspective-period-(s), for scientific-exploration, referencing ones' direct-territory-(ies), listless of article-(s), so in credibility, literature-mute-reader, not present-annually-seasonally-predicate, for no history make present, but we obituary-commemorate-pass-along different-from coroner, as is Military-Mentality, separate, at constant-twenty-four-hour-stance, for how reliance is upon routine, by social-demographic-bias, not formal by physics-method-repopulation-genetic-impasse.266¶|||01m .:51s.47m.s.|||§ sound for sociological-thought, being-naturally-defended, for no schedule, almost-preferred, yet mass miss such, each color, as red, riverbed instill confidence, from news-article-(s)-usage, when posit-(s) develop, when tectonic-plate, is not where an individual-character, is at, referencing cultural-enhance-(s), of passing-position-(s), securing land-space, when settling upon tectonic-plate, upon offset-perceptive-county-states-boundary-territory-(ies), following subjective-exchange, [under // over // compass depth-perspective-effect], [left / right /// rear-flat-forward] from method-(s), singular by offset-ecological-agenda-resourcing-schedule-supply-social-demand-dynamic;[344] religiously going to pass an book on press-shelves, for at one-impasse, or another, developing-surrounding-area-(s), erect by century-trust, guiding the Holy-County-Bible-Land, continued from communion-of-mass-aspect, apose literate-writer-capability, contrapose political-historical-congressional-archiving, where terrain, implement sign-posit-(s), juxtaposition unconjured evocation-tense-interpretation-(s), presenting idiom-feature, requisite over-preference-prelate being-mentality-perspective, focusing density-bouancy-pressure-compositional-posterior-infer-ior-anterior-superior-exterior-compositional-posture, fleshless-interior-composition, as common-reflection by solar-earth-density-tension-inversion-(s), offset person-(s)-personality-perspective-(s), based on elemental-resourcing, (from an) indefinite-prelate-unconscious-historical-national-trading-complex-(es), basically lived, by overtly-limited interaction-(s), due-to common-means'-level-interaction-(s), unclassified when time pertain-(s) [being-(s) / person-(s) aging-process] upon calendar-blank-day-extraspective-event-dating, in-an non-statistical-common-year-month-day-proglamation-(s) as basic-human-right-(s), embrace-mechanism-repair, impassing inhabitant-remain-(s), completely-inclined, because of their reluctant-impasse-(s), of daily-reoccurrence, naturally-numb, those inclination-(s), where blood-pulse-tension-inhabitant-interposed-conver-sation-(s), from whom conceive, a-the first-suffix-Article-(s)-present-reset-subjective-query?... are any individual-(s') to judge other inhabitant-(s') when cases-document-psychological-medical-offset.[267]¶||||03m.:41s.[04m.s.|||]§ If the season pass… would one interject? CanDO!! First-Alert-fifty-subterior-half-median-mutual-man (men), forget to ask how to affect will with word-(s), theirs' an bunch of [hoc-male-jargon // bias-feminine-rhetoric..] not by incredibly faulted means' of possession-(s), upon natural-circumstance-(s), swearing lifestyle-sacred-life-thinking, not formal by claim of debt-objector-(s)? I prefer to observe.. Wait… pause.. Eat. Vision is numb from exterior-wave-action-in-fluctuation-(s), (from an) spectator-listening-perspective stance, [lay / sit], in idle-fashion, microcosm-understanding, indigenous-individual-bias, physiological-hoc-belief, indifferent of other-(s') existence, while relying on community-effort-(s), enhancing our direction-(s) of action-(s), because of how developing is pertained (as an) invigoration, of serious-implication-(s) for affecting our perspective upon other-(s'), during impasse of tension-(s), evoked for-an numerous rates of those [participating-being-(s) / person-(s)] that understand accepting an part of how any one would be waged, when placed objective-(s),56L. are served at a-the right-time-data, from labor-monetary-workload-wage-rate-(s), upon common-impasse-interaction-(s)... ¶|||04m.36s.22m.s.|||§

§..Preposition-Count-[20 / Twenty]-of.. ..Article-Essay-[21 / Twenty-One].. ..Book-Essay-[193 / One-Hundred-Ninety-Three].. ..Lines-[56.61 / Fifty-Six.1.six-tenths-2.one-hundredths].. ..transit is not teaching any one thought by silence, due to those tasks required by objective- subjective-compositional-being-perspective-(s'), practicing how to serial-residential-retain, [1.hygiene / 2.clothing / 3.accessories / 4.books / 5.bed-(s) / 6.desk-(s) / 7.backpack-(s) / 8.printer-(s) / 9.bookshelf / 10.table].¶|||19s.37m.s.|||§

194c¶§How are person-(s), post-applicating greater-10-tenth-column-denominator-total-recip-rocal, apose lesser-invariable-(s)-adjacent-perspective-motion-(s)-reflexes'/offset/ variable-(s)-base-property-aspect-offset-seventh-limit-maxim-land-tectonic-posit-(s)-numerator-reset-fun-ction-(s), by-at entity-metropolitan-operation-(s), from agriculture-yield-count-(s), comprehen-sively (apprehensively-perceived), while in an formation of characteristic-(s), managing basic-human-need-(s'), where considered what of population-perceptive-assumption, for credit of rights, signifying read-era-competency, without yearly-lifestyle-recreating, for how independent vary communion-families-civilization, by where family-(ies) are to cooperate, land, but not have any written-confidence, to vocalize anything other than squabble-perplexed-intuitive-reactions, as how mathematical-magnetic-offspring-reproduction, not happen in American-wilting, (as an) professional-eight-hour-environment-(s), posterior-stance-(ing-(s)), organization-(s), by profit-confidence, non-for-profit-retardation-illiterate-mass-(es), sitting by stoic-interior-work-settings, as the Myers-Briggs-cabinet-personality-Intuitive-Introvert-Feeling-dichotomy-beings, by Federal-national-international-ambassador-trade-reliances, puppet-(s) for protestant-treasury-safe-keeping, proglamated-Federal-issue-(s), resorting all thirdths-balancing, reset by Executive-cabinet-senate-congressional-bills-meetings, intertwining assortments of mass, impassed or implored neutral-blank-negative-longitude-pacifist-protestant-recidivist-independent-posterior-human-mentality-governement-legislational-creativity-inhabitant-interpreter-(s'), colloquial-intermittent various-parcel-(s), permitting survival where shade-(d)-development, hue comprehension, considering impasse-purpose-(s), occurring from posit-posterior-manual-D.M.V.-test-referencing, because how data assist work-place referencing, by form-documenting, implored by government, during national-international-clause-(German / Alemannia / ßavaria-Deutschland);[345] for family, does not equivalent position of family-heir-mansion-house-hold-inaudible-explanation of origin-observation-(s)-power, but yet an entire-globe of trade, rely on vehicle-resourced-metal-collection-(s), being evident of an superior-civilization-(s), but when work-wage, has not been contradicted formally, developed relied, interior-comfort-effort-(s), sweatless for gauging morale, at appreciation, making clog-cognition-interpretation-(s), when-where, placing-(s) of juxtaposition-aging, remain an inclusive-part of daily-impasse-tension-(s), by-an median of independent-characteristic-(s), upon group-coordinated-activities, evoking timbre-posit-(s), comprehensively-coordinating-notion-(s), where objective-article-(s), discourse subjective-referencing,| 610 |crescendo-crescent-experience, as visual-tone-vernacular-term-definition-count-(s), repeat pertinent subjective-article-(s), (from an) consensus-succession of momentary-period-(s), denominating, [moment-(s) / elapse-(s)] per tense-perceptive-notion-effect-grade-elapse-(ing-(s)), If for how mentality applied at tense-impasse, only evoke, an singular blood-pulse-evocations', temperate interpretation-confirmation-(s), when registering indifferent granted-fashion, where individual-(s') seldomly accept Truth & Fact-(or-(s)), intermittent other inhabitant-(s), aware-confirmation-(s) discourse, subjecting directive-article-(s), from book-routine, for monthly-article-equivalent-statistical-perceptive-impasse-extension-(s), intricating subject-studies, by-an daily-period-subjection-intrication-(s'), conversive by individual-perspective-juxtaposition-(s'), when fundamental-resource-(s), require, furthering population-still-vernacular-interposed-count-production, specifically-denoting, item-object-(s), during impasse-(s) at stationary-trade, where timbre of extension-data, compile prevalent-thought, Written & Interject-(ed), an continuum of think-(ing), aspecting-theory, from intellectual-visual-observation-inquiry, by mathematical-integer-American-measure-method, from sociological-mode-aging, surveying Inhabitant & Territory-(ies), by weekend-experience where interactive-affair-(s), affect family-repopulation-discourse, where-when-why what illiterate-means', matter from commercial-enabling, of executive-branch, cabinet-extent, as American-Exponent, verifying commercial-functioning-citizen-(s), pertinent by subjective-personal-commercial-library-refer-encing, juxtaposing local-library-reasoning, apose congressional-library-reasoning, to affirm how, day to day impasse-(s), elicit corporate-operation-(s), where dawn work-attained-object-(s), not retain from, patent-dimension-concept-object-(s), which matter by usage of directive-(s), confluencing [[1]act-(s) / [2]effort-(s) / [3]object-(s) / [4]tool-(s) / [5]instrument-(s) / [6]mechanism-(s) / [7]machine-(s) / et cetera...] showing how common-humanity, cognate book-article-(s), not recognizing, awhile commencing of perfect-observation-perspective-occur, pronoun-voluntary-inquiry, would upon installation-(s) for calendar-year-(s), place an natural-impasse-(s) of routine, affecting hyper-tense-(s)-observation-(s), verify-(ing) hour-impasse, in continuum-fashion, because elapse-imagery-blood-pulse-reverberational-sound-vibrational-hyper-tensive-functional-organism-(s),| 611 |aging along an synapse-(s)-parallel-fission-force-(s)-offset-(s)-(gravity), as we hover afloat tectonic-plate-(s)-density-solidification-(s)-offset-(s), during daily-period-(s), rotating-circulation, apposed lunar-total-density-mass-(es), positioning juxtaposed celestial-ambient, because daily-voluntary-impasse-(s), year-revolution, circular-oval-north-south-ambient-rotation, in an unknown aspect of force, from mathematical-community-observation-(s), denoting mile-(s) of tectonic-plate-(s)-offset, as how county-city-universe-campus, revolve around an continuum-energy, where daily-impasse-(s), affect trade, by-an magma-combustion-fission, never ceasing, voluntary-motion-(s), simply-as,| 612 |an unconscious-reasoning-(s) of Visual-Perspective & Grandiose-Juxtaposition-(s), peripheral constant-independent-offset-motion-(s), not requiring monetary-enabling, but when each individual continue forward-action-(s),63L. because how daily-motion-(s), intercede upon Centripetal-Force & Physics...¶|||04m.:43s.27m.s.|||§

§..Preposition-Count-[21 / Twenty-One]-of.. ..Article-Essay-[22 / Twenty-Two].. ..Book-Essay-[194 / One-Hundred-Ninety-Four].. ..Lines-[63.26 / Sixty-Three-1.two-tenths-2.six-hundredths].. ..taking time to interact, may serve dues by how each moment is configured by when dating suggest where reset-motion-(s) exist, by an introverted-default-means, for survival.¶|||15s.19m.s.|||§

195d¶§Count can be read upon this book, by-an numerous-ways, while time direct supplied-order-(s), subjecting weekly-course-work, where wire-(s) affect able-effort, referring data, by-an [action / monitoring-system] when notion-(s) remaining an part for how daily-impasse-(s), remain an part of objective-item-(s), put in circulation, from Federal-reserve-confidence, applied through circulated Treasury-safe-military-protection, apose registered-Federal-withdrawal-confidence, not informed, but egalitarian-controlled, where direction-(s) at centripetal-force, move by data-confidence, formulating work-week-form-documenting-(manuscript), put in trade-(d)-confidence, from daily-effort-(s), wanted by trust-(ed)-affair-(s), remaining monitor-(ed) by diplomacy, conduct-from-academia-affair-(s), trusted when registered-corporation-(s), affirm dynamic of operation-(s), by metropolitan-reset-entity-(ies), not creating an reproductive-hospital-population, adjunct constructed-purchase-entity-(ies), (as an) superlative-joke, placing physical-basic human-right-(s), apose conducted traded-objective-product-circulation-effort-(s), without either commer-cial-independent-grandiose-purchase-(s), or minuté impasse-(s), equinox-offset winter-central-centripetal-force-cycle-(ing), equator-median-centripetal-force-influctuation, by north-south-centripetal-latitude-degree-(s)-force-wobble-continuum, denominating proportion-refer-encing, as method of, remain an part of purchased-objects, in preparation of commercial-article-(s), when considering objective-article-(s)-sort-(s), because general-application, remain an part, by perceptive-compartmentalized-interpretation-(s), sectoring decease-word-(s), transferred as we age, through wage-hourly-payment-(s), apose salary, managing each task-(s), subjunct (un)-shift-(ed), effective-application-(s), considering conscious-livelihood, along generation by-per generat-ion, (un)-lesson-(ed)- article-(s)-extension-(s), resetting by international-documenting, an subversive-context, per individual-(s')-nation-(s), to-an, nth-interject, by what grounds of origin, will logic matter, if not ever evocated, when using common-written-interposed-redact-article-(s)-referencing, offset parental-institution-trading, inept of letter-whole-five-percentile-population-classification-(s) Interpreting & Shifting, Language & Literature, when innuendo imbue language, sustem from where independent-preference, base language, upon an dead-aging, from moral-cordiality-(ies), rather scientific-inquiry, or literary-objective-observation-(s), not visualizing impassed-routine-(s), modify-(ing) article-(s), by term-mutual-co-worker-neighbor-religious-member-comradery, comprehensively-intertwining, religious-listener-morale, when dating has been of-an season-disambiguous, factoring fatigue, because how human-conduct, remain issued behavioral-affair-(s), where psychological-count-(s), not commonly-count-track, manual-vehicle-discourse, supplementing tectonic-plate-(s)-land-parcel-cultivating, during regular-routine-(s)-trading, partaken by-an order of operation-(s), influencing how impasse-(s)-outcome-(s), fortify daily-impasse-continuum-focus, mundanely-suggesting, we are required to [think // write // read] because [condition-(s) // clause-(s)] affect literal-tangible-existence-documenting, while duration concert focus of dated-article-lesson-(s), amid mentality, religious-meander of European-language-(s), when orientation, demonstrate an morale, not action-(s), affecting outcome by conducted-motion-(s)-timbre, as impasses of separate-individual-(s'), element-grandiose, required inquiry mass-function-(s), accounting for concepted-patent-object-(s), affirmed by mutual-location-interaction-(s), tasking reproductive-elemental-resources, upon cultivated land-surface-area-space, by-an dynamic of topography-condition-(s), affecting each daily-awake-perspective-effect, coherently-interpreting, interval incorporating interaction-(s), where colloquial-interaction-(s), routine land-space, apose general-genetic-identifying, from where masses-offset, not internationally-communing, family-interaction-(s), when interpretation not origin, from populat-ion congesting, in an social-simultaneous-objective-collection, affecting how humanity persuade traded-technique-(s), from congressional-bill-commercial-objective-product-mean-(s), impassing monetary-currency-method, awhile thoroughfare-routine affect how perspective of visual-perspective-impasse,| 613 |not tone-pitch-vernacular-key-signature, government-bastardized-resourcing, from how individual-(s), naturally negate segregation-identity, motivated by *mass-genetic-voluntary-resourcing-effort-(s), influctuating market-trade, from our weighted-compos-ition-(s), apart from natural-intersection-transgression, affirming currency-trade, by [which // what], posit-(s), pertain data-hour-date-recollecting, fragment-(s) of product-instructed-informat-ion, directing [person-(s) / being-(s)] whom survive through an indifferent confluence of opinion-(s), by-where optional-bias, not indicate, defined-inclination-(s), [juxtaposed // apposed] formally-apprehend-(ed)-subdue, for constraint-(s) of mass-communal-population, indicative interior-perspective-reclusion, as how act pertain upon each fact of individualism-belief, not involved upon interaction of inhabitant-(s)-interpretation, involved upon land-act-(s), serving action-(s), intermittent centripetal-force-revolving-progression, inclinate-(d) through daily-rotation-notion-(s), discoursing revolution-offset-impasse-(s), confluencing perceptive-opinion-(s'), acting upon appropriated-pronoun-property-land-terrain-responsibilities', implementing-objective-(s), for establishing an self-desired-motion-(s), awhile, daily-impasse-(s), influctuate-notion-(s), by-of mass personal-metropolitan-interposed-impasse-progression-(s), for congressionally-document-ing from valedictorian-denominative-observation-(s), subterior 5%-percentile-classification-(s), remaining independent-individual-interpretation-(s), at presumption of traded-fiction-article-(s), naturally pre-suppose-(d), for not serving an implemented-standard-(s), focusing on post-humorous-literary-interposed-read-action-enhancement-(s), due to their being an purpose of standard-(s), emphasize-(ing) inhabitant-(s) article-perspective-(s), knowledgeable when letter-word-(s)-interpretation-(s), juxtapose sentence-phrase-elapse-article-syntax-impasse-conjugation, deviating how thought is particulated, from personal-bias-idle-felt-evocation-(s), apose extraspective-interaction-(s), which convey an written-literary-standpoint-(s), rather normal-visual-observation, when vernacular-tone, has yet existed, for describing how to conjugate, each article-(s)-syntax-directive, juxtaposing how [voluntary-bias // personal-voluntary-observation / physical-motion-action-(s) // constitutional-foot-hold] affect how focus by elder-lineage-life-passing-(s), have no direct emphasis of present-commercial-debt-effort-(s), sufficing market-shelf-life-free-will-demand-purchases, by-an diligence, upon educational-interpretation-(s), then when liturgical-reading-impasse-(s), deviate common-human-standard-(s), brought-upon personal-bias-hoc-interpretation-behavior, as when linear-visual-perspective-progression, not think how sound-tone-pitch-key-signature-observation, juxtapose sighted-comprehension, forgetting that an decimal-fraction-incremental-observation-continuum, is how from physiological-reading-writing-vernacular-conversive-perspective, conjure continuum-motion-(s), interposed individual-perspective, juxtaposed, introverted-personalities, referring article-(s)-issue-(s), alluded natural-mass-population-resourcing-error, tolerating commercial-trade-impasse, as governmental-economic-type-font-manuscript, not reference memo-slant-economic-production-blue-print-dimension-outcome-(s), perplexed, by how family is suppose to cover-tectonic-plate-(s)-land-territory, when physical-vernacular-implement, interpose-result human-fatigue, focusing thought amid concerted observational surrounding-(s), where [terrain / territory] are not physically-impasse-covered, alluding to how unpopulated, human-existence remain, without an present religious-commercial-feminine-power-repopulation-dialogue, for creating new book-(s), settling on mansion-family-living,| 614 |(from an) total-masculine-feminine-offspring-exponent-population, communally-offset-reproducing, apose those individuals' whom comprehend solar-kinetic-energy-continuum-fatigue, applying rates of physical-impasse, for subjective-under-standing, interposing conjecture of vernacular-vocal-timbre, decibelized volume-tempo-scaling, driver amid [1.objective / 2.subjective / 3.construction / 4.engineering-production / 5.currency / 6.population-motivation / 7.genetic-historical-lineage], whom only ever conjure-shift-language, formally considering self-family-posit-observational-inclination-(s), without an reproductive-exponent, physiological-mass-individual-(s')-(ism)-timbre-(s'), considering literature apose [sport-(s)-mass-trading-method / theatre-trading-method / university-educational-trading-method / film-mass-(es)-trading-method / Friday-Saturday-Restaurant-trading-method / any traverse implication-(s) of off-time-dormant-action-(s)] relying confidence from exterior-perceptive-efforts', as individuals' idle, appreciating boring-still-idle-thoughts', apply circulation of international-currency-(ies), when implying how process-(es) of instructed-data, interpret service-duty-action-(s), meant through quality-expression-(s'), cohesive as-when-at, daily-impasse-distance-subterior-moment-(s), because general-inquiry, lose focus afloat-amid, centripetal-combustion-force, not singularly-calculated, upon grandiose-impasse-(s) of secondths, qualifying thesis-modification-(s), requiring extraspective-interpretation-(s), inputting visual-observation-(s), by tectonic-temporal-condition-(s), evaluating thoughts, formally-conjured-perceptive-conceive-(ing-(s)), interposed throughout, every blood-pulse-voluntary-motions-impasse-interval-(s), where land emphasize distance-(s) upon such time-frame of second-(th-(s)), to ever consider how think-(ing), interval-pre-progressive-action-(s), consecutive concluded-interpretation-(s), aforemention-(ing), visual-sounded-vernacular-key-signature-volume-reference-interpretation-(s)-impasse-(s), rather word-(ed)-memory, sustemming from-an historical-adage-denoting, articlizing, obituary date-time-hour-means', understood only if independent-individual-(s),| 615 | are to conjure thought upon an sequential-time-calculated-progressive-formation-(s), tiering schedule apose solar-kinetic-centripetal-density-dimension-velocity-pressure, inert living-comprehension, because how notions are intermittent-between, celestial-impact-confluence, juxtaposed Common-Individualism & Exterior-Grandiose-Continuum-Celestial-centripetal-Force-Impact...¶|||06m.:59s.25m.s.|||§

¢If [American-middle-class // ultra-wealth-class], not interact amicable by religious-genetic-city-housing-neighbor-interaction-(s), then their enacting, totalitarian-mandate-superior-ity by-of House-Representative-Legislational-Act-(s), {State // Federal} deemed not as valid, grammatically-punctuation, by superiority, rendering complete voluntary-freedom, of-by-a-the equal those whom repopulate in [low-rise // mid-rise] annexing, virtuing Perspective & Lifestyle.¶|||19s.47m.s.|||§

..¶§196e§¶.. Physiological-recognition, not content national-state-county-orders, however any one-individual-perspective, see-visualize-hyper-tense-(s), evoking from blood-pulse-sweat-ligament-muscle-skin-tension-stress, apose bone-posit-contact, offset-composite-weight-(s), not seen from other individual-perspective-recovery, apose a-the dynamic of physics, as we blood-pulse-vibrate, hertz-frequency-millisecondths, counter-applicating, natural-tense-(s), voluntary when free-will-motion-(s), fuse mute-vernacular-dialogue, intermittent Rating & Raising posit-(s), as Birth & Aging, Imply-(ing) effort-(s) (as an) natural-way of thinking, presumed where individual-(s')-character-mode-cycling, evocate superimposed deductive-means, of trade-(ing), thought from-an conceptive-governmental-impasse-(s) of thought, interpreted, awhile notion-(s) of mass-motion-(s)-impasse-(s), continue awhile dating permit understanding of thought, because of elemental-concept-compositional-modification-mass-development, understood, from when collecting data, use mass-product-objective-article-(s), where routing repetitive-comprehension, interval-intermittent, daily-hour-moment-(s')-act-impasse-(s), comprehending numerative-means, interpreting, exterior-visual-distance-tangible-quality-(ies), lessoning from extraspective-competent-mass, article-(s)-impasse-observation-extent-(s), exaggerating of Self-Introverted-Discipline & Extraspective-Genetic-Inquiry, formulating-count-(s) of instruction-data, numerative visual-observation-impasse, apose composition-cycling-denominating, by-of elemental-resource-(ing-(s)), during when posit-(s) of routine, Fatigue & Revive, tectonic-plate-rotation-revolution-impasse-(s), comparing product-objective-(s), as how, we decide what significant self-basic-mean-(s'), which remain in order from (inter)-national-state-constitution-confirmation-(s), which pertain constitutional-statehood-order-(s),20L. amid those empirical-dynamics of Federal-Nationalship-...¶|||01m.:33s.91m.s.|||§

{Article:} {August}

§..Preposition-Count-[12 / Twelve]-of.. ..Article-Essay-[24 / Twenty-Four].. ..Book-Essay-[196 / One-Hundred-Ninety-six].. ..Lines-[20.51 / Twenty.1.five-tenths-2.one-hundredths].. ..what data matter if illiterate-survey-genetic-population remain intro-vert-ed.¶|||09s.97m.s.|||§

{January-(31)-February-(28 / 9)-March-(31)-April-(30)-May(31)-June(30)-July-(31)-August-(31)-Month-(s)-Day-Hour-Minute-(s)-Secondths-Count}-{23,587,200 D.H.M.S. = 60S. × 60M. × 24H. × D.}-{6,552 D.H.}-{393,120 D.M.}-{273 D.}

[Article:] [September]

..¶§197f§¶..q As time elapse-(s), intimacy between an opposite-sex-partner, not consider more than an marriage-high-school-religious-lover, conceding monetary-extraspective-impact, for an adequate-reclusive-introverted-independent-individual-apprehension-thinking, evoking similar mass-individual-(s')-family-consensus-emotion-(s), simply without passion for physical-tense-intimacy, during those mode-(s) of work-place, thinking it as the end-all-be-all, when literary-labor-effort, scarcely exist, upon centripetal-force-ambient-mathematical-proglamation-analyzation-observation-discourse-(s), comprehend-perceiving, differently, those rolling kinetic-wave-ripple-(s), of solid-density-tectonic-plate-(s)-density, from-an [vapor-air-infinitive-space-ambient /// solar-Exomolten /// earth-molten-liquid-density-magma-centralized-pressure-density-force], influencing interior-entity-environment-setting-(s), as-an place for postulate-sitting, not bothering to affirm-reasoning, relying on political-richer-person-(s), for patent-mass-product-production-effort-(s), not reasoning from literary-prose & Ingenuity, hyper-tense-observation-(s), evaluating amid those facet-(s) in culmination of circulated-ordering-(s), apart by friction, from of how tension (like sweat from a the human-physique), is put upon an disposition of fatigue-deteriorating-condition-(s), requiring an extraspective-worker-labor-demand-effort-(s), creating emotional-religious-hostilities, from request, alluding as why payment-(s), comfort those lives, when-whom would exist upon daily-local-entity-scheduling, furthering an (inter)-national-Pilgrim-Protestant-Scientific-Techno-metallurgy-Mission, from various Gestalts, encompassing-progress from-an, Tectonic-Terrestrial-Stance, implored by career-aged-expertise-paid-hour-request-(s), as-how visual-sound-tone-reverberation-(s), transmit commands, municipally-permitting, each instructed-modification-(s), per [construction-site-(s) / engineering-plant-(s) // site-(s)] afflicting condition-(s), through ambient motion-(s), alone, not fully defining-human-character-(istic-(s)) from posture-anterior-visual-being-(s')-impasse-(s), from where environment, fatigue whom remain undeclared, defining (auto)-biological-entry-(s), or celebrity-live-appreciation, when method of payment-(s), suffice-independent-family-race-ethnic-individual-nation-(s), for how money-trade, posit Protestantism, by American-Humanity-Future-Mentality-Year-Season-(s)-3000-Unconscious-Belief, having false-ideal-(s'), notioning-from-extent of existence, meander each [Marine-Archaic-Mentality // Navy-Ancient-Personality-Input-Impasse], not documenting each individual-(s), that create from patent-trade-work-one-offset-career-brotherly-mechanized-systems, of particularly 15-Executive-cabinets, in lieu of [Transit-Food-culture-social-farmer-trade-singular-cycle-sleeping /616/ C.E.O-Microcosm-Composite-Colloquial-Expansion-Governmental-Commercial-Religious-Maintaining] alluding no extras-pection, without monetary-guidance, each and every monitor-(ed)-impasse-(s), sustemming from educational-background, during vocal-commercial-moment-(s), imploring metropolitan-choles-terol-storage, peripheral by [race // ethnic inbred-nation-habit-(s)], uninquisitive, to join or falter, by search of books, mundane-fact-location-entity-finding-(s), finding stadiums more often, than an private-library, outside Governmental-Neutrality, for illiteracy-retardation, conjured by physical-impasse-reaction-(s), not counting from those origin-(s) of posit-(s), exterior, apose interior-visual-medium-article-(s)-interpretation-(s), for from history, not monthly-1000-pages-read-writer-(s')-non-fiction-continuum-redact-paper-documenting-national-paper-usage-proportion of objective-(s)-extent-(s), I think many shelves remain dormant, when trading, premise subjunct-nation-(s)-market-shelf-(ves), not morale by impasse, because of illiterate-ignorance, upon book-shelf-observation-(s) from past-present, future-present-impasse-application-contin-uum, sedentary hypothetical-cross-reference-trading, haphazardly-evoked, from present-to-future-evolution, (from an) present-requisite-live-past-awareness, recollecting under governmental-mundane-article-(s)-collection-(s), as-from [being-objective-usage / person-objective-article-(s)-usage] uninterested by-of mutual-extraspective-communion-labor-church-territory-constitution-revival-interaction-(s), extension their-of university-commercial-literary-comprehension, thinking not to conquer-territory, but expand by repopulation-expectancy-present-thousandths-|years-(months //of/or// week-(s))|, interposing schedule-periodical-moment-(s), for understanding how each determinate-fact, pertain factorial, upon referenced-interpretation-(s), not considering this book from Supermarket-Season-Shelf-Life-2018-expectancy, an season from International-Publishing, subjective-lesson-intensive-focus, annotating exterior-value-(s), of various-facet-(s) [[1]of / [2]upon / [3]at / [4]in / [5]or / [6]to / [7]from / [8]out / [9]during / [10]while / [11]when / [12]a / [13]the / [14]for / [15]their / [16]there / [17]as / [18]is / [19]why / [20]how] through-existence, Paragraph-Sub-Section-(s) & Article-(s), equivalent by notions reading, interposed-interpreted, while I interpret as juxtaposed-apose-contra-posit-(s)-position from any individual-writing-society, trading from military-infancy-handling of physio-logical-80-Years-Retirement-2-4-Offspring-opposite-Sex-Reproduction-(s), defining [generation-(s) // legion-(s)], under such imperial-traded-belief, not premising other [individual-(s') /// objective-product-item-(s) /// Etymological-Tier-Perfect-Article-(s)-Pertinent-Thousandth-pre-sent // apose // visual-tonal-sounded-physical-impase-(s)-mode-location-offset-posit-(s) // Ten-Thousands-One-Hundred-Twenty-Year-Hundredth-Millionths-personal-dollars-year-decade-book-(s)-denominat-ive/////\\\\\ Ten-Billionths-Age-57{Fifty-Seven // Thirty-Year-Ambition-personal-social-extra-spection-analysis-survey-review-Extraspective-common-personal-genetic-non-technical-reset-national-10-Millionths-dollars-reset-commercial-enterprise-territory-redeveloping-process] are free to believe that their be an difference, but (from an) constitutional-median, upon press-editing, modifying-interpretation-(s), embedded in-by self-perspective governmental-religiously-belief-traditions',| 617 |offset-distance-metropolitan-complex-measures-(s), confiding non-preference-being-(s'), illiterate to an common-population-written-paper-non-fiction-live-non-auto-biography-documenting-genre, annotation-redacting, of article-(s), offset understanding, how fluent-extent, [intercontort-vernacular // written-data-mode-time-period-(s)-scale-level-(s)-extra-spective-confirming],| 618 |requiring common-hourly-live-awake-twenty-four-hour-week-placing-location-interaction-(s), intermittent natural-ambient-posit-(s), juxtaposing [1 objective-product-purchase-place task-(s) / 2 errand-(s) / 3 order-(s) / 4 listing-(s) / 5 motion-(s) / 6 contracted-period-(s)] aging awhile evaluation-(s) of [physiological-being-exper-ience / person-subjective-objective-directive-physiological-experience] determines' how one will sweat, juxtaposed Subordinate & Superior, none of those human-impasse-(s) contingent article-usage-application-(s), thinking from self-inanimate-bias, from dense-object-(s) apose tense-physiological-posterior-form, process proceed location-federal-regional-climate-operation-(s), through movement-(s), influencing article-function-(s), when mass-(es) implicate-motion-(s), when data require an constant-etymological, auto-biographical-offset-implored, [Aspect-(s) // Word-(s) // Definition-(s)-Tier-Offset-Chronological-Historical-Dating] as how term-(s)-reset-interpretation-(s), formally-resetting, by three-non-fiction-pronoun-action-(s)-impasse-(s) refer-ence-subjective-deductive-conjunction-(s), physiological-impasses, from plotted-data-value-(s), modernly-applying, upon designated-location-land-surface-area-space-(s), [governmental-com-mercial // private-social-residential-planning-commune-offset-(s)] not by military, (inter)-national-regional-economic-Establishment-proportion-center-(s),| 619 |religious-national-resourced-county-city-establishment-(s), mundane for judiciary-review, as how economic-circulation, be socially-requisite-free, militarily-historical, politically-economical, religiously-observed, perceptively-evaluated, psychologically-fatigued, when-where, eight-hours-human-day-work-cycle, deteriorate those posterior-distance-(s), cutoff by limit-(s), bounded where physical-effort-(s), are waged, to calm ego of objective-technical-placings-payment-method-(s), not understanding artistically-practiced-payment-procedure-(s), [slouch // posture] ones'-self-affirmed, family-ingestion-prerogative-(s), when fanatics', can not normally challenge readers', but where those readers-limit-(s), retain introverted-commercial-book-reading-(s), not lifestyle-private-housing-book-clubbing, by weekend constant hour house-(s), pre-planned-territory-population-proportion-trail-planned-event-(s), juxtaposing city-chambers, for weekend-two-day-10-houses-14-hours, community-excursion-(s), without an house-grounds, handling families, upon each location-(s), for an-thousand-person-(s), [flyer // invitation-event-(s)] from those primary-observation-(s), of housing-lifestyle, not having yearly-work-week-retaining, focus an genetic-finding, amid 1,440-city-house-(s)-event-(s)-locational-interaction-(s), apposing-per religious-work-neighbor-community, by 100,000-persons-housing-invitation-flyer-passing-pract-ice, practicing self-house-hosting, by an discovering of what literary-talent-(s) exist, when human-wrist-cursive, not invited-by-known-writers, reluctant extenuated-dating, when theological-dating, be how religious-Gregorian-Calendar-Mysticism-Observance, politically-enumerating-mass-(es), not incorporate morale-housing-effort-(s), devising from [housing-tradition // participation-(s)], an plan for observing perceived-circumstance-(s), where natural-belief, remain idle-common-sweat upon physics, mild when impassing boundary-road-reset-routine-mile-(s)-directive-work-routine, at terrain-conjunction-(s), from how mass-repopulating-reproduction, understand from whom, what product-item-influence-(s) affect interior-self-posterior-posit-(s'), assign-affirming, motion-(s), for count-monitoring, force-movement-(s), during [interior / exterior effort-balance-(ing-(s))], through duration-instance-(s), intermittent visual-sound-perceptive-reverberated-wave-vibration-(s), scaling from algebraic-exponent-count-(s), counting by offset-perceptive-conjunct-visual-sight-posit-(s)-item-count-(s), hundredths-whole-integer-numeral-ration-scaling-(s), awhile impasse affect intention of communal-state-territory-community-(ies)-obligation, not learning from pastor, how to have literary-house-group-article-(s)-sect-reading-(s), at weekly-housing-shift-site-(s), redacting-point-(s), by one-day-brainstorm, one-day-written-annotating, for per lesson, learning-party-(ies), not lecturing, but referencing by an steady-group-reading-(s), to interpose year-(s), by [bimonthly-book-article-reading-weekend-(s) // one-weekend-mild-festivities /// one-weekend-group-events-house-circulation-fund-raised-festivities], year-seasonally-house-family-fundraise-planning, practicing human-citizen-extraspective-coordinating, from various-action-(s), attempting an persuasion-method-(s), for city-seasonal-subterior-paperwork-documenting, citizens'-redact-article-(s)-interpretation-(s), identifying those condition-(s) of [population-community-eating // clothing // hosting // working // resourcing-product-requirement-production-recreation-(s)] recollecting by physical-perceptive-motivated-extraspective-action-(s), for appreciating-power, by [water // electricity // food-person-count-food-(s)-expectancy-coordinating-effort-(s'), delving (from an) juxtaposed A /// B /// C-class-(es)-citizen-inhabitant-occupant-interaction-(s), amid natural-independent-city // county-community-impact, when from church-house], monthly-city-government-open-constitutional-grounds-day-(s), (have not been) capable for relay, due to obedience of currency-exchange, believing in political-states-nation-senate-executive-cabinet-military-(s)-labor-monetary-economy-humanity-mass-(s), ab extra population, practicing an etymological-usage-deductive-repetitive-exponent-progression-cycle-rate-count-(s), from book-(s), having paper as trees, ink (as an) soluble refinement of oil, plastic (as an) fusion of sand-Mojave-Aloe-Regeneration-plasma-mild-heating-fusion-technique, et cetera.. as-when physics, is really an over-hoced-subject, not mathematically-meticulating, celestial-terrestrial-rotation-revolutionary-lunar-revolving-offset-mile-(s)-meter-(s)-feet-distancing-(s) of meters-impasse-(s), deducting from individual-populant-mass-(es)-offset-literary-competency, perceptive-self-extraspective-observation-(s) of basic-human-rights-currency-algebraic-proportional-objective-(s) per [month /// season /// year /// decade /// extension-(s) // eating-portion-(s) per day // exponent-dynamic] not crossing in to other [ethnic // racial bastardized-genetic-gene-pools] when attempting an newer-genetic-classification-(s), from when requiring an emphasis on housing-families,33L. as I attempt to alleviate my Fathers'-duplex-lifestyle, for beyond Schloss-Neuschwastein-lifestyle-literary-housing-development-(s')...¶|||10m.:18s.85m.s.|||§

§..Preposition-Count-[32 / Thirty-One]-of.. ..Article-Essay-[1 / One].. ..Book-Essay-[197 / One-Hundred-Ninety-Seven].. ..Lines-[33.575 / Thirty-Three.1.five-tenths-2.seven-hundredths-3.five-thousandths].. ..Proportional-ecological-terrain-family-lineage-population-lifestyle-thinking, this be how to consider [where / when] whom transpose those reasonings of wage-work-place-circulating, noting how past-requisite-human-history, be from an antithesis from present-dating, what timing signify purpose by tier of daily-secondths-denominative-impasse, apose those terrain-land-parcel-space-(s) making how develop-mental-moment-(s), juxtapose lifestyle-moment-(s), as does thought of a the physics-around, yet exist.¶|||33s.07m.s.|||§

..¶§198g§¶..Think! White-Americans'-(white-collar), prefer to win-(dow), rather than (Sch)-lose(-e+s), because those loss-(es), show an hurt-insane-fanatic-extraspective-monetary-sardines-expenditure-(s), not fundraising, by an applied-subjective-prose, when uninspired religious-pastor-housing-interactive-hoc, not communion, discussing how national-outsourcing affect community labor-lower-economic-genetic-morale, from-an denominative-means, juxtaposed Black-Americans'-physiological-fun-product-shelf-life-effort-(s)-(blue-collar // white-collar // playing-collar-(s)-{N.F.L // N.B.A.}), simply-repopulating, without an political-religious-intensive-purpose, non-chalânt about North-Hemisphere-Tectonic-Land-Territory-Canadian-militarial-(Madison's'-war)-intertwining,| 620 |lazy as sweatless-cool-friendly-introverted-preference-(s), adjust various-skin-colors, [1.taupe-yellow // 2.white-pink // 3.white-brown // 4.grey-white-pink // 5.primaries taupe /// 6.yellow /// 7.white] basically recidivist-(clowns'), against pink-skin-nature, lightly-white, from how self and offspring, can be, working on oneself-independent-economy, by addressing those inadequacies of National-Genetic-Territory-(ies)-Extraspective-Religious-Communion-Obligation,| 621 |affecting how political-literary-religious-interpretation-(s), over-pastor-one-hour, an transit, physiologically-product-objecting-embodying, Chinese-outsourcing-method, relying resourcing-trade-intelligence,| 622 |from other-tectonic-elemental-resources, rather sweating out those primary-labor-(s), for then working singular-writer-seasonal-late-twenties-literary-interposed-continuum-non-fiction-statistical-literary-observation-(s), out from under an limited-patent-creativity, for subjective-housing-church-written-document-(s)-segregation-lifestyle-method-(s), because no psychological-investigation, has ever been, not premised by-an [pink-white-lifestyle-practice // being] (I currently tan-brown-pink // green-orange-day-(light-(mirror-night)) // yellow-brown-mild-light-distancing-night);346 working on an every-child [blonde-hair /// blue-eyed /// pink-skin-customs-family-society-(ies')-civilizations] rather the American-Generation-Three-Post-War-Eras-Continuum-Four-Treasury-Congressional-Literary-Stagnation, as-how primordial-present, continue work-presence, durating posture-(s) by physical-motion-(s), practicing Federal-circulation-method-(s), (from an) technical-practicing-method, not [subjective-documenting-practicing-(s) /// physical-impasse-market-trading-method-(s)] expensed mega-vehicle-usage-(s), from how national-factory-dynamic, affect product-social-interactive-environment-(s),27L. enabling [common-census-voting-citizen-(s) // deziden-(s) //// being-senate-executive-person-(s)]...¶|||02m .07s.20m.s.|||§

§..Preposition-Count-[1 / one]-of.. ..Article-Essay-[2 / Two].. ...Book-Essay-[198 / One-Hundred-Ninety-Eight].. ..Lines-[27.6125 / Twenty-Seven.1.six-tenths-2.one-hundredths-3.two-thousandths-4.five-tenths-thousandths].. ..strong-individual-(s), do not comprehend how weak each individual-remain, by voluntary-perceptive-reclusion, not interacting, for general-timbre for lifestyle.¶|||13s.03m.s.|||§

..§¶199h¶§..September;347 by this time of year, commencing this months' [Two-Hundred & Fourty-Three-Day(s){243D.} / Five-Thousand & Thirty-Two-Hour-(s){5,032H.} / Three-Hundred-Fourty-Nine-Thousand-Nine-Hundred-Twenty-Minute(s)-{349,920M.} /623/ Twenty-Million-Nine-Hundred-Ninety-Five-Thousand-Two-Hundred-Secondths {20,995,200 Sths.} / Two-Hundred-Fourty-Three-Sleep-(s)-{243S.} / One-Hundred-Fifty-Work-Day-(s)-{173-Wd.} / Ninety-Three-Week-End-Day(s)-{70W.E..} / statistical et cetera...] individual-effort-(s'), proglamate in an conjunction of human-mass-population-effort-(s), deviated from those impasse-variation-(s), partaking in formal-person-manuscript-bill-text-inferencing, referencing from basic-human-right-(s)-impasse-transgressions, how event-(s), because where methodical-logic, prefer senate-common-congressional-factoring, denoted mass-effort-(s), along count-(s) of data, intended to affirm elemental-resource-product-(s), apposed material-item-usage, at an location-directive, generalizing governmental-territory-(ial)-impasses', affecting construe-(s') of infor-mation, where reference by-of human-discourse, perceive through social-interactive-vocal-event-(s), not understanding the purpose of no-child-left-behind, because, each inhabitant, have an usage of data, that interpret how information is applied upon those ruses at time-frame-interval-(s), alluding through misgiving-(s), how information motivate individual-(s'), remaining reclusively-introverted, (as an) part of archaic-inferencing, unreferenced of Germanic-Vowel-Modification, pertaining an National-Transition, because how subjective-article-(s) of data, transgress daily-event-(s), because how physiological-timbre, affect determination-(s) of day,| 624 |gathering food, later intermittent daily [cultivating / collecting / preparing / cooking / serving] each pre-count-(ed)-individual-(s), whom understand the significance of genetic-cultural-development, impassing amongst individual-mass-nation-population-work-routine-cycling, influctuating pop-ulation-interactive-event-(s), transgressing national-state-position, formulating how effort-(s), put an desire of national-ambiguous-effort-(s), developing-lifestyle time-impasse-customs, genetic-ally modifying cultural-developments', by-an means of logical-structural-impasse-(s), understanding how visual-perceptive-observation-interpretation-(s'), are pertinent for modifying-element-(s), by-an use of article-(s), surrounding tectonic-perceptive-posit-(s), where land is from subjective-physiological-impasse-constitution-review, for developing-era-(s), during those historic-fatigue-(s'), upon daily-interpretations', as when of solar-combustion-fission-impact-continuum, define [where / when] those celestial-posit-(s), offset tectonic-location-(s), for [subjective / objective / entertainment-invigoration-(s)], during colloquial-impasse-(s), as physiological-act-(s'), affirm perceived hyper-tense-(s'), visualizing-heard-blood-pulse-tense-(ing-(s)), observing grandeur by-of minute-macro-interpretation-(s), impassing location-(s), in an remote-operating-memory-pattern, subversive for thinking, how act-(ed)-(s), by muscle-memory-action-(s), layer upon our hyper-tense-(s), in-an subtle-way, because generally-though, notion-(s) of mass-population-(s), Etymology & Lexicon, not affirm how colloquial-language, affirm literary-remark-(s), apose an [rubric /// benchmark // subjective / article-(s)-basis] interposing how interpretation-affirmation-(s), conjure government-work-mode-influence-(s), per theatrical-persuaded-motion-(s), dating time-file-(d) & dispose-(d), [daily / yearly-elapse-method-(s)] of at-will-employment, along those means of trade, reoccurring at remote distance measure-reaction-(s), considered from when at momentary-impasse of-an extraspective-moment-present-reset-impasse-premise-interpretation-(s),| 625 |relevant upon what requisite-past-information, can be interval at digestion, while abating interactions', for furthering impasse-motion-(s), acting an part, at participant-confidence, conjuring later-date-interaction-(s), inclinating elaborated logic, from colloquial, active-assessing thought, verifying notion-(s), with visual-vocal-vibration-(s)-reverberational-conjuring, scaled as [0 = 10,000-notions-common-hourly-wage-deviation / 1 = 100,000-notion-salary-gamma /a/ 2 = 1,000,000-notions-unit-production-offset-gamma //b/ 3 = 10,000,000-C.E.O.-Salary-notions-gamma ///c/ 4 = 100,000,000-author-wealth / historical-units-production-notions-gamma ////d/ 5 = 1,000,000,000-corporate-(league)-revenue-notions-gamma ////e/ 5 = 10,000,000,000-richest-billionaires-notions-gamma /////f/ 6 = 100,000,000,000-degree-mode-median-capital-composites-notions-gamma //////g/ 7 = 1,000,000,000,000-trillions-recent-thousands-government-bailout-post-2010-notions-gamma ///////h/ 8 = 10,000,000,000,000-Federal-Government-Total-Accumulated-denominative-historical-Debt-notion-Gamma ////////i/ 9 = 100,000,000,000,000-Treasury-Yearly-Federal-Bailout-currency-human-rights-savings-notion-gamma], (as an) example, for how notion, compare by unit-(s), an superlative amounts by capable interpretation-(s), enumerating unit-(s)-box-(s), factory-national-demand-productions, transiting elemental-weight-compound-ordering, from census-population-count-(s), affirmed from entity-(ies), where Federal-government-currency,| 626 |bailout, hosting population-event-(s), when relevant-commercial-production, apply pertinent Past-Expense-Purchase-Deviation-(s) & Future-Referendum, scheduling-inhabitants', where program of patent-concept-motivation-product-(s), affirm issue-(s) as-an problem from [state / ethnic / race-population-action-(s)-(academic-effort-conduct-issue-(s))] ascertain-(ed), by graded-review, confirming methodical-analysis, partaking liaison-(s) of (wo)-men, whom are prudent competent-comprehensive-article-(s)-tier apose subjective-lesson-repopulation-cycle-ordering, daily-directive-focus-(es), upon lifestyle-timed-moments, as mass-objective-cooperative-continuity-concept-(s), consist [local-fatigue / state-fatigue], free-participate-scheduling (un)-active communication-(s), steadfast sacrament [1.ritual-(s) / 2.tradition-(s) / 3.habit-(s) / 4.custom-(s) / 5.inclination-(s)] culminating awhile daily-hour-enumerative-impasse-(s), ration, secondths-minute-(s)-visual-tone-elapse-memory-(ies), ruling consideration, for why subjective-article-(s), remain fundamental for exertion, then shifting interpretation-(s), along conjured-impasse-formality-(ies), directing locational-ecological-agenda-(s), intercoordinating data, pursued by voluntary-participation, in agreeable-fashion, where choice-decision-(s), remain nth-(s)-minute-(s)-inclination-(s), [entering / exiting] from place-posit-time-degree-latitude-longitude-time-offset-location-land-mass-micro-secondths-two-dimensional-terrain, comprehensive developmental-resources-territory-land-parcel-population-exertion-posit-(s)¶|||05m.:16s.50m.s.|||§

> §¢¶$°Time-Signature-Enumerated-Non-Fiction-6/3-timing: Key Signature-Rhetoric-Slight-B-flat-Tempo-Cursive-105°-words-beat-per-minute-vernacular-reading-timbre-sustain.. 1.75-words-per-secondths-rate, apose 0.5701-breath-rest-vocal-deviation-per-minute.°$¶¢§

..§¶200|¶§..If undirect-(ed), how does individual-perspective, find oneself, upon those condition-(s) of existence, influctuated [(in)-dependent // constant-variable-(s)] visually-meandering, those impasse-(s) by-of [action-(s) / motion-(s)] for how our primordial-vocal-usage, impasse [day-grandiose-revolving-tropics-Cancer // Capricorn-day-wave-influctuation-vibrations], equinox-transiting solstice-deviation-range-maxims, juxtaposed our physiological-psychological-subject-ive-translucent-compos-itional-being-(s')-offset, as centripetal-force-inertia, identify earth-central-molten-kinetic-vapor, surging from solar-exomolten-fission-parallel-iron-combust-ion-influctuation-core, between an Tectonic-Earth & Solar-Grandiose-Combustion-bounds, inanimate conscious-recognition, examining expanse, remained upon airless-height-clearance-suspense-black-spacing-bounds, incomparable formulated impassed-deed-(s), by metropolitan-common-effort-reclusion, intersecting inhabitant-(s'), lied study-(ies)-credential-(s), for no government-monetary-economy, first literary-envision, an lifestyle, for one to fixate, an currented-trading-moment-means, mute from those male-domineering-sexual-encounters of youth, listless intimate-attractions', unthought for how each facet of exchange, apply upon conditioned-interaction-(s), by Male & Feminine-basic-physical-function-(s), directing agenda, from analytical-psychological-observation-(s), upon daily-developmental-grounds, impassing produce before intercourse, understanding [where / when], each motion sequenced, throughout each [day-{Summer} / evening-{Autumn} / night-{Winter} / morning-{Spring}] from where at an posit of perspective, only if literate.268¶|||01m.:29s.47m.s.|||§ Rather study-(ies), men have executive-enumerated-congress-ional-senate-legal-power, serving physiological-population-tangent-capacity-(ies).269¶|||01m.:36s. 72m.s.|||§ Women are left remissive upon creative-meander, relative genetic-identity, when comprehension, is mattered due to favor-(s) acted from male-discourse, (not particularly sexual-favor), when land-favors, remain [seldom / scarce], among instituted-intuition, displaying command brotherhood, inept desired-pursuit, elongate-(d)-elaborate-(d), thinking by-an simultaneous timbre-tempo-visual-color-shape-form-weight-measure-(s)-sound-(ed)-reverber-ational-wave-vibration-(s), amid a-the impasse of dead-language, lineage-words-transitions, (un)-sequential [Birth & Era / dynasty / Totalitarian-generation] as-for review, of an population by regeneratable-adult-productions, prepared for consequence-(s) of action-(s), defining impasses, through fluid-motion-(s), remarking each facet-(s) of fact-(or-(ing-(s))), observed-by visual-posit-weight-comparison, marking interim-mean-(s'), per motion-(s)-notion-expectancy-extent, influc-tuated communal-guarantee, from how Shelf-Life & Back-Stock, collect by an overage of goods, in demand-preparation, as supplied-goods, influence each moment awhile daily thought is suppose to be intricate-evoke-(d),| 627 |as-an mathematical-grammatical-elemental-scale-(s)-measure-(s)-sort-order-comparing, of perceptive-carried-weights, still out of an (scientific)-grading-context, reminded from youth, constant comparative [independent-variable / depen-

dent-variable-documentation-observation-impasse-momentary-fo-
cus] from those plotted-hypothesis, per experiment, which pertain
where significant thought, for order-demand-deviation-pric-
ing-funds-balancing, fluent from when whom adjust formal-sequen-
tial-reference-reset-ordering-(s), observed from how human-desire,
partake by [objective // substantial-means'], for survival, but un co-
ordinated, by how temporal elemental-scheduled-location-im-
passe-(s), remain, by an celestial-county-degree-hundredths-thou-
sandths-miles apose visual-meters-distancing-impasse-personal-con-
tinuum, perceiving individual-free-effort-limits, upon extreme-(s) of
centripetal-circular-circumference-parallel, annotated through kinet-
ic-combustion-exomolten-deviation-(s), inertia inanimate-dis-
tance-time-period-tectonic-mass-analysis, convening at national-fed-
eral-rule, offset-territorial-observation-(s), sustemming from [[1]ba-
sic-human-right-(s)-{perception} / [2]tectonic-plate-state-bound-
ary-range-encampment-(s)-{maximum-extent of human-limit-(s)} /
[3]local-jurisdictional-population-(s) {macro-extent of human-limit} /
[4]individual-character {micro-minute-base-extent-(s') {intro-
vert-(ed)-morale-examination} / [5]community-interaction-(s) of enti-
ty-posit-(s) (not transit-passing-(s')) {extraspective-morale-interac-
tive-examination-(s)} / [6]genetic-bastardization-depth-infini-
tive-(pre-black-offset) {nationalism-standard-(s)} / [7]elemental-prima-
ry-quasi-characteristic-(s) [7.1]{Sun & [7.2]Earth / [7.3]Atlantic-Ocean / [7.4]Ar-
tic-Ocean / [7.5]Pacific-Ocean / [7.6]Indian-Ocean / [7.7]Europe / [7.8]Asia /
[7.9]North America / [7.01]South America / [7.02]Australia / [7.03]Africa / [7.04]Ant-
arctica / [7.05]American-Militarial-Air-Zone-Stealth-hyper-veloci-
ty-pressure-space-(ing-(s)) / et cetera of [7.06]Sea-(s) / [7.07]River-(s) /
[7.08]Lake-(s) / [7.09]Canal-(s) / [7.001]Island-(s) / [7.002]Peninsula-(s) / [7.003]Archipela-
go-(s)] for how constitutional-inquiry, fail from diplomacy of hu-
man-character-competency, affecting how each interaction-(s), be
reprimanded, by national-creation, for understanding how to direct
an genetic-community-resource-(s), always pertaining physical-ef-
fort-(s), in direct-association, by subjective-documenting, upon an

frequency of decibel-volume-hertz-notion-offset-reasoning-(s), working literary-article-(s), into singular-subjective-books, exploring how to reset those fundamental-progressive-qualities of population-effort-(s)-observation-(s), apose those elemental-patent-commercial-product-(s)-production-(s), signifying weather individual-character-(s'), are implement-(able) upon an currency-mass-communion-dating-rate-(s), offset globally-sharing-repopulation-tradition, hospitalized, deliber-ately-ignorant-illiterate-hoc-bias,| 627 |spend-pertaining [sport-(s)-ticket-(s) / television-entertainment-depreciation], through idolized existence, unaware of physic-(s), for how mass-debilitation, characterize spectator-interpretation-(s), viewed from how to avoid [sweating / deterioration / fatigue], amid affecting individual-(s')-impact,| 628 |swearing concurrent-power, when concurrent be vocal-physical-vernacular-interactive, apose enumerated-literary-obser-vation-(s), not yet practiced from writer-observation-(s), an [book-editor-(s) // agent-(s) // publisher-dynamic] for terrestrial-maximum, would require writer-author, to spend on an book-editor-cabinet-fashion, an [transit-effort / bicycle-transit-effort / vehicle-transit-capacity-gamma] intercoordinating from Men & Women, structure-orchestrated-task-(s), for juxtaposed [complementing // supplementing], land-perimeter-scheduled-cultivation-(s), tabling those time-frame-(s), of each perimeter-planned-parcel, per individual-person-substantial-objective-resourcing-(s), affecting individual-(s'), or proportion of mass, from Gregorian-Calendar-Mentality, persistently desiring, (as an) method of thinking, reworking process-(es), for arranging agenda-duty-(ies), apose fatigue-effects, upon ac self-impasse-progression-(s), subsequent daily-routine-primary-continuum-effort-(s), as how psychology, require an extraspective-communi-cating, from independent-diary-documenting, for surrounding perimeter-parameter-(s), consent relied monetary-schedule-payment-work-inhabitant-dynamic-basis, [I.R.S. // Congressional-Federal-Taxes // Commercial-State-Taxes parameter-(s)] for after-day-envisioning, from genetic-trust, an relied square-surface-area-perimeter-surrounding-(s), factoring visual-conjunction-(s), sort-deducting, each particular-tangible-quality-(ies), an label-word-thought, for subsequent applying, physical-composition, through moment-(s) of presence, around Impasse & Routine,82L. amid Parameter-surface-area, maintaining sustained-territorial-judgements, in lieu of Federal-Note-Congressional-Documenting...¶|||06m.:57s.39m.s.|||§

§..Preposition-Count-[26 / Twenty-Six]-of.. ..Article-Essay-[4 / Four].. ..Book-Essay-[200 / Two-Hundred].. ..Lines-[82.9375 / Eighty-Two.1.nine-tenths-2.three-hundredths-3.seven-thousandths- 4.five-tenths-thousandths].. ..arranging how those thoughts of prose incur, remaining intended on configuring where Entity & Objective, elapse task-period-(s), directing how period-(s) of skilled-effort-(s), make where dating increments of minuté voluntary-thinking, succession how pick up per date, have hourly-interval-(s) concert an method per individual-operator, inclinating where elapse-impasse, configure what to instrument by worship of land-territory-developing of claimed-tectonic-territory.¶|||30s.75m.s.|||§

..§¶201|¶§..At this moment {3:11 P.M. Hollywood Beach, Florida} I am in an International-Truck, on my way back to a-the headquarters, of {Luke's-Landscape}, by Interstate-95, transition Interstate-595, exit 24, North-ramp-west, returning intermission, to {Trojan Labor}, to fill out my time-ticket, and return to Trojan-Labor.270¶|||15s.93m.s.|||§ In light of my daily-payment, I anticipate deviating my [Fed. /// Med. /// Social Security] learning of a-the various payment-parameter-(s), per manual-labor, with intention, of comprehending exchange of an honest time-ticket-signature-objective-(s), at practice by service through premise-(s), resetting awhile day to day rejection or repeat, have an physical-process, verifying v.i.n.-number, intersection-repetition, rejecting from petroleum, juxtaposed-citizen-book-bibliography-reference-premise-(s), upon contracted-trust-(s'), arranging personal-bias, partial by ones' hoc, where at-will, minimal-sweat-effort-(s), pass visualized-effort-confirmation-(s), for how retained seasonal-fatigue, increment data-receipt-confirmation-(s), transaction managerial-shelf-stock-product-demand-confirmation-(s), when budget is perceived from moderate interactions', waging self-bias-personal-influence-(s), as how [task-(s) / project-(s) / effort-(s) / agenda-(s) / customer-service], duration-act-(s), superseded relayed-process-(es), mattering morale-appreciation, upon perimeter-parameter-surface-area-(s), (as an) outline for judiciary-territory-constitutional-space-(s), specifying independent-character-trait-(s), in lieu per passing-hour-(s), conceiving Impasse & Routine, abound parameter-surface-area-(s)-space-(s), surrounding physical-impasse-boundary, by impasse-imagery-experience, motioning where independent-population-posit-(s),19L. evoking motion-moment-(s), through [duration-elapse-(s) / synapse-(s) / syntax-(s)]...¶|||01m.:34s.99m.s.|||§

§Preposition-Count-[6 / six]-of.. ..Article-Essay-[5 / five].. ..Book-Essay-[201 / Two-Hundred-One].. ..Lines-[19.63 / Nineteen.1.six-tenths-2.three-hundredths].. ..Counting in mathematical-mode-fashion, should matter appose height-altitude-clearance, impasse-zone-(s), for considering extraspective-interaction-(s), with other individual-(s) awhile date be considered from an point of view, for [elaborating / efforting / considering], where point-of-view, documenting how millimeter-spacing-(s), affect how learned interpretation-(s), are suppose to proportion by visual-tonal-vocal-voice-understanding, what object-(s), fulfill those purpose-(s) for daily-aging-survival, when arrangement, increment order-(ing-(s)), for developing how periodical-notion-(s), have an intermediate-documenting-goal-(s), not understood by most illiterate-individual-(s) from commercial-industrial-product-(s)-market-shelf-trading, how territory require an emphasis on population-genetic-apprecia-tion-adoring, exterior-compositional-being-(s), of an similar-class, for appreciating why we are offset here, and alive.¶|||45s.90m.s.|||§

[Article:] [September]

..§¶202k¶§..Now I am at an service-center-headquarter-(s), alleviated waiting earlier-work-impasse-(s), as when each moment, remain in-an temporal-mode, impassing an active presence of action-(s), slipping from absent-memory, evoking along an parallel of posit-position-(s), influencing parameter-(s), where either exterior-ambient-land-territory-surface-area, apose interior-wall-(ed)-surface-area-hall-corridor-offset-(s), as [[a]East / [b]North / [c]South / [d]West compass-rotation-horizontal-revolution-vertical-revolving-axis] deviate apose interior-isolation, losing thought amid active-thinking-process-(es), because populant-(s') have blood-pulse-reflexive-bias-hoc, not ordered by-an sequential-succession of segment-(ed)-portion-(s), confluenced from introverted-interior-perspective-bias, [when / where] focus of perimeter-parameter-(s), are under an technical-sexual-effort-(s), erroneous by dishonest [cursive / sweat / interactive-effort-(s)], effective at duration of mass-product-patent-process-(es), supposed subjective-application, where supple-mental meager-means', attain symposium-dialogue, after educational-study-(ies), legit-justify classroom-hour-subject-impasse-effort-(s), clearly determined where homework-effort-(s), are logically lied, impassing an G.P.A. average word-rate-cycle, per 0.5-tenths, equivalent 10,000-word-(s)-day-rate-impasse, juxtaposed logically incline-(d)-passing, each day evoking hour-year-cycling-(s), rather maximum-continuum-follow-through, per at secondths-denominative-day, as conscious-awake-hour-(s), voluntary-impasse, through two-measure-(s), as letter-word-syntax-cycling-(s), extent [repetitive-word-(s) / Term-definition-intrication-(s)] upon influctuating indeterminant unscheduled time-physical-tension-impasse-(s), where [managerial / general-perspective], comprehend an dynamic of time-impasse-cooperation-(s), hence an thought-process, affecting [secondths-absolute-constant-inferencing / minute-(s)-sixtieth-denominative-elapse-point-posit-(s)-inferencing / hour upon day denominative three-thousand-six-hundredths-second-ths] verifying secondths along operation-(sum-continuum-cut-off-count-(s), where task-(s), (have not been) layer-(ed) simultaneous per mass, (as like individualism ('mass-superlative-conjunct-ion)));348 due to comprehension-capability-(ies), not year-season-layered, employee-mass-population-thought-process, alluding technical-inadequate-thinking-process, Uncreative & Stagnant of Appreciation, because ideology affect influenced-moment-(s), present [when / where], daily thought, can not fully be interval-(ed), from-an through-interactive-process, because count-(s) in tier of [calendar-interval-(s) / time-impasse-count-(s)] think from eureka-moment-(s), simultaneous satire of human-disposal-gore, as illiterate-motivation, [1(re)-fashion-(ed) / 2tool-(ed) / 3calibrate-(d) / 4dispose-(d) / 5itinerant / 6discuss-(ed) / 7inform-(ed) / 8interpret-(ed) / 9assign-(ed) / 10validate-(d) / 11affirm-(ed) / 12assure-(d) / 13+et cetera] beyond an contingent-totalitarian-oblige-objection of genetic-bastardization-national-currency-product-trade-conjunction, attaining youth-age-biological-anatomical-phychological-rate-mode-interpretation-focus, apose motion-notion-(s), as life remain (un)-interpret-(ed) fraction-variable-whole-percentage-(s),| 629 | juxtaposing feasible-fear-(s), synonym Holocaust-impact!!!! So now I pause for an moment, after 15-minute-(s) of typing, to wait for affirmation, weather my payment from Time-Slip-Transaction-(s) & Vehicle-Drive, remain in anticipation for Trojan-Labor-(premise-(s)), returning to Trojan-Labor-payment-formation-premise-(s), by payment from time-slip-(s)-disposal, gathering illiter-ate-motivation, referenced awhile daily-revolving-centripetal-force, drift along an ritardando-latitude-axis-momentum-aspect...¶|||03m.:08s.31m.s.|||§

§..Preposition-Count-[11 / Eleven]-of.. ..Article-Essay-[6 / Six].. ..Book-Essay-[202 / Two-Hundred-Two].. ..Lines-[37.53 / Thirty-Seven.1.five-tenths-2.three-hundredths].. ..Community development-(s), require planned-documented-activities, from when dating each piece of matter, remain significant by lifestyle, learning why thought invigorate why we exist, and documenting an collective-territory, important to understand, why we are suppose to Exist & Genetic-Lineage-Plan-Schedule-develop, tectonic-plate-(s) formal [silt / metal-(s)] existing.¶|||20s.46m.s.|||§

[Article:] [September]

..§¶203|¶§..Later I pick up, at 4:52 P.M. in an Chevrolet-Impala-(black) on a the Interstate-Highway-{I-95}, where I am in transit from Broward-Dania-Beach, Florida, as county-jurisdiction, transgress enigmatic-ties', amongst commercial-impasse, align-injuncting along Atlantic-Ocean-revolving-parallel, signify an slur,| 630 |by brief-mode-moment,-(s') impassing, centripetal-force, while awaiting to return my (group)-time-slip, confirming-trade, as commercial-operation-(s), servicing an government-park-center, via {Luke's Landscaping}, when an brief day-(s)-cycle, emit by such company interaction-(s'), sent off elsewhere, to continue a-the pattern of payment, meandering a the daily $80.00-Gross-Payment, for an $64.30-Net-Payment-cycle throughout a the Trojan-Labor-Premise-(s)-(primarily Hire Quest), curb tension by year, which I am in a the middle to attempt composing an book, non-fiction-constitutional-observational-prose, through human-condition-(s), by-an Totalitarian-implication-(s), yet persuade-(d) [1.lector // 2.leader // 3.victim // 4.order // 5.reader // 6.entrepreneur // 7.writer-(s) // 8.congressional-redactor // 9.citizen // 10.independent // 11.member-moded-motion-(s)-perfect-major-day-micro-season-subter-ior-observational-presence], through notion-(s)-annotation-(s), considering how literature at hand be convey-(ed), from individual-purchase-initial-consideration, through of an Gestalt-process, of sequential-property-(ies), with identity, at those individual-(s)-impasse-(s), ignoring perspective, evocating sensation, awhile force-density-revolving-combustion, not be seen through solar-exo-molten-fission, believing an core of a the solar-universe, when earth core revolving around an solar-exo-molten-magma-pressure-fission, offset iron-aquatic-tectonic-cement-dust-density-matter, for then having [state-boundary-(ies) // state-(s) of matter-composition // stated-vocal-tonal-congressional-sworn-rhetoric], without executive-cabinet-branch-literary-redactive-quality-context, as-for whom can contend-comprehend, simultaneous-thinking-offset, adjusted [vocal-timbre-mode / silent-thought-process / past-reference per literature of purchase-tangent / present-commercial-literature-origin-(s')-presence-production-age-offset-continuum-(formal-regard-(s))] as how I attempt to personify my context, apose an presence of educational-commercial-literary-perspective, (yet asserted, aspectual-facet-(s), that define-(s') what is utmost faulted, difficulty of-those interpretation-(s), as [youth / generation-(s) / decade-(s) / era-(s)] homogenize-present-comprehension, from humanity-mentality-requisite-future-perspective remaining unscheduled upon a-the effort-(s) of individual-(s'), not premising basis of living, conceded interaction-(s), which implore, nothing, sat idle in loop-(s) per independent-routine, while not pushing for-an further-emphasis of living, implemented, casual-impact-(s), meandering formal-role, per objective-dynamic, by America-bastardized-citizen-(s), composited-perspective, not influenced, from lordship, subservient-land, when present, not offset, or perspective not perceive, through how interpretation, constant-influctuate, remissions', that impasse for of locational-development, requiring gated-barrier, affirming what belonging-(s'), are pertinent, to an individuals'-self-worth, serving extraspective-requirement-(s'), in duty to view, how process-progression-(s), operate objective-usage, apose scheduled-physiological-interaction-(s), yet using (T)-time & (S)-schedule, persisting inclined Nationalism & Religious-Bastardized-Belief-System, requiring sweat of other-(s)-collar-(s), for enforcing-work, upon timed-condition-(s), degree per daily-kinetic-impasse, when an circulation of [Birth / Death] apose present-physiological-stage-(s)-ingestion-cycling, (as an) adjusted terrible-stain, lacking constitutional-standard-(s'), upon block-housing-title-discourse, conversing-event-(s), idle Setting & Surrounding-(s'), which pertain extraspective-regard, per resource, throughout usage of such instrument-objective-(s), by-an discourse of judgement-(s), balancing article-variable-(s), intermittent human-kind, typical manuscript-type-ethnic-genetic-variation-(s) (*not elaborating cursive*), before imploring through boundaries, how other personal-land, affect territory-(ies), of human-inhabitant-(s), whom gain from personal-satisfaction, through discoursed-action-(s), rather supplementing visual-impasse-(s), when daily-thought, pertain thinking-process-(es), influencing perspective, to apprehend-data, as does government police political economic-affair-(s), due to, a-the militarial-origin-(s), resourcing, an stain of blood, from historical-action-(s), intertwined, deviation by-of historical-review, swiftly evaporated, apose solar-combustion-impact, as terminological-review, not be an {English} part of history, isolating island-control, through ruled-order, enigmatic, rotary-revolving-water, Filter & Shared, from government-resourcing-(s), acquiring adequate basic-human-right-(s'), at constant consensus, awhile free-will, yield ten-year-(s)-default-fashion, interacting millennium-nature, sustaining substantial-culturing, cultivating terrain, in an fashion of ingestible-supplementing, along an visual-maintain-observational-hoarded-housing-corridor-(s'), that not encompass majestic-grandiose-ambient, [when / where...] dating has not been understood accompanying, mass-population-(s), for bias-proglamation-(s), not study, by youth commercial-fiat-impasse-(s), because how each moment be perceived, from-an apprehensible point of view, default all birth, extreme-moderated-mannered, educational-structure-(ing-(s)), yet affirming Balanced-Block-Terrain, & Population-Formal-Operational-Routine-Impasse-Schedule-Dating-Development, documenting daily-periodical-(s), from-of-an discourse per review, mathematically inclined visual-notion-(s), weather [weight / compositional-count / length-(s) / width-(s) / height-(s) / oblong-circumference-ring-(s) / shape-(s) per-of subterior-form-posit-(s) / compositional-six-side-(d)-plate-(s) / depth-perspective-distance-(s) / percentile-integer-(s), conjunct-perceivable-sight-(ing-(s))] affirming numerative-fashion of [exterior / self-interior-anterior-denominative-fashion-(s) / color-(s)-deviation-(s) / vapor-pressure-(s) / kinetic-solar-terrestrial-mile-exponent-combustion-offset-deviation-rate-continuum-infinitive-cycling-(s) / count-(s) per-of population // objective-(s) // entity-(ies) // infrasturctural-electrical-outlet-(s)-current-(s)-pressure-payment-ration-offset // water-current-median-transit-point-pressure-usage // et cetera...] along an monetary-governmental-means', offset measured distanced-impact, by-our basic-human-effort-(s), receiving newer update-(s) of data-information, from when educational-subjective-article-(s), require adept-thinking,| 631 |upon self-amidst-genetic-mass-deviation-(s)-impasse-influctuation-continuum-living-existence,70L. as perspective base from [black-pupil-spacing /// red-interior // blue-exterior-skin-ventricle-blood-pulse /// white-skin-plasma-composition interior-perceive-synapsing, exterior-wave-vibration-(s)-offset-(s)] per gamma of visual /632/ tonal-space-tension-reverberational-offset]...¶|||06m.:04s.47m.s.|||§

§..Preposition-Count-[27 / Twenty-Seven]-of.. ..Article-Essay-[7 / Seven].. ..Book-Essay- [203 / Two-Hundred-Three].. ..Lines-[72.48 / Seventy-Two.1.four-tenths-2.eight-hundredths].. ..I am working on refining authorship, to reduce those wastes of trees, papering human-living, not understanding why author-competition, work along nature, an means of intensive-reasoning, constantly-considering where we live, proportional to why we conduct, an means of Literary-Residential-Commercial-Entity-reference-geological-proportion-(s), understanding residential-entity-(ies), to proportion an one-book-(author) per year, per nation sales development, apposing an Global-{1}-Tectonic-Plate-(s)-{7}-Geological-{75}-ecological-{14,610}-centralized-historical-governing-library-entity-(s), that comprehend why existence requires referencing, for understanding genetic-allure-purpose, for spacing time lovingly, amongst those groups of 10-35 classed-citizen-individual-(s), by an means of dating-impasse, apose those short-timings, that occur each centripetal-force-revolutionary-drift-129,000°-Geological-Physics-designation-(s), via perpendicular-revolving 360° × 360° by an [Global-sperical-circumference-square-miles-1,219,965,184-mile-(s) // 2,098,240,116,480sq. meters].*geological-estimate ¶|||01m.:09s. 28m.s.|||6,928k.s.§

[Article:] [September]

..§¶204m¶§..Moving on, everyday remains' increasingly evident, [interval-portion-fraction // percentile // scale-rating], not simultaneous by integer-impasse, offset [sleep-future // work-present // off-time-future-present] ranging historical-response, apose an form of literary-documenting, yet morally-responsive, from prior-day personality-mass-proportioning, cycling matter, thorough an letters-words-national-obituary-humanity-conjunction, apose militarial-mentality,| 632 |when personality, stretch an pleather of intervoluntary-cue-(s), resounded meander-impasse-intertwined, momentary active-motion-(s), different-from hyper-tense-(s)-synapse-(s), visual-blood-pulse-unwritten-physical-perspective-(s'), when condition from air, offset apose shelf-life-fact, not cluing how to remerchandise market-(s), from-an active-present-motion-(s), effecting when [product // human // site-resourcing-fact-(s)], at location per central-locked-operation-(s)-responsibility-(ies), presumed, (as an) required-trade, based upon resources, for common-mass-mutual-currency-appreciation,| 633 |designated-offset, locational-directive-objective-article-(s), awhile dating-impasse, remain in part of an year-(s)-posit-designating, sworn-parameter-(s), at working attempt-(s), for considering whom would preside, per process, along limit-(s) of boundary, at non-mathematical-measure-inch-scaling, apose range-boundary-offset-proportion-distancing, as why effort-(s) of individual-(s), are significant throughout daily physical-communal-purpose, for dating & counting, schedule intermittent, physiological-sociological-perspective, listless for how conversation of how resource-(s) are attained, revealing how, we are supposed to work upon instance-(s), as experience-count-(s), not numerically-proportioned, apose grammati-cal-punctuation, as each daily-truth-(s) plan for retirement,| 634 |with [neighbor-(s) / peer-(s) / co-worker-(s) / genetic-community-member-(s) / other interaction-(s)] from when an base-national-perspective of inhabitant-(s), not subject-interact, whom consider a-the Einstein-Max-Planck-Berlin-Brother-perspective, mundane those mutual-interaction-(s),| 635

|through anatomical-physiological-mathematical-gramatical-subjective-visual-hearing-being-(s), whom comprehend how diplomacy pertain an pleather of those controversies-(s) of [diction / colloquial-tongue / rhetoric / vernacular-tone-syntax / condition / term / definition], through-upon [expository / narrative-lessoning-(s)], actively-considering how paper-ink-visual-documenting, affect perspective, as basis whom-of blood-pulse-perspective, matter from mass-objective-development, where difficulty-(ies) of development,| 636 |lye-in, singular-reduced-mass-development-(s), because terrain, matter only if an mass-population, agree to Interpret & Impasse, throughout-upon those stride-(s), that remain implemented, force of [action-(s) / motion-(s)], to be pertain-(ed), when of-an usage of terrain, from industrial land-posit-(s)-retrograde, (as like metropolitan-city-(ies)), cycling interposed, intermittent-observation-(s), juxtaposing, independent-individual-(s'), as citizen-(s), whom remain on an self-pertinent-focus-(es), where-at surrounding-area-(s), because perspective be that amazing, if visual-effect, would confluence sleep, though those surrounding-momentary-individual-(s), offset American-Colloquial-Ambassador-monetary-interactive-trading, not harmonized original-central-topic-(s), but from how [Hittite / Muslim-land / Assyrian / Ageptian-civilizations] aposetier,| 637 |West-Holy-Roman /apose/ East-Holy-Roman-Empire-(s), Divert State-politics, where [intersection // strait & direction-(s)] contrast juxtaposition-(s), physically subjecting, empirical-mode, to matter upon interpretation-(s), of [refer-(ed)-subject-(s) / location-(s) / objective-(s)] continuing progressive past-day-thought, through primordial-impasse-(s), at-an present-day-furthering, every week-month-season-(s), by-an form of schedule, only if extraspective-pertinence, embody-mass-state-expanse-territory-(ies), which matter juxtaposed present aging, when primordial-motion-(s), enumerate denominative-fashion, at procession of effort-succession-(s), not understanding why hardcover-English-literature-article-(s)-interpretation-(s), is not rhetoric, naturally-focusing, apposing relative-output,

as perspective-(s), not commonly-exist, while upon Personality & Physics, pay-effort-(s), determine those embodied-motion-(s), acting as how, one is supposed to inquiry cultural-(s), state-history-genetic-background, proceeding-step, intermittent interval-impasse, as extraspective-extent, be-of, paid-commercial-output, through entity-due-(s), as well self-physique-cleansing, (as an) part of thought, requisite-reset, constant-offset-conjuring, if person-comprehension, filter moral-act-(s), as mathematical-grammatical-hypothesis-concept-comprehension, for-then experimenting constitutional-mass-ordering-(s), [1.district-(s) / 2.village-(s) / 3.city-(ies) / 4.corridor-(s) / 5.town-(s) / 6.section-(s) / 7.et cetera], learning how ecological-boundary-(ies), offset motion-(s)-conjunction, not by-an formal national-maxim-boundary-limit-(s) (for which Americas;[349] (9,293,851 billion square-mile-(s)), remain superfluous eqivalent-reproductive-poplation-space-parcel-(s)-conjuction-method-proportional-relation-(s)), assessing each and every piece of matter, Clear & Decisive, (from an) fashion, comprehending how Handling & Service, have yet been fluently considered, because [when / where] each citizen-being-inhabitant-(s'), properly-ensemble,| 638 |perforated-sectional-ordering-(s), incrementing as visual-perspective, tonal-interpret population, upon physiological-morale, not educationally-obeying vocalized-command-(s), (for an) lack of legislative-default-commencement-quality-(ies), physical-objecting, to free-will-belief, reconsidering totalitarianism, as genetic-valedictorian-summa-cum-laude-magna-cum-laude-cum-laude-tier-percentile-method, [1.interpreting / 2.considering / 3.understanding / 4.perceiving / 5.receiving / 6.apprehending / 7.conveying / 8.lessoning / 9.comprehending / 10.reviewing / 11.interposing / 12.redacting / 13.rewriting / 14.ordering / 15.referencing / 16.inducting / 17.differencing / 18.multiplying / 19.adding / 20.subtracting / 21.dividing / 22.hypothesizing / 23.tabling / 24.collecting / 25.testing / 26.examining / 27.counting / 28.intervalizing / 29.placing / 30.extracting / 31.digging / 32.excavating / 33.developing / 34.engineer / 35.constructing / 36.anticipating / 37.accelerating / 38.enhan-cing, continuum-learning], as-an fault-error of individual-(ism), clout difficult-basis per idea (not thinking), to re-modify-language, upon an mass-voluntary-interpretation-basis, or else gibberish be how sense, motion modern-currency, sustemming (from an) ancient-dead-passing-(s) where worded-subject-article-(s), attain an micro-interpretation-(s)-article-perceptive-conjunctive-interpretation-(s), allotting experience of perceptive-interpretational-vision, apose tonal-vernacular-vocal-letter-word-intonated-survey, perceptive-impasse-conjugation, understanding how article-(s) tier apose Physiological-Analysis & Subjective-Deviation-Intrication...¶|||05m.:03s.47m.s.|||§

§..Count-Preposition-of-[23 / Twenty-Three].. ..Article-Essay-[8 / Eight].. ..Book-Essay-[204 / Two-Hundred-Four].. ..Lines-[71.59 / Seventy-One.1.five-tenths-2.nine-hundredths].. ..cooperative-development, should show how developing territorial-space, require an patience for comprehending an subtle inclination-(s), for simple comprehension, making where location would define how dating identify where each pace by individualism, make how comprehension mandate where thoughts should have an moment of interactive-discussion-(s), making where thinking should motivate an interrogative-discussing per individual-bias, where woman make life amazing, affecting where locating item-(s), make how development of tectonic, an goal for instinctual-peace, when conforming by simple-prerogative-measure, by locale per interrogative-reaction-(s).¶|||35s.03m.s.|||§

..§¶205n¶§..Dimension-(s) are throughout every-single-impasse-(s) per anterior-perspective-(s), indicating an premise of data, which pertain recipient-perspective, at-an impasse of sociological-development, when consideration per thought is reset-(ed)-generationally, because how having object, affect mentality, unconceptive those development-(s) per-of territory, by-an push-(ed)-impasse, during birth-sequenced-metropolitan-remaining-order-(s), from serviced location-(s), authorized, consistent location dialect-year-denominative-aspect, apose presence-mode-(s)-hour-(s)-command-(s), sequenced from past-day-thought-process-(es), by Sphere & Horizon-impasse-(s), celestially-revolving, due formational-physics, because perspective individuate {myself}, by-an reality of illiteracy, envisioning routine, confluence mass-extraspective-(unconscious)-ignorance, as natural-focus-application at land-parcel-(s), interval, data-date-reference-time-physiological-perspective-application, [when / where] [one / two / et cetera...] are how inhabitant-vision, meander-past-upon-future-present-presence, at (not to) an impasse of article-objective-mass-subjective-visual-blood-pulse-vibrational-tense-(s), juxtaposed invigorated ideological-tense-(s), where application of thought, sustain physiological-compositional-posture, when-where aging-decade-archaic-interpretation-tensing, impasse presence of millennium spherical-circum-ference measuring, as how Sphere & Dimension-(s), be upon an deviation of matter-(s), (due to an) rationale of mass, during inhibit-(ed)-impasse-(s), because birth-to-presence-locational-posit-(s), intermittent citizen-inhabitant-(s), whom conjure-data, by-an emphasis of basic-human-need-(s), as when reaction-(s) to thought-comprehension, survey economic-mean-(s), (from an) focus of observational-interpretation-(s), per Self & Prefer-(ed)-genetic-means', Reasoning Mass-Population-Being-(s) Superior-Person-(s), confluencing rated impasse time-frame-moment-(s), as hour-elapse, juxtapose secondths-continuum-effect, upon day-denominative-measure-(s), considering visual-difference, as paper-ink-letter-word-sentence-syntax-synapse-paragraph-liturgical-grammatical-redactive-applicat-ion-thinking, served an dues to tangible-quality-(ies) from surrounding-inferior-weight-(s), as {gravity upon earth apose solar-earth-double-magma-fission-pressure-impact-continuum} Weight-(s) & Measure-(s), an posit-location balance, from land-territory-property-(ies), surrounding-matter-dynamic, for [Dry-Substantial-(solid-(s)) /639/ Wet-Substantial-(s)-(liquid-(s)) / Centripetal-Force-Rotational-Revolution-Pressure-Time-Deviation-Equivalent-(gas-(eous))] upon visual a tonal-scaling, balancing objective-quality-(ies), for matter [Density-(ies) // Buoncy-(ies) // Vapor-Pressure] accommodate depth-perspective [measure-(s) / Weight-(s) / Potting Dimension-(s) // Mile-(s)-tectonic-(radius // diameter)-Distance-(s) / Count-(s)], as depth-perspective-qualify-(ing-(s)), document-signifying hyper-tense-(s)-cycle-(ing-(s)), during only those by-past-impasse-hyper-tense-observation-imagery-recollection-(s), affluent oneself by conducted condition-(s), which matter, as physiological-primary-application, first incurring reference (or no other logic be consider-(ed) of enabled-article-(s) can be, due to Memory & Human-Idle-Sensation-(s)), attain-retaining-data, by-an commercial-governmental-notion-(s)-extent, that remain an part of perspective-visual-tonal-interpretation, not incorporating nasal-smell-mouth-taste-tension-interpretation-(s), for intertwining written-ordeal, not to leave an clear-basis of secondary-place-(ing)-(s), as developmental-logic, leave an continuum-pleather-infinitive of belief, to actually be quite Limit-(ed) & Restrict-(ed), due to predicate-preparation of effort-(s), anticipated composite-matter-(s), in view of motion-(s), awhile our elapse-(s) continue [to // an] fatigue-cycling, as year-impasse, elapses from moment-(s) per interval-time-posit-(s), upon an notion-(s) of calendar-document-dating, affirming how each notion-(s) of mass, require an posit of perspective, to then assess weather their can be-an value to validify information, during qualification-(s) of perceptive-character-(istic-(s))-trait-(s), affirming how Constitution & Commercial-Term-(s) & Condition-(s), continuum cycle-(s) of mass-blank-thought-process-differencing & Hoc-Archaic-Trust-Census-Grammatical, (non-statistical-percentage-count-(s))-Person-Reference-(ing-(s)), working upon presence, at future-retirement-impasse-(s), yet evoked, where relative youth and old thought-process, father an neglected, political-stand-point, upon racial-standpoint, unaware of genetic-standpoint, for genetic-antiquity, not posit-(s), an tier of racial-genetic-character-(istic-(s)), ordering an particular-nasal-lip-jaw-cheek-head-chin-ear-size-style-(s), of genetic-intrication-(s), where at an developmental-impasse-conjuring-interpretation-focus, per celestial-inhabitant-being-(s')...¶|||04m. :06s.25m.s.|||§

§..Count-Preposition-[25 / Twenty-Five]-of.. ..Article-Essay-[9 / Nine].. ..Book-Essay-[205 / Two-Hundred-Five].. ..Lines-[50.5125 / Fifty.1.five-tenths-2.one-hundredths-3.two-hundredths-4.five- thousandths].. ..Humans voluntary-interactive-conscious, shows why developing tectonic-plate-(s), makes how timing require regeneration from the initial-celestial-combustion, due to how creative-development means that cultural-ties, make an emphasis for comprehending why dating from time, should proportion mass-population-surrounding-(s), that identify why each individual notion-(s) why comprehensive-data, suggest where placed-objective, circulate those fact- (s), for understanding why item-production, would serve how extraspective-interaction, is important for living by an means of celestial-dating, apose human-compositional-timing.¶|||34s.41m.s.|||§

..§¶2060¶§..How has passing time, been conceived, if we are intermittent amongst an voluntary routine-superfluous, concernless of other inhabitant-(s')-interaction-(s), throughout various-placement-mode-(s), applied upon, exerted practice, to affirm-tangent-(s), of plot-parcel-territorial-land-surface-area-space-(s), because how posit-instructive-formation-(s), matter from when daily-interpretation-(s), have had an comprehensive logical-denotative-understanding, applying visual-tonal-perceptive-observation-(s), written on to paper, from pen, awhile day-impasse, range objective-posit-(s) at an place, due to efforts that concert an ambiguous fact-(or-(ing-(s))), because family-dynamic, implore moral-human-thinking, consistently affecting those outcome-(s) of interaction-(s), for individualism, continued personal-introverted-behavior, without opening up independent-visual-tonal-interpretation-(s), thinking from-an sociological-commer-cial-educational-posit-(s), belief ignorance, overcome such intolerance-(s) of mass, enabling free-observation-(s), proving how sweat not accumulate gram-(s), to even consider ounce-(s), upon those dating-impasse-(s), per pound-(s), formally applicating logical-perspective, not believing (not to be construed with [memory / idea-(s)]), [where / when] data continue upon those settled reset-(s) of syntax, elapsing-(s), kinetic-imagery, that remain implicated, due to the pressure of [liquid / molten / air-after-effect-motion-(s)] at ambiguous mass misperspective, where I impasse continuum-observation, per vibration-(s), upon an hyper-tense-(s), consortium pupil-retina-direct-blood-pulse-vibration-eye-socket-impasse, visualizing, reverberated intonated sound-(s), upon organ-(s), of ingestion {[1]mouth / [2]esophagus / [3]larynx / [4]throat / [5]stomach / [6]intestine-tract}-{ear-left-lateral / ear-right-lateral-canal};[350] reverberating throughout the interior-skull, live-anatomical of visual-imagery, which not be vocally affirmed, unless currency-exchange, persuade pertained individual-objective-(s), believing freedom, from-an false-sequential-progressive-parameter-(s) of action-(s), instilled sequential-application-thinking,| 640 |mathematically-sectoring, book logical-identifiable-interpretation-(s), through motion-(s) from self, upon ambiguous notion-(s)-extent-(s), statistically-referring by either [base-posit-fraction // percentile // integer // dimension // count-numerative-offset], juxtaposed an denominative-concurrent-singular-topic-count, how extraspective-interpretation-(s), [in / at / of], physiological-blood-pulse, not naturally-focus, how-to persuade other genetic-inhabitant-(s), through an resourcing-cultural-literary-effort-(s), individuating existence primary-impact-(s), which fundamentally, requisite preluded present-tense, because Each & every matter, sustem from how mass-individual-human-(s), continue an natural-impasse, through physiological-compositional-subjunction, affirmed due to faction-(s) of governmental-currency-trade, not viewed, awhile direct count-(s), confirm census-interpretation-(s), because how every inhabitant-soul, not remain contingent, from observed-fact-(s), ancillary national-county-commercial-currency-structural-development-(s), accrued as governmental-perimeter-personal-private-residence-parameter-(s)-continuum-impasse, juxtaposed individual-hyper-tense-(s), by-an creative-technical-methodical-practice,| 641 |impassing past-age-dating-obituary-interpretation-data, when aging affect, an distance-sustaining-practice, developing experience, by impasse-(s) from-of [subjective / objective / genetic-populational-territorial-exchange-practice-method / population-aging-activity-(ies)] handling-article-(s) per mass-population-trade, tracing individual-responsibility-(ies), conversive in comprehending what [basis / application / longevity / genetic-direction / motivation-(s) / desire-(s) / handling-(s)] are interpretable, by exerted-concertion-focus, per individual-application, amid notion-(s) identified-extent, where-of mass-impasse-effort-(s), without an natural-impasse for logical-focus, to be at an thinking-perpetual-rationing, affirm article-directive-imagery, upon live-pulse-hyper-tense-(s), configuring how vision, pertain, natural-tension-(s)-focus, for survival, evoked from [1.desire / 2.pulse / 3.intention / 4.default / 5.want / 6.notion-(s) / 7.motion-(s)-of-living] implored at compositional-posit-(s), of physiological-deterioration-fatigue, amid physics-tectonic-plate-impasse, kymograph physiological-impasse, extenuating [elapse-syntax-rate-(s) / syntax-elapse-rate-(s)] which matter from-an numerative-value-(s), incurring, logical-conceiving, from those inhabitant-interpre-tational-comprehension-(s),| 642 |at-an dormant-impasse, for-of mass-subjective-comprehensive-bailout-extortion-ignorance, remaining stubborn from federally-guided lifestyle-impasse-(s), manuscript-reviewing daily-logic, ascertaining how daily-thinking require physiological-substantial-objective-recollecting, or else non-other-reasoning will incur, for how independent-individual-(s'), believe-in fictional-lie-(s), sustemming from educational-study-(ies), interpreting understudy bias-socialism, rather-than critical-comprehensive-brainstorming-thinking, to pertain perceptive-individual-(s), because how data incur through exterior-elemental-product-entity-(ies)-constructive-impasse, placing item-(s), for thought-(s), from-an mathe-matical-grammatical-moral-impasse, as human-conduct, proportion how thinking is supposed to interpret other-(s) repetitive-impasse-(s), as daily-presence, physiological-posit, when daily-thought-process, remain in an continuum, upon physical-temperate-condition-(s), composition-ally-perceiving, what is not personally-interpreted-conveyed, extraspective-subjective-understanding, different upon millimeter-offset, measuring [feet / inch-(es) / centimeter-(s)] compre-hending from dimension-(s), how distance-denominative-boundary-offset, rely mathematical-algebraic-impasse, throughout an thinking of grammatical-primary-basis-letter-word-definition-syntax-term-process-impassing, aligning focus at historical-factoring, interposing fluent-observational-interpretation-(s), because how [effort / academia-standards / formal-unbiased-fact] conduct perspective, due to feature-(s) of humanity, moding progression-logic, sustemming from-an unconscious-ideological-process, not interpreting, locational-objective, sustaining-an influctuation of motion-(s), perforated, visual-annotation-(s), at visual-perceptive-imagery, awhile daily-physiological-thought, construe processed currency-economic-mean-(s'), affluent where individualism, not segregate totalitarian-burden, upon tensable-effort-(s), as easy-physiological-observation-(s), serve an natural blood-pulse-respiration-cardiovascular-tension-(s), apose centripetal-force, awhile [rotational / revolutionary spin] not revolve motion-degrees°, because how sweat irritate every-perceivable-blood-pulse-being-(s), friction-tension-(s), where dating at, entity-thought-process-(s)-offset, flagrantly-forget, how [high-school-dynamic / University-school-dynamic / Commercial-Co-Worker-impasse / Neighbor-interaction-(s)] are tier-aging, religiously-purchased-held-objective-(s), judging thought-(s), because how currency-exchange, when-where interpretation-(s) have yet impassed, serve an error of honesty, erroneously-conceiving, where celestial-spin is posited, by an hour-dating-year-revolving-impact-motion, where kinetic-impasse-continuum-energy, remain completely-incapable tangible logical-understanding, as human-(s'), influctuate subservient-deposition, describing order-(s), rather blindly-objecting, loyal-visual-genetic-faith, for understanding how thought process, because physiological-perspective, matter upon written-constitutional-proportional-equivaent-sum-inter-pretation-(s), sufficing daily-impasse-(s) of real-estate, from historical-action-(s) of lineage-effort-(s), inept by family-genetic-ethnic-ritual-racial-affair-(s), unburdened daily-American-effort-(s), unspecific because of natural-reliance on Federal-currency, internationally repressing, European-root-(s')-logical-origin-(s)-inter-action-(s), forgetting what have ever conceive, white-black-racial-brown-Asian-ethnic-interpretation-(s), from how language, upon tectonic-plate-impasse, remain individualized, obsolete, extraspective-community-reaction-(s), continuum-applying, visual-tonal-motion-(s)-interpretation-(s), amidst tectonic-ambient-grandeur, sustemming desired-Ephe-sian-emphatic-functioning, pertained locational Mass & Desired-want-(s), coming from-an perceptive-comprehension, per article-(s), which continue an influctuation of variable-blood-pulse-matter-(s), retaining-responsibility, per objective-article-(s)-usage, for how daily-logic, remain mathematically inclined, for currency-impasse-(s), intermittent communal-interaction-(s), where physiological-effort, influence aptitude-point-(s), affirming each and every-notion-(s) of motion-(s), from when exponent-action-(s)-effort-(s)-deviation-(s), affect hyper-tense-conductive-academic-understanding observing effort-quality-(ies), for how mass-family-birth-existence-opinion, matter, due to each facility of constructive-solitude, due to territorial-impasse-(s), consider-(ed) when dating is not exactly-ordered, sequential-schedule-impasse-formal-review-motion-(ing-(s)), evaluating how each movement,| 643 |remain numerative-enumerated, from celestial-denominative-interpretation-(s), awhile, centripetal-force-revolve, an subterior-grand-iose-subversive-impact, from human-presence-perspective, amid those extenuating impact-(s) of Centripetal-Force & Human-Perspective...¶|||06m.:37s.82m.s.|||§

§..Count-Preposition-[26 / Twenty-Six]-of.. ..Article-Essay-[10 / Ten].. ..Book-Essay-[206 / Two-Hundred-Six].. ..Lines-[95.625 / Ninety-Five.1.six-tenths-2.two-hundredths-3.five-thousandths].. ..continuum-perpetual-centripetal-force, rotate around where conjuring human-inhabitant-(s), psychological-observation-(s), emit an celestial-illiteracy, by natural-mammal-human-physiological-introversion, not preferred to think, from how raising your children, apply intermit tent, commercial-wage-trade-scheduling, from why dynamic per individual-notion-(s), matter as when dating is grandiose, from ecological-deposit-sociological-developing, because when timing impasse through celestial-notion-(s), have not formal-direct-concertion, of particular-direct-literary-documenting-celestial-proportion, per individual-introverted-bias-hoc-opinion.¶|||33s.22m.s.|||§

..§¶207p¶§..Psychological-Literary-Characteristic-(s)-inclination-(s), are how human-(s') object to an infinitive-superfluous of motion-(s), configured by Mass-Being-(s) & Notion-(s)-Extent-Count-Referencing-(s), affirming data-information, upon sequenced-logic, thus dating, objective-article-(s), for-an maximum of [empirical-year / season / month-(s)-tier-referencing] assuring that we sleep for period-(s) of dating (not time), from how sleep, blank, visual-thought, upon our offset visualized presence-(s'), mode-deviation-intervalize, affirmed objective-article-inclination-(s) from self, pertinent upon rational-scheduled-motion-(s)-progression-(s), intermittent [mass-motivation / existence-requirement-(s) / person-(s)-method of payment-handle-maintaining] throughout how time is upon [Age / Dating / Weight / Height / Motion-(s)-fatigue / population-(s)-sum-(s)] unconscious those type-(s) of weight-(s), [moved / eaten / drinked / placed / disposed / constructed / engineered / et cetera...] to-be balanced from mass-particulate-consensus, deviating time, that allude through an serie-(s)-impasse-(s), per-of distance-(s), scaling arc-circumference-acre-surrounding-surface-area-range-(s), distanced an further-scheme, amid denominative-balance-(s), when their are those pattern-(s) of repetitive-logic, incapable from voluntary-movement-(s), for discovering how [mind-(head) / body-(torso) / soul-(genitalia-offset)] affect group-(s), by nation-default, maintaining those inconsistencies of collection-(s), factoring each notion-(s), incompetent, under Cum-Laude-offset-individualized-education, from-an historical-crescendo, up until forty-year-(s) of age, not dating by written-documenting, solar-exo-molten-circulating-revolving-central-spaced-kinetic-pressure-constant-propulsion, apose horizontal-earth-magma-propulsion-offset, as tectonic-plates-offset, define from solar-light-conductive-iron-kinetic-aquatic-current-pressure, an east-west-centripetal-force-revolving-force-impact, without literature-mathematical-bore, perceiving how confluence of median-self-composure, perceiving moment-(s), indicates what define-(s) human-visual-tonal-observation-(s), as how thinking equivalent nature, for not being creative, amongst those, similar-character-(istic-(s)), when not preferred, to consider why warranted-opinion, suffice whom, comprehend What & When, conjure by cognitive-thinking, an preference from ethnicity or city, to have restoration, live in an nation of idle-individualism, because train of thought, instill an primary-basis, upon mouth, through inhabitant inconclusive extra-spective-reasoning, mattered through an count-progression-scale-table-deviation-(s), that are intentional from thought upon physical-application-(s), where-when concertion-(s) of distance-(s), have remained ignorant of logical-impasse-(s), because what clue-(s), are to cue any purpose,| 644 |where developmental-product-(s), serve commercial-governmental-process-(es), uninclined, psychological-prose, diverting calendar-date-posit-impasse-(s), not timed yet, at intermediate natural-grade-impasse-posit-(s), because Objective-Usage & Subjective-Lesson-Article-Paragraph-Application, monitor those progressed genetic-mass, in an technical-manner, not able to have an purpose to give, deducing unreceived think-(ing), which continue-purpose of living, by Hunger & Survival;[351] Trade & Storage;[352] Supply & Demand;[353] Family & Community;[354]

Mass & Self;355 uninfluential, present-modern-motion-(s), without prepubescence, hourly-wage-aged-morbid-impassing-thought-process, [taking // trading word-(s)], from Government-Education & Sustaining-Past-Death, subconsciously-sustaining, Inclination-(s) & Indication-(s) of Grandeur,37L. upon those Process of Self & Extraspective-Extent...¶|||03m.:06s.72m.s.|||§

§..Preposition-Count-[17 / Seventeen]-of.. ..Article-Essay-[11 / Eleven].. ..Book-Essay-[207 / Two-Hundred-Seven].. ..Lines-[37.38 / Thirty-Seven.1.three-tenths-2.eight-hundredths].. ..Consideration for how others exist, is significant, for developing how each dating-hour-period-(s), de nominate through physiological-motion-(s'), action-(s) that matter from when an 483-years his tory-ancestry, showing why earth-molten-core-centripetal-force-drift, rotate-revolve at an velocity which impact how our daily existing pertain how documenting, remain for identifying where excavated-abyss-(es), should be more ecological-state, from where local-resourcing should bring an purpose for sociological-intellectual-interaction-(s), per individual-moment, when thinking should be enjoyed, attempting where proportioning of elemental-extraction-(s), meander why timing is incredibly fast, for how mode remains at proportion per mathematical-inclination-(s) at capacity per interior-entity-construct, due to elemental-dynamic-existing.¶|||38s.84m.s.|||§

..§¶208q¶§..Translucent-Dating, apose say Common-Age-Stage-(s)-impasse-Auto-Biography, is naturally-retarded, by enlistment-elemental-resource-composition-collecting, for forging-kinship, yet existing, by technocratic-development, democratized through our mean-(s'), for-of reason-ing-data, as how formal-objective-impasse-(s), subservient-location-(s), when visual-posit-(s), confirm an Visual-Range & Isolated, for when during cubicle-space-(ing-(s)), function-(s) of motion-(s), feel-review, intermittent colloquial-reader-customer-worker-mode-(s), apposed sleep, casing daily-review, [mingled-hoc // bias-mode-subterior-emotion-(s)], (ill)-logical-(ly) by mathematical-consideration, from a the effort-level-applicate-(d), during each inference-(d)-effort-(s), that matter awhile focus of moment allot-experience, but without Formal-Documenting & An Loyal-Genetic-Population-Inference-Reference-Society, what impasse-(s) at possessive-generation, can object physiological-mode-(s), various-(in)-variable-notion-(s), occur, yet mass-independent-introverted-state-motion-(s), premising what can occur, where at physiological-stance-stature-posture-implication-(s), synapse-reference-visual-tonal-review-function-(s), effer-vescence-(tly)-affirm-(ing), micro-compositional-function-(s), apose [physical-motion-(s) // action-(s) // act-(s)], intermittent present-formal, personal-individual-effort-(s), at site-(s) per-of action-(s), having each motion-(s), storage, from collective-effort-(s), at natural-mechanical-construe, documenting-numerical-mathematical-compact-data, where-at effort-(s), transiting academic-(a)-system-(s), offset coordinated conducted-affair-(s), when information not instruct contributed daily-life, comprehending how logic, thought because-of voluntary-impasses', self-mediated, constant-continuum-update, can only be from referencing latter-presence, as how to interpose at-sectional-land-surface-area-space-(s), maintaining from physical-presence, an objective-retaining, surfaced, when holding from toiletries, an logical-shelf-process, by-an bill-subjective-belief, not Valedictorian-diploma, directable those five-percent-deviation-directive-tier-effort-(s), from developmental-literary-Dewey-decimal-referencing-activity, apose religious-daily-impasse-(s), as President Barack Hussein Obama, vocal-talk-carpe-tunnel-rhetoric, conversive-microphone-speech, as how AARP-retirement-method, affect zen-youth, through [1.issue-(s) // 2.affair-(s) // 3.topic-(s) // 4.article-(s) // 5.matter-(s) // 6.talk-(s) // 7interceding-(s)-intrication-entity-impasse-(s)] evidently unartistic, for Associates'-comprehending-time, as hour-university, for calendar-sitting, affect outcome by perspective-(s), slow-largo-mediated, still-blank-sensational-moment-(s), as mathematical-calculation, have commonality by manuscript-context, apose cursive-leadership, (as an) expertise, during throughout, present-work-cycle-thinking, [incline-(d) / indicate-(d)] as how movement, supersede through at-site, voluntary-formal-action-(s), where solar-etch-(es), bypass exterior-physical-objective-use-(s), gamut from how the nature of tectonic-plate, modify agricultural-soil-grass-growth, not exponent-tenths-scaling, decimal-offset-(s), as like influctuated-motion-(s), tangibly-calculated,| 645 |plugged-in-integer-mode-(s), to-an decimal-tenth-numerating, by whole-one-hundredths-category-maxim, not by-an motion-action-(s)-Dewey-decimal-directive, hence an ambiguous-unconscious-unknowing, supersedes ignorance, as molly-coddle-habius-corpus, yet with state-federal-legislating-comprehension-deteriorating, from 17th-Century-Country-Origin-(s),| 646 |an 18th-Century-National-Founding, modified by 19th-century-industrialism, amid 20th-century-patent-resourcing, through modern-swearing, by those uninquisitive-library-(ies), from past-knowledge, of subjunct-(s) of book-subject-(s), not interposing diction-vernacular-order, for by-an lack of common-reading-interpretation,| 647 |whole-percentile-fractions-percentage-(s) of independent adult and child related article-per-week-dynamic, as how adult balance-(s) an day-seventeenth-twentieth-cabinet-range-gamma, affirming senate-sub-committee-bill-article-rhetoric, as how without cum-laude-crowd,| 648 |who from education, actually read school-year-proportionally, an 0.959-g.p.a.-youth-article-(s)-comprehending,| 649 |as how per-class-article-(s)-understand, impasse-(s) calendar-(con)-census-chapter-documenting-method, holiday insol-uble, total-denominating dimension-elemental-compound-development-(s), from how tectonic-rotational-default-posit-(s), reset numerical-ordering by [dimension-(s) // 360°-circumference-interior-point-radius-percentage-ordering /// weight-count-(s) //// unit-count-(s) ///// reading-grammatical-punctuation-redact-offset-symbol-(s) ////// interior-room-storage-collection-count-(s) /////// exterior-general-conjunct-compositional-matter-depth-perspective-count-(s)—(these prior seven-numerical-count-order-(s), are article-(s), for an mathematical-book-analyzation-tier-extenuating-article-offset)] alluded fortitude intervoluntary-development, sustained at focus, vocally-reverberating, what word-term-interpretation, adjust, juxtaposed those [scientific-density-(ies) // Substantial-Buoyancy-viscosity-(ies) // vapor-centripetal-force-tectonic-plate-wave-drift-water-pressure-deviation] proportional by common-output, that is meant only when effort is concerted, at point, for intellectual-property-(ies), because how data affirm applicated-precedent, per decade, as millennium is moving how common-effort-(s), are pertained upon| 650 | [women-repopulating // land-territory-constitutional-family-merit-continuum-documenting] amid mass-commercial-effort-(s), weak where week-(s) slip away, from formal-interposed-reference-locational-geographical-document-conjuring,| 651 |as men are politically-upheld through-an Military-Mass-Resourcing-Millennium-Offset-Sacrifice & Common-Idle-Expenditure-Game-(s), revealing an idiotic-overburdening, by systemized-white-collar-lazy-effort-(s), at blue-collar-locational-development-(s),| 652 |implicating each thought-process, from-an past-history-manner-(s), through subjective-book-article-(s)-paragraph-subject-tier-impasse-deviating, present day, engineering-effort-(s), faulted revolution-dynamic, because comforted-motion-(s), trust Family and Ethnicity-(ies), through commercial-product-public-object-(s), not concerting where an visual-tonal-vernacular-genetic-lineage-state-geographical-territory-mass, would be how living is more perfect, when collected such an extraspective-identity, envisioning how global-development, would be while alive, and in an offspring-lineage-directive-comprehension-focus.271¶|||04m.:58s.71m.s.|||§ Why is engineering not considered from inhabitant-impasse-(s), when daily-mass, rely self-life-conveyer-plant-shipped-purchase-(s), evident an cultural-locational-land-territory-development, hand-maintained, when vision is formulated, by self-interpretational-imagery, at impasse per hyper-tense-mode-moment-(s), affirming-past-data, by count-(s) through present-exponent-reading-reset-(s)-interpretation-(s), from when religious-language-(s), serve an ideological-conceiving, that will only ever apply, when method-(s) of trade, not extreme-effort-(s), candor national-counter, existing at impasse of state-constitutional-taxed-bill-policy, proficiently chaptered, analyzing sociological-population-citizen-inhabitant-(s)-national-duty-(ies), when affirming tectonic-plate-denominative-method-(s), for moding how perspective is to then redact those impasse-mode-count-(s), for then an environment-ten-miles-acreage-title-handling-per-person-visual-physical-cardiovascular-hour-day-impasse-(s)-continuum-balance-scaling,| 653 |where-when patent-conceptive-objecting, delve how population, if for thought, pertain an constant-formulating, for [lifestyle / afterlife / interval-elapse-syntax-moment-(s)] upkeeping-inquiry upon tangible-coverage, where loyalty is required upon physics, to then formally-consider [why / who / what], matters, surrounding-vocal-truths, by-an particular-visual-personal-book-reference-guidance, that yet commonly-exist, for how my book not be globally-sold, and yet much fact, require creative-funding, to construe how trade can become refined, by those general-transaction-(s), working toward an perfect-genetic-national-constant-reformation-cause-reasoning, succession-purpose, through numerical-thinking, offset population-applicating mathematical-enumerated-development-(s), currently newspaper-obituary-offset-county-counted, by manuscript-manual-written-documenting-(s), reset affirm, each individual-notion-(s)-extent, where data require conjugated-data-information remain pertinent to considering logic, upon an means' of trade, that is regardless of thought, understanding when date can be counted upon hour, affirming minute-secondths-interpretation-(s), matter-time-inverted-proportional, weight-mea-sure-count-incremental-proportioning, measuring through vapor-pressure, on top of tectonic-plate-handling, weighing those [solid // substantial-density-(ies)], for then time-distance-denominative-territory-weight-person-usage-reset-count-(s), by when incremental-focusing, suffer an storage-security-dynamic, that define how national-allegiance, affect interacting reference moment-(s), from English-liturgical-article-(s), observational their-of [physics / tectonic-plate-(s)-impasse / objective-engineering-development-(s) / land-property-engineering-development-(s) / et cetera], by-an creative-engineering-impasse-(s), developing, affirmed perspective-posit-(s), when significant-order of operation-(s), amuse impasse of tectonic-plate-(s), rather then denominate juxtaposition cartography-centripetal-force-time-offset-count-(s), which can only be perceived, by-at visual-resonance-wave-vibration-(s), detected through visual-tonal-perspective-impasse-(s), ordering week-day-offset-(s), understanding when to motion-(s)-effort-(s), by-an locational-posit-(s), pertain tectonic-significance, because when Wear & Tear of objective-(s), account reasoning, amid kinetic-friction-action-(s)-energy-(ies), remaining an part of [tangible-psychological-perceptive-solid-object // substantial-intake // electric-wattage-usage-(s)], surpassing linear-development, as grandiose-entity-façade-construction-site-(s), journey an anticipated-entity-dynamic, longer than dictated-objective-subject-focus-schedule-agenda, fixat-ing tectonic-constitutional-elemental-compound-article-development-(s), formally-incurring, from when dating is suppose to affirm why notion-(s), are acted upon, an [geological-national-territory // geological-state-territory // geological-county-territory // geological-city-territory // geological-city-district-land] impassed at land-subjective-harmonizing-visual-tonal-circulating-smell-taste-breathing-respiration-perspective-(s'), desiring objective-want-(s), Expository & Explained, how each moment upon hyper-tense-impasse-(s), pertain fluent extraspective-operation-interaction-(s), fatigued through hourly, aging-limit-(s), understanding we all have visual-tonal-competency-perspective, to adjust how [physical-cardiovascular-territory-coverage // objective-function // auxiliary-function // subjective-documenting] cycling various acted-product-resources, that forego understanding at-an subjective-intensive, interpreting-competency, [where / when] data initially-find, present-article-act-usage, deviating geological-territory-coverage, converging from self, pertained later-interpretational-impasse-(s), when perceptive-studies-referencing, not conjure why article-(s), matter during revolutionary-rotational-impasse, apose interior-introverted-independent-individual-perspective, which incur by population-influence, only abiding from money, or currency, by wage-per-hour-dynamic, apose those salaries, whom predict an fluid-ordering-operation-(s), trusting demand, from citizen-desire-unit-characteristic-(s), implored from physical-offset-geological-territory-impasse-usage-(s), as article-(s) upon self-visual-reverberational-interpretation-(s), suspend grandiose-motion-(s), motioning tension-(s), upon individual-offset-conjuring, without incremental reference-logic, scheduling-review, because ignorance by sweat-motion-(s), continue during timbre-character-genetic-impasse-(s), analyzing geological-territory, by how to sustain [water / electric-metal-wire-currenting-state-national-resourcing-review], when attempting how to predict repopulation-development,124L.|654| by-an mass-human-population-developmental-conjugation-developmental-excavating-impact-(s)...¶|||09m.:13s.02m.s.|||§

§..Count-Preposition-[20 / Twenty].. ..Article-Essay-[12 / Twelve].. ..Book-Essay-[208 / Two-Hundred-Eight].. ..Lines-[124.64 / One-Hundred-Twenty-Four.1.six-tenths-2.four-hundredths].. ..Attempting to find elemental-resourcing-refinement, shows how Voice & Context; demonstrate an purpose for lifestyle, where Women are naturally superior of male, for reproductive-lineage-extenuating, from how those stomach-reproductive-breast-reproductive-vaginal-outlets, are intended to propose why women have an dating-tempo-half-timing, from how posit of individual-(s) imbue where locational-impasse, have an requirement of limited-restriction, when 10-miles-circumference-anterior-crust-posit, be how no [man / male / child / woman / et cetera...] affect any impasse, except their two-dimensional-meter-(s) in compositional-posit-(s'), because present-date, is behind an normal-observational-effort-(s) by when individualism not get that extraspective-voice, should be timed by subconscious-perspective, when [secondths / millisecondths], are in turn by natural-position, because how environment, continue reasoning upkeep, apose those foundations of nature, by how identity is not an means of conscious-interaction-(s), when community work slow from incremental-progress, at cause for an global-lineage-repopulating-position-principality, instilling woman as charge over settling, from male subservience to locational-motion-(s), per motion-(s), developmental-bones-lineage-celestial-deposit-collecting, from how layering juxtapositions of matter, know an sedimentary-distance-limit-(s), working on collecting resourcing-(s), of elemental-resourcing-tier-trade-(s), by [origin / offset dynamic] due to an calendar-dating-progression, sub-par, natural feminine appreciation when worn by celestial-deterioration, for why women are to be so appreciate-(ing / ed) when aging has yet recovered those elemental-circulated-resource-(s), by [deposit / circulation // Centripetal-force], per moment at sedentary-stance .¶|||01m.:18s..82m.s.|||§

..§¶209r¶§..Work is required, by each and every citizen, for fluent-process-progression-operation-action-practice-(s), offset parameter-parallel-development-(s), while organizing square-surface-area-spacing-(s), controlling surrounding-parameter-(s), for-an reasoning upon mutual-inhabitant-(s), whom intend to give trust, at national-effort-(s), ordering progressive-procession of data, [measuring-(s) / kymograph-time / weighing-(s)], apose distance-territory-offset-grandiose-wave-vibration-(s), as each minor-minute-compositional-piecing-(s) of matter, then be organized, while an basis of analytical-order-progression-enhancement-practice-applicating, deviate Population & Constitution, required when term-(s) of trade, comprehend competent-motion-(s) direct-instructing, comprehensive-mass, orchestrating-progress, by-an means' of interposed-inflicted-interpreted-impasse-(s), meandering plots of populant-(s), as-an piece of nature, which matter to an, whole celestial-ordering, deviating thought-process-(es), where death-remain-(s) subsequent, circumference-radius-anticipation of life-existent-impasse-(s), because location-development-(s), not fully-function, perceptive-ordering, referencing notion-(s), apose physical-action-(s), when dating is not exact, of a-the thought-process-referencing, for how to affirm, where-when-word-initial-thought-(s), be evoked, understanding vernacular-context, from vocalized-effort-(s), by competent-individual-(s), whom interject progressive-motion-(s)-dynamic, where routine is sworn, at longevity of action-(s) {Laborer / Doctor / Lawyer}, monitoring monetary-exchange, at monitored-impasse-(s), tone-key-signature-staff-note-suggesting, indicated [note // rest] interpretation-notion-(s), referencing manual-transit-dashboard-nation-reliance, for an default-manuscript-manual-monetary-exchange, devising how hyper-tense-impasse-(s), rely auxiliary-dependency, not considering regulated individual-inhabitant-(s')-effort-(s), because how parallel-perimeter-surface-area, is [visualized // envisioned] from, non-inhibited-conception-(s)-form-(s), unless characterized, active-interpretation-(s), conjure how parameter-(s) become covered, because logical-interpretation-(s), serve order-(ed)-orchestration-(s), at currency-generated-open-agency-location-(s), formal by, commercial-practice-(s), identified by-an offset-developmental-juxtaposition-(s), pertained by nation-state-tax-territories, acting upon dating, referenced [impasse-(s) / action-(s) / motion-(s) / interpreted-method-logic], where visual-interpretation-(s), sustem an actual-impasse-extent, as respiration-vapor-circulation, depict because, extraspective-confirmation-(s), conjure required comprehension-data-fragment-piece-(s)-recovering, by pertin-ent actual-factor-(ing-(s)), furthering our genetic-obligated-pursuit-(s), upon those [momentary condition-(s) / circumstance-(s) / temperature-locational-impasse-(s) / term-(s) / method-(s) / motion-(s)] thus coherently-comprehending logic, self-motivated-action-(s), upon an religious-mass, communal-mean-(s), signifying purpose, through those visual-tonal-perspective-(s) of will, during physiological-impact, upon section-(s) of tectonic-plate-boundary-offset, revising bill-location-(s), by article-depth-interpretation-(s), because of [methodical-practice

/ effort-(s)], achieve objective-point-(s)-parameter-(s), for repopulation-development, only from when regional-resource-(ing-(s)), accumulate tangible-data, by agenda-inclination-(s), for-an confluence acceding-developmental-spacing-(s), because tectonic-plate-development-(s), techno-autocrat, common-intrigue of character-preference-(s),37L. by-an calendar-schedule, living upon population-mass-engineering-constructive-development-(s) amid parallel-surrounding-tectonic-boundary-development-(s)...¶|||02m.: 46s.54m.s.|||§

§..Preposition-Count-[14 / Fourteen]-of.. ..Article-Essay-[13 / Thirteen].. ..Book-Essay-[209 / Two-Hundred-Nine].. ..Lines-[38.08 / Thirty-Eight.1.zero-tenths-2.eight-hundredths].. ..Directive of paper paging upon tectonic-ecological-impasse-proportioning.¶|||09s.28m.s.|||§

..§¶210s¶§..Cognitive-Thinking have seismic-ways', balancing how subconscious-thinking-grade-point-average-grade-(s), influctuate relative, vehicle-function, enabling as shelf-methodical-action-(s), caution-intuitive-mundane, acts upon when-where evocation is dull, lackluster for negated-inhabitant-physiological-interaction-(s), as [omission / ignorance] of debt, remain without an clue for how interpretation-(s) of trade, remain confluenced confidence-reliance, based upon mass-governmental-commercial-production-trust, affirming tradition-good-(s), from-when observation of mass-(es)-means', relied article-(s) adaptation-(s), as purchase-(s) from governmental-currency, formulate-accounting census teeter-along, voluntary-effort-(s), accumulated, during impasse-formation-(s) impassing from when event-(s) are in ordered-action-(s), because how dating impasse-(s), through time, intertwining-mode-(s), account for each formulated type-matter-(s), continuum through reasoning, visual-vocal-opinion, emphasized stringent-formal-reference-article-counting, free-will-bias-partial-belief, conjuring how any individual-pertain-s'-thought, awhile statehood-territorial-mass-population-deviations, space off major-objective-production-(s), as commercial-development, relate from historical-impasse-(s), common-offset-reset-comfort-(s), settled at formally-centralized-designated-parameter-(s), for population-educational-social-hoc-extortion, not commercially-counting, literary-focus, common-ly accepted, amidst consideration-cycling-(s), for reasoning [Essex // annexing] personal-controlled-development-(s), in circumspection, by shared-engineering-objective-(s), building an payable-cycled-entity, when only housing can be directive so far, yet by an common acreage-family-housing-expansion-constituting, because how manuscript-application, interpret count-(s) of [ounce-(s) // feet], or [gram-(s) // inch-(es)], proportioning how objective-rating, affirm an space, as like wallpaper, by an room, fluttering differentiated variable-(s)-time-mode-impasse-(s), which matter from what controlled-that-corporate-purchased-object-(s) may not be from how per upon self-this-object-possession-(s), for how community creatively recycle, action-(s), by objective-motion-(s)-effort-(s), from how basic-human-right-(s), persuade American-Funda-mental-Issue-(s), because methodical-objective-storage-apprehension, yet be communicated, by an tool-function-interpose-usage-wear-n-tear-deterioration-count-(s)-cycling, under an month or day, of usage, relevant to those condition-(s) of environment, not formally-implicating, time-secondths-interval-factoring, upon deviation-(s) of letter-word-inductive-data, that remain upon [sentence-deductive / sentence-inductive point-(s)], referencing paragraph-extension-tangent-ordering-(s), for visualizing why an book, should calculate from Mundane-Time-Impasse-(s) & Calendar-Repopulation-Location-Expectancy-(ies), that matter, when those impasse-(s) of life, configure weight-volume-composition-comparison, influctuating mode-median-moment-inclin-ation-(s), that sustem accounted-order-(s), because how self-identify-data, gamma an historical-back-wash, that is not formally-referenced, due to when inhabitant-(s) are preferenced by physical-action-(s), because where dating slip pass physiological-self, by those mean-s', article-concert, subjective-select-reference-periodical-book-ordering, practicing year-season-schedule-posit-(s), pre-set-intentions', matter where data in presence of time, impasse affirm instructed-information, by-an peer-aging-proportional-interpretation-(s), not indicatable, as bias remain opinionated, from visual-tonal-vernacular-perspective, directing mode, ranged at no distance, scale key-reset-measure-gauging, self-weight-usage, juxtaposed exterior-densities-buoyancy-force-wind-press-ure-(s), where stillness appears upon [direct-pupil-perspective // eye-color-peripheral-perspective], mental-capacity-handle-notion-(s), by how interior-distancing, affect tectonic-furnishing, by entity-entrance, isolating-visual-distance-(s), measured by an cubic-feet-interior-rectangular-prism-constructive-economic-cycling-application, not hosting by home, intermittent governmental-political-state-county-territory-hour-yearly-interaction-county-maxim-spiral-street-circulating Day & Transit, comprehending how repopulation, lacking yearly-mass-repopulation-offset, thirty-year-generation-age-tiering, tangible-matter-(s), upon regular-citizen-focus, not competent-vernacularly-common-book-referencing, by personality-formal-directive-acceptance-enabling, where person-(s) have objective-entity-responsibilities, not layering object-(s) in range-open-distance-(s), because how dry-interior-objective-development, an [entity // hall dynamic], not make public-enclosed-hallway-(s), upon tectonic-plate-developmental-impasse, identifying how to balance developments, by [pasture-grass-soil-agricultural-crop-(s) // horti-cultural-pasteurizing] physiological-subjective-article-dating-documenting, along calendar-hour-(s)-impasse-capacity, or bleak-blank-wilted-optimistic-fatigue, work per hour, not advancing visual-physical-vocal-vernacular-subjective-article-paragraph-reading-extents,| 655 |per environment-entity-territory-compositional-currency-circulation-dynamic, per constructed [resid-ential // commercial // governmental // militarial // religious-artistic-cement-conceptive-construct-ive-entity-planning], by how an denominative-governmental-dynamic, not be shifted yet, (from an) [library // government // university-library // religious-family-citizen-library-(ies)], because corporate-commercial-objective, have an patent-instructed-manual-clause, not referencing-data, because of off-time-moral-literary-documenting-guidance, as mathematical-inclination-(s), remain not specified, furthering what notion-(s) not get described, amid [{11ft.W. × 13ft.L × 8ft.H.}-1,144-sq.cu.ft.-personal-room-space, /// {5ft.W. × 5ft.L. × 8ft.H.} -200-sq.cu.ft.-storage-space //// {4 × 3 × 2.5}-30-sq.cu-ft.-commercial-tool-box-space-(s)///// {8.5ft.W. × 53ft.L × 9.167ft.H.}-4,129.58334sq. cu.ft. ////// et cetera..]{8ins.L × 10ins.L. × 2.1269-insH.}{435-context-page-(s)};[356] confluencing how our physical-extraspective-competency-mode-effort-(s), cycle-experience, reset by how transit of pavement-clearance-thoroughfare-impasse-(s), gamma existence, by tectonic-plate-density-water-vapor-pressure-exosphere-impasse-complexity, above solid-friction-magma-density, occurring as constitutional-observation-(s), which matter in-an judiciary-review-process, from effort-(s), by written-literary-extent-(s), require, micro-inter-pretation-intrication, for understanding how to impasse-terrain, from housing-restriction, upon public-transit-commercial-area-impasse-(s), factoring each of those article-(s) of information, by either individual-observation-(s'), (bias-personal-opinion-hoc), parameter-ordering, by an person-al-possession, of territory, not transit, elucidated, common-mass-incompetency-ignorant-instructed-data, banked at work, referencing object-(s), but without an possessive-claim, when neutral-entity-usage-operation-(s), cycle any genetic-operations, at an given-day-working-time, for how resourcing, not interpose night-time-action-(s), while in use of millennium-differentiation-shift, because [topic-(s) / issue-(s)], not matter mathematical-grammatical-punctuation-matter, through citizen-manual-library-referencing, how each facet of instruction, not direct mathe-matical-comprehension, or literary-remark-(s), for redacting along an reference-competition, that can dutifully-lack-interposed-presence-competition-(s), much to do with any family-individual, sharing passionless-possession-(s) (grey-matter) of life, because resourcing requires weather in excavating-mechanism, or walking around acreage-territory, (non)-sweat-effort-(s), having no indication of progress, because how modification of spacing-(s), pertain through an national-default-requirement-(s), unless focusing element-(s), transition an storage-factory-resourcing-process, non interpretable, by saving-(s), to be applied upon, an territorial-claims', for population-methodical-land-development-progression, differentiated, because how each mode is seized in an imbalance of basic-human-right-(s')-enabling, not grammatically-counting, physical-psycho-logical-motion-action-(s), at actual-visual-tonal-vernacular-interpretation-competency-percept-ion, working on communal-notion-(s), from motion-(s)-extent, applied when progression of action-(s), succinct pertinent numerical-mathematical-enumerated-numerative-deviation-(s), requiring denominative-enumerated-higher-tenths-column-offset, apose an Dewey-Decimal-One-Hundred-whole-(s), by decimal-extent-interpretation-ration-proportioning, [action-(s) // motion-(s) // directive-document-(s)], reformulating object-unit-factor-(s), that affirm governmental-official-(s), not spinning and killing, because suppressed-empathy, apply by [black-(s) / white-(s) national-racial-dynamic (holocaust-victim-(s) / gypsy-(s) / et cetera)], that have government-position-(s), in an eighty-three-million-population-mass, deviating [racial / ethnic qualities], differed by three-million-military-servicemen, across serviced-branch-(s), breeding sociological-trust, by-an national-archaic-enlisting-cycling, archaic-stillness-Treasury-development, as politician-(s), fleet person-maintaining, casting territory-mass-development, in an estimated population county-mayor-councilmen-personel-state-government-governer-personel-federal-president-house-member-(s)-senator-(s)-bill-handling, where agent-(s) construe basic-inalienable-right-(s), under military-impunity, lacking inventive-patent-writer-documenting-labor-effort-(s), competing literature, by an collection of [paper // ink // currency-note // commercial-objective-units-per-individual-effort-impasse], documenting not initiated by free-lance-effort-(s), amid logical [free-press // religion-sect // speech] affecting how work-motion-(s), measure by an quality-scaling, from more repetitive, through [scarce-resourcing // modifying], monetary-balanced-awareness, yet refuting how tangible-grandiose-scale-(s), are required for literal [macro-literary-hardcover-sales-extent // micro-context], thousandths-word-intrication-(s), not confirmed, through how education, simply commence, our competency, deviating commercial-payment-(s), when motion-(s) are not expected, exceeding experience, from genetic-lineage, from one through another, to experience-matter, not fulfilling mathematical-grammatical-annotation-observation-(s) by [still // motion-(s)], [visual-interpretation // tonal-interpretation] for how much time of consumption, any object require, avoiding how saturation, affect how over-produced-anticipated-shelf-stock-life-product-(s), clutter our entire trading system, as where to collect, (from an) history of [mute-avid-thinking-contingent-visual // tonal-perceptive-observation-(s)], incorporating how Environment & Entity, persuade communal-unity, offset by national-elemental-resource-trading, scarce per county-development, for even where an [American-city-county-resourcing // plant-operation-(s) // Shipping & Handling-unit-(s)-pertain], dimensional-mathematics, has yet subterior, how human-exo-tectonic-visual-tonal-clearance-physiological-cardiovascular-square-miles-population-dynamic, not think per decade, how to advance an modern-refresh-repopulation-civilization, as we wait an entire millennium, to aspire weather any mass can sit-read-redact-listen, for fundamental-findings, offset (from an) [reader // writer dynamic], requiring reader to comprehend personality, to avoid being mentality, as for now personality, guard-off, from civil-decorum, slow by millennium-dating-method, rather time in secondths-documenting-dynamic, upon scaling by [walking-feet-distance-measuring // trotting-meters-distance-measuring // vehicle-driving-mile-(s)-distance-measuring] not due to payment-(s), but comprehension-regard, of physiological-self, upon exterior-mass-extraspective-genetic-mean-(s)-existence, influctuating self-prerogative-(s), from exterior-observation-(s), by communion-other-(s), select an physio-logical-stillness, upon practiced-action-(s), from value-(s) of data, mattering where memorandum-information, only apply during work-major-singular-focus-impasse-(s), as does Commercial & Government-premise-(s), competently-comprehend, legislational-literary-inquisition, inhibited at an lack of intensive-focus, from subjective-medium, from where usage of objective offset-environment-priorities, that not communicate through population-genetic-faith-church-seminar-(s), an period of time (under an month) for live-ink-written-calendar-practice-(s), for first documenting practice, then coming into daily-locational-offset-distance-measure-extent, fractioning all those objective-subjunct-entity-work-variable-(s), by-an documenting of visual-perceptive-compositional-tangible-mass-cubic-area, for then offsetting measured-object-(s), by an distance-harmony-transposing, each and every singular-matter, because of land-space, where magic not exist, modifying tectonic-plate-(s), by first practicing from physical-tangible-cardiovascular-public-terrain, what impasse is by an daily-monthly-county-location-maxim-transit-work-entity-(s)-extent, conjuring what mass-population is required for developmental-deviating-physical-effort-(s), equating wage by those entity-denominative-resource-circulating-process-(s), that would inevitably, require an literary-guideline, for what [[1]tool-(s) // [2]interior-décor // [3]appliance-(s) // [4]book-shelf-(ves) // [5]desk-(s) // [6]table-(s) // [7]infrastructure // [8]food // [9+]et cetera to purchase..], ignoring present-gifting, for saving money, by generating by labor-literary-economic-disparity-effort-(s), amid white-collar-citizen-(s) & blue-collar-citizen-(s)-composite-effort-(s), together in an lack of developmental-spaced-effort-(s), because idle-nature, fear sweat, from understanding when article-(s) would pertain, upon the environment, linguistics-literary-competency-grading-interpretation, so as write, I 100%-book-article-(s)-interpretative-directive, but without an first-book-editor-interrogative-interaction-(s), affirm, as is like literary-agent, an continuum-book-literary-determining-referencing, when geological-environment-distance-(s), are required human-perceptive-population-clearance-transit-impasse-coverage, compared by yearly-birth-rate-(s), for covering by repopulation, an formal adult-contingent-transition-generation, by those inhabitant-surrounding-effort-(s), pinned by [physical-respiration-rectangular-prism-12 // 14-feet-compositional-fatigue-visual-tonal-vernacular-term-perceptive-competent-(s')] when from reader-dynamic, the common-socialism, may bias by personality-economic-handling, an belief in basic-human-rights, being shared by humanity-economic-disparity, when in the middle of present-election, tragedy of Western-Treasury-Idle-Behavior, display the lector-competent, an error of common-reader-discussion-(s), premised-by offset-mathematical-dimension-observation-detail-analyzation-(s), commonly-unappreciated, by central-denominative-figures, for whom to worship, by honorable-deity-effort-(s), construing interposed-mode-interval-time-period-(s), from range-hour-mode-minute-(s)-block-(s), calculating how [sitting // typing // pause // posture-forward-direction // glancing // looking // considering // motioning // objective-acting // et cetera…] influencing an total enumerated, denominative-observational-capacity, that require [spectacle // awe // fantasy] to influence how consideration at directable-objective-unit-handle, offset at perimeter-outlined-spacing-(s), typically [fenced // hedged // demarcated] interval-(s) by human-physical-family-voting-perspective, affirming government-usage-rate-(s), impassing [1perspect-ive-primarily / 2territory-secondary / 3land-spacing-(s)-thirtiary / 4subjective-reference-inferencing-quaternary / 5population-agreement-quintary / 6objective-construe-sextupuary / 7physiological-motion-(s)-practice-documenting-septuary / 8time-interval-(s)-of-secondths-upon minute-(s) (motion-(s)) // hour-(s)-deviation-(s)-octonary / 9calendar-input-effort-(s)-documenting-(s))-dating-age-century-juxtaposition-(s)-nonagonary / 10reference-article-(s)-subjective-written-redactive-schedule-focus-pattern-action-(s)-pattern-(s)-decatiary / 11+et cetera….] upon resetting-offset-(s), not affirming extraspective-dynamic-(s) community-thinking, rather motions-prerogative, which beset, how belief, matter by subtle-mute-national-shelf-life-eating-enabling with one another, without an accurate-land-acre-development-transit-timed-commercial-oper-at-ions-calculating, for affirming data-information, shifting letter-participle-word-(s), by-an vernacular-diction-vocal-pitch-time-signature-frequency-key-signature-note-tone-accent-inter-pretation-cycling, validating-action-(s) at vernacular-vocal-impasse-(s), to affirm how date is confirmed upon [hour-elapse / syntax-cycling-(s)] (from an) mass-perceptive-consensus, not deciding what mass-genetic, signify an greater-purpose, than another, for while in the groove of documenting, self-analyzation, rule prior, exterior-voluntary, but when it is difficult from hoc and bias, to direct off-time-action-(s), from when commercial-communal-development, swear on biblical-hoc-reset-data, through manual-transit, commercial-instruction-(s), when militarial-command-(s), not Wall-Street-Floor-Scramble-Scream, apose how senate-congressional-forum, variate house-lector-speaker, by those [vernacular // vocal mode-(s)], which differentiate interposed, perceptive-unit-fact-(s), filtered upon perceptive-identified-inhabitant-objective-unit-usage, as how we calculate-data, while upon bill-acted-motion-(s), not thought by-an survey-literate-act-(s)-notion-motion-(s), conferred before ratified-legislation, an thousandths-mass-signature-document,| 656 |to draft into notion-(s)-documenting, when topical-direction-(s), serve duty-(ies), transfiguring elemental-matter, by-an national-objective-commercial-patent-transit-directive-(s), identifying those geological-commercial-flaw-(s) of sociological-medical-legal-trading, when not abiding subjective-psychological-communion, at those impasse-(s), by residential-commercial-transit-lifestyle-(s), moded by free-will-belief, not documenting from literary-book, an juxtaposition of annotated-objective-engraved-context-(s), for an labor-wage-earning-process-(s), pre-deliberated, by concerting an legislative-conceptive-program, for then an commercial-stock-market-funding-mechanism-dynamic-(s), pertaining an forum for denomin-ating an open-operation-(s), when an influctuation of malfunctional-literary-youth-competency, not document from subjective-lessoning, but context, an general-means', for comprehending how to trade, by an adult requirement,| 657 |to specify grammatical-punctuation-geometrical-sociological-political-commercial-communal-observational-count-(s), omitting how psychology, remain in limbo, weather to use or not, (as an) basis of legal-documenting, rather than judiciary-legal-paper-chaptered-ramification-(s), yearly-wasting paper, for their remain an incapable reading-listening, for hardcover direct personal-prerogative, from how perspective is individual by fatigue-deterioration-merit, requiring extraspective-interaction-(s), to affirm weather one-another, [fit // suit // configure], how accumulation-(s) of data, are to be denoted-relayed, (from an) literary-year-reset-week-per-book-referencing, that would suit how to recollect, such vast extent-(s) of Geology, the true error of physics, not engineering by off-time-creative-schedule-resource-collective-strict-effort-(s), affirming how deposited-developmental-mass, whom believe over-riding a-the impasse of natural-effort-(s), when effervescent produce, under-produce, capable competition-(s), because a-the metropolitan-disposition, remain without an inquiry of population-(s), for-how agenda, dissolve-from, [racial / ethnic national-background-(s)-wage-effort-(s)], not understanding why literary-purpose, matter upon natural historical-development-(s), due-to visual-understanding of data, from when fact-(s) instructed-information, [direct-(commercial) // command-(military)]| 658 |individual-obligation-(s) by-of mass-genetic-resource-engineering-construction-currency-(method of elemental-resourcing-production)-production-(s), serving methodical-comprehension of article-(s), during data-verification-(s), through sparse-spacial-trade, intermittent national-corporate-reliance, to those militarial-development-(s), rather-than commercial-development-(s), which emphasize why belief, not meticulated mathematical-velocity-motion-(s)-impasse-mile-(s)-clearance-deviation-(s), that are impassed, by physiological-perceptive-being-(s), yet attuned for why such observation is pertinent upon life-style-living-observation-(s)...¶|||16m.:58s.66m.s.|||§

§..Preposition-Count-[50 / Fifty].. ..Article-Essay-[14 / Fourteen].. ..Book-Essay-[210 / Two-Hundred-Ten].. ..Lines-[202.7125 Two-Hundred-Two.1.seven-tenths-2.one-hundredths-3.two-thousand-4.five-tenths-thousandths].. ..This book is contextual.¶|||09s.31m.s.|||§

..§¶211t¶§..Numerator-operation-inducting, influctuate physiological-visual-tonal-interpretation-off-set-comprehension, by grammatical-letter-word-syntax-referencing, from what unit-item-(s), are pertained, at-an population-agreement-deviation-(s), disproportioning developmental-product-(s), because how thought [reference / deductive clues], cycle-comprehension, upon our individual-compositional-physiological-structure-(s'), offset without an understanding of extraspective-visual-interpretation-(s), for common-identify by sensation, under schooling top 15%-educational-classification-tier-(s), by governmental-mundane-regulated-guidance, for how objective-article-(s) subject reciprocal-enumerated-impact-(s), from how physical-psychological-social-subjective-perceptive-dynamic, oversee overall-mass-secular-state-enforcing-governing, awhile independ-ent-family-perspective-(s), creative-living, without creative-artistic-assisting, from-an extra-spective-group of cooperating-individual-(s'), whom abide by-an means' of trading, because their remain an lack of genetic-elemental-compound-extraction-(s), from population-concept-schedule-article-identifying, supposed-impasse, at placed tectonic-constitutional-territory-(ies), cheated by our founding-father's-complex-superego, not expecting future-capable-commercial-industrial-objective-item-personal-usage, havoc by-an exceedingly grammatical-punctuational-etymolog-ical-ontological-reference-highlight-emphasis-reasoning, present-modern, not evoking hyper-tense-(s), through physiological-perspective-conjuring, affirming lifestyle-impasse-(s), as logical-interpretation-(s), remain dormant, amid denominative-relation-(s) of self, upon simultaneous-year-aging, juxtaposed calendar-dating-deviation-(s), as per-year [associate-(s')-years / bache-lors'-years / graduates'-years / masters'-years / philosophical-years-(census-count-birth-offset-deviation-(s), per year] apose how individual-(s), compare [written-subjective-standards / cardiovascular-standards' / muscular-standards'] through whom matter upon an [physical-territorial-development // written-subjective standards by community], because how daily-impasse, develop territory, upon tectonic-perspective, because visual-tonal-perceptive-focus, remain inputted by rationale, because where one-individual, would remain [stanced / sat / laid] as schedule cannot as of now, verify population, for mathematical-statistical-dating-offset-document-verifying,| 659 |each inhabitant, in an conclusive-juxtaposition of article-neutral-data, for literary-unconscious-copyright-competition, to rid of any childishness-ambiguity, that date-matter, throughout an formal-impasse, asserting what matters, particulate-moment-(s), only if vocal-action, impose upon-an surrounding, when setting transient-perspective, focusing from self, an uncultivated-tectonic-posit-(s), offset mattering architectural-aspect, upon self-housing-construct-ion, gathering engineering-construction-center-(s), affirming schedule-subjective-juxtaposition-system-ordering-(s), when dating impasse division-deviation, numerical-integer-visual-impasse-constant-influctuation-(s), for mathematical-timbre, yet conceived by common-religious-letter-word-belief, using colloquial-language-soly, from life-expectancy-denominative-governmental-census-retirement-aging, applicating (not pre-literary-calculating), numerical-offset-integer-rate-(s)-offset, when impasse of day-(s), interval-accede, an impact of simultaneous-twenty-nine-thousand-three-hundred-eighty-three-day-denominative-jutaposition-family-race-genetic-mass-population-increment-offset-(s), as existence matter, from only where developmental-terrain, matter because how tectonic-crust-resource-surface-area-article-(s), remain upon, assisted impasse-(s), per action-(s'), that are intended, at influential-moments, to glorify cultural-development-(s), because how daily-dynamic, have each hour, per fifteen-minute-(s), have an simultaneous-offset-rate-integer-purpose-focus, relevant juxtaposed grammatical-punctuational-observation-(s), an referencing-method, pertaining objective-genetic-direction, not by-an national-enabling, which have forgotten, purpose of education, by prose-pursuit, enhancing from historical-prose, present-date-reflective-denotative-history, directing article-(s)-usage-(s), because how developmental-impasse, pertain practice-existence, amid [action-(s) / motion-(s) / et cetera..] where surface-area, require at all time-(s), an monitor, for developmental-impact-(s), directing efficient-comprehension, by-of engineering, cursive-context-document-writing, [text-ambiguous-perspective-identification-(s) / text-constitutional-notion-extent / two-dimensional-graphic-elaboration-(s) / three-dimensional-modeling] defining an mutual-standard-(s), for maintaining national-independent-currency, remained-reliant, from saturated-mass-subjective-objective-maintaining-focus-(es), when scheduled-agenda-operation-(s), develop an offset-visual-tonal-perspective-(s), as hourly date-impasse-(s), so frequent, by self-extraspective-juxtaposition-(s), when no natural-individual-interpretation-(s) of human interaction exist, resolve individual-family-independent-impasse-(s), counting (inter)-national-observation-(s), at regional-state-colloquial-tou nge-usage, interpreting from cardiovascular-musculature-day-night-week-awake-cycling-effort-impasse-(s), throughout centripetal-force, by water-magma-tectonic-density-buoyancy-untangible-impact, sufficing [natural-parallel-plane-air-volume-space-exterior-sensat-ion-(s) // interior-respiration-push-impasse-blood-pulse-respiration-(s)] from cardiovascular-affect, because explanation, not be finalized-governmental-observation-(s), for literary-article-(s), supposed to be extended on mathematical-grammatical-punctuation-live-Federal-civilization-subterior-offset-observation-(s), but when government-bill-bailout, suffice, those inhabitant-(s), which modify each fact of office-work-class-time-impasse, not implementing homework, by written-typed-text-identification-(s), lacking pertinent focus, from birth, to affect later-age-growing-development, set upon an retirement of two-children-maximum, where no planning has been commonly-completed, amid physiological-impasse-(s), when metropolitan-basis, further assert factoring of mass-population-bill-(s)-schedule-subjective-objective-impasse-dating-time-interval-(s)-verifying, acceding municipal-union, by-of bastardized-genetic-retardation-effort-comprehension-(s), because how impasse-(s), offset how metropolitan-citizen-schedule-count-(s), wage-mundane-develop, surrounding premise-(s'), above such basis, being from pupil-ventricle-blood-pulse-synapse-tonal-vocal-vernacular-impasse-interpretation-(s), meriting monetary-appreciated-development-(s), rather than concerting presence, where visual-vocal-genetic-inhabitant-interpretation-(s), concert throughout, an watch-count-duration, at approximate [locational-developed-enter // exit-impasse-(s)], conferring elemental-origin-resourcing, from-an direct-emphasis-control, as those exterior-introverted-family-citizen-(s), influence notion-(s), (from an) birth-regeneration-offset, as genetic-hyper-tense-(s)-function-(s), have an reset compartmentalized-subjective-article-(s)-essay-paragraph-word-vernacular-ordering-concept-compre-hension, apose entity-statured-formation-(s), when-where logical-vocal-conversational-impasse-(s), are intended to intercontort, word-term-definition-(s), affirming how to complete, [manual // book-read-redactive // local-communion-survey-task-(s)], accumulating surrounding-terrain, to demonstrate an inculpable-error, for how reference-chapters, have been library-book-shelves-posit-accumulated, an sporadic, deductive-reasoning, from-an presence-prior-observat-ions'-{requisite}, [where / when] live-presence, suppose future-interpretation-(s), awhile document-dating, interval those focuses at Population & Dating-(s), existing unassured by-at comprehension-(s') of parameter-sequence-data, [where / when / what], cycle monetary-wage-resource-objective-instructive-focuses, requiring an commercial-literary-book-progressive-order-ing,| 660 |by-an physical-logical-focus, directing why hard-data-interpretation-(s), matter upon select-focus-(es), from-by-an extraspective-vocal-visual-impasse-(s), pre-cycling-inter-pretation-(s) from [information / data / logic], applied by-those incompetent-subordinate-(s') of commercial-at-will-work-place-dynamic, simply attending, blank by whatever disposition-direction, post-education-selethisize, not retaining subjective-book-(s), for genre-interposing, how geological-existence, persuade an numerous perceptive-focus-(es), while homage, [complete / finish / finalize-developmental-enterprise-(s)] upon those condition-(s) of monetary-currency-policy, handling compositional-conjunct-population-commercial-resourcing-developmental-mass, according through sequential-direct-visual-location-ecological-offsets, by county-commercial-objective-entity-acquisition-offset, by those posit-(s')-mundane-subjunction-format-ion-(s), establishing involuntary-tracking, by-of those belief-(s) from superlative-population-mass, not indicating which climate-environment, is most suitable-adequate, for geological-elemental-extraction-(s), for an state-national-elemental-commercial-patent-product-schedule-impasse-sociological-cycle-practicing, whom should impasse an entity-objective-instructed-structural-designation-(s), amid centripetal-force-day-revolving-year-revolutionary-physiological-impasse, apose perceptive-sleep-pattern-week-month-season-impasse-(s), counting by paper-page-half-blank-sheet, [objective / subjective-article-(s) / book-(s) / population-interaction-(s)-extent-(s)], extrapolation-(s) of hereditary-surrounding-inhabitant-(s), whom delve Cause & Reason, informing why motivation-(s) of grandeur, are so significant, for voluntary-cooperation-(s), by-an focus when genetic-mass-civilization-desires', have yet [explained // expositorialized], how Territory & Offset-Sexual-Genetic-Repopulation,| 661 |implore Feminine & Masculine-Physique-(s'), during-throughout, daily apprehension-impasses', instruct-pertaining, location-entity-product-objective-article-(s), without an subjective-chapter-lesson-article-(s)-comprehen-sion, instructed by commercial-entity-handling-practice-(s), amid subterior-syntax-isolation-(s) of logic, not demanding an effective-direction-(s) of tabled-information, simultaneous at group-mass-communion-equivalent-focus-(es), identifying density-measure-(s), apose substantial-weight-(s), for how measuring mile-(s), scale distance-(s), through supratangible-mass-impasse-(s), patent-variablized, visualized logical demanded presence, by sequential-action-(s), attempting schedule-impasse-affirmation-(s'), for-of objective-entity-instructed-collection-(s), which matter through-out self-mass-physiological-data-impasse-(s), [1.denoting / 2.connoting / 3.implying / 4.differing / 5.affirming / 6.conjuring / 7.conceiving / 8.referencing / 9.complicating, logical-subjective-interpre-tation-(s)], aware by linguistic-(s')-literature-shifting-(s), attaining personal-substantial-posit-(s), for visual-observation-tonal-vocal-vernacular-perspective-(s), juxtaposed linear-centralized-concerted-hyper-tense-(s)-documenting-impasse-function-(s), amid duration-(s) of focus, sustem from self, upon extraspective-directive-presence-youth-aging-surrounding-(s'), circumstancing territorial-none-claim, by work-objective-collective-instructive-pre-presence-demand, apose retirement, throughout an militarial-commercial-political-economical-sociological-effort-(s)-impasse-(s'), voluntarily disbanded, from union-legislational-constitutional-grandiose-populat-ional-effort-(s), by-an methodical-religious-unions', where formal-pertinent-practice, have-an singular-state-national-agenda, not by, tectonic-plate, American-monetary-objective-trade-value-(s), for handling way-(s), of off-time, blind-objective-continuum-trade-trust-(s')...¶|||09m.:00s. 31m.s.|||§

§..Preposition-Count-[28 / Twenty-Eight]-of.. ..Article-Essay-[15 / Fifteen].. ..Book-Essay-[211 / Two-Hundred-Eleven]. .Lines-[19.56 / Nineteen.1.five-tenths-2.six-hundredths].. ..$45.59¢¶|||11s.35m.s.|||§

..§¶212w¶§..For how many inclination-(s) by tension-(s), exist, our core-basic-compositional-posit-(s)-blood-pulse-skin-bone-tension-(s), logically adapt scheduled-activity-(ies)-impasse-(s), during our feature-(s) of finger-(s)-tension-(s)-reflexive-motion-(s), fixed from how prose matters, through transitional minute-particulated-force-(s), inertia upon hyper-tense-(s)-visual-reverber-ational-function-(s), sequenced simultaneous, impasse-(s) at palm-centralized-posit-(s), balancing locational-hold, for objective-article-(s)-documenting, by-an wrist-ligament-muscle-nervous-reaction-(s), simultaneous visual-vocal-letter-word-term-definition-syntax-tensible-perspective-interpretation-(s), mattered-upon, contacted-effort-(s), because when impasse-(s) by physical-tension-(s), afflict how presence of pre-conjured-interpretation-(s), annotate through perceptive-visual-peripheral-notion-(s), affirming independent-individual-sight-(s), during patterned-presence, formulating an routine, while repetitive-serie-(s), collect an art postulate-emphasis-found-observation-(s), defined by monetary-document-(s), from elder-default-finding-(s), at offset-posit-location-(s), as lineage-mass-cooperative-effort-(s), allude validified-observation-(s), through articlizing-(s) daily-offset-(s)-impasse-(s), in-an continuum-voluntary [substantial-ingestion / domestic-tranquility / objective-purpose-(s)], by-an common-means dating-(s), identifying logical-visual-interpretation-reasoning-(s), envoy monetary-guidance, amid move-ment, that-is simply evoked, as how any of us specify thought, for enveloping experience per [event-(s) / lesson-(s) / study-(ies) / seminar-(s) / outings / et cetera of mutual-communion] tallying-experience, through discourse by-of mundane-motion-(s), affirming impassed-inclination-(s), thought by objective-scheduled-designated-experience-(s'), intermittent voluntary-impasse-(s), to post-educate, commercial-realm, determination-(s), visual-surpassing, surface-area-spacing-(s)-routine-impasse, from-an accord of factor-(s), suited upon debated-proglamation-(s), proclaiming-observation-(s), by entity-scheduled-agenda-(s), visualized from cognitive-relapse-(s), where populant-citizen-(s), conceive educational-background, deductive-reference-reasoning-skill-(s), supplied-developmental, serie-(s) governmental-currency-resourcing-(s'), | 662 |upon-an vast-territory-(ies) of tectonic-plate-(s), acquiring refined-commercial-corporation-product-concept-purchase-(s), packaging-consumer-product-(s), bounded because metropolitan-progression-(s) balance modern-federal-locational-objective-(s), not prevalent, pre-written-conceptive-observational-impasse-observation-(s), where commercially-viable-trade, matter as their remain retirement-fashion, trading personal-objective-(s), juxtaposing deviated-dating, by-an segregated-communion-developmental-producing, pertinence from of peer-(s(')), as territory duraté various-conscious-moment-(s), where-when-why, population-mass-elemental-agenda-resource-schedule-requirement-(s'), throughout perpetual-centripetal-force-rotational-revolutionary-shift-drift-grandiose-default-impasse-(s), defining means' of determined-formal-survival, as for why, comfort-(s)-surround, an non Perceived & Desired, studying of how consumer-construe, settle-routine-(s), awhile each fact-(s) by-an extraspective-inquiry, array [1.task-(s) / 2.effort-(s) / 3.driving-(s) / 4.competency-(ies) / 5.interpretation-(s) / 6.morale / 7.capability-(ies) / 8.inclination-(s) / 9.objective-interest-(s) / 10.subjective-review / 11.scheduling-cycle-(s)], in fruition of aging, thus comprehending, each value-(s) of logical-prose, from self-motivation, deviating redactive-reproduction-(s), by-of mechanical-force-patent-compositional-objective-(s), in lieu of national-sociological-dependence, juxtaposed family-lineage-particular-production-(s), from how objective-thought, juxtapose territorial-elemental-developmental-claim, upon population-objective-schedule-usage-(s'), planning objective-article-savings', in-an indication-of retirement-fashion, living through ambervision commercial-labor-engineering-constructive-quality-(ies)-development, from when relative-mass-quantity-(ies), offset obliged-voluntary-citizen-(s), whom swear statehood-constitutional-development-(s), apposed default-national-currency-exchange-method-payment-progressive-modern-cycle-era, as when, [who / where / when / what] we appreciate, matter by-an intent of directive-agenda, articlizing an subjective-parallel-matter-(s), to pertain upon motion-(s)-impasse-(s) per movement-(s), Cause & Reasoning, why our presence-mean of significance, characterize physics-elemental-impasses', 47L. throughout self-individualism-ego-offset-(s), directing instructive daily-rotation-(s), amid subterior revolutionary-continuum-impact-infinitive...¶|||03m.:33s. 91m.s.|||§

§..Preposition-Count-[17 / Seventeen]-of.. ..Article-Essay-[16 / Sixteen].. ..Book-Essay-[212 / Two-Hundred-Twelve].. ..Lines-[48.05 / Fourty-Eight.1.zero-tenths-2five-hundredths].. ..counting date-(s), matters for how hour-(s), tier, subversive, those sedentary-value-(s), practicing on an tectonic-plate-location-(s), that matter for Genetic-Body-Experience-(s), by visual-vocal-vernacular-interpretation-(s) inducting apose [secondths-inclination-increment / millisecondths-moding] for proportioning where numerical-inclination-(s), state how those values pertain, by each indication from grammatical-understanding practicing physical-effort-(s), onto paper-tree-trade-refinement, balancing those mass-populants, apose an timing to contradict how other ethnicities, may be able to document each value, under constraints to determine why morale pertain at an difficult physics terrain developing, when fact-(s), are only temporal, commercial-impasse-trade-effort-(s) from residential-family-dynamic, present historical-view, from an juxtaposition to each physiological-mass-compositional-proportion-perspective, [under / inbetween / above], contingent thought, abstracted awhile mathematics-subversive-scale-mode-documenting, refine from ambiguous, who can not cooperate, without those literacies by calendar-dating time-posit-inclination-incremental-indication-(s) document-ordering, by an cohesive-working under strict-diligent-formal-focus-stress, per [secondths / minutes / hours], under day thought, affecting why thoughts have yet been soly alone those causes from civilization-sociological-introverted-individualism, pre-Cambrian Commercial-Industrial-Developmental-Constructive-Architecture-documenting, by thought from Genetic-population, what resources, make those modification-(s) from calendar-deposit-(s), matter by celestial-observation-(s), pertained order of thinking being [1.sustain / 2.retain / 3.maintain / 4.pertain / 5.attain] tectonic-compositional-matter, at making longevity matter by geological-terrain, (lukewarm / cool / Cold) by territory-claim, deeming how hot-climate-territory-extreme, offset apose cool, and cold-extreme, because how conduct from inhabitant-dipitipational-compositional-being-(s), exist per individual as [*Female* / ***Male*** / Children], remain [limited-(city) / restricted-(county) / boundary-(state) / contained-(Federal) / Off-set-Border-Dispute-(s)], demonstrating why area-individual-coverage-act-(s), are suppose to be paper-counted, by an scaling from circulated-bank-entity-dynamic-(s), as when timing from Automatic-Teller-Machine-(s), would have how schedule, would then be intricated, along those fact-(s), for arranging what formal territory-schedule-population-survey-wage-trade-trait-(s), happen by positioning what payment per individual, notion-(s), how act affect from off time, how morale, is suppose to interpose with hourly-wage-labor-composition-concertion-(s), practicing from work-wage-mode, an consistency of subconscious-reasoning, for think-considering, what thoughts define for-mal necessary-action-(s), upon ones stance per day along centripetal-force-revolving-drift¶|||02m.18s.56m.s.|||§

..§¶213x¶§..September-transit, revolves in to view-(s'), as primordial-base.273¶|||02s.91m.s.|||§ Banning Cartography-Gregorian-imagery, temperate physiological-character-composition-ventricle-blood-pulse-tension-liquid-substancial-plasma-ventricle-eye-motions'-reflexes, as celestial-ambient-peripheral-vision-(s), visualize-sound-tonal-thought-(s'), amid mode-(s), at perceptive-depth-perspective-distance-(s), that are massive by physiological-character-(istic-(s)), from how visual-offset-perceptive-spacing-(s), not reverberate [hot / cold temperate-reflexes] yet only upon physiological-being-structures', at state-county-maxim, for whom visualize-ten-mile-radius-maxim-fatigue, not in an acre-per-juxtaposition of [physical / psychological-documenting / housing-objective-collective-storage-space / exterior-commercial-objective (*non subjective)-entity-(ies), without inclination-(s) / indication-(s)] what meaningful hyper-tense-(s), revolve, an evocation of-an emotion, voluntary amid intersected-repetitive-planned-development-construct-ion-(s)-impasse-(s'), afflicting dialogue-interpretation-(s), for where dating, affect word-(s), works' by wage-hour-subterior-scaling, not measure how literary-legislational-matter-(s), enumerate visual-letters-words-terminological-documenting, from whom-(s')-signature, interpret-confirmation-(s'), where-at article-note-receipt-particle-piece-data, identify median visual-logical-reasoning-secondths-watch-bit-(s),| 663 |against-apose-juxtaposed, five-minutes-mode-deviation-numerative, juxtaposed life-hour-expectancy-perceptive-mode-offset-(s), Bold-letter-type-face, where peripheral-perspective, abstract periodical visual-reverberational-impact-(s), underlying what remark-(s), remain significant where determinant-visual-impact-extent-(s), upon where physiological-vibration-(s)-distance-(ing-(s)), juxtapose how human-beings' measure acreage-altitude-existence, rectangular-prism, person-(s)-ignorant-mathematical-political-handl-ing, to apose how vision conceives', an logical-value, for interior-vision-interpretation-(s')-recollecting, as modification-(s) of elemental-data, matter when imagery-mode, offset perceptive-formalities', described in-kluxed, as for [who / where / when-mutual-neighbor-importance] signify-means', factoring Warning & Caution, when physiological-impasses' affect oneself-timbre, adjunct time, because calendar relay, physiological-blood-pulse, upon muscular-mundane-simultaneous-equivalent-impact-(s), when millimeter-vapor-pounds-buoyancy-ounce-grams-viscosity-exponent-tons-density-exterior-wind-pressure-supratangible-impact, offset internal-part-(s) of numerical-enumerated-data-identification-(s), affirming where one apose natural-physics-visual-tonal-offset-variable-adjunction-impasse-(s), interpreting by-an lifetime-timbre, amid visual-incandesant-solar-exo-molten-current-friction-glow, where daily-reverberational-idle-thought-(s), envelope into intersected collective-by-pass-(es), trunk-factoring, product-receipt-data-information, numerically-deviated, by an ten-dollars, thousandths-cent-deviation, not comprehended, from how ten-thousandths-dollars, one-millionth-cent-deviation, executive-senate-congressional-legislated, where-upon climate-environment-tectonic-state-impact-(s), lived by universal-signed-limit-(s)-location-(s'), defining our physiological-impact-continuum-presen-ce, because how identifying date, is how data, exomolten-wave-reverberate, visual-current-resonance, [new / aged], amid solar-earthly-terrestrial-lunar-offset-impact, in supratangible-belief, rather physiological-motion-(s), affirming how data, past-storage-weight-collection-(s),| 664 | apposed referenced-upon, relied-library-station-market-commercial-governmental-public-private-dynamic, furthering why investigation-(s) of debate, slip affirmed-literary-thought, before their be an conscious-perspective, by what piece of an objective-data-bit-(s), increment matter, by how our visual-dynamic, tense those weight-(s)-physical-tension-measure-(s), upon circulated-weight-perspective-motion-(s), identifying [artistical-currency // technical-currency] amid logical-location-objective-shelf-time-life-thought-(s), when present-interpretation-(s), cycle-(s) taxonom-ic-cent-mode-Protista-offset-(s),| 665 |limited at motion-(s)-limit-mass-motion-(s)-understanding, when physical-education, omit communion-written-education-mass-matter-intersecting, when impassing-continuum, [1.receive / 2.store / 3.repopulate / 4.extenuate / 5.comprehend / 6.actively-7.interpret] when celestial-morbid-statistics, remain an part for counting, sociological-mass-count-interpretation-(s), impassing vapor-impasse-friction-(s), from kinetic-celestial-centripetal-force-physics-offset, due in part, inconsistent human-bias-factoring not scheduling human-voluntary-will, but Bachelors'-wage-luring, from how still-maintained-bias-generational-transitional-handling, verify those idle-I.R.S.-receipts, inconsistently-deviating, geological-excavational-development, as where local-Federal-matter, hospital-populate, national-state-county-city-library-reference-(s), for grammatical-non-statistical-algebraic-sum-integer-articlizat-ion-(s),| 666 |mundane-dictating, free-affair-(s), in depth-perspective-congruent-subjunction-(s), locational-perceptive-observation-based, yet not by confirmed (affirmed-belief), dated-document-(s), juxtaposing distance upon [measure-impasse // visual-mathematical-introverted-posit-offset-(s)], for what formal-repopulational-matter, signify an described-competent-document-impasse-(s), during implicated legal-judiciary-executive-reset-subjective-origin-(s), carefully prosecuting, fraction-denominative, integer-numerative-continuum-enumerating, how physical-mass,| 667 |yet commonly-auto-biographical-audience-write-type-produce, at-county-ten-mile-north-south-one-hundred-square-mile-(s)-maxim, variated because how [state // count-ies], are by an water-ecological-resourcing, variating how densities are refined, where engineering, is not yet an total-common-perceptive-subjective-dynamic,| 668 |ordering organic-substantial-land-territory-family-matters', due pertinent upon territorial-tectonic-impasse-(s), defining environment-location-(s), juxtaposed entity-reasoning-(s) as state-sexual-affair-civilization-(s), modify per [thirty // sixty year-generation-(s)], apose decade-aging-intervalization-(s), identifying individualistic-modern-presence, interposed independent-male-voter-dating, ancestral-present-time, amid calendar-elongated-speech, requiring more to reverberate a-the formal-developmental-conquering by segregational-conscious-reasoning, through pragmatic-presence, auxiliary-enhanced, per marvel, during ambient-existence, but cannot inhabitant-perceivable-affect-(s), living through an past-conceptive-objective-direction-(s), yet influenced from technical-creation, for how creative-economic-note-(s), dictate Manual-Documenting-Obituary-Belief, upon Nine-Five-Nine-Wars, which could, or not, understand military-number-one-front-line-theory, when during impasse-(s) of thought, receded at presence, far-distance, not measured, by how thinking-work-rate-(s), affect what acre-timbre, remain seeded, in those [1.Marriott's / 2.Inns / 3.Western / 4.Hiltons' / 5.Trump] juxtaposed motel-incendiary-peripheral-perspective, amid township-potential-fatigue-impasse-maxim-repopulating-dynamic, settled by under-seeing an total of physics, before retirement-obituary-senile-thinking, can under-cut subterior concurrent-legislational-literary-remark-(s), prebubesent,| 669 |common-aging-presence.$_{274}$¶|||05m.:55s. $_{34m.s.|||}$§ Procedure-operating, past working-age, from how youth-subjective-contextual-perspective, affect exterior-perceptive-value-(s), amid grandiose-impasse-(s), existing as untangible-element-(s), in place of celestial-abstract-grasp-(s), signifying why information help define micro-hyper-tensional-primary-base-data, requiring like dollar-weight-value-deviation-(s), an direction of documenting primary any objective-resourced-auxiliary-used-control-interposed-maintaining, as how credit verify different temperate-caste-observation-(s), when mode have an life-age-median-presence-comprehension-competency-time-maxim-interposing, shifting those perceptive-impasse-weight-scale-(s), (from an) historical-rationale, commonly not sustain-affirming, what tangible-explanation, verify all of life, amid statistical-variable-increment-deviation-(s), from supratangible-appropriate-denominat-ing, by repopulation-mass-perceptive-documenting-dynamic,| 670 |intermittent monetary-checking-savings, reclusive-approximate-land-parcel-surface-area-physiological-coverage, de-velopmental at territorial-governance, permitting impassed-operation-(s), (from an) proper-etiquette-trade-circulating,| 671 |yet fan-writing-entailing, redacted-literary-observation-(s), but when personality-reliance, remain common-law, unethical, by subjective-perspective-timbre-scale-analyzing, as in my prolegomena, an instant-note, of those literary-work-(s), which bibliographical-reference, by self-book-shelf, before those other-book-(s), not shelfed, at library-labyrinth-aisle-Dewey-decimal-commercial-shelf-ordering, compared by commercial-prior, Congressional-Library-Documenting, as for how the national-effort-(s), require referencing, in an contorted-comprehensive-competent-reference-applicating, by how resourcing, apply by Geological-Enumerating, where physics be the denominative-comparative-density-exponent-weight-(s)-extract-(s), [verifying // validifying], which competent-active-outlook, intercede-monetary-ordering, affirming where data-matter, by whom measure-scaling {14 Cubic-Square-Feet}, juxtapose those entity-(s)-documenting [{10,000-Square // 100,000-plus-cubic-Feet} /// housing-corridors {100-Square // 10,000-cubic-Feet}] ascertaining volume-article-(s)-referencing, for directing self-discipline-schedule-impassing, during conceptive-tangible-identifying, what time-mode, requires what substantial-objective-compositional-development-(s), apply at proport-ion-ration-deviation-(s), spaced construed natural-article-(s)-objective-(s), defining through physical-motion-(s)-impasse-(s), by-an commercial-wage-residential-title-property-spacing-common-acreage-adequacy, maintaining responsibilities, as duties by free-will, by not coordinating along militarial-timing, apose human-sociological-scheduling, scale-measured, by-an rate-perceptive-deviation-variable-independent-tasking, along an extraspective-mass-commun-ion-population-impasse-being-(s), lacking an titular-principle, for-how maintaining proprietary-objective-storage-lock-safekeeping, upon written-dating, for as whom matter, for-an product-(s) while celestial-grandiose-ambient, kinetic-solar-exo-molten-wave-centripetal-rotation-galaxy-propulsion-continuum, juxtaposed sedentary-tectonic-posit-(s), where magnetic-friction-metal, does not technically-stable-magma-core, as how centripetal-force-magma, impact-apose, exterior-physics-parallel-(s), because we only would transit by-an singular-solar-life-form-cycling, or solar-distance-(s) Kinetic-Thrust & Tectonic-Plate-(s)-Kinetic-Reverberation-(s)-offset-(s), mattering when distance guide, why information insight, where cited-referencing, accumulate-allot-data, (from an) youth-subjective-comprehension-effort-(s), only when value, complement-comprehensive-competency, amid commercialism being Inquired & Trade-Exist, where their be-an excessive-amount of incompetency, portrayed-by disguised-reasonists', left unread of consistent literary-mass-communion-motions'-effort-(s), apose formal-genetic-inquiry, apose modern-standards', to-an conventional-genetic-reformation-continuum-impact, imbued grandiose-celestial-ambient, honoring existence, from effective-motions'-impasse-(s), where developmental-population-debate-confirmation-confirm-affirming, how population persperate-perceive, through an interior-physical-components'-offset-(s), only relevant through [1.races /// 2.ethnicities //// 3.mammals ///// 4.animals ///// 5.maricultural ////// 6.aviculture //////// 7.apiculture ///////// 8.horticulture ///////// 9.perceptive-visual-muscle-organ-tension-tangible-compositions'];[357] juxtaposed (in)-capable construe, by comprehensive-comprehension order-orchestrating, how existence direct from, impasse-construe-evaluation-(s), by an limited-physical, subjective-minor [language / literature / vernacular-vocal-expression / perceptive-product-objective-document-scheduled-perspective-(s')] their at, environment-nature, factoring-traded-industrial-material-matter-(s), term-temperate-conditioned, clause-(s)-item-location-usage, factoring amid perceptive-mass-competency-observation-offset, not yet pertaining, subjective-sustained-library-mute-etiquette, not by-an formal-presence, presented-by-hyper-tense-(s),| 672 |upon an balance of subjective-Land-Title-Historical-Hierarchical-Lineage-Tier-Age-Generation-Classifications-Season-Year-Motioning, from-an European-global-item-unit-(s)-trade-concept-projection-(s), custom-ancient, present-literary-influence-{2018}, by continuum-movement-(s)-offset, mattering from-an pri-mary-formation-family-origins'-reset-(s'), through independent-variable-thinking, apose object-ive-weight-balance-(s), scale-visual-decimal-numerical-gauging, objective-subjective-ink-paper-letter-character-word-term-sentence-definition-paragraph-essay-visual-observation-analy-sis-documenting, not compared-analyzed,| 673 |commercial-university-debt-American-article-(s)-usage, expecting physical-commercial-wage-effort-(s) watt-water-pressure-matter-scale-tiering, historical-human-disposition-(s), centripetal-velocity-depositing, east-west-motion-volume-gram-ingestion-supplementary-unit-(s), as substantial-comestible-(s), which pertain in juxtaposition by genetic-background, upon an ethnic-segregated-refined-mass-impact-culture-(s), apposing-juxta-positions, by [their / then] of family-subjective-reliance, paid entity-labor-wage-interpretation-(s), not balancing entities-monetary-influctuation-trade-family-reliance, as general-distance-(s), ambiguate through posited-transit-dynamic, for family-history, generalize lineage-background, not auto-biographical-referencing, in fruition Federal-Geological-resourcing, bastardized-conceptive-Chinese-Outsourcing,| 674 |an modern-generated-population-activity-(ies), default-common-characteristic-repopulation-effort-(s), upon an structured-entity-environment, comprehending by [dimension-length-width-two-dimensional-ground-posit-rotation-rectangular // square-layer-expanse-identifying /// three-dimensional-cubic-Length-Width-Height-rectang-ular-maxim-square-integer-enumerating], what data-bit, pertain visual-perspective, by time in mode, upon hour-(s), but secondths-rate-count-(s), incurring minute-(s)-mode-impasse-(s)-location-sleep-offset-effort-(s)-respiration-reset-(s), dating locational-act-(s), in-an timely-fashion, for reliance government-currency-trading, I.R.S.-Note-Paper-Statement-Documeting, those tensable-affirmation-(s'), by-what notion-(s), range an human-population-trading-morale-offset, simultaneously-swear-dating, by 2018-Gregorian-Calendar-Further, rather from {1536}-Solar-Earth-Magma-Lunar-tectonic-Plate-(s)-Mundane-Default-Perceptive-Offset-(s), [ranging // gamma / misinterpreting], whom matter between an [1. supervisor / 2. manager / 3. superior / 4. Boss & Employee-interactive-extraspective-work-dynamic]| 675 |still for confirming action-(s) amid valid-mutual-requirement-trading, through-an family-market-interactions-background-ethical-confirmation-(s), illiterate for mass-documenting, not by auto-biographical-present-date-cycle-period-(s), affirming vigilant-traded-means, collecting article-paragraph-syntax-affirmation-(s), relevant by, currency-note-(s), trading by-an fair-rationale, verifying by Goods & Services, how physical-conduct-academic-tone, only matter if an vernacular-subconscious-visual-perspective-observation-exist, or no count, verify-offset, which literary-enumerated-denominative-rate-extents, amid those redacts of physical-perceptive-observation-interposed-preference, (as an) activity by hopefully, an 10,000,000-white-pink-maxim, variative, yearly-mute-subconscious-book-documenting, rather book-shelf-collection-(s), in early-earlier-edition-(s), of [prolegomena /// Epilegomena //// article-(s)-body-paragraph-essay-etiquette-visual-vernacular-observing], comparing by reference, where those topography-environment-condition-(s), can place literary-non-manned-stable-competent-objective-(s), amid physical-14-cubic-feet-life-style-mode-(s')-maxim, presenting from interior-self, cause from monotheistic-observations', rather polytheistic-symmetry-exterior-aspect-(s).$_{275}$¶|||13m.:07s.$_{33m.s.|||}$§| 676 |From abstract-ambient-celestial-infinitive-spacing-(s), remain where numerous-superlative dating, construe [1.distance-(s) / 2.timing-(s) / 3.variable-(s) / 4.count-(s) / 5.weight-(s) / 6.article-(s) / 7.word-(s) / 8.syntax / 9.sentence-(s) / 10.paragraph-(s) / 11.essay-(s) / 12.book-(s) / 13.article-(s) / 14.agenda-(s) / 15.hour-(s) / 16.schedule-calendar-hour-dating-ordering, et cetera…] for how thought remains' an part of an daily-routine, recollecting present-modern-fact-(s), durating secondths-tic-continuum-simultaneous, where [percentage / fractions-denominative-calculating-(s)], remain capable numerative-operation-(s), pertaining from visual-tonal-observational-perspective-(s), when dating objective upon location-(s), require [paper-documenting / cement-infrastructure-storage], vilifying why daily-timely-impasse-fashion, assist article-(s)-documenting-rationale, calculating numerical-data, when [letter-explanation // expository-thinking], apose awhile population-educated-inhabitant-(s), suffice from rubric-benchmark-etymological-observation-(s), expressing where extraspection at communal-lineage, by Family & Ethnic-Genetic-Background-(s), serve how relied Goods & Service-(s), transpose industrial-transportation-mode-cycle-(ing-(s)), at-an parameter of singular-function-concentration, factoring objective-collection-(s) of product-(s), by ordering process-interger-mass-mode-box-unit-patent-tier-usage-activity-(ies)-impasse, when international-resourcing, have an succession, progressing date-three-hours-motion-cycling, apose [six-hours-compartmentalized-seasonal day-summer // night-winter //// morning-spring // evening-autumn translucent-rotation-count-(s)] historical, by identified land-territory-perspective-role-(s), conver sing percentage-(s), omitting cause-(s) by being-(s) apposed person-(s)-mass-superlative-commercial-documenting-handling-(s), forfeited from literary-thinking, due to labor-physical-effort-(s), [sweating // persperating] calmly, for Federal-excavating, of restricted-tectonic-terrain, at centralized-focused-conjunction-(s), at state-boundary-central-city-point-(s)-operation-(s), time-calendar-dating, constitutional-congressional-currency-concurrent-chain-operation-(s), cycled where, planned indeterminant-dating, proceed-succession, locational-objective-physical-perceptive-factor-(s), never constituting-land-square-mile-(s), outside a-the family-population-masse-dynamic-maxim, from how [divorce /// single // married opposite-sex-repopulating-dynamics], affect surface-area-perimeter-length-width-height-mobility, for-an clearance-tectonic-plate-crust-dynamic, perceiving how distance, have an [ground-land-parameter-parallel // air-space-vapor-velocity-parallel // tunnel-subway-mine-tectonic-density-excavating-parallel] exten-uate each [visual / tonal-observational-reverberation-(s)], from-an focus-emphasis of mouth-throat-larynax-esophagus-stomach-digestive-tract-air-passage-vernacular-perspective-modificat-ion-(s), from-when-an visual-physiological-motion-(s), learn-by tallied-indent-mark-(s), cunei-form-letter-repetitive-competent-comprehended-matters, when an concertion of emphasis, dynam-ic extraspective-pertinence, due-to logical-inquiry, affirming when data, peripheral-perplex visual-hyper-tense-physiological-tension-observation-(s'), simply posited as "is", reducing mortem-memory-(observation), alluded aforemention unconscious-untimely-death, from lineage-wage-payment-recollection-(s), not learning why subjective-analysis, denominate-congressional-obser-vational-documenting, for referencing apose reader-bookshelf-rate-count-(s), comparing by-an architecture-geological-engineering-sociological-psychological-schedule-documenting-purpose, for reasoning desired-purpose, amid visual-liquid-wave-breath-vibration-(s)-perceptive-existence, by extraspective-peripheral-interpretation-(s), pertaining physiological-inclination-(s), for inter-preting, upon an lifestyle-deviation-(s), matter-(ed) (from an) derived-logical-interpretational-functions' juxtaposed thinking-inclinations, as personal-visual-tone-reverber-ational-pulse-tension-(s)-interpretation-(s)-grade-point-average-subjective-interpretation-rate-(s), apposing each article-fact-(s), through living matter-(s'), premised due-to logical-prose, identifying temperate-tense-volume-respiration-impasse-(s), practicing how to document-title-claim-peri-meter-planned-expansion-land, coordinating those motion-(s) upon an tangible-dimension-(s)-schematic-documenting, surrounding-cubic-clearance-terrain, developing awhile hypothesized-planning, attempt to collect the formal-organism-taxonomic-city-ecological-dynamic, (as an) artistic-state-federal-balance, not yet confirming-subjective-article-(s)-interpretation-(s), when mutual-perceptive-impasse-(s), factor logical-interpretation-(s), impasse revolving-impact, exter-ior, by personal-reclusive-extraspection, identifying individualism-land-space-existence, for interposed-conceiving, suffering an largo-tempo-morale, upon duration at constructed-entity-environment-cubic-interior-space-(s), for handling maintained-factor-(s), juxtaposing policies by inhabitant-(s')-bias-perspective-view, requiring survey-geological-sociological-psychological-physical-commercial-product-inquiry-exponent-mass-document-affirming, what views of mass-quantity-shelf-cycle-usage-reproduction-rate-(s), variate whom understand fair-economical-practice, pertained swiss-knife-10-tool-inquiry, not yet to be understood, by mundane-practice-(s), amid centripetal-force-physics-supratangible-mega-states, boundary-state-maxim, human-physical-limit-(s), requiring federal-petroleum, by interstate-county-city-limit-(s), relevant to ones'-housing-annexing-dynamic, yet expanding in metro-politan-city-sewer-system-electrical-commercial-transformer-industrial-circuit-wiring-water-pressure-current-cement-pavement-rebar-ecological-infrastructure, identified through denominat-ive-variable-interjections', variating scientific-term-reasoning, from when developmental-engineering-excavating-posit-(s), deposit [kinetic / sedentary-lunar-mass-rotation / solar-kinetic-revolutionary-kinetic-infinitive-continuum / terrestrial-algea-forestral-central-core-combustion-magma-infinitive-interpretation-(s)] physics-offset, taxonomic-elemental-matter, different from product-observation-intrication, through an historical-outsourcing-patent-dynamic, observable by plastic-metal-resource-wage-payment-dynamic, deviating logical-entity-subjective-primary-school-impasse-purpose-(s), when (de)-posit-(s) of individual-perspective, be-an-part, of surrounding-feature-(s), conferring extra-spective-interpretation-(s), intermittent natural-visual-tonal-respiration-nasal-physical-perspective, tasting-mode-rate-deviating, numerative [percentage-(s) / fraction-(s)] interposing when time-line-reference, have-an key-map-scale-directive, various enumerated-measured-reasoning, impasse-deviation, by secondths-centripetal-force-revolving-movement-(s), motioning juxtaposed logical-clearance-impasse-(s), throughout-existence-voluntary-movement-(s), inhibit-ing life, at thought of surrounding-parameter-(s), factoring by competent-mutual-reference-interaction-impasse-(s), collecting an genetic-motive, for elemental-referencing, or continue an millennium-continuum-lifestyle-(s), repopulating under an common-exponent-extent, never to enhance from marriage, as for how long an lineage can continue an writer-creative-documenting-verifying, when state-Federal-Currency-reset-(s), forget how dollar reset by one-hundred-cent-(s), per thousandths-dollar-(s), cubic-land-space-impasse-(s)-disparity-dynamic, conceiving-data, by factual-existence, amid consecrated constant-interpretation-(s), adaptation-modify-formulating, (from an) objective-conceptive-schedule-operation-(s)-moment-mode-continuum-impasse-(s), where daily-living, apprehend at-will-scheduling, by-an monetary-extraverted-means', trading amongst an mutual-communion, yet ordering communication-(s), for an understanding numerical-incremental-interpretation-subjective-coordinating, understanding per serie-(s) of means, conceived from, blood-pulse-liquid-volume-air-pocket-(s)-tension-(s), intermittent ventricle-muscle-active-ligament-sedentary-inlay, friction where respirational-volume-reverberation-(s), as how our fatigue-characteristic-(s), balance upon sworn-comprehended-observed-conducted-operation-(s), why labor-physical-effort-(s), require an feminine-literary-objective-repopulation-control-dynamic, for ignoring past-historical-trading, by-an live-historical-redacting-review, juxtaposed morbid-parental-passing-(s), from how monetary-trade, dynamic in lieu of daily-activity-(ies), from Cause & Reason, affirming by tension-(s) how exterior-visual-blood-pulse-white-plasma-sedentary-receptor, juxtapose organ-passage-way-(s), (as for) what happen to American-Idol-University-Reception-Retaining-Ticket-Retardation, by contextual-data-infor-mation-(s), reverberating sound-wave-vibration-(s), from muscle-blood-pulse-visual-mark-inter-pretation-(s), explaining vernacular-vocal-tonal-pitch-vibration-(s), in an control of linear-canal-passage-(s), location-perceptive-limit-boundary-offset, continuing locational-objective-designat-ed-auxiliary-usage-(s), relative by personal-experience-interpretation-(s), requiring reference-geological-impasse-documentation, for identifying-tangible-data, dating trusted document-(s), by-an ecological-survey, denoting-treasury-repopulation-reproduction-development-(s), upon Federal-commercial-coordination-(s), proglamated, when in part, ascertaining why referenced-article-(s)-comparative-paragraph-context, contort an core-common-applicated-reference, when interaction-(s), do not relate an book, by-an communal-bible-effect, understanding from reader an writer, an dynamic of superiority, directing how conceived vision-tone-pitch-vernacular-audible-interpretation-(s'), not save memory,| 677 |from presence-command-usurpation-(s),| 678 |by-an relative-blood-pulse-physical-compulsion-introverted-dynamic, interpreting awhile exterior-peripheral-impasse-(s), can only require direct-visual-tonal-volume-pitch-subjective-reference-comparative-observing-extenuating, which Federal-default-commercial-placing-(s), input from self-perceptive-formality-logical-reasoning, not debated from family, when house-representative-interpretation-(s), ratio-proportiated-maxim, why marine-barrack-sociological-annex-developing, affirm how population-census-measure-(s), juxtapose senate-commercial-regulating, from committee-confirmation-(s), conjuring by dating, how centripetal-force-time-impasse-mode-expectation-(s), regular-basis repetitive-logical-formulation-(s), controlling by citizen-free-will-voluntary-impasse-(s), an sway of schedule-(s), basis-through physical-interpretation-(s), voting subjective-matter-(s), not yet implementing psychological-congressional-entity-(ies), by-an state-territory-library-denomin-ating-basis, apposing ecological-sociological-geological-constructive-development-(s), balancing physical-citizen-being-(s), person-citizen-(s), whom serve diligent-comprehension-(s), progressing by-an work-off-time-input-extraspective-dynamic, binding diploma-(tic)-comprehension-(s), fluently-considering, why product-objective-matter-(s), pertain to citizen-observation-(s), of elemental-environment-interior-measure-(s), apose objective-measure-(s), comparing usage-wear-and-tear, cycling human-being-voluntary-perspective-(s'), at Land & Commercial-Technique, because how dating continuum comprehensive-multi-facet-understanding-(s), per whole-tangent-ration-rating, from zero-reset-continuum-offset, as when we only have one-perspective, but what extension of location, remain mute for labor-swearing, reclusively-motioning-away, from extraverted-subservient-comprehension-interpretation-(s), enhancing free-merit-(s), neither absolute or fixed, but more, free judgement as guideline-(s), for inhabitant-(s), behaving as citizen-(s), amid conducted-state-trade-operation-(s), remaining in close-friction-(s) by glycemic-index-tangible-eating, under governmental-aging-bias, where America, is directed through racial-ethnic-collective-background-(s), engineering-commercial-technique-judgement-(s), at point-(s) trade-(d)-commercial-product-(s), in an regard of govern-mental-patent-(s), through biological-person-(s), requiring reference-non-fiction-symposium, factifying historical-figure-(s), per significant-product-(s)-bill-policy-documenting, apose commercial-C.E.O.-histori cal-headquarter-figure-head-(s), in close estimated-approximation-(s), dating scheduled-dating-free-motion-(s), by how voluntary-will, (as I have alluded earlier) have an difficulty, for persuading an extraspective-genetic-resource-engineering-geological-political national-state, understood through mundane-judgement-(s), delving referenced-credible-written-cursive-manuscript-context-(s), ascertaining why article-(s)-reference-usage, remain an part of interposed visual-tonal-vernacular-letter-marks-subjective-perspective, is suppose to youth-premise, how adulthood-commercial-trade, interpose retained-person-(s)-sociological-national-documenting-dynamic, awhile fundamental-government, forget to use properly, time-mass-exponent-geological-numerical-count-per-cubic-feet-interior-storage-extent, at hand to [[1]influence / [2]persuade / [3]collect an genetic-specie-(s) / [4]mechanically-engineer (Maryland technical dictionary note) / [5]Geological-engineering / [6]construction-architect-planning-method / [7]product-schedule-population-impact-agenda] their at judgement-capacities, where executive-force remain dormant, amid mass-belief, fearing mass-population-(s), positing unilateral-politics, not in an focus of imposing ethical-guidance, co-axis contextual-trade-impasse-(s), using power of locational-position-(s), along with literary-inquiry, for persuading why [civilians' / citizen-(s')], should inquiry house-representative-legislational-congressional-literary-member-reform, every ten-book-(s) per representative-(s'), from House-Representative-{114th session}-population-census-impact-(future), by-an means' of credential-inquiries, affirming why information, remain pertinent, upon traded-influence, historically following, senate-congressional-proceeding-(s)-executive-cabinet-member-inquiries, where federal-government-treasury-funding, remain the responsibility-(ies), for analyzing-humanity-trade-deadline-(s), believing how [product-registered-demand-anticipation-currency-note //

change-circulation] enable population into land-settlement-coverage, outlining trade-policy, swaying product-trade-demand-obligation-(s), amid an mass-congregation-genetic-offset-impact, where public-commercial-development,| 679 |shy from, ambitiously-judged, communal-developmental-dynamic-reasoning, from factual-written-non-fiction-tangent-subjecting, apposed in position of power, not working on suppressing those lower-percentile-work-classification-(s), building on ones'-intention-(s), upon collective-communion, not understanding from logical-intelligent-thinker-(s'), how geological-historical-sociological-psychological-thought, never have infused such an genre, of non-fiction, analyzing from natural-truth, those error-(s) by introverted-independent-bias, from superego, an international-trade-outsourcing-impact, that is federally-Interior-inclined, by American-civilization-(s), not offset-comprehending how we can nationally-federally-trade, but offset from national-shipping, to those sociological-center-(s), because prior not being heed-(ed) warning of elemental-geological-resourcing, commercial-wage-tangible-monetary-appropriating-production-(s),| 680 | balance from, [English (Macmillan) / German (Holtzbrinck)-publishing-house-(s)] are yet infused logical-sequential-trading, furthering citizen-governmental-copyright-control-premise-basis, but where interpretation-(s) of legislational-observation-(s'), remain disproportioned common-law, to comprehend those psychological-executive-congressional-human-error-(s),| 681 |amid mass-population-lifestyle-trade-impact, neutrally-demanded, commercial-insurance-quota-expectation-(s), along Gregorian-Calendar-belief-system-{2018}-count-pattern, blindly-visualized, at solar-historical-initial-exomolten-offset, deriving tectonic-plate-landing-(s), apose aquatic-swashes, | 682 |remaining in anomaly by human-common-manual-reasoning, from political-person-bill-policy-(ies), estimating how to sway free-will, when monetary-policy-(ies), have yet had an literary-free-will-directive, emphasizing those error-(s) of educational-youth-sociological-retire-ment-expectation-dynamic, cycling year-count-(s), apose Four-

Hundred-Eighty-Year-(s)-contin-uum, suggesting an initial-origin-calendar-year, of {1536}, as the initial-basis of living-existence, default from organism-offset-development, not evolving from mammal-offset, for animal-specie-(s), juxtapose physiological-impasse-(s), presence amid-apose solar-exo-molten-hollow-kinetic-friction-central-impasse-halogen-effect-impact, only comprehended from those whom actively visual-tone-pitch-phenomena-perceive, an default-physics-impact, prose geological-sociological-organism-(s), proceeding civilization-(s), at harmony for land-title-maintaining, while Gregorian-Calendar, direct us (from an) false-historical-calendar-count-(s), as {2018}, economic-trade-year-(s), somewhat simultaneous historical-tangible-season-count-(s), from my earlier years-origin-assertion, when mass-educational-offset, not train [formal-ambervision-physical-written // read-subjective-effort-(s)], prosing how Thinking & Physiological-directive-effort-(s), influctuate-extraspective-interpretation-(s), not in an fluent-cognitive-perspective, guiding hyper-tense-(s), where formal-training, succinct from how physiological-direction-(s), perceive by mass-ethnic-racial-history, human-limit-(s')-capacity-judgement-(s), from extraspective-introverted-deviation-(s), by daily-trade-production-(s), from how human-consumption, overlook inhabitant-motion-effort-purpose-(s), working on elemental-impact, from geological-elemental-compositional-extraction, apose government-objective-patent-(s)-method, in lieu of commercial-conceptive-product-(s), acceding by scheduled-wage-directive-(s), configuring what government-currency-trading-policy-(ies), (which could be in due of an massive legislational-executive-overhaul, using common-law deviations of mass apose statehood-legislation-(s) impacting [national / international commerce], from populational-sociological-judgement-(s), of educational-testing, and repopulate-ion-rate-(s), factify analyzing, formal-identified-possession-(s), from cycled-shelf-waste-demand, | 683 |trading from substantial-product-schedule-agenda-population-off-time-demand-cycling, however dating-impasse, remain ignorantly-forgotten, lacking an self-educational-confidence, because of how Existence & Literature, juxtapose dating, by those extraspective-dependency-(ies), flagrantly-omitted, awhile evident consensus of American-bastardized-trading, not conclude, an conclusive-counting for each and every populate-inhabitant,| 684 |from formal-disposition of [House-Representative // Senate-ethical-practice-(s)], for-an suppression of sociological-introversion, not show in studies, what geological-ecological-economical-rationale, timbre free-will-bias, through-an natural-repression, of executive-handled-trade-affair-(s), similar to our National-Congressional-Library spending, which have Interior-Geological, premise what abominable-minimal-impact of Six-Hundred-Million-Dollar-(s)-insured-year-maxim-spent, affect patent-objective-merchant-billionaires', unregulated from political-prose, to call out those travesties of mass-population-sweatless-effort-(s), from-an historical-impact, juxtaposing how trade-retention, sustem an oblivion of means', not foreseen, by those eye-(s'), of birth-hospital-inhabitant-lifestyle-county-existence, not measuring impasse, as analytical-logical-mathematical-juxtaposition-scaling, apose how [clearance-height // ground-tectonic-plate-revolving-revolution-ary-reset-limit-(s)], are to affect repopulation-measure, yet not evoked, for the droves of human-illiteracy, not considering why literary-youth-subjective-studies, have commercial-book-entertain-ment, sustem geological-ecological-economical-aspiration-psychological-sociological-survey-analysis-literary-congres-sional-house-bill-approval-measure-effort-(s)-militarial-religious-genetic-communion-goal-directive-genre, attempting to advance an voluntary-participating-geological-insourcing-genetic-society, apose our present introverted-repressive-ideology-(ies), not in literary-control of factual-observation-(s), requiring cited-geological-interior-book-(s)-referencing, for interposing scale-integer-data, upon an [percentage / statistical-measure-(s) / fraction-(s)-hundredths-percentile-population-deviation-(s)] accruing from historical-observation-al-data, supreme-period-(s) of territorial-evaluation, litigating those drove-(s) of incompetent, to be an non-fiction-goal, by mere-simplicity of Repopulation-Observation & Documenting-Potential-method-(s),| 685 | inherently producing, inert [1.respect / 2.honor / 3.loyalty / 4.duty / 5.service / 6.integrity / 7.prosperity] rather swearing much, to suffice idle-means, "in one ear, out the other"-interaction-(s), by daily centripetal-force-perpetual-revolution-year-deviation-(s)-rotational-dating-deviation-(s), incul-pable by common-suspicion, where Captivated & Impressed-mass-(es), "require entertainment", for [reading // sharing // hearing], how routine of daily-offset-(s), are intertwined, for formal-extraspective-interactive-regard, time hour-interval-(s), upon calendar-dating, for reviewing how each impasse of mass, offset an schedule-time-frame-evaluation-period-(s)-method,| 686 | interval-concluding by cited-reference-(s), independent-observation-focus, expressing why, this is why I suggest that their be-an extraspective-human-introversion-evaluation, yet confirmed, by affirmed-inquisition, for context, not manuscript-effort, document-subversive-cursive-letter-appreciation, commonly not inspired by car-leisure, taxed by prerogative, as when geological opposite, (not &) Metropolitan-Documenting, light-spacing, how density, reflexes as substantial-tension-perspective-motion-(s), composed-against bone-still-settling, physiological-fatigue, reflexes by [1.work // 2.effort-(s) // 3.timed-hour-(s) / 4.incremental integers / 5.grammar / 6.letter / 7.word / 8.syntax / 9.timbre-subjective-perspective & calendar-hour-(s) / (posit-hour-(s)-mode-date-after-effect-(s)-continuum-observat-ion-(s')], from presence-past-collection, future-accounting, remaining possible in an judgement of factoring-(s), yet birth-comprehending, for-of reclusive-study-(ies), affirm-(ed) by [The Macmillan / Library of Congress-standard-(s)] barring by seasonal-upkeep, interior-statistical-sapomadic-study, against survey-university-document-comprehension-(s), as debate enlighten, subjective-post-humorous-present-documenting, as geological-count-(s), are yet diploma-administer-(ed), apose Geological-fund-questing, as how interior affirm-bond-stock-state-development-(s), in grandiose-introverted-regard, requiring Personal-Cartography-dimensional-satrap-linear-plot-point-bends, amid simultaneous-continuum, family-repopulation-factoring, not though pose whom collect and eat, or then apprehend idle-feverous-tendency-(ies), count-blank-elapsing, visual-tonal-volume-reference-interactions,| 687 |for how common-purchase, fragment-climate-purchase-identification-(s), yawn-yearn-earn, amid sleep from persperous-sweat, affable introverted-mute-vernacular-presence-voice, when Personal-Regard & Thinking, affluctuate what-one-individual-handle-(s)-believe-(s), monotheistic-through-self, or where library-isolate-books-pre-deterior-placings, where stealing suggest an criminal-legal-mental-system, for how coroner-date-to-year-obituary-pay, where-would recidivistic-purchase-value, belief from tangible-introverted-tense, possessed by market-purchase-advertising-limit-(s) aspect, from interior-converse, speaking from annotation-moment-(s), ruling how mail-is-sent, without an common-sending-regard, returning from departure-house-dull-leisure-comfort-(s), impoverished of care, affirmed by exterior-intuitive-idleness, walking for short-period-(s) of time, yet vehicle-bicycling-interval, city-county-extent-(s), apose district-city-extent-(s), going to place-(s), as when identity, affect view, when by-an impact, through null-effort-impact, fighting-fatigue, rudely-offensive, at time-(s), like-calendar-historical-belief, present at hour-siture-(s), silent by point-(s) of direct-fatigue-wear, paying-off-pursuit-measure-(s), offset spacial-land-surface-area-planned-measure-(s), to be moved by machine, in transit of radio-wave-idle-motivation-(s), rather when age-hour-range-gamma-mode-offset-(s), apose particulated-empirical-cabinet-Geological-subjective-measure-(s),| 688 |for third-dynamic-perceptive-balancing, relevant from Grammatical-punctuation-Mathematical-conjecture, but (from an) reference of Geological-Barren-Limit-(s), apose metropolitan-literary-deficiencies, per vehicle-written-type-font-perspective, losing cause for year-offspring-reproduction-(s), careless upon measure-(s), we overlook through peer-purpose, and then disregard sociological-survey-analyzation, by given-perspective, opinionated, not through an strict-rigid-repopulation-handling, as is like-human-physical-sociological-steading, another break to "the-beshied", interior-self-bias-system, by timed-operation-(s), [believed // reference-injunction], citing by depth-presence-convincing-mode-operation-(s) supposed an adequate-reflexive-non-auto-biographical-common-means', from auto-biographical-denominating, present-historical, future-allusion, through purchase-independent-mental-vision-individuals', deviating timed-limit-(s) upon physiological-space-coverage, through "ehh", mathematical-prose, amid an historical-inquiry-proglamation, because motivated-count-(ing)-(s), continuum-offset, masculine-legal-independent-individual-(s'), impasse-effort-(s)-affecting, from wage-restriction-(s), exterior-measures, at pressurized [skin-exterior // intestine-tract-muscular-transmutation], eating-etiquette-perspective, durating at self-motions'-impasse-(s), impassing Physics & Geological-distance-(s), under metropolitan-city-housing-road-regard, where state-boundary, variate, whom place motion-(s)-dynamic, from continuum-present, not historically-reflected, for early-physics-dating-populating, from sexual-mute-skill, practice as-at entity-legislational-comfort-intuition, as compositional-interior-substance-(s), influctuating exterior-mundane-municipal-signage-movement-(s), because when case oblige legal-eating, but disrespect post-currency-purpose, without Calculating & Considering, when off-time, compare an notion-tier-purpose, upon physical-self, for documenting oneself-amid-exterior-driving-petroleum-documenting, as-an motive for oneself, when-what matter-(s) upon formality of activity-(ies), are lied of educational-homework, but then socializing, where-it's "all being ok".₂₇₆¶|||34m.:39s.₄₅ₘ.ₛ.|||§ Convinced of [custom-(credit) / mysticism-(money) / tradition-(trade) / behavior-(bill-(s))-(credit-conduct)] paper-holding, sustain-symbolism-note-(s), whatever is said, and then being upfronted when [1.patent / 2.bill-(s) / 3.register / 4.copyright / 5.Trademark] affect why person-(s) can not accept humanity-sub-par-effort-(s), but can register an premise-basis, without Trademark-Copyright-(s), because of whom produce serie-(s), apose encyclopedia-geometrical-documenting-offset,| 689 |by Geometrical-State-Dictionary-Documenting & Literary-Sociological-Live-Extenuating, when Dictionary-{Calendar-Hour-mode-offset} & Encyclopedia-{Time in Secondths-perspective-deviation-continuum-offset};[358] offset because all personal-lie-(s'), all awhile lying (from an) basis of merits, adept default-impact, because each dating-period, matter, by-an cause of repercussion-(s), intermittent mega-heavy-auxiliary-verb-perceiving, as subterior-articlizing, periodicalized by [1. magazine / 2. newspaper / 3. book / 4. flyer-business-card-advertisement-(s)] again omitted by statistical-reference, then ignored, and mocked, for of flagrant-sociological-sub-par-ignorance, amid locational-limit-(s)-false-effort-(s), valued-evaluated, by-an visual-tension-perspective (*tonal-tension-Perspective) when liquid-substantial-friction, not [sense / tense /// friction-ventricle-plasma-coronary-perceive],| 690 |which impasse-(s) at falsified-effort-(s), easily-seen sporting [cross-country /// track & field-three-miles-purpose-effort-(s)], as when geological-projection, develop at impasse under-par-extraspection, for literary-extraspection, fall far under one-percent-(ile)-(age), by tenth-one-hundred-thousandth-percentage, juxtaposition hundredth-one-millionth because Baron-(Von Stuben)-belief, afflict personal-property, for blue-book-conjecture-value-(s), remaining, in-an false-attainment, when product-objective-impact, apose auxiliary-force-(s), interceding-inter-action, for demand-market-trading, but when turn around, free-will, can wilt abstract-care, by geometrical-handling, where library-cabinet-documenting, have Interior-Geological-Govern-mental-Library-Dewey-Decimal-System-Genre-Documenting, an aggressive negative-inclin-ation-(s), for sake of clearance-depth-perspective-length-motion-pressure, visualize-homogen-izing, perceptive in view-(s), rather than purposing, sitting lavishly, but fighting standing to then fulfill the purpose of laying, callous upon merit-(ed)-interaction-(s), unqualified from Grade-Point-Average-repression-rate-interpretive-moding, argued agreement-(s), for sociological-objective-thought, as where to place objective, hoarding past youth-impasse-(s) of property-objective-monetary-theft, planning where population, offset an overload of needy-human-inhabitant-habit-reliance-(s), unconventional, literary-circumventable, land-title-spacing-purpose-(s),|690|where primarily-introverted-self-perceptive-physical-blood-pulse-inhabitants, have no care for handling, amid objective-purchase-conceptive, so traditional-skill-technique, learn how reluctantly abide family-purpose, guided by commercial-cause, apose residential-interaction-(s), upon communion-(s) of conversive-house-(s), then in future, working on-an genetic-constitutional-communion, reducing extraspective-behavior, at state-center-(s)-distance-geological-coverage-aspect, when [1.forest // 2.(city)-park / 3.state-park //// 4.excavation // 5.plains /// 6.swamp ///// 7.desert /////// 8.mountain ///////// 9.ocean / 10.canal // 11.river /// 13.stream //// 14.straight ///// 15.lake ////// 16.bay ////// 17.metropolitan-lawn-dynamic]| 691 |where-amid introversion-conclusive-family-rivalry-dichotomy-deviation-denominative-percentage-(s), expose from self-introversion, extraspection in Zero-whole-percent-decimal-point-zero-tenths-nine-hundredths-percentage-deviation apose Ninety-Nine-decimal-point-zero-tenths-one-hundredths-five-thousandths-four-ten-thousandths-percentage-deviation, due-to-an excessive-superfluous, of responded-short-sighted-interpretation-(s), redundant from natural-transit-currency-energy, when calendar-dating, not appear as-an quandary of expectation-(s), in lieu historical-book-shelf, not in room by geometrical-book-shelf-room-purpose,| 692 |but when state-house-entity-development, library-house,| 693 |demonstrate [state-county-entity-state-central-boundary // county-boundary-economic-transition-schematic], not span much geological-dating, as belief (one of a the primary-reason-(s) for common-law, (from religion-perspective)), impasse amid [drift / brisk / refer / impasse / immense upon-one-another], by those other-(s) in-between substantial-survival, to mutely-impasse, motion-(s) of physiological-stance, by-an belief in thought, avidly seclusive, voluntary-action-(s), where sweat bother one-another, during day-(or equator-proximate climate)-vocal-pitch, not metronome-toning, vocal-harmony from vernacular-redacted-syntax (book-editor-query), affable presence-festive-book-fair-focus, remaining in line of lineage-memory, court-settled, down friction-energies, mute-physiological-effort-(s), swearful throughout constructive-consistency-(ies), obituary-hypotenuse-period-(s), saw by blatant-disregard, for maintained [1. land / 2. territory / 3. mutual-inhabitants / 4. subjective-documenting // 5. debate // 6. population-psychological-mathematical-mass-offset-studies-inquiry] evident entire psychological-lie-(s),| 694 |(from an) [youth-(girl) / (boy)-friend-rape-aging-retirement-high-school-cycling], by male-youth-peer-introverted-pressure, not inclined, formal educational-purpose, by cycling how entity-maintained-handled-response, formulate early repopulated-motion-(s), (I at twenty-five, probably late-twenties, expect to repopulate), seeing an influence from book-(s), retain why morbidly-grave-lifestyle-motion-(s), remain upon memory-influenced, immature-sexual-offset-offspring-product-effect-(s), to intermingle, girlfriend-effect, per relationship, as how common-marriage, juxtapose merited-traded-posit-value-(s), by mass-population-Federal-impasse-(s), when at ageless-(es)-mode-(s), affirming how interference, affect how relationship-matter-(s'), [influencing / persuading reference-book-cycling-(s)], as [etymological-grammatical // punctuational & mathematical-function-(s)] verify formal sequential-serie-(s)-progressive-calendar-operation-(s), oblige-abiding-by, [addition / subtraction / multiplication / division-processing-(s)] apose general-serie-(s) of [fraction-(s) / percentage-(s) / greatest // least-common-denominator-integer-plotting] apose-deviation of [greatest // least-common-numerator-integer-plotting] directing-referencing, as-an juxtaposition of integral-(s)-perceive-integer-information-interpretation-(s), apose algebraic-exponent-measure-reset-subject-ive-perceiving, in lieu of statistical-article-count-(s), predicate calculus-extenuating, algebraic-statistical-equation-(s), deviating-from literary-press-encyclopedia-dictionary-article-(s)-refer-ence-(s), which date-hour, plug informal-equation-(s) of thought, juxtaposed monetary-notion-(s), from-how population-extraspective-etiquette, apose independent-self-observation-(s), self-conducted, proportional, physiological-action-motion-(s)-(sweat-effort-(s) (not physique)), juxtaposed subjective-interpre-tation-(s) presence-ignoring, subjective-physical-psychological-sociological-geometrical-mathematical-grammatical-punctuation-political-redactive-literary-synthesis-studies, amid homework-recollection-(s), from where, peer-subjective-sociological-progress, settle in-by, opposite-sexual-relationship, rather those various-form-(s) of common-relationship, focusing subjective-studies, as intensive-group, be out of question, for blank-assertions, define physical-characteristics, without an clue of cartography for topographic-subjective-insight-(s), aposing Article-(s) & Subject-(s), upon Physics & Geological-Observation, with what Mathematical & Grammatical-Punctuation affirm Science & Religion, where-when Schedule, require Calendar & Date-Hour-Time-Interval-(s), atlas-global-physio-logical-impasse-(s), to then learn-visual-mode-deviation-impasse, for anatomy-practice-transition, which method of biological-post-usage-decay, pattern pathological-sociological-Lineage-Ancestry-inquiry from national-cursive-constitutional-impasse-reference, constituting two-dimensional-perimeter-surface-area, per Longitude & Latitude-square-mile-(s), where every thousand-word-minimum, upon population-proceeding-(s)-tier-congressional-offset-denotative-documenting, checked-in, on every inhabitant, by an minimum of ten-thousand-word-(s) & currency-denominative-counting-method-tier-influctuation-hard-objective-counting-paper-subjective-counting-(s) of each cent-piece, of matter, register-trademark-influctuating, patent-commercial-conceptive-developmental-material, (where headquarter-Chief-Executive-Officer) populate-articlize, human-objective-location-entity-offset-usage, when arc of practice, identify [1.issue-(s) //// 2.debate-(s) //// 3.interpretation-(s) //// 4.affair-(s) //// 5.practice-(s) //// [6]impasse-(s) //// [7]focused-action-(s)-count-(s) //// [8]entity-(ies) //// [9]method-(s) per commercial-dynamic //// [10]objective-(s) //// [11]primary-empirical-tier-deductive-subjective-(s) //// [12]mission-(s) //// [13]ecological-sociological-dynamic //// [14]elemental-posit-(s) //// [15]elemental-composition-extraction-method-(s) /// [16]transit-developmental-impasse-document-pattern-(s) //// [17]mode-time-interval-(s)-sequential-denominative-matter-of-counting-(s)-(independent-motion-(s)) juxtaposed calendar-period-(s)-decimal-denominative-engineering-construction-production-impasse-extraction-method-(s)- counting abstract fragment-(s)-process-(extraspective-developmental-impasse-(s) //// et cetera upon inquisition of logical-methodical-prose…] that define how our organ-breathing-respirational-passage-(s) & blood-pulse-ventricle-mobius-influctuation-tension-(s), by-from visual-tonal-pitch-reverberational-wave-vibration-(s), as-when, etymological-grammatical-punctuational-mathematical-scientific-reason-ing-(s), remain under an pressure of visual-extraspective-confirmation-(s), from introverted-subjective-affirmation-(s), alluded each tangible-fact-(s) from inhabitant-cause-(s') upon terrain-land-property-spacing-(s), judiciary-reviewing, executive-independent-due-(s'),| 695 |where factual-geometric-metropolitan-impasse-(s), rely agricultural-horticultural-substantial-ingestion-usage-(s), for tangible-compositional-perspective, which matter for how each secondths-retrograding-redactive-processing, space how clearly-comprehended impasse-(s), time Minute-(s) & Hour-(s)-Documenting-Omitted-Elapse-(s), for syntax-regard, not adapt-focus of product-objective-dating-usage-period-(s)-impasse-(s) & subjective-dating-interpretation-period-(s)-impasse-(s), related through genetic-population upon an posit-terrain, understanding why are independent-impasse-(s')-inhabitants', affecting an juxtaposition by-at-of perceptive-existence, from-an mundane-impasse, from celestial-ambient-grandiose-moderate-dialysis.[277]¶|||44m.:57s.[258]|||§ Incipient-condition-(s) influence an effervescent-incandesant-peripheral-numerical-exponent-variable-calculating-sequential-operation-(s)-(ing), executive-influencing, sociological-economical-trade, offset local-geological-metropolitan-interpretation-(s), interior log-counting, practicing ticket-memo-bill-docent-documenting, from where our default obligated-requirement-(s), document mass-impasses', dully accepting historical [tradition-(s) / custom-(s) / ancestral-lineage-offspring-impasse-interpretation-(s)] avoided when strict-focus, has yet implored how to extract, an geological-sociological-psychological-physiological-literary-subjective-effort-(s), fulfilling political-living-standard-(s), amid breathing, in attempt to counter-offset, post-retirement (I do not believe in retirement), by-an extraspective-lifestyle, for cultivating geological-resource-(s), after-life-lineage-method-appreciation-process-(es), construing how other-inhabitant-physical-perspective-(s), can be persuaded, by-when live-article-(s) at entity-functional-interest-(s), extenuate beyond retirement-aging, when visual-tone-pitch-reverberational-vibration-(s)-impasse, serve those fine-astute-political-person-(s),| 696 | where an open-reclusive-singular-genetic-territorial-developmental-engineered-concepted-extracted-resourced-effort-(s), reluctantly-direct, each awake moment, to matter by-an family-altruistic-tense, surrounding-privileged-presence, per independent family-{primary-4 being-(s) / Mother // Father /// Children-(Girl // Boy)}, which identifies an elder-error, from sedentary-surrounding-location-(s), for how to enhance an repopulating-family-lineage, greater than common-basic-human-trading-needs'.[278]¶|||46m.:22s.|||§ Maintaining-offset, (I at twenty-Six, probably late-twenties, expect to repopulate, while yet ever opposite-sexually interacting), influence perspective by-of book-(s), commonly-remaining, morbidly-grave, lector-reader-intelligent-retaining, an first legislational-congressional-piece, requiring citizen-independent-creative-application-interaction-perspective, to commence an common-write, for how I see an past & future documenting-retaining-observation-(s), for how memory influence-(d)-(s), immature-effect-(s'), to have an girlfriend-relationship-repopulation-timbre, be how marriage is merit-less, by mass-population-impasse, when literary-trade, remain age-(s)-(less)-mode-(s), affirming how literary-reference matters', as [grammatical // punctuational & mathematical-function-(s)], superlative how subjective-focus-origin, not require scientific-scaling, for how American-political [measure-(s) // weight-(s) // scale-(ing)-(s)]| 697 |sequential-quandary-serie-(s), operating [addition / subtraction / multiplication / division process-(es)] apose serie-(s) of [fraction-(s) / percentage-(s) / greatest // least-common-denominator-integer-plotting] apose-deviation of [greatest // least-common-numerator-integer-plotting] referencing where-when, juxtaposition of fragmented-incremental-integral-(s), an traded-receipt-(s)-information, apose algebraic-geological-exponent-suppose-(ing), in lieu of statistical-article-count-(s), predicate calculus-extenuating, algebraic-statistical-equation-(s), from literguri-cal-press-encyclopedia-dictionary-article-(s)-reference-(s),| 698 |plugging-(additional)-(multiplicit) formal-equation-(s) by-of thought, juxtaposed perceptive-notion-(s), from how population-extraspective-etiquette, apose upon self-observation-(s), proportional physiological-action-motion-(s) (sweat-effort-(s) (not physique)), juxtapose subjective-observational-interpretation-(s), ignored by common demand redactive-studies, as homework-recollection-(s), subject-interpret, peer-subjective-sociological-progress-(es), settling in by, sex-relationship, rather various-form-(s) of relationship, which focus subjective-studies, by-an intensive-group-study-focus, out of question, for blank-assertion, without an clue of cartographic, topography-insight-(s), where global-physiological-impasse-(s), view anatomy-practice-(s), for-an method of biological-post-usage-decay, sort-pattering, pathological-sociological-inquiry, as national-cursive-constitutional-impasse-(s), prelate-affirm, constitutional surface-area, per square-mile or every thousand-word-minimum upon population-proceeding-(s)-tier-denotative-documenting.279¶|||45m.:08s. 32m.s.|||§ Awhile every inhabitant, suffice an minimum per ten-thousand-word-(s) & currency-denominative-counting-method-tier-influctuation-hard-objective-counting-paper-subjective-counting-(s), of each piece per matter, that for which, pertain, patent-commercial-conceptive-developmental-material-(s),[605L.] for collecting, population-article-(izing) when using practicing [issue-(s) //// debate-(s) //// interpretation-(s) //// affair-(s) //// practice-(s) //// impasse-(s) //// focused-action-(s)-count-(s) //// entity-(ies) //// method-(s) of commercial-dynamic //// objective-(s) //// primary-empirical-subjective-(s) //// mission-(s) //// ecological-sociological-dynamic //// elemental-posit-(s) //// extraction-method-(s) elemental-composition-(s) //// transit-developmental-pattern-(s) //// time-interval-(s)-sequential-denominative-matter-of-counting-(s)-(independent-motion-(s) juxtaposed calendar-periodical-(s)-denominative-engineering-construction-production-extraction-method-(s)-counting-fragment-(s)-process-(es), (extraspective-developmental-conscious-interaction-vocal-impasse-(s) //// et cetera upon inquisition by-of logical-methodical-prose...] defining our organ-breathing-respirational-passage-(s) & blood-pulse-ventricle-systems...¶|||45m.:58s.29m.s.|||§

§..Preposition-Count-[28 / Twenty-Eight]-of.. ..Article-Essay-[17 / Seventeen].. ..Book-Essay-[213 / Two-Hundred-Thirteen].. ..Lines-[613.965 / Six-Hundred-Thirteen.1.nine-tenths-2.six-hundredths-3.five-thousandths].. ..what an extent, with reading, an final-production-(s), how would an Genetic-Geological-Terrain, build from barren-development-(s), concerting an surveyed mass-genetic-separated-population, through those year-(s), eval-uating literacy, for how if ever capable, an competent-mass, would affect those factor-(s), per individualized-perspective-(s), practicing to an Genetic-Mass-Population-Extraspect-ive-Wage-Thought-Quality-Activities-(s), avid through various-fashion-(s), from offset-celestial-genetic-population-deviation-activity-(s), proportioning how [Women-Seminar-Yearly-Repopulation-Cycling-Seminar-Daily-Extra- spective-Geological-Inquiry-Dis-cussion // Subservient-Male-Labor-Subjective-Notes-Personal-Memoir], clarify how while aging, thought of how underdeveloped those celestial-fields-remain rotating in pro-portion of a the sun-combustion-cycling, trying to find ones place amid those terrestrial-boundary-(ies), that go through life, in proportion of mass-population-consensus-genetic-documenting, thinking until our bones are out, from our physiological-compositional-muscle -eye-sight-reflexive-dynamic, where to deposit with longevity, our place amid Ancestral-Pre- sence-Review.¶|||01m.:03s.75m.s.|||§

..§¶214y¶§..Formal-extraspective-interactive-regard, time calendar-dating, in review by traded impasse-(s) of mass, when-an offset-schedule-time-frame-evaluation-period-(s)-method-(s), calculate concluded-reference-(s)-(ing), concerting independent-perceptive-focus, sealed as-when period-(s) of dating- This is why I suggest that-their be-an, extraspective-evaluation-conundrum judgmental-processing, yet (State // Attorney) practicing from reclusive-study-(ies), an Macmillan & Library of Congress-literary-legislational-standard-(s) temporal simultaneous-continuum-factoring, subjective-perceptive-observing, entertaining idle-mass-(es), feverously slept, from idle-sweating, affable introverted-presence, recidivistic from converse, then speak when hunger-moment, frustrate dull-comfort-(s), impoverished by care, affirmed by-when exterior-intuition-idleness, walk for short-period-(s) at-of aging-lifestyle-calendar-time-elapse-attention-span-timbre-pacing, Going & Placing, where null-impact, offset spacial-land-surface-area-planned-measure-(s), to-be moved by machine, in transit of radio-wave-idle-motivation-(s), rather careless upon measure, not perceiving overlooked peer-purpose, when given another break, to be supposed applause is adequate, deviating time-period-(s), upon physiological-space-coverage, "ehh" by mathematical-prose, as subconscious-idle-observational-inquiry, omit to count, through [exterior-observational-measure-(s) // weight-(s) // supratangible-exponent-latitude-longitude-acre-mile-(s)-meter-depth-perspective-documenting] affirming when perspective durate through physio-logical-impasse-(s), at-an compositional-interior-perspective, upon exterior-geological-inferior-exponent-movement-(s), because when any inhabitant, are obligated to eat, primordial-humanity, may disrespect currency-purpose, without Calculating & Considering, what motive, direct oneself, when matter-(s) upon formality of activity-(ies), remain lied by educational-homework, which may sight where socializing, interact whom deem which item-(s), as for why perceptive-thinking, signify intrinsic-value, politically-mattered, when at posit-visual-tonal-perspective, amid collective-objective-trading, their of, centripetal-force-revolving-latitude-molten-tectonic-press-ure-force-bend.281¶|||01m.48s.46m.s.|||§ [Convincing // persuading // dissuading] incurred upfronted, because subconscious-repression-conscious-vocal-interactive-lie-(s), form an basis of merit-(s), as default-impact-(s), affecting how each dating-interval-period, vary when matter by size, require an proportional-attention, by-an cause of repercussion-(s), intermittent traded-articlizing, periodicalized by serialized-magazine, again omitted by interior-geological-commerce-agri-culture-labor-education-energy-urban-housing-sociological-health-and-human-services-trans-portation-defense-political-congressional-literary-statistical-reference-(s), then mocked, socio-metrical-commercial-territory-trading, accounting from space, as how to develop from Housing & Storage, an personal-traded-subjective-values, for-an folly of perspective, emitted when during impassing, falsified-traded-effort-(s), easily-seen, as-an projection of development, for whom interchange-impasse under-par, belief of personal-property, remaining in an false-possessive-objective-collective-attainments, swearing under government, when surrounding impact force-(s), elapse around, common inhabitant, negative-subconscious-inclination-(s), for sake of homogen-izing & fortified perspective-(s), rather purporting, lavish standing purpose, layering merit-(ed)-interaction-(s), qualified at agreement by-of objective-thought, where place-(d) traded-objective-(s), hoard amid, voluntary-impasse-(s), comprehended by [1Steve // 2John // 3Mike // 4Bill // 5Darrel // 6Hank // 7Tom // 8Joseph // 9Paul // 10Peter // 11Richard // 12Michael // 13Robert // 14William // 15Thomas // 16George // 17Jon // 18Adam // 19Louis // 20Russell // 21Douglas // 22Timothy // 23Jeffery // 24Rudolph //

[25]Scott // [26]Dennis // [27]Henry // [28]Martin // [29]James // [30]Walter // [31]Patrick // [32]Gregory // [33]Greg // [34]Nick // [35]Nicolas // [36]Norman // [37]Andrew // [38]David // [39]Marc // [40]Mark // [41]Donald // [42]Arthur // [43]Ron // [44]Charles // [45]Eric // [46]Mitch // [47]Charlie // [48]Chuck // [49]Allen // [50]Jose // [51]Juan // [52]Christian // [53]Chris // [54]Allah // [55]Jesus // [56]Jèsus // [57]Lorenzo // [58]Ezekiel // [59]Abe // [60]Alfredo // [61]Nelson // [62]Rafal // [63]Rafael // [64]Victor // [65]Drake // [66]Abel // [67]Allan // [68]Alphonso // [69]Alton // [70]Alonso // [71]Ulysses // [72]Benjamin // [73]Xavier // [74]Zachariah // [75]Zachary /// male-first-name-pronoun et cetera.,], as male-political-responsibility of objective-subjective-planning, is-at visual-population, by-an overload from needy-incessant-inhabitant-reliance-(s), uncircumvential by-of, Title-land-Owner-City-County-Planned-Perimeter-space-purpose-(s), for perceptive-focus, primarily relied self, then reluctant for-an family-purpose, guiding live-documenting-interaction-(s), upon an communion, per conversational-housing-church-genetic-separation-rule-dynamic, afterwards working an genetic-principality-constitutional-communion-document-ordering, reducing from present-continuum-mass-population-impoverished-extraspective-behavior, an state of [principality-mass-population-introversion-(33%) // extraversion-(66%)-dynamic],| 699 |conclusive when-where, whom perceptive-dichotomy-deviation-denominative-percentage, amid extraspection in Zero-whole-percent-decimal-point-zero-tenths-nine-hundreths-denominative-compartmentalized-percentage-deviation range-apose Ninety-Nine-decimal-point-zero-tenths-one-hundreths-five-thousanths-four-ten-thousandths-percentage-deviation, excessive-superfluous, repressed-short-sighted-inter-pretation-(s), redundant from natural-energy, when calendar-dating, not appear (as an) quandary of interval-documenting-expectation-(s), in brief-lieu, an history, never spanning, even as much for, hour-dating, crescent-aging, believed (one of a-the primary-reason-(s) for common-law), as where each impasse, would [[1]drift / [2]brisk / [3]incur / [4]impasse / [5]immense-upon-one-another], reclusive by those mass-consensus-other-(s), in-between substantial-survival, mutely-impassing, exteri-

or-consensus-mass-motion-(s), individualized, as-when physiological-stance, by-an belief in thought, avidly seclusive through action-(s), for human-nature, learn only in youth, rather through Dictionary & Literary-Periodicalizing, sweating in-between, bounded-purchase-reading, as for how evident, their be, dating during (or equator-proximate climate), vocal-pitch, not toning harmonious redacted-syntax, (book-editor-query), affable for focus, remaining in lineage-(s) from memory, courted through city-municipal-settlement-(s), solar-down-tectonic-plate-(s)-lunar-friction-energies, evident amid mute-physiological-(less)-effort-(s), politically-swearing, throughout monetary-traded-consistency-(ies), believing to shroud by-when-where-offset, mass-population-commercial-traded-interaction-(s), retain shelf-life-product-(s), not auxiliary, from how massive each of those hour-period-(s), ignoring blatant disregard, for [1. land / 2. territory / 3. mutual-inhabitants / 4. subjective-documenting // 5. debate // 6. population-psychological-mathematical-mass-offset-studies-inquiry] evident of an entire psychological-lie, from male & female girlfriend-boyfriend-youth-age-high-(school)-raping, idling by University-maturity, to then omit commercial-common-literature.282¶||| 05m.:33s.63m.s.|||§ Male-youth-sexual-reclusive-press-ure, not be sensed, yet where Federal-Judiciary-review, identify educational-purpose, mundane decade-Gregorian-Census-(es), identify-apprehending, teens'-twenties-early-repopulating, to then believe in an family-racial-ethnic-international-planetary-tectonic-political-boundary-universe-(s)-city-district-realization-government-(s)-monetary-bill-policy-code-mobius-influctuation-tension-(s), as-when, visual-tonal-pitch-reverberational-vibration-(s), derive per etymological-grammatical-punctuational-mathematical-scientific-term-definition-(s), under an pressure by visual-tonal-perceptive-affirmation-(s), alluding each present-fact-(s), for voluntary interposed inhabitant-discourse-(s), upon perimeter-terrain-land-property-space-(ing-(s)), identifying factual-global-impasse-(s), as agricultural-horticultural-substantial-ingestion-usage-(s), Construct & Object,| 700 |Tangible-Matter-Development-(s) & Culture-(s), when each secondths-retrograding-redactive-process-(es), derive clearly juxtaposed comprehensive-reference-recollect-ion-comprehension, from-an present-historical-impasse-(s), elapsing Minute-(s) & Hour-(s) documenting, adapting focus of Objective-Dating-Usage-Period-(s)-impasse-(s) & Subjective-Dating-Interpretation-period-(s)-impasse-(s), at an relation by-of genetic-population, upon perimeter-terrain, understanding why we are, by-of-an impasse of existence, from when mundane-impasse-(s), intermittent celestial-ambient-grandiose-moderate-dialysis, [condition // temperate // environment // exponent-density-(ies)] affect intervoluntary-scheduled-trade, (as an) cause for primitive-inhabitant-colloquial-family-tribal-communion-interpretaion-(s), focus log-counting, practicing ticket-memo-bill-docent-documenting, as when Default-Requirement-(s) & Respons-ibilities, document awhile mass-impasse-(s)-offset, dully intercede [tradition-(s) / custom-(s) / ancestral-lineage-offspring-impasse-interpretation-(s)] avoided strict-focus by eminent effort-(s), fulfilling living-standard-(s), while breathing, [1.be // 2.at // 3.to // 4.an // 5.of-exist and 6..attempt] hosting post-retirement (I personally, do not believe in retirement), by-an after-life-lineage-method-appreciation-process, persuading live-article-(s), amid functional-interest-(s),| 701 | mattering beyond age-visual-tone-pitch-reverberational-vibration-(s)-impasse-(s), by those fine-astute-person-(s) open through reclusive-singular-genetic-territorial-developmental-engineered-concepted-extracted-resourced-effort-(s) per-of mass-populus, affecting sake awake moment-(s), to matter in an altruistic-tense, upon environmental-metropolitan-population-Federal-surrounding-(s), live at present,-commer-cial-residential-developmental-fash-ion...¶|||07m.:53s.35m.s.|||§

§..Preposition-Count-[26 / Twenty-Six]-of.. ..Article-Essay-[18 / Eighteen].. ..Book-Essay-[214 / Two-Hundred-Fourteen].. ..Lines-[105.37 / One-Hundred-Five.1.three-tenths-2.seven-hundredths].. ..American-English-Letter-Word-Count-proportion-deviation, upon lines-count-reading-deviation, example how physiological-cardiovascular-muscle-liga-ment-intestine-brain-hip-bones-tension-perspective-release, exterior-grandiose-ambient, not learning for why to document each factor-(s) of existence, practicing by physiological-motion-(s), an path for etymological-ontological-grammatical-mathematical-literary-cabinet-proof-(s), for layering an centralized-death-location, to live alive to, from how desolate those Global-Geological-deposit-(s) of protestant-human-beings, survive, not coordinating-educational-competency from those year-(S) of [2017-Gregorian-Contin-uum© /// 483-Offset-Year-Live-Physics-Documenting-Count©].¶|||48s.$_{97m.s.|||}$§ [Literary-Year / Physics-year]¶|||01s.$_{87m.s.|||}$§

{January-(31)-February-(28 / 9)-March-(31)-April-(30)-May(31)-June(30)-July-(31)-August-(31)-September-(30)-Month-(s)-Day-Hour-Minute-(s)-Secondths-Count}-{23,587,200 D.H.M.S. = 60S. × 60M. × 24H. × D.}-{6,552 D.H.}-{393,120 D.M.}-{273 D.}

{Article:} [October]

..§¶215z¶§..I presume that with literature, if other inhabitant-(s), would participate intermittent algebraic-reasoning, interposed extension-(s) by-of literary-redacting, upon etymological-grammatical-punctuational-referencing, calculating each lined-interpretation-(s), when redacted-isolated-pertinent-interpretation-(s), tier along an calculus-equation, interaction-(s) logic, would help affirm how interpretation-(s) of introvert-(ed)-bias, go along an product-shelf-means', amid extraspective-dialogue, cause denominating pertinent condition-locational-geographical-metropolitan-planned-existence, benign-feature, due in part of sweated-effort-(s), upon resource-(ing), by elemental-product-(s), afflicting how personal-ideals', steadfast how part-(s) of production stand-(s), while millennium-belief beyond day, through year-(s) unarticulated passings', in-an Stance & Sitting, traded-extortion-method-(s), complicity collecting product-(s)-trade, due for an storage-means, at daily-factoring, without an formal-factoring of designate-(d)-period-(s), affirming intermittent an push for genetic-purpose-(s), putting how totality by-of human-behavior, be in-an, remissive-fandom-ticket-retained-trading, throwing out garbage, implicit through ignorance, without an family-loyalty, identifying reproduction, because how daily-living, is without an flagrant-regard, by personal-accountability-effort-(s), working at intermittent locational-two-dimensional-planning-posit-(s), as other-(s) in sincere-fashion, remain in an view of act-(ed)-reference, while dated impasse-(s), idle from significance, where common-interpretation-(s), can not [go / recede / proceed / affirm / conjure] without an extraspective-directive-guidance-(s'), to progress each voluntary-motion-(s), respirating upon timely-matter-(s), because input-presence, affect an output per afterlife, viewed in line of documented-article-(s), or in-an form of strict-diligence, even when I can not sustain an singular one-hundred - five-hundred-page-(s)-extension-(s), when apparent-paragraph-deviation-(s), block total-book-flush-inter-pretative-testing, where latter paragraph-(s), reference-article-(s), at helm where, an six.seven-nine-point-extension, micro-minuté-focus, (un)-applicated, simultaneous, communicative-terrain-presence-(s'), transgressing conscious-factoring by-of vocal-tension, affirming, rather than, technological-idiomatic-functioning, that is tensionless by-of factoring, for-of purchase Wait & See, normalized by, momentum-period-(s)-interpretation-(s), unaffected, by how, extraction-effort-(s), is much a-the process of expert-(s) among life, from nationalism-posit-valedictorian-period-posit-(s), apposed what has not ever been Communicated, or Documented & Proportionally-Rationalized, due to feign ideal-(s), at an impasse of progression-(s), never being controlled by freedom-interpretation-(s), for how still-idle-belief, muster an anomaly at-of time-(s), alluded tragedy, forgotten when after-life, discard bone-after-life-retention, because how planning is not by human-figure, physiological-tension-(s)-decay, and if presence not be detect-(ed), no factor-(ing) can be viewed to ascertain bone-(s), in afterlife, dating upon year, ancestral-entity-directive-retention, when living-existence, figure upon article-pertinence, never to have yet exist-(ed), for how literary-agent, not redact book-article-(s), book-editor merely-impasse-(s)-traded-objective, amid viewing ontological-etymological-grammatical-punctuation-proof-(s), when reader may or may not object, to an longevity traded-purpose, at government-library, default by-of debt-purposed-purchase-(s), uninclined to further how ticketed-event-(s), can be schedule-transgressed, upon an Psychological & Sleep-Measure, dating being, in part of an minute-(s)-hourly-count-equivalent-presence-(s'), infixed from [perceptive-vocal-vernacular-visual-tonal // objective-placing // subjective-interjection-tiering-interaction-(s)] at matters when, none-presence, characterize what is output-(ed), where continuum centripetal-force-revolving, impact Time & Date, influctuating inhabitant-(s)-prerogative-activity-(ies), prior paragraph-document-obser-vation-(s), as human-youth-sex, before eighteen, euthanize how thought is juxtaposed, upon interactive-direct-subjective-metre, when intuitive-commercial-scamming, be a-the method, aging when alluded death, by obituary-grace, existing by-an mean-(s) of negative-human-timbre, based upon traded-retention-lie-(s), from youth-subjective-interpretation-(s), rather than honest inquisition, through aging deducting population-repopulation-reproduction-(s), amid those physical-impasse-(s) of humanity to decay, not partaking an passionate-interpretative-impasse-(s'), upon such educational-background, due at an timed-interactive-presence-(s), as how non-fiction-interpretation-(s), remain by-an lack of focus as Time-Interval-(s) & Dating, introvertly-eating, by ethical-behavior, believe silverware, etiquette while no focus of mass-ingestion, interpose interchange-point-(s), at subjective-background, upon latter-day-continuum, by national-trade, never partaking intermittent, Human-Behavior-(humanity) & Perceptive-Person-Visual-Vocal-Vernacular-Tonal-Conduct-(personality), for how each [objective // subjective // locational-focus], by-of logical-impasse-(s), revolve amid perpetual centripetal-force, because how factoring date, is without an conclusive-data-interposed-presence, because when mathematical-inclination-(s) are yet to have preceded family-interposed-interaction-(s), because egalitarian-elect-processing, remaining in an view of life, negated basic-effort-(s), [1.eating / 2.dispelling / 3.moving / 4.urspurating / 5.fornicating / 6.driving / 7.mechanizing / 8.working / 9.impassing-presence of interpretational-substantial-(s), from objective-traded-assistance], affluent ignorance, (as an) means' to sensate, propose-less, at placings in pleather, moving directionless, at daily, common-however-perspective-(s), placed offset, revolving-combusted-kinetic-existence, none voluntary, intermittent, progressive-motion-(s), meandering feature-(s) of surface-posit-(s), by-an calculating-method, dissent-(ed) during an basis of living, interposed awhile presence, calculate interval time-hourly-dating, minute-(s)-incurring currency, when motion-(s) of matter-(s), task an efficient-conjure-presence, intervoluntary common-conjecture, as how subterior-perceptive-literary-subjective-youth-studies-thought, decrescendo-present, perspective-subconscious-inclin-ation-(s), juxtaposed conscious-dialogue-impasse-(s'), all awhile centripetal-force, is without an formal-denominative-counting, yet verified grandiose-central-topic-(s), inclinating interactive-examination-(s), for how voluntary-impasse-(s), subdue (from an) unopen interacting of data, throughout mechanized-focus, affirming why traded matter, is important to oneself, awhile feature of dating, not apprehend-interpretive-emphasis of (two)-(Length × Width-perimeter)-(three)-(Length × Width × Height)-dimension-(s), for redaction, at [1.length-(s) /// 2.width-(s) /// 3.height-(s) cubic-measuring] not by [1.hypotenuse // 2.arc // 3.adjacent // et cetera..] diameter-radius-circumference-rectangular-prism-square-measure, for constitutional-surrounding-mile consider-ation, by-an Totalitarian-focus, upon reference, Commonly-inducting & Visually-Tonal-Vocal-Vernacular-Referencing, [1.metre / 2.measure / 3.verify / 4.tense / 5.extract (pick / mechanize)] ascertaining physical-voluntary-impasse-(s), by-an means' through intermittent-counting, because how an part of self, offspring only if verifiable-proof, present juxtaposition of visual-tonal-perspective-(s), affirming idle visual-perspective, in case by area-prose, validify-(ing)-traded-article-(s), that only permit-interpretation-(s) from past day, formally qualified, upon when example-(s) of verifiable-perspective-(s'), accrue data, past date data, as lineage-qualification-(s), affirming how viable-data, can be used as-an dissemination of death, or have still-routine, superceding American-routine, quotient elemental-resource-(s), lacking an national-spacing-(s), posited in an juxtaposition of perceptive-territorial-boundary-miles-exponent-climate-(s), temporarily affirming Federal-action-(s) of territory, while dating is perceived, hourly at, involuntary-impasse-(s), upon terrestrial-tectonic-space-(ing-(s)), because how dating-perspect-ive, is upon an reactive-motion-(s) of indignation, while mutual-interpretation-(s), formally-conjure, linguístical-literary-letter-participle-word-form-(s), derive-defining existence, at presen-ce, pertain-(ed)-(ing) in lieu, statehood city-county-dynamic-place-(ing-(s)), due in part of pragmatic-emphasis, as individualism, be the common-belief, intermittent time, their at existing-moment-(s), for idle-freedom, awhile filtered-action-(s), demonstrate how practicing-data, progress awhile class-(ed)-citizen-(s), remain unconstitutional by, archaic-execution, mandated from Federal-bill-policy-sociological-commercial-trade-regulation-(s)-impasse-(s), by Mode & Outcome, where Genetic-Mass & Age, Date & Time, while Being & Sleeping, because mentality-personnel, understand-common-error-(s), false conjuring adequate-communion-traded-impasse-(s), through individualism-ego-error, for being-(s) rationalized, never consistent, document-dating-(s), in an voluntary-consistency, for dating human-inhabitant-(s), require-(ing) legislation-al-interval-(s), as mass-population-(s), remain in part of, default-enabling, surpassing formal-extraspective-interpretational-thinking, awhile perceptive-locational-two-dimensional-planning-formation-(s) of stance, cease to exist, for how involuntary-motion-(s), pertain under politics, amidst undignified-perspective-(s), throughout an monetary-basis, imperial-influctuating, how mere humanity prefer to live, by-an means' of physical-wage-entity-commercial-objective-(s), indifferent perpetual-revolving-molten-centripetal-force, as when, solar-earthly-friction-contin-uum, irrelevant visual-place-(ing-(s)), referenced by visual-tonal-vocal-vernacular-perspective, historical-context, never perceiving before an Spanish or English-language-consistency-(ies), dormant awhile surface-area-miles-grandiose-exponent-degrees, of rotation-revolution-kinetic-magma-pressure-fission, mobius human-distance-(s), without an clear exemplification-(s), for how organism-(s) are at tangible-prevalence-form-distance-(ing-(s)), at an compositional-charisma, stationary for at non intuitive-reasoning-(s), idle-fatigue-impasse-extents, determined extraspective-commercial-development, not harmonious at restricted introverted-perspectives', fatigued by auxiliary-mechanical-effort-(s), without an outstanding-conclusive-interpretation-(s), individualizing Characterized-Physique & Dictate-(ed)-effort-(s), subtly alluded, without an muse of reliance, because an lack by-of community-centralized-genetic-effort-(s), remained involun-tarily-attended, interpose-(ing-(s)), human-driven-impasse-(s), awhile featured-function-(s), can not stand longer than an hour per, without requiring an form of rest, from influctuated-effort-(s)-dynamic, pertinently describing an reluctant-imposing by-of those conditional-environment-feature-(s), intervoluntary awhile, physique-fashion intertwine factual-data, superceding central-aquatic-terrain, for-an populational-inductive-impasse, amidst water-vapor-revolving-drift, ink-paper-press-defining, micro-physical-perceptive-characteristic-(s), requiring repopulation by women, for moment-(s), where life is suppose to balance effort-(s), at genetic-individual-objective-resourcing, metre perceptive-total-means', pertinent for [maintaining land / reproducing agricultural // horticultural-yield-(s) / repopulating in juxtaposition of neighbor-(s)-date-aging] for an tier of population-perfection, by-an consistency of physiological-impasse-(s), intermittent constitutional-territory-development, retaining objective-(s), at locational-environment-entity-(ies), awhile an aggregate superfluously surpass, common-perceptive-recognition, intermittent functional-means', upheld where feature-(s) of elaboration, are meant to signify hyper-tense-(s), in an unideological-progression, featuring aspectual-handle-(ing), meaning that, proper-function of article-practice-(s), serve where perspective is maintained, awhile fortification of entity-(ies), compile closer in approximate-distance-(ing-(s)), where empty-entity-(ies), are remain-(ed)-dormant, without an conclusive-contextual-pragmatic-program, for full-function-feature-(s) of paper-trade-data, denominating along enumerated-displayed-perspective, where at an county-colloquial-dialogue-impassing, methodical-procedure, fulfilling territorial-purpose-(s), as land-task-(s), monetary-balancing, hour-date data, for an various pattern-(s) of subject-(s), where-when, congruent-function-(ing-(s)), equivalent symmetrical elapsed-reference-(s), as dated-hour-(s), have been ignored presenced-minute-(s)-deviation-interval-(s), between surface-area-minute-(s)-impasse-(s), for how in outer-space, our tectonic-yearly-365-degree-drifting, have an grandiose-new-continuum-offset, resetting influctuated from kinetic-molten-pressure, presence-impasse-(s), pertained in secondths-psychological-interview, recollecting each of those tectonic-visual-tonal-perceptive-synapse-(s), impassing where intervoluntary-perceptive-focus, remain by physio-logical-motion-(s), when secondths, are rate-continuum-influctuate-counting, intonation by-of vibrational-frequency-(ies), by-an rate of elapse-(s), noting mode, where kinetic-reverberational-friction-heat-resonance, affect how our individual-imagery-offset-(s), climate-latitude-longitude-posit, through [imaged-sightings ///// reverberational-vibration-(s)-hearing ///// pitched-sound] from of-an linear-exterior-parallel-characteristic-(s), as for how volume-(s) of unit-(s), contain common-ignorant-awe-appreciation, politically-mattering, rather than education-subjective-interpretive-meriting, thus conveying, that [objective-(s) / entity-(ies) / person-(s)] are waged only, from when surprised suspicion crop up, intervalizing at impassed-interaction-(s), during-simultaneous [[1]preposition / [2]civilization-(s) / [3]culture-(s) / [4]space-(ing-(s)) / [5]subterior-anterior-interior-impasse / [6]infinitive-quantitive-physiological-impasse-limit-(s)-(mechanical-application not applicable) / [7]exterior-ecological-chain-cycle-systemizing-(Ecology) / [8]commerce / [9]residence / [10]physiology / [11]anatomy / [12]biology / [13]debate {I / II} / [14]Geometry / [15]Statistic-(s) / [16]Calculus / [17]Trigonometry / [18]Language-Art-(s)-{I / II / III / IIII} / [19]Civics-Studies / [20]World-History / [21]National-History / [22]Florida-History / [23]Police-Training / [24]Military-Training / [25]Electrician-Trade / [26]Plumber-Trade / [27]Criminology / [28]Marketing / [29]Geology / [30]Geography-(atlas) / [31]Engineering / [32]Merit / [33]literature / [34]Psychology / [35]Pathology / [36]Phrenology / [37]Gastronomy / [38]Carpentry / [39]Contractor / [40]Supervisor-training / [41]Engineering-school / [42]nursing-school / [43]adjective-(s) / [44]adverb-(s) / [45]Noun-(s) / [46]Pronoun-(s) / [47]Verb-(s) / [48](In)-Transitive-Verb-(s) / [49]Feminine / [50]Masculine / [51]Plural / [52]Place / [53]Theology

/ [54]Political / [55]Scientific / [56]Physic-Theory / [57]Earth-Space-Science-(s) offset-directives of survival] all are an part, of a-the daily-eight-thousand-six-hundred-five-hundred-secondths, matter-(ed) upon an nightly-interposed locational-offset, that are awake-measure, Legal & Interior, in an gentle-transition millennium mean-(s), mattering because how secondths-millimeter-equivalent-beat-ventricle-blood-flow-rate-interval-(s), (what can millen-ium-pronoun timed-presence-numerical-offset-function-factoring, draw from conclusion-(s), amid ventricle-blood-pulse-friction-muscular-wave-detection-tension-(s), component-(ed)) compart-mentalizing, moded-hours-entity-work-dating, commercial Age 17 through (dash / -) Sixty-Two, pertaining in part by-of handle-(ing)-(s), from physiological-reset-griping, pressure depth-perspective-tangible-contact-impasse-moment-point-(s), rotating-revolution (revolving) physique, as Irrigation & Military, partake in an variational conditional-activity-(ies), amid-discourse-of, conventional communion, through Military-Mechanized-Religious-Reliance-Engineering, conjunction historical-morale-tension-(s), untraded by, discourse-schedule-conduct, traditional as congunctive-cognitive-conjunction, outlet-purchase-(re)-create, an cause for conceptive-product-(s), in lieu from when perceptive-tension-(s) offset hyper-tension-(s), (Four-independent-tangible-aspect-(s)-detecting, deviated by wind-pressure, (skin) apose Ear-Canal-(Skin-Ligament-Bone-tension) / Nasal-(ligament-Bone) / Tounge-taste-(muscle-exo-interior-teeth-pressure) trade, that matter for considering whom is meaningful to one another, in an realm of context, impassing mode-(s) by locational-three-dimensional-square-feet-competency, from how those direction-(s) of study-(ing), not reemphasize, an governmental-commercial-treasury-militarial-trade-process, during agricultural & horticultural design, at common-impasse-(s), developmental upon each factor-(ing-(s)) intermittent traded secular-handling, directing common-mass-objective-land-space-(ing-(s))-debt-purchase-item-(s), scheduling routine-confidence, as simultaneous-factoring-(s), procedure management-protocol, indifferent subjective-objective-means', for how ecological-economic-resourcing-chain-operation-(s), quite tedious, recalling weather, dialogue by-of Goods & Services, fulfill sociological-commercial-stock-shelf-life-trading-deadline-purpose-(s),| 702 | by-an local-market-creativity-competency-rationale-ration-(s), as product-(s), affirm how method-ical juxtaposed free-will, at point-(s) physiological-trade, have an merchant provide service-(s), from those inhabitant-(s) offset, working parameter-perimeter-pattern-time-(ing-(s)), inter-voluntary, Maintained & Proposed, due to a-the consistency of traded-colloquial-limit-matter-(s), having an interior-primary-surface-area-ground-space-(ing)-(s)), upon mass-(es) of item-unit-shelf-(ves), propping stock-product-(s), in-an commercial-trade-tier-space-(ing-(s)), where governmental-basic-human-right-(s), influence directive process-(es), for how outcome-(s) of merchant-acquisition-(s), taken, amidst locational-conditional-environment-surrounding-(s), further shelf-surface-area-space-(s), [interior // exterior], an direct comparison of [product-count-(s) / weight-(s)-measure-(s)] to then heighten dimensional-cubical-spherical-rectangular-prism-complexities, identifying posit-subjective-lesson-word-term-paragraph-interpretational-matter-(s), upon hour-time-frame, consistent-developmental-trading, in regard by-of medium-trade-impasse-(s), where monthly-store-shop-stock-saving-(s), matter upon temporal-figure-(s) inter-posing, paper-ink-year-series-count-currency-dollar-cent-unit-rate-(s), per ounce, complexing trading, to be always by an advanced, indelible, weekly-shelf-unit-compositional-consistency-(ies), balancing weights of [1.product packaging // 2.substantial / 3.objective /// 4.auxiliary // 5.paper-ink-objective-subjective communal-economic-commercial-residential-governemental-socio-log-ial-religious-impasse-(s)], continually aforementioned, for how complete hour-time-line, genetic-statehood-reformation-application-schedule-(d)-agenda-design-(s), emit during a-the vague-arid-belief-(s), by-of government-commerce-cabinet-trade, identifying why Factory-Plant-Product-Offset-Objective & Monetary-Note-Count-Currency, land no national-juxtaposition-(s), per [state // county / city] amid physiological-visual-perceptive-matter-(s), for how independent-individual-ism, balance by an extraspective-work-social-educational-festive-impasse-(s), when identifying present moment-(s), engaging from Romantic-Affair-(s) & Militarial-International-Defense, self-rested, or idle-waste, rather than, proactive-pertinent-impasse-enjoyment per Life & Decease-Aging-Lineage-Legacy-Objective-Tradition-Meaning-(s),179L. deriving value-(s) for substantial-luxurious-existence...¶|||15m.:29s.55m.s.|||§

§..Preposition-Count-[71 / Seventy-One]-of.. ..Article-Essay-[1 / One].. ..Book-Essay-[215 / Two-Hundred-Fifteen].. ..Lines-[179.50125 / One-Hundred-Seventy-Nine.1.five.tenths-2.zero.hundredths-3.one.thousandths-4.two.tenths-thousandths-5.hundredths-thousandths].. ..Conduct reading this book aloud, voicing each letter-word-fricative-dynamic, for then subconsciously-posit-envisioning, each character-(s) of data, for thinking how to refine each context, by proportion of Geological-thought, of barren-terrain.¶|||20s.28m.s.|||§

..§¶216¶§..October is an period again, repeated an estimate of {1 / 12 // 31 / 365}, for monthly-reset-percentage-(s), of {0.08333333 // 0.0849315068}, (as an) starting-point-(s), per year, sustemmed from individual-independent-impasse-(s'), from unconscious-vocal-dating, inputting an calculation per locational-dating-reasoning, for calculating logical-deviation-(s), where [city-district // city /// county //// state-population-(s)], referendum legal-matter-(s), upon an surface-square-mile-(s), counting time-momentary-impasse-(s) of dating, by those dimensional-distance-(s), offset from what movement-(s), juxtapose tangible-posit-(s), apposed weather where, book-referencing, think, how can location-three-dimensional-emphasis, direct numerical-inference-(ing)-(s), by square-mile-height-gender-step-(s)-denominative-count-(s)-impasse-(s'), considering calculated numerical-output, proportional formal-reasoning, intermittent voluntary-reasoning-function-(s), due in part, by-of factory-mechanized-labyrinth-meander-infrastructure-engineering-(s), install-(ing)-(ed) demarcated-mark-tally-(ies), inclusive-Objective-(s), interconnected by hand, for-an mechanical-advantage, taken in to an sequential-progression-(s), as physical-action-(s), secure schedule-(d)-voluntary-impasse-(s), because voluntary-at-will-place-(ing-(s)), due an traded-inclination-(s), verifying action-(s), to be count-(ed), with common-consideration-(s), of a-the, -denominative-monetary-wage // objective // land-title-note-method-(s)] affirming inclinated-notion-extent-(s), alluding weather objective-article-(s), pertain denominative-mode-impasse-(s), mathematical-grammatical-punctuation-paragraph-line-syntax-subversive-hour-(s)-impasse-count-extent-(s), subterior-incremental-day-impasse-(s), collecting by storage-designation-(s), objective-elemental-reset-(s), for particular-creative-compositional-dimensional-patent-docu-menting, how subjective-copyright, intercede thought, by-how consideration by-of physiological-motion-(s), follow an progressive-calendar-succession-(s), at progression-(s), by-an [group-count-denominative-population / group-task-(s) / stroke-(s)], per objective-usage, at an surface, of Space-(d)-Length-Width-Depth & Height-Impact-(s), requiring precision-function-(s), for their to maintain furthering from perspective, technical-governmental-count-(s), subject-articlizing, an impression of dating, considering how hour-(s)-revolving, impact dating upon yearly-outtake, which is for how Season-(s) & Month-(s)-interval juxtapose purpose function-(s), dating-impasse-(s), as like hour-(s)-comparison-(s), develop cultural-focus, by custom-(s) & cultural-intent-(s), considering Vision & Tonal intercoordinated-examining, note how an entity can have an listed-superlative-documenting-practice-(s), directing subject-(s), by wage-method-(s), influctuated independent-personal-belief, as when an entire-comprehensive-article-(s)-book, simplify methodical-conversion-count-(s), written from those pronoun-(s)-participant-(s), upon meticulat-ing act-(ed)-precession, along an basis through stroke-count-(s), as Objective-Count-(s) & Weight-(s), date-physical-impasse-period-(s), as payment-(s) per Hour & Salary, developing cement-panel-(s) onsite, order-deviating objective-purchase-count-(s), handling object-(s), where two-dimensional-perimeter-surface-area-floor-count-(s), room-(s)-count-(s), counting [1.door-(s)-(handle-(s))-(bolt-(s) // 2.hinge-(s) // 3.screw-(s) // 4.minute-interior-layer-(ed)-fitting-(s) // 5.schedule-40 // 6.pvc // 7.cement-rectangular-prism-brick-(s) // 8.eighty-pound-cement-bag-(s) // 9.stucco // 10.dry-wall-panel-(gypsum) // 11.crane-operating-lift-(s) // 12.ice-cube-scooper-stroke-(s)-per-cube-(s)-count-offset-reset // 13.water per pound time /// 14.weight /// 15.action-(s) of onsite-function-(s) // 16.wiring // 17.lighting // 18.outlet-(s) // 19.drilling // 20.sawing // 21.paneling // 22.painting // 23.installing // 24.metal-paneling // 25.jackhammer-usage-(s) // 26.material-cart-(s) // 27.trash-disposal-cart-(s) // et cetera] counting first, developmental-objective-shelf-life-shipping-storage-back-stock-item-(s)-cycle-production-(s), juxtaposing population-count-(s), when voluntary-at-will-employment, tier those weight-(s), under dating-impasse-(s), enveloping hour-minute-(s)-motion-(s)-impasse-(s), acceding process-formal-function-(s), processing Subjective & Literary-redacting, as those formality-(ies) by-of logical-inquiry, believe how, outcome-(s') of proceeding-(s), remaining in part, through directive planed-agenda-(s), construing documented onsite count-(ing), all there when coordination per locational-development, would exist in-an continuity-dating, better-conveying, how each product-item-objective-article-(s)-piece-(s), factor objective-interpretation-(s), persisting sustained-interpose-(d)-action-demand, encapsulate Schedule & Budget, imploring on how locational-modification-(s), interact workplace, conducting, by-an astute-function-(s), when time is important, upon developmental-continuity, because weather, consistent-ordering-(s), work an open-door, trading-policy-(ies), through various-impasse-mean-(s'), intermittent through dating, aiding how conceived-logic, matter only if count-(ed), during an furloughs of count-(s), influctuated by denominative-affirmation-(s), apposed those numerical-impasse-(s) from individual-(s)-effort-(s), (due to an) consistency by-of product-ordering, impacting governmental-extraction-(s), expecting elemental-compound-component-(s)-tally, inclusive, for commercial-inclination-(s), by-of commercial-factory-development-(s), defense-patent-integrated, techno-logical-thought, meant to signify human-behavior, at posit-(s), impassing formulated-inclination-(s), constructive concrete-consistency-(ies), operating-ordering-(s), as when, what would enhance congressional-library-genre-(s), (from an) non-fiction-observational-orchestration-(s) of thought-(s), input indent-tally-mark-(s), conceiving inclusive fashion-dating, for emphasizing weather each mode-(s) of physical-action-(s), remain construed upon an formal-transit-impasse-(s), dating each [person // objective // floor // project-interval] for denominative-comparison-(s), integrating an grading system-function-(s), upon those juxtaposition-(s) by-of matter-(s), enabling Vision & Vernacular-Vocal-Interpretation-(s), due to vivacious-perspective, mattering sake of existence, rather than feign-belief, when visual-tonal-vocal-vernacular-interpretation-(s),31L. define market-place-interaction-(s') per offset individual-(s)... ¶|||05m.: 25s.46m.s.|||§

§..Preposition-Count-[18 / Eighteen].. ..Article-Essay-[2 / Two].. ..Book-Essay-[216 / Two-Hundred-Sixteen].. ..Lines-[31.4 / Thirty-One.1.four-tenths].. ..Ontological-Vocal-Proof, adjust how each fact by timing, layer under deductive-dating, calendarized, when each notion-(s) of thinking, remain adjusted from calorie-thinking-digestive-procession, apposed an entity of safe-sleeping, juxtaposing our transit-market-entity-circulation-dynamic, timed while impassing from elemental-product-resource-(s), packaged by clean-liness, to document etymological-existence, by numerical-inclination-(s), proportioning those count-(s) of physics, upon an juxtaposition of physiological-mass-(es), understanding why lineage-ancestral-documenting, has yet to exist, due to those inconsistencies of youth-study, transgression along adulthood, where daily-fervor, operate in an circulated-function-(s), by entity-policing, observing how from Genetic-Tectonic-celestial-offset, an religious-commercial-wage-effort-(s), should be made, when juxtaposing how balanced-item-(s), variate why ordered-cognitive-purpose, weigh on top of an alternative-pane, for proportioning by principalities, how Carta-Designation-(s), should tier objective-entity-commercially-planned-development-(s), by religious-centralized-celestial-genetic-entity-house-(s), that work apose those fact-(s) of alternating context, for then saving money, offset individual-(ism), practicing on an Commercial-Genetic-Montana-Canadian-Geological-Ethnic-Group, surveying in my lifetime, how weather dating by hourly-mode-set-impasse, de terminal an inclination of thought-(s), that conjure why each individual-sustain-(s), an instinctual-effort-(s), blankly-seeing surrounding-terrain, then simply sitting an forgetting what to secure, for maintaining an constant understanding for lifestyle-wage-context-repopulating. ¶|||01m.:20s.79m.s.|||§

..§¶217¶§..Would grammatical-punctuation-ontological-etymological-text, or mathematical-procedure-(s), operate-visual-perceptive-tonal-interpretation-function-(s) consisting primary subsequent-understanding, through subjective-impasse-rate-deviation-interpretation-quality-(ies), coherently-comprehending from how depth-tangible-perspective, motion through exterior-impasse-(s), interpreting grandeur by visual-tonal-denotative-impasse-(s), not without, fixated-vocal-interpretation-(s), acceding which product-shelf-piece-(s), mark-(s)-tally, in-an inclusive understanding, for how we are in & of-an, posit-(s)-perceptive-offset, sent on our way, to live upon directive-location-operation-function-(s), existing for no past-tier-lineage-article-(s), achieving begotten-fact-(s), without incurring (from an) historical-vernacular-terminological-reference-conjunction, ad hoc, personal-possessive-property-data, improvised through motion-(s), yet reconjured, by interpretative-emphasis, meandering how letter-(s)-word-(s), are to be practiced, by referred-literary-highlight-monetary-bookmark-data, by limited-mark-(s), modified by participle-repetitive-offset, interdialouging from equivalent-book-article-(s)-reference, an title-owner-written-cursive-directive-documenting-interaction-(s),| 703 |solidifying those exo-developmental-three-dimension-establishment-(s), due to comprehensive-consideration, supple-menting calculated-estimated-tectonic-plate-top-crust-ground-self-being-height-clearance-socio-logical-circumvention-impasse-(s'), awhile focus from physiological-posterior-stature, provide oneself an compositional-system, maintaining elongated mark-(s)-instructed-daily-managing-product-objective-mean-(s') item-matter-(s), proceeded though [location-reset-(s) / product-objective-placed-offset-(s) / bookshelf-library /704/ personal-title-collective-ontological-subjective-interpretation-(s)], for etymological-mutual-comprehension, perceived by Ontological-Historical-Present-Observation-(s') & Etymological-Enumerated-Documenting, on an millionths-word-interval-deviation-merit-basis, thinking through [[1.]product-trade-method-(s) // [2.]work-place-title-instructive-maintained-directive-operation-(s) /// [3.]process-(es) /// [4.]transaction-(s) /// [5.]proceeding-(s) // [6.]customary-petroleum-vehicle-transit-homage // [7.]unison-reenergizing-sleep-pattern // [8.]commercial-educational-adult-literary-subjective-comprehensive-{8in.W. × 10in.L. × 1.36in.H.}], by-of living, proportional, intercoordinated, elemental-resourced-mean-(s), clearly by-an legal-congressional, extraspective-third-person & First-person-moderated-denoting-dynamic, from when concepted-patent-product-filed-objective-(s), not be copyright-literary-inquired, which formal-schedule, interval-impasse-location-human-interactive-contact-commun-ication-(s), (presently reclusive under 10-communion-city-living-work-place-licensed-commun-ications, transiting in believed requirement-(s), where cognitive basic-human-right-(s), somehow ignore literary-prose, post constitutional-revising-(s), because peer-pressure, swear sociological-neglect, Evocating & Emotionally-Existing, when conjuring by-an subjective-contortion, proportional physiological-effort-(s), large in part when, revolving-centripetal-force, perpetually rotate, intermittent mint-currency-production-(s), for how Cent-(s)-(*Canadian 100-Note // American-Penny-Cent) & Secondths, posit-(s) an total-denomination-trade-process, but when secondths-total-denominative-factoring, variate from cent-(s)-dollar-deviation-(s), in an common-colloquial-error dollar-subjective-inflation, upon hardware-trade, superceding numerative-mathematical-integer-operation-continuum-offset-(s), working through an total-note-denom-in-ative-factoring, proportions of implemented formal-contortion-mean-(s'), awhile interceding fatigue, meander physiological-action-(s), when subjective-contortion-perceptive-merit-(s), validify note-currency-inflation-(s), (over)-transport-(ed), voluntary-effort-(s), due in part objective-requirement-aspect-(s), taken out of proportion, upon finite-detail-(s), at account-(s), implying which configuration of organ-component-(s), deserve cash-monetary-fiat, surfacing false-labor-effort-(s), which remain an part of interposed-examination, during sociological-impasse-(s) of human-behavior, to remain completely in examination-(s) of purpose, intended to insight, minimal-purchase-motion-(s)-purpose, of-a-the basis by individualism, being method-ically-configurated, where procedural-proceeding-(s), are order-(ed), as particular-minute-complex-(es), identifying at which, extenuated-circumstance-(s), further impasse, trade-logic, compartmentalizing, impassed-process-proceeding-(s), tiering historical-progression-(s), to then arrange formal-recollection-(s), influencing a-the formal-propaganda, per setting of schedule-(d)-impasse-(s), because how monitoring location, mean objective-factor-(s), can only be upon an guise by traded [factor-(ing-(s)) / justice-truth-(s) / founding-(s') / schedule-agenda-imple-mentation-(s)] throughout-an live-recollected-familiarity-(ies), presence-(t / -ce)-impasse-perceptive-vocal-vernacular-interpretation-(s), conjured at an form-of past-date-(ing)-year-unchronological-reference-(ing-(s)), while fate incur Closer & closer, by-an offset-mortality-decai-impasse-military-line-impact-(s), where western Ideological-Treasury, effect Federal-Mass-Population-international-land-extortion, not literary-congressionally-regulating, inland politician-(s), (whom have never understood why poverty is important, and have an post two-thousand-(s)-year-(season-(s))-trade-inflation-error, affecting how exponent-note-track-observation-(s), affirm inhabitant-(s')-fate, remaining in an approximate-century-lineage-military-fatality-dynamic, Easter-Hemisphere & Western-Hemisphere;[359] for how formal-ethics apply, upon subjective-metre, locational-construct-(s), interposed-subjective-impasse-(s), upon a-the hardware-technological-presence, swearing modern-presence-allegiance, when death suppose an moded evocation, by perceptive-personality-individualism and natural-effort-(s), awhile exponent-decade-census-repopulating, has yet yearly taken focus, by exponent-genetic-reformation-(s), because how each inhabitant, prefer-(s'), side to side political-system-process-(es), uninclined, balanced livelihood, in an strict-manner, upon mathematical-count-(s)-impasse-(s), for how legislational-review, by each objective-impasse-(s), remain in line by-an progressive-usage-ordering-(s), commended outline-two-dimension-length-width-clearance-planning-function-(s),| 705 |at visual-perceptive-terrain, (un)-incorporate-(d), constant-constitutional-derived-define-(ing)-(s), because how written-population-implementation-(s), manuscript extenuate-(d)-(ing) formal-traded-impact-(s), from logical-interpretation-(s), serving major-role-feature-(s'), at presence through matter, awhile live-dating provide an mechanism, for functioning, how objective-usage, by subjective-data, pertain intermittent visual-tonal-voluntary-perspective-(s), providing-insight, from-of-a-those facet-(s), capable for translating subjective-article-(s)-application-(s), making live-data, become an work of Physical-Usage & Elemental-Refining, when-where-whom can apply such verified-point-(s), by systematic-information, proportioning-refined-production-package-means', working applicational-function-(s), because how meander of introvert-(ed)-mass, is without an respectful-nature for post-life-appreciative-observing, as physiological-input-(s) of presence, continually-location-offset, regardless monetary-fashion, at point-(s) of sale, interacting subjective-dialogue, affirming what product-process, rely what objective-(s), are preferred for documenting, when mode pertain, elongated-echeloned, by either panel-interior-bordering, past date-infrastructure, devising excavating-elemental-compounds, directing refined-patent-objective-(s), for modification by-of surrounding-location-title-report-terrain, signifying empirical-subjective-date-impasse-populational-function-(s), delving deeper upon past-year-dated-article-(s), numerically deviating, tangible Mode-time-minute-(s)-Substantial & Objective-usage, comparing dry-interpretation-(s) by progressive-matter-(s), influctuating throughout normal-perspective, conducted work-load-(s), where population-simultaneous-impact, affect patent-item-product-production, preoccupying how location-market-shelf-life, access an uphill-battle, by bias-individualism-purchase-interpretation-(s), not aware by common-municipal-restraint, practicing|706|populational-mass-developmental-impasse-elemental-resource-(ing)-conceptive-factory-product-compos-itional-developmental-alloting,| 707 |allowed by pre-population-action-(s)-motion-(s)-measure-(s), attaining objective mutual-genetic-mass-achieve-ment-(s), in fruition beyond [year / season / month / week / day temporal-material-(s)] proportioning how collective-location-(s), have generalized & Specified-Objective-usage-effect-(s), as empty-space, apose objective-imploring, remained by-an relative-proportional-decai-impact-(s), outcoming what data upon presence, and an past-objective-item-usage, identify item-(s)-extent & populant-experience, for which define article-integrity, where progressive objective-handling-(s), remain pertinent from physiological-blood-pulse-ventricle-muscle-ligament-bone-tension-impasse-(s) & visual-tonal-pitch-vocal-vernacular-vibration-(s)-reverberational-sound-impasse-(s), compositional, location-accumulating, as how each logical-comprehensive-aspect, abide-by confluential-processionary-function-(s), from-an morbid-past-day-influence-inter-pretation-(s), for which matter as-when-during tangible-presence, can affirm how progress be affluent upon conjuring, each topical-traded-matter-(s), on an merit-basis, for article-paragraph-topic-reiteration, sequencing periodical population-simultaneous-genetic-comprehension-(s), reforming state-operational-agenda-(s), scheduling each piece of [matter // populant // dimensional-subjective-verification-(s) // objective-usage-experience-impasse // et cetera…] denote-defining how isolated-territorial-surround-(ing-(s)), can be audited (from an) particular-location-impasse-focus-(es), through common-notion Action-(s) & Motion-(s), defining where accomplished-serie-(s) of land-parcel-function-(s), discourse impasse-(s), awhile physiological-effort-(s), concert subjective-fixation, identifying scheduled-routine, per inhabitant,109L. locational-voluntary-interactive-inhabitant-article-(s)-subjective-adult-engaged-participation-(s)...¶|||08m.: 28s.24m.s.|||§

§..Preposition-Count-[20 / Twenty]-of.. ..Article-Essay-[3 / Three].. ..Book-Essay-[217 / Two-Hundred-Seventeen].. ..Lines-[109.19 / One-Hundred-Nine.1one.tenths-2.nine.hundredths].. ..where will you be as an reader, offset from this book-(s) purchase-observation-(s), incrementing apose an referencing-medium-constant-article-reference-usage-(s); how timing affect live-visual-effort-(s), lackluster from historical-physics-year-count-juxtaposition-(s), apose those Gregorian-Year-(s)-offset, as physiological-compos-ition-(s), settle interior qualities, spacing where visual-composition-effort-(s), define-subconscious realization-(s), determined by whom be referring etymological-ontological-vocal-person-(s')-literary-Book-data-referencing-(s)', [highlighting // underlining], which moment-(s), would surface, beyond calculation-(s), from historical-function-(s), when program-(s) per thought, not direct [human / physiological-feminine-offset-male-children-reproductive-being-(s)] inhibiting offtime-sitting-(s), from where time-frame-(s), can be partaken upon, literal-offset wage-work-dynamic, largo present 0.compositional-physiolog-ical-form-(s), due to each surface-area, planned amongst centripetal-revolving-revolution-ary-force-(s), defining our default-limited-storage-resource-(s), from continuum compos-itional-elemental-resourcing-act-(s).¶|||56s.94m.s.|||§

..§¶218¶§..Halloween is an day, amidst month of October, when American-tradition, develop, an influence through those secretarial-societies, by-an interposed sociological-impasse-(s), comprehending how festive-nature, is suppose to balance every individual-inhabitant-(s), when manager-(s) recluse themselves-(s'), hoarding monetary-fund-(s), rather than mass genetic-sociological-religious-interpretation-(s), affecting how each moment-(s), through daily-interposed-impasse-(s), await dating tangible-matter, by what descriptive-emphasis, elaborate conceptive-model-(s)-objective-emphasis-(es), removed electrical-outlet-usage-(s), where mass is supposed to subjectively-centralize, and then disperse, upon territorial-impasse-(s), by-when title-land-directive-(s), indicate when [caution // warning // beware // attention-sign-information], remain an warning of American-bastardized-genocide, from our various Eastern-Outlet-(s) of monetary-information, for how Americans' premise introverted-money before mass-pure-genetic-population-(s), amid idle-bastardization, will-less by-of emphatic housing-room-effort-(s)-concertion-(s'), through momentary-emphasis, Proglamating self-impasse-interval-(s), progress-ive by wage-monetary-individual-belief-(s), spent where Physiological & (dry)-(wet)-substantial-(s), interpreting Conjured & Formulated, from past day word-obituary-decai-article-finding-(s), in-an historical-perspective, not considered, for why insight can only be used, when asserting what data can pertain, where territory-(ies)-land-occupancy-(ies), move throughout occupied-vacancy-(ies), as why developmental-constructive-conception-(s), interpose-intermittent, existence-extent, as monetary-national-inclination-(s), imprison metropolitan Lock & Key, for-of an intuitive-awareness, sweatless at location-compositional-effort-(s), because inhabitant-logical-impasse-(s), have an denoting [1.method-(s) // 2.practice-(s) // 3.operation-(s) // 4.process-(es) // 5.transaction-(s) // 6.proceeding-(s) // 7.cycle-(s) // 8.homage-(s) // 9.pattern-(s) // 10.comprehensive-(s)] by-of information, for how off-time-interaction-(s), have yet interpreted, a-the purpose for prevalent-interactive-informative-action-(s), when directive-objective-subjective-sociological-physiological-psycho-logical-perceptive-visual-tonal-formulation-(s), direct dialogue, by geological-physics-dynamics, for how no formal-purpose, exert by-an off-time-perspective-(s), alluding five-nine-five-military-international-line-ultimate-American-Objective, concepting militarial-egalitarianism, imposed omitted-technological-inclination-(s), false-referencing, biblical-guidance, due their, not to ever exist an juxtaposed-construe, of letter-participle-word-sentence-syntax-paragraph-article-(s)-essay-format-(s), where-at-when, romantic-Germanic-prevalence, maintaining quantified-balanced-dishonesty, permitting wasted-effort-(s), because of an lack by white-collar-subjective-interpretation-(s), apose blue-collar-concert-(ed)-effort-(s), premise basis, title-owner-perimeter-parameter-(s), influctuating dynamic of purchase-parity, inconsistent volume-count-(ing-(s)), to relation dry-pound-(s), by isolated-matter, counting amid juxtaposition, where-when time-elapse-rate-impasse-effect-(s), simultaneous vernacular-vocal-letter-word-sentence-syntax-articulation-emphasis, pertaining isolated-fixation-(s), where physical-exterior-hyper-tense-(s)-observation-(s), impasse each Visual-Blood-Pulse-Plasma & Tonal-Pitch-Vocal-Vernacular-Reverberational-Vibration-(s), for-an numerical-scale-measuring, apposed distance-macro-pertinence, micro-posit-emphasizing, an sequential-retentive-operation-(s), referencing integer-whole-posit-(s), through decimal-denominative-numerative-function-place-(ing-(s)), by-an verification per total-denominative-article-(s)-explanation-(s), tangible-objective-usage, affluent subjective-reason-(ing)-(s), from person-debate-confirmation-explanation-(s), apose interpretive-impasse-(s), through volume-mean-(s), that matter because how secluded-sequential-perspective-trade-impasse-(s), pertain periodical-mean-(s), articulated at confidence, based upon locational-objective-function-(s), where territorial-land, permit pertinent population-objective-trusted-impasse-(s), trade accumulating, elemental-composite-matter, by-an numerical-calculated-cause-(s), from sociological-inquiry, adjusted from-an subjective-paragraph-article-(s)-inquisition, per action-(s), determining an accumulation by-of instructed-informative-competent-hearing-(s), awhile focus at daily fixated-article-(s)-understanding, remain an constant-reminder, for how Motion-(s) & Action-(s), presence prevalence by exemplificated, formal feudal-identification-(s), reluctant-subservient, actual-article-comprehensive-retention-(s), due in part for how, context centralize-visual-tonal-perspective-(s), by-an conclusive-focus through surrounding-surface-area-impasse-sphereical-oblong-oval-comprehensional-tangible-apprehension-vision-familiarity, when-where-whom denote-observation-(s), vary Physiological-Inductive-Perspective-Visual-Tonal-Verbal-Reverberated-sound-Resonance & Subjective-Reference, pertinent inhabitant-mutual-application-(s), when population-offset, have no local constitutional-prevalence-place-(ing-(s)), elementally extracting, compound-comprehension-method-(s), where subjective-fixational-focus, develop, temporal-modern-impasse-(s), when setting upon surrounding-moment-(s), interject, subjective-emphasis of momentary-action-(s), while influctuated-variable-(s), remain relevant, from observational-reference-standpoint, where American-colloquial-tounge, not interpret consideration from self, upon distance-(s), juxtaposing population-impasse-(s), deducting observational visual-tonal-reverberational-vibration-(s)-resonance, applicated by-an neglect of featured-focus, amounting an merit through-of vernacular-vocal-impasse-(s), alluded future-dating-expectation-(s), without an inquiry of reason, from each and every single inhabitant-(s), due to intuitive-male-female-youth-consensual-stauatory-raping-economy, inclinating morbid-memory, unconscious of Eastern-Treasury-genocide-pragmatic-method, for how Western-Treasury-dishonesty, have an youth-sincerity, that rely American-empire, ideal at, blank-memory, to reflect subjective-interpretation-(s) of presence, never conceding weather Cause & Reason, | 708

|meet at helm for meticulating awareness at direct-practiced-location-al-article-(s)-comprehension, for conjecture, not rebuttal existence, for where their remain an measure by-of Distance-(s) & Weight-(s), compulsive by ideological-behavior, male-hoc-dominant, over female-repopulation-reproduction-(s), as male complexify-error-(s), where physiological-effort-(s), rather suffering, physical-trial-(s), domineering each other ethnic-group-(s), reducing egalitarianism, through whom bully feminine-existence, remain an male-preference, amid male-sport-(s)-premise-(s), because difficult subjective-article-(s)-analysis, insist an substantial-transport-requirement-(s), interposed where denotative variational-task-(s), assign common-demand by-of physical-effort-(s), lacking an consistency, measuring vehement-ignorance of behavior, balancing person-(s)-conduct, juxtaposing mass-being-(s)-population-physiological-action-(s), forgetting articulated political-blue-collar-effort-(s), because preference through dull-idle-upbringing,76L. premise those mass-continuum-terrain-supra-tangible-commercial-objective-offset...¶|||05m.:43s.14m.s.|||§

§...Preposition-Count-[19 / Nineteen]-of.. ..Article-Essay-[4 / Four].. ..Book-Essay-[218 / Two-Hundred-Eighteen].. ..Lines-[76.467 / Seventy-Six.1.four-tenths-2.six-hundredths-3.seven-thousandths].. ..Verifying how counting each variable per data variation, can only be valid, if extraspective-first-second-person-juxtaposition, observe how dormant physio-logical-perspective-remain.¶|||17s.53m.s.|||§

..§¶219¶§..Seldomly does mute-inhabitant, vocally-sustain, an active-analysis by-of daily-location-impasse-(s'), forgetting mathematical-inquiry, for story-(ied)-tradition-(s), participated by-an reluctant-preference, affirming idle-social-purchase-method-(s),| 709 |not ordered by communion-voluntary-observation, not evaluating community-purchase-selection-method-(s), (as an) Nice-Relaxing-Working & Activity-(ies) environment-(s), not pertained ecological-cycle-processing, intervoluntary substantial-Comestible-(s) &

Compost-(s), repetitive daily-cycling-(s), regarded from constitutional-background, permitting youth-age-basis, in an mental-hour-location-effort-(s)-perceptive-reflex-(es)-motion-(s), layering rheytouricaily-{rhâ - tor - is - əl - ee}, statehood-debate-(s), interacting only upon mention, from ordained-orated-development-(s), for how conjecture of synapse-posit-(s), not vocalize-extenuate, while reading youth-educational-subjective-comprehension-basis, as when reader not pick-up the personal-reading-characteristic-educational-point-(s), for reverberating by Sound & Subconscious-Visualizing, instilling from how human-characteristic-(s), embody from [fashion // eating-method-(s) // showering-cycle-(s) // electrical-charge-handling-period-(s) // skin-tone // eye-color // hair-color // minutè-pocket-objective-(s) // population-census-denominating], through an promotion of aging-(ed)-political-cabinet-policy-commercial-bill-matter-(s),| 710 |contrasting computerized-denominative-read-memoranda-documenting-(s), reminding why posit-(s) through Physical-Labor & Subjective-Book-Articlizing, balance from conjectured-affair-(s), amassed by emphasis at municipal-basic-human-right-(s), not theorizing practice-(s), by-of Federal-Legal-Entity-(ies), existence, because how managing, maintain yearly-one-work-character-routine-(s), empirical-characteristic-(s), isolated visual-sensational-common-motion-(s), as how equal-politics, show an lack of physiological-subjective-common-congressional-literature, for why no inhabitant, cares for an national-genetic-feminine-exponent-decade-repopulating, awhile Employee & General-Laborer, adequate subordinate, such deplorable-work-condition-(s), when male-physical-idle-sufficing-capability, sustem (from an) fear of superego, unhumble at presence, Attaining & Sustaining, dimensional-function-(s), through [tectonic-rock-lava-density-(ies) // dry-terrestrial-excavating-geological / wet-liquid-substance-(s) // air-wind-pressure-revolving-aquatic-blue-vapor-solar-light-reverberation-effect], as objective-function-(s), register developmental-contract-location-(s), reiterating mundane-indication-(s), because perspective, verify processed-commercial-industrial-instructed-factory-manufacturing-posit-(s), made an common-union-neutral-production-(s), as how wage moniker, human-mass-morale-evaluation-(s), when physiological-effort-(s) alone, can not proceed an industrical-competitive-population-extraspective-interactive-effort-(s), by when literary-commercial-subjective-etiquette, price how ethnic-genetic-tier-effort-(s), serve Natural Tectonic-Plate-(s) & Physics, because how objective-determination-(s), wear colloquial-physical-preferenced-human-visual-perceptive-being-(s), past article-(s)-usage-(s), better helping, term-defining, by how then location-(s) of terrain, permit an Three-Dimensional-Average-Exponent-Square-Feet & Meter-(s)-Dimensional-Component-Perceived-Aspect-(s), for how during intermingled-voluntary-effort-(s), (an conundrum of decision-(s)), confluence conception by developmental-product-(s), matter-(ed) in presence (not part-(s) particularly) for pertaining Wood & Gold objective-(s), apose book-visual-perceptive-tonal-article-(s), depict-envisioning, Mathematical-Operation-(s) & Grammatical-Punctuation-Etymological-Proofing, upon a-the population-impasse-dynamic when crust-surfaceable-surrounding-walk-dialogue-impasse-area-(s), encapture visual-imagery, by-an perceptive-focus-epiphany-visual-mark-(s), annotating pertinence-observation-(s), then re-incurring upon formal-prerogative-parameter-(s), commenced by two-dimensional-perimeter-basis, for then intervalizing, each Cubic-Feet & Meter-Dimensional-Objective-Retention-Maintaince-Progress, by date-impasse-scheduled-agenda-planning, for how motion-effort-(s), intertwine from past-day-set-up-action-(s), layering per Hour & Day, upon perceptive-presence, an imbue upon present-confidence-storage, for impassing objective-procedure-(s), as objective-relay, hold whom may matter, by-an basic-visual-vernacular-hyper-tense-(s), an simple draw, for way why, (an) moment, can Mean & Signify, such dignified-(con)-verse-(ive)-(-e)-value-(s), interacting supplemented awake-conscious-impasse-(s), volumiz-ing dictionary-vocal-definition-(s), vernacularizing our perceptive-preference-(s), because person before place, place before subjective-rigor, amid scheduled-cabinet-commercial-agricultural-interior-developmental-defense-operational-agenda, directing objective-production-effort-(s), at an post further, impasse-synapse-(s)-period-(s), time-hour-minute-(s)-aging-intervalizing, sequen-tial segmented, dating-monetary-purchase-agenda-monitoring-checkpoint-(s), day per day, hour per hour, where-when, purchase has been an default-family-reclusive-behavior, defined from conducted-affair-(s), from various [physical // subjective // acted-voluntary-motion-(s)], visualizing requisite-impasse-(s), familiarizing oneself by those sequence-(s) of parameter-impasse-(s), causing progressive-thought-pattern-(s), meandering existing-period-(s), during when, volume-respiration-circulation-interior-ventricle-blood-flow, vitalize physiological-visual-tonal-perceptive-hyper-tense-(s)-interior-compositional-function-(s), compulsive by-an reflexive-motion-(s)-implication-(s), apose [policy // bill // referendum // amendment // article-(s)-act-(s)] variated (due to an) effect of policy-agenda, directing various human-sensational-motive-(s), reverberating upon skin, when articulation-(s) from exterior-visual-perceptive-value-(s), intercontort-pertain an basis of observational-documented-count-(s), collecting from standard-(s) by act-(s), an pertinent interactive continuity, progressing logical-outcome-(s), linguistics-vernacular-literary-reference-derived-etymological-ontological-physio-logical-impasse-dynamic-heart-(esphogus-ventricle-blood-pulse-beat)-counting, through [1. Geological // 2. Economical // 3. Ecological], traded-pattern-(s), when repetitive-confirmation-count-(s), tension at objective-applying, for then subjective-investigation, (un)-populate-(d) at common-parameter-(s), for Resource-(s) & Function-(s), by human-impasse-(s), awhile letter-(musical)-note-nth-factoring, suppose an complacent-complementary, surrounding-setting-(s), in centralized-subjunct-fashion-(s),| 711 |impassing through emitting common-visual-perceptive-understanding-(s'), from developmental-complex-comprehension-(s), impassing human-inter-active interpretation-(s'), in lieu for enhancing ones' collective-extraspective-others'-method, for ascertaining visual-tonal-perceptive-subjective-geological-title-physiological-aging, by grandiose-exterior-monotheistic-impasse-cutoff-point-(s), interposed present-developmental-commer-cial-circulating-traded-impa ct-(s)...¶|||06m.:17s.59m.s.|||§

§..Preposition-Count-[11 / Eleven]-of.. ..Article-Essay-[5 / Five].. ..Book-Essay-[219 / Two-Hundred-Nineteen].. ..Lines-[73.2625 / Seventy-Tree.1.two-tenths-2.six-hundredths-3.two-thousandths-4.five-tenths-thousandths].. ..For how those standpoints of perspective go, the year we document in, may actually be faulted, from how Trial & Error, not sequence an coordinated-logical-function-(s), in purpose per operational-parameter-(s), amidst those fact-(s) per individual-(s'), whom comprehend where developmental-location-(s), affirm why global-politics, resolve issue-(s), by those perspective-(s), at physiological-tangible-distance-(s), apose human-bias-hoc-behavior, not incrementing formally, why develop-ment of geological-terrain, would matter for interactive-interpreting, whom think clearly why etymological-ontological-grammatical-mathematical-physic-existence, would be. ¶|||40s.75m.s.|||§

..§¶220¶§..[1.Interpretation-(s) / 2.indication-(s) / 3.implication-(s) / 4.incurrence-(s) / 5.impasse-(s) / 6.writing-(s) / 7.motion-(s) / 8.apprehending-(s) / 9.understanding-(s) / 10.comprehending-(s) / 12.study-(ing-(s)) / 13.finding-(s) / 14.reading-(s) / 15.calculating-(s) / 16.sustaining / 17.maintaining / 18.obtaining / 19.retaining / 20.measuring / 21.weighing / 22.balancing / 23.collecting / 24.grazing / 25.cultivating / 26.germinating / 27.cleaning / 28.hypothesis], are all basis-(ed) from micro-matter-infetesimal-particle-content-incremental-crescendo-referencing, apose those feature-(s) by-of independent-manual-hyper-tense-perceptive-motion-(s), apart from physiological-impasse-(s'), while physiological-component-anatomical-interior-mass-encompassing, remain for how any one of us, exist upon an parallel of parallel-spherical-revolving-compass-depth-perspective-centripetal-perpetual-force-height-clearance-plane-(s), visually-perceptive by physiological-components-fatigue-limit-(s), revolving amid revolutionary-terrestrial-tectonic-impact,| 712 |from those state-(s) of primary-physics-supratensible-condition-(s)-grandiose-form-(s), either in direct or peripheral-vision, impassed from when, [1.commercial-man-agerial-ordering // 2.personal-residential-purchase-(s) // 3.sociological-work // 4.off-time-religious-cursive-writing-practice-(s) /// 5.executive-cabinet-government-policy /// 6.senate-congressional-commercial-government-bill-(s) /// 7.sociological-decade-census-mass-act-(s) /// 8.philosophical-literary-perceptive-physiological-vocal-hyper-tense-(s)-common-being-attribute-(s)-ethnic-group-origin-decade-act-(s)-currency-cycling-analyzation-percentage-mass-fraction-deviation-senate-bill-(s)-socio-logical-geological-interior-document-(s)-commercial-denominative-market-(s)-currency-exchange-military-treasury-enlistment-standards-population-percentage-weapon-regulating-executive-policy-economic-stimulus-offset-amending-(ment-(s))-(-ing)], conducting affluent mutual-cooperative-effort-(s), interpreted-interposed free-will-monetary-cycling-analysis, along an simultaneous (vernacular)-vocal-resonance, reverberated upon ear-canal-organ-vibration-(s)-rate-(ing-(s)), receiving data rated upon an perceptive graded-studying, identifying how literary-article-(s) direct an concerted evident-sociological-mass-proof, be all human-being-(s)-living-life-style-(s), weather objective-component-(s), or subjective-literary-observation-analyzation, would imply upon tonal-usage, an conjuring-grade-point-average-semester-whole-point-millionth-words-subconscious-thinking-hundredths-thousandths-words-written, survey human-mass-exponent-competency, by empirical-logical-formal-impasse-progression-(s), juxtaposing literary-book-(s), to apply by-an method of location-purchase-action-(s), determining how weather, pre-architectural-blue-print-engineering-consideration,| 713 |is ever denominating-simultaneous intelligent-voluntary-genetic-reformation-analysis, geometrically-appraising, sociological-metropolitan-repopulation-practice-(s),| 714 |visually-perceiving, affirming commercial-market-inter-national-standard-book-number-collecting-personal-library-reader-practice-(s), balancing by Conventional-Self-Vocal-Limit-Lecturing, communicating by directive highlight-(ed)-redactive-reading-(s), ascertain referenced [line-(s) / paragraph-(s) / syntax-observation-(s)] thought by-an registering through dating, considering apose hour-time-method-impasse-(s), inclinated work-position-(s)-payment-function-(s)-dynamic, gandering whether monetary-currency-savings Note-(s) & Coin-(s), partake in progress, by-an retirement-purchasing-method, mid-age-working-monetary-parity-investing, for an juxtaposing commercial-bank-saving-(s), for retaining those objective-(s), required by trading-parity-burden-(s), in lieu alleviating plant-constant-over-reproduction-factoring, from Treasury-superlative-commercial-factory-product-objective-production-(s), impacting each (in)-dependent-variable-(s), amid mass-constant-offset, aware when community-inhabitant-(s'), voluntarily interact, intermittent fatigue-constant-offset-effect, upon tectonic-plate-elemental-national-resourcing, state-retained-objective-wage-inhabitant-maintained-congressional-sustain-ed-item-impasse-(s), from-an dispersion of [topic-(s) // matter-(s) // issue-(s)], covered by offset-locational-newspresse, progressing lifestyle, uncertain, incomprehensible bias-perspective-(s'), affecting deadline of agenda-(s), between juxtapose-(d) corporation-agenda-(s), developing how voluntary-function-(s), value or deem, in lieu at reclusive-management-(s), nationally-filed, not capable for locale-operation-(s), because an lack by religious-university-genetic-scheduling, separate apart, independent-being-mass-progress, bi-partisan-unilaterally-proportional, for proceeding date-impasse-act-(s), intercorrelated by religious-communion-circumstance-(s), conditioned by environment-temperate-impact, effecting read-legible-conjuring, by-an sequence-(d)-subjective-reading-timbre, apposed redactive-subjective-timbre, being formulated from-by monetary-substantial-cause-motion-(s), alternating past-requisite-progression-(s), determining auxiliary-function-(s), in review by act-(s)-policy-traded-decade-impasse-(s),| 715 |due from constitutional-national-state-land-population-ration-conjuring-restriction-(s), unperceived, from-by-when educational-academia, not filter personal-independent-moral-subjective-interpretation-(s), declining-decrescendo competency-reference-age-rating-comprehension, during aging-year-month-juxtaposition-(s), yet to consider seasonal or monthly-dating, layering hour-(s) by day or week-(s)-extent-period-(s), [analyzing from modern-production-{time-live-year-dating-analysis} // past-collective-development-(s)-{Calendar-month-week-past-historical-objective // subjective-proportion-documenting-analysis}], for furthering juxtaposition-(s) by human-focus, in light of agricultural or horticultural temperate-production-(s), harmonized apose-intermittent, basic human-right-(s), executively-senate-congressionally-flawed, for how deplorable the total-human-genome, [1.literary-mathematical-numerical-ordering / 2.scaling / 3.counting / 4.calculating / 5.operating / 6.budgeting // 7.dimension-(s)-measuring /// 8.objective-production-weighing-Grammatical-term-word-(s)-paragraph-definition-(s) // 9.paragraph-syntax-essay-article-(s) // 10.personal-document-(s) // 10.Federal-form-application-memo-operation-(s)-documenting-genre-method-(s)] ordering predicate-action-(s), from monetary-communion-practice-(s), directing resourced-temperate-action-basis, aligned [cause-(s) / want-(s) / desire-(s) / requirement-(s)-by-data] referencing-objective or subjective-impasse-usage-(s), avoiding elicited-effort-(s), because genetic reliable-livelihood, affect each of those legal-governmental-binded-case-(s)-human-volume-(s),| 716 |handling denominative-population-(s), by county-state-federal-issue-(s), as moment-(s) exist through an scenic-perceptive-imagery, alluded only if vocal-vernacular-subject-conjuring, attain-sustain at an personage-impasse, or tend to dissolve an formal-legislational-congressional-literary-balance,72L. by total-mass-population-executive-municipal-police-code-unaware-ambiguous-history...¶|||05m.:08s.46m.s.|||§

{Article:} [October]

§..Preposition-Count-[9 / Nine]-of.. ..Article-Essay-[6 / Six].. ..Book-Essay-[220 / Two-Hundred-Twenty].. ..Lines-[72.57 / Seventy-Two.1.five-tenths-2.seven-hundredths].. ..If the Physics-Year is 483, while we conduct political-year 2017, what continuum view do we have from tomorrow, for living as today, during an inconsistency-(ies), from physiological-human-bias-hoc-position-(s), juxtaposing an negative-outcomes, from physics-tectonic-standpoint-(s), due to an early-offset of matter, yet formally proportioned, from linear-visual-vocal-tonal-focus.¶|||24s.53m.s.|||§

..§¶221¶§..Focus, intensifies visual-imagery-vocal-tonal-vernacular-perceptive-conjuring, by-an method of listening-imagery-retention-perspective, based upon unit-tangible-(s) for fixation by-of thought-(s), relevant during processing-presence, for human-presence only signify when at locational-environment, an time-period-surrounding, not (as an) live-calendar-range or estate-geological-posit-(s), exemplifying how applied monetary-unit-point-(s), date economical-trade-cycle-(s), due in part purchase-effort-(s), emanating, attained objective-article-(s), that fruition trade-basis for monetary-article-circumvention, expressing derived notion-(s), from [motion-(s) // action-(s) // inclination-(s) // indication-(s)] implicit unless, extraspective-communion-communication-effort-(s), devise an formal-extent for purchase-purpose-(s), in exponent-purchase-communion-communication-(s)-extent-(s), working from minor-longevity-object-purchase-(s), through [auxiliary // tool-(s) // horticulture // agriculture-documented-cycling-(s)], to be-an actual-purpose, for common-effort-(s), for how methodical-hour-time-dating-operation-(s), act awhile presence-effort-(s) are upon implication-(s) by-of standard-(s), unmet because how commercial-impasse-effort-(s), are consider-(ed) in grade by academic-impasse or conduct-impasse-(s), observing an influctuated-common-interpretation-behavior, more prevalent from [musical // film // novel // magazine-serial-present-continuum-copyright-commercial-market-retention-(s)], rather than conducting modern-conventional-book-copyright-purchasing, ignoring the remaining copyright-function-(s), until monthly-musical-festival-(s) /balance/ work-week-philharmonic-performance-(s) / juxtaposing/ seasonal-historical-film-act-(s)-documenting alluding how presence by posterior-usage, have implication-(s), for reviewing attain-(ed)-trade-property, relative to mass-function-(s), directing how transit amid [work-place-payment-method-operation-(s) // home-sleep-eating-recuperation-(reading)-lifestyle // church-congregating] are pre-historically-instilled, for how an interactive-crowd-setting, has yet existed, for operating church-house-(s), twenty-four-hour-(s) an day, simultaneous by literary-communion-geological-terrain-metropolitan-planning-population-schedule-cycling, as how monetary-work-wage, can extent the greater lifestyle-(s'), from whom matter property-storage-trading, by an comprehension of how [[1]eating-(stove-(electricity / propane) / [2]substantial /// [3]liquid-diet-scheduling) // [4]sleeping-(charging-device-(s)) // [5]writing-(Ink & Paper) // [6]typing-(electricity) // [7]showering-(projected-water-pressure) // [8]walking // [9]placing // [10]calculating-(mason / carpentry-dimensional-spacing)] thus jested irrelevant-notion-(s)-offset, distanced by physics, where mass-collective-(s'), apose required-directive-motion-(s), for individualism be concentrically-inclined, pertinent component-part-(s), for how each daily-focus, be in (re)-view, by locational-title-objective-storage-responsibilities, because when-where reference, (from an) historical-material-data-base-(s), [indicate // imply // interpret // suggest], an part of data enhance house-legislational-argument-(s), for pronoun-presence-stint, reveal their needing an presence-mass-(es)-impact, when (re)-action-(s) during voluntary focus from inhabitant-(s)-point-of-view, remain Secluded & Reclusive, without concepting hour-dating /apose/ mode-dating-schedule-(s) because how depth-perspective-forward-linear-common-mass, forward-continuum-remain, from when diplomatic-person-(s), handle an majority-preference-presence, at an posterior physique-view, apose individualism-monetary-budget-handling, for how interpretation-(s) are confluenced amid year-millennium-aging, not considering how we age through month or season range-count-(s), moding through [hourly /sleeping/ daily / (work)-week-(end) / weekly-time-reset-(s)], (from an) idle-disposition-offset, sleeping relevant through voluntary-impasse-action-(s), persuaded from subjective-interpretation-(s)-central-curb, defining how definite-interpretation-(s), juxtapose ambiguous-circulating-observation-(s), requiring review, due to perspective-discourse-(s), servicing duty-(ies), by colloquial-common-affair-(s), by-of title-land-space-(ing-(s)), attaining spacing, when physique is suppose to impasse-(s)-presence, in fashion at directed-market-(s), retaining-commercial-market-item-unit-(s)-objective, from parental-family-lineage-transgression-(s), upon racial-ethnic-intuitive-diversion-(s), following how color-shape-translucent-effect-(s), emit consideration, where imagery-movement-(s)-presence, influence-locational-posit-(s)-outcome-(s), conducting traded unit-(ed)-product-(s), defining product-ideal-(s'), (from an) extraspective-vision amid existence, rather than individualistic-ideal-(s), twist-(ed) /and/ bent, on ravaging normal-commercial-good-(s), implicated where voluntary-imagery-perceptive-physical-motion-(s), [1.territory // 2.location // 3.entity // 4.city 5.block // 6.acre // 7.estate // 8.restricted-area-offset], for how individualistic-behavior, define underpar-grade-point-average-(s), by-when educational-study, apose each beneficial-year-millennium-aging, not integrating unit-shelf-item-(s), from purchase-cycle-dispersion, separating each unit-(s), amongst federal-production-yield-(s), offset by city-county-commercial-factory-warehouse-production-site-(s), for where sociological-market-(s), register monetary-dollar-note-currency-exchange-(s), functioning residential-spacing-purchase-(s), interacting by mutual-effort-(s), at present-mass, consensually-inclined, for an lack of respect, not preferring conduct-etiquette-impasse-purport, for reasoning by-an literary-means', circumventive-yearly-feminine-offspring-repopulation-literary-method-(s), constant-settling, community-feminine-genetic-scheduling, bored by study, because sweat-maximum-effort-(s), indicate-fatigue, when present-rotation, slip past comprehensive-interpretation-(s),60L. for what remain by Lock & Key-(s), under putative-law...¶|||04m.:45s.69m.s.|||§

§..Preposition-Count-[8 / Eight]-of.. ..Article-Essay-[7 / Seven].. ..Book-Essay-[221 / Two-Hundred-Twenty-One].. ..Lines-[60.47 / Sixty.1.four-tenths-2.seven-hundredths].. ..I intend to create book-(s), that have my reader comprehend how to live, while thinking what mode of living one supersedes, by visual-observation, from those offset-Global-Tectonic-Geological-State-County-Territory-Boundary-(ies), that serve an dispersion of Genetic-ethnicities, not in an fulfilled National-State-Functional-Living.¶|||22s.38m.s.|||§

..§¶222¶§..Citizen-(s) rampart past auxiliary-objective, swearing by family-vehicle-transportation, that all is well, by ethical-objective-retention-(s), from how retirement, phase out, elder work-place-effort-(s), when substantial-effect-(s), are taken for granted, by imperial-presence, because how duty is impose-(d) (rather than voluntarily-participate-(d)), signifying through disseminated-truth-(s), referenced focus matter-(s), most discreet by means, communicated by elemental-extraction-(s), affluent from work-place-dialogue, discrepant for-of personal-title-concealment, incapable for individualistic-extremes, allured at limit-boundary-maxim, by individualistic-reclusion, forgetting how each product of trade, incorporate life-style-method-(s), relevant implemented extraspective-impasse-voluntary-effort-(s), never totalitarian-capable, individualistic-rationale, for how legislational-bill-(s), require-house-decade-census-deadline-bill-act-(s),| 717 |evaluating executive-policies, by commerce-dollar-cent-(s)-circulation-subcommittee-analysis, reducing those bill-(s), by decade-effort-(s)-evaluation, diminuendo-decrescent Federal-bailout-proglamation-(s), upon commercial-sociological-circulation-two-hundred-forty-year-educational-accountability-process-(es), examining agricultural-product-con-ception-(s), by government-bus-transportation-(s), amid hertz-vehicle-special-occasion-usage-(s), when re-cycling those repossessed-vehicle-(s), may serve better if formal-repopulation from political-commercial-power-(s), ever balance hundredths-thousandth-annual-dollar-debit-rate-(s), per child 0 - 18 $21,000-raising, resetting the balance of common-classification-(s), for evaluating why offspring, has been mishandled by political-power, when religious-belief, corrupt the vocal-vernacular-educational-inter-pretation, relied after educational-intervention, and interior-cabinet, physiological-subjectively-guide, [citizen-(s) // inhabitant-(s)] amongst geological-state-county-city-territory, but when bus would serve an county or city-state-driven-usage-function, rather then personal-reclusive-preference, as how bus could then transfer cause after an decade-bill-act-sociological-payment-evaluation-(s), guideline for political-monetary-ethical-salary-shift-(s), compiling literary-review, for citing by word-count-(s), those dollar-purpose-(s), for circulating monetary-economic-stimulus.283¶|||02m.:00s.19m.s.|||§ As rash non-syntax-comprehensive-under-standing-(s) continue personally, shelf-stock-life-maintaince, serve mass-local-population-deviation-consensus, unsegregated by bastardized-communities, moving as like cattle-drove-(s), through vehicle-transit-day-(s),| 718 |not vernacularizing, conclusive-extraspective-transaction-(s), as existence up until now, interpose elemental-state-tectonic-constitutional-land-word-population-syntax-proportional-demand-(s),| 719 |simultaneously-mutual, massive-at-will-work-place-schedule-effort-(s), for how daily-circumvention, omit why educational-subjective-purpose-(s), signify formal-methodical-literary-subjunctive-trading, resetting at decade-census-impasse-(s), primary-subjective-literary-analyzation-(s), religiously intended, to test-(ing), extenuating each furthering, by-of inanimate-cultural-impasse-(s), observing those pitfalls, for developing-terrain, compensating (as an) role from custom-(s), refer customary-ruse, per method-(s), combined consumer development-(s), progressing geological-crust-surface-area-proportion-(s) of acreage-estate-terrain, [cultivating // developing // adding], infrastructure-continuation-(s), extenuating selective-infrastructure (abandon-(ed) or inhabit-(ed)), arousing motion-(s) by Taste & Smell rather visual-mark-(s) of sight, predicate of vocal-vernacular-term-glossary-dictionary-definition-hearing-interpretation-(s), then (re)-affirming influctuated-logic, to deviate by Hearing & Sight, for then later incurring, physical-presence, relied by truth-(s) by-of cited-article-centralized-referencing-(s), to convey how fatigued-presence, require article-directive-(s), avoiding personal-story-pacifism, confluencing from individualism, an concealment at [door-entrance /apose/ exit-opening-(s')], intuitively-suggested, but ignored-simultaneously, overtly-ambiguated, not bothered enough, from central-ventricle-nervous-blood-interior-respirating-friction-pressure, an pre-eminent-temporal-forecast of fate, remain dulled in car-cubicle, contained from self-glory, to remind oneself, what was never communicated, by-an preference for-of reclusive-clamorous-nervous-retardation, reclusive for talking amid youth, and continued through adulthood, temporal-imagery-retained-perceived, because how discretion of article-(s), maintain an possessive-trading-piece-(s), only relevant from family-interaction-(s), from self, through community-objective-trade, naturalized by neutral-religious-political-view, defining human-quality-(ies), in an limited-locational-mode-(s), to soon deceased, by physiological-compositional-aging, when century-mark, have yet reached two-century-capacity-(ies), motivated by literary-subjective-trading, enhancing treasury-trading-method-(s), physically-aging into obituary-articlizing, intervalizing interior-recipe-consumption-presence, incapable for belief in potential-quality-(ies), never commonly-convey-(ed), for literary-non-fiction-conventional-presence, drifting wiltfully, between [tectonic-plate-earth-molten-circulating-magma-rotating-revolving-core // Lunar-tectonic-rotating-drift // Solar-central-exo-molten-revolutionary-core-celestial-universe // Satellites], as afterlife-enigma, due to live-presence-influence, construct non competent-competitive-copyright-consideration-(s), (where black-(s) compete physiologically, not feeling an ancestral-angst, where dull-perspective, rule over subjective-effort-(s)), when upon each development-cause, in regard while-in /moving/ at-progression-(s), intermittent centripetal-force, perpetual-interval-underling, acceding where-dating personal-possession, not be unit-(ed)-registered-material-(s), yet required for cycling objective-literary-subjective-application, for determining which topic-(s), by conducted academic-development-(s), revealing relevant momentary-means', at land-space-purpose-scheduled-impasse-(s), indicative-presence, never assenting from bias or hoc-progressive-grandeur, repopulation-literary-planning, unrestraint by self-internal-discipline, served an locational-objective-place-(ing-(s)), accepting voluntary-bias, to remain in an false-transit-monetary-construe, quick at youth, to forget subjective-methodical-subjective-contortion, freely unreview-(ed), because how sleep-pattern-(s), space physiological-fatigue-sequence-(s), as subjective-will, not yet exist, mark-indent-indicating, procedural-instructed-directive-(s'), serving commercial-wage-act-(s), (not) meticulated from monetary-circulation-purchase-motion-(s), when each superlative moment, have an data-exponent-interpretation-offset, like physics, population-personal-dating, superfluous processed-methods of payment-(s), in lieu as used numerical-numerator-calculated-operation-(s), vigor various-independent-ethnic-being-(s), attesting, from how conjure-practice-(s), pertain systemized method-(s) of payment-(s), in lieu from daily-parameter-depth-perspective-focus, deriving possessive-common-shelf-objective-(s), at market-commercial-operation-(s), instilling an scheduled-operational-functional-act-(s), cycling from metropolitan-locational-conjunction-(s), objective-monetary-generated-production-(s), sustaining an market conceptive-action-(s), coordinating from when duration of motion-(s), supersede conjuring monetary-at-will-employment, at consideration-(s) for developmental-constructive-premise-(s), in according sequence-(s) per hour, wage-work-week-impassing, by an proportional-sleeping-process, when daily-after-effect-(s) of population-entity-function-(s), abide at [open // close-shop-(s) // market-(s)-premise-(s)] awhile each dating-period-(s), incept, an required agenda, for governmental-incorporation-filing-(s), for then commercially-operating, each of those factor-(s) of trade, by-an routine for free-will-attraction-(s), valuing how circulated-routine, can formally supplement, Supply & Demand, because remnant-(s) [to / from] transit-county /local/ city-minute-(s) of inhabitant-intersected-impasse-(s), factor objective-product-(s), upon an documented-tax-(es)-filing-(s), for maintaining traded-trade-mark-objective-registered-receipt-operations, at commerc-ial-post-(s), by commercial-prowess, progressing Lifestyle & Existence, when we are under constrained-distress, not having enough self-motivation, pertain oneself upon thought at empirical-interceded-environment-surrounding-(s), hall subjective-ethical-book-volume-copyright-serializ-ing-(s), (due to an) dishonesty by mutual-interaction-(s), which maintain an perceptive denotative-documenting, by [person-(s) /// empirical-subject-(s) /// objective-(s)-time-frame-fatigue-usage /// plan-(s) of ownership-ten-year-serializing /// monetary-incremental-spending, until self-adult-architectural-constructive-plan-(s)-exposure, match city-perimeter-plan-(s)-guideline-(s) /// et cetera…] for how inhabitant sociological presence, physiologically-compose, visual-peripheral-subjunct-notion-(s)-extent-(s'), visually-rationalizing, surrounding environment-factor-(s), volumeized by monetary-periodical-sequence-(s), [determining plan-(s) / receipt-(s) / consumer-commercial-production-dynamic-(s)] circulating working-lifestyle-inhabitant-(s'), motivating directed-environment-effort-(s), untheatrical voluntary-creative-impasse-(s), accumulating point-(s), amid blank vast territory-(ies), offset metropolitan-existence, where perimeter-terrain-denominating, not have an monthly-seasonally-yearly-decade-century-population, at colloquially-tongue, identifying grade-points-average-illiterate-claim-(s), juxtaposed perimeter-land-offset, sufficing an lack of cursive-contextual-redactive-capabilities, without an national-masculine-feminine-repopulation-method-schedule-agenda-dynamic, through fervor by-of study-(ies), regardless by mass-mutual-intelligence-competent-(s), forcing breeding from youth, intermittent mutual-range-age-demands, axis upon reformation from human-transit-transgression-(s'), which remain a-the primary-ultimate-conundrum, from-at human-trade, extraspectively-stifled, from elemental-topographic-resourcing-(s), affirming an marines-Americas-tectonic-subjunction, formed in sacrifice by-of humans'-eight-hours-idle-thinking-impasse-effort-(s), forgetting from western-weakness-(es), how non developed free-will-prerogative, vast-range, visual-pick-up-standard-(s), by-an system of product-(s)-objective-sociological-twenty-four-hour-cycle-(ing-(s)), maintaining functional population-count-(s), in order by review, from when objective-entity /used/ site-usage, serve an functional-procedure-mean-(s), where-at daily-impasse-numerative-function-(s), [year-(s) / season-(s) / month-(s) / week-(s)-denominative-impact-(s)], assisting in directive physiological-perceptive-effort-(s), as mass-minute-hundredth-community-consensus-(es), extenuate vernacular-conversation-(s), from political-forum, formulating methodical objective-schedule-spacing-(s)-sophistication-(s) elaborate managerial-comprehension-subjunction-(s), by way at mass-incompetent-(cies), inconsistent for how attention-span, consequent-succession, linage from aging-discourse, not fully embodying how existence has been (from an) perpetual-centripetal-force-latitude-degrees-revolving-solar-lunar-terrestrial-core-parallel, by those linear-surface-plane-(s), composing mass-negotiated-impasse-(s'), fatigued through work-place-eight-hour-(s)-per-day-forty-hours-per-week-pay per-iod-(yearly)-scheduling, at wage per hour, or salary payment-method-(s), derived by conclusive-registered-reference-data, conjured by past-commercial-historical-receipt-dating-impasse-inter-pretative-transgression-(s), by how competency is not an common-visual-vocal-listening-perceptive-function, amid considerable literary-reference-(s), confirming which parcel-(s) of planned-geology, have ever been documented, or pre-hypothesized, from genetic-repopulation-dynamic, by such case, from inhabitant-routine-location-motion-(s), assured, signed voluntary-at-will-contract-employment-payment-(s), by temporal mutual-commonism, reclusive by religious-monetary-interaction-(s), when considering calculated-context-interpretation-(s), abiding objective-article-(s)-act-(ion-(s)), in perimeter-parameter-limit-boundary-sequence-(s), because how physical-voluntary-motion-(s), grandeur factored information, awhile each daily-thought, be in an realm by-of existence, sharing how data is suppose-(d) to [guide // direct], physiological-impasse-(s), (from an) schedule-standpoint, eating [food // deli // canned-good-(s) // et cetera] by commercial-packaging-means', that derive-potential axiom offset-(s), by inhabitant-progressive-action-(s), learning from sociological-impasse-(s), why interpretation-(s) direct weather physical [motion-(s) // action-(s) // deductive-reference-(s)] remain pertinent upon religious-inquiry, through-of-a-those family-genetic-development-(s), when durated-hourly-time, sustem from secondths denominative-calculating, for then conjuring how simultaneous-exponent-(s)-mass, occur at primordial-decision-(s), motioning where land-territory, remain-planed, because [what // who // where], article-(s) for primary-dating, remain an part by continental-land-territory-(ies), apposed residential /code/ commercial-volatility, aforementioned at predicate, location-presence, awhile self-life-style-birth-youth-commercial-aging-inter-pretation-(s), tier hourly-daily-living-impasse-extent-(s), past primary-perceived, adult-confidence-trade-(s), living only validified, objective-conscientious-thinking-trait-(s), intriguing communion-genetic-objective-trade-dynam-ic, serving an, mass-population-schedule-impasse-monetary-circulation-dynamic, reviewing consideration, by what be fact-(or-(ed)), (from an) listless-scheduled-trade-history, yet to assert beyond physical-ventricle-blood-pulse-muscular-ligament-tense-voluntary-presence-(s'), because for when idle-inhabitant-motion-(s), extent-limit, educational-sociological-building-(s'), asserted-through, county-documenting-periodical-(s), unspecific by city-district-block-housing-document-ing, for dating-record-(s),| 720 |referring governmental-commercial-planning-trusted-interaction-(s), while judgement-(s) of [county // state // Federal // International-means'], be too grandiose to affirm-total-individual-impasse-perspective-(s'), from who signify before, state-constitutional-sworn-allegiance, because how momentary-monetary-wage-method-(s),48L. derive at relevant-physiological-inhabitant-interposed-objective-interaction-(s)...¶|||12m.:04s.68m.s.|||§

§..Preposition-Count-[28 / Twenty-Eight]-of.. ..Article-Essay-[8 / Eight].. ..Book-Essay-[222 / Two-Hundred-Twenty-Two].. ..Lines-[48.59 / Fourty-Eight.1.five-tenths-2.nine-hundred-ths]..

{January-(31)-February-(28 / 9)-March-(31)-April-(30)-May (31)-June (30)-July-(31)-August-(31)-September-(30) October-(31)-Month-(s)-Day-Hour-Minute-(s)-Secondths-Count}-{26,265,600 D.H.M.S. = 60S. × 60M. × 24H. × D.}-{7,296 D.H.}-{437,760 D.M.}-{304-(5) D.}

[Article:] {November}

..§¶223¶§..0**Poorism**;360 a-the disease of introverted-intervoluntary-blue-collar-common-inhabitant-effort-(s),| 721 |swearing statehood-entity-objective-instructed-development-(s), omitting subjective-geological-inquiry, by-an proportion of basic-human-right-(s')-mean-(s), considered through common-motion-effort-(s), determining (how an) situated-site, state commercial-Federal-monetary-development-(s), imbued by logical-interpretational-impasse-(s), due to placed [objective // dry / wet-substantial-food-(s)], to [eat // cloth], populated domestic-tranquility, while perspective remain juxtaposed, from rich-person-means', to suffice temporal-interactive-motion-(s), when living, where at an impasse for rich-soil-land-belief, in poor-metropolitan-inhabitant-(s)-standard-(s), inept comprehensive-competent-(s), per sermon or meeting, complemented-physical-effort-(s) from rich-wealthy-industrial-production-monetary-handling-intervention, Interval by such is, [1]RICHISM, or [2]*Wealthy-ism*;361 a-the disease of introverted-uncertain-white-collar-managerial-industrial-investment-effort-(s), bastardizing furt-her, selethisized commercial-means', inconclusive common-poorism-introversion, (as for how) one-(s')-power, use persuaded-interpretation-(s), from sensational-poor-thinking, not considering legislational-constitutional-house-statehoodship, alleviating present-modern-mass-reproductive-burden-(s), when monetary-government, tax American-Defense, (un)-attain-(ed) by-an mass-population-civilization-resourcing-mysticism-offset, from state-Federal-elemental-conceptive-product-(s)-entity-planning-comprehensive-population-denominative-relevant-proportional-equating-cycling-(s), where count-(s) of poorism-purchase-sale-(s), influence proportional monetary-method-of-payment-retention-dynamic, amid conscious-interactive-moment-(s), inter-facing act-(s), by voluntary-trade-impasse-(s), alleviating-burden-(s) from poorism,| 722 | by-an genetic-territorial-commercial-conceptive-objective-product-population-relevent-proportion-al-physiological-psychological-schedule-motion-(s)-impasse-(s)-re-formational-mean-(s'), relying G Relationship-(s') metre per human-behavioral-action-(s), (as an) extent of historical-impasse-(s), relevant to those industrial-purpose-(s), factor intermittent, communion-transit-traded-action-(s), persuading how common-trade-fact-(s), pertain favored-delight, delect taste-nutritional-relationship-(s), intermittent [1. sexual-opposite-(s) // 2. friendship // 3. co-worker-(s) // 4. colleague-(s) // 5. neighbor-(s) // 6. religious-member-(s) // 7. sequestrated-interactive-off-work-time-inhabitant-(s)] interposed daily-monetary-denominative-diligence, due to, relative-commercial-factoring, at Federal-national-census-population-(s).284¶|||02m.:21 s.96m.s.|||§ Because how individual-interpretation-(s), reside calculated identified natural-basis, when conjuring indicated presence before interpretation-(s), derive when usage of colloquial-dialect, affirm vocal-objective-(s), intermittent vocal-voluntary-discussion-(s), as tangible-universal-surrounding-(s), remain relevant, at an mode-interpretation-(s), applying voluntary-developmental-function-(s), from daily-voluntary-location-objective-impasse-(s), pertaining human-physiological-effort-(s), as tectonic-continental-environment-surrounding-(s), continuum ideological-impasse-belief-(s), fun-ctioning metal-steel-(coal)-developmental-part-(s), as commercial-conceptive-interpretation-(s), defining how Male & Feminine-interaction-(s), derive an racial-ethnic-background-interpretation-(s), colloquially-comfortable, esteemed statehoodship, at mass collective-group-effort-(s)-action-(s)-perspective-(s), disseminating common-extraspective-interaction-(s), characteristic-(s)-flaw, upon title-two /amid/ three-dimensional-impasse-trade-develop-offspring-affair-(s), because how identity resort (from an) perpetual-central-force-presence of characteristic-(s), based in at, influctuated validated-information, when during historical-guidance-impasse-trading, direct why human-(s), should implement, metal-pole-municipal-signage, to redirect hourly-time-physical-voluntary-discourse-placings, where product-(ed)-good-(s), resemble how Elemental-Geograph-ical-scale-composition-retaining, have yet verified, topical-rule, incendiary physical-perceptive-motive-action-(s), peripheral by, linguístical-interpretation-(s)-(subconscious), apose direct-letter-word-literary-letter-participle-word-marking-(s)-(conscious), affirming weather, term-subject-glossary-point-article-(s), can use how, exterior-developmental-force-(s'), articlize followed progression-(s), sequential Centripetal-Force-momentum-action-(s), deriving-notion-(s),| 723 | physically-unconscious, intervoluntary-Elemental-Geometrical-extraction-(s), retaining from, Composite-Clay-Metal-clump-solid-(s), segregated-refined, resourcing-good-(s), by an Govern-ment-patent-commercial-conceptive-documentation-(s), reserved from balanced of human-exchange, Military-Treasury-conceal-saving-(s), protected-gold-armament-(s), by those title-Terrain-inclination-(s), from basic-human-right-(s), incapable of civil-obliteration-(s), from humanity, because sub-par-human-basis-effort-(s)-perpendicular-plane, guide various cautionary-impasse-(s), preferred introverted-basic-human-rights-needs-individualism, before directive-concertion, by group-location-directive-effort-(s), for-when daily-revolutionary-periods, have an increment-percentage-(s), accumulating colloquial-experience, set at objective-article-(s)-locational-development-(s), perceiving, impact-trace-(ing-(s)), as-when weighted-volume-(s), increment-crescendo, vocal-timbre-volume-impasse-(s), amidst various-constant-sustain-outcome-(s) from where population-cultural-ideal-(s), not conceive, idle-begotten-inclination-(s), | 724 |grandeur upon surrounding-outcome-(s), testament-(s)-examination-(s)-mass-(es), at corporate-conception-(s), at centripetal-point-(s) by-of reference-visual-tonal-vibrational-reverberation-(s), extenuating solar-impasse-(s), conclusive between, momentum apose human-effort-(s), considering mechanized-appeal, validating each construe by article-datum-form, because their exist upon relation-(s'), an male-superiority of physique, greater-than feminine-appeal, where work-lifestyle-duration-illiterate-common-subjective-12th-century-future-not-exist, amid Gregorian-National-Calendar-existence, remaining in an time-constraint-(s) of methodical-payment-(s), yet interpreted by-an better measure for maintaining-inter-relational-affair-(s), through sociological-trade, for-how-I-presume their to be an Balance from physiological-reflexes-visual-blood-pulse-air-pocket-pupil-ear-organ-tonal-canal-perspective, those [1.modern-literary-article-subjective // 2.Composi-tional-Geological-Elemental-compositional-formation-commerce-purchase-rate-(s) // 3.commercial-register-entity-work-wage-schedule-activity-motion-(s)-dynamic /// 4.commercial-residential-consumer-waste // 5.Water & Power-department-(s) // 6.residential-housing-consumer-lifestyle-religious-communion-socialism // 7.Commercial-patent-production-plant-(s) // 8.Shipping-Warehouse-(s) // 9. Hotel-(s) // 10.CDL Class A / B / C-load-road-task-handling // 11.constructed paved-roadway-(s) // 12.constructed-low / mid / high-rise-composite-entity-(ies) // 13.municipal-sign-(s) // 14.auxiliary-road-cognitive-mega-transit-device-(s) // 15.Excavator // 16.Bulldozer // 17.Crane // 18.Forklift // 19.Deer533 // 20.Gates // 21.traffic-light-(s) // 22.Electric-Industrial-0.002-rubber-current-wiring // 23.Train // 24.Bus // 25.F12963-Air-Force-Delta-Military-Engineered-Jet // 25.Door-(s)-(entrance / / / exit) // 26.elemental-Geological-particular-tangible-(s)-article-(s) // 27.monetary-ink-cotton-ream-paper-constitutional-article-observation-impasse-(s)], performing subjective-cultural-development-(s), where viewing placed-product-objective-trade-(s), affect each and every-outcome, by human-circumstance-(s), rule while proceeding-(s) [1.measure // 2.weight // 3.applicate // 4.ingest // 5.consume // 6.meter // 7.mile // 8.appreciate] supersede conjuring, at those commercial-developed-entities, where offspring extenuate from women-blood-trimester-fashion-(s), transfusing [1.bone // 2.muscle // 3.skin // 4.ligament-(s) // 5.organ-(s) // 6.nail-(s)-composition-(s) // 7.hair // 8.blood // 9.ventricle-system] for administrating-existence, intermittent an political-background, that conjunct each compositional-inhabitant, amid existence-recipient-remain-(s'), at particulate-(d)-developmental-moment-(s), proceeding voluntary-work-mode-function-(s), pertaining an [1.application // 2.instruction-(s) // 3.workbook-guide // 4.memo

// [5]shrink // [6]unit-ordering-(s)-volume-page-(s) // [7]Corporate-Legal-Binding-Term-(s) & Condition-(s) // [8]seasonal-corporate-unit-(s)-produced / [9]store-shipped / [10]shelf-life-purchase-statistical-deviation / [11]dime-out-(s) /// [12]restaurant-order-ring-out-margin-budget-(in)-corporated-corporate-volume-statistic-(s)-materials] suppose schedule-(s) from men, upon offspring, harmonize millennium facts-interpretation, suspended by calendar-aging-offspring, where moment-(s), sequence women with children from men, by only two-three-children, per thirty-year-monetary-social-communion-rate-(s), not literary-economical-planning, natural-moded-sequence-(s), learning how patent-objective-component-(s), adjust in counter-compositional-unit-posit-(s), like individual-objective-project-purchasing, (from an) level of resourcing by mutual-colleague-co-worker-competency-function-(question // inquiry), faceting communion-sociological-interested-mass, that make out how each impasse from birth, pertain yearly-monthly-reset-(s), considering what object-(s), are temporally-orchestrated, by-an century-schedule-trade /or/ resourcing-demand-anticipation-(s), supplying enumerated-continuum-object-(s), at volume-unit-growth, boxed into consideration-(s), where-when-why, focus when surrounding interior-visual-tonal-perspective, commonly-reflexive-notion-(s), offset hour-celestial-time-analyzation, through youth, before [self // elder-existence-impasse-(s)], age due to a-the continual-growing-extenuation-(s), from elder-offspring-will-objective-article-(s), mustered between vin-verification-(s), for surrounding-county-transit-lifestyle-perspective, surviving from physical-instinctual-housing-means', not teaching through comprehensive-literary-reader-redac-ting, formal-methodical-trade-practice-(s), from population-mass-county-offset-civilization-(s), from whom attain construe by voluntary-manner-motion-consideration-(s), by free-will, to [examine // observe // redact // review // highlight] what article-(s) can commonly-assist, visual-vocal-tonal-mathematical-numerical-grammatical-letter-mental-communion-pattern-consider-ation-(s), posited throughout existing [intersection / public-commercial-square / work-trait-season-year-handling-swearing / labor-literary-dynamic], amid proglamation-(s) at off-time-work-effort-(s), insistent on common-currency-exchange-ignorance, rather parted present-focus, for materialized data, by muscular-ventricle-blood-pulse-circulation-physiological-visual-plasma-space-ear-canal-ligament-organ-symmettry-kinetic-pulse-receptive-vibration-(s)-hyper-tense-(s)-perceptive-interior-components'-being, dating sequential-inter-sexual, punctuation-word-term-definition-syntax-competent-comprehensive-subjective-article-(s)-perceptive-momentum-time-hour-mode-motion-(s)-comprehending, juxtaposed when dialect-impasse-(s), reverberate colloquial-historical-tone-(s), affirming [eating // clothing // housing-material-(s) soly] unaware how exquisite-extent, particle posited-elemental-governmental-patent-commercial-conceptive-monetary-currency-dollar-circulation-exchange-commercial-unit-sale-(s)-validfication-(s), requisite-historical-year-to-date-modern-cycle-referencing, amid presence, before their even exist an direct-national-genetic-isolated-continental-focus, when counted-commercial-unilateral-cash-store-market-trade-dependent-variable-(s), not acceded between daily-hour-ly-aging-determinat-ion-(s), at interior-case-(s) for human-life, mattering because assent from present-disposition-(s), exist in how trade verify daily-effort-(s), claiming [substantial // objective tangible-piece-(s)] where time-political-cronyism, enabling from common-mass-illiterate-default-politician-(s), educationally-timid, from human-physical-composition-(s), undetermined at suburban-ego-capable-retirement-commercial-work-consideration-(s), accommodating action-(s), listless, lazy-sweatless-effort-(s), illiterately-extracting-monetary-wage-effort-(s), from those general-class-ification-(s), acting upon an blue-collar-mutual-male-work-career-method-(s), then redistributing feminine-youth-interpretation-(s), for objective-population-directive-schedule-impasse-(s), by-an non-collar-effort-(s), [worshipping / appreciating], monetary-wage-demand-reactive-action-(s), not intervoluntary, an vocal-tonal-appreciation-(s), conjuring from hour-week /or/ month-day-impasse-(s), moral-act-(s), documenting hour-occurrence-(s), fulfilling an literary-theoretical-schedule-time-observing, when year-denominative-dating-activities, supplement with-out an concisive-application-(s), human-location-offset-faith-timbre, because lineage-generation-(s)-colloquial-linguistic-family-modification-(s)-vocal-moment-(s), seldomly juxtapose co-worker-instructed-at-will-schedule-lingüístics-tounge, fraternizing National-Genetic-objective-tounge, verified in discretion an political-tier, by educational-subject-(s)-chapter-lesson-(s)-question-redacting, an differentiation from national-centralized-global-historical-comprehension-(s), believing in human-effort-(s), construing intermittent, family-population-count-(s), then concertion-cultural-balance-trade-action-(s), relying particular-cultivation-techniques, undone, where written-constitutional-impasse, has stopped at State-Constitutional-Maxim-Extent, not having County-Constitutional-Valedictorian-Effort-(s), hence far as to convey, men as moronic from sport-ticket-ball-goal-mass-entertainment-function-(s), amid adult-feminine-repopulation-consideration-(s), requiring obligated moment of sweat, cycling such delicate-opposite-human-timbre, juxtaposed each facet-(s) at existence, yet ever fruitful, by sexual-desire, beefed-up by idle-prestige, comforting by where-ever-climate-environment-(s), change by city /or/ county-constructed-development-fatigue-capacity-hour-extent, not thinking from fatigue-limit-impact, how subjective-interpretation, influence an medium-ink-paper-creative-platform, for communion-protestant-genetic-church-house-constitutional-article-(s)-documenting,| 725 |those local-customary-standards, for literary-analyzation, of Common-Exponent-Repopulating, or not.285¶|||12m.:01s.18m.s.|||§ Indifferent population-genetic-characteristic-(s), centralized-constitutional-article-(s)-political-party-perception-documenting-method-climate-(s), clearly-convey, evident Thomas Jefferson-document-reliance, for over the past 240 Year-(s), by what cultural-custom-(s), largo society developmental-documenting-constitutional-abbreviated-considered-communicat-ion-(s);[362] awhile independent-fatigue not realize how old daily-planning, exemplify an massive-error-(s) in human-transit-extenuation-(s), comforted non scheduled-entity-function-(s), upon origins-extraction from raw-resource-(s), not hosting constitutional-genetic-mass-Developmental-effort-(s), [1.Convention // 2.Church house // 3.personal-estate // 4.mansion // 5.stadium // 6.arena // 7.center // 8.Hall // 9.Field // 10.Interstate-Rural-Undeveloped-Exit-(s) // et cetera..] due to a-the nature of Physics & Populational-Prerogative-Existence, why Metropolitan-aforemention-Civilization-function,|726|when restricted-Geological-Military-interval-example-grounds, exemplify no-mention-(s), from human-illiterate-perspective-(s), when magazine /or/ news-presse, not suffice calendar-hour-mode-time-setting-(s), arranged by government, to make an basis of basic-human-living-exist, where no fact-(s) exist, where drove-(s) common-bias-perceptive-fate-(s), idle through oblivion, sanctified housing-objective-cycled-means, habiting accustomed tradition-(s), at no-preference through scheduling, anticipating intersexual-reproduction-(s) /or/ moment-(s), when [1.Genetic-vocal-comradery // 2.Elemental-extracting // 3.Farming // 4.Agricultural-inlaying /// 5.picking-up // 6.comestible-(s)-cleansing-(s) // 7.recipe-conjuring // cooking // 8.product-patent-conceptive-Population-(s)-dating-schedule-manufacturing-(s) /727/ 9.Elemental-Composite-transit-effort-(s) // 10.subjective-reading-(s) /// 11.redacting-(s) /// 12.reformulating-(s) /// 13.mathematical-denoting-(s) //// 14.measuring-(s) /// 15.annotation-(s), non-fiction-pertinent-context-(s) /// 16.conjuring-(s) /// 17.Theatre-performance-(s) /// 18.orchestra-ensemble-performance-(s) //// 19.theatre-accompanying-(s) // 20.family-housing-Architectural-Constitutional-planning-(s)];[363] intertwine implicated natural-trade, thinking how Constitutional-Merit & Mass-Population-Genetic-Competent-Effort-(s), suppose monetary-civilization-city-county-communion-mass-(es), yet proportion-population, upon an [sociological 1.Village // 2.Town // 3.City // 4.County ECOLOGical-BOUNDary-(ies)-Geological-Incremental-metropolitan-site-(s)], vary, restricted-Geological-outer-barren-state-mid-west-rocky-terrain, title-spacing, human-title-perspective, serving greater-custom-(s), transgressing-incremental-road-intransitive-impasse-(s), for affirming what placed-objective-(s), theatrically-develop, compliment human-effort-(s)-denoting-express-ion-(s) rather than paper-currency-circulation-wage-purchase-monetary-savings-disparity, noting how each product, per shelf-market-trade, should be amulet, in an exponent-housing-unit-item-personal-purchase-market-object-trade-collection-(s)-deviation, compared monetary-note-(s)-grade-point-average-deviation-(s), retrograde an formal-constituting, human-comfort-qualities, compared fair-consistent-genetic-trading-effort-(s)-confidence, until mass-population-genetic-conception-(s), superseding Federal-Bank-series-note /and/ cent-monetary-count-holding-inflat-ing, sworn by basic-human-rights, not identifying visual-vocal-tonal-perspective, denominating, simply clearly-communicating intermittent, voluntary-inhabitant-motivation-interaction-(s), generally understanding a-the basic-human-fact-(s), that exist, when physiological-composition, remain consistently primary, focused from existence, before Substantial-Comestible-(s) & Patent-Conceptive-Objective-(s), supposing supplement directive, existing only from perceptive-observation-vernacular-reference-interposing, much of a-the rural-undeveloped-continental-tectonic-plate-(s), displaying an reclusive-idle-introverted-retarded-mass-independent-individual-istic-effort-(s), where lifestyle-progress, require interval-mode-count-(s), interposing auxiliary-facet-(s), remaining secluded from perceptive-observation-feature-(s), deriving family-self-effort-(s),| 728 |unaware diction-discoveries, existing by-an word-ambiguous-documenting-denominative-counting weekly-dating-reset-offset-pattern-(s), extenuating through retirement, working by mid-age-educational-traded-position-(s) conclude in daily-obituary-newspresse-precession-(s), undaunted by idle-nature & Auto-Biographical-Common-Book-Rate-Experience-History-Deviation, not observing [future-conceptive-grammatical-context-inquisitioning // mathematical-schematic // percentage-(s)-fraction-(s)-integer-exponent-deviation /729/ Ten-Column-Integer-multiple-numerical-sum-perceptive-deviation] because an lack of Presence & Directive-Effort-(s), not constitute common-article-observation-lifestyle-denominating-context-deviation,| 730 |variablizing independent-voluntary-sharing, interpreting how those claimed-possessive-objective-(s), modify State-Continental-constriction-engineering-developmental-Territory,185L. by-an preference-bias / or/ hoc-perspective, intermittent genetic-reproductive-threatrical-constitutional-sociological-method-living.286¶||||16m.: 49s.43m.s.|||§

..§¶224¶§..Intonation has been hugely overheard, from musical-rubric-principal-staff-coor-dinating-practice, omitted for informing relationship-(s), forgetting how vocal-vibrational-tonal-pitch-visual-letter-mark-(s)-syntax-reverberation-(s), mode-simultaneous, conjuring living-motion-(s), for visual-tonal-subjective-perspective-(s), for how sexual-hetero-relationship-(s), are supposed to inter-dialogue /or/ document, as many article-(s), requiring [Federal // National Article-(s)-Constitutional-Parchment-Tier-Perceptive-Observation-Staff-Key // Time-Signature-Predicate-Planning], plausible for trace by hyper-tense-(s)-observation-(s), amidst physiological-impasse-(s), not learn by psychological-perspective, for-of forum-psychology, scheduling religious-congregation-confessional-dynamic-focus, yet designated from compelled, comprehend-ed intimacy, between female-interaction-(s), by human-consideration-(s) from feminine-sexual-interaction-(s)-act-opinion, ever understanding time-frame-calendar-event-(s), understanding why no sexual-interaction-(s)-happen, comprehending how subjective-comprehension, not been conclusively constitutional, amid continental-tectonic-plate /or/ shelf-confirming-(s), by-an proportional-worshipping-population-(s), tiering from nation /or/ state constitution-(s), apose local-population-constitution-(s), affirming denominative-confirmation-(s), (from an) humanistic-morale-schedule-wage-savings-purchase-dynamic-(s), compart time-mode-interval-(s), subterior under dating, human-morale, because how factifying ation, participle-predicate-school-reference-subjective-practice, forgetting how homework, not guide common-literary-treasury-physical-human-visual-tonal-perspective, pertinent impassed humanity, not because of idle-behavior, due in part, factual-comprehension-(s), lessoning how physiological-subjective-mode-genetic-interaction-(s), are suppose to commercially-proportion city-community-dimension-documenting-literary-survey-sustained-agricultural-acre-com-munity-yield-(s), responsible by subjective-article-(s)-usage-(s), interacting how human-physique, embody-anticipating, longevity-discourse, from visual-isolated-color-shape-form-letter /or/ numerical-mark-reset-notion-secondth-method-ical-word per second, [definition // paragraph // essay // article-(s)-interpretation-(s)] juxtaposing, | 732 |ear-organ-isolated-vibration-(s)-decibel-reverberational-resonance-volume-(s), apose year-deviation-schedule-aging-reflexes-continuum-dating, marking vocal-inter-pretation-(s), intend-(ed) for conceiving-article-(s), intersected, transit-mode-considering, conservative-application-(s)-process-genetic-complication-(s), intermittent throughout each synapse-(physical) of syntax-(subjective), conjuring notion-(s) of [mass-motion-(s) // subjective-article-schedule-literary-observation-(s) /// monetary-bank-exponent-confidence-series-note-deviation //// location-spaced-meter-feet-measure-(s) /apose/ interior-objective-(s)-collection-(s)-calendar-period-aging-expectation-(s)], as literary-mean-(s), term each inhabitant, whom characterize illiterate-observation, pre-reader, pre-redacting-citizen, pre-auto-biographical or pre-lifestyle-non-fiction-documenting-observation-(s), because how competency, is relevant upon those whom expenditure, apose an historical-purchase-deviation of trade, where thinking has no realm of those practice-(s), at helm, with perceptive-explanation-(s), where this such book, factor thought, can /and/ will, tiered apposed family-notion-(s)-limit-perspective, not understanding notion-motion-(s)-population-civilization-intercounty-universal-offset, because how metropolitan-development, apply, from national-Federal-constitution-(s), incapable where each traded-effort-(s), by cursive-contextual-documenting-(s), apose how local-development-(s), conjure custom-(s), affirming how denoting is suppose to validate, what object-(s) cycle at residential-entity, different from commercial-sociological-impasse-entity-cycling, variating how [enter-(in) // exit-(out)],| 733 | pertain our community [fluent-language-vernacular-word-term-perceptive-time-secondths-interpretation-proportion-aging-day // week /// month //// season-continuum-motion-(s) male // feminine-dynamic-location-objective-developmental-literary-lifestyle-planning-usage-(s)], because article-(s)-factifying determination-(s'), require intense-inquisition-(s), in to conscious-human-interpretation-(s)-interacting-(s), having how I would suggest, why not teach ones' own-children, or have religious-bible-redact-disciple-subjective-practice-(s),| 734 |including how city-county-sociological-community-Federal-Commercial-development-(s), limit by dating, an impact upon those decision-(s), remain by basic-human-right-(s), when inclusive-physics-extreme-condition-aspect-(s), think, as how consideration-(s), proportion from secondths-deviation-hundredth /or/ thousands-tic-deviation-inquiry, appealed through extent-impasse-tics-deviation-definition-mode-rate-paragraph-interpret, reverberating as whom redact / and/ conjure-article-(s)-defining, intended to persuade, sociological-anatomical-physiological-collective-mean-(s), (reminding how common-sociological-population-inhabitant-illiteracy, sound-entertained-sustain, formal-colloquial-vocal-sound-appeal), due to commercial-component-part-(s), cycling outcome-impasse-effect-(s), outlined by-an various-mutual-interaction-(s), proceeding when purchase-action-(s), should learn from work-place-operation-(s)-subterior-conducted-affairs, occurring where set-entity-(ies)-structure /or/ object-(s)-placed, [1.objective // 2.mission // 3.agenda // 4.schedule // 5.itinerant // 6.list-(s) // 7.subjective-topic-(s) // 8.issue-(s)], articlizing-tier-functioning-(s), for as why, individual-(s) /or/ extraspective-perceptive-mass-population-(s), can only be from when rebirthing from self, matter by-an variation from [1.self / 2.spouse // 3.intimate-partner / 4.neighbor-(s) / 5.co-worker-(s) / 6.colleague-(s) / 7.managerial-corporate-operation-(s)-tier / 8.church-member-impasse-(s)-congregation-lifestyle], assenting collective-purchase-(s), crescent inanimate-objective-usage-placing-(s), where church should serve headquarter-(s)-offset-interaction-(s), from corporate-terrestrial-inquiry-dating-development-constructive-curb, apposed proceeded-impasse-operation-(s), directing agenda-coordination-(s), upon constitutional-grafting-(s), by collecting-article-(s), for amendment-repopulation-period-(s)-cycling, substantial-reproduction-(s), allotting interposed-ingestion-(s), with communion, to affirm-dating-(s), tangible when product-interaction-(s), pertaining past-communion-consensus, confirming [weight-(s) // measure-(s) //dimensional-product-(s)-density-occupying], upon mass-usage-schedule-count-(s), apposed independent-subjective-spacial-distance-(s)-competency-observation-examination, from word-term-propor-tion-count-(s), as well [physiological-block // acre-(s) // estate-25-acre-family-development-practicing-lineage] by either millennium or century-decade-repopulation-literary-dynamic-document-practicing, amid expanse, because how great parallel-plane, revolving-centripetal-force-earth, upon those transit-environment-climate-base-condition-(s), living where Impasse & Development-(s), develop-logical-impasse-motion-(s)-reset-limit-(s), requiring purpose for lineage-literary-population-count-(s)-documenting, perpetual circumference mile-(s)-metric-distance-gap-(s), either where-when metropolitan-inhabitant-(s), differentiate rural-title-owner-inhabitant-prerogative-abstract-individualism-developmental-commercial-debt-dynamic, when citizen-(s) whom impasse, vacant-barren-territory-(ies), consider because how mark-(s) of objective-observation-(s), not serve data, identifying common-civilian-inhabitants'-subjective-article-(s)-redact-competency-limit-(s), entertained, unless repetitively-common, inhabitant-enter-tain-ment-affair-(s),| 735 |bemused conclusive document-order-orchestration-delving, Exterior-Compart-mentalized-Date-Aging-Categorical-Decimal-Enumerated-Value-(s)-Juxta-positional-Article-Column-Order-Classification-(s) upon topic-(s) /or/ issue-(s), relying Amer-ican-International-Bailout-Trade-enabling, when misinterpreted-information, forget why waste require recycling /and/ decomposing-waste-receptacle-(s), awhile cause of action-(s), follow commercial-production, balance taxes, with Electricity & Water, commercial-expenditure-(s), learning why sociological-international-population-(s), (have not been) formally-regulated by congressional-government, affirming-international-ambassador-impasse, exterior-existence, from those variat-ion-(s) of ethnicity-(ies), intermittent by commercial-Federal-settling-(s), proceeded from individualism, not existing by purpose, from lineage-standard-(s), intertwined constitutional-written-merit-(s), as how ideological-compulsive-belief, remain an intricate-part, for-how thought is conjured by each verifiable-inclination-(s), existing by legal-rule-case-leather-bound-historical-executive-judiciary-review-filing-order-method,| 736 |fixating-objective-grounds-order-demand, from those characteristic-(s), input-description-(s), because how ideological-figure, postpone schedule-location-motion-(s)-progression-(s), as in youth-education, teaching sociological-hallway-transitive-method-repetition, in either social-traffic-inhabitant-(s), not understanding classroom-subjective-studying grandeur, from grey-like-thinking-imagery-existence, insistent on driving to commercial-entity-(ies), mile-(s) apart, rather than, conscious-block-conceptive-acreage-family-planning-surrounding-(s), proceeding, walking-meter through mile-(s)-exper-ience-civilization-offset-locational-visual-tonal-volume-depth-perspective-distance-determinat-ion-(s), trading based from religious-population-mutual-currency-count-(s), assented by military-historical-morale-determination-(s), sociological-lifestyle, family-communion-literary-competen-cy, [banking-cent-percentage-credit // debit-commercial-debt-inflating], commercial-work-mode-(s), congruent during eight-hour-work-mode-(s), requiring-transit-to-arrival-destination /or/ from-departure-mode-point-(s), deeming-effort-(s), at location-(s), totaling general-work-week, at seventy-two-hour-(s)-work-routine, retrieving Substantial-Comestible-Commercial-Relied-Acre-Title-Lifestyle-Personal-Power & Water-Geological-Engineering-Difficulty, appraising from exterior-trade, how extraspective-interaction-(s), are what make our introverted-objective-possession-(s), trafficking intersected outcome-(s), continuing daily-dynamic, by-an police-twenty-four-hours-county-city-road-circulating, superimposed-domestic-trade-process-(es), for then sleeping intermittent work-week-daily-impasse-(s), affirming how purchase-action-(s), influence market-monetary-supply-trade, supplying purchase-participating-citizen-(s), maintain-ing repetitive-work-function-(s), at consumer, placed-objective, Time-(s) & Dating-period-(s), year-month-periodical-etymological-word-interpretation-origins-tiering, awake visual-tonal-vocal-perceptive-lifestyle, presently-continuum, performing meager survival-low-rise-mid-rise-customary-purchase-developing, past-structured-development-(s), accustomed by one-hundred-sixty-eight-hour-weekly-hour-(s)-cycle-reset-(s), yet never measuring hour-minute-(s)-secondths-mode-counts-deviation-time-calendar-week-month-season-dating-interval-(s), [month-(s)-{Seven-Hundred-Twenty-Hour-(s) -/- Forty-Four-Hour-(s)} /extent/ pay-period-(s)-{Five-Hundred-Hour-(s) – Two-Thousand-Hour-(s)} // season-(s)-{Two-Thousand-one-hundred-sixty-hour-(s)} // et cetera…], referencing by focusing dating /or/ hour-period-(s)-interval-Commercial-production-Geological-Compositional-mass-proportion-progression-deviation-documenting, proportioning self-observation, juxtaposed community-personal-bias-mute-lifestyle, by subject-ive-physiological-observation-(s), Documenting & Conjuring, traded-factor-(s), for population-method-routine-consideration-(s), from when thought-process, remain by-an portion of group-at-will-wage-payment-method-focusing (only method of human practiced-effort-(s)) unto group-rationale-simultaneous-observation-first-motion-still-reactive-effort-(s), inclined complaining, when task-(s)-work-wage-objective-attention-span-observation, dating, time-denominative-hour-(s)-interval-(s), with an rule instilled per [1.process // 2.action-(s) // 3.motion-(s) // 4.maneuver-(s) // 5.modification-(s) // 6.impasse-(s) // 7.interpretation-(s)]| 737 |from-of observation-(s), administer-(ed) from product-(s)-commercial-conceptive-notion-(s), concerted, where presence not exactly, perceive fluent-vocal-term-script-syntax-extent-vernacular,| 738 |from letter-posit-word-transitive-noun-offset-adjective-empirical-definition-tier-posit-article-(s), wristwatch /or/ watch-time-frame-minute-(s)-impasse-progressive-function-(s), reserved when declined help from other-(s), due to-an natural-human-fact-(s), at an reliance by [1.wage-site // 2.market // 3.headquarters // 4.shipping // 5.payment-federal-commercial-operation-effort-(s)], amid metropolitan-municipal-literary-lock, without common-pen-key-temporality, adjusting rate-competency, schedule-orchestrating, by oneself-preference, constraint by elemental-continental-statehood-federal-interior-geological-industrial-location-agrarian-extraction-offset-requirement-(s), commercial-factory-production-(s), merited by Federal-(Inter)-national-premise-(s) for trading,| 739 |dispersed throughout the North-American-tectonic-plate-(Free-Trade-Agreement), plant-commercial-industrial-trading, as well soil-food-groups-compositional-tangible-quality-offset-quantity-celest-ial-grid-agricultural-control, without creative-comprehension of such soluble-substantial-recipe-density-pressure-(s), guiding packaged-trading-method-(s), not understanding metropolitan-limit-work-spacing-offset, how [family // introversion // county-communion-transit-working-dynamic], occur when Agricultural & Horticultural Food-Group-(s)-Count-(s), inter-coordinated by mundane-trading, relied upon directive-agenda-(s), inclusive when action-(s), affirm from labor, national-currency-confidence, apose university-legislational-government, never developing further extenuation-(s), by-those national-state-quality-(ies), that would encourage genetic-purity-repopulation-expediting, from communication-(s)-youth-perceptive-live-visual-presence-appre-cia-

tion-(s)-retention-offset-family-capcity-mentality-dynamic, by ba-
sic-human-need-(s), inter-posed exquisite refined-sophistication-re-
tirement-lifestyle-purchase-commercial-savings-histor-ical-so-
cial-security-reclusive, when open-consider-ation-(s) not Micro-
soft-excel-graph, periodical-population-name-orchestration-numeri-
cal-hun-dredths-tenths-thousandths-population-operation-(s)-dy-
namic, lifetime-decade-millennium-cal-endar-Federal-repopula-
tion-mass-pop-ulation-non-calculated-mode,| 740 | secondths-year-re-
set /or/ age-offset-impasse-juxtaposition-perspective-scaling, yet
fragment educational-subjective-metre-(s), upon proportioning hy-
per-tense-(s)-compartmentalized-singular-competency-offset-ex-
tent-reasoning, visual-tonal-letter-word-syntax-subjective-lesson-ar-
ticle-(s)-perspective, per subjective-contortion-time-numerical-math-
ematical-dating-grammatical-punctuation-syntax-staff-key-time-sig-
nature-physical-aorta-pulmonary-ventrical-muscle-response, requiring still-inquisitive-perspective-(s), reminding age-secondths-thou-
sandths-secondths-deviation /or/ hundredths-secondths-deviation,
juxtaposing how value of presence,| 741 | function through an series
of grandiose-offset-(effect)-(effort)-impact-(s), Nationally-demo-
graphic-race /or/ ethnic-category-(ies), between metropolitan-civili-
zation-(s), where homogenized-trade, has never cycled-out from
weekly-reset-offset-(s), routine an uncreative-chain-of-operation-(s),
sustemming from-an history, neglected by vocal-constant-vowel-col-
loquial-vernacular-method-(s), for how dating /or/ hour-impasse, remain without common-critical-tier-operating, how action-(s), reset
per [year // season // month], amid week-day-hour-(s), conjuring-prac-
tice-review, by secondths-tic-lifetime-deviation-point-inter-
val-mode-impasse-conducting, when scaling how [1.time // 2.tempo ///
3.syntax //// 4.rubric ///// 5.benchmark ////// 6.liquid-substantial-acid / 7.wa-
ter / 8.bleach-tangible-scaling /////// 9.dry-substantial-food-group-(s)
//////// 10.wind-vapor-pressure-millibar-respiration-count-(s) /////////
11.blood-pulse ////////// 12.motion-reflexes-compul-sion ////////// 13.sit ///////////
14.stand-at-rest //////////// 15.sleep ///////////// 16.eating-mode-deviation-pos-

ture-general-means-cycle-(s)], intimate, by spouse-opposite-sex-lifestyle-presence, with self-sex-zygote, thinking how agenda affect-schedule, as in labor-pool, when one is assigned an location by site-designated-task-(s), working at means from daily-article-(s)-marking-denominating, furthering how we reference at present-posture-vernacular-composure-impasse-(s), extending at present-dating, geological-terrain, offset Metropolitan-inhabitant-communal-county-block-parcel-dynamic,| 742 |not realized, by-an focus of extenuated topical-issue-matter-circumstance-(s), beyond present-day-requisite-monetary-communion-exchange-confluence-(s), family-personal-reclusive, historical lineage-ancestry-(ies), which remain past, creative-active-sociological-sleep-day-night-hours-lifestyle-method-(s), eroded through commercial-day-working, night-sleep-recuperating, apose retirement-extenuating, life-sleep-routine-preferenced-impasse-pattern-(s), from present-vernacular-shifting, inclined tiering [action-(s) /or/ motion-(s)-by-of-effort-(s)], (not) desiring sexual-juxtaposition-(s), juxtaposed children-year-century-rate-(s), amid sleep-three /and/ six-hour-thirty-six-hour-seven-day-week-cycling,| 743 |remaining organically-different, from how human-compositional-perceptive-subjective-offset,|744|suffer an animal-subjective-adult-commercial-evident-under-common-par-standards, façade fundamental-(s) of constitutional-article-(s)-thinking, where metropolitan-development-(s), not measure soil-ground-underneath, counting dynamic of restricted under-ground, air-space, for how we clearance through perspective, nominative throughout timed-present-count-(s), for how male is suppose to [behave // conduct], amongst woman-(en), where logical-focus, is pertained, from dollar-currency-inquisition-(s), [surveying sighted physique // composition-(s) // object-(s) // book-(s) // schedule-(s) // subject-(s) // et cetera...], that would serve an purpose by monthly-work-routine-schedule-(s), upon an market-currency-circulating-generating-Ecological-checking /or/ saving-(s)-calculation-offset, personal by individualism, when subjective-written-application-(s), fulfill interval-checkpoint-period-(s), (from an) basis by-of [developmental-constructive // deconstructive // Engineering-Geology // auxiliary-(mega)-vehicle-Geological-Metropolitan-transit-impasse], function-(s) of trading, meriting upon formal-conjuring-(s), sustemmed factory-commercial-elemental-refined-plate-measure-component-(s), over-weighed at initial-cause, but proportional Geological-Metropolitan-bill-debt-payment-(s), mass-patent-factory-operation-conceptive-product-(s), discoursing trade-(s), apposing at all moment-(s), an family-Sociological-communion-county-circulation-dynamic, never thinking how Literary-Article-(s) & Congressional-Legislation-categorical-decimal-column-ordering, acceded youth-relationship-(s), squabally-seclusive-reclusive, commercial-transit-trade-note-(s), current-ing denominative-trade-purchase-(s), by commercial-good-(s), in lieu from Cause & Effect, titling block-housing-parcel-land-spacing-(s),| 745 |suburban, futures-stock-market-dating, commercial-housing-investment-(s), because how rule be implemented, maintaining governmental-agenda, national-pertinence, by population-count-(s), dating-rule-(s), for interpose-intervalizing, time-hour-code-(s), upon such rule-(s), perceptively-gauging, subjected-objective-material-directive-(s), simultaneous bastardized-national-state-function-process-(es), in parameter-(s) by-of action-(s), while juxtaposition, be-as Physics-supratangible-quality-(ies)-impact-(s)-continuum-climate-condition-effect-(s) effort, have yet found a the maximized-potential, from human-effort-(s), [serialized-magazine-(s) / county-newspaper // musical-ensemble / band / artist-piece-(s) // theatre-play-(s) / show-(s) / Broadway // literary-book-(s)], as how bias amongst youth-Peer-(s) & Colleague-(s), distort still enough Sit & Write-contextual-dynamic, in tier by article-(s)-pertinent-focus, when their at intermediate-function, difficulted, when conscientious attempt from education-context, not yet commercially-intricate, enumerated-monetary-concurrent-congress-ional-executive-commercial-factory-production, from literary-non-fiction-natural-offset-ob-ser-vation-(s), conceding rhetoric, from mass-wage-effort-offset, interpreting data, at-will, by thought of dating-theoretical-conception, rather than scientific-method-literary-transgression-hypothesis, through intermediate-mega-capacity, for handling, commercial-sociological-survey-inquiry-(ies), of [population-genetic-religious-mass // statehood-climate-topographic-territory // commercial-objective-religious-bulk-purchase-(s) // calendar-scheduling], an retirement-planning-national-break, apposing how perspective is premised, from wage-purchase-dynamic-(thought-(s)), inquired in line of parcel-block-commercial-market-resourcing, yearly-aging, calculated cultivated-comestible-(s) /and/ element-(s), by Tectonic-Geographical-Continental-Development-(s) & Object-(s), from| 746 |[tool-calibration // mechanism-(s) per land-excavation // population-land-housing-acreage-developmental-planning-dispertion-dynamic /// personal-handling-(maintaining) // stadium-(s) gathering consensus-constituting // identifying sociological-denominative-timing-sworn-possibility-common-deviation-competency], by survey-sociological-subjective-inquiry-interpretation-(s) per Subjective-Tier & Formal-Theatrical-Repopulating, confluencing [[1]theatrical-art-(s) // [2]philharmonic-ceremony // [3]subjective-inquiry /747/ [4]housing-capable-maximum-of-article-(s)-per-of-possession-agenda-offset-means' // [5]festival // [6]mural-(s) // [7]painting-(s) // [8]t cetera...], pertaining [capitia-visual-ora-perspective-location-impasse transit // major-governmental /// commercial // theater-house-(s)], confirming hour-(s) by-of thought, scheduled by-an totalitarian-government-developing, without a-the waste from citizen-commercial-product-continuum-outputting-waste, off-work-time-creatively-saving [cardboard / glass / aluminum / plastics / wood / scrap-metal] for projecting architecture-creative-kiln-pottery-clay, metal-Bessemer-Activities, spending [idle /748/ hobby-time-(s)], through ecological-geological-literary-congressional-book-payment-(s), (as an) example for live-requisite-shopping-eras, seasonally transitioned by objective-product-(s), but when space have [governmental / commercial / residential-resources-secular-limit-(s)], mundanely-directing, human-sedentary-sociological-population, required by extraspective-wage-obligation, showing how off-time-recuperation, largo physiological-nutrition-needs, in fashion, rather unifying an genetic-wardrobe, for monthly-installment-(s), or method of industrial-capabilities-proportioning, city-district-singer-clothing-abridgement-(s), from either stitching or hemming, but not beyond an common-fashion, for confluential-adaptive-trade-basis, upon personal-family-locale-pattern-(s), eating consumable-comestible-(s), yet balancing [eating / sleeping / transit] to [location-objective-developing / instructive-location-operation-monetary-circulation-physical-motion-(s)-function-ing], determining a-the capitia of-an surrounding-constitutional-minor-community, (as how I example, without any other writer-(s), whom proportion the total-mass-lifestyle-denominative-impasse-motion-(s),| 749 |or exemplify, how conveyer-belt-factory-commercial-head-quarter-Federal-warehouse-unit-product-trademark-shipping-market-registering, human-civilization-(s)-metropolitan-offset-dynamic,| 750 | from food-developmental-article-(s)-paragraph-(s)-syntax-intrication-(s)-maximum, resetting from syntax,| 751 |upon Grammatical-Tier-Numerical-Word-Term-Abbreviation-qualities-individual-term-definition-usage-offset,comparing exponent-syntax-work-method-(s), along an [sociological-communion // sociological-family-work-place-individualism // sociological-retirement-aging-method-physical-subjective-obligated-effort-(s)], continuously by at-will-effort-(s), high on retirement-commercial-syncratic-district-city-communities-(s)-repopulating, too technical, for their to assert, an #10,753,247-women, comparative #3,963,150-men, or so, apposing Horticulture & Agriculture-livestock-planning-(s), | 752 |in an analysis of title-land-excavating-subjective-physical-estate-maintaince-practice, proportioning, uncharted open-mass-rural-community-population-documenting-(s), juxtaposed other genetic-committee-(s)-commitment-(s), for whom comprehend development-(s) of land, by offspring-yearly-repopulation-sociological-conversive-method, abstract, sociological-query, pre-historic or requisite, from national-militarial-constitutional-currency-merit, per taxed [district /// city /// county //// state-constitution-inquiry], transitioning currency by-of nation, in-to an merit-factory-production-(s)-system, planning an thirty-year-offspring-period-hypothesis,| 753 |from extenuating-observation-(s), interceding how community-maintaince, coincide with populating communicative-competence-interactive-mean-(s), estimating an hundredths-millionths-political-party, by-an denominative-congressional-tolerable-agenda, reverberate-frictioning-congressional-ethic-(s), from mathematical-literary-schematic-incremental-impasse-coverage,| 754 |for validify-ing how Title-Land-Appropriated-Insurance-Governmental-County-Address-Limit-(s), define how perspective of transit, may intercede, like R.V. & Writer, [Surface-Area-Project-Infrastructure // City / County-Inspection-procedure // Commercial-Entity-Operation-(s) // Transit-Municipal-Compliance], may always never be redacted, by common-subjective-feudalism-yearly-monthly-weekly-impasse-(s),249L. swearing by literary-historical /or/ presence-calendar-observation, to retract those common error-(s) from common-physical-effort-(s)...¶||| 21m.:37s.97m.s.|||§

§..Preposition-Count-[35 / Thirty-Five]-of.. ..Article-Essay-[2 / Two].. ..Book-Essay-[224 / Two-Hundred-Twenty-Four].. ..Lines-[249.6125 / Two-Hundred-Fourty-Nine.1.six-tenths-2.one-hundredths-3.two-thousandths-4.five-tenths-thousandths].. ..Can other individual-(s), agree on an basis for comfortable-lifestyle-living, or does inhabitants have an introverted-ego, incapable from family-historical-practice, to accede how their be an requirement for scheduling, dated objective-dependent-variable-(s), apose entity-independent-variable-(s), from how physiological-anatomical-compositional-human-being-function-(s), matter, with when differentiated moment-(s), mode why developmental-labor, occur primary due to educational-progress, from youth-book-directive-lessoning-(s), practicing every day, during offtime-parameter-(s), how to conduct what operation-(s), for Ancestral-Lineage-LONGevity, how to conduct an utopian-lineage, from those confides of present-living.¶|||38s.68m.s.|||§

..§¶225¶§..Had for all of American-history, not preferred to write with ink, on paper, thinking perceptive, for how an total-population should practice why (all! / some?) function-(s), proportional labor, upon literary-labor-engineering, or common-non-literary-inquiry, not foretell extent-compromise, how past-dating count from expert-handling, finding an transit-sedentary-work-complex-routine, identifying market-(s), where Toilet-Paper & Paper-Towel-(s)-become-produced, Federal-Congressional-Senate-Executive-Cabinet-Member-(s) candor numerous drove-(s) of [box-(es) / plastic-(s) / glass-(es) / aluminum-can-(s) / paper-(s)], producing from primordial-present-year-offset-competency, without visual-inclination-(s), noting tonal-recognition-(s), registering circumventional-market-(s), apose conventional-theatre-book-space-(s), staging routine, from civilized-conjunct-(s), from form where illiterate-perspective, still magic-mythicize-perceptive-production-maintain, produced-from, resourced, to idle by market, in confidence of Paper & Plastic-Trading, maintained or sustained, by-an numerical-account-verifying, losing vision at bold-face-Times-new-Roman, processing [bill-(s) / typed-manuscripts / memorandum-(s) / television-cursory / signatures' / identification],| 756 |early-human-physic-observation, transitive-psychological-effort-(s), denominating an singular-state-sociological-climate-topographic-terrain, as each article-(s), await individual-numerical-populant-(s')-impasse-(s), extending family, reliant blank circulating, characterized-pronoun-interactive-basis, based upon legible-fact-(s), historically inclined, personal-lifestyle, relevant community-faith-interaction-(s), [spending / saving], as how we [extenuate // extemporize // examine], by our personal hyper-tense-(s)-impasse-reflex-(s), not truthfully-(judiciary-fact), conversed, how youth-open-book-pre-examination-policy, not exist, identifying which parameter, to direct bookshelf, from classroom-complex-dynamic.287¶|||01m.:55s.02m.s.|||§ Creative-observation from "here", can only be for executive-mass, legislate bill-bailout-policy-commercial-industrial-sociological-specimen-affair-(s), Federal-International-Total-resourcing-reliance, left in interior-geological-vehicle-(auxiliary-mechanism)-cartographic-physical-coverage-boundary-restriction-extent, far too conclusive, common-family-independent-individualism, posture no character-direction, supplementing method of received-payment-(s), Scheduling & Considering, [what / which], [object-(s) / article-(s) / physical-perceptive-posture-(s)], matter as Ecological-civilization-(s), past-developed, present-continuum, by-an paper-shortage, where inhabitant-competency, directly-collect, in ethical-revise, state-denominative-enumerated-territory-(ies), affecting circumstance-(s) of living, without much of an focus of dating, regardless of conduct, as for how it can affect blank-centralized-independent-perspective, complicated, from story-(ied)-low-mid-high-rise-county-civilization-(s), in low-rise-distance-competencies,| 757 |illiterate / written?-{Percentage-offset-deviation}-{Fraction-offset-deviation}-{Factorial-15-calculator-trillionths-exponent-cut-off}, effort-(s) by Sedentary-Academia, when pupil-eye-space, fragment an cerebral-air-pocket-exterior-depth-perspective-tangible-focus-impasse, temporal, [1.décor / 2.cursive / 3.fashion / 4.alcoholic-taste / 5.vocal-vernacular-syntax-literary-scaling-mathematical-letter-word-participle-primary-thousands-sociological-impasse-hundredths // 6.thousandths // 7.ten-thousandths /// 8.hundred-thousandths-daily-word-conjunction-(s)], not offset, any particular-layer of thinking, primarily mute, from Person & Being, temporal-circumstance-impasse-living, quickly-deter-mining, how dating has / is-(maybe live / presence-appeared), gracing perceptive-qualities, from tentative-denominative-awareness, greatly-impacted through juxtaposition-(s), amongst one-another, for then having hour-three-thousand-six-hundred-denominative-secondths, demonstrate-pressure, awhile moment-(s), superlatively-impasse, moded-invigoration, where [visual / tonal / taste // nasal-observation-(s)], by natural-perspective-(s), only if their is focus (from an) inhabitant-present-physiological-compositional-past-subjective-lesson-(s)-perceptive-conduct-purport-timbre, representative by how population-perceptive-dynamic, hedonize-free-democracy, while swearing Republican-mass-person-party-(ies)-offset, for guarantee-reliance, of {$}senate-executive-fundamental-legal-action-(s), yet pertained, from visual-tonal-vocal-hyper-tense-reflex-es-perceptive population /and/ land-constitutional-continental-statehood-subterior-focus-impact, indicative by those minor-minute-(s), mattering house by block-community, from those title-deed-attainment-civilization-method-(s), accumulating documented [scheduled-entities // agenda // itinerant // method // effort-(s)], administering minor-field-ecology-focus-(-aspect), common-legal-thought amid present-literary-upkeep, juxtaposing how life commemorate millennium, or practice for an century, working an few decade-(s), to travel an decade, and realize population-perceptive-literary-purpose, or per se, upon an [jargon / rhetoric / colloquial-tounge], serve voluntary-motion-(s), equal, interpreted formal-impasse, concentric, from inhabitant independent-visual-perspective-(s'), physically-trading because how article-(s), suffice monetary-appetite, at parcel-title-work-objective-sociological-conceptive-parameter-(s)-activities-schedule-humanis-tic-flaw, either disregarded again and again, by [[1.]religious-sect-constitutional-communion / [2.]geology / [3.]cartography / [4.]biology / [5.]anatomy / [6.]physiology / [7.]psychology / [8.]political-person-population-legal-science-(s) / [9.]dentistry / [10.]mathematical American Weight-(s) & Measure-(s) / [11.]congressional-executive-cabinet-(s)-(20 committees)-house-article-(s)-literature-constitutional-amending-article-(s)-politics], pertained, yet from U.S.D.A.-Resourcing-solubility-edible-prob-ing, where capable-sociological-active-usage, by hard /or/ soft-product-tangible-mean-(s), may amongst physiological-genetic-compositional-visual-perspective-(s), determine less then human-physical-accounting-one-by-one, whom understand why stillness is important upon distress of listening /and/ observing, yet (from an) minute-vocal-tonal-vibrational-reverberational-tonal-focus, implicated [1.person-Nation-industrial-populant-count / 2.land-parcel-room-exponent-offset-living-quarterths-deviation-production / 3.furniture / 4.appliance-(s) / 5.desk-(s) / 6.bookshelf-(ves) / 7.bed-(s) / 8.closet / 9.bathroom-quarterths / 10.kitchen-dining-quarterths / 11.living room-communion-conjoining], through visual-physical-perceptive-formality-(ies), supposed applicated-interaction-(s), by-an ruse of visual-physical-perceptive-observation-(s), different at family-offset, by generalized-conjunction, not perceiving why writing-article-(s) may differ from {8in × 10in-layed // shelfed-articles-literary-extent-(s)}, discoursing first, primary-term-tier-article-premise-visual-tonal-existing-perceptive-inhabitant-characteristic-(s), detail-elaborating, apposed, climate-living-condition-(s), implicit for determining numerative-subterior-posit-numerical-variable-(s), thinking how [offset-competent-experience // capabilities / entity-paperwork-procedure-(s) / objective-adjusting / product-trading], condition those thoughts, for documenting, clarifying from calendar-year-historical-scheduled-dating, yet literary-scheduling, human-behavior, for of an serious lack of common-cursive-pages-tier-documenting-extent-(s), believing that general-writers, and readers, dynamic free-will-human-capacity-(ies), yet awhile-an numerative-process-progression, tier from major-elaborated-thinking, while denominative-count-exponent-numerical-article-(s), | 759 |apose minor-intricated-visual-depth-features-calendar-tonal-vocal-numerical-grammatical-characters-day-impasse-interpretation, where hour /and/ day-impasse-denominative-motion-effect-(s), are rather insufficient of repopulation-perceptive-emphasis, for literary-circulation, which can not soly relay an singular-valedictorian-emphasis, unless an balance of [cum-laude / magna-cum-laude / summa-cum-laude], an 15th-% / 100%-denominative-tenths-millionths-populat-ion-census-political-religious-constitutional-documenting-article-(s)-party-accurate-count, reader-citizen-responsible, an industrial-Federal-subjective-Geological-century-state-county-city-block-family-district-principality-(ies)-goal-objective-(s), by auxiliary-plant-function-guidance, quite possible-action-(s), driv-er-identity, conceal introversion, rather comfortable, when visual-in-vigoration, is not capable, or emotional, from illiterate-inhabitant-be-ing-suggestion-(s), not valuable, fixating measure, by developmen-tal-location-(s)-two-dimensional-perimeter-impasse-(s), measuring objective-three-dimensional-composition-(s), where-ever distance, impacting how impasse-effect-(s), are too simply visualized, experi-encing by repetition, intrication-(s) from past-day-hour-comforted-re-flexive-muscular-subjective-visual-characters-vocal-pronuncia-tion-perceptive-(common-consideration), intertwining various-(s) juxtaposed, market-register-mone-tary-observation-(s), subtly differ-ent from how laundromat, not be housing Washer & Dryer, or dry-cleaner-(s), not tailor-(ing), as is restaurant, not be supermarket, or grocery-market, from headquarter-Federal-commercial-offset-pro-duction-(s), forgetting to elaborate though perceptive-existence, to think, or else [1.wither // 2.wilt / 3.deteriorate / 4.wisp / 5.wane / 6.wish], from at an basis of birth, global-Centripetal-simultaneous-aging-re-volving-parallel-revolutionary-impasse-perpet-ual-force-continu-um-impact, congruent-offset, physical-formation-(s), at boundary-re-population-limit-(s), rampant delible-idle-nature, commonly-suf-ficed, when at limited-traded-basis, relevant why subjective-disposi-tion-(s) exist, unused, (dry), of reference-(s), self-book-shelf-cited, for then interacting by literary-soirée, examining intermittent reli-gious-communion, for staging our housing, not apartment-(s), by-an lifestyle-appreciation-method, for referring present-living sleep, that if believed by commercial-present-youth-transition-aging-perspec-tive, remain casted in, Metropolitan-civilization-(s), for experiencing climate-Geological-Barren-lifestyle-cultivation-(s), a-the most finest content, granulized by tangible-acre-incrementation-(s), characteriz-ing product-market-value, for not sociological-independent-land-ti-tle-handling, exist without our metropolitan-work-schedule-dynam-ic-activities-offset, not [writing / typing] our innermost-activities-lit-erary-scheduling-(s), interposing newspresse-political-issue-(s), as would be permitted, in salvaging oneself, from those presumed historical-war-(s), addressing, human-diplomatic-affair-(s), sedated for such, when international-political-dynamic, not make sense, at [bachelors // associates // high school diploma-(s)],| 760 |aging-interpretations, for subtly-extracting [1.international-dry-substantial-(s) // 2.housing-objective-(s) // 3.repopulating-objective-(s) // 4.food-cycling-super-market-objective-(s) // 5.vehicle-auxiliary-excavating /// 6.transiting /// 7.hauling-objective-(s)], either by self-independent-family-effort-(s), or sociological-genetic-political-survey-party, following legislational-article-(s)-guideline-(s),| 761 |amid visual-letter-marks-conscious-vocal-tonal-subconscious-perceptive-comprehensive-competency, literate when self-writing, understand lifestyle-literary-theory, listless from droves of humanity, as motion-(s) of individual-(s), simply matter from ecological-documenting, not contingent thesis of hypothesized-validity, yet of future-anticipation-(s), but without current-literature, formulating (from an) interposed-national-literary-genre-(s)-superlative-article-physics-elemental-tier-reset-subjective-function-(s), from how presence-consideration-conduct-(extraspective) juxtaposed human-behavior-(introversion), motion-exponent-offset = notion-(s), familiar similar-effort-axiom,| 762 |yet to male-subservience, time-motion-range-barren-project-geological-interior-terrain, female-mundane characteristic-(s), validifying consequent purpose-(s), debt purchasing, though literary-folly, motioning document-(less), work-aging-mode-inquiry-survey-impasse-(s), legal-leather-bound-frat-story-board-advise,| 763 |legislational-historically, misunderstood common-congressional-Federal-Literary-Encyclopedia-Extent, different diction-ary-word-thesaurus-thousandths-ambiguous, [1.uninterpreted / 2.unthought / 3.unwritten / 4.untasked / 5.unarticlized / 6.unfamiliar / 7.uncontrolled], [Judiciary-Leather-bound-Legal-Case-Sociological-Documenting /// Legislational-Senate-Congressional-Third-Person-Federal-Bill-(s)-Observat-ion],| 764 |from-apose cabinet-executive-committee-(ies), Federal-county-sub-committee-subver-sive, bill-policy-monetary-reserve-act-timbre-denotation-(s), offset.¶|||10m.:52s.02m.s.|||§

§..Preposition-Count-[23 / Twenty-Three].. ..Article-Essay-[3 / Three].. ..Book-Essay-[225 / Two-Hundred-Twenty-Five].. ..Lines-[42.42 / Fourty-Two.1.four-tenths-2.two-hundredths].. ..Where will this book be stored for millennium-entity-preserved-utilizing, beyond lineage-reader-commercial-book-purchasing-4000 / 6000-unit-production-(s), when mass circulation, incur juxtaposed wage, laboring by day, an enjoying evening offtime interaction-(s), with community-members, whom can coherently apply, thought of general-intellectual-inquiry, serving those fundamental-circulated-substantial-nutritional-production-(s), anticipating how [Horticulture / Agriculture / Mariculture pre-nutritional-elemental-resourcing-(s)] have weights still to be practiced of balancing, by concerted-compositional-individual-weight-count-(s), apose [field-distancing-(s) / soil-level-(s) / water-condition-(s) / et cetera..] conjuring what commercial-product-(s), are required to direct an Formal Senate-Executive-Congressional-Budgeting, apose each of those Ethnic-Offset-Compiled-compositional-survey-physiological-human-being-(s), whom are in an direct-competent-awareness, evaluating from year [2018-Gregorian / 484-Physics] continuum aging-offset-progression, how voluntary inhabitant-effort-(s), have yet credited up, an Genetic-Citizen-Class, practicing from requisite continuum, past aging layering of ones' motioned-act-(s), how to verify what inclination-(s), pertain an formal practice for interpreting Housing-City-Block-Ecological-Geological-Topogra-phic-review, noting why ecological-resourcing, counterpose Global-War-Offset-Terrain-Resource-(s), for how each indication of lifestyle-living, affect what motivation-(s), individual-Citizen-(s), offset-labor-trait-circulate, why developing an Centralized-Canadian-Hoarded-Resourc-ing-Society, would be significantly-better, than living South of those Topographic-Condition-(s) far too humid, to understand why if each individual, sits idle, not thinking what to offset moniker, how would daily living be, from those bored thresholds of sedentary-wait, not making an Sociological-Literary-Society, whom comprehend Relig-ious-household-Meeting-(s), amongst communicated-Genetic-City-County-Community-Individual-(s), dialoguing from literature, how to retain those (Decade) Century-Resource-(s), for making life invigorating, from those vocal-conscious-interaction-(s), that appear wonderful, as I intend to initiate communication-(s), for building to an Blonde-National-Endeavor, through my lifetime, as each coherent individual, understand-(s) why dating; remain steadfast-longevity, apose those timing-(s)-incremental-passings, deductive by hour-cutoff-limit-(s), as how individual-(s), require concertion from attention-span, due to each of those limit-(s) of physiological-limited-capabilities, apose Feminine-Offset-Yearly-Repopulating, when Physics remain, Globally Boundaried, without an capable [individual / mass / population] inhabitants, capable for communicating, without literary-Commercial-Intelligent-Inquiry, making all those [1.action-(s) / 2.motion-(s) / 3.purchase-(s) / 4.interaction-(s) / 5work-effort-(s) / 6offtime-thinking / 7.transit-(s) / 8+et cetera..], be more worthy from quality-visual-vocal-tonal-youth-educational-rubric-benchmark-commerical-adult-literary-lifestyle, [breeding / raising / circulating / wage-conceiving], how to generate an mass-population, whom would think how an mass-offset-influencing, would push communicate to one another, those individual-(s), by family-dynamic, which motion-(s) need to be past-scheduled-documented-noted, about realizing what wage-nutrition-circulation-proceeding-(s), enable an ||5,000,000+|| requisite Total-Population over those age-(s) of Goal-100, while those proglamation-(s) of Wage & Retirement, sustem an Physic 484-Year-(s) (Gregorian-American-Literary-English 2018) error of Formal visual-vocal-tonal-attractive-existing.¶|||02m.:50s.26m.s.|||§

{January-(31d.)-February-(28d. / 9d.)-March-(31d.)-April-(30d.)-May(31d.)-June(30d.)-July-(31d.)-August-(31d.)-September-(30d.)-October-(31d.)-Novermber-(30d.)-Month-(s)-Day-Hour-Minute-(s)-Secondths-Count}-{28,857,600 D.H.M.S. = 60S.× 60M. × 24H. × D.}-{8,016 D.H.}-{480,960 D.M.}-{334 D.}

{Epilegomena}

¶Their are Twenty-four-Hours,| 765 |an day Thirty.four-tenths-one-hundredths-six-continuum-thousandths-seven-ten-thousandths-days-per-month amid seven days per week; awhile ninety-day per season; deviate upon eight-hours per day, work-place-environment; for how perspective inducts from oneself; offset family-(ies) of four, for whom impasse several-currency-circulated-place-(s), in work for two-children, three primary-aging-classification-(s)-{Birth to Thirty-(youth-young-adulthood) /// Thirty to Sixty-(sustainable-adult) /// Sixty-(grandparent-aging)}, amid four seasons occur-revolutionary, amidst those four phases of day, {1. Day //// 2. Evening //// 3Night //// 4. Morning}-celestial-rotation, (by an) pentagon of military-security, reclusive (at those) point-(s) of security, for A-Mercian-Freedom, in alternative regard, apose those Eastern-custom-(s), that are enjoyed from our lord-(s) whom hold human-stock-trading, in Germanic [Hohenstaufen // Hohenzollern // Hoehn-superlative-dynasty-castles-(s)], throughout Duetschland, pertaining no civil duty to those citizens, (whom are attained at an), savage human-lordship-nature, amid-an focus from technological-industrial-de-velopment-(s), that in my [(one)-A Book; An Series of Essay-(s)], examine how other-(s) are pertaining focus from surrounding-matter, not ecologically-international, but yet where our lords have put an dispertion-population-clause, not understood, why I would tolerate their riches, for how ignorant human-behavior react to those fact-(s) by our lord-(s), amid-an non-reactive-survival-discourse of humanity, having how they are in fact suppose to conquer the globe, amid-an longevity-discourse of subdued-event-(s), as how I try to publish my book with Macmillan-Holtzbrinck-Publishing-corporation-(s), (for an) split of currency, that would delve my neutrality on their castle-empirical-fashion, amid my repopulating-emphasis, that impede on the dynamic of human-behavior, because how war is fought, arm-in-arm, apose any righteous-line-(s) of sanity, under an construe of population-control, yet maximized (by the) age Seventy-Five higher-dynamic, (for how) cultivating each property at title-directive-control, deviate competent-individual-(s'), understanding those Primary-Six-Subject-(s)-Lessoned-Context-(s) per day to be focused on, by [1Mathematics-Primary-Function-(s)-into-Geometry; 2Language-Art-(s)-into-Grammatical-Punctuational-Rubric-Benchmark-Liter-ary-Reasoning; 3Cartography-Geographical-Topographical-political-boundary-earth-space-scien-ces-matter-biological-anatomical-reasoning; 4Economics-human-ethics-resourcing-wage-(s)-4.1industrial-factory-shipping-production-operation-(s)-4.2market-time-management-budgeting-4.3stock-market-industrial-saving-(s)-4.4(international)-exchange-trade-funds-management; 5History in [5World / 5.1National-Offset / 5.2State-National-Offset / 5.3County-Principalities-Offset-politics; 6Physical-Education or {Push-ups /// Sit-ups /// Mile-runs}] incorporating gradually other workout-stretches-routine-(s), tiering-an balance of cardiovascular-respitorial-reflexive-movement-(s). Delving back into how their are seven tectonic-plate-(s); eight years for an maximum American-Presidential-Candidate-(two)-Term-(s), apose nine-five-nine-reset-conjunction-offset-occurrence, self-pre-embody, what form-(s) matter upon an exterior-perceptive-persona-rate, uninclined death, by secretive-fashion-(s), without-an global-discourse-(s), per subjective-quality-(ies), offset historical-vernacular-custom-(s), variating how each moment is (from the) predicate-subjective-pertained-focuses, (at any) given time, hourly per date, in {1 / 24} offset each notion apose an query of open-mindedness, per datum-given-customs', {1 / 7}-datum, when other {tenths / hundredths}-denominative-continuum-influctuation, cent-dollar apparently-occur simultaneous-apparatus, per dating function-(s), awhile momentum, not yet inertia-comprehension, *for free*; amid an superlative-offset-variable-(s), that momentary-adjust, why pronoun-posit-{$47.00 per book; {2017 / 2018 / 2019 / 2020 / et cetera..}}, adjective-process, repeat befuddled modification-usage-(s); particulated-at junction per-of-an-prerogative-induced-occurrence, solar-celestial-offset, as how each continuum-moment-secondths-in-fourths, repeat amidst repetitive-process, that influctuate, why factoring datum, haven't-contracted-lingüístics', where word-secondths-letters-milli-secondths-processing, not commonly-factor, html://, in clued fashion per tier-objective-input-letter-method-(s), amid an flurry upon formulated-conjective-data, understood, when at purchase-fashion-production, or not an clue per offset common-barbarism, not fashionable upon barbaric-tradition, for how agreement-(s) repressed an subdue per trade, what grief make culture-eatable-fashion, in live-belief, yet by my truth, oor belief?!, (fur how) perceptive-undifferentiation, occur sporadic-per-contextual-clues, deciphering amid issue of grief, ooor not an radical motion, will not be made, transit boundary, yet be fully-guided, when alterior-function-(s), pertain each individual-apparatus, attaining at an grounds' for what define perimeter-boundaries', because how retained-perimeter-boundary-perceptive-grounds', swear-an particular-offset, per offset-planned-perimeter-boundary-spacing-(s), where developments of commercialism, superlative why their ethos an pact-agreement-(s), treaty-(ied)-off, when not adjected-internationally, per cultural-developmental-societies, as trades' are accepted by different-culture-(s), parameter perimeter-boundary-clarification-(s), per ecological-room-space, when in [introverted // extraverted customary-offset-(s)] person-(s) have not interposed primary-predicate, for extemporaneous-redact-(ing), per notion-(s) offset, locational-developmental-sweating-custom-(s), for how men & women, solemnly-sit, without formal-gradience from how dating (has an) seasonal-denominative-period {$^{1} / _{90}{}^{th}$}, comparative thousandths, apose whole-period-table-(d)-customs', traditionally-valued, at approximate-distance-secular-space-(s), familiarized, from whom meet one-another, at customary-value-(s), because cultivation, derive custom-(s), from physical-quality-(ies), when matter develop from horticultural-value-(s), when purity fixate an particular-Yorkshire-et cetera...-cultures-of-pink-blue-blonde-tangible-qualities, (as an) delicacy of human-eating, and sacred-rite-(s), (from how) custom-(s) believe where developmental-qualities, matter per approximate-parameter-secular-layer-(s), when municipal-code, coldly-abides (to those) condition-(s) of nature, in various topographical-contexts, not inclined juxtaposition-(s), of tectonic-plate-(s)-conjunction-conditions-complex, conclarative, each denominative declarative-value-(s), when idealizing physical-motion-(s), per payment, from self-subjective-architectural-literary-funds-compiling-bonds-acres-composition-(s)-contenting, informing why each action-(s), under Federal-Literary-Congressional-Watch, forget mathematical-timbre, as did my Epilegomena, omit Prolegomena-predicate-indication-(s), which should continuously be remodifying, each notion-(s) of tangible-function-(s), be only before hand, at slightly gripped, from the offset continuum, of unconscious-effort-(s), continuously alive, but at an bue of state-ten-mile-(s)-conjunction-offset-develop-mental-curb, as when sat-transited-function-(s), pertain only restricted-self-mile-(s), objective-usage-mechanical-quality-operation-(s), because how operational-handling, offset minor-process-ing-(s) as is [http:// / html:// tier-(s)] coding how individualism is subterior breadth-an subconscious-code, not practical, in an conclusive-fashion. What define an clause-context, for such reasoning-(s), or else dimwittedness, would subbcome any inbreed, (for how) serving our lord, is suppose to be ethically-foreseen, not hallucinogenic, (from an) forensics conundrum, historical Germanic-Barbarism, and global-economy, (that can only be had from each) position per known individual, when an common-interpretation-basis, have not other cult-family, male-elder-eating-(s); Reenter for why I see no need to type-December, when an superlative-paging-(s), not refine those word-seasonal-qualities, amidst Territory & Nature, where subtle pressing-(s) incur as when every-dating of language centripetally-impasse-offset, from one-another, when fluctuated at an circumventional-revolutionary-pattern, resetting from physiological-sleep-cycling-(s), an purpose for each of those cause-(s), that carry an burden upon reasoning, or calls into convention, an played sequences, recycled, when formal-training impression why an call for action-(s) exist, why developing an ecological-boundary, preserve what rite-(s) of an being are to be recollected, beyond an limit-(s) per daily-ontological-proof-(s), would be practiced before an consulate, practicing-evaluation-(s), of rite-(s), to council weather an practice remain suitable, or not; no other influctuation of motion-(s), could exist when exterior-boundary-(ies), are collected from still-introverted-mute-mutational-being-(s), abetting lifestyle, (from an) still classicism, embodied from when handling horticultural-value-(s), could be rather than apose purposed action-(s), that matter (from an) directive-prerogative-function-(ing), rather recidivistic-pacifistic-approach, to those matter-(s) at hand by preference-effort-limit-extent-(s), idealizing why their not become an ruse for learning, (as by what has been) instated, for decade-year-month-century-count-(s), not aware of millennium impassing, (as by what has had an) collective-grouping-gathering-(s), because how intervention per premise-interpretational-basis, before had, and yet be orchestrated, impasse-territory-function-circulating, an immense effect of labor-sourcing-methodical-effort-(s), as what I attempt to inclinate, through each word, evoking context, administered by expert-minor-subterior-analysis, upon macro-volume-stock-market-Federal-Currency-saving-(s), apose those small-business, like [food / beauty / hygiene / medicine / liquor / et cetera...] as where currency fall in through various, too all hands, at different-metric-mediums, but without an mechanical-Christianized-passage-(s), by how fighting has taken an discourse for medium-right-(s), rather cooperating on what device-(s), administer formal belief-(s), upon an item-article-manual-objective-trading-(s), because for when dating is continuum per factor-(s), without an groove for how each dating-variable, is supposed upon every procrastinated-point-(s) of exterior-dated-universal-surrounding-(s), reclusively adept, by human-mundane-morale, when dating have an circumstantial-progressive-prerogative-(s), redated per moment, when action-(s) are taken at place, for conducting amicable-interaction-(s), because how transient-motion-(s), affirm our influctuated place-(ing-(s)), due to our during motion-continuum-frontal-present-focus, alerting that without communication-(s), procession-(s) could not ever be formally met, when dating at locational-distancings, revolve at various harmonies, because how developing at an establishment, function from [male / female genitalia-activities], at any given moment, could only be had to in feminine-male-action-fashion-(s), that mechanical-advantage, impression more than their actually be calculated, relevant to an calculated-resourcing-operation-(s), upon [Agricultural /// Horticultural /// Mined /// crude-oil-Extracting /// et cetera-raw-extraction-lands] per offset civilization-(s), that incorporate-(d), from how sweat is supposed to be expected, apposed washed from civilian-effort-(s), practicing formal incorporation-legal-right-(s), apposing any advantage from pre-impending-discourse, (at any other) civilian-right-(s) (by an) incorporation-programmer-function-purpose, not fully inclinated, agitation of poverty, (from how those) isolated-terrain-(s), affect (much of a the) similar-inclination-(s), are pre-pertained, before an requisite-indication-(s), at offset-Geographical-origin-(s), (that maintain much of a the) land remained disproportional, East of a the Mississippi-River, South amidst Missouri-River-Land-expanse-(s)-{3 / 11}, from [Red-River // Colorado-River discoursing's'] inconclusive for how to handle perimeter-politics, due at equal-deviation-(s) for understanding why motion-(s), matter in contraction of muscle-pulmonary-gland-glycemic-index-nutrition-proportionating-muscle-(s), amid-an continuum-con-text-(s), equally-valued, when dating each value-(s) (from an) position of common-rambling, can only statistical-fact, reintroduce an premise of said-presiding-(s), to formulate any [subconscious / conscious-reasoning-(s)], when [type / written-cursive-font-(s)], face an script by juxtaposition of roman-fiction-material, predominant common-perspective, when metric-physics, impasse each meter, unseen from the naked-eye, offset each imagery, from visual-tonal-perspective, adjust-affirming from quality-tangible-valuing, before any fact of trade, because how Gewerbesteuer, not take local-taxes, from individual-(s), amid an contextual-profits payroll and capital, where tax-deviation-(s), have not had an Gewerbesteuer-take, on what approach of civilization-(s), are matter-(ed), but apprehended because how transit (has an) outlook, (from an) historical-contextual-dynamic-count-(ing); when each variable of matter is slipping beyond each common-feature-(s), that allude each latter-moment-(s), when circle-conferencing, signify an customary-means-(s), per traded-objective-item-(s), select by what has been put in circulation, per moment-(s), at traded-market-(s) apprehension-(s), because when dating rule from position-(s), how one-(s') role partakes' an anima from men, whom animus (from an) omitted physical-responsibilities-nature, because, historical-factor-(ing), not present an exterior-nation, away from one-(s) secular-development-(s), with an non-verbal-directing what motion-(s), would be formidable to attempt to physiologically-cover-ground, not soly based on ones' family, from past day requisite-encounter-(s), amid an periodical-supposed-process-(es), forgotten to note into the fact, as is planned certain-meeting-fact-(s), forgetting from commercial-processing-(s), how many jobs are supposed to complement one another, from each daily factoring-(s), where moment-(s) per cycle, are not in an controlled-designation-(s), for why we are free to express our self-sentiment-(s), while in contextual-deviation, not sentimental, too repair our resourced-good-(s), as well as cycle more goods, because how work is supposed per objective, at traded-resourced-market-trademark-registered-center-(s), insighting from numerical-shelf-trade-policy-(ies), an wide range of issue-(s), that consider why human-fatigue, have an customary-shelf-value-(s), when tolerance has reached an continuum subdued means', not expanding beyond 100-years of lifestyle-(s), per ordinance-rule, underscripted motive for life after 100-years-aging, when cycled-work-material, pertain an guidance per prerogative at an basis from past day customary-eating, detailing what any individual may believe in, (from an) international-motive-prerogative-(s), for how detailing land, be worked (from an) reasoning each deviation of human behavior, when at work-load, defining why land would be accounted because how covering ground, works different (from an) individual-dynamic, from why non human-history, define which grounds one determine-(s) formal-developmental-trade-practice-(s), but without common-documenting, then every-individual, (has an) duty to reset page-(s), juxtaposed thousandths-pages deviation, per offset territory, when millisecondths, do not have an letters-deviation-exact-count-(s), proportional count-process-offset-(s), per community-squared, amid those context-(s) of content-(s), reconfiguring each programmed-perimeter-parameter-distancing-(s) offset, when customs of learning, remain colloquial by personal-tounge, incapable of impartial-means', when subjective-predicate, not fully morale total-human-qualities, amid those convention-(s) of Barbarism, per traded-production-(s), as how customs have been developed, by long-withstanding-culture-(s), persuaded market trading, when dating each interpretational-function, have no truth to cultivating particulated-situated-estate-(s), offset distance by diameter-(s), of Yards & Meter-(s), as how compass affect view of distance, upon squared-box-perimeter-distance-(s), per physiological-impasse, comprehending that we attain substantial-piece-(s), but yet with an understanding for isolating an terrain, for cultivating [horticultural // maricultural // agricultural-western-state-nation, subterior ecological-rural-development-(s), emphasizing how distance is too be contoured] outside of metropolitan-domesticated-living, apose those custom-(s) of trade, practiced, (from an) repopulating-dynamic, yet fully practiced, by all living-species, or human-custom-(s), for how homage of past date-action-(s), has church bring an extraspective-revivalism, to an group of individual-(s), with an awe for those daily circumstances, offset each contrite right-(s), held in possessor of title-duty-(ies), conundrum resolve, without educational-dissolution from past abided rights, (due to) commercialism-practice-(s), absolved from sin, without taking an insight or so, to comprehend what matters remain most important, when eating on an daily-basis, not having researchers of surveyed consumption-matter, per capitia offset spaced off territory, only applicable when conveyance by socialism-lifestyle, matter from physical-lineage-customs appreciation, per lineage dynasty, unto an social-market, not fully mattered by attire, but allegiance to an ball or activity, or hobby; repetitively as weeks reset fifty-two-.45667, every year, so what draw up such quantitive, when delving those topics, for refined ecological-architectural-lifestyle-room-(s), per acre-estate, spacing land shift-proportional, impassed-metropolitan-boundary-(ies), abided when supple-menting, figurative-space-(ing-(s)), eepterischzé block-perimeter-boundary, as how timing, have subjective-quality, to decipher how interpretation-(s) by parameter-limit-(s), abide an capable-comprable-unit, of item-pieced-men, that conjure from factory-shipping-dynamic-trading, an discourse, inculpable for total-asset-(s) affirmation-(s), per item-squared-surface-area, because how divergent each company has had an configuration, by general-mean-(s), iconic to all that bequeaths before visual-tonal-vibrational-cartilage-pressure-level-(s), informational per notable-fashion-(s), that exist before an modification-(s), because how interrogative-data, interpret-(s), when fashioned to an status of men, in wilderness-conjunction, obscured what piece of voluntary-data, confirm information-callously, for deep in the untamed terraculture-terrain nothing alone, will cultivate an ecological-setting, drawing into real-question, how to sustain an ecological-growth, to delve an multitude of sequence-(s), that draw objective-inquisition-skills, in comparison of executed-power-function-(s), per task, designated at an station of sort-(s), for isolating those technical-characteristic-(s), for formal-project-usage-(s), amid an global resourcing for many of those [[1]bullets / [2]guns / [3]gasoline / [4]petroleum / [5]oil / [6]plastic-(s) / [7]et cetera...], refining those raw-resources, into an controlled-product, capable of subduing each international-market, relevant through those capable function-(s), of shipping, which has been hugely overlooked, (by an) historical relevance, referencing each piece of context, in an superlative-comparison-(s), juxtaposing data, for voluntary-mean-(s), to have an proportional-purpose, of each ruled-task, (pick up where last left off) measuring various cooridors, hallways, rooms, living-space-(s), exterior-perimeter-housing-unit-space, understanding that their are building-material-(s), that complicate the proglamated-process, of developing interior-cooridors-grounds, for facilitating each inhabitant involved from how to perform those task-(s) by rigor, vivacious when idle-self, sustain moment-(s), from our ambiguous-grasp, of our tangible-celestial-universe, mathe-matically-inclined, (on an) decreasing slope, scoping out every driven impasse, (on an) continuum-rampant-rate, (for not) entrenching oneself (into an) process for invigorating every live-awake-conscious-moment-lived, rather than live consciously-idle, by subconscious-process, enlivened because how each notion-(s) gets' molly-coddled, from male-introverted-idle-nature, amid an feminine-non-enabling, for how to repopulate metropolitan-physics, because how each and every [inclination-(s) / indication-(s)] from human-effort-(s), have not an sweat-concertion, which affect-(s) heavily from centralized-metropolitan-congestion, because how various environment-procedure-(s), would have stabilized due to cognitive-psychological-event-(s), yet elucidating why project-procedural-development, remain an fact for further considering why sloth enters the psyche of those lazy independent-work inhabitants, (due to an) lack of work-charge-morale, which identifies weakness in those inhabitant-being-(s), whom perform up to task, not furthering the job identifies (for an) inhabitant to formally go, as government have an enabling-psychological-code, municipal from when each of those right-(s), are administered, debunking sport-(s) physiological-psychological-conduct, would have an subversive-unconscious Pronoun-(s), thirty-year-scoped by aging-drive-routine-(s), when labor remind mundanely that subterior-surrounding-(s), can only gander an premise per operational-functional-inhabitant-incorporator, from dues of count-(s), by [minor / major-subterior-career-(s)], facing how each [[1]notion-(s) / [2]event-(s) / [3]annotation-(s) / [4]inclination-(s) / [5]indication-(s) / [6]inductive-evidence / [7]vocal-vernacular-scholarly-timbre / [8]English-American-Renovated-Language]; deductive-confluence, amid those figures per hourly conduct, (as why) commonality, combine total-output, where each dating-figure, can only host, by showing-(n'), how conduct intermittent other individual-activities, have workplace, retire, and wilt, always upon an structure of will, to modify why dated-work-labor-(s), can only coalesce past where information-data, cooberated [1goods / 2services / 3resources / 4raw-goods // 5industrial-mechanical-operational-patent-(s)]; continuum by what strip-mall-(s), have had an international market-wall-(s), foresteading those territorial-bounds, previous evident by collection of book-(s), where market-(s), have an usage per ordered-layered-mineral-(s), how certain-mineral-(s), you are not supposed to eat.. ..but do we eat folic-acid, or swallow thiamine, for how dosages, pertain per aging, how ingested-material-(s), affect how we digest each comestibles or materials not meaning anything, due to surrounding-cultivating, each particular-location-development-(s), requiring the other objective-(s), furthering how we concept exterior-visual-imagery-(ies), because how awake-conscious-content, matter only when prevalently aware, at which impasse mark why each location is not simply eaten-at, or still-idle-at, without recognition of independent-right-(s), supplemented-up, when each group of inhabitant-right-(s), obscure what can be interpreted, when crescendo locational-right-(s), believe in how each independent-individual-comprehension, would matter apose competent-interpretation-conjecture, intermittent other individual-practice, per offset situation, conundrum by how I am ending now, before their ultimately-end, what synthesis of syntax, verify what pages will define my incremental-inductive-piece, labeled by A Book; An Series of Essays; [authored / written / edited] by Manuel Pagan Jr.

I am working on my one-hundredth-book, to age per year, at the moment, so how each week to month variate the top ten-book-(s), (has an) rate I would like to enhance at retirement-pace-sit-down and work, without interrupting my prior context, because (as to whom it would) concern react amid motion-(s)-impasse-propulsion-movement-lapse-(ing), for ever however many chains are retorted, or dialoged (through to whom), kudos aging, not understand land-space-purpose-geological-measuring...

> ....So Euthology of Mortem, would subterior race, (as an) collective-identification-system. or no 100th book not be made for the entertainment of human-disparity, or else fiction-mystery becomes the only article-(s) [read / written / spoken], how to find from Face book, or other governmental-app, social-security-number-(s) with names, no photo, for identification, but an phone number and pronoun-book, would be the twenty-fifth-book-test, every lifetime to one hundred, because yearly aging have an assortment for how each diagram of context, would variate each mode of perceptive-individual-interpretation-(s), only (as an) blank-gesture-(s), when no sub conscious / conscious-acceptance can ever be, due to the during-continuum-impact, offsetting most of what we see of chrome-shine, can never fully in-depth the next level of Geological-Ecological-pronoun-responsibility-(ies)...

...So I left the middle of the book barren from what should be yet to come up next in context, but as of now, would agree the editor and I, do not fuss with the year-(s) timing it would take to reedit all the punctuation-mark-(s), hence I left out an perfect meticulating of context, when it would deserve the entire-book-properties-predicate, or an blog-project of literary-formal-redact-analysis...

..However so, it is with utmost Renault-value, do I present to you, (after the fact), A Book; An Serie-(s) of Essay-(s)..¶|||20m.:17s.38m.s.|||§

{Epilegomena}

¶§Final Witty Remarks; diminuendo paging per perimeter-three-dimensional-unit, not interchangeable, hardcover-interpretational-power-(s), yet instilling what paragraph-lesson-(s), interpose per paragraph, between Essay-Conventional-type-face, when alliteration, can only sound if men conduct from various-forms per context, in ritual-procession, out of line, from formal-military-conduct, where land be an part for maintaining, total-premise-operation-(s), amidst an flurry from circumventable-circumstance-(s), continuously incurred, after collecting-point-(s), for living, each day away from calculated-functional-labor-operative-task-(s), as per city centralized-development-point, an technical-living affirm where dating physiological-wage-effort-(s), pertain why land can be covered, when circulated-currency-(ies), refresh from commercial-market-manager-budgeting, inconclusive thorough community-families-breeding, because how no family, exist without an offset geographical-age-tiering, inconsistent to the breeding age-(s) per under eighteen birth-(s), that are required for progressing along tectonic-plate-geological-crust-surface-area, offset human-physiological-impasse-competency, amid those resource-(s), used as would bicycling has been, for covering land, (by an) welding-metal-pole-(s), and metal-brake-cords, used when no individual, come up soly with how to configure an mechanism, per restricted parameter-(s), due to monetary-commercial-trade, looking at [patents / copyrights / registering's / licensing's], thinking how commercial-market-(s) are suppose to balance an various commodity-resourced-need-(s), amongst human-labor-wage, for understanding what impasse-(s) (are to be) formally understood, when traveling in county-city-approximate-distance-(s), tiering family offset matter, which remain uncommunicative, per trade, not bolstering an furthering of independent-inductive-interpretation-matter-(s), bringing cited-literary-data, to administer (why data is to) interpose along our deteriorating-aging, an configuration circumstance-(s), (that continue from how) meticulated-aging, is going through-an deterioration (from how) substantial-compositional-matter, cycle through each eating of interior aorta-ventricle-muscular-knot, conjoined (by the) esophagus, for supplementing nutritional-matter, (from how our) intestine-tract, is suppose to build an daily-relevant matter-compositional-back-up, for reaching our esophagus-aorta-nutritional-supple-menting, where data can only be relevant, upon human-past-day-reading, amid present-day-reading, (for how to) consistently communicate, what resources, are required by an formal-gratitude, to the land, and those individual-(s), whom cultivate each piece of commercial-market-material-(s), traditional for how to get those [cultivation-(s) / harvest-(s) / extracting-(s) / excavating-(s) / et cetera...]; always from at least how I see circumstance-(s), an purpose to bow to each individual, upon an moment at fair gratitude, seeing that life is not simply made (from an) single-motion-(s), and learning which formal-routine-(s), adjust cultivating land, must always understand their is an ecological-purpose to etiquette, upon our commercial-circumstance-(s), (due to an) lack of physiological-capabilities, that resort all of humanity to the Minor-Principle, of commercial-engineering, agricultural-cultivating, horticultural-grazing, slaughtering-method-(s), preparation-process-(es), et cetera... ...for outside ourselves, exist an nice-individual-(s)-society-member-(s), when understanding every individual reacts at first (to their) personal-space-(ing-(s)) being protected, amid an voluntary-discourse for how interaction-(s), work (from an) consistency of interactive-dialogue, while remembering the Bible to always divert what function-(s), require an member of an civilization-mass, to formally Cultivate & Extract resources of land-(s), to accommodate our fine-sophisticated-lifestyle-(s)...¶|||03m. :06s.50m.s.|||§

Manuel.Pagan.JR.

{|1.Gregorian9.-2.Calendar17..-3.Schedule25.-4.Circulation36.-5.Survey42.-360°|}

[|Initial-Hypothesis|]

¶For how a-the discourse of this book would go, I have not an initial-firm clue, as for where those bookstore-(s) would be, or digital-sales can be circulated, so I am working on an advertisement-scheme, to hopefully intrigue my readers, for comprehending that off-time-lifestyle, is suppose to interpose work-effort-(s), for comprehending why each developmental-aspiration-(s), continue an contingent focus, premised by holdings of this book, for interacting with other Colloquial-Citizens, whom interact highlighting of such printed-production-(s), making not simply an Book, but an Generation of intelligence, working each day, practicing thinking each and every awake-secondths, to juxtapose how celestial-matter, ambient apose elemental-condition-(s), juxtaposing those physiological-compositional-being-effort-(s), for objecting in entity-(ies), an Titled-spacing-(s), that affirm why population-extraspective-interaction-(s), is truly the most important fact to existing, working on Resourcing & Cultivating Geological-Terrain, Practicing from our current offset disposition-(s), how to Centralize, an offset Elemental-edible-cultivating, for then shifting an Northern-Vacation-Lifestyle, for Retirement-Working, practicing how to resource from Tectonic-Plate-Cavern-Abyss, an mundane society that communicate through Table-Non-Fiction-Literature, concurrent fact-(s), for understanding what other functions of paper are required for denominating human-mass-population-Geological-Centralized-Terrain-Decided-Placings.¶|||01m.:03s.,84m.s.|||§ I hope this work, book sells in to weekly circulated-production-(s), shortly after publishing.¶|||01m.:07s.,49m.s.|||§

{|1.Gregorian9.-2.Calendar17..-3.Schedule25.-4.Circulation36.-5.Survey42.-360°|}

[|Initial-Hypothesis|]

[My Estimate Book Sales Circulation Production]

{|1) 4,000; Weeks of April 25, 2018 - May 14, 2018a.|}°| How does one store data |
{|2) 4,000; Weeks of May 15, 2018 - June 26, 2018b.|}°| intermittent amongst Com |
{|3) 6,000; Weeks of June 27, 2018 - July 14, 2018c.|}°| mercial-circulated- |
{|4) 6,000; Weeks of July 15, 2018 - July 24, 2018d.|}° book(s)? |
{|5) 8,000; Week July 25, 2018 - August 4, 2018e.|}° |
{||||||||[||||||||[||||||[||||[[|{||{{{|*6) 8,000; Week August 5, 2018 - August 11, 2018f.*|}}||}|]]|||||]|||||||||]|||||||||]||||||}|
{|7) 10,000; Week August 12, 2018 - August 18, 2018g.|}° |
3[2,658,606,720i.c.i.] {|8) 10,000: Week August 19, 2018 - August 25, 2018h.|}° |
1[13,564,320n.p.e] {|9) 10,000; Week August 26, 2018 - September 1, 2018i.|}°|
0Pages-up-to: {|10) 10,000; Week September 2, 2018 - September 8, 2018j.|}° |
2[40,128,000-d.p.e.] {|11) 10,000; Week September 9, 2018 - September 15,| 2018k.|}° |
4[7,865,088,000i.c.i.] {|12) 10,000; Week September 16, 2018 - September 22, 2018l.|}° |
{	*13) 10,000; Week September 23, 2018 - September 29, 2018m.*	}°
{	14) 10,000; Week September 30, 2018 - October 6, 2018n.	}°
{	15) 10,000; Week October 7, 2018 - October 13, 2018o.	}°
{	16) 10,000; Week October 14, 2018 - October 20, 2018p.	}° [Purchases:]
{	17) 10,000; Week October 21, 2018 - October 27, 2018q.	} ° [Estimated-Store
{	18) 10,000; Week October 28, 2018 - November 3, 2018r.	}° Random-interval
{	19) 10,000; Week November 4, 2018 - November 10, 2018s.	}° -numerative-offset
{	20) 10,000; Week November 11, 2018 - November 17, 2018t.	}market-merchant-holding-
{	21) 10,000; Week November 18, 2018 - November 24, 2018u.	}°	practic3]
What	{	22) 10, 000; Week November 25, 2018 - December 1, 2018v.	[45,927-n.s.]}
is [an / the / a]	{23) 10,000; Week December 2, 2018 - December 8, 2018w.[180,000
reduction of cost	{	24) 10,000; Week December 9, 2018 - December 15, 2018x.	}°
from copyright-book	{	25) 10,000; Week December 16, 2018 - December 22, 2018y.	}°
apose reader social	{	26) 10,000; Week December 23, 2018 - December 29, 2018z.	}°
sales, amid daily	{	27) 10,000; Week December 30, 2018 - January 5, 2019a.a.	}°
offset circulation, until	{	28) 10,000; Week January 6, 2019 - January 12, 2019a.b.	}°
[*Year:*]2017-(25)-{8-years}	{	29) 10,000; Week January 13, 2019 - January 19, 2019a.c.	}°
[Year:]ɷ483-500-{17-years}	{	30) 10,000; Week January 20, 2019 - January 26, 2019a.d.	}°
{	31) 10,000; Week January 27, 2019 - February 2, 2019a.e.	}°
{	32) 10,000; Week February 3, 2019 - February 9, 2019a.f.	}°

{|1.Gregorian9.-2.Calendar17..-3.Schedule25.-4.Circulation36.-5.Survey42.-360°|}

[|Initial-Hypothesis|]

| {|33) 10,000; Week February 10, 2019 - February 16, 2019a.g.|}° |

| {|34) 10,000; Week February 17, 2019 - February 23, 2019a.h.|}° |

| |Barbasol| {|35) 10,000; Week February 24, 2019 - March 2, 2019a.i.|}° |

| {|{36) 10,000; Week March 3, 2019 - March 9, 2019a.k.|}° |

| {|38) 10,000; Week March 17, 2019 - March 23, 2019a.l.|}° |

| {|*39) 10,000; Week March 24, 2019 - March 30, 2019a.m.*|}° |Slogân-Cliçĥé| ||||||||||

| {|40) 10,000; Week March 31, 2019 - April 6, 2019a.n.|}° |

| {|41) 10,000; Week April 7, 2019 - April 13, 2019a.o.|}° |

| {|42) 10,000; Week April 14, 2019 - April 20, 2019a.p.|}° |

|| {|43) 10,000; Week April 21, 2019 - April 27, 2019a.q.|}° |

||| {|44) 10,000; Week April 28, 2019 - May 4, 2019a.r.|}° |

|||| ...I am trying {|45) 10,000; Week May 5, 2019 - May 11, 2019a.s.|}° |

||||| to reduce the waste {|46) 10,000; Week May 12, 2019 - May 18, 2019a.t.|}° |

|||||| of Paper... {|47) 10,000; Week May 19, 2019 - May 25, 2019a.u.|}° |

||||||| from {|48) 10,000; Week May 26, 2019 - June 1, 2019a.v.|}°|

||||||| Shagbark-Composite {|49) 10,000; Week June 2, 2019 - June 8, 2019a.w.|}° |

| reducing warehouse shelve {|50) 10,000; Week June 9, 2019 - June 15, 2019a.x.|}° /

||||||| ply-wood-cords, {|51) 10,000; Week June 16, 2019 - June 22, 2019a.y.|}° |

|||||| comparing Hardcover-Context {|52) 10,000; Week June 23, 2019 - June 29, 2019a.z.|}°|

||||| from various locale purchase {|53) 10,000; Week June 30, 2019 - July 6, 2019b.a.|}° |

|||| rate-(s), {|54) 10,000; Week July 7, 2019 - July 13, 2019b.b.|}° |

||| making those trees {|55) 10,000; Week July 14, 2019 - July 20, 2019b.c.|}° ||

|| non-deforestation {|56) 10,000; Week July 21, 2019 - July 27, 2019b.d|}° |||

| apose {|57) 10,000; Week July 28, 2019 - August 3, 2019b.e.|}° each ◦Geological ||||

| {|58) 10,000; Week August 4, 2019 - August 10, 2019b.f.|}° ₁Ecological |||||

| {|59) 10,000; Week August 11, 2019 - August 17, 2019b.g.|}° [◦Genetic / Ethnic] ||||

| {|60) 10,000; Week August 18, 2019 - August 24, 2019b.h.|}° offset, by |||||||

\ {|61) 10,000; August 25, 2019 - August 31, 2019b.i.|}° {₁Pink / ₂Periwinkle / ₃Taupe / |

| {|62) 10,000; September 1, 2019 - September 7, 2019b.j.|}° |Bruise-{sarcastic} |||||||

| {|63) 10,000; September 8, 2019 - September 14, 2019b.k.|}° / ◦yellow / ₀white||} ||||||

| Commercial- {|64) 10,000; September 15, 2019 - September 21, 2019b.l.|}°entity |||||

| re-cement- {|65) 10,000; September 22, 2019 - September 28, 2019b.m.|}° orchestrating ||||

|| Entity-Sleeping {|66) 10,000; September 29, 2019 - October 5, 2019b.n.|}° |||

{|1.Gregorian9.-2.Calendar17..-3.Schedule25.-4.Circulation36.-5.Survey42.-360°|}

[|Initial-Hypothesis|]

||| Barrack-dynamic-(s){|67) 10,000; October 6, 2019 - October 12, 2019b.o|}° {|[p]Magenta|}||

|||| Retirement-Sexual-Repopulating {|68) 10,000; October 13, 2019 - October 19, 2019b.p.|}°|

|||| {|69) 10,000; October 20, 2019 - October 26, 2019b.q.|}° |

|||| {|70) 10,000; October 27, 2019 - November 2, 2019b.r.|}° |

||||| [Qwerty-Inquiry] {|71) 10,000; November 3, 2019 - November9,2019b.s.|}° /

||||| For every tree one wastes, {|72) 10,000; November 10, 2019 - November 16, 2019.b.t.|}° |

|||||| Do banks have dollar-(s)? {|73) 10,000; November 17, 2019 - November 23, 2019b.u.|}°|

||||| {|74) 10,000; November 24, 2019 - November 30, 2019b.v.|}° ||

|||| Year-Anamoly {|75) 10,000; December 1, 2019 - December 7, 2019b.w.|}° |||

|||| Reset:1776-{0}- {|76) 10,000; December 8, 2019 - December 14, 2019b.x.|}° ||||

||| D.M.Y.{|77) 10,000; December 15, 2019 - December 21, 2019.b.y.|}|201-(2 / 3 / 4 / 5) ||||

|| {|78) 10,000; December 22, 2019 - December 28, 2019b.z.|}° [Continuum] ||||

| {|79) 10,000; December 29, 2019 - January 4, 2020c.a.|}° *Estimate ||||

| {|80) 10,000; January 5, 2020 - January 11, 2020c.b.|}° [Terracultrual ||

\ {|81) 10,000; January 12, 2020 - January 18, 2020c.c.|}° [Cord-(s):] |

| {|82) 10,000; January 19, 2020 - January 25, 2020c.d.|}°-{|[0]430,110,000-Interior-Pages|} |

| {|83) 10,000; January 26, 2020 - February 1, 2020c.e.|}° {|[1]810,000-Books|} ||

| {|84) 10,000; February 2, 2020 - February 8, 2020c.f.|}{[2]168, 222, 150, 000-offset} |||

|| {|85) 10,000; February 9, 2020 - February 15, 2020c.g.|} |[3]23,735-inclination-(s)| ||||

||| {|86) 10,000; February 16, 2020 - February 22, 2020c.h.|}° ..Reset.. ||||||

|||| [Cement; {|{87) 10,000; February 23, 2020 - February 29, 2020c.i.|}° ||||

|||| Architectural-Structure {|88) 10,000; March 1, 2020 - March 7, 2020c.j.|}° ||||||

||||| City-Principality-County- {|89) 10,000; March 8, 2020 - March 14, 2020c.k.|} ° |||||

|||||| Conceptive-Montana- {|90) 10,000; March 15, 2020 - March 21, 2020c.l.|}° ||||

|||||||Retirement-Aging-Eating-Survival{|91) 10,000; March 22, 2020 - March 28, 2020c.m.|}° |||

||||||| Shared-Resourcing] {|92) 10,000; March 29, 2020 - April 4, 2020c.n.|}° ||

|||||||| {|93) 10,000; April 5, 2020 - April 11, 2020c.o.|}° |

||||||| {|94) 10,000; April 12, 2020 - April 18, 2020c.p.|}° ||

||||||| {|95) 10,000; April 19, 2020 - April 25, 2020c.q.|}° |||

|||||| {|96) 10,000; April 26, 2020 - May 2, 2020c.r.|}° ||||

|||||| {|97) 10,000; May 3, 2020 - May 9, 2020c.s.|}° |||||

||||| {|98) 10,000; May 10, 2020 - May 16, 2020c.t.|}° |||||

|||| {|99) 10,000; May 17, 2020 - May 23, 2020c.u.|}° ||||||

||| {|100) 10,000; May 24, 2020 - May 30, 2020c.v.|}° √966,038-purchased |||||||

{|1.Gregorian9.-2.Calendar17..-3.Schedule25.-4.Circulation36.-5.Survey42.-360°|}

[|Initial-Hypothesis|]

{|101) 10,000; May 31, 2020 - June 6, 2020c.w.|}° circulated-book
{|102) 10,000; June 7, 2020 - June 13, 2020c.x.|}°[Create-Space-Online/
{|103) 10,000; June 14, 2020 - June 20, 2020c.y.|}° Macmillan-Publishing /
{|104) 10,000; June 21, 2020 - June 27, 2020c.z.|}° or Else]
{|105) 10,000; June 28, 2020 - July 4, 2020d.a.|}° initial estimate.
{|106) 10,000; July 5, 2020 - July 11, 2020d.b.|}°
{|107) 10,000; July 12, 2020 - July18, 2020d.c.|}°
{|108) 10,000; July 19,2020 - July 25, 2020d.d.|}°
{|109) 10,000; July 26, 2020 - August 1, 2020d.e.|}°
{|110) 10,000; August 2, 2020 - August 8, 2020d.f.|}°
{|111) 10, 000; August 9, 2020 - August 15, 2020d.g.|}°
{|112) 10,000; August 16, 2020 - August 22, 2020d.h.|}°
{|113) 10,000; August 23, 2020 - August 29, 2020d.i.|}°
{|114) 10,000; August 30, 2020 - September 5, 2020d.j.|}°
{|115) 10,000; September 6, 2020 - September 12, 2020d.k.|}°
{|116) 10,000; September 13, 2020 - September 19, 2020d.l.|}°
{|117) 10,000; September 20, 2020 - September 26, 2020d.m.|}°
{|118) 10,000; September 27, 2020 - October 3, 2020d.n.|}°
{|119) 10,000; October 4, 2020 - October 10, 2020d.o.|}°
{|120) 10,000; October 11, 2020 - October 17, 2020d.p.|}°
{|121) 10,000; October 18, 2020 - October 24, 2020d.q.|}°
{|122) 10,000: October 25, 2020 - October 31, 2020d.r.|}°
{|123) 10,000; November 1, 2020 - November 7, 2020d.s.|}°
{|124) 10,000; November 8, 2020 - November 14, 2020d.t.|}°
{|125) 10,000; November 15, 2020 - November 21, 2020d.u.|}°
{|126) 10,000; November 22, 2020 - November 28, 2020d.v.|}°
{|127) 10,000; November 29, 2020 - December 5, 2020d.w.|}°
How many {|128) 10,000; November 6, 2020 - December 12, 2020d.x.|}°
Books are required {|129) 10,000; December 13, 2020 - December 19, 2020d.y.|}°
to be subjunct, from {|130) 10,000; December 20, 2020 - December 26, 2020d.z.|}°
how {[1]Physics} / {|131) 10,000; December 27, 2020 - January 2, 2021e.a.|}°|
{[2]Principality} / {[3]City} / {|132) 10,000; January 3, 2021 - January 9, 2021e.b.|}°
{[4]County} / {[5]Ecology} / {|133) 10,000; January 10, 2021 - January 16, 2021e.c.|}°
{[6]Nation} / {[7]Geology} {|134) 10,000; January 17, 2020 - January 23, 2021e.d.|}°

{|1.Gregorian9.-2.Calendar17..-3.Schedule25.-4.Circulation36.-5.Survey42.-360°|}

[|Initial-Hypothesis|]

||||| {[8]Billions Population} {|135) 10,000; January 24, 2020 - January 30, 2021e.e.|}° |

|||| {[9]Factory Unit {|136) 10,000; January 31, 2020 - February 6, 2021e.f.|}° ||

||| Production {|137) 10,000; February 7, 2021 - February 13, 2021e.g.|}° |||

|| Mechanics} {|138) 10,000; February 14, 2021 - February 20, 2021e.h.|}° ||||

| along with {|139) 10,000; February 21, 2021 - February 27, 2021e.i.|}°construction|||||

| {|140) 10,000; February 28, 2021 - March 6, 2021e.j.|}° {[10.001]Engineering - |||||

| {|141) 10,000; March 7, 2021 - March 13, 2021e.k.|}° {[10.01]Principality} / {[10.02]City} |||||

| {|142) 10,000; March 14, 2021 - March 20, 2021e.l.|}°{[10.03]County} / {[10.04]Ecology} ||||||

| {|143) 10,000; March 21, 2021 - March 27, 2021e.m.|}°{[10.05]Nation} / {[10.06]Geology} /|||||||||

|| {|144) 10,000; March 28, 2021 - April 3, 2021e.n.|}° {[10.08]Tunnel-Excavating}|||||||||

||| {|145) 10,000; April 4, 2021 - April 10, 2021e.o.|}° {[10.07]Aquatic-Parallels} ||||||||

|||| to {|146) 10,000; April 11, 2021 - April 17, 2021e.p.|}° {[10.08]Auxilary-Flights}} ||||||||

||||| circulate {|147) 10,000; April 18, 2021 - April 24, 2021e.q.|}° ||||||

|||||| PHYSICS-retrieval- {|*148) 10,000; April 25, 2021 - May 1, 2021e.r.*|}° |||||

||||||| serial-note-(s)-economy, {|149) 10,000; May 2, 2021 - May 8, 2021e.s.|}° ||||

||||||||for formal book-copyright{|150) 10,000; May 9, 2021 - May 15, 2021e.t.|}°documenting. |||

||||||| {|151) 10,000; May 16, 2021 - May 22, 2021e.u.|}° ||

|||||||| {|152) 10,000; May 23, 2021 - May 29, 2021e.v.|}° |

||||||| {|153) 10,000; May 30, 2021 - June 5, 2021e.w.|}° |

|||||| [Thought] {|154) 10,000; June 6, 2021 - June 12, 2021e.x.|}° |

||||| {Secondths Rate = words-deviation}-{|155) 10,000; June 13, 2021 - June 19, 2021e.y.|}° |

||||| .offset. {|156) 10,000; June 20, 2021 - June 26, 2021e.z.|}° |

|||| {Millisecondths Rate = characters-rate {|157) 10,000; June 27, 2021 - July 3, 2021f.a.|}° |

||| deviation} {|158) 10,000; July 4, 2021 - July 10, 2021f.b.|}° |

|| ...only in progression...{|159) 10,000; July 11, 2021 - July 17, 2021f.c.|}° ||

| by offset pursuit-survey-inquiry. {|160) 10,000; July 18, 2021 - July 24, 2021f.d.|}° |||

| {|161) 10,000; July 25, 2021 - July 31, 2021f.e.|}° How many ||||

| {|162) 10,000; August 1, 2021 - August 7, 2021f.f.|}° [0]{blank-(s)}[1] |||||

| {|163) 10,000; August 8, 2021 - August 14, 2021f.g.|}° [0]{space-(s)}[2] |||||

|| {|164) 10,000; August 15, 2021 - August 21, 2021f.h.|}° [0]{pause-(s)}[3] ||||||

||| {|165) 10,000; August 22, 2021 - August 28, 2021f.i.|}° [0]{intermission-(s)}[4] |

|||| {|166) 10,000; August 29, 2021 - September 4, 2021f.j.|}° [0]{omission-(S)}[5] ||||||

||||| {|167) 10,000; September 5, 2021 - September 11, 2021f.k.|}° [0]{vacancies}[6]|||||||||

|||||| {|168) 10,000; September 12, 2021 - September 18, 2021f.l.|}° [0]{silence-(s)}[7]||||||||||

{|1.Gregorian9.-2.Calendar17..-3.Schedule25.-4.Circulation36.-5.Survey42.-360°|}

[|Initial-Hypothesis|]

|||||| {|169) 10,000; September 19, 2021 - September 25, 2021f.m.|}° ever incur |||||||||

||||||| {|170) 10,000; September 26, 2021 - October 2, 2021f.n.|}° offset ||||||||

|||||||| {|171) 10,000; October 3, 2021 - October 9, 2021f.o.|}° population; |||||||

||||||||| {|172) 10,000; October 10, 2021 - October 16, 2021f.p.|}° Geologically ||||||

|||||||||| {|173) 10,000; October 17, 2021 - October 23, 2021f.q.|}° Zone-(d)-s. ||||

||||||||| {|174) 10,000; October 24, 2021 - October 30, 2021f.r.|}° |||||

|||||||| {|175) 10,000; October 31, 2021 - November 6, 2021f.s.|}° ||

||||||| {|176) 10,000; November 7, 2021 - November 13, 2021f.t.|} |

|||||| {|177) 10,000; November 14, 2021 - November 20, 2021f.u.|}° ||

||||| {|178) 10,000; November 21, 2021 - November 27, 2021f.v.|}° |||

|||| {|179) 10,000; November 28, 2021 - December 4, 2021f.v.|}° ||||

||| {|180) 10,000; December 5, 2021 - December 11, 2021f.w.|}° ||||

||| {|181) 10,000; December 12, 2021 - December 18, 2021f.x.|}° |||||

|| {|182) 10,000; December 19, 2021 - December 25, 2021f.y.|}° ||||||

| {|183) 10,000; December 26, 2021 - January 1, 2022f.z.|}° |||||||

| {|184) 10,000; January 2, 2022 - January 8, 2022g.a.|}° ||||||||

| {|185) 10,000; January 9, 2022 - January 15, 2022g.b.|}° canyou interpose |||||||||

| {|186) 10,000; January 16, 2022 - January 22, 2022g.c.|}° |||||||||

| {|187) 10,000; January 23, 2022 - January 29, 2022g.d.|}° |||||||

|| {|188) 10,000; January 30, 2022 - February 5, 2022g.e.|}° ||||||

||| {|189) 10,000; February 6, 2022 - February 12, 2022g.f.|}° ||||||

|||| {|190) 10,000; February 13, 2022 - February 19, 2022g.g.|}° |||||

||||| {|191) 10,000; February 20, 2022 - February 26, 2022g.h.|}° |||

|||||| {|192) 10,000; February 27, 2022 - March 5, 2022g.i.|}° ||

||||||| {|193) 10,000; March 6, 2022 - March 12, 2022g.j.|}° |

|||||||| {|194) 10,000; March 13, 2022 - March 19, 2022g.k.|}° |

|||||||| Icanseeintothefuture {|195) 10,000; March 20, 2022 - March 26, 2022g.l.|}° |

||||||| {|196) 10,000; March 27, 2022 - April 2, 2022g.m.|}° |

|||||| {|197) 10,000; April 3, 2022 - April 9, 2022g.n.|}° |

||||| {|198) 10,000; April 10, 2022 - April 16, 2022g.o.|}° ||

||||| {|199) 10,000; April 17, 2022 - April 23, 2022g.p.|}° |||

|||| {|200) 10,000; April 24, 2022 - April 30, 2022g.q.|}° ||||

||| {|201) 10,000; May 1, 2022 - May 7, 2022g.r.|}° |||||

{|1.Gregorian9.-2.Calendar17..-3.Schedule25.-4.Circulation36.-5.Survey42.-360°|}

[|Initial-Hypothesis|]

|| {|202) 10,000; May 8, 2022 - May 14, 2022g.s.|}° ||||||

| {|203) 10,000; May 15, 2022 - May 21, 2022g.t.|}° |||||||

| {|204) 10,000; May 22, 2022 - May 28, 2022g.u.|}° ||||||||

| {|205) 10,000; May 29, 2022 - June 4, 2022g.v.|}° ||||||||||

{|206) 10,000; June 5, 2022 - June, 11 2022g.v.|}° |||||||||

| {|207) 10,000; June 12, 2022 - June 18, 2022g.w.|}° ||||||||

|| {|208) 10,000; June 19, 2022 - June 25, 2022g.x.|}° |||||||

||| i {|209) 10,000; June 26, 2022 - July 2, 2022g.y.|}° ||||||

|||| Intend this {|210) 10,000; July 3, 2022 - July 9, 2022g.z.|}° |||||

||||| book, enlightening {|211) 10,000; July 10, 2022 - July 16, 2022h.a.|}° ||||

|||||| my fair avid readers {|212) 10,000; July 17, 2022 - July 23, 2022h.b.|}° |||

||||||| to impartially work away {|213) 10,000; July 24, 2022 - July 30, 2022h.c.|}° |

|||||||| from the present international {|214) 10,000; July 31, 2022 - August 6, 2022h.d.|}° |

||||||||| foreign-visa-trading of factory {|215) 10,000; August 7, 2022 - August 13, 2022h.e..|}° |

|||||||||| commercial-products, {|216) 10,000; August 14, 2022 - August 20, 2022h.f.|}° |

||||||||| for an incremental-domestic {|217) 10,000; August 21, 2022 - August 27, 2022h.g.|}° |

|||||||| reduction of products, {|218) 10,000; August 28, 2022 - September 3, 2022h.h.|}° |

||||||| for developing an retirement {|219) 10,000; September 4, 2022 - September 10, 2022h.i.|}°|

||||||new-(*Blonde hair)-nation, {|220) 10,000; September 11, 2022 - September 17, 2022h.j.|}° |

||||| for book-lifestyle-origin-{|221) 10,000; September 18, 2022 - September 24, 2022h.k.|}° |

|||| GEOlogical- {|222) 10,000; September 25, 2022 - October 1, 2022h.l.|}° *Blue ||||

||| Domestic {|223) 10,000; October 2, 2022 - October 8, 2022h.m.|}° eyes |||||

|| Copyright© {|224) 10,000; October 9,2022 - October 15, 2022h.n.|}° Patent® ||||||

| {|225) 10,000; October 16, 2022 - October 22, 2022h.o.|}° Philosophical |||||||

| {|226) 10,000; October 23, 2022 - October 29, 2022h.p.|}° Thinking Society, ||||||||

| {|227) 10,000; October 30, 2022 - November 5, 2022h.q.|}° harmonizing an ||||||||||

| {|228) 10,000; November 6, 2022 - November 12, 2022h.r.|}°centralized Lecture|||||||||

| reducing {|229) 10,000; November 13, 2022 - November 19, 2022h.s.|}° Classification||||||||

|| waste of resources {|230) 10,000; November 20, 2022 - November 26, 2022h.t.|}°Progress|||||||

||| while appreciating {|231) 10,000; November 27, 2022 - December 3, 2022h.u.|}°Ordering-(s)|

|||| those vision-(s) per {|232) 10,000; December 4, 2022 - December 10, 2022h.v.|}°71W ||||||

||||| *Pink Skin individual-(S). {|233 10,000; December 11, 2022 - December 17, 2022h.w.|}°564C|||

|~~AAAAAAAAAAAAAAAAAAAAAAAAAAAAAAAAAAAAAAAAAAAAAAAAAAAA~~|

{|1.Gregorian9.-2.Calendar17..-3.Schedule25.-4.Circulation36.-5.Survey42.-360°|}

[|Initial-Hypothesis|]

{[Pages:]} 501

{Words:} 152,537-dash-subjunct // 229,703-particular-word-(s)

{Characters no spaces:} 1,697,455

{Characters with spaces:} 1,858,601

{Paragraphs:} 917

{Lines:} 20, 448

www.ingramcontent.com/pod-product-compliance
Lightning Source LLC
Chambersburg PA
CBHW020717180526
45163CB00001B/3
9781543463187